RECENT TRENDS IN APPLIED PHYSICS AND MATERIAL SCIENCE

The book comprises selected papers presented at the Second International Conference on Recent Trends in Applied Physics and Material Science (RAM 2024), organized by Bikaner Technical University, Bikaner, during November 15–16, 2024, in hybrid mode. The central theme of the conference focused on promoting interaction between researchers working in Material Science and Applied Physics, encouraging knowledge sharing, and enhancing visibility for their research. With expert plenary talks, oral sessions, and poster presentations, the event aimed to foster collaborations across topics like nanomaterials, biomaterials, computational physics, and nuclear physics. The conference proceedings include contributions vetted for technical depth and relevance. The objective was to bring young and senior scientists together, facilitating a platform to exchange innovations, progress, and challenges in the domain.

RECENT TRENDS IN APPLIED PHYSICS AND MATERIAL SCIENCE

Conference Proceedings of 2nd nternational Conference on Recent Trends in Applied Physics & Material Science (RAM 2024)

Edited by
Sudhir Bhardwaj
Manoj Singh Shekhawat
Bhuvneshwer Suthar

CRC Press
Taylor & Francis Group
Boca Raton London New York

CRC Press is an imprint of the
Taylor & Francis Group, an **informa** business

First edition published 2026
by CRC Press

2385 NW Executive Center Drive, Suite 320, Boca Raton FL 33431

and by CRC Press
4 Park Square, Milton Park, Abingdon, Oxon, OX14 4RN

CRC Press is an imprint of Informa UK Limited

British Library Cataloguing-in-Publication Data
A catalogue record for this book is available from the British Library

ISBN: 978-1-041-16456-2 (hbk)
ISBN: 978-1-041-16452-4 (pbk)
ISBN: 978-1-003-68471-8 (ebk)

DOI: 10.1201/9781003684718

Typeset in Times New Roman
by Aditiinfosystems

Recent Trends in Applied Physics and Material Science – Sudhir Bhardwaj et al. (eds)
© 2026 Taylor & Francis Group, London, ISBN 978-1-041-16452-4

Contents

List of Figures

Recent Trends in Applied Physics and Material Science – Sudhir Bhardwaj et al. (eds)
© 2026 Taylor & Francis Group, London, ISBN 978-1-041-16452-4

List of Tables

Recent Trends in Applied Physics and Material Science – Sudhir Bhardwaj et al. (eds)
© 2026 Taylor & Francis Group, London, ISBN 978-1-041-16452-4

Preface

It gives us immense pleasure to present the **Conference Proceedings of the Second International Conference on Recent Trends in Applied Physics & Material Science (RAM 2024)**, held on **November 15–16, 2024**, at **Bikaner, Rajasthan, India**. This prestigious event was organized jointly by ***Bikaner Technical University, Bikaner*** and the ***Condensed Matter Research Society, Bikaner*** with the support of our esteemed publication partners — **CRC Press** and the **Journal of Condensed Matter**.

RAM 2024 brought together over **400 participants**, both offline and online, from across the globe, reflecting the vibrant and growing international interest in the domains of **Applied Physics** and **Material Science**. The conference featured **plenary and keynote lectures** by eminent experts, **oral presentations**, and **poster sessions**, providing a stimulating platform for the exchange of knowledge and recent advances in the field.

The primary aim of this conference has been to foster meaningful interaction among researchers, academicians, and students working on various aspects of material science and applied physics. It is our hope that this event has helped build new collaborations, encourage the exchange of innovative ideas, and inspire future directions of research in these crucial scientific domains.

The overwhelming response to the call for papers was truly encouraging. After a thorough peer-review process, **140 high-quality research papers** were selected for inclusion in this volume of the proceedings. These contributions cover a broad spectrum of cutting-edge topics, including nanomaterials, optoelectronics, semiconductor physics, thin films, quantum materials, sensors, and energy materials, among others.

We would like to express our heartfelt gratitude to **Prof. Ajay Kumar Sharma**, Hon'ble Vice-Chancellor of Bikaner Technical University, for his constant encouragement and support as **Chief Patron**, and to **Dr. Y.N. Singh**, Dean Academic, BTU, for his guidance as **Patron** of the conference.

We are thankful to all authors for their valuable contributions, to the reviewers for their critical insights, and to all delegates and participants for making RAM 2024 a memorable and intellectually enriching event. Special thanks to our publication partners, CRC Press and the Journal of Condensed Matter, for their collaboration in disseminating this important body of work.

We hope that this proceedings volume will serve as a valuable reference for researchers, educators, and practitioners and will continue to inspire further advancements in applied physics and materials science.

Sincerely,

Dr. Sudhir Bhardhwaj | *Organizing Secretary*

Dr. M. S. Shekhawat | *Joint Secretary*

Dr. Bhuvneshwer Suthar | *Convener*

Recent Trends in Applied Physics and Material Science – Sudhir Bhardwaj et al. (eds)
© 2026 Taylor & Francis Group, London, ISBN 978-1-041-16452-4

About the Editors

Dr. Sudhir Bhardwaj is a distinguished Indian physicist specializing in theoretical high-energy physics. He serves as an Assistant Professor in the Department of Humanities, English, and Applied Sciences (HEAS) at Bikaner Technical University (BTU), Rajasthan. Dr. Bhardwaj earned his Ph.D. from the University of Rajasthan in 2009. Since 2007, he has been a faculty member at BTU, contributing significantly to both teaching and research in his field.

With over 100 publications to his name, Dr. Bhardwaj has made substantial contributions to experimental high-energy physics. His work has garnered more than 21,000 citations, reflecting the impact and recognition of his research within the scientific community.

According to the AD Scientific Index 2025, Dr. Bhardwaj holds the top position among scientists at Bikaner Technical University. He ranks within the top 3% of scientists in India, Asia, and globally, based on metrics such as the h-index and citation counts.

Dr. Bhardwaj's extensive research and academic contributions continue to advance the field of experimental physics, making him a prominent figure in the scientific community.

Dr. Manoj Singh Shekhawat is a dedicated physicist and academician, currently serving as an Assistant Professor in the Department of Physics at Government Engineering College, Bikaner. He earned his Doctor of Philosophy in Physics from Maharaja Ganga Singh University, Bikaner. His research interests encompass condensed matter physics, material characterization, and the development of functional materials for advanced technological applications.

Dr. Shekhawat has an extensive publication record, with over 50 research articles published in reputed journals and conference proceedings. His work has garnered significant citations, reflecting his impact in the field. Notable publications include studies on the synthesis and characterization of neodymium-based acylpyrazolone complexes, the effect of mercerization on natural fibers, and the development of optical materials for optoelectronic devices.

In recognition of his contributions, Dr. Shekhawat has been actively involved in organizing international conferences, including serving as the Joint Secretary of the 2nd International Conference on Recent Trends in Applied Physics and Material Science (RAM 2024) and the 3rd International Conference on Condensed Matter & Applied Physics (ICC 2019). His dedication to research, education, and scientific collaboration has significantly advanced the fields of condensed matter physics and material science, inspiring and influencing the scientific community both nationally and internationally.

Dr. Bhuvneshwer Suthar is an accomplished physicist and academician, currently serving as an Assistant Professor in the Department of Physics at Government Dungar College, Bikaner, Rajasthan, India. He obtained his Ph.D. in Photonics from Maharaja Ganga Singh University, Bikaner, in 2011. His research interests encompass photonic crystals, optical sensors, metamaterials, nonlinear materials, and the design of photonic devices for biomedical and sensing applications.

With over 93 research papers, 6 book chapters and 8 books to his name, Dr. Suthar has made substantial contributions to theoretical photonics and optical sensor applications. His work has achieved about 2,027 citations and 32 h-index, reflecting his impact in the field. In recognition of his contributions, Dr. Suthar has been listed among the Top 2% Scientists globally by Stanford University and Elsevier in 2023 and 2024. According to the AD Scientific Index 2025, Dr. Suthar holds the top position among scientists at Govt. Dungar College, Bikaner. His rank is in the top 3% of scientists in India based on h-index counts.

Dr. Suthar is an associate member of Institute of Physics (IOP), London, N1 9BU, United Kingdom. He is actively involved in various professional bodies, including the Indian Society for Atomic and Molecular Physics, the International

Association of Computer Science and Information Technology (IACSIT), Singapore. He is also a reviewer for esteemed journals such as Silicon (Springer-Nature), Optik (Elsevier), Optics and Laser Technology (Elsevier), and Optical and Quantum Electronics (Springer-Nature).

Dr. Suthar's dedication to research, education, and scientific collaboration has significantly advanced the fields of photonics and applied physics, inspiring and influencing the scientific community both nationally and internationally.

Recent Trends in Applied Physics and Material Science – Sudhir Bhardwaj et al. (eds)
© 2026 Taylor & Francis Group, London, ISBN 978-1-041-16452-4

Optical Biosensors Using Photonic Crystals for Medical Applications

Anami Bhargava*,
Ankita, Sapna Dinodiya
Nanophysics Laboratory, Government Dungar College,
Bikaner, India

Abstract: One-dimensional photonic crystals offer promising avenues for optical biosensing due to their tunable photonic bandgap (PBG) properties. This study explores the application of 1D PCs with engineered defect layers for sensitive detection of biomedical parameters. The introduction of nanocomposite and metal layers enhances tunability and sensing performance by manipulating defect modes in the PBG. Optical biosensors are designed and analyzed for the detection of malaria, hemoglobin, sucrose concentration, and tuberculosis. Key performance metrics—such as sensitivity, quality factor, figure of merit and detection limit are evaluated using simulated structures. Results show high sensitivity (up to 1114 nm/RIU), ultra-fast response times, and compact designs suitable for label-free diagnostics. The influence of structural and material parameters, including refractive index, incident angle, and layer thickness, is systematically investigated. These findings demonstrate the strong potential of 1D PC-based sensors in real-time biomedical diagnostics and point-of-care applications.

Keywords: Biosensor, Photonic crystal, Malaria, Tuberculosis etc.

1. INTRODUCTION

Photonic crystals (PCs) are micro or nano-scaled artificial structures, which contain a periodically arranged high and low refractive indices. Such periodic arranged dielectrics gives forbidden bandgap, which are known as photonic bandgap (PBG). Any disturbance in periodicity just by introducing as additional layer or removing one works as a cavity, which introduce a transmission peak (defect mode) into PBG region. The geometrical characteristics of the defect layer, such as thickness and refractive index, can be tuned to affect the defect mode along with angle of incidence and number of layers on either sides of defect layer (Bhargava et al. 2009, Suthar et al. 2009, 2012, Kumar et al. 2013). In the present work, we have summarized some of our recent studies on an optical biosensor using 1-D PC with defect that attracts researchers due to the ultra-fast response optical biosensor with compactness and precise result. The acute detection of a disease is key parameter for testing of blood, stool and urine, whose RI is interesting for a pathological analysis and treatment. The performance of an optical biosensor may be evaluated with sensitivity, quality factor and defection limit. These parameters for an optical biosensor are investigated for symmetric structure as well as asymmetric structure. In this study, the material structural changes, geometrical changes, fluid density changes and concentration changes with incident wavelength range to account for enhance optical sensor parameters. The results are presented primarily for blood samples for malaria diagnosis, haemoglobin, sucrose concentration and tuberculosis, in acqueous solution (Dinodiya et al. 2018, 2021, Suthar et al. 2023A, Bhargava et al. 2013).

2. RESULTS AND ANALYSIS

2.1 Malaria Diagnosis

In this study, 1-D defect photonic crystal (DPC) is taken as an optical biosensor application to diagnose malaria. The blood sample is taken as defect layer, which control defect mode's position (transmission peak). The defect mode is characteristics by the refractive index of defect layer.

*Corresponding author: anamib6@gmail.com, bhuvneshwer@gmail.com

DOI: 10.1201/9781003684718-1

As the refractive index of blood sample changes with the concentration of hemoglobin, which can be affected by the infection of blood sample in case of malaria. Therefore, the defect mode gets shifted as the infection increases for the blood sample of infected human. The sensing parameters as sensitivity, QF, DL, and response time are achieved as 495.73 nm/RIU, 2.03×10^5, 8.07×10^{-6} RIU, and 73.8 fs. The proposed device shows good results for the diagnosis with ultra-compact size and ultra-fast response (Ankita et al. 2021, 2022, 2022A).

When two metal layers are added on the either sides of a defect in the one-dimensional PC, the performance of optical sensor is improved. The detailed study with the various parameters such as refractive index of defect, metal layers' thickness and angle of incidence were investigated to optimize the geometrical parameters to improve sensing parameters. As it was observed that the sensitivity increases with the metal layer's thickness while the transmission decreases. Apart from this, the sensitivity gets an increase along with angle of incidence and achieve maximum value at 86° in case of TM mode. The sensitivity of the proposed biosensor is achieved as 1114 nm/RIU for malaria diagnosis, which can be used to detect malaria at an early stage (Ankita et al. 2022A). Refinement of 1-D PC using a nanocomposite (silver-doped TiO_2 and silver-doped SiO_2) layers introduced on both sides of defect layers leads to augmentation of the structure's tunability, unlocking promising applications in sensors, optical filters, and switches. The study explores how the structure behaves in relation to the refractive index, thickness of the nanocomposite layer, fill factor, and incident angle of the defect layer. Importantly, the improved structure shows greater sensitivity, which is especially noticeable with bigger fill factors and thicker nanocomposite layers. A lower transmission peak results from the trade-off associated with this increased sensitivity. The fact that greatest sensitivity in TM mode occurs at an incidence angle of 83°, which corresponds to visible transmission, is an interesting finding. Silver-doped TiO_2 is clearly the best option for sensing applications in this adjustable 1-D PC, according to this study. This study contributes valuable insights into tailoring the properties of 1-D PC structures for enhanced functionality in various optical devices (Ankita et al. 2024).

2.2 Sucrose Concentration

A 1-D binary photonic crystal (BPC) structure as $(Si/SiO_2)^N$/nanocomposite/defect layer/nanocomposite $/(Si/SiO_2)^N$ is studied as an optical sensor to detect sucrose concentration. Particularly at high incidence angles and significant defect layer thickness, the suggested structure shows high sensitivity and excellent tuning with random changes in sucrose content. The relationship between the concentration and the solution's refractive index is examined. As the breadth of the nanocomposite layer increases, the transmission peak is significantly diminished and the absorption peak is significantly increased. Increasing the angle of incidence for TE polarization significantly increases the reflection peak while significantly decreasing the transmission peak. Brewster's angle makes the situation slightly different for TM polarization. The transmission peak rises as the angle of incidence increases until it reaches 60 degrees, at which point it begins to fall. Increasing the thickness of the defect layer can improve the sensitivity. As the thickness of nanocomposite layer increases, the sensitivity decreases. A nanocomposite layer thickness of 5 nm yields the highest sensitivity. It is found that the sensitivity of the BPC to sucrose concentration can be enhanced with the increase of the incidence angle (Almawgani et al. 2022).

2.3 Haemoglobin Concentration

An optical biosensor based on 1-D defect photonic crystal (DPC) was designed to measure Hb concentration. The defect layer as blood sample introduced a defect transmission mode in PBG region. The simulations are optimized using the defect thickness and the number of layers on either sides of defect layer. At the concentration (0 g/dL), the transmission mode was obtained at 640.2 nm. The wavelength shift was achieved as 15 nm, as Hb concentration change from 0 to 18 g/dL. The parameters of proposed sensor device are achieved as 439 nm/RIU for sensitivity, ~4.91 for quality factor, ~3.29 for FOM and ~ 3.03 ×10-7 RIU detection limit (Dinodiya et al. 2022).

2.4 Tuberculosis

In this study, we investigated three configurations as including two symmetric configurations as $(A/B)^N/D/(B/A)^N$ and $(B/A)^N/D/(A/B)^N$ while a single asymmetric configuration as $(A/B)^N/D/(A/B)^N$ of 1-D DPC. We used TMM approach to investigate the optical characteristic of 1-D periodic structures. The resonant defect mode appeared into PBG. Such defect mode gets shifting with refractive index values of defect, giving a good value of sensitivity for an optical biosensor. The results revealed that the defect layer thickness, the number of unit cells and incident angle gives significant change in sensing parameters such as sensitivity, FOM, and DL. The best sensing parameters are obtained as high sensitivity of 730.23 nm/RIU for asymmetric structure, whereas high FOM of 10.93×106 RIU^{-1} and low DL of 4.58×10^{-7} RIU for sym metric structure (Suthar et al. 2023).

REFERENCES

1. Almawgani, AHM, Suthar, B., Bhargava, A., Taya, S.A., Daher, M.G., Wu, F. (2022) Zeitschrift für Naturforschung A 77 (9), 909–919 (2022)
2. Ankita, Bissa, S., Suthar, B., Bhargava, A. (2022) Materials Today: Proceedings 62 (8), 5407.

3. Ankita, Bissa, S., Suthar, B., Nayak, C., Bhargava, A. (2022A) Results in Optics 9, 100304.
4. Ankita, Suthar, B., Bhargava, A. (2021) Plasmonics 16 (1), 59–63.
5. Ankita, Suthar, B., Bissa, S., Bhargava, A. (2024) Optical and Quantum Electronics 56 (7), 1116.
6. Bhargava, A., Suthar, B. (2009). Chalcogenide Letters 6 (10), 529–533.
7. Bhargava, A., Suthar, B. (2013). AIP Conference Proceedings 1536 (1), 15–18.
8. Dinodiya, S. Bhargava, A. (2021). Advancement in Materials, Manufacturing and Energy Engineering 1, 303–310.
9. Dinodiya, S., Suthar, B., Bhargava, A. (2018). AIP Conference Proceedings 1953, 060016.
10. Dinodiya, S., Suthar, B., Bhargava, A. (2022). Journal of Physics: Conference Series 2335 (1), 012014.
11. Kumar, V., Suthar, B., Kumar, A., Singh, KS., Bhargava, A. (2013). Optik 124 (16), 2527–2530.
12. Suthar, B., Bhargava, A. (2023) Appl Nanosci 13, 5399–5406.
13. Suthar, B., Bhargava, K., Bhargava, A. (2023A) Optical Engineering 62 (5), 057104–057104.
14. Suthar, B., Kumar, V., Kumar, A., Singh, K.S., Bhargava, A. (2012) Progress in Electromagnetics Research Letters 32, 81–90.
15. Suthar, B., Nagar, A.K., Bhargava, A. (2009) Chalcogenide Letters 6 (11), 623–627.

Recent Trends in Applied Physics and Material Science – Sudhir Bhardwaj et al. (eds)

Glycine Lithium Nitrate Crystals Doped with $NaNO_3$ and KNO_3: Morphological Analysis of Crystal Structure and UV Absorption Research

Nimisha S. Agrawal*, Prathmesh R. Vyas
Department of Physics, Sarvajanik University, Surat, India

Ishvar B. Patel
Department of Physics,
Veer Narmad South Gujarat University, Surat, India

Dimple V. Shah
Department of Physics,
Sardar Vallabhbhai National Institute of Technology, Surat, India

Abstract: An excellent optically transparent Glycine single crystal with the presence of A small quantity of sodium nitrate, potassium nitrate, and lithium nitrate were cultivated in a solution by slow evaporation method which have good non-linear optical (NLO) behaviour. The crystalline perfection of crystals that have formed was extremely good with no internal structural grain boundaries. Morphological study of Glycine Lithium Nitrate and doped material were observed by SEM. UV-Vis spectroscopy was used in to determine the band gap of energy of glycine lithium nitrate and doped material.

Keywords: Growth from solution, Slow evaporation, Glycine $LiNO_3$, SEM, UV-Vis spectroscopy

1. INTRODUCTION

Ionic bonding between an organic ligand and an inorganic host creates new hybrid materials with modified mechanical, chemical, and physical characteristics that are appropriate for device creation. Compared to inorganic materials, organic nonlinear materials are desirable because of their high optical threshold for laser power, enormous optical susceptibilities, and intrinsic ultra-fast response times. Materials based on a combination of ionic salts and amino acids have been recognized and investigated in the field of NLO. For any device fabrications in the electronics industry, pure and defect less single crystals, i.e., good quality crystals with perfections are needed. In this research work, some undoped and doped organic and inorganic crystals with NLO have been grown for its possible uses in the domains of laser technology, photonics, both optical computing and optical communication. Because of their possible medical uses, amino acids, glycine and their complexes are biocompatible substances that are being researched extensively. Within the group of amino acids, gamma glycine, a chiral amino acid, is an effective NLO molecule. Several authors have recently reported on the formation and other research on gamma glycine crystals. Single crystals of gamma glycine were observed to have been formed utilizing several kinds of modifications, including potassium chloride, sodium chloride, and others.

*Corresponding author: agrawalnimisha173@gmail.com

DOI: 10.1201/9781003684718-2

This paper investigates the type of crystal structure, morphological study and study of UV-Vis spectra for energy band gap of crystals of Glycine Lithium Nitrate doped with Sodium Nitrate and Potassium Nitrate, that is, crystal shape, morphological study and band gap of crystals. The growth of glycine crystals only has been studied in a few papers; bulk growth has not been covered. Glycine Lithium Nitrate crystals have been gowned and also doped with different doping concentration like 20%,30%,50% and 60%. In this paper Glycine Lithium Nitrate crystal were cultivated using the slow evaporation method. After that, added with doped sample of Sodium Nitrate and Potassium Nitrate with different concentrations. Then after, crystals of different shape and size were harvested. Morphological study of harvested crystals of Glycine Lithium Nitrate and doped material were studied by SEM which 20X to around magnification of 30,000X and a spatial resolution of 50–100 nm. A direct energy band gap of Glycine Lithium Nitrate with doping samples were calculated by using UV-Vis spectrometer. An energy band gap of different concentration wise doped crystals was calculated. Energy band gap of pure sample was found to be 2.71 eV. While energy band gap of doped samples with Sodium Nitrate and Potassium Nitrate was observed decreasing with increasing concentration of doped samples.

2. LITERATURE REVIEW

Further, the study of Semi-organic bis glycine maleate crystal for thermal, functional, and dielectric investigations were conducted by V. J. Priyadharshini and G. Meenakshi (2016). Using the slow evaporation approach, bulk single crystals of BGM were successfully produced from the aqueous solution in this work. The crystal structure of the BGM crystal was found to correspond to the hexagonal crystal system, according to a single crystal XRD investigation.

Glycine, sodium nitrate, undoped, and lithium nitrate-added single crystals: solution growth and investigations (2015). In this study, the XRD tests on powder validate the crystallinity of the developed crystals. The crystals grown study can be considered as promising NLO crystals as they have lower cut-off wavelength between 200 nm to 400 nm i.e. 321 nm and 324nm for pure and 1M% Lithium Nitrate added GSN crystals respectively.

Mechanical analysis of gamma and nonlinear optical glycine by M. Vijayalakshmi, C. Yogambal, D. Rajanbalu and R. Ezhil Vizhi researched a single crystal in the presence of lithium nitrate (2012). Powder X-ray diffraction investigations were used in this study to calculate the crystal lattice parameters, and the results showed good agreement with published values. For optical window applications, grown crystals might be advantageous, according to the UV-Vis investigations.

Lithium bromide added to gamma glycine in a nonlinear optical single crystal: structural, mechanical, optical, thermal, and electrical investigations (2017). Single crystal XRD in this work verified that the produced crystals are part of the hexagonal system. The full visible spectrum's optical transmittance, with a reduced UV cut-off at about 241 nm. SEM analysis shows that the developed crystals had a few inclusions and fissures, which could be caused by the growing conditions.

3. METHODOLOGY AND MODEL SPECIFICATIONS

The process of growing glycine lithium nitrate involves gradual evaporation, crystals by determined ratio of lithium nitrate to glycine was 4.5:1.5 respectively in 10ml double distilled water. A super saturated solution of given material was made and after that solution was filtered with what man filter paper. There was a perforated cover over the solution. The solution was then maintained in an atmosphere free of dust, for the purpose of growth. After 4 to 5 days, tiny and colourless seeds were harvested. After that a 10 ml solution of Sodium Nitrate and Potassium Nitrate in double distilled water was prepared with different concentrations like 20%, 30%, 50% and 60% respectively. Seeds of Glycine Lithium Nitrate were hanged with suitable string and doped in solution of Sodium Nitrate and Potassium Nitrate with concentration of 20%, 30%, 50% and 60% respectively. After 10 to 15 days, doped crystals were harvested with different shape and size. Now, crystals of pure Glycine Lithium Nitrate and doped with Sodium Nitrate and Potassium Nitrate with different concentrations were ready to check for different characterisations.

4. EMPIRICAL RESULTS

4.1 UV-Vis Spectroscopy

Thermo Scientific Evolution 260 B10 was used as a Photo-spectrometer. The central instrument in UV-Vis spectrometer that emits a wide range of UV or visible light through sample and measures the transmitted intensity

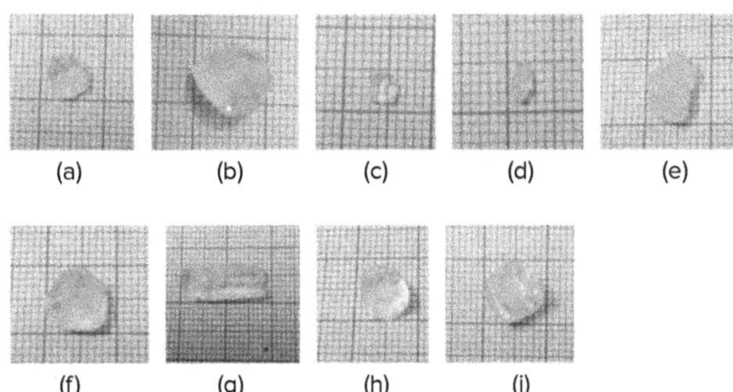

Fig. 2.1 Glycine lithium nitrate (a), Doped with 20% (b), 30% (c), 50% (d), 60% (e) $NaNO_3$, 20% (f), 30% (g), 50% (h) and 60% (i) KNO_3

or absorbed light. Main thing of a UV-Vis spectrometer includes a light source (often a tungsten lamp for UV light and a deuterium lamp for Vis), a monochromator for wavelength selection, a sample cell and a detector.

UV-Vis-NIR spectroscopy could be described as the assessment of radiation emission or absorption associated with changes in the atomic and molecular electron distribution. Optical transmission of the Glycine Lithium Nitrate and doping with Sodium Nitrate and Potassium Nitrate crystals were measured within the range of 200–1100 nm. Plot of absorbance verses wavelength of given sample was analysed and the direct band gap of given material were calculated.

Summary statistics: The summary statistics of Glycine Lithium Nitrate and dopant material energy band gap by UV Vis spectroscopy are presented in the Table 2.1.

Table 2.1 Energy band gap of various concentration of glycine lithium nitrate with doped samples

Doping Concentration	Energy Band gap in eV	
	$NaNO_3$	KNO_3
20%	3.96	4.06
30%	3.69	3.91
50%	3.65	3.70
60%	3.62	3.64

Graphs of Glycine Lithium Nitrate and doped with Sodium Nitrate and Potassium Nitrate with various concentration are shown in Fig. 2.2 and Fig. 2.3. Given Table 2.1 shows that energy band gap of different material using formula of $E_g=1240/\lambda$. Pure sample can't show any band gap. In a

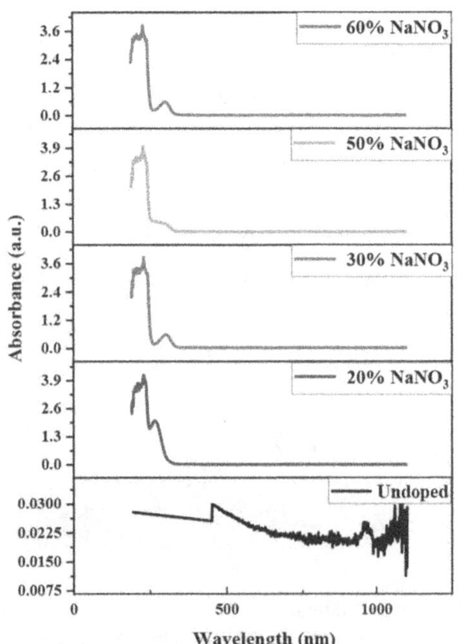

Fig. 2.2 UV-Vis spectroscopy of glycine lithium nitrate doped with sodium nitrate

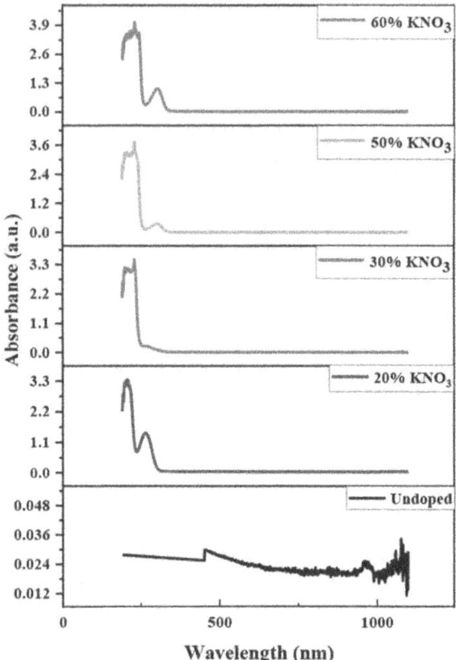

Fig. 2.3 UV-Vis spectroscopy of glycine lithium nitrate doped with potassium nitrate

doped crystal with Sodium Nitrate and Potassium Nitrate, energy gap was observed decreasing with increasing concentration.

4.2 SEM (Scanning Electron Microscopy)

The scanning electron microscope (SEM) creates a spectrum of signals at the surface of solid objects by using a concentrated stream of high-energy electrons. A two-dimensional picture showing spatial differences in these characteristics is created from the resultant signals as a result of electron-sample interactions. The crystalline structure, chemical composition, external appearance (texture), and orientation of the constituent parts are among the details about the sample that are revealed by these signals. A typical SEM technique (magnification range of 20X to about 30,000X, spatial resolution of 50 to 100 nm) can be used to capture areas on camera in scanning mode that are between 1 cm and 5 microns in width. This approach is particularly helpful for qualitative or semi-quantitative evaluations of chemical compositions (using EDS) and crystalline structure and orientations (using EBSD). The SEM can also examine individual point locations on the material.

5. CONCLUSION

At normal temperature, Glycine Lithium nitrate crystals were successfully formed by gradual evaporation. According to UV-Vis spectroscopy, it is clear that whenever the dopant was added in pure material, Energy band gap was found continuously decreasing with increasing of concentration. The morphological structure analysis of the grown crystals was verified by SEM analysis.

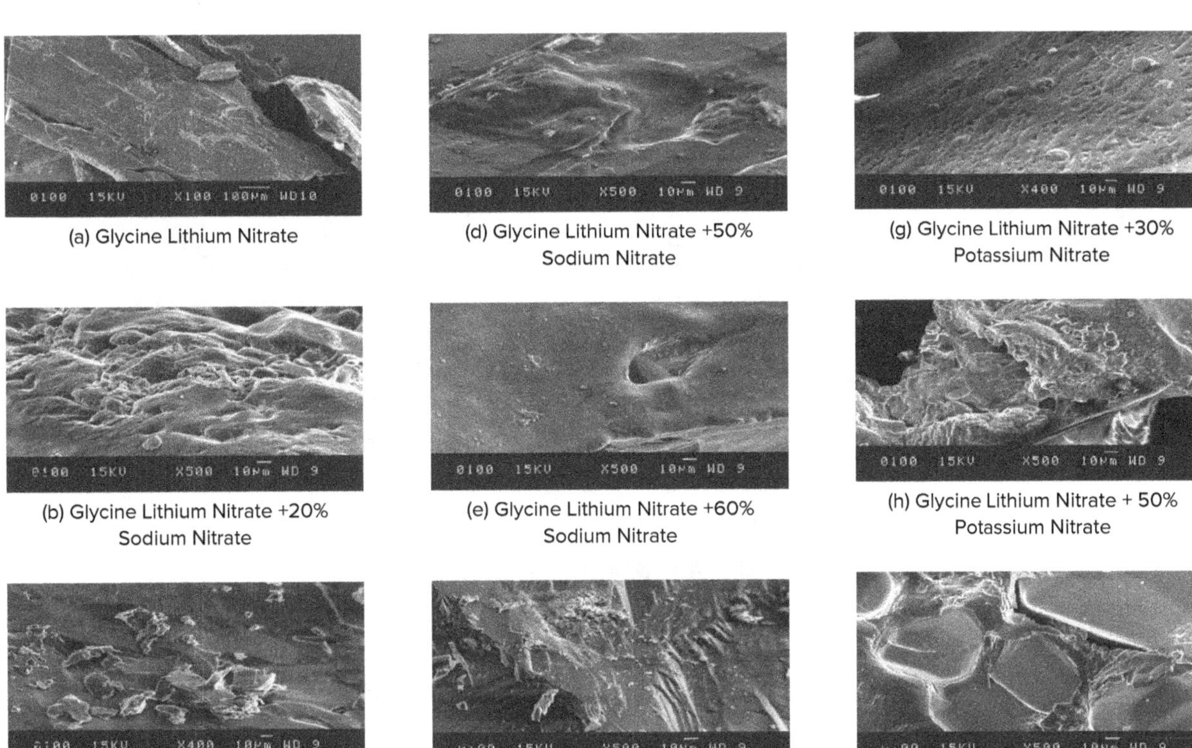

(a) Glycine Lithium Nitrate

(d) Glycine Lithium Nitrate +50% Sodium Nitrate

(g) Glycine Lithium Nitrate +30% Potassium Nitrate

(b) Glycine Lithium Nitrate +20% Sodium Nitrate

(e) Glycine Lithium Nitrate +60% Sodium Nitrate

(h) Glycine Lithium Nitrate + 50% Potassium Nitrate

(c) Glycine Lithium Nitrate +30% Sodium Nitrate

(f) Glycine Lithium Nitrate +20% Potassium Nitrate

(i) Glycine Lithium Nitrate + 60% Potassium Nitrate

Fig. 2.4 SEM analysis of pure glycine lithium nitrate doped with sodium nitrate and potassium nitrate

REFERENCES

1. K. Ambujam., S. Selvakumar, D. Prem Anand, G. Mohammad, P. Sagayaraj, (2006). Cryst. Res. Technol.41:671.
2. K. Srinivasan. J. Arumugam, (2007). Opt. Mater,30:40.
3. J. Thomas Joseph Prakash, S. Kumaraman, (2008). Physica B 403:3883.
4. C. Sekar, R. Paimaladevi, (2009). Spectrochim. Acta A 74:1160.
5. R. Parimaldevi, C. Sekar, (2010). Acta A 76:490.
6. P. Selvarajan, J. Glorium Arul Raj, S. Perumal, (2009). J. Cryst. Growth 311:3835.
7. B. Narayana Moolya, A. Jayarama, M. R. Suresh Kumar, S. M. Dharmaprakash, (2005), J. Cryst. Growth 280: 581.
8. M. Esthaku Peter, P. Ramasamy, (2010). Spectrochim. Acta A 75:1417.
9. Mohd. Shkir, Haider Abbas, (2014). Experimental and Theoretical Studies on bis (Glycine) Lithium Nitrate (BGLiN); A physico-Chemical approach.
10. V. J. Priyadarshini and G. Meenakshi (2016). Investigation on thermal, functional, dielectric studies of semi organic bis glycine maleate crystals.
11. D. Dooslin Mary, M. Mary Freeda and Gerardin Jayam (2015). Solution growth and studies of undoped and lithium nitrate added glycine sodium nitrate single crystal.
12. Ollaa M. Mailouda, Adly H. Elsayedb, A.H. Abo-Elazmb, H.A. Fetouhc (2018). Synthesis and study the structure, optical, thermal and dielectric properties of promising Glycine Copper Nitrate (GCN) single crystals.

Note: All the figures and tables in this chapter were made by the authors.

Recent Trends in Applied Physics and Material Science – Sudhir Bhardwaj et al. (eds)
© 2026 Taylor & Francis Group, London, ISBN 978-1-041-16452-4

Effective Shielding: A Study of Shielding Effectiveness of PANI/CuO Composites in the UV Region of Electromagnetic Spectrum

S.V. Tayade[1], A.P. Bangar[2], S.A. Waghuley[3]
Department of Physics Sant Gadge Baba Amravati University,
Amravati, Maharashtra, India

Abstract: Electromagnetic Interference (EMI) shielding is a critical technique designed to safeguard electronic devices from disruptive electromagnetic fields (EMFs). This paper examines the health implications of Electromagnetic field exposure, which has been linked to various adverse effects, including headaches, sleep disturbances, and potential cancer risks. Common sources of Electromagnetic fields include mobile phones, Wi-Fi routers, and household appliances. EMI shielding reduces exposure by blocking or attenuating these fields through mechanisms such as absorption, reflection, and rerouting, with effectiveness influenced by material. By mainly focusing on shielding against ultraviolet (UV) light within the electromagnetic spectrum. This paper discussed the importance of understanding both the beneficial uses and potential health risks of UV radiation. Further the analysis of shielding effectiveness (SE) through the absorption value is discussed. Through the SE value in decibel for the polyaniline (PANI)-based copper oxide (CuO) composites with varying weight percentages of CuO, incremented by 5%. it is concluded that absorption mechanism predominates, highlighted the potential of PANI/CuO nanocomposite as lightweight materials for EMI shielding applications. The total electromagnetic interference shielding effectiveness (SE) of PANI/CuO was observed to be comparatively improved for the10wt%(S2), 15wt%(S3), 20wt%(S4). The synthesized nanocomposite shows a shielding effectiveness (SE) that exceeded the minimum threshold necessary for adequate electromagnetic interference (EMI) protection.

Keywords: Electromagnetic interference shielding, UV radiation, Shielding effectiveness

1. INTRODUCTION

Electromagnetic Interference (EMI) Shielding: EMI (Electromagnetic Interference) shielding is a technique used to protect electronic devices from electromagnetic interference that can disrupt their performance (Kumar, P. et al. 2019). EMI shielding for human health focuses on reducing exposure to electromagnetic fields (EMFs) that can potentially have adverse effects. Prolonged exposure to certain levels of EMFs has been linked with various health issues, including headaches, sleep disturbances, and, in some studies, potential links to cancer (Chen, Y. et al. 2021). Common sources include mobile phones, Wi-Fi routers, power lines, and household appliances (Nazir, A. et al. 2022). Shielding can help mitigate exposure from these sources.EMI shielding involves blocking or reducing electromagnetic fields from penetrating or transmitting. Shielding works through absorption, reflection, and rerouting of electromagnetic waves (Ganguly, S. et al. 2018). The schematic representation of Electromagnetic Interference shielding mechanism is shown in Fig. 3.1.

Effective shielding materials for health protection include conductive fabrics, metals (like aluminium and copper), and specialized coatings that can block or absorb Electromagnetic field. EMI shielding is critical in various industries, including telecommunications, aerospace,

[1]surajtayade14896@gmail.com, [2]akshaybangar0101@gmail.com, [3]sandeepwaghuley@sgbau.ac.in

DOI: 10.1201/9781003684718-3

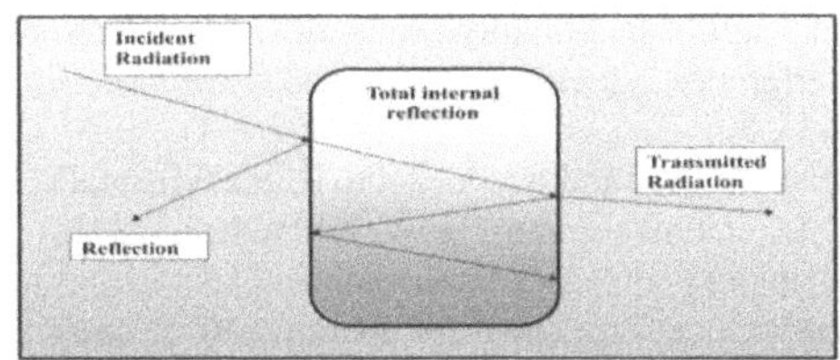

Fig. 3.1 Schematic diagram of the mechanism of Electromagnetic Interference shielding

automotive, and medical devices, where reliability is paramount. In this paper we are focusing on the shielding of UV lights from the electromagnetic spectrum which has wide application with respect to the human health, the effects of UV radiation in the electromagnetic spectrum shown in Fig. 3.2.

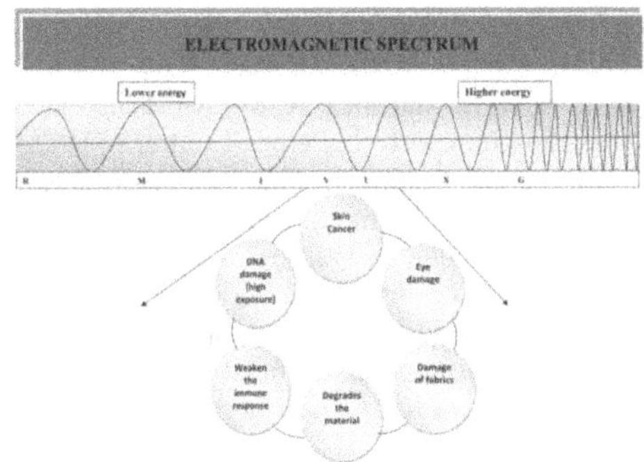

Fig. 3.2 Effects of UV radiation in the spectrum of electromagnetic spectrum

Ultraviolet (UV) Radiation: Gadgets emitting UV radiation include UV lamps, sanitizers, UV-enabled electronics, portable UV-C devices, and UV LEDs, commonly used for sterilization, lighting, and industrial processes. UV exposure poses health risks such as skin damage, eye injuries, and immune suppression, especially from UV-C and UV-B radiation. Increased exposure to UV radiation can give different health issues in humans, including skin cancer, cataracts and immune system damage. Ultraviolet (UV) light can pose several health risks to humans like skin damage, skin cancer, Eye damage, immune system suppression, photosensitivity reaction, DNA damage particularly with excessive exposure (Furukawa. et al. 2021). Excessive UV exposure can weaken the immune response, making the body less effective at fighting off certain infections and diseases.

2. METHODS

Polyaniline was synthesized by dissolving 10 ml of aniline in 150 ml of HCl and stirring the solution. Separately, 10 g of ammonium persulfate was dissolved in 50 ml of double-distilled water and added dropwise to the aniline solution over 5 hours, forming a diamond-green solution. The reaction was conducted at 25–30°C, followed by filtration and washing of the product with double-distilled water. The final polyaniline powder was obtained by drying in a vacuum oven at 40°C. CuO nanocomposite with polyaniline (PANI) were prepared using an ex situ method by varying CuO content in PANI at intervals of 5 wt% (5 wt% to 25 wt%). The mixtures were combined in acetone as the organic medium.

3. RESULT AND DISCUSSION

Figure 3.3 presents the UV-VIS spectra of PANI matrices loaded with varying CuO content (5–25 wt%, at 5 wt% intervals) recorded in the 200–600 nm range. The absorption coefficient is nearly linear between 400–600 nm, while broad peaks in the 200–225 nm range indicate π–π* transitions in the polymer matrix. (A. Kaushik et al. 2008) And more absorption in the range in the 230nm to 400 nm. With the standard result it is confirms that peaks of PANI and CuO shifts minutely to lower energy (Jundale, D et al. 2013)

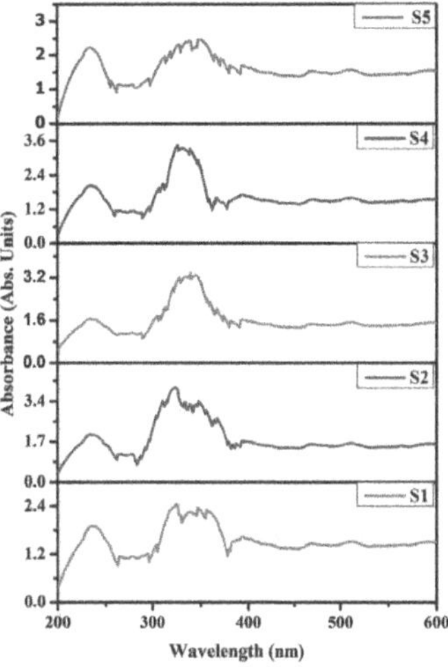

Fig. 3.3 UV–VIS spectrum of different wt% of CuO loaded PANI

Electromagnetic Interference Shielding Effectiveness: Electromagnetic interference shielding effectiveness (SE) refers to the attenuation of electromagnetic (EM) waves achieved through the use of shielding materials. Absorption serves as the primary mechanism, while reflection acts as a secondary mechanism (Lyu, L. et al. 2018). multiple internal reflections (SEM) mechanism can be neglected if shielding effectiveness due to absorption(SEA) is greater than 10 dB (Modak, P. et al. 2021). Measure Absorbance(A): This is related to how much UV light is absorbed by the nanomaterial. It's often calculated as equation 1. (A. Kaushik et al. 2008).

$$A = -\log (T) \quad (1)$$

Higher absorbance values indicate better shielding effectiveness. To express shielding effectiveness as a percentage, it's usually directly represented by the absorbance value itself describe in equation 2:

$$\text{Shielding Effectiveness} \approx A \quad (2)$$

(when expressed as a percentage)

To express shielding effectiveness in decibels (dB), need to convert the absorbance or transmittance into a decibel scale. Shielding effectiveness in dB quantifies how well a material blocks UV light, and it is commonly used in various fields to describe attenuation. It is given by equation 3.

$$\text{Shielding Effectiveness SE (dB)} = 10 \times \text{A (dB)} \quad (3)$$

Figure 3.4 shows the shielding effectiveness (SE) of PANI/CuO nanocomposite measured in the frequency range of 789.4–882.3 THz. SE increases linearly between 790–833.3 THz, with samples containing 10 wt% (S2), 15 wt% (S3), and 20 wt% (S4) showing significant changes beyond 833.3 THz. The total SE, dominated by absorption phenomenon.

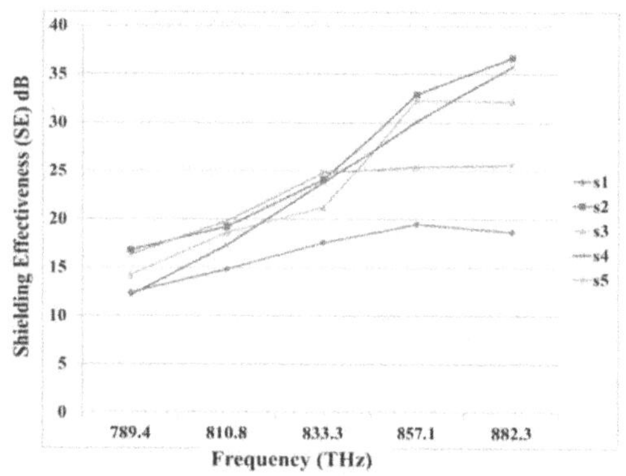

Fig. 3.4 Electromagnetic shielding effectiveness

4. CONCLUSIONS

The results indicated that the absorption mechanism predominates, highlighting the potential of PANI/CuO nanocomposite as lightweight materials for EMI shielding applications.

REFERENCES

1. Chen, Y., Li, J., Li, T., Zhang, L., & Meng, F. (2021). Recent advances in graphene-based films for electromagnetic interference shielding: Review and future prospects. *Carbon, 180*, 163–184.
2. J., Ding, T., Wang, Z., ... &Guo, Z. (2019). Electromagnetic interference shielding polymers and nanocomposites-a review. *Polymer Reviews, 59*(2), 280–337.
3. Nazir, A. (2022). A review of polyvinylidene fluoride (PVDF), polyurethane (PU), and polyaniline (PANI) composites-based materials for electromagnetic interference shielding. *Journal of Thermoplastic Composite Materials, 35*(10), 1790–1810.
4. Cheng, J., Li, C., Xiong, Y., Zhang, H., Raza, H., Ullah, S., ... &Che, R. (2022). Recent advances in design strategies and multifunctionality of flexible electromagnetic interference shielding materials. *Nano-Micro Letters, 14*(1), 80.
5. Omana, L., Chandran, A., John, R. E., Wilson, R., George, K. C., Unnikrishnan, N. V & Paul, I. (2022). Recent Advances in Polymer Nanocomposites for Electromagnetic Interference Shielding: A Review. *ACS omega, 7*(30), 25921–25947.
6. Kumar, P., Narayan Maiti, U., Sikdar, A., Kumar Das, T., Kumar, A., &Sudarsan, V. (2019). Recent advances in polymer and polymer composites for electromagnetic interference shielding: review and future prospects. *Polymer Reviews, 59*(4), 687–738.
7. .Liang, C., Gu, Z., Zhang, Y., Ma, Z., Qiu, H., &Gu, J. (2021). Structural design strategies of polymer matrix composites for electromagnetic interference shielding: a review. *Nano-Micro Letters, 13*(1), 181.
8. Ganguly, S., Bhawal, P., Ravindren, R., & Das, N. C. (2018). Polymer nanocomposites for electromagnetic interference shielding: a review. *Journal of Nanoscience and Nanotechnology, 18*(11), 7641–7669.
9. Furukawa, J. Y., Martinez, R. M., Morocho- Jácome , A. L., Castillo-Gómez, T. S., Pereda-Contreras, V. J., Rosado, C., ... & Baby, A. R. (2021). Skin impacts from exposure to ultraviolet, visible, infrared, and artificial lights–a review. *Journal of Cosmetic and Laser Therapy, 23*(1-2), 1–7.
10. Romanhole, R. C., Ataide, J. A., Moriel, P., & Mazzola, P. G. (2015). Update on ultraviolet A and B radiation generated by the sun and artificial lamps and their effects on skin. *International journal of cosmetic science, 37*(4), 366–370.
11. Modak, P., &Nandanwar, D. (2021, May). Enhanced electromagnetic interference shielding effectiveness of carbon-based conducting polymer nanocomposites. In *Journal of Physics: Conference Series* (Vol. 1913, No. 1, p. 012054). IOP Publishing.
12. Jiang, D., Murugadoss, V., Wang, Y., Lin, J., Ding, T., Wang, Z. & Guo, Z. (2019). Electromagnetic interference shielding polymers and nanocomposites-a review. *Polymer Reviews, 59*(2), 280–337.
13. Lyu, L., Liu, J., Liu, H., Liu, C., Lu, Y., Sun, K., ... &Wujcik, E. K. (2018). An overview of electrically conductive polymer nanocomposites toward electromagnetic interference shielding. *Engineered Science, 2*(59), 26–42.
14. Tayade S.V, Bangar A.P., Waghuley S.A.,(2024). “Evaluating Polyaniline-CuO composite for next generation Electromagnetic interference shielding.”. ICMPE Conference Proceeding.
15. Jundale, D. M., Navale, S. T., Khuspe, G. D., Dalavi, D. S., Patil, P. S., & Patil, V. B. (2013). Polyaniline–CuO hybrid nanocomposites: synthesis, structural, morphological, optical and electrical transport studies. *Journal of Materials Science: Materials in Electronics, 24*, 3526–3535.
16. A. Kaushik, J. Kumar, M.K. Tiwari, R. Khan, B.D. Malhotra, V. Gupta, S.P. Singh, J. Nanosci. Nanotechnol. 8, 1757 (2008)

Note: All the figures in this chapter were made by the authors.

Recent Trends in Applied Physics and Material Science – Sudhir Bhardwaj et al. (eds)

4 Slurry Coating Production of Binary Zinc Chalcogenide Film Samples for Optoelectronic Usage

Vipin Kumar*
Department of Applied Sciences, KIET Group of Institutions (Delhi NCR, Ghaziabad) India

Dhirendra Kumar Sharma, Kapil Kumar Sharma, Akansha Agrwal
Department of Applied Sciences, KIET Group of Institutions (Delhi NCR, Ghaziabad) India

Parvin Kumar
Department of Electronics and Communication Engineering, KIET Group of Institutions (Delhi NCR, Ghaziabad) India

Dilip Kumar Dwivedi
Photonics and Photovoltaic Research Lab, Department of Physics and Material Science, Madam Mohan Malaviya University of Technology, (Gorakhpur) India

Nagendra Prasad Yadav
School of Electrical and Electronics Information Engineering, Hubei Polytechnic University, Huangshi, (Hubei) China

Abstract: Over the past 50 years, the II-VI group zinc based binary chalcogenides has been considered as noteworthy materials for diverse optoelectronic usage. This article presents the production of binary zinc chalcogenide film samples via slurry coating deposition process. The deposited samples were examined for their optoelectronic attributes. X-ray diffraction (XRD) plots establish the polycrystalline character for the samples. The mode of reflection of UV-Vis spectroscopy was employed to determine the "band gap (Eg)" of the samples. Usual approach of two probes was employed to assess the DC electrical conductivity of the samples.

Keywords: Slurry-coating, XRD, Zinc chalcogenide, Conductivity

1. INTRODUCTION

The production of "II-VI" group compound semiconductors has captivated significant awareness to their unique features and extensive scope of utilization in optoelectronic tools. Among these semiconductors, "zinc chalcogenides" are of great interest due to their prospective usages from visible to UV-region (Hsu 1998). The major reason of their fame is high absorption coefficient and direct nature of band gap (Kumar et.al. 2021).

The lower cost of production, simplicity in manufacturing and superior performance in diverse conditions is the origin of practical importance of polycrystalline devices. A large number of film deposition processes such as "electrodeposition" (Gromboni et al. 2016), "thermal-evaporation" (Priya et al. 2017), "sputter-deposited" (Rakshani 2013), "chemical bath deposition (CBD)" (Sagadevan et al. 2017), "spray pyrolysis" (Al-Diabat et al. 2019) and "screen-printing" (Kumar et al. 2012) were described to yield binary "zinc chalcogenide" semiconducting films. Commonly, across all these processes, uniform, polycrystalline and stable films are yielded.

In this article, binary "zinc chalcogenide" (ZnS, ZnSe and ZnTe) film samples were fabricated on washed glass substrates applying slurry-coating deposition process. This process is comparatively cheaper and felicitous for large area deposition. The deposited film samples were portrayed for optoelectronic attributes to scrutinize the usage of film samples towards optoelectronic tools.

*Corresponding author: vipinkumar28@yahoo.co.in

DOI: 10.1201/9781003684718-4

2. EXPERIMENTAL

Regarding the fabrication of binary zinc chalcogenide films, AR grade commercially accessible powders of these "zinc chalcogenides" (ZnS, ZnSe and ZnTe) were used as the starting materials. Slurry containing either of these powders (ZnS or ZnSe or ZnTe) along with (10% wt.) of zinc chloride (adhesive) adequately mixed with ethylene glycol (binder) via an agate mortar & pestle. The acquired slurry was used for fabricating the films on the clean glass slides via a silk screen. The fabricated film samples were initially dried via the heating plate (at 120^0C for 240 minutes) and finally fired (at 450-500^0C for 10 minutes) in a furnace.

The XRD plots were used to characterize the films for structural attributes, whereas reflectance mode of UV-Vis spectroscopy was employed to determine the "band gap" (E_g). Usual "two probe" system was applied to undertake the "electrical conductivity" (DC) measurement of the film samples.

3. RESULTS AND DISCUSSION

The room temperature XRD plots of samples are depicted in "Fig. 4.1". Polycrystalline behaviour of the samples is authenticated from the presence of intense XRD peaks. Creation of well crystallized samples is confirmed from the acute and strong peaks. Further characteristics on these XRD plots of the samples can be referred elsewhere (Kumar et. al. 1998).

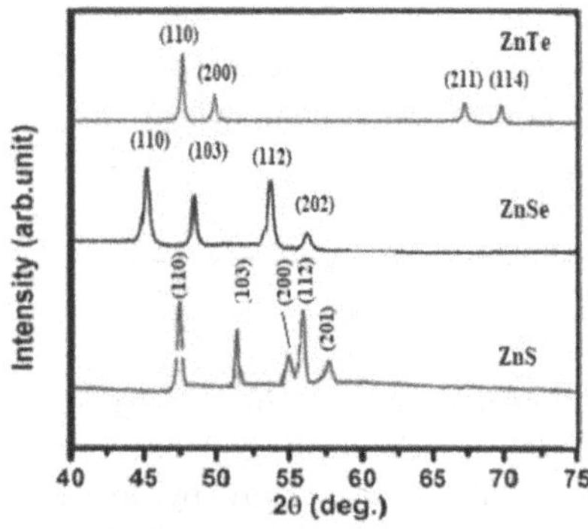

Fig. 4.1 XRD plots of binary "zinc chalcogenide" samples

It is well established that all II-VI group binary zinc chalcogenide films shows direct type transition of "band gap" (Kumar et al. 2017). To assess the "band gap" of the samples, the reflection curves were recorded. To assess the "band gap (E_g)" from the reflection spectra one must create a graph between $(\alpha h\nu)^2$ or $[h\nu \ln (R_{max} - R_{min})/(R\text{-}R_{min})]^2$ on the y-axis against ($h\nu$) on the x-axis (Kumar et.al. 2017) as presented in "Fig. 4.2".The "band gap (E_g)" is determined by extending the straight section of this plot (Fig. 4.2). The band gap (E_g) for the produced binary zinc chalcogenide films alters from 2.26 eV to 3.50 eV. Such a broad range of band gap (E_g) designates

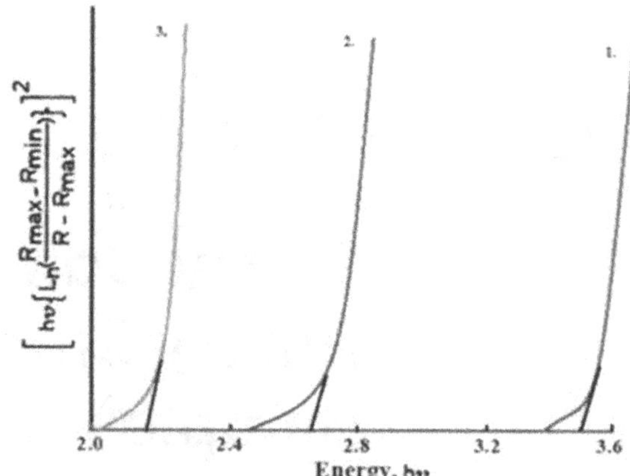

Fig. 4.2 Plot between $[h\nu \ln (R_{max} - R_{min}) / (R - R_{min})]^2$ against ($h\nu$) for binary "zinc chalcogenide" samples

the appropriateness of these binary films for diverse optoelectronic usage.

The examination of electrical transport properties in these binary zinc chalcogenide films is crucial for determining the potential utility of the material being studied. For optoelectronic usage, key characteristics encompass electrical conductivity. The direct current electrical conductivities of the produced films were assessed within the temperature range of 300 to 400 K using a two-probe approach. A direct current voltage was introduced to the sample, and the developed current was analysed using an electrometer (Keithley). A thermocouple made of (copper and constantan) was employed to assess the temperature (T) of the samples. The assessment of conductivity was conducted by applying a thin layer onto a glass substrate, which had been pre-coated with indium electrodes featuring a narrow gap. The graph depicting the relationship between the "log σ_{DC}" and "1000 / T" for the produced samples and is illustrated in "Fig. 4.3".

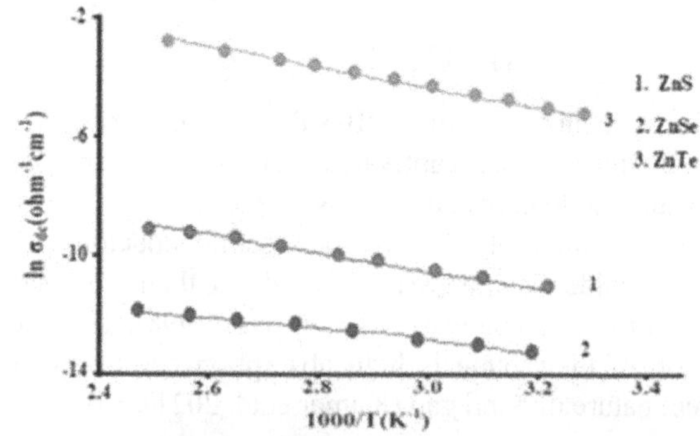

Fig. 4.3 Graph of "log σ_{DC}' against "1000 / T" for binary "zinc chalcogenide"samples

It was noted that the conductivity rises as the temperature increases, indicating the semiconducting characteristics of the samples. The observed rise in conductivity is attributed to the enlargement of crystallite-size and rise in carrier-density (Kumar et.al. 2017).The linear relationship observed in the curve of "log σ_{DC}" against "1000 / T" indicates that the conductivity (σ) of these samples conforms the "Arrhenius formulation", suggesting that the conduction attributed to the thermal stimulation of charge-carriers. The values of "activation energies" for the produced samples were determined by analysing the slope of the graph plotting "log σ_{DC}" against "1000/T" and are presented in Table 4.1.

Table 4.1 Determined values of "electrical Conductivity" and "activation energy" of binary "zinc chalcogenide"

Composition	Electrical Conductivity ($\Omega^{-1}cm^{-1}$)	Activation Energy (eV)
ZnS	2.70×10^{-5}	0.80
ZnSe	0.41×10^{-6}	0.35
ZnTe	1×10^{-3}	0.21

4. CONCLUSION

In Summary, binary zinc chalcogenide (ZnS, ZnSe and ZnTe) were produced utilizing an economical and uncomplicated slurry-coating approach. The produced samples exhibited polycrystalline characters, as verified by XRD plots. The "band gap" of the fabricated samples varies between 2.26 eV to 3.50 eV. The direct current (DC) "electrical conductivity of these binary zinc chalcogenide film samples aligns closely with the Arrhenius formulation. The findings acquired signify that these slurry-prepared samples are applicable in multiple optoelectronic-usages

REFERENCES

1. Hsu, C. T. (1998). Epitaxial growth of II-VI compounds semiconductors by atomic layer epitaxy. Thin Solid Films, 335(1-2):284–291.
2. Kumar, V. and Masih, V. G. (2021). Study on optoelectronic properties of slurry coated binary cadmium chalcogenide films. In Muzammil, M., Chandra, A., Kankar, P. K., Kumar, H. (eds) Recent advances in mechanical engineering. Lecture Notes in Mechanical Engineering. Springer, Singapore, 135–139.
3. Grombani, M. F. and Mascaro, L. H. (2016). Optical and structural study of electrodeposited zinc selenide thin films. J. Electroanal. Chem. 780:360–366.
4. Priya, K. Ashith, V. K. Rao, G. K. and Sanjeev, G. (2017). A comparative study of structural, optical and electrical properties of ZnS thin films obtained by thermal evaporation and SILAR techniques. Ceram. Int. 43(13):10487–10493.
5. Rakshani, A. E. (2013). Effect of growth temperature, thermal annealing and nitrogen doping on optoelectronic properties of sputter-deposited ZnTe films. Thin Solid Films, 536:88–93.
6. Sagadevan, S. and Das, I. (2017). Chemical bath deposition (CBD) of zinc selenide (ZnSe) thin films and characterization. Aust. J. Mech. Eng. 15 (3): 222–227.
7. Al-Diabat, A. M. Ahmed, N. M. Hasim, M. R. and Almessiere, M.A. (2019). Growth of ZnS thin films using chemical spray pyrolysis technique. Mater. Today: Proc., 17 (part 3):912–920.
8. Kumar, V. Kumar, V. and Dwivedi, D. K. (2012). Growth and characterization of zinc telluride thin films for photovoltaic applications. Phys. Scr. 86: 015604.
9. Kumar, V. and Sharma, T. P. (1998). Structural and optical properties of sintered ZnS_xSe_{1-x} films, Opt. Mater.10 (4):253–256.
10. Kumar, V. Agarwal, S. and Dwivedi, D. K. (2017). Study on optical investigations and DC conduction mechanism in polycrystalline chalcogenide (Cd, Zn) semiconductor films grown by screen-printing method. J. Mater Sci.: Mater. Electron. 28:1715–1719

Note: All the figures and tables in this chapter were made by the authors.

Recent Trends in Applied Physics and Material Science – Sudhir Bhardwaj et al. (eds)

First-Principle Approach to Understand the Electronic Structure of Actinide Nitride (AmN and CmN)

Namrata Yaduvanshi* and T. Uthayakumar
Department of Physics,
Dayananda Sagar college of engineering, Kumaraswamy Layout,
Bangalore, India

Abstract: The Actinide pnictides containing a large-radius ion are particularly significant for studying the impact of the structural properties of compounds by p-f mixing. Actinide-pnictide compounds have several potential applications due to their unique properties. This paper examines the Actinide Nitride (AmN and CmN) their phase transition under high pressure using the NaCl crystal structure. We utilize the improved interaction potential model (IIPM) approach to conduct a theoretical analysis to explore the structural characteristics of these compounds. Our results demonstrate that that NaCl (B1) undergoes a structural phase transition to CsCl (B2). Our IIPM approach yielded phase transition pressures and volume collapse that exhibit remarkable concurrence with the prevailing theoretical data.

Keywords: Rare-earth pnictides, Phase transition, Volume collapse

1. INTRODUCTION

A series of rare-earth nitrides crystallize in rock-salt structures. Transition-metal nitrides have been extensively studied, but rare-earth nitrides have received considerably less attention. Materials of this type are hard and brittle, stable but susceptible to oxidation and hydrolysis. A variety of properties of these compounds are currently being studied, including their magnetic, electronic, and optical properties (Duan et al. 2007, Larson et al. 2007, Takahasi et al. 1985, Yaduvanshi et al. 2016). The first-order phase transition between the NaCl (B1) and CsCl (B2) states has been demonstrated in many theoretical studies on CeN structural properties (Kanchana et al. 2011, Schram et al. 1997, Pataiya et al. 2016). Using the tight binding linear muffin-tin orbital (TB-LMTO) method within the local density approximation (LDA), the electronic and ground states of americium pnictides (AmY) have been calculated by Murugan et al. (2016). The study investigates the electronic, structural, mechanical, and magnetic properties of actinide nitrides (AnN) in three cubic phases: NaCl (B1), CsCl (B2), and zinc blende (B3). Under normal pressure, UN (uranium nitride) is a ferromagnetic compound, while both NaCl (B1) and CsCl (B2) are also found to be ferromagnetic. These nitrides are predicted to undergo a structural phase transition from the B1 to the B3 phase when subjected to pressure. Additionally, their electronic structure indicates a metallic character. (Zhang et al. 2017).

This paper investigates the high-pressure phase transition of Actinide Nitride (AmN and CmN), as well as its elastic properties. High pressure behavior was studied using our improved interaction potential model (IIPM). Based on Hafemeister and Flygare's (1965) approach, the present model includes a long range Columbic, a three body interaction, a short range overlap repulsive interaction, and an electronic polarizability effect.

2. PROPOSED MODEL AND CALCULATION METHOD

The Gibbs free energy (G = U + PV – TS) incorporates the effects of pressure, volume, temperature, and entropy, making it essential for modelling crystal phase transitions.

*Corresponding author: namrata123yaduvanshi@gmail.com

DOI: 10.1201/9781003684718-5

At absolute temperature T, the internal energy U denotes the lattice energy at 0 K, while the vibrational entropy S is reflected as a function of pressure P. At T = 0 K, the Gibbs free energy is equivalent to enthalpy, and it can be expressed for the NaCl (B1) and distorted CsCl (BCT) structures as follows:

$$G_{B1}(r) = U_{B1}(r) + PV_{B1} \quad (1)$$

$$G_{BCT}(r') = U_{BCT}(r') + PV_{BCT} \quad (2)$$

The lattice energies for sodium chloride (NaCl) and the distorted structures of cesium chloride (CsCl), which are represented by the initial terms in equations (1) and (2) respectively, can be articulated as a function of the unit cell volumes associated with each phase, denoted as V_{B1} and V_{BCT}.

$$U_{B1}(r) = [-(\alpha_m z^2 e^2)/r] - [(12\alpha_m z\, e^2 f(r))/r] - (e^2\alpha e)/r^4 + 6b\beta ij \exp[(ri + rj - r)/\rho] + 6b\, \beta ii \exp[(2ri - 1.41r)/\rho + 6b\beta jj \exp[(2rj - 1.41r)/\rho] \quad (3)$$

$$U_{BCT}(r') = [-(\alpha'_m z^2 e^2)/r')] - [(16\alpha'_m ze^2 f(r'))/r'] - (e^2\alpha e)/r'^4 + 8b\beta ij \exp[(ri + rj - r')/\rho] + 2b\beta ii \exp[(2ri - r')/\rho] + b\beta ii \exp[(2ri - r'')/\rho] + 2b\, \beta jj \exp[(2rj - r')/\rho] + b\beta jj \exp[(2rj - r'')/\rho] \quad (4)$$

The ionic charge is denoted by the symbol ze, where ri (rj) represents the ionic radii of the respective i(j) ions. The range parameter is indicated by ρ, while the hardness parameter is represented by b. The parameter associated with three-body interactions, denoted as f(r), is defined in relation to the nearest neighbor separation within the NaCl phase. In the tetragonal structure, the nearest neighbor distances corresponding to the lattice parameters a and c are denoted by r' and r", respectively.

3. RESULTS AND DISCUSSION

This study uses measured values of the equilibrium lattice constant (r_0) and bulk modulus (B_t) with K=2 for the B1 phase to derive model parameters presented in Table 5.1. A single set of parameters [ρ, b, f(r)] applies to both the B1 and BCT phases due to atomic rearrangement during the transition. Inter-ionic separations are determined through a minimization technique, adjusting f(r) based on the inter-ionic separation (r). This allows for the calculation of separations r and r' for the B1 and BCT phases, respectively.

Table 5.1 Input and model parameters

Compound	Input Parameter		Model Parameter		
	r_0	B_T	b (10^{-12} ergs)	ρ(Å)	f(r)
AmN	2.49[a]	189	5.32	0.524	0.101
CmN	2.51[a]	154[b]	5.02	0.453	0.043

a. (Eller et al. 1986), b. (Petit et al. 2009)

Table 5.2 Calculated transition pressure and volume collapse

Compounds	Transition	Phase Transition Pressure (GPa)	Volume Collapse (%)
AmN	B1 –» BCT	46.5	7.0
CmN		101	8.2

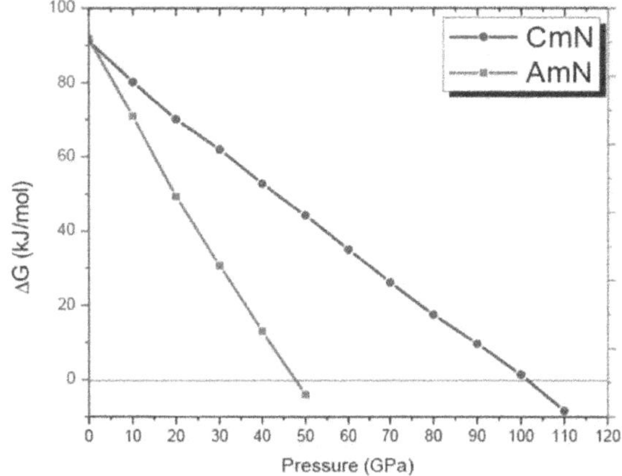

Fig. 5.1 Variation of change in gibbs free energy with pressure for AmN and CmN

We assessed the Gibbs free energies, $G_{B1}(r)$ and $G_{BCT}(r')$, finding that the phase transition pressure (Pt) occurs where ΔG approaches zero, indicating phase coexistence. The transition from B1 to BCT for AmN and CmN occurs around 37 GPa, with a 12% volume collapse, as shown in Figs. 5.1 and 5.2 and Table 5.2.

REFERENCES

1. C. G. Duan, R.F. Sabirianov, W.N. Mei, P.A. Dowben, S.S. Jaswal, E.Y. Tsymbal, (2007). Electronic, magnetic and transport properties of rare-earth monopnictides Journal of Physics: Condensed Matter 19: 315220.
2. P. Larson, W.R.L. Lambrecht, A. Chantis, M. van Schilfgaarde (2007). Electronic structure of rare-earth nitrides using the LSDA+*U* approach: Importance of allowing 4*f* orbitals to break the cubic crystal symmetry Physical Review B 75: 045114.
3. H. Takahasi, T. Kasuya (1985). J.Phys. C 18, 2695. M. De, S.K. De(1999). Electronic structure and optical properties of neodymium monopnictides. J. Phys. Chem. Solids 60: 337–346.
4. N. Yaduvanshi, S. Singh (2016). Pressure induced phase transition and elastic properties of cerium mono-nitride (CeN). AIP Conf. Proc. 1731: 030003.
5. V. Kanchana, G. Vaitheeswaran, X. X. Zhang, Y. M. Ma, A. Svane, and O. Eriksson (2011). Lattice dynamics and elastic properties of the 4[x]*f* electron system: CeN, Phys. Rev. B 84: 205135.
6. R. P. C. Schram, J. G. Boshoven, E. H. P. Cordfunke, R. J. M. Konings, and R. R. van der Laan (1997). Enthalpy increment measurements of cerium mononitride, CeN, J. Alloys Compd. 252: 20.

7. J. Pataiya, M. Aynyas, C. Makode, S. P. Sanyal (2016). Ground State and Electronic Properties of Americium (Am) Compounds. Advanced Materials Research 1141:176–179
8. A. Murugan, G. Sudha Priyanga, R. Rajeswarapalanichamy, M. Santhosh, K. Iyakutti (2016). First principles study of structural, electronic, mechanical and magnetic properties of actinide nitrides AnN (An = U, Np and Pu). Journal of Nuclear Materials, 478: 197–206.
9. Yu-Juan Zhang; Zhang-Jian Zhou; Jian-Hui Lan; Tao Bo; Chang-Chun Ge; Zhi-Fang Chai; Wei-Qun Shi J. (2017). Theoretical investigation on electronic and mechanical properties of ternary actinide (U, Np, Pu) nitrides. Appl. Phys. 122: 115109.
10. W. Hafemeister, W. H. Flygare (1965). Calculation and Interpretation of the 129I Isomer Shifts in the Alkali Iodide Lattices, Journal of Chemical Physics. 43: 795.
11. P.G. Eller, R.A. Penneman, in: J.J. Katz, G.T. Seaborg, L.R. Morss (Eds.) (1986), The Chemistry of the Actinide Elements, Chapman and Hall, NY 1: 962.
12. L. Petit, A. Svane, Z. Szotek, W.M. Temmerman, G.M. Stocks (2009), Ground-state electronic structure of actinide monocarbides and mononitrides. Phys. Rev. B 80: 045124–045131.

Note: All the figures and tables in this chapter were made by the authors.

Recent Trends in Applied Physics and Material Science – Sudhir Bhardwaj et al. (eds)
© 2026 Taylor & Francis Group, London, ISBN 978-1-041-16452-4

Investigate Photoluminescence Properties of Cerium Lanthanide Ions Doped into Cesium Cupric Chloride Nanocrystals

Harsha Sonawane[1], Ashok Sunatkari[2]
Siddharth College of Arts, Science & Commerce, Fort, Mumbai

Swarnalata Sunatkari[3]
N.G. Acharya & D.K. Marathe College, Chembur, Mumbai

Shital Sonawane[4]
K.J. Somaiya College of Science & Commerce, Vidya Vihar, Mumbai

Abstract: Researchers have focused a lot of attention on CsCuCl3 perovskite because of its great stability and environmental friendliness. However, the process of capturing a charge carrier is primarily controlled by its self-inherent trapping and surface imperfections brought about by strong excitation-phonon coupling. Therefore, the quantum efficiency of the photoluminescence is significantly decreased, making it impossible to detect the photoluminescent signal and restricting research into its optical characteristics. This work presents the Cerium elements, which were effectively synthesized by the solvent-based thermal synthesis approach, to passivate internal and surface flaws in order to address this issue. This work results in a 52% rise in Cerium doped CsCuCl3 NCs PLQY, which sets a standard for the fabrication of lead-free perovskite optoelectronic devices with good optical characteristics.

Keywords: Copper-based perovskite, Cerium doped, Photoluminescence

1. INTRODUCTION

Cerium doping CsCuCl3 perovskite nanocrystals were synthesized and their optical, structural, and environmental stability were characterized. The substance demonstrates photoluminescence at its peak, and materials that are incredibly durable also have optimal absorption characteristics that extend to the near-infrared spectrum with an appropriate band gap (Aamir, 2018). Shape modification, size management, and compositional alloying are crucial processes in PL tuning and optoelectronic components, including photodetectors, solar cells, screens, LEDs, and lasers. Additionally, one practical method for controlling the optical, electrical, and structural stability of halide perovskite NCs is doping, or adding impurity ions (Cui et al.,2020). Fast electron injection, which can lead to high efficiency, is greatly impacted by the conduction band alignment between lanthanide ions and perovskite. Concerns regarding the composites have recently grown significantly because to the lead-free halide perovskites' and lanthanide ions' improved energy conversion efficiency and charge transport (Zhang et al., 2017). However, the process of capturing a charge carrier is primarily controlled by its self-inherent trapping and surface imperfections brought about by strong excitation- phonon coupling. Lately, there have been significant worries regarding the composites due to the lead-free halide perovskites' and lanthanide ions' improved energy conversion efficiency and charge transport. This study presents a solution to the problem by introducing Cerium elements, which were effectively synthesized by the solvent-based thermal synthesis approach, to passivate internal and surface flaws. When the luminescence, photoluminescence, and crystal structure of cerium-doped CsCuCl3 were studied, the NCs PLQY increased by 52%, setting a new standard for

[1]harshasonawane24@gmail.com, [2]ashok.sunatkari@gmail.com, [3]swarnalatasunatkari@gmail.com, [4]shital.sonawane@somaiya.edu

DOI: 10.1201/9781003684718-6

the construction of lead-free perovskite optoelectronic devices with good optical properties (Dang et al.,2021).

2. RESULTS AND DISCUSSION

2.1 Properties of UV and Photoluminescence

The optical characteristics of the perovskite material were investigated using UV-visible absorption and photoluminescence spectroscopy in ethanol at room temperature. The absorption spectrum of CsCuCl3 is seen in Fig. 6.1 (a). At 410 nm, the inorganic perovskite material exhibits an absorption peak. The direct band gap of the compound is estimated to be 2.28 eV based on the Tauc plot, where the band gap was calculated using the absorption peak at 410 nm (Tsukahara et al., 1995). The compound Ce3+: CsCuCl3 NCs shows absorption peaks at 310 nm, as seen in Fig. 6.1 (b), and the Tauc plot estimates the band gap to be 2.15 eV. Figure and time-resolved spectrum of copper halide perovskite cannot be obtained, which restricts optical research on the material and its applications. However, samples of cerium doped CsCuCl3 with visible fluorescent light were created by adding cerium ions to minimize defects, as illustrated in Fig. 3.8. Its luminescence centre wavelength is 820 nm, its FWHM is 384.8 nm, and its light absorption spectrum is 310 nm. Its fluorescence spectrum has an almost Gaussian distribution. Consistent band edge absorption at indicates that band edge emission from CIE fluorescence spectra is the source of the fluorescence, seen in Fig. 6.2 (b) The CRI, CCT, and 1931 CIE (x, y) chromaticity coordinates were used to describe the colour of radiation treatments and their impact on human vision (Park et al.,2018). The CRI and 1931 CIE (x, y) chromaticity coordinates for each radiation treatment were determined by entering spectrum data into the colour Calculator software (version 7.23; OSRAM). https://www.osram.us/cb/toolsandresources/applications/ledcolorcalculator/index.jsp (Fig. 6.2) Wilmington, NC; Sylvania (Park et al.,2018). As can be seen, its colour coordinate of 0.16, 0.19 indicates that the Ce3+ doped CsCuCl3 nanocrystal is a material with strong blue luminescence (Pan et al., 2017). Figure 6.2 (a) displays the photoluminescence (PL) emission spectra of cerium doped cesium copper chloride and cesium copper chloride. Photoluminescence (PL) peaks at 273 nm were seen in the inorganic perovskite CsCuCl3 NCs excitation at 652nm. When activated at 820 nm, photoluminescence (PL) peaks were observed in Ce 3+ doped CsCuCl3 NCs at 211 nm. There is an effective energy transfer from the lanthanide ion energy levels to the CsCuCl3 NC host since the perovskite NC host's primary absorption band is situated at 310 nm (see Fig. 6.2 (b) (Pan et al., 2017). The excitonic absorption

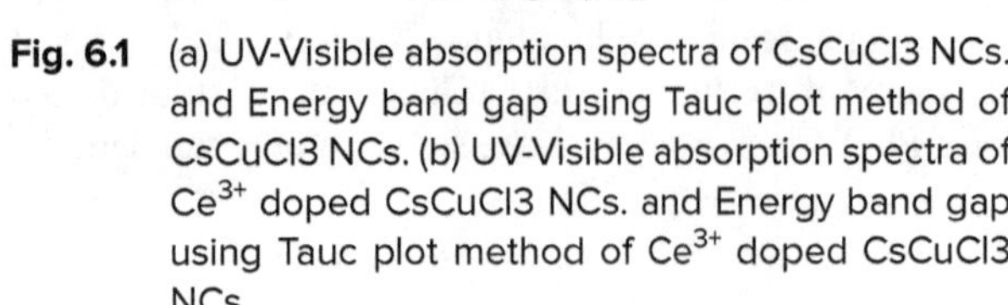

Fig. 6.1 (a) UV-Visible absorption spectra of CsCuCl3 NCs. and Energy band gap using Tauc plot method of CsCuCl3 NCs. (b) UV-Visible absorption spectra of Ce^{3+} doped CsCuCl3 NCs. and Energy band gap using Tauc plot method of Ce^{3+} doped CsCuCl3 NCs

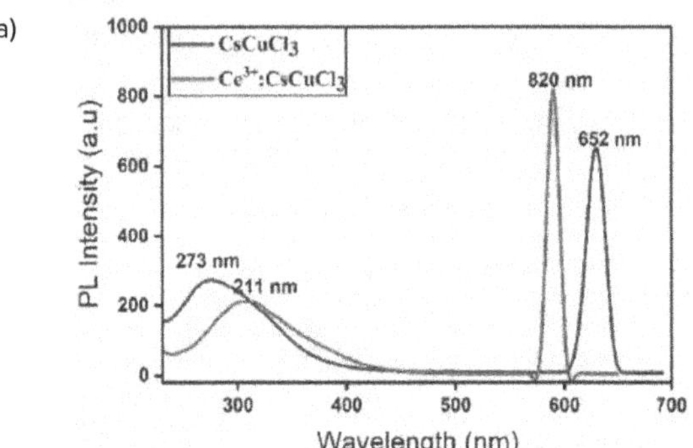

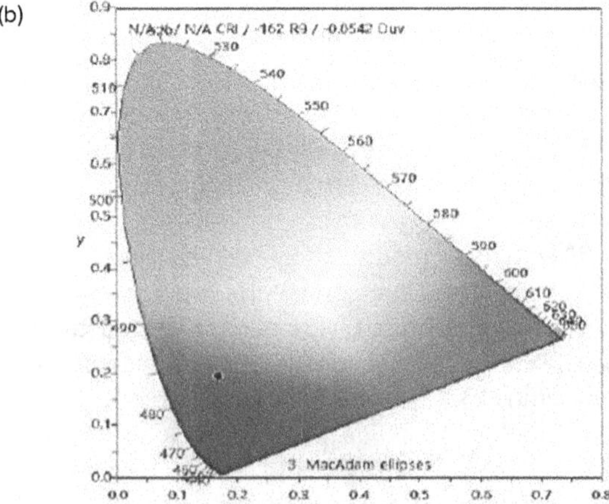

Fig. 6.2 (a) Photoluminescence spectra of CsCuCl3 NCs and Ce^{3+} doped CsCuCl3 NCs CsCuCl3 NCs and Ce^{3+} *doped CsCuCl3 NCs. (b) CIE chromaticity coordinates of Ce* $^{3+}$*doped CsCuCl3 (0.16,0.19)*

of the CsCuCl3 NC host matches the broad UV component from 220 to 450 nm in the excitation spectra, which was carefully observed at the peak position of the lanthanide ions (Pan et al., 2017). This suggests that the CsCuCl3 NC host is primarily responsible for increasing the lanthanide ion emissions. As Ce3+ rises, the PL component's relative intensity improves as well. The procedure by which Ce3+ of CsCuCl3 NCs works is clarified by the previously mentioned data. Ce3+ concentration is another significant factor in PL modification. Given the heterovalent nature of Ce3+ doping, donor states are probably formed by Ce3+ below the band of conduction. These donor states are readily excited to the conduction band upon photoexcitation, and hence may serve as a substitute for photoexcited carriers trapped in trap states. The enhanced PLQY phenomenon could perhaps stem from the enhanced electrical characteristics of CsCuCl3 NCs subsequent to Ce3+doping. The addition of CeCl3 to the precursor concurrently introduces a second Cl− source and a heterovalent dopant. The concentration of CeCl3 does not exactly correlate with the Cu-Cl coordination number, and excess Cl−around Ce3+ results in a drop in PL. It implies that PL enhancement may not always result from just increasing the Cl− concentration or passivating structural errors (Zang et al.,2020). On their crystal surface, copper halide perovskites exhibit an extremely high defect density because of their double-stranded spiral development on the axis. Strong exciton-phonon coupling and surface imperfections that regulate the charge carrier trapping process create intrinsic self- trapping as a result, which significantly lowers radiation recombination and produces a very weak photoluminescence signal. Consequently, the steady- state fluorescence spectrum.

3. CONCLUSION

In conclusion, the cesium copper chloride perovskite's luminescence was studied using a solvent-based thermal synthesis doping approach. The findings showed that a high-quality system comprising CsCuCl3 and Ce3+ doped CsCuCl3 had been developed. The amount of CsCuCl3 mixing with the cerium lanthanide ions gives uniformity and quality of CsCuCl3, which is the important factor affecting the optoelectronic devices, its PLQY improves by 52%, according to research done by comparing the PL efficiency of samples. Cerium ion doping enhances band edge absorption and luminescence performance while reducing defect density in an efficient manner. Because of its high efficiency luminescence qualities, the research mentioned above enhances CsCuCl3 and promotes its application in optoelectronic devices.

REFERENCES

1. Aamir, M., & Au840146, R. N. (2018). Synthesis Of Metal Halide Perovskite Materials for Light Harvesting Applications.
2. Cui, S., Chen, Y., Tao, S., Cui, J., Yuan, C., Yu, N., ... & Zhang, X. (2020). Synthesis, Crystal Structure and Photoelectric Response of All-Inorganic Copper Halide Salts CsCuCl3. 22, 2165–2169.
3. Dang, Y., Liu, X., Cao, B., & Tao, X. (2021). Chiral halide perovskite crystals for optoelectronic applications. 4(3), 794–820.
4. Booker, E. P., Griffiths, J. T., Eyre, L., Ducati, C., Greenham, N. C., & Davis, N. J. (2019). Synthesis, characterization, and morphological control of Cs2CuCl4 nanocrystals.123(27), 16951–16956.
5. Zhang, C., Li, T., Pu, L., Wen, W., Luo, X., & Zhao, L. (2020). Enhanced photoluminescence and stability of ZnSe microspheres/Cs4PbBr6 microcrystals/CsPbBr3 nanocrystals composites. 31(9), 2499–2502.
6. Wu, R., Bai, Z., Jiang, J., Yao, H., & Qin, S. (2021). Research on the photoluminescence properties of Cu 2+-doped perovskite CsPbCl3 quantum dots. 11(15), 8430–8436.
7. Dong, Y., Gu, Y., Zou, Y., Song, J., Xu, L., Li, J., ... & Zeng, H. (2016). Improving all-inorganic perovskite photodetectors by preferred orientation and plasmonic effect. 12(40), 5622–5632.
8. Sun, Y., Seo, J. H., Takacs, C. J., Seifter, J., & Heeger, A. J. (2011). Inverted polymer solar cells integrated with a low-temperature-annealed sol-gel-derived ZnO film as an electron transport layer. 23(14), 1679.
9. Jiang, Y., Liao, J. F., Xu, Y. F., Chen, H. Y., Wang, X. D., & Kuang, D. B. (2019). Hierarchical CsPbBr3 nanocrystal-decorated ZnO nanowire/ microporous graphene hybrids for enhancing charge separation and photocatalytic CO 2 reduction. 7(22), 13762–13769.
10. Yang, D., Yang, R., Zhang, J., Yang, Z., Liu, S. F., & Li, C. (2015). High efficiency flexible perovskite solar cells using superior low temperature TiO2. 8(11), 3208–3214.
11. Zhang, L., Yang, X., Jiang, Q., Wang, P., Yin, Z., Zhang, X., ... & You, J. (2017). Ultra-bright and highly efficient inorganic based perovskite light-emitting diodes. 8(1), 15640.
12. Stranks, S. D., & Snaith, H. J. (2015). Metal-halide perovskites for photovoltaic and light-emitting devices. 10(5), 391–402.
13. Zhao, L., Sun, C., Tian, G., & Pang, Q. (2017). Multiple-shell ZnSe core-shell spheres and their improved photocatalytic activity. 502, 1–7.
14. Li, X., Wu, Y., Zhang, S., Cai, B., Gu, Y., Song, J., & Zeng, H. (2016). CsPbX3 quantum dots for lighting and displays: room-temperature synthesis, photoluminescence superiorities, underlying origins and white light-emitting diodes. 26(15), 2435–2445.
15. Pan, G., Bai, X., Yang, D., Chen, X., Jing, P., Qu, S., ... & Song, H. (2017). Doping lanthanide into perovskite nanocrystals: highly improved and expanded optical properties. 17(12), 8005–8011.
16. Park, Y., & Runkle, E. S. (2018). Far-red radiation and photosynthetic photon flux density independently regulate seedling growth but interactively regulate flowering. 155, 206–216.
17. Tsukahara, K., Sawai, N., Koji, K., & Nakazawa, T. (1995). Photophysical properties of a new type of viologen-bridged triad: the bis (N-propyltetraphenyl porphyrinatochlorozinc (II)) complex. 246(3), 331–334.

Note: All the figures in this chapter were made by the authors.

Recent Trends in Applied Physics and Material Science – Sudhir Bhardwaj et al. (eds)
© 2026 Taylor & Francis Group, London, ISBN 978-1-041-16452-4

Hot-Injection Synthesis and Characterization of $CsPbI_3$ Nanostructures

Hari Ganesha Y, and K. Gopalakrishna Naik*
Department of Studies in Physics, Mangalore University,
Mangalagangothri, India

Abstract: Cesium lead iodide ($CsPbI_3$) perovskite nanocrystals (NCs) were prepared using the hot-injection method with cesium carbonate (Cs_2CO_3) and lead iodide (PbI_2) as precursors, 1-octadecane (1-ODE) as solvent, and oleic acid (OA) and oleylamine (OLA) ligands as surfactants. XRD pattern indicates the presence of perovskite cubic a-$CsPbI_3$ and orthorhombic black γ-$CsPbI_3$ in the synthesized $CsPbI_3$ NCs. The observed stability of the synthesized $CsPbI_3$ in cubic a-$CsPbI_3$ and black orthorhombic $CsPbI_3$ (B-γ-$CsPbI_3$) may be due to the nano-scale dimensions of the synthesized NCs. FESEM images show the growth of $CsPbI_3$ nanorods along with nanoparticles during the synthesis. The energy bandgap of the synthesized $CsPbI_3$ NCs were obtained from the UV-visible optical absorption measurements.

Keywords: Hot-injection, ODE variation, $CsPbI_3$

1. INTRODUCTION

Cesium lead iodide ($CsPbI_3$) is perovskite material belonging to family of perovskite halides of the general formula ABX_3 describes a structure in which A is a monovalent metal cation, B is a divalent metal cation, and X represents a halide. This class of materials is becoming increasingly important as a semiconductor for various optoelectronic applications, including solar cells, LEDs, lasers, and photodetectors (Zhang et al., 2021). $CsPbI_3$ has been reported to have four known structural phases; the room temperature stable non-perovskite orthorhombic d-phase or yellow phase (Y-$CsPbI_3$), and three high-temperature perovskite phases, characterized by cubic (α-phase), tetragonal (β-phase), and orthorhombic (γ-phase) crystal structures (Ke at al., 2021). The perovskite phase $CsPbI_3$ has been reported to have the energy bandgap in the range of 1.6 eV– 1.8 eV and hence, it is a promising material for solar cell applications (Ke at al., 2021). However, room temperature stability of perovskite phase of $CsPbI_3$ is the main problem for any of its applications. In the bulk form a-$CsPbI_3$ is reported to be stable only above 300 °C and as the temperature is decreased it transforms into the undesirable d-$CsPbI_3$ having a wider band gap of 2.82 eV through high-temperature metastable b-$CsPbI_3$ followed by g-$CsPbI_3$ phase (Marronnier et al., 2018, Chen et al., 2019). Hot-injection (HI) method is one of the widely used methods for the synthesizing $CsPbI_3$ NCs (Protesescu et al., 2015). The size controlled colloidal synthesis has been considered as one of the best methods for the synthesis of α-$CsPbI_3$ (Li et al., 2018).

In this study, $CsPbI_3$ nanocrystals were synthesized through the hot-injection (HI) method. Synthesized NCs were characterized using powder XRD, FESEM, and UV-visible optical absorption measurements.

2. EXPERIMENTAL METHODS

The $CsPbI_3$ NCs were prepared by HI method using cesium carbonate (Cs_2CO_3), lead iodide (PbI_2), 1-ODE, OA and OLA adopting the procedure similar to the one adopted by Protesescu et al., (2015), however, the synthesis was carried under vacuum instead of inert gas atmosphere. Cesium oleate (CsOL) solution was used as

*Corresponding author: gopal1mng@gmail.com

DOI: 10.1201/9781003684718-7

Cs source. The CsOL solution was prepared in a 3-neck 100 ml round bottom (RB) flask using 0.163 g (0.5 mmol) of Cs_2CO_3, 1.6 ml of OA, and 3.4 mL of ODE as precursor and heating the precursors under vacuum at 150 °C for 2 hr with continuous stirring. The above-mentioned quantities of Cs_2CO_3 and OA were selected so as to have the Cs: OA ratio equal to 1:5 so as the prepared CsOL precursor solution remains completely soluble even at room temperature as reported by of Lu et al., (2019). The prepared CsOL solution was cooled and stored in a vial under ambient conditions. The $CsPbI_3$ NCs were prepared using three different Pb precursor concentrations by keeping the CsOL precursor concentration and OA and OLA quantities same in all three syntheses. The OA and OLA are used as surfactants and also as reported by Protesescu et al., (2015), OA and OLA are required for the dissolution of PbI_2 in ODE. The Pb precursors were prepared in the same RB flasks using 0.0928 g (0.2 mmol) of lead iodide (PbI_2) each in 15 ml, 12 ml, and 6 ml of ODE solvents, respectively. In each of the synthesis, the Pb precursor containing flasks were heated at 150 °C under vacuum for 1 hr with constant stirring. 2 mL of OA followed by 2 mL of OLA solutions were injected into each of the hot Pb precursor solvents maintained at 150 °C. Then, 2 mL of CsOL was injected to each of the hot solutions. After the CsOL injection, the stirring was stopped and solution temperature was maintained at 150 °C for further 35 minutes and then flasks were placed in ice baths to halt the further growth of NCs. The resulting $CsPbI_3$ NCs were centrifuged under hexane solution to purify the prepared NCs. The centrifugation was repeated 3 times before collecting the synthesized NCs. The nanocrystals were dried at 150 °C for 1 hour using an open-air heater. Synthesized $CsPbI_3$ NCs were characterised by XRD using CuKa X-ray source (l= 0.15406 nm) for structural properties, FESEM for morphological study and UV-Visible optical absorption spectroscopy for optical properties.

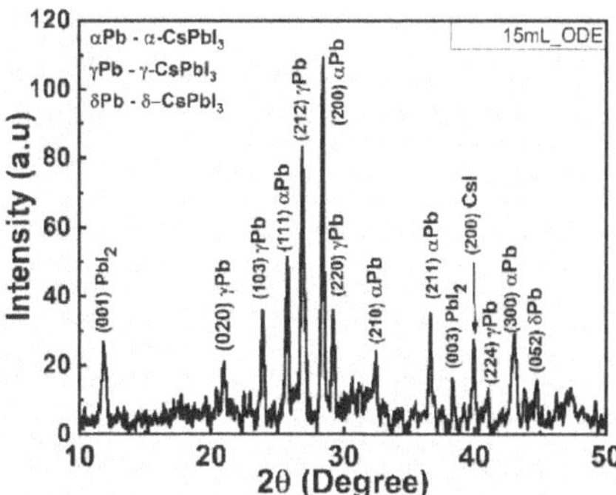

Fig. 7.1 The XRD pattern of the synthesized $CsPbI_3$ using Pb precursor in 15 ml ODE nanostructures recorded 10 days after the synthesis

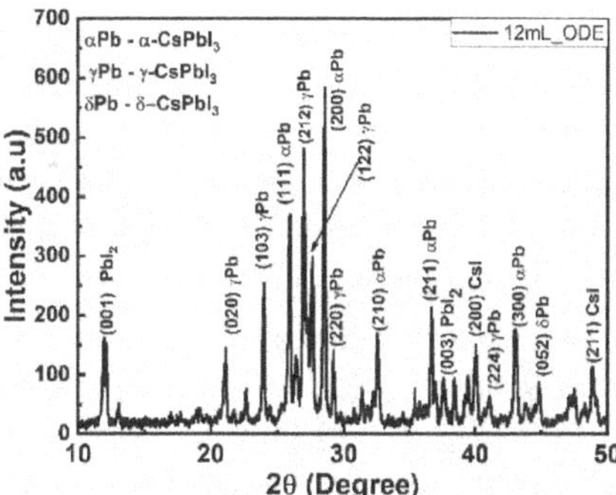

Fig. 7.2 The XRD pattern of the synthesized $CsPbI_3$ using Pb precursor in 12 ml ODE nanostructures recorded 9 days after the synthesis

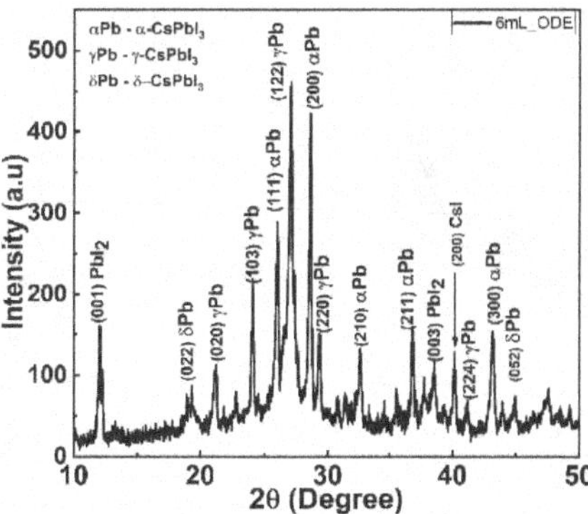

Fig. 7.3 The XRD pattern of the synthesized γ-$CsPbI_3$ using Pb precursor in 6 ml ODE nanostructures recorded 10 days after the synthesis

3. RESULTS AND DISCUSSION

Figure 7.1 to Fig. 7.3 shows the XRD patterns of the synthesized $CsPbI_3$ nanocrystals (NCs) prepared using 15 mL, 12 mL, and 6 mL of ODE solvent, respectively, recorded 10 days, 9 days, and 10 days after synthesis. XRD peaks were identified comparing with the published literatures. The XRD peaks observed at 2θ values of around 26°, 28.5° 32.5°, 36.6°, and 43° are related to (111), (200), (210), (211) and (300) planes of a-$CsPbI_3$, respectively (Shenn et al., 2017). The XRD peaks observed at 2θ values of 21°, 24°, 27°, 29°, and 41° can be assigned to (020), (121) (103), (122), (220) and (224) planes of B-γ-$CsPbI_3$, respectively (Sutton et al., 2018). The XRD peak observed at 2θ value of 19° in Fig. 7.3, and of 44.8° in Fig. 7.1 to Fig. 7.3 can be assigned to (022) and (052) planes of δ-$CsPbI_3$, respectively (JCPDS card No. 18-0376, Yang et al., 2021). The XRD peaks observed at 2θ values of 40° in Fig. 7.1 to Fig. 7.3 and 48.82° in Fig. 7.2 are associated with the (200) and (211) planes of CsI (Farzaneh et al., 2016). The XRD peaks observed at 2θ values of 11.8°, and 38.4° can be assigned to (001), and (003) planes of PbI_2, respectively (Burschka et al., 2013). The presence of CsI XRD peak may be due to the dissociation of the synthesized $CsPbI_3$ NCs. The presence of PbI_2 in the synthesized NCs may be due to incomplete conversion of PbI_2 precursor during the synthesis or the dissociation of synthesized $CsPbI_3$ NCs. The nano dimensions and the OA-OLA surfactants could contribute to the room-temperature stability of the synthesized $CsPbI_3$ perovskite nanocrystals under ambient conditions. (Swarnkar et al., 2016).

Figure 7.4 to Fig. 7.6 show the FESEM images of the $CsPbI_3$ nanocrystals using 15 ml, 12 ml, and 6 ml ODE solvents, respectively. The FESEM images show the growth of both nanorods and nanoparticles of $CsPbI_3$ in all the synthesized samples. The FESEM images synthesized from the Pb precursor prepared in 15 ml ODE solvent indicates growth nanorods with lower aspect ratio than the $CsPbI_3$ nanorod grown from that the $CsPbI_3$ nanocrystals from Pb precursor prepared in 12 ml ODE solvent. The $CsPbI_3$ nanocrystals synthesized from Pb precursor prepared in 6 ml ODE solvent shows the formation of nanoparticles as well as nanorods with relatively small in size compared to the $CsPbI_3$ NCs synthesized from 12 ml ODE solvent. The above observations indicates that the nucleation and subsequent growth of $CsPbI_3$ seems to follow the LaMer model of nucleation and growth of nanoparticles in the solutions (Pradhan et al., 2022).

The synthesized nanostructures were dispersed in toluene for the UV-visible optical absorption measurement. Figure 7.7 (a) to Fig. 7.9 (a) show UV-visible absorption spectra of the synthesized $CsPbI_3$ nanocrystals synthesized using 15 ml, 12 ml, and 6 ml ODE solvents, respectively. Figure 7.7 (b) to Fig. 7.9 (b) show the Tauc's plots of the corresponding absorption data. The bandgaps of the synthesized $CsPbI_3$ nanocrystals using 15 ml, 12 ml, and 6 ml ODE solutions were found to 1.76 eV, 2.00 eV, and 2.40 eV, respectively. The observed increase in the energy bandgap with decreasing ODE volume suggests a reduction in nanocrystal size as precursor supersaturation increases. This occurs because the reduced ODE solvent volume promotes the growth of smaller particles due to the heightened supersaturation (Haydous et al., 2021). The manifestation and increase in the quantum size effect as NC size decease with increase in precursor supersaturation may account for the measured band gap values from UV-visible absorption spectroscopy (Swarnkar et al., 2016, Pradhan et al., (2022).

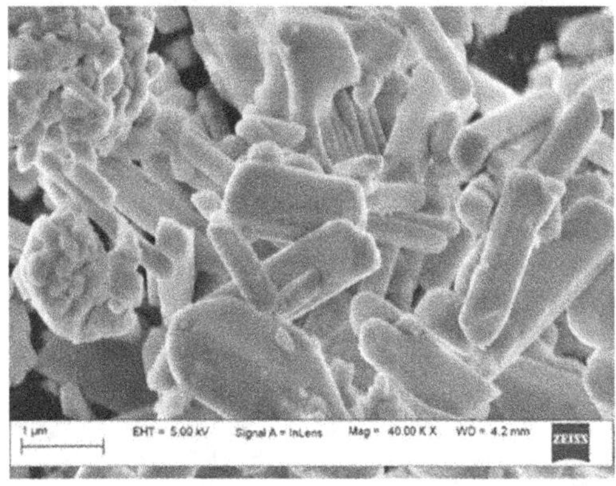

Fig. 7.4 FESEM image of $CsPbI_3$ prepared using 15 ml ODE solution

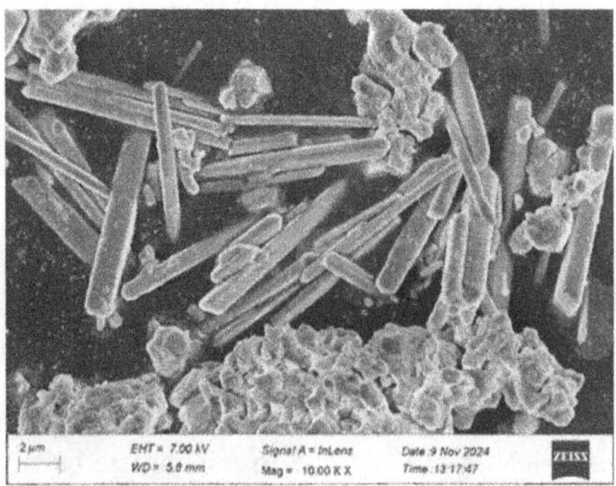

Fig. 7.5 FESEM image of $CsPbI_3$ prepared using 12 ml ODE solvent

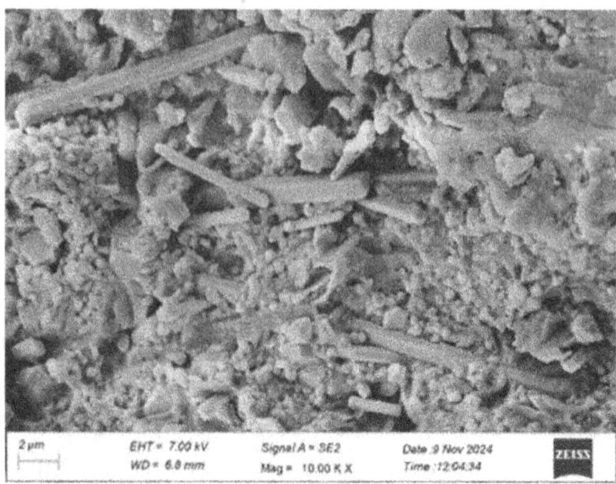

Fig. 7.6 FESEM image of $CsPbI_3$ prepared using 6 ml ODE solvent

4. CONCLUSION

The stable perovskite $CsPbI_3$ NCs were successfully prepared using the HI method. XRD analysis confirmed the presence of stable α-$CsPbI_3$ and black γ-$CsPbI_3$ phases in the synthesized NCs. The FESEM and UV-visible optical absorption studies indicate the precursor supersaturation as an effect on formation of $CsPbI_3$ NCs in during Hot-

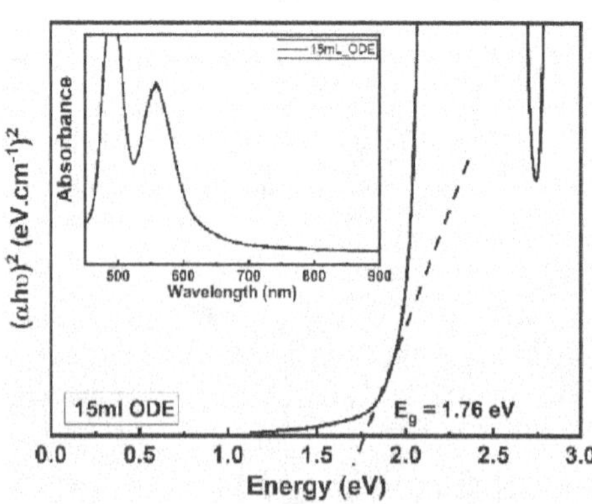

Fig. 7.7 The UV-vis optical absorption spectrum (shown in the inset) and the Tauc plot derived from the absorption spectrum of $CsPbI_3$ nanocrystals synthesised using 15 mL of ODE solvent

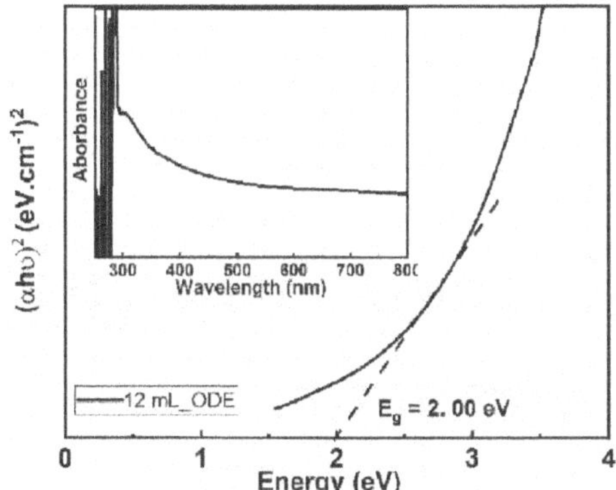

Fig. 7.8 UV-vis optical absorption spectrum shown as inset and Tauc's plot of the absorption spectrum of CsPbI$_3$ nanocrystals synthesised using 12 mL of ODE solvent

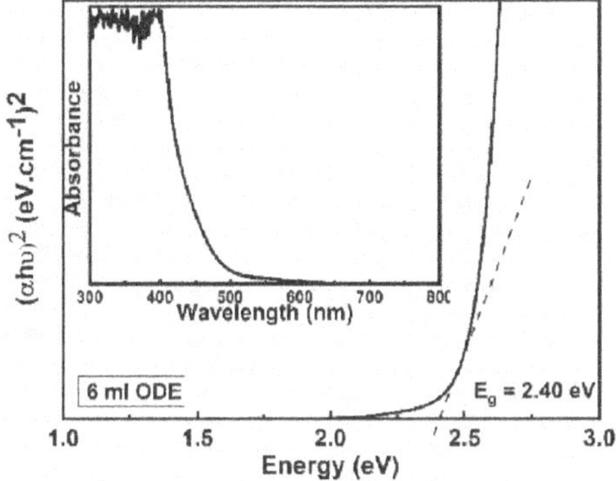

Fig. 7.9 UV-vis optical absorption spectrum shown as inset and Tauc's plot of the optical absorption spectrum of CsPbI$_3$ nanocrystals synthesised using 6 mL of ODE solvent

injection synthesis. The higher precursor saturation shows the growth of small crystal as indicated by the FESEM images and energy bandgap estimated from the UV-visible optical absorption data.

REFERENCES

1. Burschka, J., Pellet, N., Moon, S., Humphry-Baker, R., Gao, P., Nazeeruddin, M. K., and Grätzel, M. (2013). Sequential deposition as a route to high-performance perovskite-sensitized solar cells, Nature 499: 316–319.
2. Chen, Z., Dong, L., Tang, H., Yu, Y., Ye, L., and Zang, J. (2019). Direct Synthesis of Cubic Phase CsPbI$_3$ Nanowires. CrystEngComm. 21: 1389–1396.
3. Farzaneh, A., and Abdi, M. R., Saraee, K. R. E., Mostajabaldaavati,M., and Quaranta., A. (2016). The preparation of cesium-iodide thin films via sol–gel method for the detection of ionizing radiation. J Sol-Gel Sci. Technol. 78: 313–321.
4. Haydous, F., Gardner, J. M., and Cappel, U. B. (2021). The impact of ligands on the synthesis and application of metal halide perovskite nanocrystals, J. Mater. Chem. A. 9: 23419.
5. Ke, F., Wang, C., Jia, C., Wolf, N. R., Yan, J., Niu, S., Devereaux, T. P., Karunadasa, H. I., Mao, W. L., and Lin, Y. (2021). Preserving a robust CsPbI$_3$ perovskite phase via pressure-directed octahedral tilt. Nat. Commun. 12: 461.
6. Li, B., Zhang, Y., Fu, L., Yu, T., Zhou, S., Zhang, L., and Yin, L. (2018). Surface passivation engineering strategy to fully-inorganic cubic CsPbI$_3$ perovskites for high-performance solar cells. Nat. commun. 9:1076.
7. Lu, C., Wright, M. W., Ma, X., Li, H., S. Itanze, D. S., Carter, J. A., Hewitt, C. A., Donati, G. L., Carroll, D. L., Lundin, P. M., and Geyer, S. M. (2019). Cesium oleate precursor preparation for lead halide perovskite nanocrystal synthesis: The influence of excess oleic acid on achieving solubility, conversion, and reproducibility. Chem. Mater. 31: 62–67.
8. Marronnier, A., Roma, G., Boyer-Richard, S., Pedesseau, L., Jancu, J., Bonnassieux, Y., Katan, Stoumpos, C. C., Kanatzidis, M. G., and Even, J. (2018). Anharmonicity and disorder in the black phases of cesium lead iodide used for stable inorganic perovskite solar cells. ACS Nano. 12(4): 3477–3486. https://doi.org/10.1021/acsnano.8b00267.
9. Pradhan, N. Growth of lead halide perovskite nanocrystals: Still in mystery. (2022). ACS Phys. Chem Au. 2 (4): 268–276.
10. Protesescu, L., Yakunin, S., Bodnarchuk, M. I., Krieg, F., Caputo, R., Hendon, C. H., Yang, R. X., Walsh, A., and Kovalenko, M. V. (2015). Nanocrystals of Cesium Lead Halide Perovskites (CsPbX3, X = Cl, Br, and I): Novel optoelectronic materials showing bright emission with wide color gamut. Nano Lett. 15(6): 3692–3696. https://doi.org/10.1021/nl5048779.
11. Shen, Q., Ripolles, T. S., Even, J., Ogomi, Y., Nishinaka, K., Izuishi, T., Nakazawa, N., Y. Zhang, Y., Ding, C., Liu, F., Toyoda, T, Yoshino, K., Minemo, T., Katayama, K., and Hayase, S. (2017). Slow hot carrier cooling in cesium lead iodide perovskites. Appl. Phys. Lett. 111: 153903.
12. Sutton, R. J., Filip, M.R., Haghighirad, A. A., Sakai, N., Wenger, B., Giustino, F., and Snaith, H. J. (2018). Cubic or orthorhombic? Revealing the crystal structure of metastable black-phase CsPbI$_3$ by theory and experiment. ACS Energy Lett. 3: 1787–1794.
13. Swarnkar, A., Marshall, A. R., Sanehira, E. M., Chernomordik, B. D., Moore, D. T., Christians, J. A., Chakrabarti, T., and Luther, J. M. (2016). Quantum dot–induced phase stabilization of α-CsPbI$_3$ perovskite for high-efficiency photovoltaics, Science. 354: 92–95.
14. Yang, T., Zheng, Y., Chou, K., and Hou, X. (2021). Tunable fabrication of single-crystalline CsPbI$_3$ nanobelts and their application as photodetectors. Int J Miner Metal Mater 28: 1030–1037.
15. Zhang, J., Yin, C., Yang, F., Yao, Y., Yuan, F., Chen, H., Wang, R., Bai, S., Tu, G., and Hou, L. (2021). Highly luminescent and stable CsPbI$_3$ perovskite nanocrystals with sodium dodecyl sulfate ligand passivation for red-light-emitting diodes. J. Phys. Chem. Lett. 12 (9): 2437–2443.

Note: All the figures in this chapter were made by the authors.

Recent Trends in Applied Physics and Material Science – Sudhir Bhardwaj et al. (eds)

Study of Graphene Oxide, Zinc Oxide and Graphene Oxide—Zinc Oxide Composites

R. P. Reshma,
N. S. Abishek and K. Gopalakrishna Naik*
Department of Physics, Mangalore University,
Mangalagangothri, India

Abstract: Graphene oxide (GO) was produced by modified Hummer's way and the prepared GO was thermally reduced to reduced GO (rGO) by heating the prepared GO at a low temperature of about 120 °C for about 72 hours. The XRD pattern confirms the possibility of reduction of GO to rGO by prolonged heat treatment at relatively low temperatures. The FESEM images show two-dimensional sheet like structures of the synthesised GO and rGO. Sol-gel process was applied to incorporate ZnO nanostructures which resulted in the growth of nanostructured hexagonal prisms. The ZnO-GO nanocomposites were prepared by hydrothermal method using a mixture of GO prepared Hummer's method and ZnO nanostructures prepared by sol-gel method dispersed in ethanol solvent by carrying out the growth at 120 °C for 24 hours. The observation intense ZnO XRD peaks indicate the increase in the ZnO crystallite sizes after hydrothermal growth process. FESEM images shows that the hydrothermal growth resulted in growth of ZnO nanorods; it appears that the sol-gel synthesised hexagonal prisms act as the seed layers for the growth of nanorods.

Keywords: Graphite, Graphene oxide, Zinc oxide, Reduced graphene oxide, Graphene oxide-zinc oxide composites

1. INTRODUCTION

Graphene, a recognized 2D sheet of carbon atoms organized in a planar hexagonal lattice, garnered significant attention due to its remarkable mechanical, surface, electrical, and optoelectronic characteristics. (Mbayachi et al., 2021 and Urade et al., 2023). Among the synthesis procedures, chemical vapour deposition (CVD) is of the major technique employed for the large-scale production of excellent quality graphene (Liu et al., 2017). However, the extensive use of graphene has been turned out be challenging owing to the difficulty and significant cost of synthesis, the complication of solubility and agglomeration in solutions (Jirickova et al., 2022). Alternately, the chemical, thermal, and photo-irradiation based reduction of GO to rGO has been proved to be the simple and economical method to obtain graphene-like nano-sheets of carbon (Jirickova et al., 2022, Slobodian et al., 2018 and Mei et al). GO is essentially a chemically modified graphene that has high density of oxygen consisting of functional groups like carbonyl (C=O), alkoxy (C-O-C), hydroxyl (-OH), and carboxylic acid (-COOH) groups which gives it hydrophilic character (Smith et al., 2019). The existence of oxygen that contains functional groups contributes GO to have many advantages over graphene like larger solubility, possibility of surface functionalization, good interfacial interaction with polar polymers to enhance their characteristics (Smith et al., 2019 and Hidayah et al., 2017). Also, GO has been one of the widely used precursor for production of graphene (Jirickova et al., 2022 and Yu et al., 2016). Modified Hummer's way is the generally adopted approach for the amalgam of GO (Hidayah et al., 2017). Reducing GO to rGO increases its electrical and thermal conductivity

*Corresponding author: gopal1mng@gmail.com

DOI: 10.1201/9781003684718-8

due to the decreased oxygen content. However, having a significant level of defect in contrast with pure graphene due to the oxygen and structural flaws from the chemical oxidation production of GO, even after reducing considerably from GO into rGO, the properties of rGO do not attain the same level as that of graphene. (Sarkar et al., 2014). The oxygen containing polar function groups in rGO results in excellent dispersion in several solvents, and the improved electrical and thermal properties of rGO makes rGO an ideal material for many applications (Slobodian et al., 2018). Nanocomposites have drawn considerable recognition because of their exceptionally enhanced properties and its applications for catalysis, energy storage, electronics, and sensing (Bai et al., 2024). Recently, GO-ZnO composites are emerging as promising materials with enhanced properties due to the synergistic combination of GO/ZnO (Bai et al., 2024). A semiconductor material with a high band gap (~3.2 eV) and tenable electrical and optical properties, zinc oxide (ZnO) has the potential to be used in gas sensors, photocatalysis, and optoelectronic devices (Raha and Ahmaruzzaman, 2022). The unique combination of GO- ZnO in the form of nanocomposites results in materials with improved performance in various applications, including environmental remediation, energy storage, sensors, and biomedical fields (Bai et al., 2024). Graphene and composites based on it have exceptional electrical properties recently, making them very interesting prospects for widespread use in technological applications (Alamdari et al., 2019). The graphene oxide helps the zinc oxide particles spread out evenly and ZnO nanoparticle stop graphene oxide sheet from clumping together (Alamdari et al., 2019).

The present work involves the synthesis and characterisation of GO and ZnO nanoparticles, and GO-ZnO composites. The synthesis of GO was carried out using a modified Hummer's method, while ZnO nanoparticles were produced through the sol-gel process. The solvothermal approach was utilized to produce GO-ZnO nanocomposites using GO synthesised by modified Hummer's method and ZnO nanoparticle synthesised by sol-gel method. The produced nanomaterials have been assessed using EDX spectroscopy, UV visible absorption spectroscopy, FESEM, and powder XRD.

2. EXPERIMENTAL PROCEDURE

In the synthesis of GO, graphite powder (LR grade, Molychem, India) was utilized as the precursor, while sulphuric acid (H_2SO_4, AR grade) and potassium permanganate ($KMnO_4$) served as oxidizing agents. Hydrogen peroxide (H_2O_2, 30% w/v, Molychem, India) was employed to reduce any excess $KMnO_4$ present. Initiating the procedure involved putting a 250 milliliters beaker in an ice bath with 1 g of graphite powder suspended in 25 milliliters of H_2SO_4.Gradually, 3 g of $KMnO_4$ was introduced to the mixture and dissolved under continuous stirring, which continued for a duration of 3 hours. Later 50 milliliters of deionised water were added dropwise, and then another 100 milliliters of deionised water was added immediately. Subsequently, 5 milliliters of H_2O_2 were incorporated into the mixture, which was stirred continuously for an hour. The resulting mix was then centrifuged to isolate the synthesised GO. The collected GO was thoroughly washed and dried for three days at 90 °C in a hot air oven. Prior to thermal reduction of GO to rGO, 0.5 g of GO was dispersed in 250 milliliters of deionised water and subjected to ultrasonication for approximately one hour to ensure the exfoliation of any remaining oxidized graphite sheets. The solution was then centrifuged to recover the GO. The thermal reduction process involved heating the GO at 120 °C for four days using a hot air oven.

Nanostructures of ZnO were prepared utilizing a sol-gel approach, following a methodology akin to that employed by Hasnidawan et al. The synthesis involved the use of zinc acetate dihydrate (Zn(CH3COO)2·2H2O, 99.5%, Lobal Chemie, India), sodium hydroxide (NaOH), double-deionised water, and ethanol under alkaline conditions. A magnetic stirrer was used to dissolve 2g of zinc acetate dihydrate in 15 milliliters and 8 grams of NaOH in 10 milliliters of deionised water, stirring continuously for roughly 10 minutes. Subsequently, the NaOH solution was gradually incorporated into the zinc acetate solution over a duration of about 15 minutes, maintaining constant stirring. Following this, 100 milliliters of ethanol were added dropwise to the mix, which was later stirred for about another 15 minutes before centrifugation was performed to isolate the synthesised ZnO nanoparticles.

To synthesise GO-ZnO composites via the solvothermal method, 1 g of GO powder, produced through a modified Hummer's technique, was initially dispersed in 20 milliliters of ethanol. This mixture was subjected to ultrasonic treatment for 1 hour at ambient temperature to ensure uniform dispersion. Subsequently, 0.5 g of prepared ZnO was imported into the GO suspension, and the mixture went through an additional 30 minutes of ultrasonic agitation. The resulting solution was transferred to the Teflon-lined autoclave (50 milliliters), where it was heated to 120 °C for a duration of 24 hours. After the reaction, the resulting products were centrifuged and rinsed with ethanol and double-deionised water multiple times, later drying at 90 °C for 24 hours.

Powder XRD studies were conducted with Rigaku X-ray diffractometer (λ = 1.5406 Å), operating at a scanning rate of 2° per minute, across 2 theta range of 0° - 80° to assess the structural properties of all synthesised samples. For morphological and elemental analysis, Zeiss Ultra 55 field emission scanning electron microscope (FESEM) with an energy dispersive X-ray (EDX) attachment was employed. Additionally, a Shimadzu UV-1800 UV-visible spectrophotometer was used to evaluate optical absorption.

For photo degradation analysis of the synthesised GO-ZnO nanostructures using methylene blue (MB) dye was performed as follows. The methylene blue (MB) solution was stored in a dark environment for an entire day. Subsequently, 4 milliliters of this MB solution was combined with 100 milliliters of deionised water. Following this, 20 mg of the synthesised GO-ZnO nanoparticles were added to the mixture in the dark, which was then subjected to continuous stirring for approximately an hour, establishing adsorption-desorption equilibrium between MB and photocatalyst. An 8 W power, 365 nm peak wavelength emission UV light source was used for the photoirradiation of the absorption – desorption equilibrated MB and the photocatalyst mixed solution. At 10-minute intervals throughout the irradiation process, 2 milliliters samples of the photo-irradiated methylene blue (MB) solution were extracted to assess photocatalytic degradation of the MB dye utilizing the synthesised GO-ZnO nanocomposites. The total duration of the photoirradiation was 50 minutes. The photocatalytic analysis was conducted with a Shimadzu UV-1800 spectrophotometer.

3. RESULTS AND DISCUSSION

The XRD pattern for the commercial graphite powder utilized in the synthesis of GO, the synthesised GO itself, and rGO are shown in Fig. 8.1. In Fig. 8.1 (a), a characteristic peak corresponding to the (002) plane of graphite is evident at a 2θ value of 26.3° (JCPDS no. 01-075-1621). The peak observed at 2θ = 11.8° in the Fig. 8.1 (b) signifies successful formation of GO through modified Hummer's method applied to graphite (Mun et al., 2015). Furthermore, the broad peak appearing at approximately 22° in the XRD pattern depicted in Fig. 8.1 (c) suggests, reduction of GO to rGO during low-temperature reduction process, aligning with (002) plane of rGO (Hidayah et al., 2017).

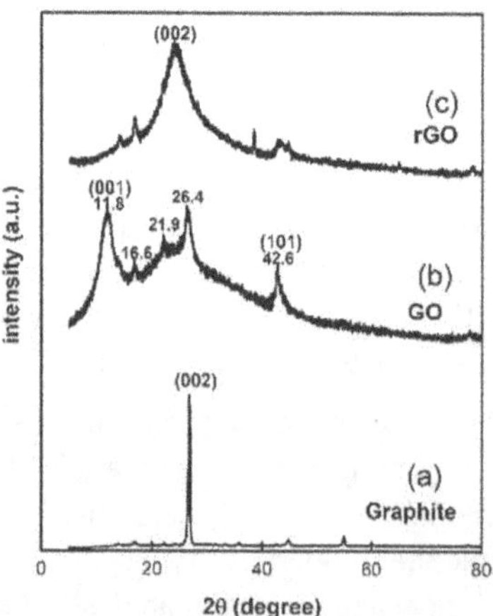

Fig. 8.1 XRD pattern of (a) graphite, (b) GO, and (c) rGO

The XRD patterns of ZnO nanoparticles synthesised by sol-gel method and GO-ZnO nanocomposites produced through hydrothermal synthesis are shown in Fig. 8.2 (a) and (b). Peaks associated with ZnO correspond closely with JCPDS no. 36-1451 for its hexagonal structure. Notably, (101) plane of ZnO is prominent in the XRD pattern. Additionally in Fig. 8.2 (b), low-intensity peak observed at 10.3° is likely indicative of the XRD signature for GO.

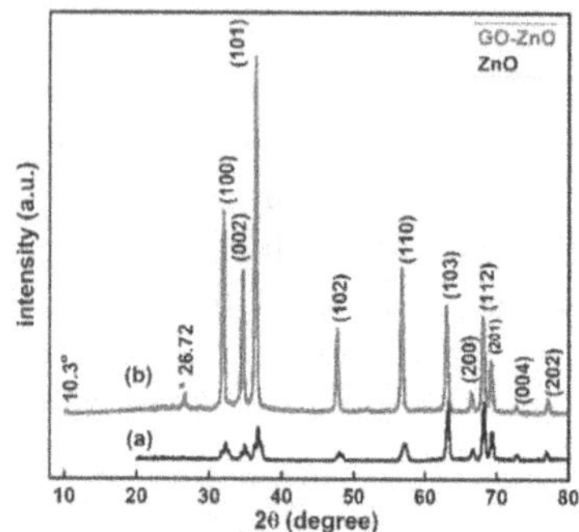

Fig. 8.2 XRD pattern of: (a) sol-gel synthesised ZnO nanomaterials, and (b) hydrothermal synthesised GO-ZnO nanocomposites

The morphological and elemental composition for GO are presented in Fig. 8.3 (a) and (b). And for rGO in Fig. 8.4 (a) and (b), which revealed crumpled, stacked sheet-like morphologies for GO and rGO. The EDX spectra for both materials confirm the presence of carbon and oxygen. The sol-gel method employed for synthesizing ZnO nanostructures led to the formation of hexagonal disk-like shapes, as depicted in Fig. 8.5 (a). The corresponding EDX spectrum in Fig. 8.5 (b) confirms that the synthesised material consists solely of zinc (Zn) and oxygen (O). Following the solvothermal treatment of GO and the hexagonal ZnO disk-like structures, ZnO nanorods were observed to grow, with lengths exceeding 10 nm, as shown in Fig. 8.6 (a). This growth of ZnO nanorods is likely facilitated by the Ostwald ripening mechanism during the solvothermal synthesis, where the hexagonal ZnO nanoplatelets serve as templates for the formation of nanorods (Singh et al., 2011).

Fig. 8.3 (a) FESEM image and (b) EDX spectrum of GO

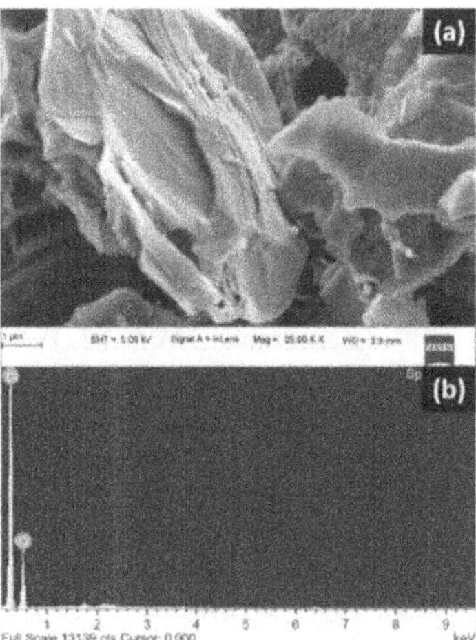

Fig. 8.4 (a) FESEM image and (b) EDX spectrum of rGO

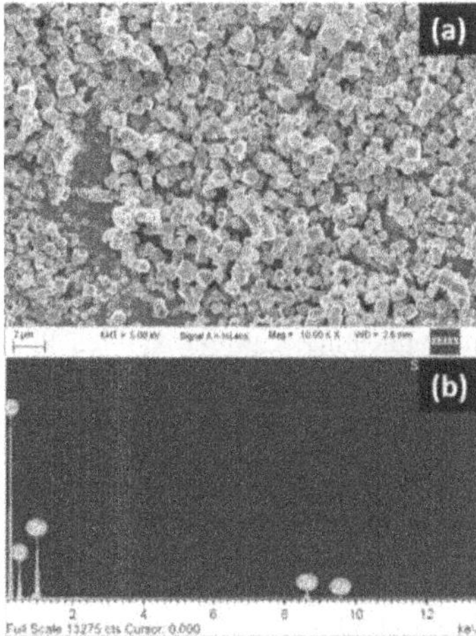

Fig. 8.5 (a) FESEM image and (b) EDX spectrum of ZnO

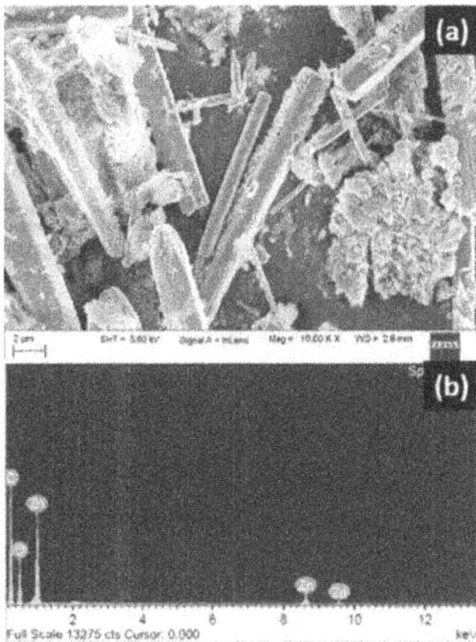

Fig. 8.6 (a) FESEM images and (b) EDX spectrum of GO-ZnO

Figure 8.7 illustrates UV-Vis spectrum of GO-ZnO nanocomposites. The optical spectrum exhibits a peak near 375 nm, which is likely attributable to the ZnO nanostructures. Additionally, a broad peak centered around 750 nm, extending from approximately 450 nm to 1100 nm, is likely associated with GO) (Singh et al., 2012).

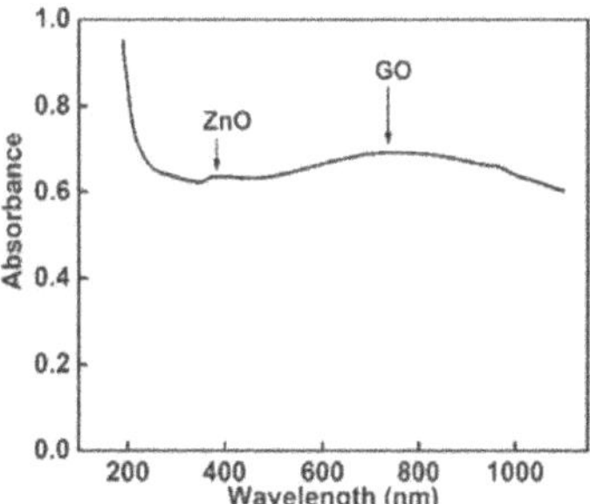

Fig. 8.7 UV-visible spectroscopy GO-ZnO composite

Photocatalytic performance of the synthesised GO-ZnO composite was analysed through degradation of MB dye under UV-light exposure. This evaluation was conducted utilizing UV-visible optical absorption spectroscopy to analyze the interaction between the GO-ZnO photocatalyst and the MB dye (Rehman et al., 2021). As illustrated in Fig. 8.8, the effectiveness of the GO-ZnO composites in degrading MB dye is presented.

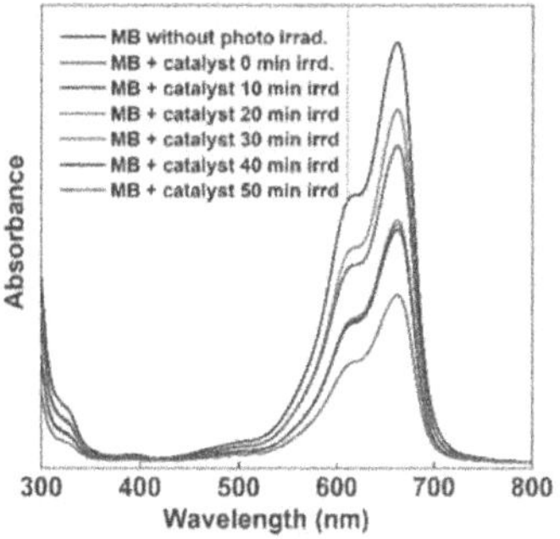

Fig. 8.8 Photocatalytic activity of GO-ZnO composites in degrading MB

4. CONCLUSION

GO was produced via a modified Hummer's method utilizing graphite powder, with XRD interpretation confirming the successful synthesis of GO. A prolonged low-temperature thermal treatment was employed to convert GO into rGO, and the XRD results indicate the transformation from GO to rGO during this process. Hexagonal ZnO nanoplatelets were fabricated through a sol-gel method, using zinc acetate and sodium hydroxide as alkaline medium. GO-ZnO nanocomposites were then developed via a solvothermal approach, incorporating the GO synthesised through the modified Hummer's way and the ZnO nanostructures obtained from the sol-gel process. This solvothermal synthesis facilitated the formation of ZnO rods characterized by a relatively high aspect ratio. The UV-visible optical spectrum of the GO-ZnO composites exhibits significant optical absorbance within the visible range, which can be attributed to the presence of GO.

REFERENCES

1. Mbayachi, V. B., Ndayiragije, E., Sammani, T., Taj, S., Mbuta, E. R. and Khan, A. U. (2021). Graphene synthesis, characterization and its applications: A review. Results in Chemistry. 3:100163.
2. Urade, A. R., Lahiri, I. and Suresh, K. S. (2023). Graphene Properties, Synthesis and Applications: A Review. JOM. 75:64–630.
3. Liu, Z., Lin, L., Ren, H. and Xiao Su. (2017). CVD synthesis of graphene. In book: Thermal Transport in Carbon-Based Nanomaterials. 19–56.
4. Jirickova, A., Jankovský, O., Sofer, Z. and Sedmidubský, D. (2022) Synthesis and applications of graphene oxide. Materials. 15:920.
5. Slobodian O. M., Lytvyn, P. M., Nikolenko A. S., Naseka, V. M., Khyzhun, O. Y., Vasin, A. V., Sevostianov, S. V. and Nazarov, A. N. (2018). Low-temperature reduction of graphene oxide: electrical conductance and scanning Kelvin probe force microscopy. Nanoscale Research Letters. 13:139.
6. Mei, X. F., Meng, X. Q. and Wu, F. M. (2015). Hydrothermal method for the production of reduced graphene oxide. Physica E. 68:81–86.
7. Smith, A. T., LaChance, A. M., Zeng, S., Liu, B. and Sun, L. (2019). Synthesis, properties, and applications of graphene oxide/reduced graphene oxide, and their nanocomposites. Nano Materials Science. 1:31–47.
8. Hidayah, N. M. S., Liu, W.-W., Lai, C.-W., Noriman, N. Z., Khe, C.-S., Hashim, U. and Lee, H. C. (2017). Comparison on graphite, graphene oxide and reduced graphene oxide: Synthesis and characterization. 1892: 150002.
9. Yu, H., Zhang, B., Bulin, C., Li, R. and Xing, R. (2016). High-efficient Synthesis Graphene Oxide Based on Improve Hummers Method. Sci. Rep. 6:36143.
10. Sarkar, S.K., Raul, K. K., Pradhan, S. S., Basu, S. and Nayak, A. (2014). Magnetic properties of graphite oxide and reduced graphene oxide. Physica E. 64: 78–82.
11. Bai, H., Li, J., Yao, J., Chen, Z., Wu, W., Zheng, S. and Zhang, P. (2024). Synthesis and characterization of zinc oxide-graphene oxide nanocomposites for electrocatalytic detection of rutin. Alexandria Engineering Journal. 91:486–493.
12. Raha, S. and Ahmaruzzaman, M. (2022). ZnO nanostructured materials and their potentialapplications: progress, challenges and perspectives. Nanoscale Adv. 4:1868.
13. Alamdari, S., Ghamsari, M. S., Afarideh, H., Mohammadi, A., Geranmayeh, S., Tafreshi, M. J., Ehsani, M. H. and Majles ara, M. H. (2019). Preparation and characterization of GO-ZnO nanocomposite for UV detection application. Optical Materials. 92:243–250.
14. Hasnidawani, J. N., Azlina, H. N., Norita, H., Bonnia, N. N., Ratim, S. and Ali, E. S. (2016). Synthesis of ZnO Nanostructures Using Sol-Gel Method. Procedia Chemistry. 19:211–216.
15. Mun, S., Kim, H. C., Yadave, M. and Kim, J. (2015). Graphene oxide–gellan gum–sodium alginate nanocomposites: synthesis, characterization, and mechanical behavior. Composite Interfaces. 22:249–263.
16. Singh, R.G., Fouran Singh., Vinod Kumar. and Mehra, R.M. (2011). Growth kinetics of ZnO nanocrystallites: Structural, optical and photoluminescence properties tuned by thermal annealing. Current Applied Physics. 11:624–630.
17. Singh, D. K., Pandey, D K., Yadav, R. R. and Devraj Singh. (2012). A study of nanosized zinc oxide and its nanofluid. Pramana – J. Phys. 78:5.
18. Rehman, H., Ali, Z., Shahzady, T. G., Zahra, A., Hussain, H., Anwar, A. and Latif, M. S. (2021). Synthesis, characterization of GO-zinc oxide nanocomposites and their use as an adsorbent for the removal of organic dyes in industrial effluents. Digest Journal of Nanomaterials and Biostructures. 16:1547–1555.

Note: All the figures in this chapter were made by the authors.

Recent Trends in Applied Physics and Material Science – Sudhir Bhardwaj et al. (eds)
 ISBN 978-1-041-16452-4

Investigation of Structural Parameters and Enhanced Photocatalytic Activity of ZnO-GO Nanocomposites

Pravin Rathod*, Vishnudas Bhosle
Department of Physics,
Government Vidarbha Institute of Science & Humanities,
Amravati, Maharastra, India

Zakir Khan
Department of Physics,
Government Vidarbha Institute of Science & Humanities,
Amravati, Maharastra, India

Ashok Ubale
Government Institute of Forensic Science,
Chhatrapti Sambhajinagar,
Maharastra, India

A. Viji
Department of Physics,
Kongunadu College of Engineering and Technology,
Thottiyam, Tamilnadu, India

Abstract: ZnO-GO nanocomposites are gaining attention for their enhanced photocatalytic applications in environmental remediation. This study explores their structural and morphological features, synthesized via a simple method, with properties analyzed through XRD and SEM. Crystallite sizes, calculated using the Debye-Scherrer equation, are 25.18 nm for ZnO and 16.18 nm for ZnO-GO. Photocatalytic performance was evaluated through organic pollutant degradation under UV and visible light, showing notable improvement over pure ZnO. The enhancement is attributed to GO's role in separating electron-hole pairs, providing active sites, and reducing recombination. These findings emphasize ZnO-GO's potential for advanced photocatalytic materials in environmental applications.

Keywords: Nanocomposite, ZnO-GO, XRD, SEM, UV-Vis, Photocatalysis

1. INTRODUCTION

Nanocomposite materials, particularly ZnO and GO, show significant potential in electronics (Senthil Kumar et al., 2023) photonics (Das, et al., 2022) photocatalysis (Ragunathan, et al., 2022), and energy storage (Bishwakarma, et al., 2023) ZnO, a semiconductor with a wide bandgap, offers excellent optical (Giri, et al., 2012), piezoelectric (Bhadwal, et al., 2023), and electrical properties (Abdel, et al., 2021), while GO provides high electrical conductivity, a large surface area (Tien, et al., 2013) and robust mechanical strength (Babak, et al., 2014), Combining these materials enhances charge mobility, surface area, and structural stability (Mututu, et al., 2019). This study synthesized ZnO nanoparticles via a chemical method and integrated them with functionalized GO through ultrasonication. Photocatalytic tests demonstrated the ZnO-GO nanocomposites superior performance in degrading MB dye under visible light, highlighting its effectiveness for environmental remediation and industrial wastewater treatment.

2. EXPERIMENTAL

2.1 Synthesis of ZnO and ZnO-GO

Zinc Acetate (99.9% Zn (Ac) $2.2H_2O$), Ethanol (99.8% EtOH), Graphite Powder (99.9%), Concentrated Sulfuric Acid (98% H_2SO_4), Hydrogen Peroxide (30% H_2O_2), Potassium Permanganate (98% KMnO4), and Hydrochloric Acid (HCl) were procured from SdFine LTD. All reactants were utilized as supplied, without further purification.

*Corresponding author: rathodpp11iyc@gmail.com

DOI: 10.1201/9781003684718-9

A straightforward chemical approach was employed to produce ZnO nanoparticles. Eight grams of zinc acetate were dissolved in 50 ml of ethanol and agitated continuously for one hour. The solution was thereafter allowed to remain undisturbed for 20 hours. Thereafter, it was rinsed 2–3 times with ethanol, and the resultant residue was subjected to heating at 400 °C in a muffle furnace for one hour. The acquired powder was meticulously pulverized utilizing an agate mortar.

Equal quantities of synthesized ZnO and GO were amalgamated and blended in ethanol. The amalgamation underwent ultrasonication for 30 minutes. The resultant product was subsequently dried at 100°C for one hour.

2.2 Material Characterization

The phase purity of ZnO, ZnO-GO, and GO nanopowder was assessed using an X-ray Diffractometer (Model: MiniFlex-II, Rigaku, Japan) with Cu Kα radiation (λ = 1.5406 Å), functioning at 40 kV and 30 mA. The surface morphology of the produced materials was analyzed using a Scanning Electron Microscope (SEM, Model: JEOL JSM-6360, Japan). The Fourier Transform Infrared (FTIR) spectra of the samples, formulated as KBr pellets, were obtained utilizing an FTIR spectrometer (Shimadzu, Japan) within the region of 4000-400 cm-1. Optical absorption spectra were acquired within the wavelength range of 300-800 nm with a UV-visible spectrometer (UV-1800 Spectrophotometer, Shimadzu, Japan).

3. RESULTS AND DISCUSSION

3.1 XRD Analysis

Powder X-ray diffraction (PXRD) examination was performed to investigate the crystalline phases and structures of the synthesized samples. Figure 9.1 depicts the PXRD patterns for the nanostructures: ZnO, GO, and ZnO-GO nanocomposites. A pronounced diffraction peak for GO is noted at 2θ = 10.62°, corresponding to the (001) plane. The peak is observed at 2θ = 12.14° for ZnO-GO, corresponding to the (001) plane. The PXRD pattern for ZnO has distinct peaks at 2θ = 31.68°, 34.39°, 36.15°, 47.48°, 56.54°, 62.88°, 66.31°, and 67.87°, corresponding to the hexagonal wurtzite crystal phase of ZnO (JCPDS card no. 36-1451).

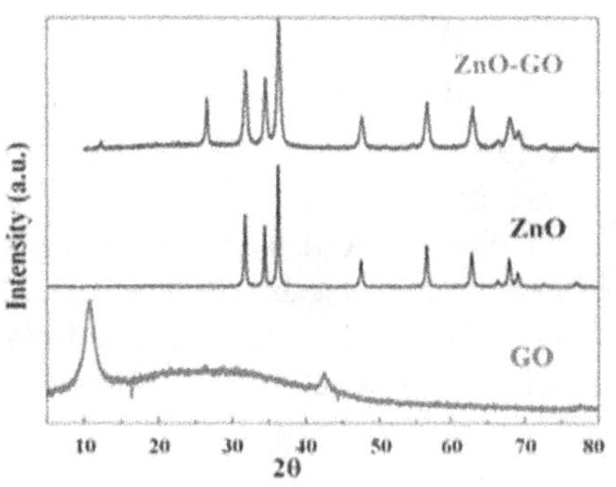

Fig. 9.1 XRD spectra of ZnO, GO and ZnO-GO nanocomposite

The diffraction pattern of ZnO-GO exhibits a distinct peak at approximately 12.14° alongside the conventional peaks of ZnO, therefore affirming the existence of GO. Furthermore, the diffraction peaks for ZnO-GO exhibit increased breadth and a minor shift towards higher angles, signifying the impact of graphene oxide on the crystalline structure of ZnO.

The crystallite dimensions (D) of virgin ZnO and ZnO-GO nanocomposites were determined utilizing Scherrer's equation:

$$D = \frac{K\lambda}{\beta \cos\theta} \tag{1}$$

In this context, λ represents the X-ray wavelength (0.15405 nm), β denotes the full width at half maximum (FWHM), and θ signifies the diffraction angle. The average crystallite diameters of pure ZnO and ZnO-GO were measured at 25.86 nm and 16.18 nm, respectively. The interplanar distance (d) for ZnO-GO (0.251 nm) was somewhat lower than that of ZnO due to a reduced quantity of oxygen-containing groups. The lattice strain (ϵ) was determined utilizing the formula:

$$\epsilon = \frac{\beta_{hkl}}{4\tan\theta} \tag{2}$$

Where βhkl and θ represent the full width at half maximum (FWHM) and the diffraction angle, respectively. The computed structural parameters and interplanar spacing are presented in the Table 9.1.

Table 9.1 Represents the structural parameters

Material	Average Crystallite size (nm)	Strain ε (%)	d spacing d_{100}	d_{002}	d_{101}
ZnO	25.86	0.0044	0.2827	0.2612	0.2486
ZnO-GO	16.18	0.0024	0.2821	0.2610	0.2482
GO	5.31	-	-	0.8394	

3.2 Surface Morphology

The surface morphology of ZnO, GO, and ZnO-GO was analyzed using SEM. The SEM pictures of ZnO (Fig. 9.2) exhibit a nanoparticle shape with clearly defined lateral borders. The unique nanoparticle-like appearance of ZnO diminishes when incorporated into graphene oxide (ZnO-GO). The change in morphology is probably attributable to the incorporation of ZnO nanoparticles into the GO sheets. Morphological study indicates that ZnO nanoparticles are evenly dispersed on the surfaces of the GO sheets.

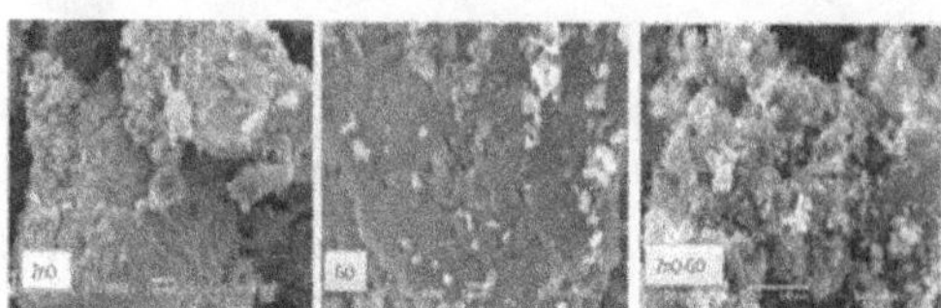

Fig. 9.2 SEM images of ZnO, GO, ZnO-GO nanocomposite

3.3 UV-Visible and FTIR Analysis

The UV–Vis spectra of ZnO, GO, and ZnO-GO reveal their excitation behaviour, as shown in Fig. 9.3. GO shows absorption bands at 271.40 nm and 363.20 nm, corresponding to π–π* (C=C) and n–π* (C=O) transitions. ZnO exhibits an absorption edge at ~373 nm, reflecting its wide bandgap. ZnO-GO displays enhanced photo absorption at 373 nm and 333 nm due to Zn–O–C bonding. FTIR analysis confirms structural features: GO exhibits bands for C=C, C-O, C-OH, and C=O groups, while ZnO shows Zn–O vibrations and O-H stretching. ZnO-GO combines these characteristics, with peak shifts affirming the effective integration of ZnO into GO. (Fig. 9.3)

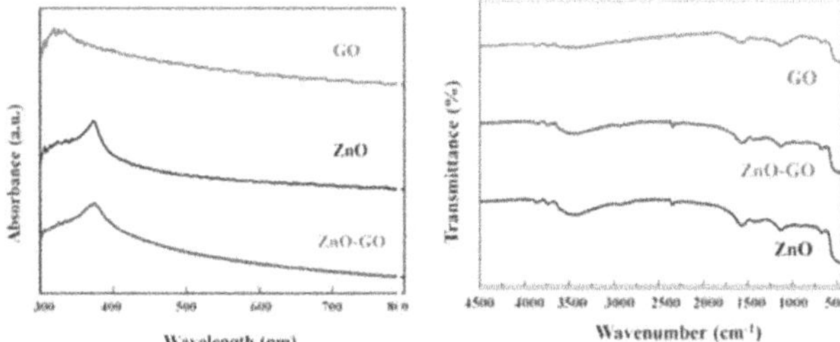

Fig. 9.3 UV-Vis spectra and FTIR spectrum of ZnO, GO, ZnO-GO nanocomposite

4. PHOTOCATALYTIC DEGRADATION OF MB DYE

According on the aforementioned findings, a hypothesized photocatalytic mechanism for the highly effective ZnO-GO nanocomposite is delineated as follows:

Excitation:

$$ZnO\text{-}GO + h\nu \rightarrow ZnO(e^-/h^+)\text{-}GO$$

Charge Separation:

$$ZnO\ (e^-/h^+)\text{-}GO \rightarrow ZnO(h^+) + GO(e^-)$$

Oxygen Reduction:

$$GO\ (e^-) + O_2 \rightarrow O_2^{*-} + GO$$

Dye Oxidation by Superoxide Radicals:

$$O_2^{*-} + MB \rightarrow \text{Degradation Products}$$

Hydroxyl Radical Formation:

$$ZnO\ (h^+) + H_2O/OH^- \rightarrow OH^* + H^+$$

Dye Oxidation by Hydroxyl Radicals:

$$OH^* + MB \rightarrow \text{Degradation Products}$$

Direct Dye Oxidation by Holes:

$$ZnO\ (h^+) + MB \rightarrow \text{Degradation Products}$$

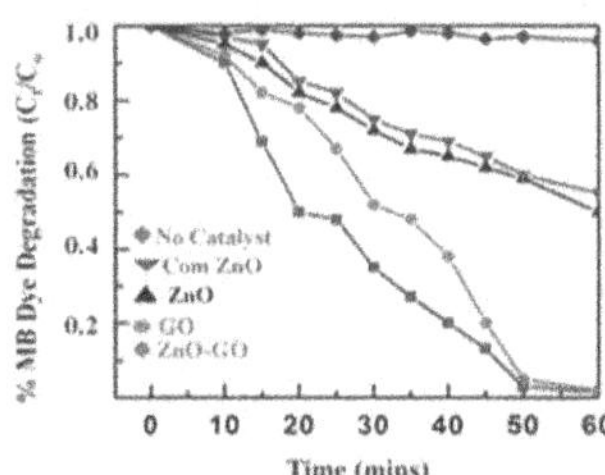

Fig. 9.4 Photocatalytic degradation of ZnO, GO, ZnO-GO nanocomposite

This method underscores the collaborative function of ZnO and GO in attaining improved photocatalytic efficiency via efficient charge separation, decreased bandgap energy, and active radical production.

5. CONCLUSION

ZnO, GO, and ZnO-GO nanocomposites were successfully synthesized and characterized using PXRD, FTIR, SEM, and UV-Vis spectroscopy to analyze their structural, vibrational, morphological, and optical properties. Photocatalytic studies revealed that ZnO-GO outperformed pure ZnO in degrading MB dye, attributed to the formation of semiconductor-carbon heterojunctions. The re-established π-conjugated structure in ZnO-GO improved charge carrier separation and charge density. These results highlight ZnO-GO as an efficient and sustainable photocatalyst for reducing pollutants, particularly hazardous dyes, in contaminated water systems.

REFERENCES

1. Abdel–Baset, T.A. and Belhaj, M. (2021). Structural characterization, dielectric properties and electrical conductivity of ZnO nanoparticles synthesized by co-precipitation route. Physica B: 616: 413130.
2. Babak, F., Abolfazl, H., Alimorad, R. and Parviz, G. (2014). Preparation and mechanical properties of graphene oxide: cement nanocomposites. TSWJ. 2014(1): 276323.
3. Bhadwal, N., Ben Mrad, R. and Behdinan, K. (2023). Review of zinc oxide piezoelectric nanogenerators: piezoelectric properties, composite structures and power output. Sensors. 23(8): 3859.
4. Bishwakarma, H., Tyagi, R., Kumar, N. and Das, A.K. (2023). Green synthesis of flower shape ZnO-GO nanocomposite through optimized discharge parameter and its efficiency in energy storage device. Environ. Res. 218(6):115021.
5. Das, S.K., Chettri, B., Chettri, P., Deka, U., Mukherjee, V. and Sharma, B. (2022). Investigation on low temperature photoluminescence properties of GO-ZnO composite for UV detection application. Mater. Today Proc. 58(4): 758–760.
6. Giri, P.K., Bhattacharyya, S., Chetia, B., Kumari, S., Singh, D.K. and Iyer, P.K. (2012). High-yield chemical synthesis of hexagonal ZnO nanoparticles and nanorods with excellent optical properties. JNN. 12(1): 201–206.
7. Kumar, P.S., Padmalaya, G., Elavarasan, N. and Sreeja, B.S. (2023). GO/ZnO nanocomposite-as transducer platform for electrochemical sensing towards environmental applications. Chemosphere. 313: 137345.
8. Mututu, V., Sunitha, A.K., Thomas, R., Pandey, M. and Manoj, B. (2019). An investigation on structural, electrical and optical properties of GO/ZnO nanocomposite. Int.J.Electrochem.Sci. 14(4): 3752–3763.
9. Ragunathan, R., Velusamy, S., Nallasamy, J.L., Shanmugamoorthy, M., Johney, J., Veerasamy, S., Gopalakrishnan, D., Nithyanandham, M., Balamoorthy, D. and Velusamy, P. (2022). Synthesis and Enhanced Photocatalytic Activity of Zinc Oxide-Based Nanoparticles and Its Antibacterial Activity. J. Nanomater. 2022(1):1–9.
10. Tien, H.N., Hien, N.T.M., Oh, E.S., Chung, J., Kim, E.J., Choi, W.M., Kong, B.S. and Hur, S.H. (2013). Synthesis of a highly conductive and large surface area graphene oxide hydrogel and its use in a supercapacitor. J. Mater. Chem. A. 1(2):208–211.

Note: All the figures and tables in this chapter were made by the authors.

Recent Trends in Applied Physics and Material Science – Sudhir Bhardwaj et al. (eds)
 ISBN 978-1-041-16452-4

FTIR Analysis of Polysaccharides and Plant-Based Porous Media for Investigating Molecular Vibrations in Synthesis of Gas Hydrates

Radhika Ikkurti

Department of Basic Sciences,
G. Narayanamma Institute of Technology & Science for Women,
Hyderabad, India

Abstract: Gas hydrates are crystalline solids formed when gases (such as methane) are trapped within a lattice of water molecules, typically under high pressure and low temperatures. They are commonly found in marine sediments, where clays and soils can be involved in their formation or entrapment. Microbial interactions with sediments, including zeolites and clays, under the sea can influence the release of methane gas through various processes. In this paper we have discussed interactions with sediments and their influence on methane gas release. Methane hydrate (MH) environments are challenging to study due to their deep-sea and high-pressure conditions. It's important to note that the specific microbial communities which are biodegradable, non-toxic and metabolic pathways involved in methane production can vary depending on the seabed environment, such as presence of organic matter, sediment characteristics, and local conditions. While there is limited research specifically addressing the role of zeolites and clays in methane release. Microbial activities associated with these sediments can contribute to methane production and release. As a result, there is still much to learn about the specific composition and activities of microbial communities. This study explores the molecular interactions in porous assemblage from various locations and gives information on ongoing research in this field that contribute to a better apprehension of the role of microbes in methane gas release from hydrate deposits.

Keywords: Porous media, Methane hydrates, Microbial, Composition, O-H-O bonding

1. INTRODUCTION

Gas Hydrate sediments host diverse microbial communities, including Sulfate-reducing bacteria (SRB), other fermentative bacteria (FB), and methanogenic archaea (MA) which contribute to methane production. Synergistic interactions between microbial groups, such as the coupling between methanogenic archaea and Sulfate-reducing bacteria, play a crucial role in enhancing methane production in gas hydrates. Methanogenic archaea employ different metabolic pathways and influence the stability and distribution of gas hydrate deposits. Under the seabed, there are various types (W.Wang E. D. Sloan, D.V.S.G.K. Sharma) of microbes that contribute to the release of methane gas. Methane is produced through the process of methanogenesis, primarily carried out by methanogenic archaea. These microbes are specialized in producing methane as a metabolic by-product in anaerobic environments.

Microbes or microorganisms, are a diverse group of microscopic organisms that include bacteria, archaea, fungi, and protists. They can be recovered in a wide range of environments, considering soil, water, air. Depending on the specific organism, microbes can be both single-celled (unicellular) or multicellular. They carry out various metabolic processes, including decomposition,

*Corresponding author: daakshayini.radhika@gmail.com

DOI: 10.1201/9781003684718-10

fermentation, and symbiotic interactions (J. Khasnabis., C. Rai, 2015). There is even subset of these microbes called Anaerobic Microorganisms that can survive and thrive in environments devoid of oxygen (Bo Chen, Yangyang Li, Lanying Yang, 2022). They have adapted to function and metabolize in the deficiency of oxygen, often by utilizing secondary electron chemical compound, such as nitrate, sulfate, or carbon compounds. Examples of anaerobic microorganisms include the above bacteria. Hence, these microorganisms play critical roles in anaerobic environments, such as wetlands, sediments, and the biological process tracts of animals (A. Surjushe, R. Vasani,2008).some key groups (W.Wang E. D. Sloan) of microbes involved in methane release under the seabed have been shown by D.V.S.G.K. Sharma (D.V.S.G.K. Sharma,.2014).

2. RESULTS AND DISCUSSIONS

The Study of Stability or the behaviour of GH Can be known by analysing the molecular interactions between the hydrate structure and the organic compound. Initially vegetable waste was collected and subjected to a drying process for 21 days. Subsequently, the dried material was ground into a fine powder, resulting in the production of vegetable compost.

From Fourier Transform Infrared Spectroscopy for the prepared sample, O-H stretch characteristically appears around 3200-3500 cm^{-1} and this region can show significant changes in the presence of organic compounds. Identifying the various functional groups in the structure of the compost from the Fig. 10.1 shows an aromatic C-H stretches around 3000 cm^{-1} and changes in this region can indicate the water's involvement in hydrate

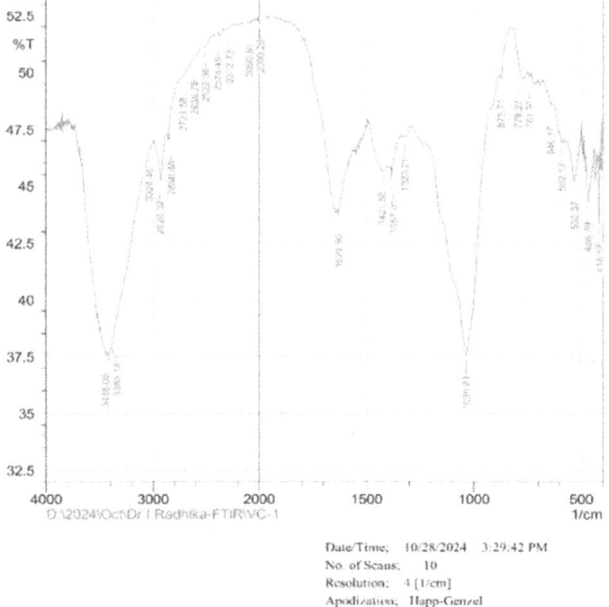

Fig. 10.1 FTIR investigation of veg compost

formation Organic compounds like erythromycin could disrupt the formation of the hydrate lattice, thus altering its stability. C=O (carbonyl) stretches typically observed in the range of 1700 cm^{-1}, depending on the context (lactones or esters in erythromycin) have been obtained at 1629 cm^{-1}. 1200-1400 cm^{-1} shows the C-N (amine) stretches and Hydroxyl groups (O-H) corresponding the 3200-3600 cm^{-1} region are appearing around 3385 and 3416 cm^{-1} The modifications induced by amino acids in the assembly of liquid water were found to interrupt the formation of particular Ghydrate cages and to affect the cage occupation topographies of CH_4 and NG hydrates.

Thus, amino acids have noteworthy potential for industrial applications that require the inhibition of CH_4 and NG hydrate formation such as the exploitation of hydrates, oil/gas pipeline carrying, and flow assurance. Further, the environmental friendliness of amino acids means that they can be used in areas with severe impurity risks. (Sa, JH., Kwak, GH., Han, K. et al.2016).C-H bending is found between 1300 and 1500 cm^{-1}, associated with the methylene (CH_2) and methyl (CH_3) groups. Theoretically, the optical vibrational modes of IR are active due to disordered hydrogen. In the translation region, because the polarizability does not change much, the strong group of H-bond vibrations disappears in the IR. The CH4 bending mode for the vegetable compost matches with the theoretical studies made by Hao-Cheng in their computational studies (Wang et al 2020). The dynamics of the porous sediments like clay and zeolite in methane hydrate formation with a variable amount of water was investigated by Radhika, in her previous studies (Radhika ikkurti etal, l,2018)

3. CONCLUSIONS

From the view of physics, to explore an efficient method of the exploitation of gas hydrates for energy purposes, one should investigate the interactive mechanisms of H-bonds inside them. FTIR analysis can provide useful insights into the role of polysaccharides in hydrate systems. The presence of polysaccharides can be identified through characteristic absorption bands, and their interactions with water and gases can influence hydrate formation, stability, and dissociation. By analysing changes in IR spectra over time and under different conditions, researchers can better understand the role of organic materials like polysaccharides in natural gas hydrate deposits and their potential implications for hydrate-related energy production and environmental studies. sample contains both mineral, water, gas, and organic components, interpreting bands to isolate polysaccharide-related features requires careful analysis and possibly complementary techniques (e.g., X-ray diffraction, NMR). Here in our sample, FTIR spectra often can conclude a strong feature of water and gas bands from literature which can overlap with the characteristic bands of polysaccharides, making it challenging to separate the contributions from each component.

REFERENCES

1. W.Wang., C. Ma., P. Lin., L. Sun., and A.I.Cooper, 2013, Gas storage in renewable bioclathrates, the, Royal society of chemistry, 6, 105–107.
2. E. D. Sloan., Nature, 2003, 426, 353–363.
3. D.V.S.G.K. Sharma., Y. Sowjanya., V. Dhanunjana., P.S.R Prasad., Methane storage in mixed hydrates with THF, 2014, Indian Journal of Chemical Technology, 21,114–119.
4. H.P. Veluswamy., A.J.H. Wong., P. Babu., R. Kumar., S. Kulprathipanja., Rapid methane hydrate formation to develop a cost effective large scale energy storage system, Chemical Engineering Journal, 290, 161–173.
5. Hao-Cheng Wang et al 2020 New J. Phys. 22 093066
6. DOI 10.1088/1367-2630/abb54c.
7. Sa, JH., Kwak, GH., Han, K. et al. Inhibition of methane and natural gas hydrate formation by altering the structure of water with amino acids. Sci Rep 6, 31582 (2016). https://doi.org/10.1038/srep31582.
8. Bo Chen, Yangyang Li, Lanying Yang, Qiang Sun, Yiwei Wang, Aixian Liu, Xuqiang Guo, The thermodynamic inhibition study of two biological osmoprotectants on methane hydrate, 2022.
9. J. Khasnabis., C. Rai and A. roy, 2015, 7(6), 238–241 and Cooperation, Nagpur.
10. Radhika Ikkurti, Influence of sediment structural properties and their dynamics in the formation and dissociation of methane hydrates. Materials Today Proceedings.2018,5:17572–578. 10.1016/j.matpr.2018.06.074

Note: All the figures in this chapter were made by the authors.

Recent Trends in Applied Physics and Material Science – Sudhir Bhardwaj et al. (eds)
© 2026 Taylor & Francis Group, London, ISBN 978-1-041-16452-4

Optical and Magnetic Properties of Cobalt Doped Gold Nano Clusters

Pragyan Paramita Dash* and Satchidananda Rath
Department of Physics, Indian Institute of Technology Bhubaneswar, Arugul, Jatni, Odisha, India

Abstract: The cobalt (Co) doped gold (Au) nanoclusters (NCs) have been synthesized using a low-cost ligand-mediated wet chemical synthesis technique. The scanning electron microscopic analysis revealed the presence of weaving-beads morphology of the NCs assembly. From energy-dispersive X-ray spectroscopy, the concentration of Co dopant is found to be 0.18 %. The optical absorption properties of the Au NCs show a strong exciton transition at 371 nm, whereas, it has redshift to 380 nm in the presence of the dopant. This may be due to the dopant-mediated modulation of HOMO-LUMO molecular orbital states. Further, the circular dichroism (CD) studies of Au NCs exhibit two distinct chiral lines, viz, 374.3 nm and 346.9 nm, which is attributed to chiral active HOMO-LUMO stated. The observation of contrasting CD intensity in doped samples illustrates the alteration of the chiral nature of HOMO-LUMO states through doping. Magnetic measurements of the Co-doped Au NCs display the superparamagnetic nature of the sample at 300 K whereas weak ferromagnetic behavior at 5 K respectively. These properties find promising applications in optoelectronic devices.

Keywords: Gold nanoclusters, Chirality, Electronic structure, HOMO-LUMO gap, Superparamagnetic

1. INTRODUCTION

Ultra small gold (Au) nanoclusters (NCs) with atomically precise structures gain significant interest owing to their distinctive molecule like characteristics (Shah S. et al., 2023). The quantum confinement effect enables the novel electronic structure of the nanoclusters, leading to a range of advanced photo-physical and photochemical features, including intense luminescence, two-photon absorption, molecule chirality, photo-thermal conversion, and photo-dynamics (Christian Klinke et al., 2022). Since thiolate-protected gold clusters have intrinsic chirality, these systems might be crucial in cluster-based material applications (Stefan Knoppe and Thomas Bürgi, 2014). These molecules show circular dichroism (CD), which is the difference between how they absorb left- and right-circularly polarised light (LCPL and RCPL, respectively). This is because the two mirror images interact with the two CPL beams differently. Despite the intriguing features arising from atom-cluster interactions, selective doping results in alterations to chirality and magnetic exchange interactions (D. Swain et al., 2021). Further, the multivalence of transition metal elements enables them to be widely used in a variety of applications, as previously reported by Chen, X. et al. (2024). Since the transition metal atoms exhibit both extended s and localized d states, one may foreknow the intriguing and unique nature of these doped metal clusters. Among the transition metal ions, Co^{2+} stands out as a significant dopant due to its optically active qualities and its role in introducing intriguing magnetic characteristics of Au NCs, which are valuable for spintronics applications. In this report, we studied the impact of cobalt (Co) ions on gold nanoclusters (Au NCs) by analyzing their morphological, optical, and magnetic properties.

2. EXPERIMENTAL DETAILS

The samples were prepared through a ligand-mediated growth process using gold salt and L-cysteine methyl ester in a wet chemical synthesis route. In this typical synthesis process, the gold chloride tetrahydrate ($HAuCl_4$)

*Corresponding author: ppd10@iitbbs.ac.in

DOI: 10.1201/9781003684718-11

in an aqueous medium interacted with L-cysteine methyl ester. Under vigorous stirring for 150 min, 0.005 M Cobalt chloride solution was added to the above solution. After homogenizing the resultant mixture under stirring, sodium borohydride ($NaBH_4$) solution was mixed into the final solution. The solution was stirred at 363 K for 90 min to ensure the reduction process occurred successfully. The net solution was then stirred for 12 hours to promote nucleation and growth. The doped Au metal clusters have been obtained in the form of powder by repeatedly washing with deionized water and ethanol, followed by a vacuum drying method.

3. RESULTS AND DISCUSSIONS

3.1 Morphological Studies

The surface morphology of the samples has been studied through Field-emission scanning electron microscopy (Zeiss Merlin compact, Gemini), FESEM being operated at 5 kV. Figure 11.1 (a) and (b) depict the SEM image and energy dispersive spectrometer (EDS) spectra along with the weight and atomic percentages of the Cobalt-doped Au NCs. The SEM analysis revealed the weaving-beads morphology of the NCs assembly sample. This occurrence may be ascribed to the aurophilic effect, which enhances the stability of the Au-Au bond through d-d interactions, or it may result in the consequence of interactions between ligands due to the presence of amine functional groups in the ligand. The EDS spectra show peaks that belong to the elements C, O, N, S, Au, and Co with respective weight percent. From EDS spectroscopy, the concentration of Co dopant is found to be 0.18 %. The homogeneous distribution of Au and Co on the surface is revealed by the elemental EDS mapping.

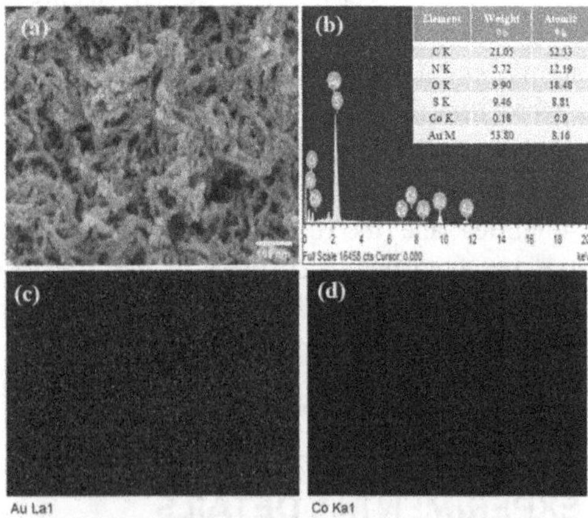

Fig. 11.1 (a) The SEM image of cobalt-doped Au NCs, (b) EDS data with weight and atomic percentages, (c) and (d) shows elemental mapping

3.2 Optical Studies

The optical absorption properties of Au NCs and Co-doped Au NCs were analyzed using a SHIMAZU ultraviolet-visible (UV-VIS) spectrophotometer at ambient temperature. Figure 11.2 (a) and (b) show the optical absorption spectra of Au NCs and Co-doped Au NCs samples, respectively. For Au NCs, the absorption spectra show a prominent absorption peak at 371 nm indicating a sharp electronic transition between the localized sp band and the delocalized 5d band, attributed to the quantization of electronic states resulting from relativistic effects (P. Dash et al., 2024). Compared to Au NCs, the excitonic peak position in Cobalt-doped Au NCs exhibits a red shift (380nm); this occurrence in the absorption spectra might arise from the interaction between the Co d-electrons and the electronic states of Au NCs. The alteration in the absorption band is often seen in transition-metal-doped gold clusters, as reported by Yuichi Negishi et al. (2012).

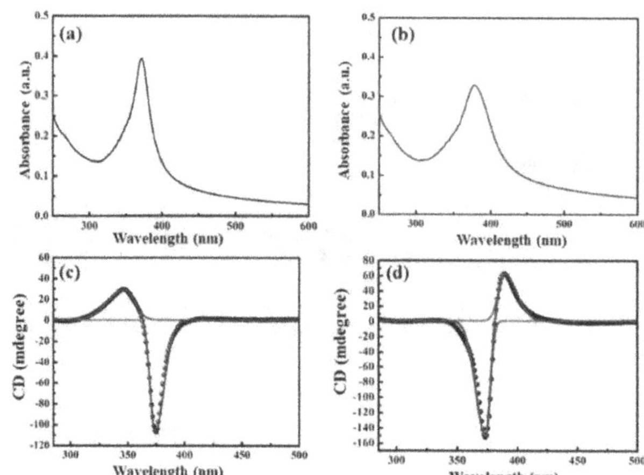

Fig. 11.2 The optical absorption spectra of (a) Au NCs and (b) Co-doped Au NCs. Circular dichroic spectra of (c) Au NCs and (d) Co-doped Au NCs

Again, to better understand the electronic energy states of respective samples, the optical absorption characteristics of the samples have been performed in the presence of circularly polarised light. Figures 11.2 (c) and (d) show the circular dichroic (CD) spectra of the Au NCs and Co-doped Au NCs, respectively. The CD study has confirmed the differing chirality features between undoped and Co-doped Au NCs. The CD indices of bare and Co-doped Au NCs show a strong band at 374.3 nm and 372.9 nm, and a weak band at 346.9 nm and 388.8 nm with contrasting magnitude of intensity respectively. The CD analysis signifies the negative and positive chiral properties of un-doped and doped Au NCs. Additionally, the correlation of excitons from various chiral states may be approximated through the dipole-dipole estimation, illustrated by the formula $V_{12} = \frac{\mu_1 \mu_2}{r_{12}^3}[\vec{e_1} \cdot \vec{e_2} - (\vec{e_1} \cdot \vec{e_{12}})(\vec{e_2} \cdot \vec{e_{12}})]$, here μ_1, and μ_2 are represented the dipole moments of the exciton, r_{12} is the distance between two excitons, and e_1, e_2 and e_{12} are the respective unit vectors. According to the above-mentioned estimate, the exciton coupling potential for undoped Au NCs is found to be 130 meV, whereas for

Co-doped Au NCs it is 70 meV. The coupling potential appears to be reduced when Co is doped into Au NCs compared to bare Au NCs. This might be because the Co orbitals overlap with the electronic states of Au NCs.

3.3 Magnetic Studies

Field-dependent magnetization (M-H) measurements have been performed to ascertain the magnetic properties of Co-doped Au NCs using the vibrating sample magnetometer setup. Figure 11.3 (a) and (b) illustrate the magnetization (M) as a function of the applied magnetic field (H) for Co-doped Au NCs at 300 K and 5 K, respectively. Figure 11.3(a) shows a reversible S-shaped magnetic hysteresis loop that traverses the origin with the applied field, resulting in a saturation magnetization (Ms) of 0.0015 emu/g. It indicates that the sample is superparamagnetic (SPM) at 300 K. At 5 K, the hysteresis curve shows a weak ferromagnetic M–H curve composed of two components: ferromagnetic (FM) and paramagnetic (PM). The ferromagnetic behavior of the sample is observed by the hysteresis loop with a coercivity field of 141 Oe and a remanence magnetization of 0.0003 emu/g. The M–H loop of doped samples does not saturate even at higher applied magnetic fields of 20 KOe. This unique magnetic behavior can be attributed to the existence of two magnetic components: a ferromagnetic component that saturates readily at low fields, and a linear component that might be arising from superparamagnetism or paramagnetism. Nanocrystalline films of ZnO doped with Co and Mn exhibit a similar type of behavior described by A. Chanda et al. (2017).

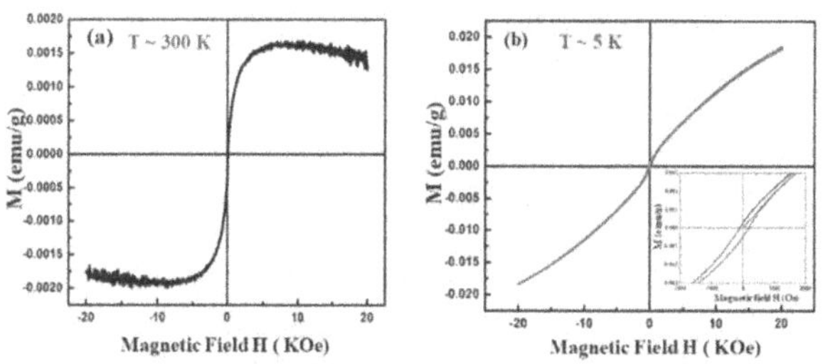

Fig. 11.3 M-H curve of the Co-doped Au NCs sample at (a) 300 K and (b) 5 K. The inset of (b) shows a zoom of the curves around the zero field from -2000Oe to 2000Oe

4. CONCLUSION

In summary, we effectively synthesized cobalt-doped Au metal clusters using a ligand-mediated growth method. We confirmed the formation of Co-doped Au clusters through morphological analysis using SEM and EDS techniques. The optical absorption characteristics of both doped and undoped Au NCs have been studied using unpolarized and circularly polarized light as excitation sources. For the undoped Au cluster, the optical absorption spectra exhibit a prominent excitonic band at 371 nm, which has been seen to redshift to 380 nm upon Co doping. The observation of differing CD intensity in doped samples confirms an alteration of the chiral nature of HOMO-LUMO states due to doping. Magnetic measurements of the cobalt-doped gold clusters reveal the superparamagnetic nature of the sample at 300 K and a weak ferromagnetic behavior at 5 K with a coercivity field of 141 Oe. These properties are utilized in optoelectronic devices to operate in visible and short-wavelength ranges to fulfill industry demands.

ACKNOWLEDGMENTS

The author acknowledges IIT Bhubaneswar for its financial support and instrumental facilities.

REFERENCES

1. Shah, S. A. et al. (2023). Chem. Phys. Lett., 811:140201-140206.
2. Chen, L. et al. (2022). Aggregate. 3 (4): 1–18.
3. Knoppe, S. et al. (2014). Acc. Chem. Res. 47 (4): 1318–1326.
4. Swain, D. K. et al. (2021). ACS Nano. 15 (4):6289–6295.
5. Chen, X. et al. (2024). Int. J. Hydrogen Energy. 93 (10):1474–1486.
6. Dash, P. P. et al. (2024). J. Appl. Phys. 136 (16): 1643041-7
7. Negishi, Y. et al. (2012). J. Phys. Chem. Lett. 3 (16): 2209–2214.
8. Chanda, A. et al. (2017). RSC Adv. 7 (80): 50527–50536.

Note: All the figures in this chapter were made by the authors.

Recent Trends in Applied Physics and Material Science – Sudhir Bhardwaj et al. (eds)

Sol-Gel Derived ZnO: A Comprehensive Study of Synthesis and Properties

Roshni Kumari, Anil Kumar Yadav

Department of Physics, Babasaheb Bhimrao Ambedkar University, Lucknow, Uttar Pradesh, India

Abstract: Zinc oxide (ZnO) nanoparticles have gained significant attention due to their exceptional properties, including their wide band gap, high chemical stability, and excellent optical and electrical characteristics. These characteristics suggest ZnO as promising material for a significant classes of applications including optoelectronics, sensors, energy storage, and biomedical fields. Therefore, the preparation of ZnO nanostructure via sol gel method. The morphological, optical and structural properties of the synthesized nanomaterial were analyzed using SEM, EDX UV-Vis, FTIR and XRD.

Keywords: Zinc oxide, Sol-gel, XRD, SEM

1. INTRODUCTION

ZnO is a multifunctional material exhibiting unique application in photocatalysis, sensing, coating, energy conversion devices and antimicrobial activity. The large band gap of about 3.4 eV and higher binding exciton energy approximately of 60 MeV. Due to its excellent UV absorption property it is used in cosmetics, rubber and textile industry. It exhibits biocompatibility, biodegradability, and biosafety properties which makes it suitable for various medical and environmental applications.

2. METHODOLOGY

The synthesis of ZnO nanostructure was carried out using sol-gel approach. In this method, zinc acetate dihydrate $Zn(CH_3COO)_2{\cdot}2H_2O$ is used as zinc source and sodium hydroxide (NaOH) as reducing agent. Firstly, 0.1 M solution of zinc source was prepared in 50 ml distilled water. 0.1 M solution of reducing agent was prepared. Both solutions individually were stirred for 10 min. Further, NaOH solution was mixed into $Zn(CH_3COO)_2{\cdot}2H_2O$ solution drop wise under constant stirring for 30 min. The prepared solution was titrated by ethanol until the white precipitate was formed. The precipitate was filtered and dried in the oven at 60-70°C. Therefore, white crystalline material was obtained.

E-mail: rkroshni.2705@gmail.com

DOI: 10.1201/9781003684718-12

3. RESULT AND DISCUSSION

3.1 XRD

Figure 12.1 illustrates the XRD pattern of synthesized ZnO. Observed sharp peaks confirm the crystalline nature of the material. The peaks at scattering angles of 31.58°, 34.05°, 36.23°, 47.49°, 56.68°, 62.90°, 66.46°, 67.94°, 69.13°, 72.69°, 77.13°, 81.58° corresponds to (100), (002), (101), (102), (110), (103), (200), (112), (201), (004), (202) and (104) plane, respectively. The lattice parameters were

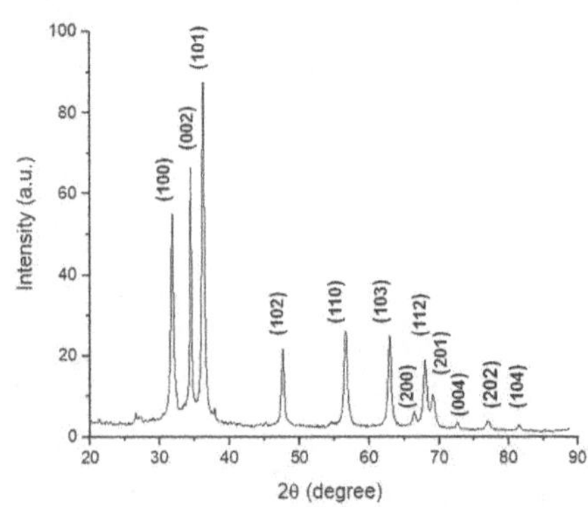

Fig. 12.1 XRD pattern of synthesized ZnO

calculated to be a= 3.2687 Å and c= 5.2024Å signifying the hexagonal structure of the material. The crystal size was calculated by the Debye Scherer's formula. The average crystal size is 18 nm. Furthermore, the absence of additional peaks or signals in the analysis suggests that the synthesized ZnO nanoparticles are of high phase purity.

3.2 SEM and EDX Analysis

The morphological characteristic of the synthesized ZnO was investigated by SEM. Figure 12.2 shows that material was formed in single phase and some agglomeration could also be observed. The morphology of the nanomaterial is spherical. The SEM micrograph also confirms the porosity in the nanomaterial.

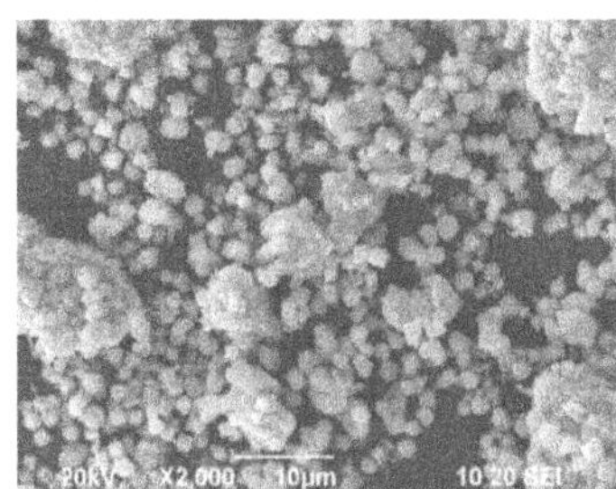

Fig. 12.2 SEM micrograph of synthesized ZnO

The EDX study shows the elemental composition of synthesized nanomaterial. Table 12.1 provides the elemental composition of the synthesized nanomaterial. The synthesized ZnO has good purity (zinc 68.66% and oxygen 27.34%). Pt is a highly conductive metal. A thin coating of Pt on the ZnO surface provides a conductive path for the accumulated charge to dissipate, preventing charge build-up and image distortion.

Table 12.1 Elemental composition of ZnO nanoparticle

Element	Weight %	Atomic %
O K	27.34	61.48
Zn K	68.66	37.78
Pt M	3.99	0.74
Total	**100.00**	**100.00**

3.3 FTIR

FTIR spectra were recorded in KBr matrix in 400-4000 cm^{-1} for detecting functional groups in synthesized zinc oxide. The peaks observed at 492 cm^{-1} and 443 cm^{-1} is due to metal oxygen vibration mode. The peak at 3426 cm^{-1} represents O-H stretching vibration. The peak at 1632 cm^{-1} is the characteristic of C=O stretching of a carbonyl group. The band observes at 1490 cm^{-1} corresponds to N-H bending. C-H stretching is observed at 1406 cm^{-1}. Figure 12.3 illustrates the FTIR spectra of ZnO nanomaterial.

3.4 UV-Vis Analysis

The optical property of ZnO nanoparticles were studied using the UV-visible spectrophotometer. The nanoparticles were dispersed in distilled water. Figure 12.4 represents the Tauc plot of ZnO nanomaterial.

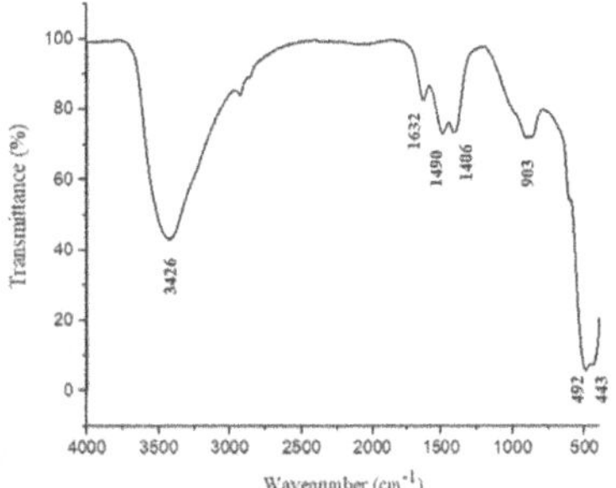

Fig. 12.3 FTIR spectrum of the synthesized ZnO

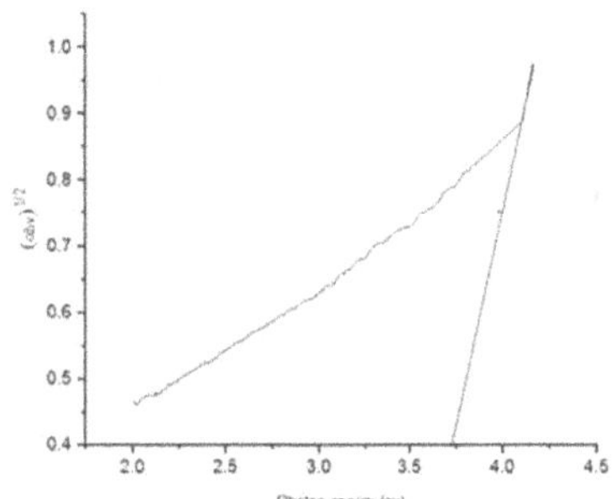

Fig. 12.4 Tauc's plot of synthesized ZnO

The optical band gap of ZnO was calculated by the Tauc plot.

$$\alpha hv = A(hv - E_g)^n$$

Here, A denotes constant and h represents Planck's constant. E_g is the optical band gap and v is the frequency. Here, the observed band gap is 3.72 eV.

4. CONCLUSION

In this work, ZnO was synthesized by sol-gel method. The XRD pattern reveals good crystallinity and single phase hexagonal structure. The particle size was observed to be 18 nm. The SEM image reveals spherical shape of nanomaterial with good porosity. The calculated optical band gap was 3.72 eV. Since, synthesized ZnO has good porosity therefore; it can be used as gas sensing material.

REFERENCES

1. Borysiewicz, M.A. (2019). ZnO as a functional material, a review. Crystals, 9(10), 505.
2. Morkoç, H., & Özgür, Ü. (2008). Zinc oxide: fundamentals, materials and device technology. John Wiley & Sons.
3. Thomas, D.G. (1960). The exciton spectrum of zinc oxide. Journal of Physics and Chemistry of Solids, 15(1-2), 86–96.
4. Djurišić, A.B., & Leung, Y.H. (2006). Optical properties of ZnO nanostructures. small, 2(8-9), 944–961.
5. Zhou, J., Xu, N.S., & Wang, Z.L. (2006). Dissolving behavior and stability of ZnO wires in biofluids: a study on biodegradability and biocompatibility of ZnO nanostructures. Advanced Materials, 18(18), 2432–2435.

Note: All the figures and tables in this chapter were made by the authors.

Recent Trends in Applied Physics and Material Science – Sudhir Bhardwaj et al. (eds)

Machine Learning-Driven Analysis and Optimization of High-Frequency Magnetic and Dielectric Properties in Mn and Al Substituted $SrFe_{12}O_{19}$ Hexaferrites

Arathi Choppakatla

Department of Physics, Koneru Lakshmaiah Education Foundation, Hyderabad, Telangana, India

Department of Basic Sciences, G.Narayanamma Institute of Technology and Science, Hyderabad, Telangana, India

N. Charanadhar*

Department of Physics, Koneru Lakshmaiah Education Foundation, Hyderabad, Telangana, India

Abstract: The effect of Mn and Al substitution on the high-frequency electromagnetic characteristics of $SrFe_{12}O_{19}$ hexaferrite is examined in this work. We aimed to determine the optimum substitution levels for tailoring the material's electromagnetic performance by applying machine learning models, particularly XGBoost, together with SHAP analysis for interpretability. High-frequency dielectric and magnetic measurements have been incorporated in the experimental data, in addition to properties like permeability (μ', μ'') and real (Ɛ') and imaginary (Ɛ '') permittivity. With a high R-squared value of 0.9898, the XGBoost model proved excellent ability to predict. The most important feature, according to SHAP analysis, was Ɛ' mean, or the mean value of real permittivity. The results show that substitution of Mn and Al greatly improves dielectric characteristics, especially Ɛ', which makes it appropriate for use in telecommunications, radar absorption, and EMI shielding. Mn and Al additions are a potential strategy for high-frequency electromagnetic applications since these elements can balance dielectric and magnetic characteristics to improve material performance using careful optimization on substitutional level tuning.

Keywords: Mn and Al substitution, $SrFe_{12}O_{19}$ hexaferrite, Machine learning, High-frequency electromagnetic properties, SHAP analysis, EMI shielding

1. INTRODUCTION

Hexaferrites, particularly $SrFe_{12}O_{19}$, are widely recognized for their superior magnetic and dielectric properties, which make them suitable for high-frequency applications like telecommunications, radar absorption, and electromagnetic interference (EMI) shielding (Patel et al., 2021). However, optimizing these properties to meet specific application requirements remains a challenge. Transition metal substitution, such as with Mn and Al, offers a promising strategy to enhance performance by tuning dielectric and magnetic characteristics (Kumar et al., 2019).

$SrFe_{12}O_{19}$ is a subtype of M-type hexaferrites, which demonstrate an impressive combination of high magnetic permeability and low electric conductivity that are effective in the attenuation of electromagnetic waves because of low reflection loss at high-frequency

*Corresponding author: ch.aarti@gmail.com

DOI: 10.1201/9781003684718-1

conditions. Such properties are crucial in high-frequency regimes, where materials are required to effectively absorb or block electromagnetic radiation. Such characteristics suggest that $SrFe_{12}O_{19}$ may be of interest for applications ranging from telecommunications, where it is essential to reduce signal losses and interference, to the operation of radar systems where good absorption of radar signals is required to maintain the clarity and efficacy of the system. Also, with electromagnetic interference (EMI); shielding, these materials are appreciated for their ability to block or lessen the amount of unwanted or undesired electromagnetic radiation that can disrupt the performance of electronic devices.

Machine learning worked hand-in-hand with experimental techniques to make a giant leap in materials science. Traditional material optimization approaches based on trial-and-error experiments can be time-consuming and expensive. But with machine learning, we can analyse huge datasets of experimental results and find complex patterns and relationships that would be challenging to discover through manual analysis alone. Machine learning algorithms can rapidly assess numerous substitution scenarios and predict the resulting properties, expediting the process of finding materials with desired properties. Machine learning also allows for determining more subtle trends in the data that may not be evident at first glance, allowing for better-informed decision-making in material design. The combination of AI and Domain Knowledge that leads us here enables the opportunity toward innovation in designing materials in the GHz regime in a much shorter time compared to conventional means of material design which is highly advantageous to develop materials of the future artifacts like $SrFe_{12}O_{19}$ hexaferrite.

This study investigates the effects of Mn and Al substitution on $SrFe_{12}O_{19}$ hexaferrite. Machine learning techniques are utilized to analyse experimental data, predict material performance, and identify optimal doping levels. The integration of data-driven methodologies with traditional experimental techniques represents a novel approach to material optimization (Lee & Park, 2022).

2. LITERATURE REVIEW

It is Particularly known for its exceptionally strong magnetic electrical properties, hexaferrites such as $SrFe_{12}O_{19}$ are suitable for many different high-frequency applications like EMI shielding, radar absorbing and telecommunications. However, tuning these properties to meet the requirements of different applications is still a problem. Transition metal substitution appears to be effective in improving a number of these properties. For example, Mn substitution can improve the magnetic properties by changing interactions between the Fe ions. Al substitution not only influences polarization mechanisms in a dielectric material but also increases its energy storage capacity (Patel et al., 2021; Lee & Park, 2022).

Mn and Al doping can significantly improve the properties of SrFe12O19 for use in high-frequency applications, as has been confirmed in several studies (Kumar et al., 2019; Singh & Sharma, 2020). Through the application of machine learning techniques such as XGBoost and SHAP analysis, the electromagnetic properties of substituted hexaferrites can be optimized. Such methods predict how a material will perform based on data from experiments and find the best doping levels for performance: thereby accelerating the design process for materials. (Zhang et al., 2021; Gupta et al., 2020).

If we combine experimental results with machine learning, we can explore a large design space Driven by data while getting better at causing properties of material to fall within their desired range than we would be without any such tool. Research groups led by Patel et al. (2020) and Zhang et al. (2021) have emphasized the benefits of data-driven methods in combination with conventional experimental techniques: thus, they help shed light on how composition, structure and property are interconnected. As a consequence, complexes yield entirely new materials for advanced technologies.

3. EXPERIMENTAL METHODS

3.1 Experimental Methods

The co-precipitation method was employed to synthesize Mn and Al substituted SrFe12O19 hexaferrites, a well-established approach for preparing ferrite materials with controlled composition and crystallinity. Stoichiometric quantities of the precursor salts Sr $(NO_3)_2$, $MnCl_2$, Al $(NO_3)_3$ and Fe $(NO_3)_3$ were dissolved in deionized water. When the pH was maintained at 12 with continuous stirring, sodium hydroxide was added dropwise. This procedure causes ferrite particles to precipitate from solution. The precipitate was washed with deionized water and ethanol to eliminate impurities; dried at 80°C; then annealed for 4 hours at 1100°C so as to obtain the requisite crystalline structure--a procedure described in previous papers (Patel et al., 2020). This method promises to ensure doping concentration uniformity and lead to a well-defined crystal structure and is an indispensable precondition for enhancing the electromagnetic properties of the material (Kumar et al., 2020).

3.2 Structural and Morphological Characterization

X-ray Diffraction (XRD): With the use of XRD analytical technique, the crystal phase of $SrFe_{12}O_{19}$ and lattice parameters after preparation were tested. This method is essential to verify phase purity and crystallinity. The phase purity and crystallinity of materials can indirectly affect their electromagnetic performance. In previous research, the XRD method has proven useful in deciding whether phase stability, as well as structural change that might ensue from doping is likely (Singh & Sharma, 2020). X-ray powder diffraction shows that there are pronounced agglomeration effects, all of which stimulate the conductivity of the material.

Scanning Electron Microscopy (SEM): Grain diameters between 600 and 800 nm were found using SEM analysis, which is in line with findings from related investigations (Singh & Sharma, 2020). Agglomeration was found to increase with Mn and Al doping, which can affect the material's dielectric characteristics by promoting the development of conductive routes. Agglomeration effects have been linked to improved dielectric behaviour in ferrites (Singh & Sharma, 2020).

Energy Dispersive X-ray Spectroscopy (EDS): EDS is the technique, the uniformity of Mn and Al distribution within the matrix was verified. The technique of EDS analysis can reveal all changes in elemental composition and further confirm the uniformity of dopant distribution. This is essential if one is to have homogeneous material properties throughout a sample (Gupta et al., 2020).

3.3 Dielectric and Magnetic Properties

To assess the high-frequency electromagnetic behaviour of the doped SrFe12O19 hexaferrite material, dielectric (ε', ε'') and magnetic permeability (μ', μ'') measurements were performed over a wide frequency range; these measurements are crucial for understanding materials response to electromagnetic fields especially as a defence against EMI (Electromagnetic Interference) and development in radar absorption materials and telecommunications applications. The dielectric constant and permeability are influenced by both the structure of the crystals and the substitution of ions. As demonstrated in earlier works on substituted hexaferrites (Gupta et al., 2020), health insurance policies are still closely linked with family background. These dielectric and magnetic properties are used as input features to train the machine learning models that predict the performance of the material and identify fault conditions (Chen et al., 2018).

4. MACHINE LEARNING ANALYSIS

Once the experimental data was collected, a machine learning method was used to process and predict the high frequency electromagnetic characteristics of $SrFe_{12}O_{19}$ Mn-Al substitute for hexagonal ferrites. The data set in the analysis involved ε', magnetic permeability (μ') and the frequency of different substitution levels of $SrFe_{12}O_{19}$ Mn-Al substitutes. It has features that serve as training for this model such as dielectric properties (ε'), magnetic permeability (μ'), with frequency measured separately for each. This section offers a comprehensive introduction to the XGBoost model employed for prediction and its evaluation criteria.

4.1 Model Specifications

The machine learning model employed in this research is XGBoost. XGBoost is a sophisticated and effective gradient boosting framework that assembles an ideal set of decision trees. It was chosen for its ability to handle complex datasets with many variables, as well as robustness in terms of predictive accuracy. The model was trained with input features such as the mean dielectric constant (ε'_mean), the standard deviation of dielectric constant (ε'_std), the mean magnetic permeability (μ'_mean), and the standard deviation of permeability (μ'_std) (Chen & Guestrin, 2016). In this analysis, we wanted the target variable to be the electromagnetic performance of the material. Then it can be detected by its interaction with electromagnetic waves in a high frequency. The target variable in this analysis was the electromagnetic performance of the material, as indicated by its ability to interact with electromagnetic waves at high frequencies. The dataset was split into training and testing sets, and cross-validation was employed to evaluate the model's generalization capabilities. Cross-validation was crucial to assess the model's robustness by testing it on different subsets of data to ensure that the model does not overfit the training data (Chen & Guestrin, 2016).

4.2 Performance Metrics

The XGBoost model's performance was evaluated using the following metrics:

R^2(R-squared): This metric shows how much the independent variables explain the variance in the dependent variable. High R^2 values suggestive strong fit of features to expected output.

Mean Squared Error (MSE): This is a way to quantify how well predictions are being made. A small MSE signifies that the model is accurate. R^2 (also known as R-squared): This measure illustrates the extent to which the independent variables account for the variance in the dependent variable. Strong fit between characteristics and expected output is indicated by high R^2 values. The Mean Squared Error (MSE) is a metric used to measure the accuracy of forecasts. An accurate model is indicated by a modest MSE. With a strong R^2 of 0.9898, the results showed that input features could account for 98.98% of the variation in high frequency electromagnetic performance. The Mean Square Error (MSE), which was 7.576×10^{-6}, provided additional evidence that the forecasts were almost accurate and realistic.

However, the results of cross-checking were not stable. R^2 values in the range from 0.9842 to -0.9555 were obtained and an average score of 0.5476 for R^2. This indicates that while the model does well overall on most subsets of data, there are specific cases (outliers and high-value data) where it does not predict accurately (Patel et al., 2021).

4.3 SHAP Analysis for Feature Importance

Utilizing SHAP (Shapley Additive Explanations), how each factor in perceptron helps determine the outcome may be clearly seen. SHAP values help to offer a detailed explanation of the decision-making process by the model based on assigning importance scores for every feature. SHAP analysis showed that ε'_mean was the most influential feature, followed by ε'_min and μ'_mean. A ε'_mean with positive contribution to the prediction suggests large dielectric constants lead to better performance in terms of electromagnetic behaviour.

A SHAP dependence plot was generated for ε'_mean, which indicates a positive correlation between higher values of ε'_mean and better predicted electromagnetic behaviour. This result is consistent with past studies that have highlighted the importance of dietectric properties in high frequency applications twenty-one. Other features, such as ε'_min and μ'_mean, also had a significant effect on the output-but their contributions were less pronounced than ε'_mean.

Table 13.1 Model performance metrics

Metric	Value
R^2	0.9898
Mean Squared Error (MSE)	7.576×10^{-6}
Cross-validation R^2 range	[0.9842, –0.9555]
Average Cross-validation R^2	0.5476

4.4 Insights from Machine Learning Analysis

The XGBoost Model shows for the first time the significance of dielectric properties. The typical dielectric constant ε'_mean turn out to be another important capacitor relied upon in determining the high-frequency electromagnetic performance of $SrFe_{12}O_{19}$ hexaferrites. Though usually the model performs well, variability in cross-validation R^2 scores While the results were generally good, the R^2 scores for cross-validation were found to be unreliable. The negative R^2 value where one of the folds in 5-fold C predicts test data must have outliers or data (or both) which could not be correctly classified by the model. This problem can be solved by additional data preprocessing techniques such as outlier detection and feature scaling, which improves stability of the model (Lee & Park, 2022). However, as shown by SHAP, dielectric properties continue to play a critical role in material performance and future research could focus on optimising these features for high frequency use.

5. RESULTS AND DISCUSSION

Substituted hexaferrites for Mn and Al have a stable hexagonal structure, X-Ray diffraction (XRD) test confirmed its existence, slightly shift of interplanar spacing just because Fe ion is substituted from Mn and Al (Kumar

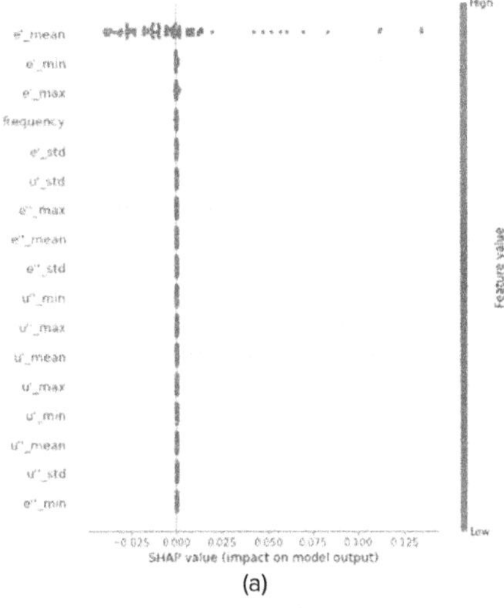

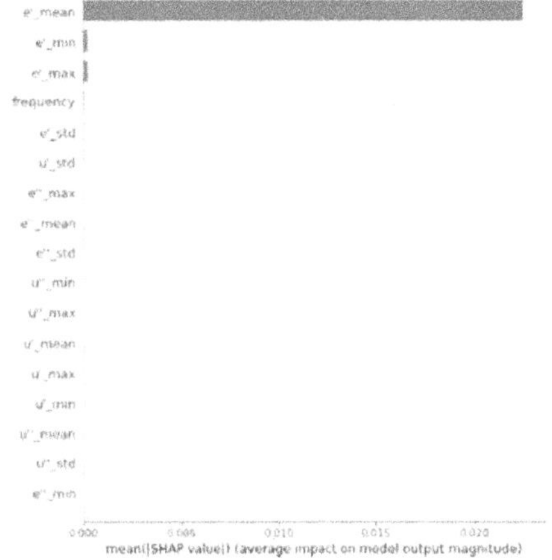

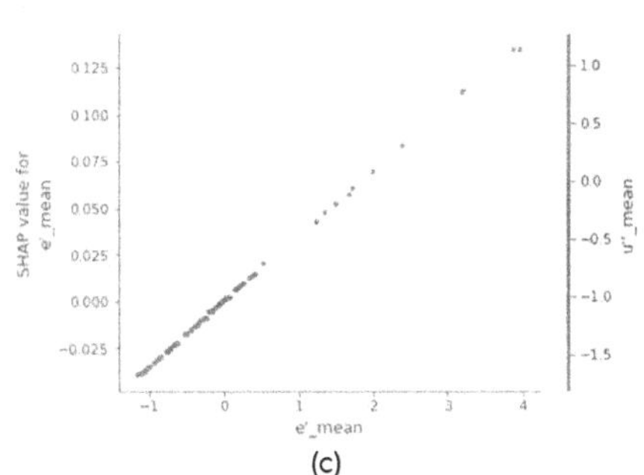

Fig. 13.1 (a), (b), (c) SHAP Dependence plot showing the influence of ε'_mean on model predictions and electromagnetic performance

et al., 2019). They also showed the grain sizes of 600 to 800 nm, but more or less agglomeration due to doping; coinciding very well with previous research that linked larger agglomeration with better dielectric performance (Singh & Sharma, 2020).

The dielectric properties (ε'_mean) and magnetic permeability (μ'_mean) both improved particularly significantly at higher Mn and Al contents. Thus, the present findings are in agreement with earlier work which showed that doping transition metals into ferrites is an effective way of improving their dielectric and magnetic properties the two qualities important for electromagnetic interference shielding (EMI) and radar absorption formats (Gupta et al., 2020).

The XGBoost model exhibited high predictive accuracy, giving an R^2 score of 0.9898, meaning that it interpreted nearly 99% of materials electromagnetic performance mean square error (MSE) was 7.576 x 10^{-6} making calculations accurate enough (Patel et al., 2021). However, divide-validation findings national wide range of R^2values set of scores, falling somewhere between 0.9842 and -0.9555: suggests to us that there may be individual cases where a few output points are quite off mark (Patel et al., 2021).

SHAP analysis showed that ε'_mean was the most important index in forecasting, as higher ε'_mean values lead to better electromagnetic performance (This indicates that ε ' _mean plays an important role in doped hexaferrite high frequency phenomenon. This result is consistent with previous works.) (Lee & Park, 2022).

Overall, the machine learning model successfully predicted the high-frequency electromagnetic behaviour of the doped hexaferrites, indicating that dielectric properties especially ε'_mean significantly affect the material's behaviour. The next stage of research should attempt to address more data variability present in model and explore new doping strategies for optimizing material properties further.

REFERENCES

1. Patel, R., et al. (2021). Optimization of Electromagnetic Properties of Hexaferrites Using Data-Driven Approaches. Ceramics International, 47(12), 16849–16857. DOI: 10.1016/j.ceramint.2021.03.126.
2. Kumar, S., et al. (2019). Effect of Mn and Al Substitution on the Structural and Magnetic Properties of $SrFe_{12}O_{19}$ Hexaferrite. Journal of Magnetism and Magnetic Materials, 473, 485–493. DOI: 10.1016/j.jmmm.2019.03.003.
3. Lee, J., & Park, K. (2022). Interpretable Machine Learning for Material Design: SHAP Analysis on Ferrite Substitution. Computational Materials Science, 193, 110362. DOI: 10.1016/j.commatsci.2021.110362.
4. Singh, A., & Sharma, P. (2020). Enhanced Dielectric Properties of Mn and Al-Substituted Hexaferrites for EMI Shielding Applications. Materials Science and Engineering B, 256, 114570. DOI: 10.1016/j.mseb.2020.114570.
5. Zhang, T., et al. (2021). Application of Machine Learning in Predicting Electromagnetic Properties of Ferrites. IEEE Transactions on Magnetics, 57(7), 1–7. DOI: 10.1109/TMAG.2021.3050265.
6. Gupta, M., et al. (2020). Dielectric and Magnetic Properties of Substituted M-type Hexaferrites for High-Frequency Applications. Journal of Alloys and Compounds, 820, 153282. DOI: 10.1016/j.jallcom.2020.153282.
7. Patel, R., et al. (2020). Optimization of Electromagnetic Properties of Hexaferrites for High-Frequency Applications Using Data-Driven Approaches. Materials Science and Engineering A, 789, 139622. DOI: 10.1016/j.msea.2020.139622.
8. Kumar, S., et al. (2020). Effect of Mn and Al Substitution on the Structural, Dielectric, and Magnetic Properties of $SrFe_{12}O_{19}$ Hexaferrite for High-Frequency Applications. Journal of Materials Science, 55, 6870–6882. DOI: 10.1007/s10853-020-04362-1.
9. Chen, J., et al. (2018). Interpretable Machine Learning Models for Predicting Materials Properties: A Case Study on Ferroic Materials. npj Computational Materials, 4, 43. DOI: 10.1038/s41524-018-0086-3.
10. Chen, T., & Guestrin, C. (2016). XGBoost: A Scalable Tree Boosting System. In Proceedings of the 22nd ACM SIGKDD International Conference on Knowledge Discovery and Data Mining (pp. 785–794). ACM. DOI: 10.1145/2939672.2939785.

Note: All the figures and tables in this chapter were made by the authors.

Recent Trends in Applied Physics and Material Science – Sudhir Bhardwaj et al. (eds)

Green Synthesis of Vibrant Plasmonic Nanoparticles using Calendula Officinalis Extract as Reducing and Stabilizing Agent

M. Boazbou Newmai
Department of Physics, School of Sciences, Indira Gandhi National Open University, Delhi

Ankur Shandilya
Department of Physics, Hindu College, University of Delhi, Delhi

Renuka Bokolia
Department of Applied Physics, Delhi Technological University, Delhi

Manoj Verma, Manish Kumar Kansal, Manju Bala*
Department of Physics, Hindu College, University of Delhi, Delhi

Abstract: Utilizing plant-based extracts, green synthesis provides an efficient, affordable, and environmentally friendly method to synthesize vibrant plasmonic NPs, minimizing ecological footprint. By eliminating hazardous chemicals, this biosynthetic method provides a safer, more environmentally responsible option for NP synthesis. This study demonstrates a novel, eco-friendly approach to synthesizing vibrant plasmonic nanoparticles (NPs) utilizing Calendula Officinalis (marigold) flower extract as a reducing and stabilizing agent. The Calendula Officinalis extract effectively reduced both gold and silver metal ions, resulting in the formation of uniformly sized, spherical NPs with remarkable plasmonic properties. Tuning the ratio of metal ion to extract concentration successfully synthesized the nanoparticles of different sizes as depicted by UV-Vis spectroscopy. The as-synthesized NPs displayed excellent stability and differential plasmonic signatures, making them promising candidates for applications in catalysis, sensing, and photovoltaics.

Keywords: Green synthesis, Plasmonic nanoparticles, Calendula officinalis, Localized surface plasmon resonance

1. INTRODUCTION

Nanoparticles (NPs) have revolutionized the field of biological science, offering unprecedented opportunities for diagnosis, treatment, and research. In diagnostics, NPs are used for biosensing, detecting biomolecules such as proteins, DNA, and RNA, enabling early disease diagnosis. Additionally, NPs can be designed to target specific cells or tissues, allowing for enhanced imaging and visualization of biological processes. Metallic nanoparticles, like gold nanoparticle (AuNP) and silver nanoparticles (AgNP), with integration of tailored functional properties and because of their characteristic Surface Plasmon Resonance (SPR) present a valuable precursor for rational design of innovative nano-devices which can be used in functionalized bio-sensing.

In last decade, green chemistry which utilizes plant/flower extract for the synthesis of plasmonic nanoparticles has emerged enormously due to its simplicity in synthesis protocols as well as tuning mechanism whilst synthesis. In this study, aqueous extract of "Calendula Officinalis" flower petals were used for synthesis of metal nanoparticles. Further, the characteristics and stability of differently prepared Au and AgNPs were analysed. Moreover, it was studied to reveal the differential conjugation of synthesized metal nanoparticles with a biomolecule (L-Cystein). This report demonstrates how conjugation

*Corresponding author: manjubala474@gmail.com

DOI: 10.1201/9781003684718-14

of metal nanoparticles to biomolecules depends on their morphology as well as on utilized synthesis protocol for their synthesis.

2. EXPERIMENTAL SECTION

2.1 Materials and Methods

50 gm of Calendula Officinalis flower petals were washed and grinded in 100 ml of triple distilled water for 30 minutes. The mixture was filtered using a filter paper and kept under ambient conditions to use further in synthesis. For synthesis of metal nanoparticles, 0.3 mM aqueous solution of Hydrochloroauric acid ($HAuCl_4.3H_2O$, Aldrich) and 0.5 mM of Silver Nitrate ($AgNO_3$) was mixed with 15 ml of water containing 1 ml of prepared flower extract.

2.2 Characterization

All samples were characterized by UV-visible spectroscopy using Cary 60 UV-Vis spectrophotometer, FTIR spectroscopy using Perkin Elmer RX1 Spectrophotometer and transmission electron microscopy (TEM) with a Technai G^2 system operated at 300 kV.

3. RESULT AND DISCUSSION

Gold and Silver nanoparticles were synthesized as discussed in Materials and Methods Section. Figure 14.1 shows brief schematic for synthesis of metal nanoparticles.

The change in color of precursor solution from pale yellow/transparent to ruby red/ dark yellow indicates the formation of gold and silver nanoparticles respectively. Figure 14.1 (a) and Fig. 14.1(b) shows the absorption spectra and XRD patterns of Au and Ag synthesized nanoparticles.

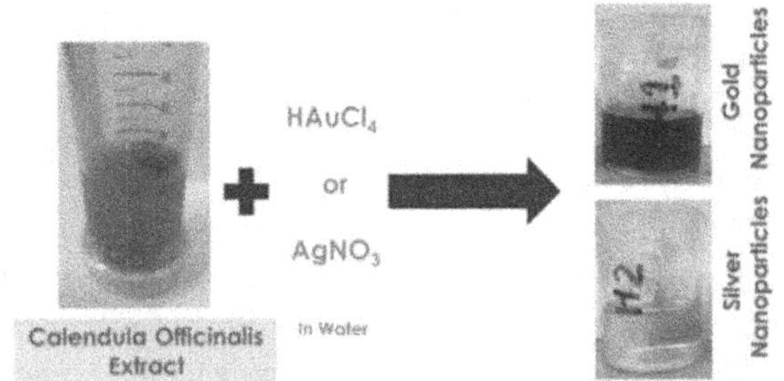

Fig. 14.1 Schematic showing procedure of green synthesis for gold and silver nanoparticles

In case of Gold nanoparticles single LSPR peak centred around 530 nm indicates the formation of pseudospherical nanoparticles of size around 20-30 nm as confirmed by TEM images. In case of Silver nanoparticle single LSPR peak around 410 nm indicates the formation of silver spherical nanoparticles of size around 20-30 nm as confirmed by TEM images. The XRD patterns of synthesized nanoparticles shows characteristic diffraction peaks of planes <111>, <200>, <220> and <311> depicting the fcc lattice formation of gold and silver. Figure 14.1(c) and 14.1(d) shows the TEM images of synthesized nanoparticles.

Biosensing of synthesized nanoparticles was evaluated using their conjugation studies with L-Cystein. 1 mM of L-Cystein was injected in the nanoparticle solution at three different pH 7, 7.6, 8.2 and kept on magnetic stirrer at 600 rpm for 6 hours to allow the complete sorption (Biosensing schematic shown in Fig. 14.2).

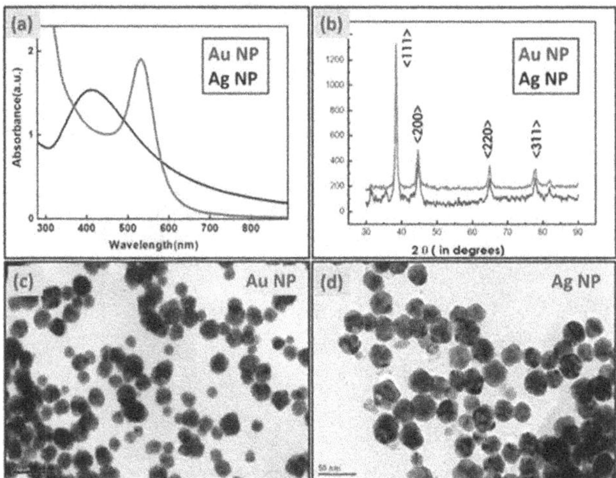

Fig. 14.2 (a) Absorption spectra of metal nanoparticles, (b) XRD patterns (c) & (d) TEM images of Au and Ag Nanoparticles

FTIR studies shows the successful functionalization of L-Cystein onto metal nanoparticles. Pure L-Cystein shows vibrational peaks for wv C-S, SH bending, mv C-NH_2, NH_3 Asymmetric stretch, sv C=O as shown in Fig. 14.4.

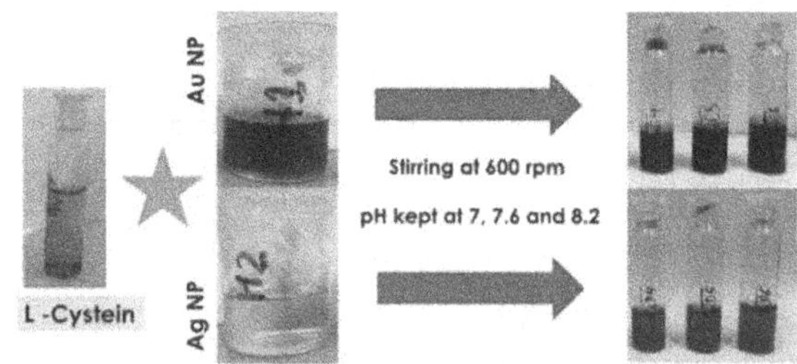

Fig. 14.3 Schematic showing functionalization of L-cystein onto synthesized metal nanoaprticles

Peak Assignment	Peak Position (cm^{-1}) (Pure L Cystein)	(L Cystein functionalized on Gold Nanoparticles)	(L Cystein functionalized on Silver Nanoparticles)
wv C-S	450	454	450
	529	529	529
	570	570	570
	596	596	596
	662	662	662
	796	802	796
SH Bending	876	880	876
mv C-NH_2	1018	1018	1018
NH_3 Asymmetric Stretch	1488	1494	1488
	1582	1588	1582
sv C=O	1674	1674	1682

Fig. 14.4 FTIR peaks of pure L-Cystein and functionalized L-cystein onto gold and silver nanoparticles

In case of functionalization onto silver nanoparticles, NH_3 asymmetric stretch were most affected vibrations so it can be safely concluded that L-Cystein is conjugated through NH_3 bond in this case. While in case of gold nanoparticles, C = O bond peak of L-Cystein was red shifted considerably accentuating the conjugation of L-Cystein through C=O bond.

4. CONCLUSIONS

Gold and silver nanoparticles were green synthesized using the reducing and capping potential of Calendula Officinalis. The synthesized nanoparticles were characterized by UV – Vis Absorption spectroscopy, XRD and TEM analysis. Biosensing of anaylte L-cystein was carried out onto synthesized nanoparticles and differential conjugation was confirmed using FTIR studies.

REFERENCES

1. Haleem, A., et al., Applications of nanotechnology in medical field: a brief review. Global Health Journal, 2023. 7(2): p. 70–77.
2. Zhou, M., et al., Green synthesis of gold nanoparticles using Sargassum carpophyllum extract and its application in visual detection of melamine. Colloids and Surfaces A: Physicochemical and Engineering Aspects, 2020. 603: p. 125293.

Note: All the figures in this chapter were made by the authors.

Recent Trends in Applied Physics and Material Science – Sudhir Bhardwaj et al. (eds)
© 2026 Taylor & Francis Group, London, ISBN 978-1-041-16452-4

Electrochemical Behavior and Theoretical Modeling of the SSCF Cathode in Intermediate-Temperature Solid Oxide Fuel Cells

Pranay R. Kautkar*, Anchal B. Sahu, Roshan P. Pustode, Sureshkumar R. Choubey
Department of Physics, VMV Commerce, JMT Arts & JJP Science College, Nagpur, India

Abstract: In the present attempt $Sm_{0.5}Sr_{0.5}Co_{0.8}Fe_{0.2}O_{3-\delta}$ (SSCF), a perovskite-type oxide was prepared as a potential cathode material through an ethylene glycol-citrate sol-gel combustion route. X-ray diffraction analysis reveals that the synthesized sample exhibits an orthorhombic crystal structure. By integrating Rietveld's analysis of XRD patterns with theoretical modeling and visualization tools, we gain a profound understanding of SSCF's crystal structure and electronic properties. This multi-faceted approach confirms the crystal structure belongs to the Pnma space group, exhibiting an orthorhombic symmetry. and identifies regions of high electron density around Sm/Sr, Co, Fe, and oxygen ions. A symmetric cell was fabricated using SSCF on an SDC15 electrolyte system to evaluate its electrochemical performance. The results of complex impedance spectroscopy and SEM analysis highlight the material's potential as a cathode for IT-SOFCs, with its low Area Specific Resistance (ASR) of 0.181 $\Omega.cm^2$ and activation energy (E_a) of 0.78 eV making it an attractive option for further research and development.

Keywords: SSCF, Rietveld, CIS study

1. INTRODUCTION

As the world transitions towards a decentralized hydrogen economy, SOFC technology stands out as a promising, environmentally responsible solution, offering flexibility in utilizing various hydrocarbon fuels (Adler., 2004; Doshi et al., 1999). To expand the applicability of SOFCs, it is crucial to reduce their operating temperatures. This requires overcoming the limitations associated with electrolyte conductivity and cathode polarization. The discovery of novel perovskite-type oxide materials with exceptional mixed ionic and electronic conductivity is a significant step towards enabling SOFC operation at temperatures below 1000°C. Lowering the functional temperature range offers advantages in terms of reduced component degradation and cost savings. However, it also leads to impaired electrode kinetics and increased polarization resistance at the interface. Furthermore, maintained highly efficient oxygen reduction reaction kinetics (Sun., 2010). The strategic selection of constituent elements enables the cathode material system to achieve lower polarization resistance while maintaining chemical compatibility with ceria-based electrolytes (Kindermann., 1997).

The current study investigates the potential of Sr-doped complex perovskite oxide SSCF as a cathode material for IT-SOFC applications, in combination with SDC15 ($Ce_{0.85}Sm_{0.15}O_{2-\delta}$) electrolyte, focusing on structural theoretical modeling and the Area of specific resistance (ASR) and cathode/electrolyte interface properties.

2. RESULTS AND DISCUSSION

2.1 Crystal Structure and Phase Composition of SSCF

The Rietveld method, as implemented in Full-Prof software, was employed to conduct a quantitative

*Corresponding author: pranay.kautkar@gmail.com

DOI: 10.1201/9781003684718-15

crystallographic phase analysis of SSCF, with the results presented in Fig. 15.1(a). The refined XRD pattern confirms the presence of an Orthorhombic phase with a Pnma space group, consistent with the reference pattern JCPDS file no. 53-0112, and indicates the absence of impurity phases. The three-dimensional crystal structure of SSCF, visualized using VESTA software, is displayed in Fig. 15.1(b), revealing the central positioning of Co/Fe ions within an octahedral environment and the distorted oxygen coordination surrounding Sm/Sr ions.

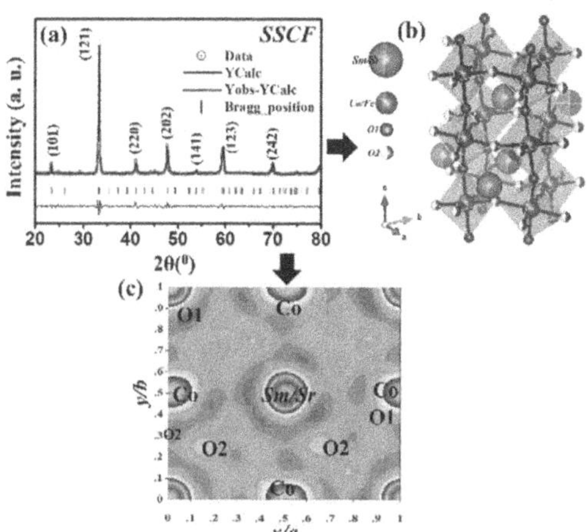

Fig. 15.1 (a) Rietveld refinement of XRD pattern of SSCF cathode oxide (b) Unit cell representation by VESTA structure model (c) EDD map of SSCF with the position of Oxygen around cations

Replacing Sm^{3+} with Sr^{2+} increases the average cation size at the A-site, leading to an expanded unit cell through longer A-O bonds. We analyzed the electron density distribution (EDD) within the unit cell by examining the reflection of x-ray waves from parallel electron density planes. This analysis utilized Gfourier mapping and Rietveld refinement, with the resulting electron density distribution contour shown in Fig. 15.1(c). The EDD map displays the probability of finding electrons near Sm/Sr, Co, Fe, and oxygen ions, offering insights into the material's structural and bonding features. This map enables a detailed understanding of the electron distribution, which can be conceptualized as a cloud of negative charge, in this visualization, sky-blue and green colors denote high and low electron density regions, respectively. The high electron density observed near cations and anions may be linked to oxygen vacancies.

2.2 Temperature Dependent Impedance Study

Complex impedance analysis (CIS) was performed in air under zero DC bias conditions to elucidate the electrode-electrolyte interface dynamics between SSCF and SDC15.

Impedance measurements on the SSCF-SDC15 symmetric cell, performed across an extensive range of frequencies (0.01 Hz to 1 MHz) and temperatures (300°C to 700°C), yielded complex plane plots [Fig. 15.2(a)] and equivalent circuit models [Fig. 15.2(b)]. The plots reveal two semi-circular arcs, corresponding to two-time constants. The impedance spectrum shows a high-frequency arc corresponding to the SDC15 electrolyte's ionic conductivity, and a distorted low-frequency arc, indicative of electrode polarization and electron charge transfer (Adler et al., 2007). The electrode resistance is significantly influenced by the cumulative effect of R3 and R4 resistances. Notably, the SEM micrograph of SSCF oxide reveals a highly porous structure, ideal for SOFCs [Fig. 15.2(c)]. This study demonstrates that substituting Sr^{2+} for Sm^{3+} significantly reduces the area-specific resistance (ASR) to 0.181 $\Omega.cm^2$ at 700°C. The temperature-dependent area-specific resistance (ASR) of the SSCF cathode is visually represented in the Arrhenius plots displayed in Fig. 15.2(d). The calculated activation energy (E_a) of 0.78 eV confirms the cathode's outstanding performance, making it an excellent candidate for IT-SOFC applications.

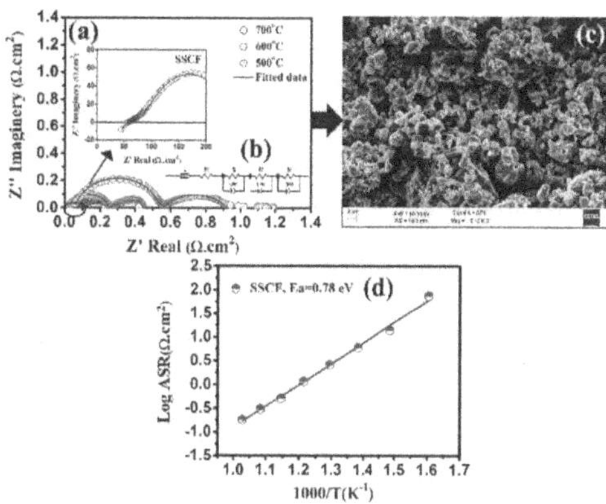

Fig. 15.2 (a) Impedance plot of SSCF cathode on SDC15 electrolyte at 500 to 700°C (b) equivalent circuit (c) SEM micrograph of powder sample (d) Arrhenius plot of temperature-dependent ASR

3. CONCLUSION

In this work, Utilizing the sol-gel combustion technique, the complex perovskite oxide SSCF is successfully synthesized and subsequently explored as a promising cathode material for IT-SOFC applications, with SDC15 employed as the electrolyte. Theoretical modeling of XRD patterns using Rietveld analysis confirms the orthorhombic crystal structure of SSCF, belonging to the Pnma space group. A detailed visualization of the crystal structure is achieved using VESTA software. Additionally, electron density distribution mapping provides a comprehensive understanding of the material's electronic structure, highlighting the spatial distribution of electrons around Sm/Sr, Co, Fe, and oxygen ions. A significant observation from the SEM micrograph is the highly porous nature of the SSCF oxide, which is a desirable characteristic for IT-SOFCs. The electrochemical assessment of the SSCF cathode yielded encouraging results, including a low area-specific resistance of 0.181 $\Omega.cm^2$ at 700°C and a relatively moderate activation energy (E_a) of 0.78 eV. The substitution of Sr^{2+} for Sm^{3+} led to the lowest ASR, suggesting enhanced electrochemical properties. This material is expected to offer enhanced oxygen reduction activity, making it a suitable candidate for (IT-SOFCs).

REFERENCES

1. Adler, S. B., (2004). Factors governing oxygen reduction in solid oxide fuel cell cathodes, *Chemical Reviews*, 104, 4791–4843.
2. Doshi, R., Richards, V. L., Carter, J. D., Wang, X., and Krumpelt, M., (1999). High-performance solid-oxide fuel cells, *Journal of the Electrochemical Society*, 146(4), 1273–1278
3. Sun, C., Hui, R., Roller, J., (2010). Advances in solid oxide fuel cells, *Journal of Solid State Electrochemistry*, 14, 1125–1144.
4. Kindermann, L., Dos, D., Nickel, H., Hilpert, K., Appel, C. C., and Poulsen, F. W., (1997). Chemical Compatibility of $(La_{0.6}Ca_{0.4})Fe_{0.8}M_{0.2}O_3$ with Yttria-Stabilized Zirconia, *Journal of the Electrochemical Society*, 144, 717.
5. Adler, S. B., Chen, X. Y., Wilson, J. R., (2007). Mechanisms and rate laws for oxygen exchange on mixed-conducting oxide surfaces, *Journal of Catalysis*, 245, 91–109.

Note: All the figures in this chapter were made by the authors.

Recent Trends in Applied Physics and Material Science – Sudhir Bhardwaj et al. (eds)

Prediction of Thermophysical Properties of Nanomaterials: A Theoretical Study

Shivam Srivastava*, Prachi Singh
Department of Physics,
Dr. Shakuntala Misra National Rehabilitation University,
Lucknow, Uttar Pradesh

Anjani K. Pandey
Institute of Engineering and Technology,
Dr. Shakuntala Misra National Rehabilitation University,
Lucknow, Uttar Pradesh, India

Chandra K. Dixit
Department of Physics,
Dr. Shakuntala Misra National Rehabilitation University,
Lucknow, Uttar Pradesh

Abstract: The present study incorporates high pressure equations of state (EOSs) to study some Thermophysical properties of nanomaterials. The Thermophysical properties under study are pressure dependence of volume, pressure dependence of isothermal bulk modulus at high pressure.

Keywords: High pressure, Bulk modulus

1. INTRODUCTION

3C-SiC (Cubic Silicon Carbide) and CdSe (Cadmium Selenide) in their rock salt phases are pivotal nanomaterials due to their unique structural, electronic, and mechanical properties, which are critical for advanced technological applications. On the other hand, CdSe rock salt phase nanomaterials have garnered attention for their tunable optical properties and high quantum efficiency, particularly in photovoltaic and optoelectronic applications.

2. METHOD OF ANALYSIS

There are number of EOS in which Birch-Murnaghan EOS gives good results (Pandey AK et al. 2023, Srivastava S et al. 2024). Birch-Murnaghan EOS which is basically based on finite strain theory is given below

$$P = \frac{3}{2}B_0[t^{-7} - t^{-5}][1 + \frac{3}{4}(B_0' - 4)(t^{-2} - 4)]$$

Where $t = \left(\frac{V}{V_0}\right)^{\frac{1}{3}}$

The bulk modulus at constant temperature can be calculated as fallows $B_T = -V\left(\frac{\partial P}{\partial V}\right)_T$

$$B_T = \frac{B_0}{2}[7t^{-7} - 5t^{-5}] + \frac{3}{8}B_0(B_0' - 4)(9t^{-9} - 14t^{-7} + 5t^{-5})$$

The differentiation of B_T w. r. t P is denoted by

$$B_T' = \frac{B_0}{8B_T}[(B_0' - 4)(81t^{-9} - 98t^{-7} + 25t^{-5}) + \frac{4}{3}(49t^{-7} - 25t^{-5})]$$

3. RESULTS AND DISCUSSION

The required parameters B_0 (GPa), B_0' of 3C-SiC(30nm) (Liu H et al. 2004) and CdSe(rock salt phase) (Tolbert S H et al 1995) are 245, 2.9 and 74, 4. We can see that 3C-SiC exhibits higher compressive resistance and a more gradual change in bulk modulus with pressure compared to CdSe. The behavior of B_T' in both materials suggests that the response of the bulk modulus to compression becomes less sensitive as the materials are further compressed.

*Corresponding author: ss_phyphd2021@dsmnru.ac.in

DOI: 10.1201/9781003684718-16

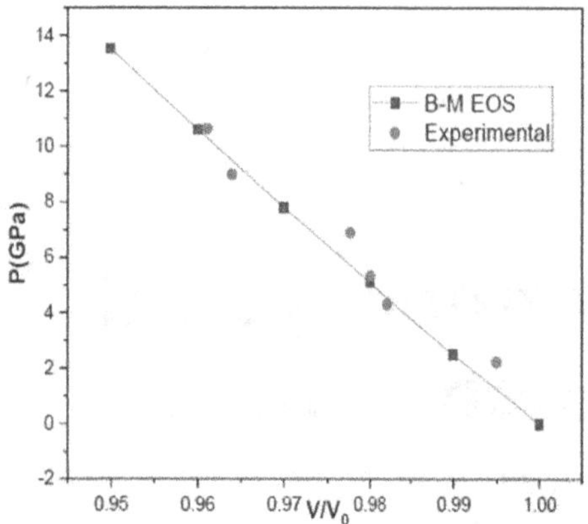

Fig. 16.1 Volume dependency of pressure for 3C-SiC(30nm) from B-MEOS

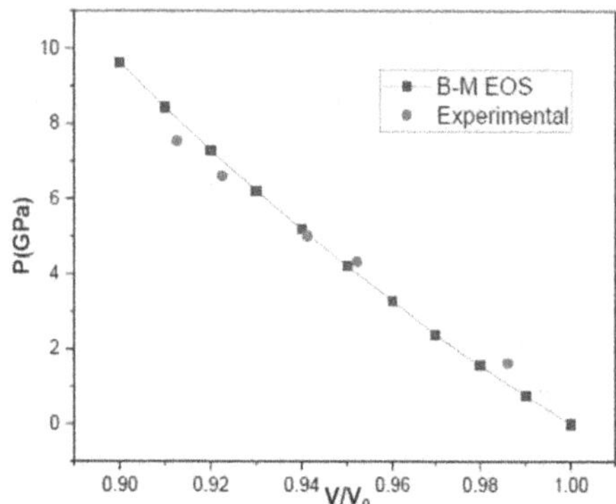

Fig. 16.2 Volume dependency of pressure for CdSe (Rock Salt) from B-MEOS

Table 16.1 Calculated values other thermophysical parmeter at different compressions for 3C-SiC (30nm) and CdSe (Rock Salt Phase)

3C-SiC (30nm)			CdSe (Rock Salt Phase)		
V/V_0	B_T(GPa)	B_T'	V/V_0	B_T(GPa)	B_T'
1	245	2.9	0	74	4
0.99	252.1952	2.860147	0.758827	77.02058	3.961686
0.98	259.5733	2.820536	1.556655	80.16629	3.924486
0.97	267.138	2.781121	2.3956	83.44336	3.888353
0.96	274.8929	2.741858	3.277909	86.85839	3.853239
0.95	282.8416	2.702698	4.205971	90.41839	3.819101
0.94	290.9874	2.663596	5.182327	94.13077	3.785899
0.93	299.3335	2.624502	6.209683	98.00343	3.753593
0.92	307.883	2.585367	7.290919	102.0447	3.722146
0.91	316.6385	2.546139	8.429103	106.2636	3.691523

REFERENCES

1. Liu H, Jin C, Chen J, Hu J, (2004) J. Am. Ceram. Soc. 87, 2291
2. Pandey A K, Srivastava S, Dixit C K, et al. (2023)Shape and Size Dependent Thermophysical Properties of Nanomaterials. Iran J Sci , 47,1861–1875
3. Pandey A.K, Dixit C.K &, Srivastava S, (2023) Theoretical model for the prediction of lattice energy of diatomic metal halides. J Math Chem ,62, 269–274
4. Pandey A, Srivastava S, & Dixit C K, (2023) A Paradigm Shift in High-Pressure Equation of State Modeling: Unveiling the Pressure–Bulk Modulus Relationship. Iran J Sci 47, 1877–1882
5. Pandey AK, Srivastava S, Singh P, Tripathi S, Dixit CK (2023) Theoretical prediction of Gru¨neisen parameter for nano Lead Sulfide at different compressions. https://doi.org/10.21203/rs.3. rs-3159558/v1
6. Srivastava S., Pandey A.K, Dixit C K, (2023).Theoretical prediction for thermoelastic properties of carbon nanotubes (CNTs) at different pressure or compression using equation of states. J Math Chem, 61, 2098–2104
7. Srivastava S.,. Pandey A K, Dixit C K, (2023)*Theoretical prediction of Gruneisen Parameter for Ɣ- Fe_2O_3,* computational condensed matter, 35, e00801
8. Srivastava S, Dixit C K and Pandey A K, Comparative Study of Elastic Properties of Some Inorganic and Organic molecular Crystals by using Isothermal EOS Available at SSRN: http://ssrn.com/abstract=4427891 or *http://dx.doi.org/10.2139/ssrn.4427891,(2023)*
9. Srivastava Shivam et al.(2023), Equation of states at extreme compression ranges: Pressure and Bulk modulus as an example, Materials Open, doi: 10.1142/S2811086223500073
10. Srivastava S, Singh P, Pandey A, Dixit C K, (2024) Solid State communication 377, 115387
11. Pandey A K, Dixit C K, Srivastava S, Singh P, Tripathi S, (2023) Theoretical Prediction of Thermo-Elastic Properties of TiO_2 (Rutile Phase), Natl. Acad. Sci. Lett.
12. Srivastava S, Singh P & Pandey A K and Dixit C K, (2023) Theoretical Prediction for Thermo Elastic Properties of Nano CdSe (Rock salt Phase), AIJR Proceedings, 1–5, https://doi.org/10.21467/proceedings.161.1
13. Tolbert S H, Alivisatos A P, (1995). J. Chem. Phys. 102, 1

Note: All the figures and tables in this chapter were made by the authors.

Recent Trends in Applied Physics and Material Science – Sudhir Bhardwaj et al. (eds)

Thermal Investigations on LSCF Cathode Materials to Boost IT-SOFC Performance

Pranay R. Kautkar*,
Sejal S. Shahu, Roshan P. Pustode,
Apurba M. Ghosh and Sureshkumar R. Choubey
Department of Physics, VMV Commerce, JMT Arts & JJP Science College, Nagpur, India

Abstract: In this study, we synthesized the perovskite oxide $La_{0.5}Sr_{0.5}Co_{0.8}Fe_{0.2}O_{3-\delta}$ (LSCF) via the sol-gel combustion method. We evaluated its suitability as a potential cathode material for solid oxide fuel cells (SOFCs) operating at intermediate temperatures. The thermal properties of LSCF were investigated using TG-DTA analysis, revealing a calcination temperature of 1100°C. SEM micrographs confirmed the highly porous nature of LSCF, with a particle size of approximately 509 nm. Electrochemical impedance spectroscopy (EIS) analysis was performed on symmetrical cell configurations to assess the ionic conductivity of LSCF. The outcome showed that the particle size substantially influences the symmetric cell's efficiency, with the lowest area-specific resistance (ASR) of 0.29 $\Omega.cm^2$ at 700°C featuring an activation energy of 0.71 eV. The data suggest that LSCF is an attractive option for use as a cathode material in IT-SOFCs, exhibiting enhanced electrochemical performance due to A-site substitution.

Keywords: LSCF, TG-DTA, EIS study

1. INTRODUCTION

The potential of solid oxide fuel cells to deliver high-efficiency energy conversion and accommodate diverse fuel sources has sparked widespread interest in their development for sustainable energy applications (Steele et al., 2001). However, traditional SOFCs operate at high temperatures (800-1000 °C), posing challenges related to materials compatibility, thermal stresses, and degradation (Sadykov et al., 2012; Gao et al., 2016). To address these challenges, The development of solid oxide fuel cells with operating temperatures between 600-800 °C has become a major focus area for researchers (Sadykov et al., 2012). A critical component of IT-SOFCs is the cathode material, which significantly impacts the cell's electrochemical performance and long-term stability. Conventional cathode materials, such as lanthanum strontium manganite, exhibit limited ionic conductivity and poor performance at lower temperatures, prompting the exploration of alternative materials (Seabaugh, 2003). Mixed ionic-electronic conducting cathodes have emerged as promising alternatives.

In this study, we fabricated a $La_{0.5}Sr_{0.5}Co_{0.8}Fe_{0.2}O_{3-\delta}$ (LSCF) cathode; examined its thermal behavior and morphological features of symmetric cells incorporating $Ce_{0.85}Sm_{0.15}O_{2-\delta}$ (SDC15) electrolyte. This analysis uncovers the interplay between microstructural characteristics, thermal properties, and electrochemical performance. The results provide valuable insights into optimizing thermal management in LSCF-based cathodes, ultimately enhancing overall cell efficiency.

2. RESULTS AND DISCUSSION

2.1 Thermal Analysis (TG-DTA Study)

As-synthesized-LSCF sample by sol-gel combustion route is generally in amorphous form or otherwise very low degree of crystallite. An XRD study was performed to reveal the structural status of the as-synthesized LSCF system. It clearly shows Fig. 17.1 (a), a very low

*Corresponding author: pranay.kautkar@gmail.com

DOI: 10.1201/9781003684718-17

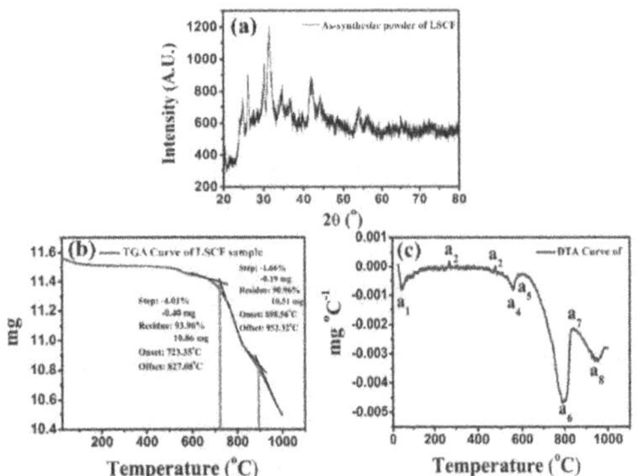

Fig. 17.1 (a) XRD pattern of as-prepared LSCF cathode. Thermal analysis of LSCF raw powder (b) TGA (c) DTA

degree of crystallinity of the as-prepared LSCF sample. To increase to degree of crystallinity, commonly high-temperature calcinations or annealing process is used. To estimate the crystallinity temperature, the thermal analysis technique especially the Thermogravimetry (TG) study is highly suitable. TG-DTA study of the as-prepared LSCF cathode system was performed. The TG curve of the as-synthesized LSCF powder is displayed in Fig. 17.1 (b). The TG curve shows that the thermal decomposition of LSCF occurs in three distinct stages, with the removal of organic material being completed at temperatures ranging from 650 to 1000°C. The first stage, at approximately 200°C, entails the loss of surface-adsorbed water. The second stage, detected between 600-830°C, corresponds to the degradation of combustion residues. The third stage, occurring at 850-1000°C (Tao et al., 2002), marks the complete dissociation of combustion-derived carbonates and the onset of LSCF phase formation.

The DTA curve [Fig. 17.1(c)] of the as-synthesized LSCF cathode powder, prepared via combustion with citric acid, reveals several thermal events. Small endothermic peaks [a1, a4] are observed, consistent with the TGA analysis, as well as more pronounced peaks at 797°C [a6] and 951.6°C [a8], attributed to the decomposition of organic residues.

A distinct exothermic peak at 839.6°C suggests the combustion of residual organic matter not fully burned during the initial combustion process. Due to instrumental limitations, the analysis was restricted to 1000°C. Consequently, the raw cathode powder was calcined at 1100°C, resulting in the formation of a pure phase.

2.2 Microstructural Analysis

Figures 17.2 (a) and (b) illustrate the particle size distribution and surface morphology of the LSCF powder, providing insight into its physical characteristics. The results indicate an average particle size of approximately 509 nm, with a narrow size distribution. Scanning electron microscopy (SEM) imaging reveals a highly porous structure, The significance of this characteristic lies in its impact on the fabrication of IT-SOFCs with enhanced efficiency.

2.3 Temperature Dependent EIS

The electrochemical performance of the LSCF/SDC15/LSCF symmetric cell was evaluated using impedance spectroscopy in air, with results presented in Fig. 17.3. The complex impedance spectra reveal a distorted semicircular arc at low frequencies, attributed to cathode resistance, which increases as temperature decreases (Fig. 17.3a). The incomplete high-frequency semicircular arc is associated with the migration of oxygen ions through the SDC15 electrolyte, limited by the frequency range ($\leq$ 1 MHz) of the Autolab potentiostat + FRA model 204. Notably, the ASR of the electrode is reflected in the distorted semicircular arc observed at low frequencies.

As depicted in Fig. 17.3(b), the Arrhenius plot of temperature-dependent ASR reveals an optimal value of 0.29 $\Omega.cm^2$ at 700°, accompanied by an activation energy (Ea) of 0.71 eV. The observed Arrhenius behavior indicates a thermally activated charge transfer process, where mobile oxygen ions and polarons (electronic defects) overcome a potential energy barrier (Ea) to jump from occupied sites to adjacent vacant sites, facilitating ionic conduction.

3. CONCLUSION

The feasibility of LSCF ($La_{0.5}Sr_{0.5}Co_{0.8}Fe_{0.2}O_{3-\delta}$) as a cathode material for IT-SOFCs was explored, with the material being synthesized through a sol-gel combustion

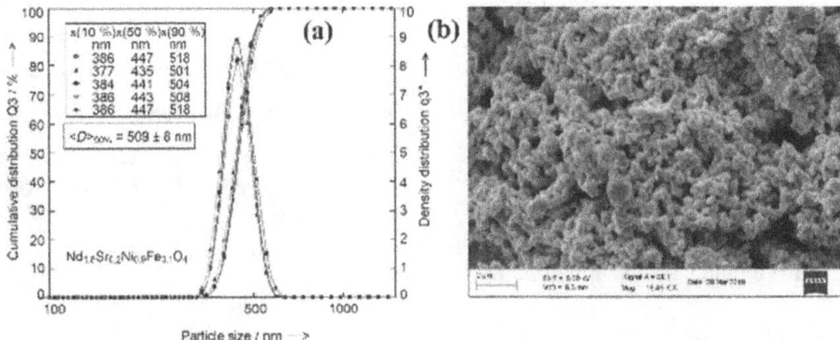

Fig. 17.2 (a) The particle size distribution of the LSCF cathode (b) SEM micrograph of LSCF powder

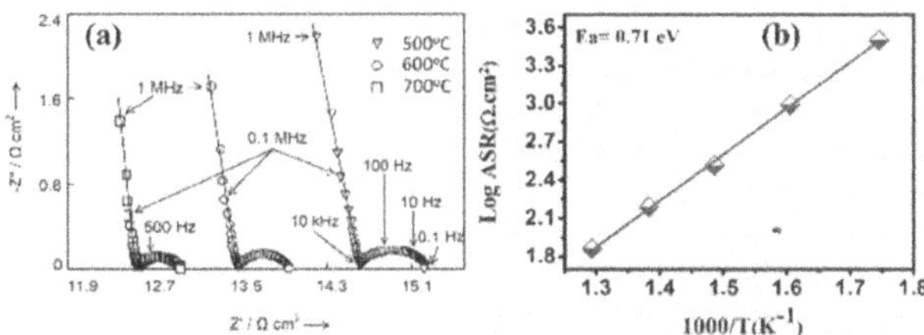

Fig. 17.3 (a) Impedance plot of LSCF cathode on SDC15 electrolyte at 500, 600, and 700°C (b) Arrhenius plot of Area specific resistance (ASR)

process. Thermal analysis revealed the optimal calcination temperature while scanning electron microscopy (SEM) confirmed the material's highly porous structure with a particle size of 509 nm. EIS was performed to assess the ionic transport properties within a symmetric cell configuration, revealing a strong correlation between particle size and performance. The electrical performance of the LSCF cathode was distinguished by a low ASR value of 0.29 Ω.cm² at 700°C coupled with an activation energy (E_a) of 0.71 eV. A-site substitution in LSCF cathodes significantly enhances their electrochemical performance, making them a promising material for IT-SOFCs.

REFERENCES

1. Steele, B. C. H., & Heinzel, A. (2001). Materials for fuel-cell technologies. Nature, 414(6861), 345–352.
2. Sadykov, V. A., Bobrikov, I., Lukashevich, A., Pavlova, S., Lukyanov, D., and Alikina, G. (2012). Perovskite-related oxides and nanocomposites as promising materials for intermediate-temperature solid oxide fuel cells, *Integrated Ferroelectrics*, 130(1), 166–178.
3. Seabaugh, M. M. (2003), Solid oxide fuel cell performance limitations associated with mixed conducting cathodes, *Journal of Power Sources*, 118(1–2), 362–374.
4. Tao, S., & Irvine, J. T. S. (2002). Synthesis and characterization of $(La_{0.75}Sr_{0.25})_{0.95}Cr_{0.5}Fe_{0.5}O_{3-\delta}$. Journal of Solid-State Chemistry, 165(2), 259–266.

Note: All the figures in this chapter were made by the authors.

Low Temperature Magnetic Studies of $NdFeO_3$ Synthesized by Starch Assisted Sol-Gel Combustion Method

P. Ramesh Babu
Department of Electronics and Communication System,
Sri Krishna Arts & Science College,
Coimbatore, Tamil Nadu, India

B. Srimathy*
PG & Research Department of Physics,
Seethalakshmi Ramaswami College
(Affiliated to Bharathidasan University),
Tiruchirappalli, Tamil Nadu, India

T. Veeramanikandasamy,
S. Devendiran
Department of Electronics and Communication System,
Sri Krishna Arts & Science College,
Coimbatore, Tamil Nadu, India

Abstract: The widespread utilization of rare earth orthoferrites in solid oxide fuel cells, magneto-optical materials, catalysts and gas sensors has garnered considerable interest in these materials. In the present study, neodymium orthoferrite was synthesized by starch assisted sol-gel combustion method. The presence of pure perovskite phases was confirmed and lattice parameters were calculated from powder x-ray diffraction analysis. Low temperature dependence of magnetization under zero field cooled and field cooled conditions showed the presence of antiferromagnetic ordering and spin reorientation phenomenon at 170 K was evident from these measurements. Saturation magnetization decreased with increase in temperature which was apparent from the hysteresis loops.

Keywords: Sol-gel combustion, Magnetism, Hysteresis, Ferrites

1. INTRODUCTION

Currently, perovskite rare earth oxides have been found to exhibit high and exceptional multiferroic and magnetoelectric properties. The potential of multiferroic materials in data storage devices, sensors, microwave and spintronic devices (Srimathy et al., 2021) are noteworthy in modern science and technology. Recently, orthoferrites comprising of rare earth elements like yttrium (Y), Gadolinium (Gd), Samarium (Sm), neodymium (Nd) are known to have a temperature-dependent magnetization, which has attracted the attention of these materials in diverse fields. Orthoferrites are very good for studying the relationship of electric and magnetic degrees of freedom, a relationship that largely depends on rare-earth elements (Srimathy et al., 2014). The magnetic phenomenon exhibited by $RFeO_3$ engage two magnetic sublattices, R^{3+} and Fe^{3+} thus creating complex magnetic phases and influence magnetization. Orthorhombic disorder in orthoferrite increases with growth of rare earth ions controlling its magnetic phenomenon, mainly due to strong switching over through Fe-O-Fe bonds.

Among these rare earth orthoferrites, $NdFeO_3$ is well known for its high magnetic order and spin reorientation in antiferromagnetic systems. According to recent research, they are said to possess ferroelectricity at room temperature. $NdFeO_3$ has orthorhombic structure and space group *Pbnm*. Generally, solid state reaction method is employed to produce $NdFeO_3$ powders. However in the current work, sol-gel combustion process was employed which offer advantages such as uniform grain size, good electrical, magnetic, and surface properties.

2. EXPERIMENT

Polycrystalline specimens of $NdFeO_3$ were produced through sol-gel combustion technique, employing starch as both a combustion fuel and binding agent. The initial reagents comprise analytical grade starch, $Nd(NO_3)_2 \cdot 4H_2O$ and $Fe(NO_3)_2.6H_2O$. A mixture of 50 ml of 1M neodymium nitrate and 50 ml of 1M ferric nitrate was prepared followed by the gradual incorporation of 50 ml of 2M starch solution into metal nitrate mixture. The resultant clear solution is ignited from ambient

*Corresponding author: bsrimathy@gmail.com

DOI: 10.1201/9781003684718-18

atmosphere to 393 K while being continuously stirred to yield a brown coloured gel. The obtained gel is maintained at 623 K for half an hour in a furnace which is preheated at a rate of 101°C/min. Finally, a capacious fluffy brown mass of precursor is obtained. Subsequently, the resultant precursor is subjected to thermal treatment for 2 hours at each of the temperatures 873 K, 1123 K and 1400 K. Upon additional heating to 1600 K for 12 hours, a single-phase, carbon-free $NdFeO_3$ is attained.

3. RESULTS AND DISCUSSION

3.1 X-Ray Diffraction Analysis

A Rigaku Geigerflex instrument with Cu-Kα radiation source was engaged to record the XRD pattern at room temperature. The measurement was taken at values ranging from 20° to 40° and with interval of 0.01° (Fig. 18.1). $NdFeO_3$ has a single phase growth pattern as indicated by the XRD plot, with no evidence of reactivity in the secondary phase. By using the JPCDS file 88-0477, the peaks were indexed and was determined that the crystal growth occurred in the orthorhombic crystal system. The lattice parameters a = 5.467 Å, b = 7.682 Å, c = 5.550 Å, α = β = γ = 90 ° are well matched with the reported values.

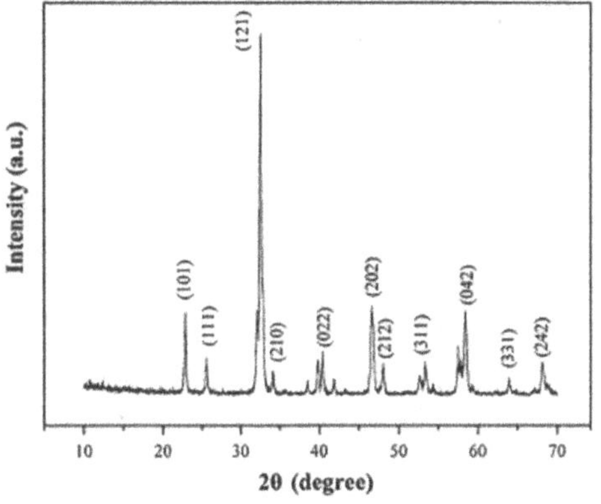

Fig. 18.1 XRD pattern of $NdFeO_3$

3.2 Magnetic Measurements

Figure 18.2 shows the temperature dependent magnetic transition curves of $NdFeO_3$ under zero field-cooled (ZFC) and field-cooled (FCC) conditions. The applied magnetic field was 100 Oe and temperature was varied

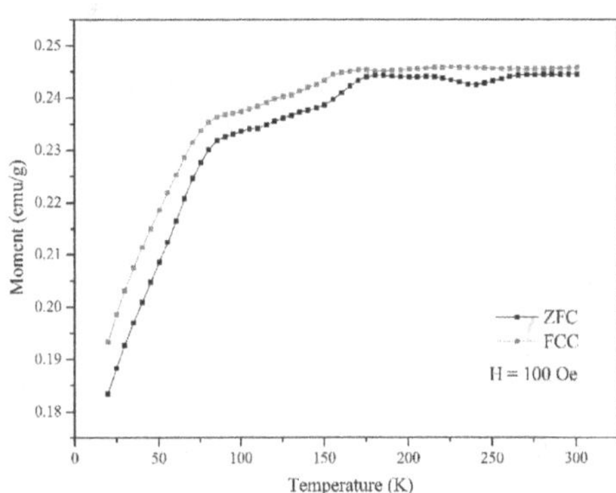

Fig. 18.2 Low temperature dependence of magnetization of $NdFeO_3$

between 50 K and 300 K during the measurements. The Fe^{3+}-Fe^{3+} interactions and Nd^{3+}-Fe^{3+} interactions result in spin reorientation at a temperature around 170 K, leading to a small curve. The magnetization increases at low temperatures in the field-cooled (FC) measurements, which are a cause of interactions between Fe-O-Fe and Nd-O-Fe ions suggesting the existence of antiferromagnetic nature. The separation of the ZFC-FCC curves before 170 K indicates that the sample exhibits magnetic irreversibility and the convergence of the curves after that indicates the onset of the paramagnetic ordering.

The M–H spectrum of $NdFeO_3$ at low temperature is depicted in the Fig. 18.3. Below the Neel temperature, all hysteresis cycles of $NdFeO_3$ exhibit antiferromagnetic feature which is consistent with the weak ferromagnetic nature of the rare-earth orthoferrites (Babu et al., 2015). In the region of spin reorientation above 200 K, the shape of the loop becomes narrow with small coercivity which is the generic character of orthoferrites.

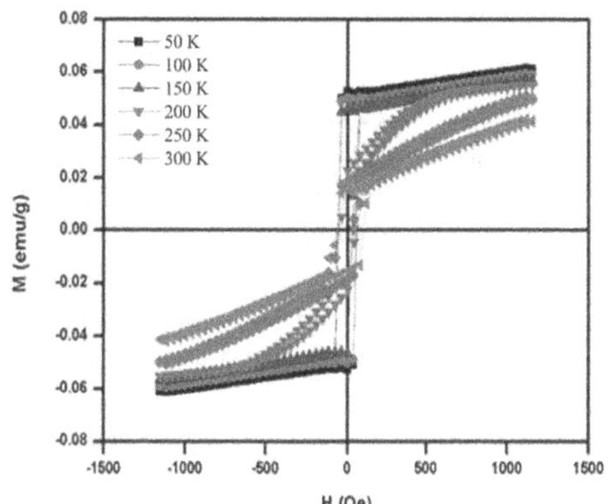

Fig. 18.3 M vs. H measurements at low temperature

4. CONCLUSION

Starch assisted sol-gel combustion technique was employed to synthesize pure perovskite phase $NdFeO_3$. The absence of any secondary phase was confirmed from powder x-ray diffraction analysis. Presence of antiferromangetic ordering at low temperature was confirmed from magnetization measurements. Occurrence of spin reorientation at 170 K was apparent from M-T curves and the same was confirmed from hysteresis loop measurements.

REFERENCES

1. Srimathy, B. and Kumar, J. (2021). Investigations on the Growth and microhardness studies of $Ga_{2-x}Fe_xO_3$ single crystals. Phys. Met. Metallogr. 122(12):1234–1240.
2. Srimathy. B, Indranil, Bhaumik, Ganesamoorthy, S, Bhatt, R, Karnal, A. K. and Kumar, J. (2014). On the Neel temperature and magnetic domain wall movements of $Ga_{2-x}Fe_xO_3$ single crystals grown by floating-zone technique. J. Alloys Compd. 590:459–464.
3. Babu, P. R, Bhaumik, I, Ganesamoorthy, S, Kalainathan, S, Bhatt, R, Karnal, A. K and Gupta, P. K. (2015) Investigation of magnetic property of $GdFeO_3$ single crystal grown in air by optical floating zone technique, J. Alloys. Compd., 631: 232–6.

Note: All the figures in this chapter were made by the authors.

Effect of Plasmon Resonance of Metal Nitrides in Semiconductor Plasmonic Devices

Pratima Rajput
Department of Physics, JSS Academy of Technical Education,Noida, India

Soniya Juneja
Department of Applied Science, KIET Group of Institutions, Ghaziabad, India

Alok Singh
Department of Physics, JSS Academy of Technical Education, Noida, India

Manmohan Singh Shishodia*
Department of Applied Physics, Gautam Buddha University, Greater Noida, India

Abstract: The dipole-dipole interaction between Donor (D)–Acceptor (A) fluorophore pair leads to the Förster Resonance Energy Transfer (FRET) from donor to acceptor. This non-radiative transfer of energy has found applications in detecting protein folding and biosensors for clinical, pharmaceutical, toxicological, environmental monitoring and agri-food analysis. Quantum dots (QDs) has emerged as versatile materials that can be used as both donors and acceptors in FRET provided absorption spectra of acceptor QD overlaps with the emission spectra of donor QD. As it is well established, Localized Surface Plasmons of metallic nano system placed in the vicinity of QD fluorophores can greatly enhance FRET, and subsequently the performance of devices like LEDs, Solar Cells, Luminescent Solar Concentrators and Biosensors. In view of high bio compability, low intrinsic loss, low work function, high melting point etc. of ZrN and TiN, the role of ZrN@SiO_2, TiN@SiO_2, ZrN@TiO_2 and TiN@TiO_2 core-shell nanoparticles in modifying FRET parameters e.g., FRET rate, FRET efficiency and Förster radius for CdSe QDs have been investigated.

Keywords: FRET, Förster radius, Quantum dot

1. INTRODUCTION

Surface plasmon enhanced FRET is likely to play pivotal role in the development of energy efficient technologies for solar harvesting (Guang et al., 2012, Durach et al., 2008). Quantum dots (QDs) such as cadmium selenide (CdSe), Si QDs are widely used in field of photonics. FRET process between QDs in presence of plasmonic nanoparticle can lead to enhanced performance of QDs for semiconducting plasmonic devices (Faucheaux et al., 2014). When metallic nano systems are placed near QD fluorophores, they amplify the electromagnetic field, leading to improved FRET performance. This enhancement can greatly benefit devices like light-emitting diodes (LEDs), solar cells, luminescent solar concentrators, and biosensors. This energy transfer happens through dipole-dipole interactions and is highly dependent on the distance between the donor and acceptor.

Zirconium nitride (ZrN) and titanium nitride (TiN) are gaining attention as better alternatives to conventional plasmonic materials because of their high biocompatibility, lower optical losses, and high thermal stability. These materials, when combined with silica (SiO_2) or titanium dioxide (TiO_2) in core-shell nanoparticle structures, offer even better control over FRET parameters like transfer efficiency, rate, and Förster radius. This research investigates how ZrN and TiN core-shell nanoparticles (ZrN@SiO_2, TiN@SiO_2, ZrN@TiO_2, and TiN@TiO_2) influence FRET parameters in systems using CdSe QDs. The findings provide insights into developing advanced energy transfer platforms for optoelectronic and sensing technologies.

2. THEORETICAL MODEL

The schematic of D-A system near core-shell nanoparticle is shown as insert in Fig. 19.1 (a). Energy transfer rate enhancement factor (ETREF) between D-A (dipole moments p_a= 25 Debye and p_d=4.4 Debye (Durach et al., 2008), parallel to + z-axis) near core (ZrN/TiN, radius-R_1)

*Corresponding author: manmohan@gbu.ac.in, pratimarajput@jssaten.ac.in

DOI: 10.1201/9781003684718-19

–shell (SiO_2/TiO_2, radius-R_2) embedded in medium with dielectric constant, ε_d=1 is (Rajput and Shishodia, 2020),

$$\eta = \left|\frac{J(\omega)}{J_0}\right|^2 \quad (1)$$

Where $j(\omega)$ is the D-A interaction energy in the presence of nanoparticle and J_0 is the direct D-A interaction energy. FRET rate for the system under the consideration is defined as (Rajput and Shishodia, 2020)-

$$\gamma_{FRET} = \frac{9}{8\pi}\frac{\sigma_A(\omega)\gamma_0}{k^4}|J(\omega)|^2 \quad (2)$$

Where $\sigma_A(\omega)$ is the absorption cross section of the acceptor molecule and γ_0 is the emission rate per unit frequency of the donor treated as a free molecule.

FRET efficiency is defined as (Rajput and Shishodia, 2021)-

$$E_{FRET} = \frac{\gamma_{FRET}}{\gamma_{FRET} + \gamma_R + \gamma_{NR}} \quad (3)$$

Where γ_R and γ_{NR} are radiative and non-radiative decay rate, respectively.

3. RESULT AND DISCUSSION

The effect of presence of plasmonic nanoshell on energy transfer between CdSe QDs is discussed. Distance between plasmonic nanoparticles and QDs is an important factor for energy transfer rate and efficiency of FRET process. Compared to the structure without presence of nano particle ~8-9 order enhancement is achieved for the FRET rate with resonance wavelength ~500nm for ZrN nanoshell and ~600nm for TiN nanoshell (Fig. 19.1(a)) and ~4 order enhancement achieved for ETREF (Fig. 19.1(b)) between QD's. Multiple resonance peaks (range 500-700 nm) are observed for ZrN nanoshell, which provide the possibility to tune the structure at various resonance wavelengths. Figure 19.1(c) shows the distance dependence of FRET efficiency. Calculated Förster radius and FRET efficiency is shown in Table 19.1. It is observed that when ZrN nanoshell is used instead of TiN nanoshell, R_0 increases by ~4nm. Increased R_0 results in increased FRET efficiency. This enhanced energy transfer interaction certainly improve the performance of QD's for photovoltaic devices. Effect of distance on energy transfer by varying shell thickness is also examined. As shown is Table 19.2, as the distance increases, Förster radius, FRET efficiency increases and FRET rate decreases. This variation of FRET rate can be shown through equation

$$\gamma_{FRET} = ax^b \quad (4)$$

Where a = 941.206, b = –2.611 are constants and x is the shell thickness.

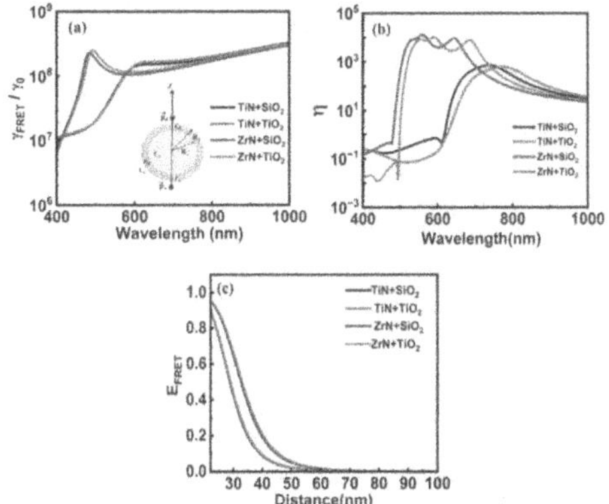

Fig. 19.1 Spectral variation of (a) Normalized FRET rate (b) ETREF,betweenCdSeD-ApairneararananoshellofZrN/TiN (core)-SiO_2/TiO_2shellwithaspectratio0.85(c)distance dependence of FRET efficiency

Table 19.1 Resonance wavelength λ_R, Förster radius (R_0) and FRET efficiency of different nanoshell of aspect ratio 0.85

Nanoshell	λ_R(nm)	R_0 (nm)	E_{FRET}
ZrN/SiO_2	645.833	32.5	0.92
ZrN/TiO_2	685.083	33.0	0.91
TiN/SiO_2	751.515	28.8	0.84
TiN/TiO_2	789.809	29.0	0.83

Table 19.2 Förster radius (R_0), FRET efficiency and normalized FRET rate of ZrN@SiO_2 nanoshell of different shell thickness with R_1 = 8.5 nm at resonance wavelength 645.833 nm

Shell radius R_2 (nm)	R_0 (nm)	E_{FRET}	$(\gamma_{FRET}/\gamma_0) * 10^8$
10	32.5	0.92	1.13
12	37.9	0.99	0.593
14	41.0	0.998	0.319

4. CONCLUSION

In conclusion, the effect of plasmon mediated FRET between CdSe QD's is studied. Significant enhancement can be achieved in energy transfer interactions between quantum dots for semiconductor applications and also the efficiency of FRET is increased with presence of the system under investigation. Overall, this study suggests the possibilities of new structure/ design based on the CdSe QDs and Plasmonic nanoparticle for improved performance of semiconductor devices.

REFERENCES

1. Durach M., Rusina A., Klimov V. I., Stockman M. I. (2008). Nanoplasmonics renormalization and enhancement of coulomb interactions. New Journal of Physics. 10:105011.
2. Faucheaux A. J., Stanton L.D., Jain P.K. (2014). Plasmon Resonances of Semiconductor Nanocrystal: Physical Principles and New Opportunities. J. Phys. Chem. Lett. 5:976–985.
3. Guang Qi Li, Shishodia M. S., Fainberg B. D., Boris Apter, Michal Oren, Abraham Nitzan, and Mark A. Ratner (2012). Compensation of Coulomb blocking and energy transfer in the current voltage characteristic of molecular conduction junctions. Nano Lett. 12:2228–2232.
4. Rajput P., Shishodia M. S. (2020). Förster Resonance Energy Transfer and Molecular Fluorescence near Gain Assisted Refractory Nitrides based Plasmonic Core-Shell Nanoparticle. Plasmonics 15:2081–2093.
5. Rajput P., Shishodia M. S. (2021). Energy Transfer Interactions and Sensing Characteristics of Gain Assisted and Graphene Coated Plasmonic Nanomatryoshka. Plasmonics 16(1):2277–2290.

Note: All the figures and tables in this chapter were made by the authors.

Recent Trends in Applied Physics and Material Science – Sudhir Bhardwaj et al. (eds)
 ISBN 978-1-041-16452-4

Electronic Characteristics of Multinary Chalcogenide Semiconductors for Energy Applications

Anima Ghosh*
Department of Physics,
School of Sciences and Humanities, SR University,
Warangal, Telangana, India

S.K. Ghorui
Department of Physics,
Vaagdevi Engineering College,
Warangal, Telangana, India

Abstract: In the field of energy applications, particularly in thin-film solar cell technology, multinary chalcogenide semiconductors are of significant interest due to their tunable and favorable optoelectronic properties. The multinary compound $CuZn_2AlS_xSe_{4-x}$ (x=0 and 2) represents a vital area of research, offering unique characteristics that can be harnessed for next-generation electronic and optoelectronic applications. In this work, we utilized density functional theory through the Wien2k code to investigate the electronic properties of $CuZn_2AlS_xSe_{4-x}$ (x=0 and 2) materials. The density of states was calculated using the Perdew-Burke-Ernzerhof (PBE) functional. Our findings indicate that this compound behaves as direct band gap materials, with transitions occurring at the gamma point between the conduction and valence levels. The bandgap increases when two selenium (Se) anions are replaced with two sulfurs (S), however, the overall band characteristics remain consistent for both S and Se. This compositional tunability significantly enhances light absorption and charge carrier mobility, which are critical factors for improving energy application efficiency.

Keywords: Chalcogenide, Density functional theory, Density of states

1. INTRODUCTION

In the rapidly advancing field of materials science, the demand for innovative semiconducting materials with tailored properties for energy applications has surged. Multinary material choice in energy devices, one of significantly important thin film solar technologies, could influence in performance. The foremost appeals of multinary semiconductors are in their tunable electronic properties, e.g., mobility of charge carrier, band gap, thermal stability, and optical characteristics, which are very crucial for improving device performance in any kind of energy technology. In particular, for solar energy harvesting, materials with optimal band gap that enables the light absorption of a broad range of sunlight wavelength are required. Multinary chalcogenides are basically consists by chalcogen atoms (S, Se or Te) with metal cations from II, III, IV and V group. The flexibility in crystal structure by altering concentration of constituent elements makes tenability of the electronic properties of these semiconductors, which could helps to absorb specific wavelengths of light to maximize solar-to-electricity conversions efficiency. A familiar and notable example of multinary chalcogenide is copper indium gallium selenide (CIGS) thin-film semiconductors (Lopes et al, 2023), which have been extensively researched for use in photovoltaic devices due to optimal band gap from constituent elements in CIGS.

A series of cross-substituted chalcogenide multinary semiconductor materials have been reported to enhance the stability and performance of semiconducting materials

*Corresponding author: animaghosh10@gmail.com

DOI: 10.1201/9781003684718-1

(Ghosh et al. 2017, Wencong et al, 2020). However, few studies have focused on multinary systems that include aluminum (Al) on the cation side. One such system, the stannite phase $CuZn_2AlS_4$, (Ghosh et al. 2017) has been reported to have a bandgap of 1.6 eV and a high absorption coefficient. Recently, Wencong et al.(2020) reported the phase stability of a series of quaternary chalcogenides with the chemical composition $I-II_2-III-VI_4$ and found that the $CuZn_2AlSe_4$ (CZASe) structure is energetically stable in the stannite phase.

Here in present report, we have analysis the density of states (DOS) of $CuZn_2AlSe_4$ (CZASe) and $CuZn_2AlS_2Se_2$ (CZASSe) using the Perdew-Burke-Ernzerhof (PBE) functional in generalized gradient approximation. The total DOS of CZASe and CZASSe were analyzed from optimized lattice parameter values. Our investigation aims to enhance the understanding of these multinary chalcogenide semiconductors and could facilitate comparisons with other extensively studied quaternary groups.

2. COMPUTATIONAL DETAILS

The scalar-relativistic full-potential augmented plane wave (FP-LAPW) method was employed using the Wien2k code (Blaha et al, 2023) to analyze the density of states (DOS) of $CuZn_2AlSe_4$ (CZASe) and $CuZn_2AlS_2Se_2$ (CZASSe). In this calculation, the core and valence subsets of the basis set are separated, however they located within the Muffin-tin (MT) spheres. The core level contributions primarily arise from the spherical component of the potential, where the density of charge exhibits spherical symmetry. The orbital configuration of the valence states are Se ($3d^{10}$, $4s^2$, $4p^4$), S ($3s^2$, $3p^4$), Al ($3s^2$, $3p^1$), Cu ($3p^6$, $4s^2$, $3d^9$) and Zn ($3d^{10}$, $4s^2$). All these orbitals are expanded within a potential using spherical harmonics. The Perdew-Burke-Ernzerhof (PBE) potential in the generalized-gradient approximation (GGA) (Perdew et al, 1996) is applied for crystal structure optimization, atomic position refinement, and evaluation total energies. The other parameters, such as cut-off energy, self-consistent convergence, basis expansion, and charge density values in the Fourier expansion, are kept consistent with previous reports (Ghosh et al, 2024). Further, the tetrahedral method is used for Brillouin zone integrations with a 7×7×3 k-point mesh.

3. RESULTS AND DISCUSSION

The density of states (DOS) in semiconducting materials describes the behavior of electronic states at various energy levels for electrons to occupy. This is important for investigation the semiconductor's electronic properties, including its conductivity, optical absorption, and overall performance in electronic and optoelectronic devices. Figure 20.1 and Fig. 20.2 show the DOS of $CuZn_2AlSe_4$ (CZASe) and $CuZn_2AlS_2Se_2$ (CZASSe) respectively.

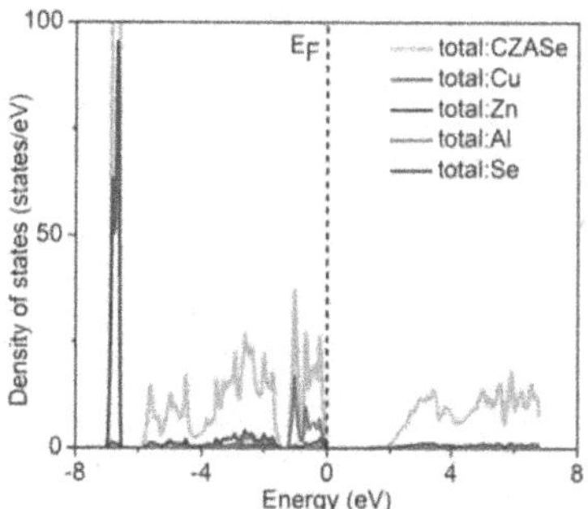

Fig. 20.1 Density of states of $CuZn_2AlSe_4$ (CZASe)

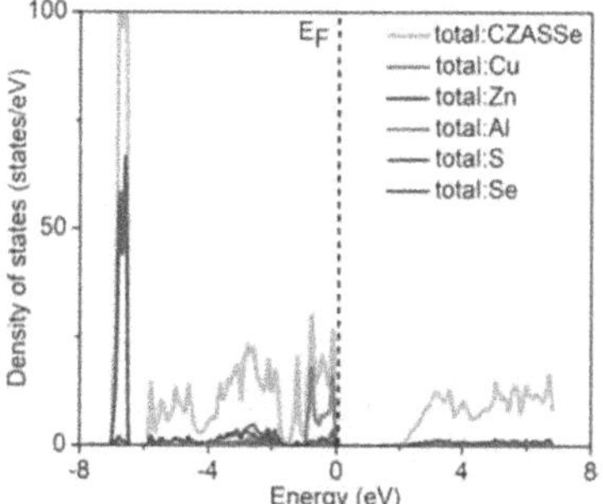

Fig. 20.2 Density of states of $CuZn_2AlS_2Se_2$ (CZASSe)

Here, E_F represents the Fermi energy. Both the DOS profiles (shown in green colour) are similar trends, but the band gap is increases from CZASe to CZASSe structure due to partial replacement of Se to S atoms. The calculated band gap values for CZASe is 1.29 eV and for CZASSe 1.46 eV which are good trends with literature reported values (Palchoudhury et al, 2024). While all atomic sub-levels contribute to the overall DOS, the primary dominating contribution for conduction level DOS comes from hybridization of *s*-level of cation with anion's *p*-level. Specifically, for valence level DOS of CZASe, the interaction involves Cu-*d* and Se-*p,* while for CZASSe, it includes both *S-p* and *Se-p*.

4. CONCLUSION

The computational study using the Wien2k code has provided valuable insights into the total contribution of density of states (DOS) of multinary chalcogenides. The calculated bandgap value of CZASe and CZASSe closely align with reported literature data. The Wien2k simulations emphasize the critical role of tuning the bandgap by varying the percentages of S and Se, which is key to controlling and optimizing charge transport within the compound.

REFERENCES

1. Blaha, P., Schwarz, K., Madsen, G. K. H., Kvasnicka, D., Luitz, J., Laskowski, R., Tran, F., Marks, L.D. (2023). WIEN2k an Augmented Plane Wave + Local Orbitals Program for Calculating Crystal Properties, Karlheinz

Schwarz, Techn. University at Wien, Austria, ISBN 3-9501031-1-2.

2. Perdew, J.P., Burke, K., Ernzerhof, M. (1996). Generalized Gradient Approximation Made Simple, Phys Rev Lett. 77 (18): 3865–3868.
3. Lopes, T.S., Teixeira, J.P., Curado, M.A. et al. (2023). Cu(In,Ga)Se_2 based ultrathin solar cells the pathway from lab rigid to large scale flexible technology. npj Flex Electron **7**: 4.
4. Ghosh, A., Thangavel, R. (2024). Structural stability, electronic band structure, and optoelectronic properties of quaternary chalcogenide $CuZn_2MS_4$ (M=In and Ga) compounds via first principles. Indian J Phys **98**: 3959–3966.
5. Palchoudhury, S, Diroll, B.T., Ganesh, P., Cobos, J., Sengupta, S., Huang, J. (2024). Multinary light absorbing semiconductor nanocrystals with diversified electronic and optical properties. Nanoscale Adv. 6: 3785–3793.
6. Wencong, S.; Khabibullin, A. R.; Woods, L. M. (2020). Exploring Phase Stability and Properties of I-II_2-III-VI_4 Quaternary Chalcogenides. Adv. Theory Simulation 3 (8): 2000041.
7. Ghosh, A.; Thangavel, R.; Gupta, A. (2017). Chemical synthesis, characterization and theoretical investigations of stannite phase $CuZn_2AlS_4$ Nanocrystals. New J. Chem. 40 (2): 1149–1154.

Note: All the figures in this chapter were made by the authors.

Recent Trends in Applied Physics and Material Science – Sudhir Bhardwaj et al. (eds)

Numerical Optimization of TiN and ZrN Nanorods based Plasmonic Nanosystems for Enhanced Solar Absorption

Alok Singh*
Department of Physics,
JSS Academy of Technical Education,
Noida, India

Manmohan Singh Shishodia
Department of Applied Physics, Gautam Buddha University,
Greater Noida, India

Pratima Rajput
Department of Physics,
JSS Academy of Technical Education,
Noida, India

Soniya Juneja
Department of Applied Sciences,
Krishna Institute of Engineering and Technology,
Ghaziabad, India

Abstract: The global photovoltaic capacity, now at 1600 GW (India: 89.4 GW), underscores the demand for efficient solar cell technologies. This study explores Titanium Nitride (TiN) and Zirconium Nitride (ZrN) as cost-effective, thermally stable plasmonic materials for enhancing light absorption in thin-film solar cells. These refractory nitrides exhibit strong plasmonic activity across the UV-IR spectrum, aligning with solar wavelengths. Nanorod-based designs are optimized for solar absorption, with nitrides showing superior scalability and efficiency compared to gold. This work highlights metal nitrides as sustainable alternatives for improved photovoltaic performance.

Keywords: Photovoltaics, Light absorption, Nanorods, PV, Titanium nitride, Zirconium nitride, Plasmonics, Nanostructures

1. INTRODUCTION

Photovoltaics is a clean and sustainable technology crucial for future energy demands. Thin-film solar cells, with thicknesses of 1-2 μm, reduce costs but struggle to absorb photons efficiently, particularly in the near-infrared spectrum (A. Hasan, 2018). Traditional light-trapping methods face challenges like recombination losses and fabrication complexity, necessitating innovative solutions (A. G. Waketola, 2024). Plasmonic nanostructures, which confine electromagnetic energy at subwavelength scales, have shown promise for boosting light absorption. Noble metals like gold (Au) and silver (Ag) are widely studied but are costly and less scalable. This study explores titanium nitride (TiN) and zirconium nitride (ZrN) as alternative plasmonic materials (A. Singh, 2023). These refractory nitrides offer UV-IR plasmonic responses, low cost, thermal stability, and fabrication compatibility. By optimizing TiN and ZrN nanorod dimensions and configurations, we aim to enhance light absorption in thin-film photovoltaics, particularly in the underutilized near-infrared range (D. F. Carvalho, 2024). This research highlights their potential to improve efficiency and scalability in solar technology.

2. SYSTEM DESCRIPTION

The schematic diagram of nanorod embedded in a medium of dielectric constant ε_m is shown in Fig. 21.1.a. The nanorod having dielectric permittivity ε_1, length L, radius r, embedded in a medium of permittivity ε_m (refractive index, n). The plasmonic system is exposed to uniform electric field $\vec{E} = E_0\hat{z}$ directed along z direction. In the present study, COMSOL Multiphysics has been used for nanorod system. The mesh setting chosen is a fine mesh to discretize the geometry. The governing electromagnetic

*Corresponding author: alok.singh@jssaten.ac.in

DOI: 10.1201/9781003684718-21

equations for the mesh elements are defined, and these elements are then assembled to solve the problem. Numerical simulations are carried out using the RF module within COMSOL Multiphysics.

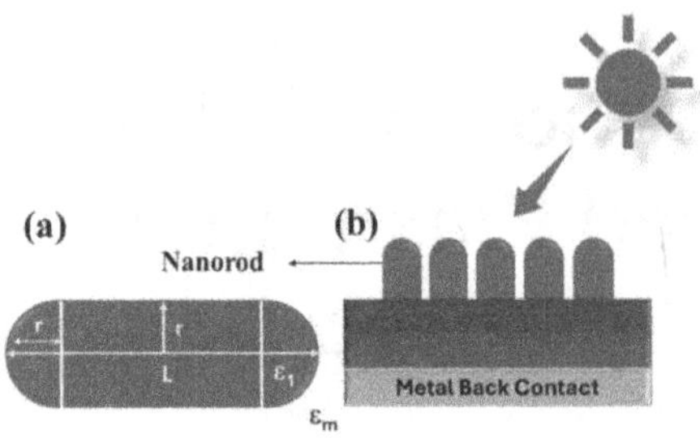

Fig. 21.1 (a). The schematic diagram of nanorod with permittivity ε_1, radius 'r' and length 'L' surrounded by a medium of permittivity ε_m, (refractive index, n), (b). photovoltaic cell exposed to typical solar spectrum

The photovoltaic cell exposed to typical solar spectrum shown in Fig. 21.1.b.

3. RESULTS AND DISCUSSION

This section presents the computational investigation of Au, TiN, and ZrN nanorods, focusing on their optical properties. The calculated spectral variation of the scattering cross-section (C_{sca}) for ZrN nanorods with different lengths (L = 105 nm, 120 nm, 135 nm, and 240 nm) and a fixed radius (r = 30 nm) is shown in Fig. 21.2. As the nanorod length increases, the resonance peak shifts toward longer wavelengths. For instance, when the length of the ZrN nanorod is increased from 105 nm to 240 nm, the resonance peak shifts from 650 nm to 1100 nm, highlighting the tunability of the plasmonic response.

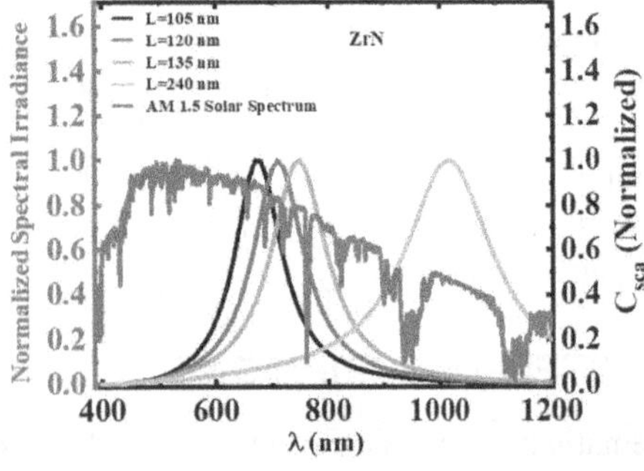

Fig. 21.2 Calculated normalized scattering spectra of ZrN nanorod nanoparticle with different particle sizes (L=105 nm, 120 nm, 135 nm, 240 nm and r = 30 nm, embedded in air, and AM 1.5 solar spectra (λ = 400 nm – 1200 nm)

Figure 21.3 depicts the calculated electric field distribution for a ZrN nanorod of length L = 240 nm and radius r = 30 nm at three specific wavelengths: (a) λ = 551 nm, (b) resonance λ_R = 1042 nm, and (c) λ = 1188 nm.

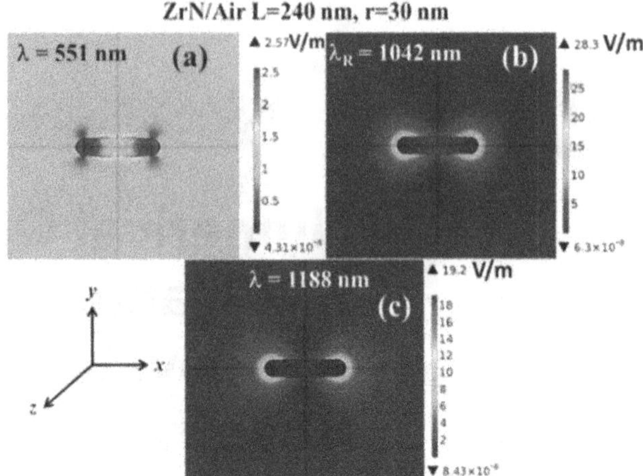

Fig. 21.3 Calculated electric field distribution (a). at l=551 nm, (b). at resonance l_R=1042 nm, (c). at l=1188 nm for ZrN nanorod of length, L= 240 nm, and r =30 nm. Other parameters are: Δλ = 1 nm, and n=1

The strong field enhancement at the resonance wavelength demonstrates the efficient light confinement capabilities of the ZrN nanorods, making them promising for photovoltaic applications.

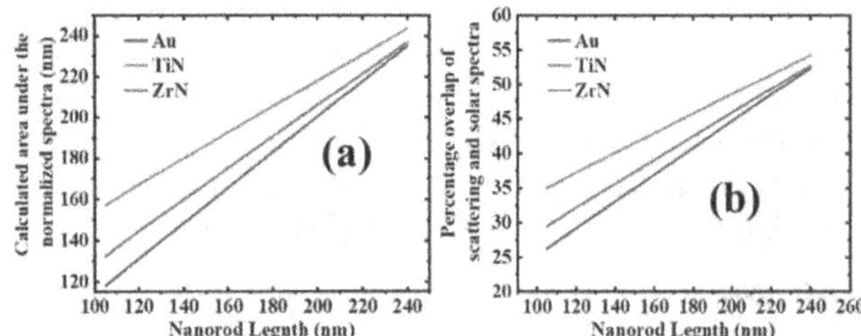

Fig. 21.4 (a) Calculated area under the normalized spectra (nm) for Au, TiN, and ZrN nanorods with lengths tuned from 100 nm to 240 nm. (b). Calculated percentage overlap between the scattering spectra and solar spectra for Au, TiN, and ZrN nanorods with lengths tuned from 100 nm to 240 nm. The spectral range is λ = 400–1200 nm, with a nanorod radius of r = 30 nm

The parameters used in this analysis include Δλ = 1 nm and refractive index n = 1. In Fig. 21.4(a-b), the effect of length tuning (100–240 nm) on the optical performance of Au, TiN, and ZrN nanorods is presented. TiN nanorods exhibit the highest overlap with the solar spectrum, indicating superior light absorption. However, ZrN and Au nanorods, while showing slightly lower solar spectrum overlap, offer a balance between optical efficiency and thermal stability, making them suitable for high-temperature plasmonic photovoltaic applications. This demonstrates the potential of material and geometry optimization for enhancing light absorption across a broad solar spectrum.

4. CONCLUSION

This study demonstrates the influence of nanorod length and material selection on the plasmonic properties of Au, TiN, and ZrN nanorods for photovoltaic applications. Length variations shift the resonance peak toward longer wavelengths, allowing tunable light absorption across the solar spectrum. TiN nanorods exhibit the highest solar spectrum overlap, while Au and ZrN offer a balance of optical performance and thermal stability, making them ideal for high-temperature conditions. These results highlight the potential of TiN and ZrN as cost-effective and durable alternatives to noble metals, advancing the development of efficient plasmonic photovoltaic devices.

ACKNOWLEDGMENT

The author acknowledges Dr. S. M. Tripathi (IIT Kanpur, India) for his assistance with COMSOL.

REFERENCES

1. Husain, A. A., Hasan, W. Z. W., Shafie, S., Hamidon, M. N., & Pandey, S. S. (2018). A review of transparent solar photovoltaic technologies. Renewable and sustainable energy reviews, 94: 779–791.
2. Singh, A., Kumar, P., & Shishodia, M. S. (2018, November). Spherical and Cylindrical Plasmonic Nanoparticles for Enhanced Light Trapping in Photovoltaic Cells. In 2018 5th IEEE Uttar Pradesh Section International Conference on Electrical, Electronics and Computer Engineering (UPCON) (pp. 1–5). IEEE.
3. Waketola, A. G., Hone, F. G., Geldasa, F. T., Genene, Z., Mammo, W., & Tegegne, N. A. (2024). Enhancing the Performance of Wide-Bandgap Polymer-Based Organic Solar Cells through Silver Nanorod Integration. ACS omega, 9(7): 8082–8091.
4. Singh, A., Shishodia, M. S., Agarwal, D., & Kumar, P. (2023). Fano resonances in graphene coated refractory nitride nanoshell and nanomatryoshka for sensing food adulteration. Applied Physics A, 129(5), 366.
5. Carvalho, D. F., Teixeira, J. P., Salomé, P. M., Fernandes, P. A., & Correia, M. R. P. (2024). Gold nanorods as performance enhancers in planar perovskite solar cells: A numerical study. Solar Energy Materials and Solar Cells, 277, 113112.

Note: All the figures in this chapter were made by the authors.

 ISBN 978-1-041-16452-4

Melting Temperature of Nano Germanium-Ge, Tellurium-Te at Different Shape and Size

Prachi Singh, Shivam Srivastava
Department of Physics,
Dr. Shakuntala Misra National Rehabilitation University,
Lucknow, Uttar Pradesh

Anjani Kumar Pandey
Institute of Engineering and Technology,
Dr. Shakuntala Misra National Rehabilitation University,
Lucknow, Uttar Pradesh

Chandra Kumar Dixit
Department of Physics,
Dr. Shakuntala Misra National Rehabilitation University,
Lucknow, Uttar Pradesh

Abstract: In our present work, we are investing several nanosolids Viz. (Germanium-Ge, Tellurium-Te, Rhenium-Re, Osmium-Os and polonium-Po) with the aim of theoretically predicting their meting temperature at different shapes and sizes, by using the equation of W.H.Q.

Keywords: Melting temperature, Spherical nanosolids, Nanowires, Nanofilms

1. INTRODUCTION

W.H. Qi developed a model known as the energy model to determine the cohesive energy of some nanomaterials, helpful to understand behavior of materials at nanoscale where their bonds become more prominent, the combine combination of atoms present within the inner structure and those on the surface (Kumari T et al 2023, Srivastava S et al 2023, Pandey A.K et al 2023, Srivastava S et al 2023, Srivastava S et al 2023). In our present work, we are investing several nanosolids Viz. (Germanium-Ge, Tellurium-Te, Rhenium-Re, Osmium-Os and polonium-Po) (National library of medicine) with the aim of theoretically predicting their meting temperature at different shapes and sizes, by using the equation of W.H. Qi (QiW. H.2005) for calculating the melting temperature of some nanosolids.

2. METHOD OF ANALYSIS

The cohesive energy is a fundamental that helps us understand other physical properties of nanosolids such as melting temperature (Kittel C 1996,Seiz F. et al 1964). The cohesive energy regulates the thermodynamic behavior of the nanosolids (QiW. H. 2016,Srivastava Shivam et al 2023, Pandey A. K et al 2023, Pandey A. K et al 2023, RoseJ. H.et al 1981, Singh Met al 2017).

$$E_p = E_b\left(1-\frac{N}{2n}\right) \tag{1}$$

W.H. Qi formula is given as [Singh Met al2017] as:

$$T_p = T_b\left(1-\frac{N}{2n}\right) \tag{2}$$

3. RESULT AND DISCUSSION

The graphs plotted between melting temperature (Tp) and D(nm). The graph shown in Fig. (22.1–22.2). From Fig. (22.1–22.2) it is clear that with an increase in the size and shape of nanosolids, the melting temperature of nanosolids for different shapes first increases; however, at a point, the melting temperatures show a linear trend with each other continuing consistently. The graph illustrates that when the size of nanomaterials is less than 5 nm for all shapes, the melting temperature decreases as the size of the nanomaterials decreases. Sizes below 5 nm are favorable for achieving lower melting temperatures (Table 22.1).

[*]Corresponding author: psc_phyphd2021@dsmnru.ac.in

DOI: 10.1201/9781003684718-22

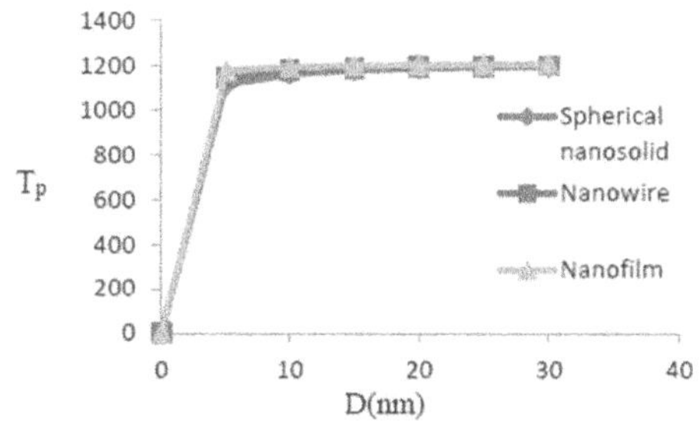

Fig. 22.1 Size-dependent melting temperature of Germanium (Ge)

Table 22.1 Value of input data Tb and d(nm)

Materials	Melting temperature of bulk material (T_b) K	Size d(nm)
Germanium (Ge)	1211.4 K	0.211 nm
Tellurium (Te)	1261 K	0.206 nm

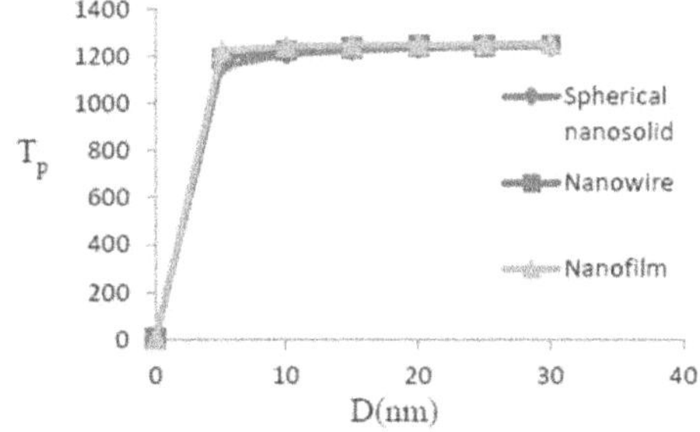

Fig. 22.2 Size- dependent melting temperature of Tellurium (Te)

From Fig. (22.1-22.2) it is clear that with an increase in the size of nanosolids, the melting temperature of nanosolids for different shapes first increases; however, at a point, the melting temperatures show a linear trend with each other continuing consistently. The graph illustrates that when the size of nanomaterials is less than 5 nm for all shapes, the melting temperature decreases as the size of the nanomaterials decreases. Sizes below 5 nm are favorable for achieving lower melting temperatures.

REFERENCES

1. Kumari T, Pandey B.K., Gupta J., Jaiswal R. L. and Shuka S, (2023), Solid State Communication, 371, 115254
2. Kittel C, (1996), Introduction to Solid State Physics, seventh ed., John & Sons Inc., New York
3. Parodic table of element, National library of medicine, https://pubchem.ncbi.nlm.nih.gov/ periodic-table/
4. Pandey A.K., Dixit C.K., Srivastava s, Singh P andTripathi S, (2023), Theoretical Prediction of Thermo-Elastic Properties of TiO_2 (Rutile Phase), Natl. Acad. Sci. Lett.
5. Pandey A. K., Dixit C. K. and Srivastava S, (2023), Theoretical model for the prediction of lattice energy of diatomic metal halides. J Math Chem
6. Pandey A. K., Srivastava S and Dixit C. K., *et al.*, (2023), Shape and Size Dependent Thermophysical Properties of Nanomaterials. *Iran J Sci*
7. RoseJ. H., Ferrante J, and Smith J. R., (1981) "Universal binding energy curves for metals and bimetallic interfaces," Physical Review Letters, 47, 9, 675
8. Seiz F. and Turnbull D, (1964), Solid State Phys., Academic Press, New York, 16
9. Singh M, Tlali S and Chandra K, (2017) Advanced Materials Proceedings, 2(2), 72–75
10. Srivastava S, Singh P,Pandey A and Dixit C.K., (2023), Solid State communication 377, 115387
11. Srivastava S, Singh P,Pandey A and Dixit C.K., (2023), Nano Structure & Nano objects, 36, 101067
12. Srivastava S, Pandey A.K. and Dixit C.K, (2023), Comparative study of elastic properties of some inorganic and organic molecular crystals from EOS. *J Math Chem.* https://doi.org/10.1007/s10910-023-01546-9
13. Srivastava Shivamet al., Equation of states at extreme compression ranges: Pressure and Bulk modulus as an example, Materials Open, doi: 10.1142/S2811086223500073
14. Srivastava A P,. Pandey B K, Gupta A K, (2025) Computational Condensed Matter 42, e00986
15. Srivastava A P,. Pandey B K, Gupta A K, (2024) Computational Condensed Matter 40, e00952
16. QiW. H., (2005), physica B, 368, 46–50
17. QiW. H., (2016), Nanoscopic thermodynamics, Acc. Chem. Res. 49(9), 1587–1595

Note: All the figures and tables in this chapter were made by the authors.

Optimizing Solar Cell Efficiency via Downshifting /Downconversion: Current Trends and Future

Sonal P. Ghawade*,
Anjali A. Mahajan, Khushi G. Gupta
VMV Commerce, JMT Arts & JJP Science College, Wardhaman Nagar, Nagpur, Maharashtra, India

Abhay D. Deshmukh
Dept. of Physics, Energy Materials and Devices Laboratory, RTM Nagpur University, Nagpur, India

Abstract: The optimization of solar cell efficiency is a vital area of research as the demand for renewable energy source. This review focuses on downconversion techniques, which convert high-energy photons from the solar spectrum are transformed into photons with lower energy. That can be more effectively harnessed by photovoltaic cells. We analyze current trends in downconversion materials, including organic phosphors, inorganic nanocrystals, and quantum dots, highlighting their potential to enhance light absorption and minimize thermal losses. Key strategies for integrating these materials into various solar cell architectures are discussed, showcasing notable improvements in power conversion efficiency in recent studies. Additionally, we address the challenges associated with material stability, integration, and manufacturing scalability, while identifying promising avenues for future research, such as novel nanostructured materials and advanced design methodologies. By synthesizing recent advancements and proposing future directions. This review aims to illuminate the transformative potential of downconversion technologies in the ongoing pursuit of highly efficient solar energy solutions.

Keywords: Solar cell, Luminescence, Downshifting

1. INTRODUCTION

Sunlight is a readily available and free energy source in many areas of the world, which can be harnessed by innovative technologies to produce electricity (Scholes, Fleming, Olaya-Castro & van Grondelle,2011). According to recent findings, available solar energy at the Earth's surface vastly surpasses worldwide energy requirements, with estimates suggesting it's roughly 10,000 times more abundant than our total energy usage. Therefore, solar energy is anticipated to effectively address a significant portion of future energy consumption needs. Although the photovoltaic (PV) industry has advanced considerably finished the last few years, achieving an efficient and economically viable change of solar energy into electricity using photovoltaic cells continues to be a challenging endeavor (Nozik, and Miller, 2010). Presently, the annual utilization of solar energy is significantly less than 1% of the overall energy utilization, whereas fossil fuels represent over 90% of total energy use. To enable widespread adoption of solar energy, it is essential to develop more efficient photovoltaic systems at lower costs (Wong & Ho, 2010). Consequently, solar energy is anticipated to fulfill a considerable share of future energy demands. Although photovoltaic technology has advanced significantly, efficiently and affordably converting solar energy into electricity remains a major challenge. Solar power still accounts for only a small portion of global energy use, with fossil fuels continuing to dominate, meeting over 90% of worldwide energy demands. For solar energy to be widely adopted, there is a need for the growth of more efficient and lower-price PV System.

A significant challenge hindering the change productivity of photovoltaic cells is their inability to effectively

*Corresponding author: sonalghawade25@gmail.com

DOI: 10.1201/9781003684718-23

respond to the complete solar spectrum. The spectral supply of sunlight at Air Mass 1.5 global (AM 1.5G) includes wavelengths that span from ultraviolet to infrared, specifically ranging from 280 to 2500 nm (0.5 to 4.4 eV). However, existing photovoltaic cells capture only a minor portion of these solar photons, as shown in Fig. 23.1 (Richards, 2006). This limitation arises because each photovoltaic material is sensitive to a specific range of solar photon energies that correspond to its characteristic bandgap. In theory, a photon with energy exceeding the bandgap is absorbed; Surplus energy is wasted as heat due to the inability to harness photons with energies above or below the material's optimal range, leads to losses of roughly 50% of the incoming solar energy during the conversion to electricity in silicon-based solar cells. A crucial limitation is that crystalline silicon, with a bandgap energy of 1.1 eV, has a theoretical maximum efficiency of around 31-41%, depending on concentration factors, as dictated by the Shockley-Queisser limit, a fundamental constraint established in 1961.

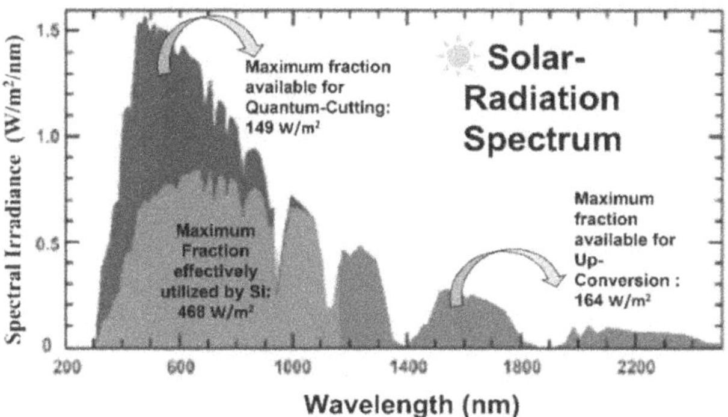

Fig. 23.1 Displays the AM 1.5G spectrum, with green indicating the portion absorbed by a typical silicon-based solar cell. The purple and red regions show the spectral areas that can be harnessed using quantum-cutting and upconversion techniques, respectively

Using luminescent materials as spectral converters can help minimize inherent thermalization and losses due to non-absorption. This approach, referred to as third-generation solar photon conversion, incorporates a passive luminescent layer into photovoltaic cell (Shalav, Richards, and Green, 2007). This technology offers a significant benefit: spectral converters can be seamlessly integrated into existing solar cells with minimal adjustments, as their optimization can be done independently of the solar cells themselves, thereby enhancing the productivity of single-junction solar cells, researchers are currently investigating three luminescent processes: upconversion, quantum-cutting, and down-shifting, in order to develop more effective photovoltaic devices. Trivalent lanthanide ions are strong applicants for effective spectral conversion due to their sophisticated energy-level architecture, called the Dieke diagram, which enables efficient management of photons.

1.1 Down conversion/ Quantum Cutting

In the down conversion process, One high-energy photon is split into several photons with lower energy levels, potentially leading to a quantum efficiency greater than 100%. This phenomenon is commonly known as "quantum cutting," where one photon is effectively "cut" into multiple photons.

1.2 Upconversion

In upconversion (UC), multiple low-energy photons (below the bandgap) are converted into one high-energy photon. The phenomenon of upconversion has been extensively researched in solids doped with lanthanide and transition metal ions since Auzel's discovery of upconversion in the 1960s. The concept of utilizing upconversion materials to enhance photovoltaic system efficiency was first introduced by Gibart's research team in 1996. They successfully demonstrated the potential of the Er^{3+}–Yb^{3+} duo in bifacial GaAs solar cells.

1.3 Downshifting

Downshifting, also referred to as photoluminescence, is a property exhibited by certain materials, similar to downconversion. However, it involves the emission of a single photon, accompanied by energy loss through non-radiative decay, ultimately yielding a quantum efficiency of less than unity. Downshifting can mitigate the weakened blue response in solar cells, a limitation often caused by inadequate front surface passivation in silicon-based devices. By tailoring the incident spectrum to wavelengths with enhanced internal quantum efficiency, overall conversion efficiency can be boosted by approximately 10% (Van Sark et al. 2005). Enhanced front passivation techniques can diminish the need for downshifting layers, which can also obstruct the absorption of high-energy photons in certain heterojunction window layers, such as CdS on CdTe solar cells. A thorough review of this topic has been provided by Klampaftis et al. (2010), offering valuable insights into this complex issue.

2. CONCLUSION

This review explores the basic principles of luminescent materials that act as spectral converters aiming to enhance efficiency solar cells. Optimizing solar cell design requires mitigating energy losses stemming from the discrepancy between the solar cell's absorption spectrum and incoming sunlight. Emerging techniques leveraging upconversion, quantum-cutting, and luminescent down-shifting materials hold promise for reducing these losses and boosting solar cell efficiency, although further development is needed. Down-shifting technology increases solar cells' responsiveness to short-wavelength sunlight, leading to enhanced energy harvesting. By incorporating specialized materials like lanthanide-based phosphors, glasses, quantum dots, and organo-lanthanide complexes into solar concentrators and planar coatings, the power conversion efficiency of solar cells can be significantly improved.

REFERENCES

1. Scholes, G. D., Fleming, G. R., Olaya-Castro, A., & van Grondelle, R. (2011). Lessons from nature about solar light harvesting. Nature Chemistry, 3, 763–774.
2. Nozik, A. J., & Miller, J. (2010). Introduction to solar photon conversion. Chemical Reviews, 110(11).
3. Wong, W.-Y., & Ho, C.-L. (2010). Organometallic photovoltaics: A new and versatile approach for harvesting solar energy using conjugated polymetallaynes. Accounts of Chemical Research, 43(9).
4. Richards, B. S. (2006). Enhancing the performance of silicon solar cells via the application of passive luminescence conversion layers. Solar Energy Materials and Solar Cells, 90(15), 2329–2337.
5. Shockley, W., & Queisser, H. J. (1961). Detailed balance limit of efficiency of p-n junction solar cells. Journal of Applied Physics, 32(3), 510–519.
6. Shockley, W., & Queisser.
7. Shalav, A., Richards, B. S., & Green, M. A. (2007). Luminescent layers for enhanced silicon solar cell performance: Up-conversion. Solar Energy Materials and Solar Cells, 91(9), 829–842.
8. Gibart, P., Auzel, F., Guillaume, J.-C., & Zahraman, K. (1996). Below band-gap IR response of substrate-free GaAs solar cells using two-photon up-conversion. Japanese Journal of Applied Physics, 35(8R), 4401.

Note: All the figures in this chapter were made by the authors.

Structural, Optical, and Transport Properties of CZTS Semiconductor Nanoparticles Synthesized via Hydrothermal Route

Dhruti Bhagora and P. H. Soni*

Department of Physics, Faculty of Science,The Maharaja Sayajirao University of Baroda, Vadodara, Gujarat, India

Abstract: Semiconductor nanoparticles find useful applications in organic and inorganic solar cells. Moreover, wide band-gap nanoparticles prepared from group I-VI elements have attracted potential applications in the emerging energy alternatives. Photovoltaics are currently a promising technology producing sustainable energy resources. The quaternary compounds, mainly chalcogenides composed of earth abundant and non-toxic elements have gained tremendous interest in photovoltaics. Among them, Cu_2ZnSnS_4 (CZTS) is a promising candidate for photovoltaic applications and various eco-friendly methods are being explored for synthesizing CZTS nanoparticles. The powder X-ray diffraction (XRD) reveals characteristic peaks of (101), (112), (200), (220), (312) and (332) diffraction planes corresponding to the Kesterite phase of CZTS. Scanning Electron Microscopy (SEM) reveals prominent agglomerated nanosphere-like shapes and particle size of nearly 50-80 *nm*. Energy Dispersive X-ray spectroscopy (EDAX) shows the elemental composition of nanoparticles. The optical properties were obtained using UV-visible spectroscopy and evaluated optical band gap of 1.88 *eV* which is favourable for photovoltaic applications. The transport properties were measured using Hall Effect at room temperature.

Keywords: Semiconductor nanoparticles, Hydrothermal, Photovoltaic

1. INTRODUCTION

Rapid growth in population and consequent enormous increase in energy consumption require for renewable energy sources. In the chase of sustainable energy solutions, solar photovoltaic technology has emerged to play a major role in harnessing renewable resources of energy with the potential to cover the energy need of the world (Moiz et al., 2020). The traditional silicon-based solar cells have dominated the market, but the quest for alternative materials with improved efficiency, cost-effectiveness and environmental compatibility remains in focus for research (Patrick et al., 2013). One promising candidate in this endeavour is Copper Zinc Tin Sulphur (CZTS) that has gained considerable attention in recent years. CZTS exists in three mineral structural forms, *viz.*, wurtzite, stannite, and kesterite. A kesterite structure is favored for photovoltaic applications because to its strong electron affinity and configurable direct band gap, which are both important for PV cells. CZTS, a quaternary semiconductor compound, has gained significant attention due to its earth-abundant and non-toxic constituents, setting it apart from more conventional solar cell materials. The remarkable combination of copper, zinc, tin and sulphur presents an interesting prospect for enhancing solar cell performance (Vipul et al., 2013). Morden advancement in nanotechnology has further opened possibility for manipulating CZTS properties at the nanoscale, leading to the requirement of better synthesizing routes providing precision of the repeatability and cost effectiveness. Additionally, CZTS nanoparticles exhibit a tunable bandgap, allowing for efficient absorption of sunlight across a broad spectrum.

*Corresponding author: phsoni-phy@msubaroda.ac.in

DOI: 10.1201/9781003684718-24

There are different reports in the literature on vacuum and non-vacuum methods for the synthesis of quaternary CZTS nanoparticles. Other synthesis methods include sol-gel (Nattee et al., 2021), chemical co-precipitation method (K. Pal et al., 2021), solvothermal (Xinlong et al., 2015) and hydrothermal (Sonali et al., 2018) techniques. However, precise control over particle size, morphology, and crystallinity remains a major challenge in the above synthesis routes.

CZTS is used as an absorber material in solar cells, and the microstructure, crystallinity, and optical characteristics of the absorber material all have a significant impact on the solar cell's performance. In addition, optimized composition is vital when application is inclined towards thin film solar cells fabrication where binary and ternary phases arises causing thermal stress in the films. In the present work we report a hydrothermal synthesis process for pure CZTS nanoparticles with improved microstructure, optical properties andtransport property of CZTS nanoparticles. The present process provides precise control with repeatability for processing of CZTS nanoparticles along with tunability of the microstructure and the optical properties.

2. EXPERIMENTAL

2.1 Materials Used

The chemicals of analytical reagent grade used for the reaction were cupric chloride extra pure ($CuCl_2.2H_2O$ ≥99.0%), Zinc chloride ($ZnCl_2$ ≥ 98.0%), stannous chloride ($SnCl_2 2H_2O$ ≥ 98.0%), Thiourea extra pure ($NH.CS.NH_2$≥ 99.0%) and Ethanol (C_2H_5OH). All the starting materials were used without further purification. Deionized water has been used throughout the synthesis.

2.2 Synthesis of CZTS Nanoparticles

In this work, hydrothermal synthesis method was followed for the preparation of Cu_2ZnSnS_4 nanoparticles. 0.196 *M* Cupric chloride, 0.07 *M* zinc chloride and 0.07 *M* stannous chloride were dissolved in 30 *ml* of deionized water and 0.42 *M* thiourea as sulphur source was added to it under constant stirring to make the precursor solution. The precursor had been stirred for 1 hour till the color of the solution turns clear, slightly whiteish. Then the precursor solution was transferred into a 100 *ml* capacity Teflon lined stainless steel autoclave and was then kept inside a furnace for 24 hours at 230°C temperature. After the hydrothermal reaction the autoclave was allowed to cool down to room temperature. The obtained precipitate was collected carefully and washed several times with deionized water and ethanol, followed by drying process at 60°C for 4 hours in ambient air. The as-dried CZTS powder was characterized through various characterization techniques.

2.3 Materials Characterization

The as-prepared CZTS nanoparticles were characterized using analytical techniques. X-ray diffraction (XRD) was used to study the structural properties. XRD measurements were carried out over the 2θ range 20-80° using X-ray diffractometer (Rigaku Smart Lab SE) with $Cu\ K_\alpha$ radiation (λ = 1.5406 Å). The morphology and elemental composition of the synthesized nanoparticles were examined using a JEOL (JSM-6380LV) Scanning electron microscope with an accelerating voltage of 20KV and an energy dispersive spectrometer attached to it. A Perkin-Elmer Lambda 20UV-*vis* spectrometer was used for the optical measurement of the sample dispersed in water and Dimethyle sulfoxide (DMSO) with 1:1 ratio.

3. RESULTS AND DISCUSSIONS

3.1 X-ray Diffraction (XRD) Analysis

XRD pattern of the CZTS nanoparticles synthesized by simple hydrothermal method is shown in Fig. 24.1. The Kesterite nature and crystallinity were confirmed by the presence of sharp diffraction peaks associated with the (101), (112), (200), (220), (312) and (332) diffraction planes with preferential orientation in the (112) plane corresponding to the Kesterite CZTS structure (JCPDS no. 26-0575) (Sonali et al., 2018).

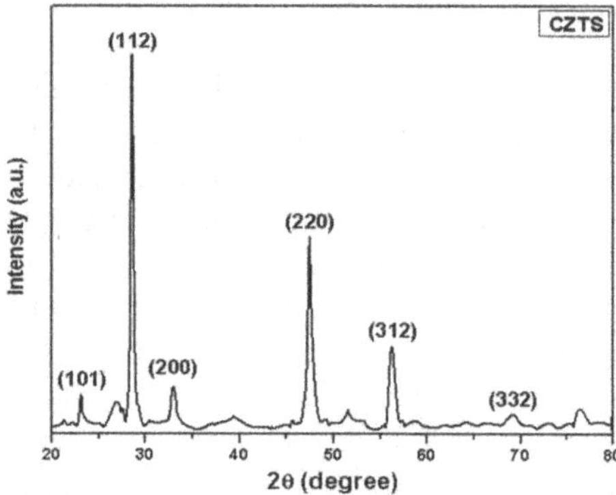

Fig. 24.1 XRD spectra of CZTS nanoparticles

Figure 24.2 shows the Rietveld refinement of the XRD pattern obtained. It yielded the lattice constant values to be a=b= 5.42 Å and c = 10.75 Å, with tetragonal phase. Similar lattice parameters have been reported by various authors (Sergio et al., 2019).

The average crystallite size was calculated using Debye Scherer formula,

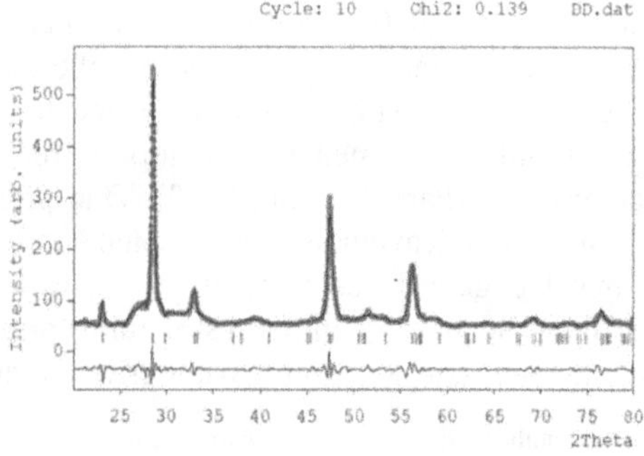

Fig. 24.2 Rietveld refinement of CZTS nanoparticles

$$D = \frac{k\lambda}{\beta Cos\theta} \quad (1)$$

Where k is a constant of approximate 0.94 shape factor, λ is the wavelength (1.5406Å) of the XRD used, β is the full width at half maximum (FWHM), θ is the Bragg Diffraction angle and D is Crystallite size. The average crystallite size of the CZTS nanoparticles was found to be 16.49 nm from the prominent (112), (220) and (312) peaks.

3.2 Scanning Electron Microscopy (SEM)

Figure 24.3 shows the SEM images of CZTS nanoparticles prepared by simple hydrothermal method. These were recorded under resolutions of scanning, mainly 200 *nm*. From Fig. 24.3, it is observed that the nanoparticles are slightly agglomerated. It is also observed that the assemblies of nanoparticles are loosely joined together. It may be due to the hydrophilic nature of nanoparticles and there are no capping agents used in the synthesis. It is also observed that the nanoparticles were present in the size range of 50-80 *nm*. This kind of larger nanoparticles can reduce the recombination rates in the solar cells due to lesser grain boundaries (Sahaya et al., 2017).

Fig. 24.3 SEM images of the CZTS nanoparticles

3.3 EDAX Spectra

Figure 24.4 depicts the chemical composition analysis performed using EDAX measurements on nanoparticles. The calculated elemental ratio of Cu: Zn: Sn: S_4 compound clearly indicates that they were present in ratio of 2.8:1:1:6 ratios. For the fabrication of highly efficient nanoparticles based solar cell, the stoichiometry of the compound is an essential feature (Xinlong et al., 2015).

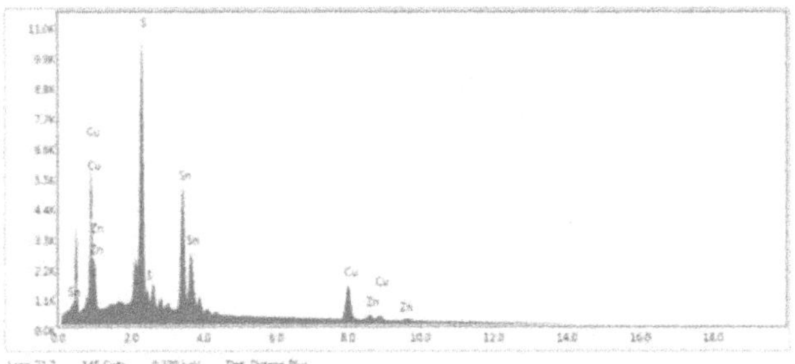

Fig. 24.4 EDAX spectra of CZTS nanoparticles

In sequence to infer the distribution of elements, elemental mapping of Cu, Zn, Sn, and S_4 was performed on nanoparticles and was obtained as shown in Fig. 24.5. It is also observed that the elements are homogeneously distributed in the prepared nanoparticles.

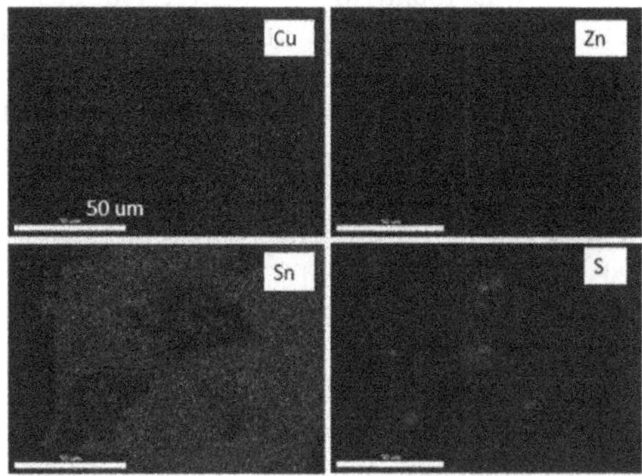

Fig. 24.5 Elemental distribution of Cu, Zn, Sn and S_4

3.4 UV-vis Analysis

The optical properties of the synthesized CZTS nanoparticles were studied using UV-Vis spectrometer. Figure 24.6 shows the spectra of optical absorbance and Tauc's plot (inset) to study the optical band gap of CZTS nanoparticles. Using the UV absorbance spectra, it can be observed that the CZTS nanoparticles are capable of good absorption of light in the visible region to UV region. This strong absorption spectrum of light can convey its possible application in solar cells which require absorption of a greater number of light. Figure 24.6 shows the inset Tauc's plot of $(\alpha hv)^2$ vs hv to determine the optical band gap.

$$(\alpha h\upsilon)^2 = \beta (h\upsilon - E_g) \quad (2)$$

Where β is a constant, υ is incident photon frequency, h is Planck's constant, and E_g is the band gap energy. The obtained band gap values for the sample is about 1.88 eV. The band gap here is much larger than those reported which can be attributed to the quantum confinement and the decreasing crystallite size (Sonali et al., 2018). The optical studies revealed that the prepared CZTS nanoparticles can find its application as a promising UV-visible light absorber material for solar cell applications.

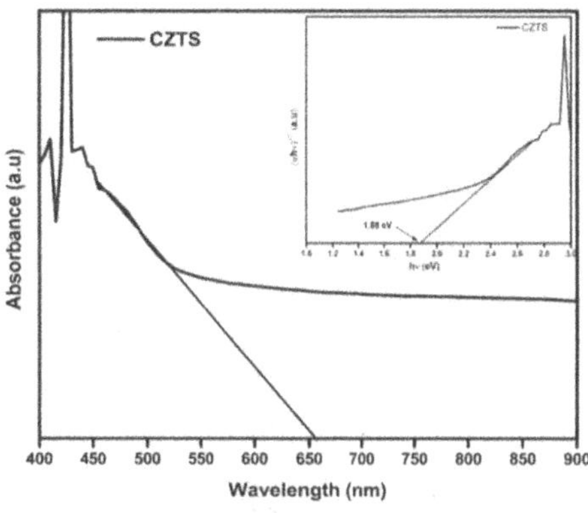

Fig. 24.6 UV-vis absorbance spectrum and (inset) Tauc's plot of CZTS nanoparticles

4. TRANSPORT PROPERTIES

Hall effect measurement: Hall effect measurements at room temperature were performed to analyze the charge carrier type, charge concentration, and mobility in Cu_2ZnSnS_4 nanoparticles. Hall coefficient of CZTS nanoparticles was found to be positive, suggesting the hole-induced transport in CZTS nanoparticles and in turn the p-type conductivity in CZTS nanoparticles (Zhang et al., 2006).The charge carrier concentration was calculated to be 3.81×10^{20} cm^{-3}, and the carrier mobility was approximately 2.59 $cm^2V^{-1}S^{-1}$.

Electrical Conductivity: The temperature-dependent electrical conductivity (σ), of the CZTS nanoparticles, is presented here. The electrical conductivity of CZTS nanoparticles was measured over the temperature range of Room temperature to 410 K, as shown in Fig. 24.7. The electrical conductivity of the samples increases with increase of temperature, showing the semiconductor like behaviour (Yu et al., 2024).

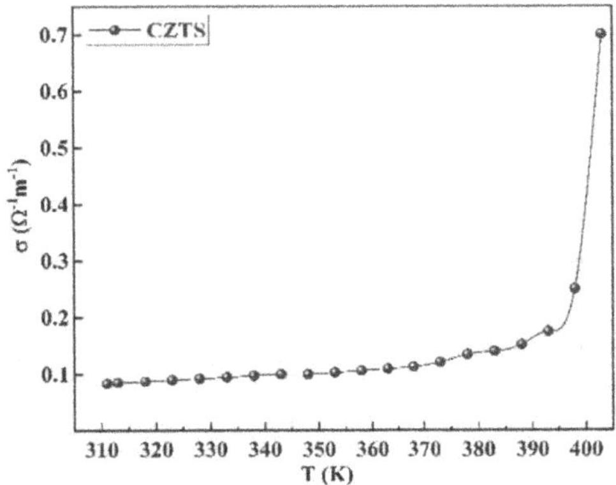

Fig. 24.7 Electrical conductivity of CZTS nanoparticles

5. CONCLUSIONS

In this paper, we have synthesized the quaternary semiconductor Cu_2ZnSnS_4 nanoparticles by a simple hydrothermal method. The X-ray diffraction (XRD) profile reveals the diffraction planes correspond to the Kesterite phase of CZTS nanoparticles. Using Rietveld refinement the lattice parameter values obtained indicates that the CZTS nanoparticles are crystalline in nature having tetragonal crystal structure with $I\bar{4}$ 2m space group. SEM andEDAX revealed prominent agglomerated nanosphere-like shape with particle size of nearly 50-80 *nm* and the elemental composition of the CZTS nanoparticles follows proper stoichiometry. The band gap of the synthesized particles is 1.88 eV which is suitable for an absorber layer in solar cells. Additionally, a Hall effect study at room temperature revealed that the CZTS nanoparticles conduction type, mobility, and hole concentration were p-type, 2.59 $cm^2V^{-1}S^{-1}$ and 3.81×10^{20} cm^{-3}, respectively. The electrical conductivity showing the semiconductor like behaviour. These characteristics show that the CZTS nanoparticles are very ideal for the use as the absorber of the low-cost and environment friendly solar cell.

ACKNOWLEDGEMENT

The authors sincerely express thanks to UGC-DAE Consortium for Scientific Research, Indore for XRD facilities.

REFERENCES

1. Balakrishna, A., Jeotikanta M., D, Bahadura., N, Medhe-kara.,and M. Aslama., (2018). Influence of the Cu2ZnSnS4 nanoparticles size on solar cell performance,Solar Energy Materials and Solar Cells.189,125–132.
2. K. pal, Dheeraj, M., Priyanka, C., Khem, T., and Bal, Y., (2021). Co-precipitation Synthesis with a Variation of the Sulphur Composition of Kesterite Phase Cu2ZnSnS4 (CZSS) without Annealing Process, Journal of Physical science. 32(2), 27–39.
3. Min, Z., Yanmei, G., Jian, X., Gang, F., Qingbo, X., and Jianfeng, D. (2013). Colloidal CZTS nanoparticles and films: preparation and characterization, Journal of Alloys and Compounds. 574, 272–277.
4. Moiz, M., (2020). Size-Controlling of Cu2ZnSnS4 Nanoparticles: Effects of Stabilizing/Reducing Agents on Material Properties, Results in Physics.19,103407.
5. Nattee, K., Theerawut, S., and Tosawat, S., (2021). Investigation of Cu2ZnSnS4 film by Simple Flow-Coating Technique, Journal of Physics: Conference Series. 012018.
6. Patrick, R., Stephan, B., andA. Tiwari, (2013). Technological status of Cu (In, Ga) (Se, S) 2-based photovoltaics, Sol. Energy Mater. Sol. Cells. 119, 287–290.
7. Sahaya, D. B.G., and Sahaya, S., (2017). Colloidal chemical synthesis of pentanary semiconductor nanoparticles – A versatile functional material for low-cost, high efficient photovoltaic applications, Materials Today Proceedings. 4(14), 12592–12599.
8. Sonali, D., Kadambinee. Sa., Prakash. M., Jagatpati, R., Injamul, A., B.V.R.S., Subramanyam., and P. Mahanandia, (2018). synthesis of quaternary chalcogenide CZTS nanoparticles by a hydrothermal route, Materials Science and Engineering. 338, 012062.
9. Sergio, G., Zachari, J., Marcel, P., V. I. Roca, Alenjandro, R., Edgardo, S., (2019). Progress and perspectives of thin film kesterite photovoltaic technology: a critical review, Adv Mater. 31:1806692.
10. Vipul, K., Kinjal, P., Sanjay, P., and Dimple, S., (2013). Synthesis and characterisation of Copper Zinc Tin Sulphide (CZTS) compound for absorber material in solar-cells, Journal of Crystal Growth. 362(1), 174–177.
11. Xinlong, Y., and Xiaoyan, H., (2015). Solvothermal Synthesis of CZTS Nanoparticles in Ethanol: Preparation and Characterization, Journal of the Korean Physical Society. 66, 1511–1515.
12. Yu, L., Paul, M., Xiaodong, L., Andrey, K., Jonathan, S., Feridoon, A., David, L., and Robert, F., (2024). Exceptional Thermoelectric Performance of Cu2(Zn,Fe,Cd)SnS4 Thin Films, ACS Appl. Mater. Interfaces.16, 11516–11527.
13. Zhang, J., Shao, L., Fu, Y., and Xie, E., (2006). Cu_2ZnSnS, thin films prepared by sulfurization of ion beam sputtered precursor and their electrical and optical properties, Rare Metals. 25,315.

Note: All the figures in this chapter were made by the authors.

Recent Trends in Applied Physics and Material Science – Sudhir Bhardwaj et al. (eds)
 ISBN 978-1-041-16452-4

Synthesis and Characterization of Surface Modified Metal/Metal Oxide (Ni/NiO) Core Shell Nanoparticles

Bhasker Soni*, P. H. Soni
Department of Physics, Faculty of Science,
The Maharaja Sayajirao University of Baroda, Vadodara, Gujarat, India

Abstract: Among various core shell nanoparticles, metal-metal oxide core-shell nanoparticles (M-MCN) have gained wider interests due to their unique optical, electrical, and magnetic properties and finds pertinent applications in the field of catalysis, magnetic and electrical devices, sensors, drug delivery, and bio-imagining. Various efforts are being made towards development of the process providing tunability over core-shell materials (organic and inorganic), shape and size of such M- MCN nanoparticles with special emphasis on surface functionalization. Here, we report synthesis of Ni: NiO nanoparticles using microwave assisted sol-gel type chemical process. The reported process permits precise control over the morphology and the surface group functionalization with and without graphitic type carbon layer over the surface. The structural, compositional, and optical properties of the derived Ni based M-MCN were examined using X-ray diffractogram (XRD), Raman spectroscopy, and X-ray photoelectron spectroscopy (XPS).

Keywords: Metal-metal oxide core-shell, XPS, Raman spectroscopy

1. INTRODUCTION

Metal-metal oxide core-shell nanoparticles M-MCN are the novel class of highly functional material allowing tunability over their thermal and chemical stability along with surface functionalization [Mahesh *et al* 2015, Jones *et al* 2024]. Their structure consists of core, shell, and the interface between them. However, the major advantage of the structure lies with the surface functionalization using different ligands. Different M-MCN can be achieved with the variable composition of similar or different elements along with possible combination of inorganic: inorganic, organic: inorganic compounds. Based on the end user applications materials can be tuned for their electrical, optical, thermal, magnetic, and other rheological properties. Although the material has gained high attraction there lies the challenge for controlling the properties with repeatability for production at industrial scale. Hence, a reliable and cost-effective synthesis route is desired. Many efforts have been reported in literature for developing these M-MCN particles which may have a limitation in some aspects of either cost effectiveness, bulk scalability, repeatability etc. [Viola *et al* 2020, Chen *et al* 2021]. Different methodology to synthesize M-MCN such as solid-state reaction, chemical route, sol-gel technique etc are used based on the product. When well dispersed nanomaterial/ powder form is required, chemical route is well accepted with its easiness and facile process (Viola *et al* 2020). More efforts are being put forward to optimize the reaction mechanism to tune the morphology, structural, thermal, optical, and other physical properties. In the present work we report the characterization of Ni:NiO M-MCN derived microwave assisted sol-gel technique.

2. EXPERIMENTAL DETAIL

The experiment was carried using precursor materials of maximum purity. The materials were poly-vinyl alcohol (PVA) as encapsulating agent, 25% ammonia solution, and nickel nitrate hexahydrate. A fresh PVA solution was prepared with required concentration and solution for Nickel nitrate of required concentration was added

*Corresponding author: bhasker.soni-phy@msubaroda.ac.in

DOI: 10.1201/9781003684718-25

dropwise in the PVA solution in the pH range of 9-10. The mixed solution was further provided microwave heating for 30 min under constant stirring inside the microwave oven and a temperature of the range 85-90°C was maintained during reaction. The reacted solution was then further cooled to room temperature and after 2h sol-gel type solution was formed. The derived sol-gel was further microwaves to dry and the derived reacted precursor power was further theat treated at required temperature of 400°C and 700°C. The derived M-MCN were characterized and analyzed with a Panalytical X`Pert Pro X-ray diffractometer using CuKα radiation of wavelength 0.1545 nm, employing at a scanning rate of 0.02/s. X-ray photoelectron spectroscopy (XPS) was employed to examine the surface oxidation state along with elemental compositions. Raman measurements were carried out with Horiba Jobin Yvon Lab RAMHR800 spectrometer with a He-Ne laser source (632.8 nm).

3. RESULTS AND DISCUSSION

Figure 25.1 shows the XRD patterns of Ni based M-MCN heat treated at 400°C and 700°C for 2h in ambient air. The derived Ni based M-MCN shows respective XRD reflections peaks matching with JCPDS card 04-0850. The (200) and (220) planes belongs to Ni which are small in intensity whereas, broad characteristic peaks corresponding to (111), (200), (220), (311), and (222) crystalline planes of NiO (JCPDS card 04-0835).

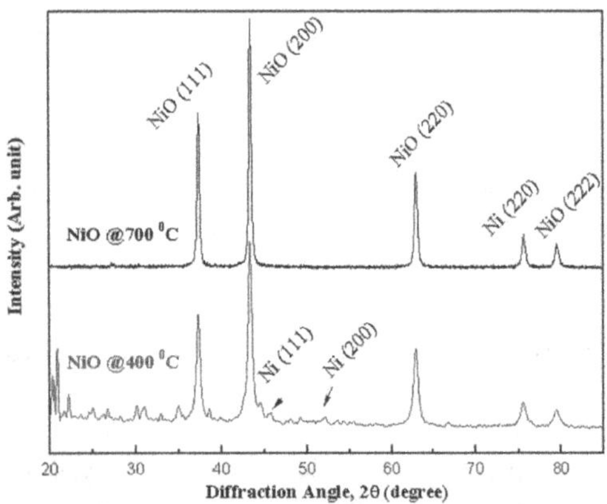

Fig. 25.1 XRD patterns of derived Ni based M-MCN after heat treating the precursor powder at (a) 400°C and (b) 700°C for 2h in ambient air

The XRD graphs in Fig. 25.1 confirmed the highly crystalline nature of the calcined samples. In both the samples, Ni and NiO phases have face-centered cubic (fcc) crystal structure. The average crystallite size as calculated using Debye Scherrer's formula is 18 nm and 22 nm in the derived nanoparticles. With increase in the heat treatment temperature from 400°C and 700°C the peak width decreased, and the oxide content was further enhanced, respectively. The shifting of (200) peak towards higher diffraction angle with increase in temperature indicates the increase in lattice constant value.

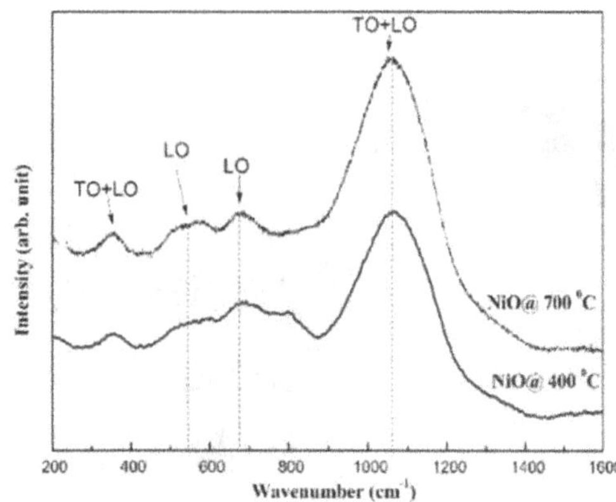

Fig. 25.2 Raman spectra of the derived Ni:NiOM-MCN derived at (a) 400°C and (b) 700°C calcination temperature for 2h in ambient air

Raman and XPS analyses were used to study the surface structure of the Ni/NiO M-MCN nanoparticles along with elemental composition. The peaks appearing at 520 and 678 cm^{-1} belongs to NiO phase in the sample. The diminished but broad peak near 520 cm^{-1} belongs to the first order longitudinal optical (LO) mode of NiO. In addition, the second order longitudinal optical (2LO) mode present at ~ 685 cm^{-1} in the sample derived increased with temperature. However, the signature transverse optical (TO) or mixed TO + LO modes (near 380 cm^{-1} and 1080 cm^{-1}, respectively) of NiO are well defined in these derived M-MCN which confirms the purity of NiO on the surface.

XPS is a extremely precise and profound technique to obtain detailed analyses of the material elemental composition and surface properties up to a depth of few nanometers from surface of the particles. The full survey spectrum of the Ni:NiO M-MCN heat treated at 400°C is presented in Fig. 25.3 in the range of 0-1100 eV binding energy. The presence of C, Ni, and O is confirmed using XPS.

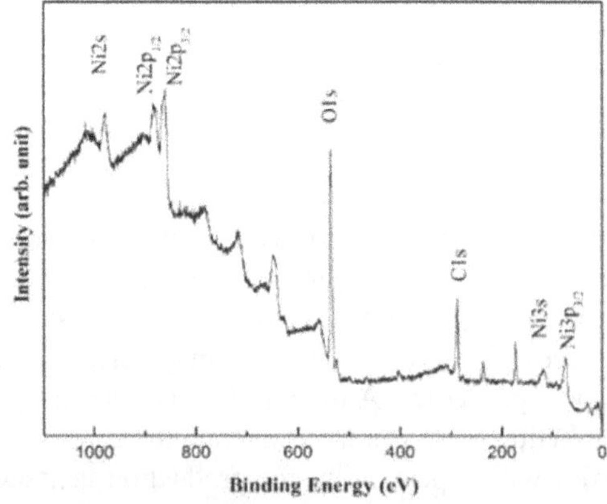

Fig. 25.3 XPS survey spectrum of Ni:NiO CSN derived at 400°C for 2h in ambient air

The derived sample showed no other elemental presence other than C, Ni, O which confirms the purity of the derived samples. The deconvoluted curves for the individual element would be performed for better understanding of the surface oxidation state.

4. CONCLUSION

A novel microwave assisted chemical route has been explored for deriving Ni based M-MCN. Structural analysis using XRD confirmed that the crystalline nature with stable fcc crystal structure. The Raman spectra justified the formation of NiO on the surface with core as Ni. XPS analysis gave a very clear confirmation for the purity of the derived samples.

REFERENCES

1. Viola, C. and Filippo R. (2020). Inorganic–organic core/shell nanoparticles: progress and applications, RSC Adv. 2: 5090–105.
2. Mahesh, P., Somnath, B., Yadav, A. K., Jha, S. and Bhattacharyya, D. (2015). Morphology-controlled synthesis of monodispersed graphitic carbon coated core/shell structured Ni/NiO nanoparticles with enhanced magnetoresistance, Phys. Chem. Chem. Phys., 17(48):32398–412.
3. Jones, C. S., Resina, L., Ferreira, F. C., Alberte, P. and Esteves, T. (2024). Conductive core–shell nanoparticles: synthesis and applications, J. Phys. Chem. C, 128(27) 11083–100.
4. Chen, Y., Zhan, X., Bueno, S., Shafei, H., Ashberry H., Kaustav, C., Lin, X., Yawen, T. and Sara E. (2021) Synthesis of monodisperse high entropy alloy nanocatalysts from core@shell nanoparticles, 6: 231–237.

Note: All the figures in this chapter were made by the authors.

Recent Trends in Applied Physics and Material Science – Sudhir Bhardwaj et al. (eds)
© 2026 Taylor & Francis Group, London, ISBN 978-1-041-16452-4

26 Transport Properties of Bismuth Telluride and MoS_2 Composites

Poonam Deshmukh*
Govt. College Shahpur, Dist. Betul, Madhyapradesh, India

Toshi Bagwaiya
Department of Physics, Barkatullah University, Bhopal, Madhyapradesh, India

Harsha Pawar
VeerRengu Korku Govt College Khakner, Dist. Burhanpur, Madhyapradesh, India

Shovit Bhattacharya
BARC, Mumbai, Maharashtra, India

Vilas Shelke
Department of Physics, Barkatullah University, Bhopal, Madhyapradesh, India

Abstract: Molybdenum disulfide (MoS_2) is belonging to a class of transition metal dichalcogenide (TMD) compounds with layered structure. In this paper, we have synthesized MoS_2 by ball milling followed by sulfurization process. This synthesis method is simple and reliable to prepare the nano/micro particles of MoS_2. The composite of Bi_2Te_3 and MoS_2 materials prepared by varying MoS_2 concentration in weight percent ratio of 2%, 4% and 10 % and followed by mechanical alloying and vacuum hot press sintering technique. The structural analysis and thermoelectric properties of bulk composites were investigated. The XRD pattern of molybdenum sulfide showed phase formation and strengthened diffraction peaks are indicating improved crystallinity in MoS_2 due to annealing. Morphology of the composites shows the MoS_2 homogenously and randomly distributed in composite samples. The electrical resistivity decreased with increasing temperature indicating semiconducting character of composites. The Seebeck-coefficient as a function of temperature confirms the n-type nature of composites.

Keywords: Dichalcogenide, Composite, XRD, Transport properties

1. INTRODUCTION

Molybdenum disulfide (MoS_2) is belongs to a class of transition metal dichalcogenide (TMD) compounds with layered structure. Transition metal dichalcogenides family has general formula MX_2, where M = Mo, W, Ti, Nb etc. and X = S, Se, Te. Over the past few years, the layered transition metal dichalcogenide has received more attention due to their useful functions in micro-electronics, energy storage, and sensing devices. The traditional metal dichalcogenide is potential thermoelectric material due to extensive properties. The two-dimensional (2D) molybdenum disulfide has high electron mobility and low band gap. Recently the monolayer of MoS_2 has shown a large thermo power value posing TMDs potential importance in the field of thermoelectric. The research on thermoelectric performance of MoS_2 mainly focuses on theoretical calculation studies. The multivalley bands and 2-D electronic structure of transition metal dichalcogenide are known to enhance thermoelectric performance. The ultrathin 2D MoS_2 showed the high seebeck coefficient value in order of ~10 m.V/K. Theoretically, MoS_2 nano-ribbons ZT value could be optimized to 3.0 at room temperature. Accordingly, MoS_2 is good material for thermoelectric applications.

In the present work, MoS_2 materials were prepared by mechanical alloying followed by sulfurization process. The composite of Bi_2Te_3 and MoS_2 materials were prepared by mechanical alloying and vacuum hot press sintering. Then the structural analysis and thermoelectric properties of bulk composites were investigated.

2. EXPERIMENTAL TECHNIQUES

In the synthesis of MoS_2, analytical pure regents were used. The weighed mass ratio of 1:2 of MoO_3 and S

*Corresponding author: poonamkhade06@gmail.com

DOI: 10.1201/9781003684718-26

powders were mixed in ball milling at 350 rpm for 20 hrs. The ball milled powder was annealed at 600 °C for 2 hour in a tube furnace in the presence of sulfur vapor for sulfurization in an Ar atmosphere. The furnace was cooled to room temperature and the black Powder of MoS_2 was obtained. The composites of Bi_2Te_3 and MoS_2 were prepared by mechanical alloying. Bi_2Te_3 and MoS_2 powders were mixed and subjected to the Ball milling at 350 rpm for 5 hour to obtain the composite powder. Then the mixed composite powders were consolidated using hot press sintering technique at 673 K temperature for 60 min. The MoS_2 concentration is varied in weight percent ratio of 2%, 4% and 10 %.

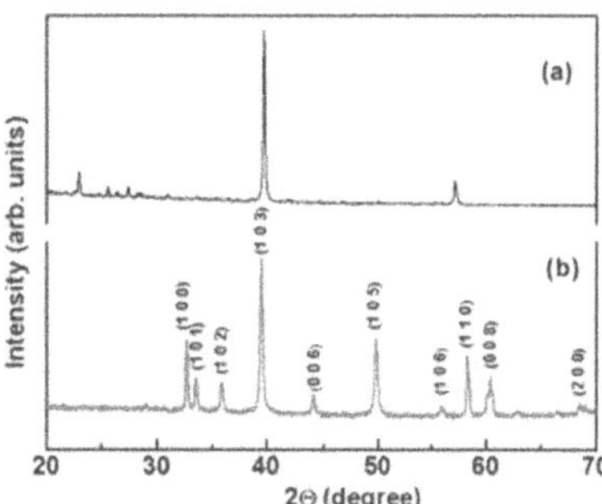

Fig. 26.1 XRD of MoS_2 (a) ball milled powder, and (b) after sulfurization

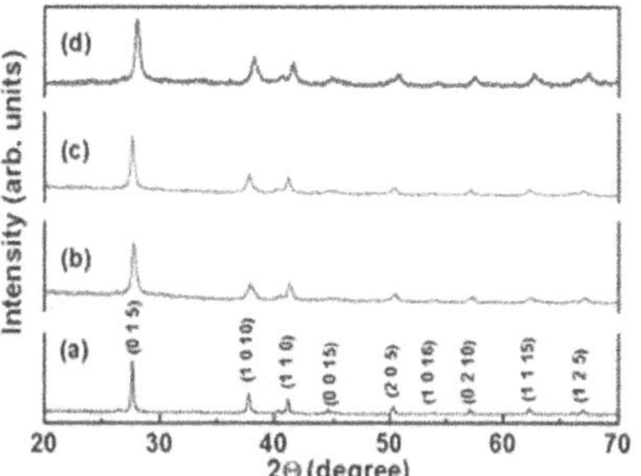

Fig. 26.2 XRD of (a) pure Bi_2Te_3 and composite with MoS_2 (b) 2%, (c) 4%, and (d) 10%

3. RESULTS AND DISCUSSION

The XRD pattern of molybdenum disulfide (MoS_2) sample of ball milled powder shows in Fig. 26.1(a) and the Fig. 26.1(b) the pure MoS_2 phase is generated after sulfurization process. The XRD peaks are well indexed with JCPDS file no. 37-1492. The diffraction peaks are strengthened slightly indicating improved crystallinity due to annealing. The calculated lattice parameters were a = b = 3.15 Å and c = 12.30 Å. The composite powders of all samples were characterized by XRD are shown in Fig. 26.2. The diffraction patterns of Bi_2Te_3 correspond to rhombohedral structure. The XRD pattern of composites samples indicate the Bi_2Te_3 phase present in all the samples. There is no indication of peak shifting in composite samples. In XRD pattern of composites, MoS_2 feature may not be observed due to small amount of MoS_2 concentration and may be high cystallinity of Bi_2Te_3 compound.

Figure 26.3 presents the SEM images of fracture surface of composites samples. The addition of MoS_2 in different concentration leads to reduction of grain size. It is observed from images that the MoS_2 homogenously and randomly distributed in composite samples. The grain size is in submicron range.

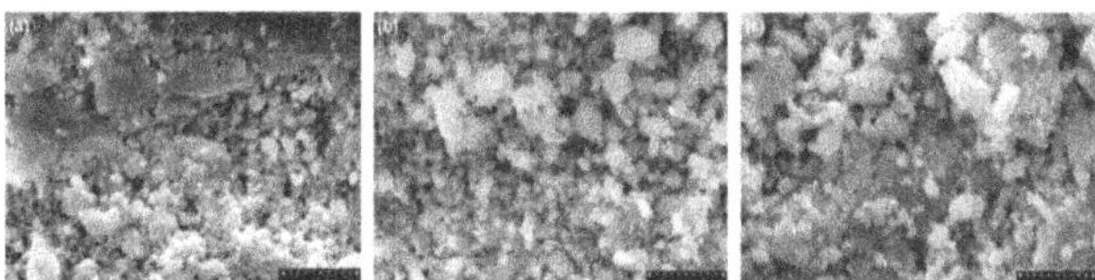

Fig. 26.3 SEM images of BT+MoS_2 composites (a) 2%, (b) 4%, and (c) 10%

The temperature dependent transport properties of Bi_2Te_3 and MoS_2 composites are shown in Fig. 26.4 (a, b). The electrical resistivity of composite samples is shown in Fig. 26.4 (a) decreased with increasing temperature for all samples indicating semiconducting behaviour. The electrical resistivity increases as the MoS_2 concentration increase in the Bi_2Te_3 material. The 4% MoS_2 added in Bi_2Te_3 showed higher resistivity with increasing temperature. The MoS_2 melting point is 1185 °C high than the sintering temperature so that the MoS_2 stay on the same size before and later than the hot pressing resulting in high resistivity. The change in electrical resistivity as compared to pristine sample is directly associated with the occurrence of interfaces.

The Seebeck coefficient as a function of temperature is shown in Fig. 26.4 (b). The negative value of Seebeck coefficient is observed that verify the n-type behaviour of composites. The addition of MoS_2 shows slight change in the behaviour. It is mainly due to the low value of Seebeck coefficient of MoS_2. The Seebeck coefficient behaviour decreases with increasing temperature. The maximum

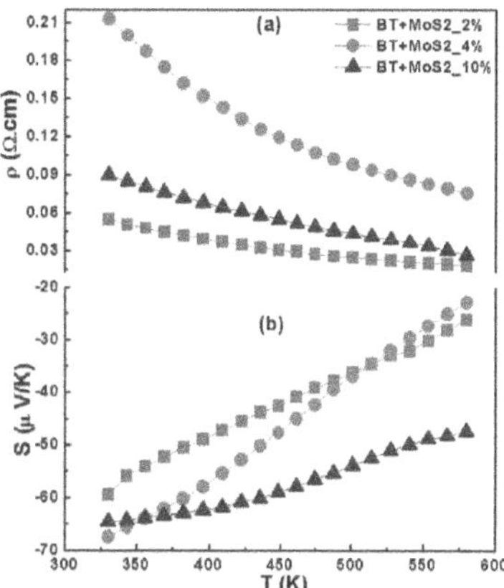

Fig. 26.4 Transport measurements of BT+MoS_2 composites (a) electrical resistivity, and (b) Seebeck coefficient

value of Seebeck coefficient is around 69 μV/K at room temperature.

REFERENCES

1. Hong, X., Liu, J., Zheng, B., Huang, X., Zhang, X., Tan, C., Chen, J., Fan, Z., Zhang, H. (2014). A universal method for preparation of noble metal nanoparticles-decorated transition metal dichalcogenide nanobelts. Adv. Mater. 26(36):6250–6254.
2. Chhowalla, M., Shin, H. S., Eda, G., Li, L-J., Loh, K. P., & Zhang, H. (2013). The chemistry of two-dimensional layered transition metal dichalcogenide nanosheets. Nat. Chem. 5: 263–275.
3. Wang, T., Liu, C., Xu, J., Zhu, Z., Liu, E., Hu, Y., Li, C., and Jiang, F. (2016). Thermoelectric performance of restacked MoS2 nanosheets thin film. Nanotechnol. 27:285703.
4. Fan, D. D., Liu, H. J., Jiang, P. H., Shi, J., Tang, X. F. (2014). MoS2 nanoribbons as promising thermoelectric materials. Appl. Phys. Lett. 105:133113.
5. Ahmad, M., Agarwal, K., Kumari, Navnita., Mehta, B. R. (2017). KPFM based investigation on the nature of Sb2Te3: MoS2 and Bi2Te3: MoS2 2D interfaces and its effect on the electrical and thermoelectric properties. Appl. Phys. Lett. 111:023904.
6. Keshavarz, M. K., Vasilevskiy, D., Masut, R. A., Turenne, S. (2014). Synthesis and characterization of bismuth telluride-based thermoelectric nanocomposites containing MoS2 nano-inclusions. Mater. Charact. 95:44–49.
7. Tang, G., Cai, K., Cui, J., Yin, Junlin., Shen, S. (2016). Preparation and thermoelectric properties of MoS2/Bi2Te3 nanocomposites. Ceram. Int. 42(16):17972–17977.

Note: All the figures in this chapter were made by the authors.

Recent Trends in Applied Physics and Material Science – Sudhir Bhardwaj et al. (eds)
© 2026 Taylor & Francis Group, London, ISBN 978-1-041-16452-4

Study of Vanadium Pentoxide (V_2O_5) Nanostructured Thin Films

Sujit K. Saini*
Department of Physics and Materials Sciences and Engineering, Jaypee Institute of Information and Technology, Noida, India
Formerly-Thin Film Laboratory, Department of Physics, Indian Institute of Technology Delhi, Delhi, India

Megha Singh
CSIR-National Physical Laboratory, Dr KS Krishnan Marg, Pusa, New Delhi, Delhi India
Academy of Scientific and Innovative Research (AcSIR), Ghaziabad, India
Formerly-Thin Film Laboratory, Department of Physics, Indian Institute of Technology Delhi, Delhi, India

Rabindar K. Sharma
Department of Physics D. D. U. Govt. P. G. College, Sitapur, Lucknow, India
Formerly-Thin Film Laboratory, Department of Physics, Indian Institute of Technology Delhi, Delhi, India

G. B. Reddy
Formerly-Thin Film Laboratory, Department of Physics, Indian Institute of Technology Delhi, Delhi, India

Abstract: This study investigates the impact of growth temperature on the structural and optical properties of V_2O_5 nanostructured thin films. The films were deposited for 45 minutes at temperatures of 350, 450, 550, and 650 °C. Structural analysis using XRD, Raman spectroscopy, and SEM revealed temperature-dependent changes, including enhanced crystallinity, phase transitions, and morphological evolution. XPS confirmed the chemical composition and oxidation states. Optical studies via UV-Vis spectroscopy showed shifts in reflectance and band gap with temperature variation. These findings highlight the potential for tailoring V_2O_5 thin films for optoelectronic and energy applications through precise temperature control.

Keywords: Vanadium Pentoxide, Thin Films, Nanograss

1. INTRODUCTION

Vanadium oxide (V_2O_5) has garnered significant attention due to its diverse applications in gas sensors, energy storage devices, catalysts, lithium battery cathodes, and solar cell windows (Seng et al. 2011, Schneider et al. 2016). These applications arise from its unique physical, chemical, and mechanical properties, which are further enhanced in nanostructured forms. The properties of V_2O_5 nanostructures—such as crystal structure, morphology, orientation, and defect states—are strongly influenced by synthesis methods and parameters. While both chemical and physical routes have been employed for V_2O_5 synthesis, chemical methods often face challenges like toxic byproducts, precursor impurities, amorphous phases, and limited control over nanostructure alignment. In this study, we investigate the effect of plasma annealing on the growth and optical properties of V_2O_5 nanostructured thin films (NSTs). Plasma annealing offers a controlled and efficient approach to fine-tune the structural and optical characteristics of V_2O_5, paving the way for enhanced performance in advanced applications.

2. EXPERIMENTAL DETAILS

Nanostructured V_2O_5 thin films were deposited on nickel-coated glass substrates using thermal evaporation followed by O_2 plasma annealing (Fig. 27.1). The plasma

*Corresponding Author" sujitsaini95@gmail.com

DOI: 10.1201/9781003684718-27

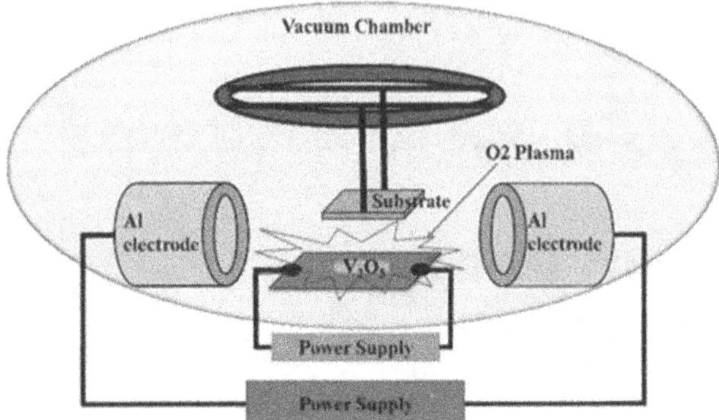

Fig. 27.1 Schematic diagram of setup

was generated using two aluminum electrodes placed 7.5 cm apart with a 2500 V potential applied. V_2O_5 pellets, placed on a heated tungsten strip, were evaporated and deposited onto substrates, which were then annealed at 350 °C, 450 °C, 550 °C, and 650 °C for 45 minutes under an oxygen partial pressure of 1.2×10^{-1} Torr (Sharma et al. 2015). Structural characterization was performed using a Rigaku Ultima IV X-ray diffractometer (Cu-Kα, λ ~ 1.54 Å). Surface morphology was analyzed with a ZEISS EVO 50 scanning electron microscope (SEM). Raman spectroscopy, using a Horiba LabRAM HR Evolution system with a 514.5 nm Ar^+ laser, provided insights into vibrational and rotational modes. Optical properties, including transmittance and reflectance, were measured with a Perkin-Elmer-900 UV–VIS–NIR spectrophotometer.

3. RESULTS AND DISCUSSION

Figure 27.2 (a-d) presents the X-ray diffractograms of V_2O_5 thin films annealed at 350, 450, 550, and 650 °C, demonstrating their polycrystalline nature. At 350–550 °C, the XRD patterns confirm the orthorhombic phase of V_2O_5 (JCPDS card no. 89-0611). Films annealed at 450 °C show preferential alignment along the (010) plane, while those at 550 °C exhibit alignment along both (010) and (020) planes. At 650 °C, peaks marked with # in the diffractogram suggest a phase transition from V_2O_5 to V_3O_7, an intermediate state between V_2O_5 and VO_2, consistent with the monoclinic phase (JCPDS card no. 71-0453). The lattice parameters calculated are a = 9.9461 Å, b = 3.5852 Å, and c = 10.0423 Å, respectively. Sharp and intense peaks in the diffractograms confirm the crystallinity of the films. Figure 27.2 (a-d) also shows the Raman spectra of V_2O_5 films deposited at varying temperatures. The high-frequency peak at 994 cm^{-1} corresponds to the terminal oxygen stretching mode (unshared oxygen). Peaks at 696 cm^{-1} and 527 cm^{-1} are attributed to doubly coordinated (V_2–O) and triply coordinated (V_3–O) oxygen stretching modes, respectively.

Peaks at 405 cm^{-1} and 283 cm^{-1} represent terminal oxygen bond bending vibrations, while peaks at 487 cm^{-1} and 303 cm^{-1} correspond to bending vibrations of triply coordinated oxygen (V_3–O) and bridging V–O–V bonds, respectively (Sharma et al. 2015, Singh et al. 2017). Peaks marked with * suggest the presence of V_3O_7 phase, a common secondary phase in V_2O_5 growth. These Raman findings align with the XRD results, further supporting the structural analysis.

Figure 27.3 shows surface micrographs of V_2O_5 thin films deposited on Ni-coated glass at 350, 450, 550, and 650 °C with a constant 45-minute duration. At 350 °C, vertically aligned blade-like nanostructures (~200 nm width) are observed. At 450 °C, randomly oriented nanograss with sharp tips (~300 nm width) forms. Higher temperatures (550 °C and 650 °C) result in wider, randomly oriented nanograss (~700 nm and ~800 nm widths, respectively). Figure 27.4 (left section) shows the XPS spectra of the V_2O_5 film annealed at 550°C. The survey scan reveals three intense bands corresponding to V (2p), O (1s), and C. (1s) core levels. The V (2p) peaks at 516.8 eV and 524.4 eV correspond to the V^{5+} ($2p_{3/2}$) and V^{5+} ($2p_{1/2}$) states with a spin-orbit splitting of 7.4 eV, aligning with reported

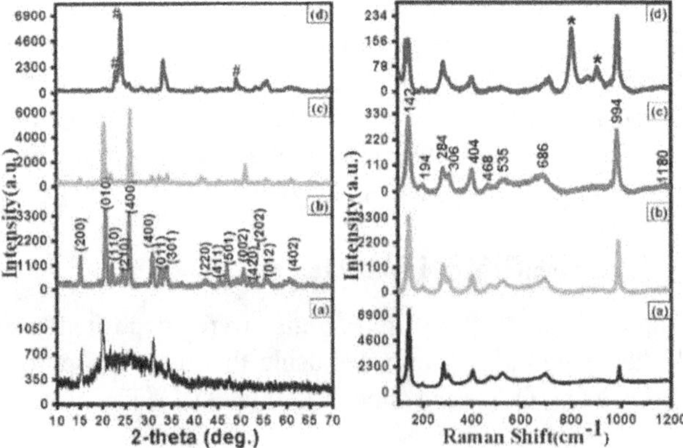

Fig. 27.2 X-ray diffractograms (left) and Raman spectra (right) of the films annealed at (a) 350 °C, (b) 450 °C, (c) 550 °C, and (d) 650 °C respectively

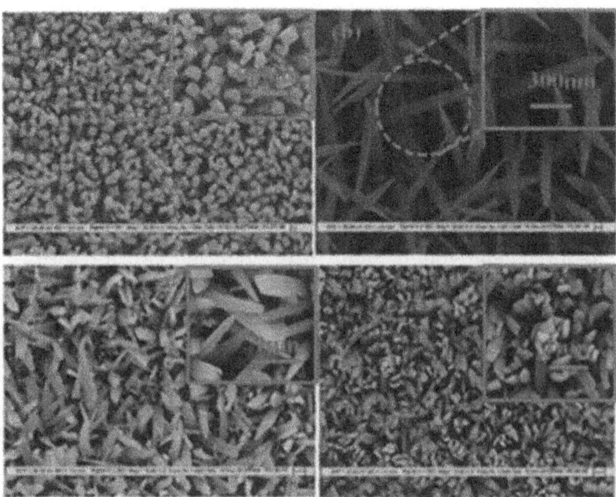

Fig. 27.3 Surface micrographs of V_2O_5 thin films deposited on Ni-coated glass at (a) 350 °C, (b) 450 °C, (c) 550 °C, and (d) 650 °C respectively

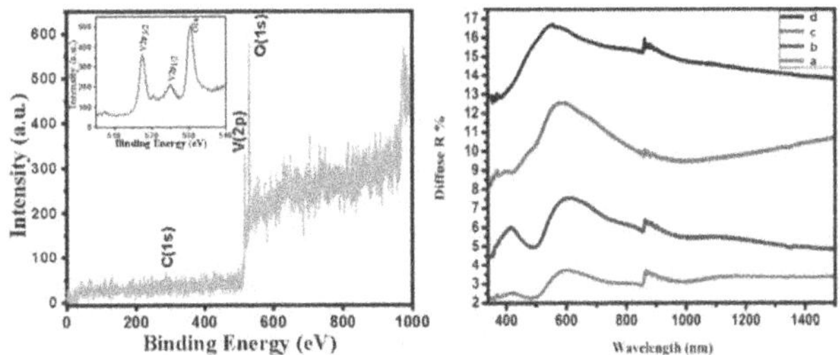

Fig. 27.4 XPS spectra (left) of the V_2O_5 film annealed at 550°C section of the figure and diffused reflectance (right) of the films

values (Singh et al. 2017). XPS analysis confirms a +5 oxidation state of vanadium, consistent with the presence of V_2O_5.

The optical properties of the as-grown V_2O_5 thin films were characterized through diffuse reflectance measurements from 350 to 1500 nm. Figure 27.4 (right section) shows the diffuse reflectance spectra for films deposited at various temperatures ranging from: (a) 350 to 650 °C, results show a gradual increase in reflectance from 3% to 17% in the visible range as the temperature rises beyond 450°C. A sharp absorption edge near 440 nm indicates improved crystallinity and low density of defects near the band edge in the V_2O_5 films.

4. CONCLUSION

In conclusion, nanostructured thin films of V_2O_5 are synthesized through plasma annealing of thermally evaporated films. At growth temperatures of 350°C to 550°C, the films predominantly consist of the orthorhombic phase of V_2O_5, evidenced by sharp and intense XRD peaks. SEM micrographs show the films are influenced by the annealing temperature. Raman spectra indicate that at temperatures from 350°C to 550°C, V_2O_5 is the dominant phase, but at 650°C, the V_3O_7 phase is more prevalent. XPS confirms the +5 oxidation state of vanadium, supporting the presence of V_2O_5. UV-VIS spectroscopy reveals an absorption edge near 440 nm, indicating good crystallinity and low defect density in V_2O_5 NSTs.

REFERENCES

1. K. H. Seng, J. Liu, Z. P. Guo, Z. X. Chen, D. Jia, H. K. Liu (2011). Free-standing V_2O_5 electrode for flexible lithium ion batteries. Electrochemistry Communications, 13, 5, Pages 383–386.
2. K. Schneider, M. Lubecka, and A. Czapla, (2016). V_2O_5 thin films for gas sensor applications. Sensors Actuators B Chemical 236, Pages 970–977.
3. R. K. Sharma, P. Kumar, G. B. Reddy (2015) Synthesis of vanadium pentoxide (V_2O_5) nanobelts with high coverage using plasma assisted PVD approach. Journal of Alloys and Compounds, 638 Pages 289–297.
4. M. Singh, P. Kumar, R. K. Sharma, G. B. Reddy (2017) Plasma assisted synthesis and growth mechanism of rare V_2O_5 nanostructured thin films. Journal of Alloys and Compounds, 690, 532e541.

Note: All the figures in this chapter were made by the authors.

Recent Trends in Applied Physics and Material Science – Sudhir Bhardwaj et al. (eds)

Growth and Characterization of V_2O_5 Nanoflowers on Cobalt Coated Silicon Substrate

Rabindar K. Sharma*
Department of Physics D. D. U. Govt. P. G. College, Sitapur, Lucknow, India

Megha Singh
CSIR-National Physical Laboratory, Dr KS Krishnan Marg, Pusa, New Delhi, India
Academy of Scientific and Innovative Research (AcSIR), Ghaziabad, India

Sujit Kumar Saini
Department of Physics and Mat. Scien. & Engg., Jaypee Institute of Information and Technology, Noida, India

G. B. Reddy
Formerly:- Thin Film Laboratory, Indian Institute of Technology, Delhi, New Delhi, India

Abstract: This report explains the growth of vanadium pentoxide (V_2O_5) nanoflowers (NF_s) through two consecutive growth steps using Plasma Assisted Sublimation Process (PASP). In first step, vertically aligned nanoplates (NP_s) of MoO_3 is deposited on Co-coated Si substrate employing molybdenum metal as sublimation source. The second step involves the growth of vanadium oxide NF_s on already grown MoO_3 NP_s by repeating PASP for Vanadium (V) metal. X-ray diffractogram (XRD) of film deposited after step II, divulges that only orthorhombic phase of vanadium oxide (i.e. α-V_2O_5) is present. The intensity count profile of peaks ascertains that the direction of preferential crystal growth is long b-axis. The surface micrographs indicates that after step I vertically aligned MoO_3 NP_s are formed whereas, a uniform growth of α-V_2O_5 NF_s on pre-deposited MoO_3 NP_s is occurred after step II with excellent coverage. The Raman spectra of NF_s show various absorption bands and all are assigned according to α-V_2O_5. A red shift in Raman peaks in the range 3 to 15 cm^{-1} confirms the existence of low oxidation state of vanadium (i.e. V^{4+} & V^{3+}). The fringe and the dot patterns in HRTEM and SAED, respectively verify the single crystalline nature of NF_s. The interplanar spacing of 0.34 nm recorded in fringe pattern infers that NF_s grows preferentially along [010] direction.

Keywords: Vanadium pentoxide, Nanoflowers, PASP, XRD

1. INTRODUCTION

Many research groups have been made a considerable effort for synthesizing variety of nanostructures of inorganic materials in remarkably controlled way. Now a days, the growth of three-dimensional (3D) nanostructures by the interpenetration of 2D nanostructures like nanoplates, nanoflakes, nanorods with random alignments employing bottom-up approaches has been an exciting field for new technological applications (Sharma and Reddy 2014) To date, variety of 3D nanostructures especially of transition metal oxides have been synthesized. Among all metal oxides, vanadium oxides of distinct morphologies have been used in diverse applications such as gas sensors, storage device, solar energy devices (Hu Peng et al. 2023). Vanadium oxide has been occurred along with several stoichiometries so having different phases as V_2O_3, VO_2, and V_2O_5 (Hu Peng et al. 2023). Among all the phases orthorhombic phase (vanadium pentoxide i.e. α-V_2O_5) is the most stable phase because vanadium atom occupies in the highest oxidation state in orthorhombic phase and categories to the P_{mnm} space group (Beke S. et al. 2008). In the present work, we discuss the growth of α-V_2O_5 NF_s into two consecutive steps. The crystallographic and surface morphological studies of nanoflowers are carried out to study the crystal structure and surface modifications

*Corresponding author: rkrksharma6@gmail.com

DOI: 10.1201/9781003684718-28

of films. The vibrational analysis is performed to study the distinct vibrational modes in NF_S. Further, the deeper morphological analysis of NF_S is done by recording HRTEM image along with SEAD pattern. Before start the growth, the substrate is prepared by depositing cobalt film (thickness ~ 100 nm) on silicon substrate using thermal evaporation of Co-powder (99.99% Aldrich). In this growth, Co-coated Si substrate is used as a base substrate.

2. EXPERIMENTAL DETAILS

The schematic diagram of both the growth steps is shown is Fig. 28.1.

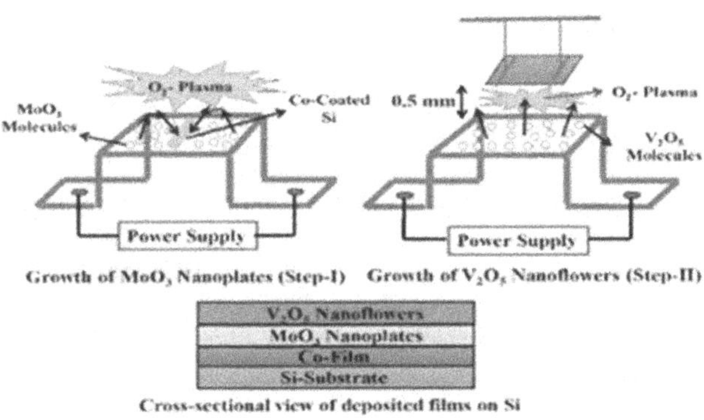

Fig. 28.1 Schematic diagram: (a) Step I: Growth of MoO_3 NPs (b) Step II: the growth of V_2O_5 NF_s (c) The corresponding view of all deposited films

In first step, MoO_3 NP_s are deposited over the base substrate through PASP at 500°C (Fig. 28.1a) whereas, second step involves the deposition of vanadium oxide on MoO_3 NP_s using the same PASP (Fig. 28.1b). The base substrate in first step is placed on Mo-metal strip at the center whereas, in second step the substrate with MoO_3 NP_s is placed at 0.5 mm above the V-metal strip (see in Fig. 28.1b). The temperature of substrate in step II is measured as 350°C. The ratio of substrate area (1.5 × 1.5) cm^2 to that of Mo/V-strips (8 × 3) cm^2 is optimized and kept at 1/10. The temperature of substrate is measured at the time of growth using thermocouple arrangement. The optimized values plasma parameters viz. plasma voltage, electrode separation and oxygen pressure are 2500 Volt, 7.5 cm and 7.5 × 10^{-2} Torr, respectively for both the growth steps. The cross-sectional view of all the deposited films is shown in Fig. 28.1c.

3. RESULTS AND DISCUSSION

X-ray diffractogram (XRD) is recorded to investigate the composition and phase of vanadium oxide film deposited in steps II. The sharp and intense diffraction peaks in diffractogram endorse the polycrystalline nature of film and also ensuring the existence of nanostructures in films.

The pattern of x-ray diffraction peaks of vanadium oxide film deposited over pre-deposited MoO_3 film is shown in Fig. 28.2a. All the recorded peaks are precisely indexed

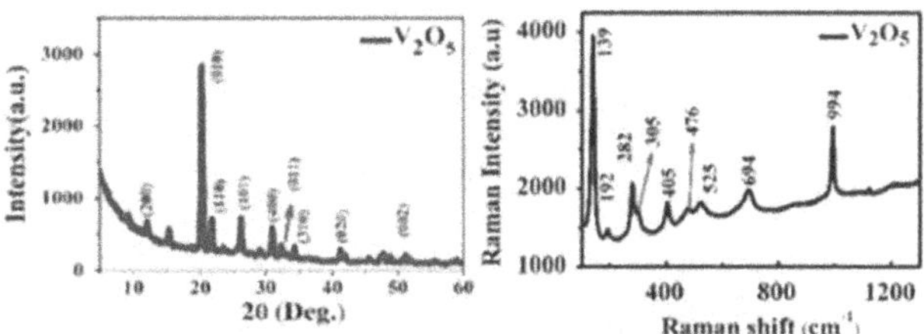

Fig. 28.2 X-ray diffractograms of V_2O_5 film after step II (left) and Raman spectrum (Right)

as per pure orthorhombic phase of vanadium oxide (i.e. α-V_2O_5) as reported in the standard JCPDS file (89-0612) with fundamental lattice parameters (a = 11.48 Å, b = 3.55 Å, and c = 4.36 Å). The peaks related to other sub-oxide phases are not obtained under the limit of accuracy. The intensity patten as recorded in Fig. 28.2 evidences that preferred crystalline growth takes place along b - crystallographic direction. The estimated crystallite size of V_2O_5 film is measured as ~ 30 nm using Debye–Scherrer formula for the diffraction peak of maximum intensity. The Raman spectrum in Fig. 28.2b shows the absorption peaks are located at 139, 192, 282, 305, 405, 476, 525, 694, and 994 cm^{-1} and all are matched with the characteristic vibrational modes of α-V_2O_5 (Meng Jian Li et. al. 2006). Most of the recorded Raman peaks show red shift of 3 to 15 cm^{-1} owing to the presence of low oxidation states of vanadium (i.e. V^{4+} & V^{3+}) and also ensuring the deviation of stoichiometry ratio (V/O) in NF_s from its perfect value (i.e. 1: 2.5).

In Fig. 28.3, the first micrograph (Fig. 28.3a) depicts the growth of vertically aligned MoO_3 NP_s with excellent and uniform coverage. The width and thickness of NP_s are 1.5 µm and 100 nm, respectively (insight of Fig. 28.3a). The V_2O_5 films deposited after step II (see 3b & c) reveals the formation like sort of nanoflowers like structures through systematic merging of nanopetals. The length and the width of a single petal of a flower are 1.5 µm and 35 nm (insight of Fig. 28.3c). The voids among vertically aligned NP_s with their sharp edges offer an excellent exposing area and furnish good number density of active centers or nucleation sites for the systematic growth V_2O_5 molecules in second step (see Fig. 28.3b). Here, the systematic interconnection among petals probably due to the unsaturated oxygen atoms having incomplete bonding with the V atoms in proximity. Such oxygen atoms are always tended to saturate their electronic configuration by forming bond with the further impinging molecules.

The ratio of vertical and the horizontal dimensions of the NF_s is nearly equal to one ~ 1 (insight of Fig. 28.3c), which is indicating that NF_s have uniform size along all directions. The fringe patten recorded by HRTEM with interplanar spacing ~ 0.34 nm and the spot pattern in SEAD image attribute the single crystalline nature of each petal of NF_s and assures that the growth is preferentially aligned along the b-axis.

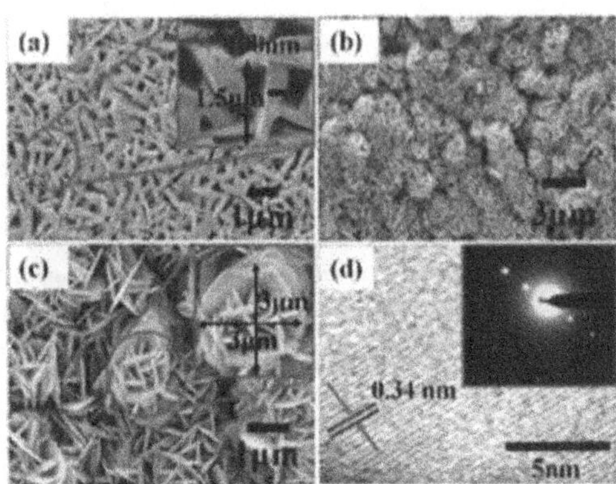

Fig. 28.3 SEM: (a) MoO_3 NP_s after step I (b-c) Growth of $V_2O_5NF_s$atlowandhighmagnification(d)HRTEMimage with SAED patten

4. CONCLUSION

In summary, this paper describes the growth of α-V_2O_5 NF_s into two consecutive growth steps using PASP. In step I, vertically aligned α-MoO_3 NP_s are deposited and then use these nanoplates as the base for the further growth of α-V_2O_5 nanoflowers in step II. The structural analysis of vanadium oxide film after step II, explores the presence of orthorhombic phase with excellent crystallinity. The SEM image after step I confirms the growth of vertically aligned MoO_3 NP_s with deep voids whereas after step II, V_2O_5 nanoflowers are grown on entire substrate with uniform dimensions. The vibrational bands in Raman spectrum among V and O atoms again ascertains the presence of only orthorhombic phase of NF_s and the red shift (3-15 cm^{-1}) in Raman bands, indicates a little deviation of stoichiometric ratio (i.e. V/O in V_2O_5) to that of its perfect value (i.e. 2/5) in NF_s.

REFERENCES

1. Sharma K. Rabindar, Reddy B. G. (2014) Synthesis and characterization of α-MoO3 microspheres packed with nanoflakes J. Phys. D: Appl. Phys. 47: 065305.
2. Hu Peng, Hu Ping, Vu Duc Tuan, Li Ming, Wang Shancheng, Ke Yujie, Zeng Xianting, Mai Liqiang, Long Yi (2023) Vanadium Oxide: Phase Diagrams, Structures, Synthesis, and Applications, Chemical Reviews 123: 4353–4415
3. Beke S., Giorgio S., Korosi L., Nanai L., Marine W. (2008) Structural and optical properties of pulsed laser deposited V_2O_5 thin films, Thin Solid Films 516: 4659–4664.
4. Meng Jian Li, Silva A. Rui, Cui Ning Hain, Teixeira V., M.P. dos Santos, Xu Zheng (2006) Optical and structural properties of vanadium pentoxide films prepared by d.c. reactive magnetron sputtering, Thin Solid Films 515:195–200.

Note: All the figures in this chapter were made by the authors.

Recent Trends in Applied Physics and Material Science – Sudhir Bhardwaj et al. (eds)
© 2026 Taylor & Francis Group, London, ISBN 978-1-041-16452-4

Structural and Optical Properties of Mn-Doped Cu Ferrite Nanoparticles

V.S. Noorjahan Begum
Government College (A), Anantapur,
Anantapur, AP., India
Department of Physics, JNTUA College of Engineering,
Anantapur, AP., India

D. Zarena*
Department of Physics,
JNTUA College of Engineering,
Ananthapur, AP., India

Abstract: The Mn-doped copper ferrite nanoparticles, a novel material of significant interest in green chemistry, stand out for their exceptional properties, including their biocompatibility and minimal environmental impact. This study is pivotal for analyzing structural and optical properties of nanoparticles (MCFNPS). X-ray diffraction and Fourier transform infrared spectroscopy studies affirmed the remarkable highlights of these nanoparticles. The conventional auto-combustion method was used to synthesize MCFNPS, and the results showed that the nanoparticles exhibit a cubic spinel structure, with particle sizes ranging from 50 to 60 nanometers, and no secondary phase was observed. The FTIR study disclosed tetrahedral and octahedral bond vibrations, providing proof for spinel structure. In conclusion, the synthesized MCFNPS possess a single-phase spinel cubic structure, making them potentially suitable for medical applications.

Keywords: Magnetic materials, Ferrites, XRD, FTIR

1. INTRODUCTION

Ferrites are ceramic-like materials made of iron oxides (Fe_2O_3)combinedwithmetallicelementssuchasmanganese, zinc, nickel, or cobalt. Ferrites can be broadly categorized into **soft ferrites** and **hard ferrites.** Nanoferrites can also be classified based on their chemical composition and crystal structure as **spinel ferrites**, **hexaferrites,** and **garnet ferrites.** Copper ferrite (CuFe2O4) has a **spinel structure** with the general formula AB_2O_4. In **the cubic spinel** lattice, Cu^{2+} ions possess the tetrahedral A-site, Fe^{3+} particles possess the octahedral B-site, and oxygen particles structure a cubic close-packed course of action.

Mn-doped copper **ferrite** is a modified form of copper ferrite ($MnCuFe_2O_4$) in which a portion of the copper (Cu^{2+}) ions are replaced with manganese (Mn^{2+} or Mn^{3+}) ions. This substitution significantly alters the physical, chemical, and magnetic properties. The chemical formula for Mn-doped copper ferrite is represented as $\mathbf{Cu_{1-X}Mn_xFe_2O_4}$, where x denotes the amount of manganese substitution. The structure is based on **a spinel crystal lattice,** characterized by two types of cation sites: **Tetrahedral (A) Sites**: filled by Cu^{2+} ions. **Octahedral (B) Sites**: filled by Fe^{3+} ions.

2. SYNTHESIS TECHNIQUE

The auto combustion method is utilized for the preparation of nanocrystalline manganese-doped copper ferrite particles. The stoichiometric calculations were carried out with the chemical formula $Mn_{0.5}\ Cu_{0.5}\ Fe_2\ 0_4$. Manganese nitrate ($Mn(NO_3)_2{\cdot}6H_2O$), copper nitrate ($Cu(NO_3)_2{\cdot}6H_2O$), and iron nitrate ($Fe(NO_3)_3{\cdot}H_2O$). This solution was heated to 160°C until self-combustion occurred, then it was cooled to room temperature and the pH level was continuously monitored. During this procedure significant gases were evolved. As a result, a dry, loose ferrite powder was created.

3. RESULTS AND DISCUSSION

3.1 XRD Analysis

The structural properties of the spinel ferrite system Mn0.5Cu0.5Fe2O4 were analyzed by X-ray diffraction

*Corresponding author: zareenajntua@gmail.com

DOI: 10.1201/9781003684718-29

technique. The X-ray diffraction pattern of Mn0.5 Cu0.5Fe2O4 is shown in the Fig. 29.1. The obtained pattern confirms the single-phase cubic spinel structure with reflected intensity peaks noted as (111), (220), (311), (222), (400), (422), (511), (440), (533), and (444).

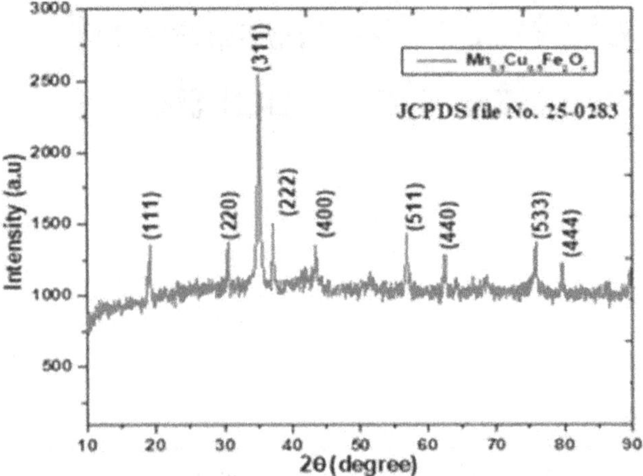

Fig. 29.1 XRD spectra of Mn0.5Cu0.5Fe2O4 Nanoparticles

These peaks show that materials were manufactured in a crystalline form. The obtained spectra in XRD observed very intensive reflection peaks. From this the lattice constant was found to be 8.6 Å, and the average crystalline size can be found to be 18.92 nm.

3.2 FTIR Analysis

The Fourier Transform Infrared Spectroscopy (FTIR) produced mixed-phase and single-phase MCF nanoparticles in the range of 2500-400 cm^{-1}, as shown in the Fig. 29.2. The cations that are dispersed among the two sites (octahedral and tetrahedral) can establish bonds with oxygen atoms, which may result in the creation of spinal structure (AB2O4).

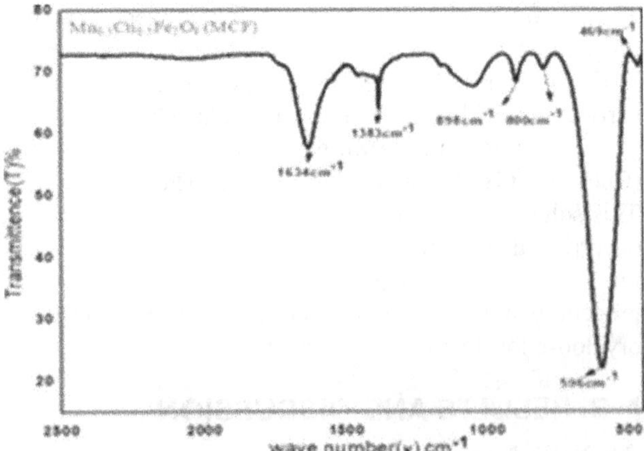

Fig. 29.2 FTIR spectrum of Mn0.5Cu0.5Fe2O4 nanoparticles

One peak is ϑ_1 at 469 cm^{-1}, and another one is ϑ_2 at 596 cm^{-1}, indicating the tetrahedral spinal absorption band and the octahedral spinal absorption band, confirming the cubic spinal structure of MCF nanoparticles. It was also observed that the extra absorption bands surrounded the octahedral site. This was the evidence for the diffraction pattern in the cubic $MnCuFe_2O_4$ single-phase structure.

4. CONCLUSION

The auto-combustion technique was utilized to make Mn-doped copper ferrite with an identical concentration. The X-ray diffraction (XRD) analysis confirmed that they have a cubic spinal structure. After more analysis, it was found that its average crystalline dimension (D) and lattice parameter (A) fell between 18.62 and 8.61 nm, respectively. The cubic spinal structure of MCF nanoparticles was validated by the Fourier Transform Infrared (FTIR) spectra. These materials may find use in biomedical applications in the future.

REFERENCES

1. Ibrahim F. Waheed, Nijat A. Dhham, Tikrit University, Synthesis and Characterization of Copper Ferrite Nanoparticles, Nov 2020, doi:10.1088/1757-899X/928/7/072125.
2. Synthesis and characterization of copper ferrite nanoparticles, Faten Haithum Mulud et al. 2020 IOP Conf. Ser: Mater. Sci. Eng., https://iopscience.iop.org/article/10.1088/1757-899X/928/7/072125.
3. Mirzaee, Sharareh, Yashar Azizian-Kalandaragh, and Parisa Rahimzadeh. "Modifiedco - precipitation process. Effects on the structural and magnetic properties of Mn-doped nickel ferrite nanoparticles." Solid State Sciences 99:106052, 2020. https://doi.org/10.1016/j.solidstatesciences.2019.106052
4. Ma. Mosabberul Haque, Asifur Rahman, Md. Shafiul Islam Shahin, Md. Ahsan Habib, Md. Abu Rayhan Khan, Md. Mahouddin, Minhaj Uddin Monir, Kaykobad Md. Razaul Karim. Manganese-doped Copper Ferrite Nanoparticles: A Promising Approach for Organic Dye Elimination under Light Irradiation, Volume 7, January 2024, 101509.
5. Abubakar Wada, Aliyu Mohammad Aliyu, Dauda Abubakar, Ibrahim Maina, Muhammad Abbas, Salisu,Investigations of the structural properties of Chromium-Zinc Spinel Ferrite by Sol-Gel Technique", Journal of Physics and Chemistry of Materials, Vol. 10, Issue 2, pp. 1–6, 2023.

Note: All the figures in this chapter were made by the authors.

Recent Trends in Applied Physics and Material Science – Sudhir Bhardwaj et al. (eds)
 ISBN 978-1-041-16452-4

30

Chemical Decomposition Method of Designing and Development of a Nano Antenna for Energy Harvesting in Internet of Things Dependent Nanotechnology Systems using Nanopowders

Manjunath T.C.
Department of Computer Science & Engineering,
Rajarajeswari College of Engineering,
Bengaluru, Karnataka

Pavithra G., Swapnil S. Ninawe
Department of Electronics & Communication Engg.,
Dayananda Sagar College of Engg.,
Bengaluru, Karnataka

Sandeep K.V., Iffath Fawad
Department of Electronics & Telecommunication Engg.,
Dayananda Sagar College of Engg.,
Bengaluru, Karnataka

Abstract: This research article presents the development of nano antennas for energy harvesting. With the increasing prevalence of wirelessly interconnected devices, the Internet of Things (IoT) is becoming a significant part of modern society. These devices, which often operate autonomously, are continually scaling down to millimeter and even smaller dimensions, creating substantial challenges for powering them. To address this, various energy harvesting approaches have been developed, includes RF, opticals, mechanicals, thermals, nuclears, chemicals, & biologicals modality. This article provides a comprehensive survey of these methods, discussing their potential to scale down to small dimensions within the contexts of current technologie & future nanosciences advancements. Since the number of untethered, wirelessly interconnected devices grows, the IoT has seen widespread adoption. These autonomous devices, which range from millimeter to sub-millimeter sizes, present significant power supply challenges. In this article, we conduct a thorough survey of current energy harvesting methods, covering modalities that as radio-frequencies, opticals, mechanicals, thermals, nuclears, chemicals, & biologicals source. These methods enable the generation of electricals powers for micros- & nano-system. We explore the potential for scalings these energies conversions techniques to smaller dimension, taking into account existing technologies and future developments in nanoscience. Additionally, the article provides an outlook on necessary advancements to overcome the challenges of powering small-scale devices and systems.

Keywords: Photovoltaic devices, Infrared, Communications, Nano antennas, Lithography, Chemical

1. INTRODUCTION

A nano antenna is a novel solar collection device that utilizes rectifying antennas. Energy harvesting for nano antennas refers to the processess of capturings & convertings ambients energies into useble electricel energies using miniature antenna structures at the nanoscale (Seren et al. 2019). Nano antennas are designed to resonate at specific frequencies and efficiently collect electromagnetic waves or other forms of energy from the environment. These harvestad energys could then be utilized for powers low-power electronical device, sensor, for wireless communication systems, especially in IoT (Internet of Things) applications (El Tantawy et.al. 2020).

*Corresponding author: tcmanju@iitbombay.org

DOI: 10.1201/9781003684718-30

Nano antennas are made using several methods, with the most common being Electron Beam Lithography (EBL), Focused Ion Beam Lithography (FIB), and Nanoimprinting Lithography (NIL) (Mohamed et.al. 2021). Both EBL and FIB are known for being costly, slow, and less efficient for large-scale production. In response, Nanoimprinting Lithography has emerged as a more economical and faster alternative for creating nano antennas. Unlike the traditional beam-based methods that use photons, electrons, or ions to create nano patterns, NIL uses a hard mold with the desired patterns on its surface (Luo and Chan, 2013). This mold is pressed into a thin polymer layer under specific temperature and pressure conditions, creating patterns due to the differences in thickness. This technique has achieved resolutions as small as 10 nm over ten years ago. An advancement in this area is UV-NIL, which involves pressing a transparent mold into a liquid precursor at room temperature.

This is then hardened using UV light. Additionally, soft nanoimprinting methods have been developed using flexible polymeric stamps that can be made from a single master mold. This adaptation helps lower the cost of mold production and facilitates patterning over larger areas with less pressure.

2. EXAMPLES OF NANO-ANTENNA

The Fig. 30.1 gives the experimental work carried out & the results observed in the multimeter. The working principle of a nano antenna relies on the interaction between light and its physical structure, typically composed of metallic nanoparticles arranged in a specific pattern. When light interacts with the nano antenna, it can excite the electrons within the metallic nanoparticles, resulting in surface plasmon resonance (Nguyen and Halvorsen, 2018). This resonance causes the nanoparticles to oscillate, generating a localized electromagnetic field that can interact with nearby materials such as molecules or other nanoparticles. The design of a nano antenna can be optimized to enhance desired properties, such as the intensity and directionality of the electromagnetic field. Consequently, nano antennas find utility in various applications, including sensing, imaging, and data communication.

The working principle is centered on the ability of metallic nanoparticles to interact with light and produce a localized electromagnetic field, which can be exploited for diverse applications (Sze and Ng, 2006). However, material selection in nano antenna fabrication presents challenges, particularly with golds (Au's) and silvers (Ag's), as they exhibit skinned effects @ high frequency. This effect lowers the efficiency of nano antennas by shrinking the wire's effective cross-sectional area and raising its resistance. To address this issue, alternative materials like graphene and carbon nanotubes (CNT) are being explored. These materials do not display skin effect

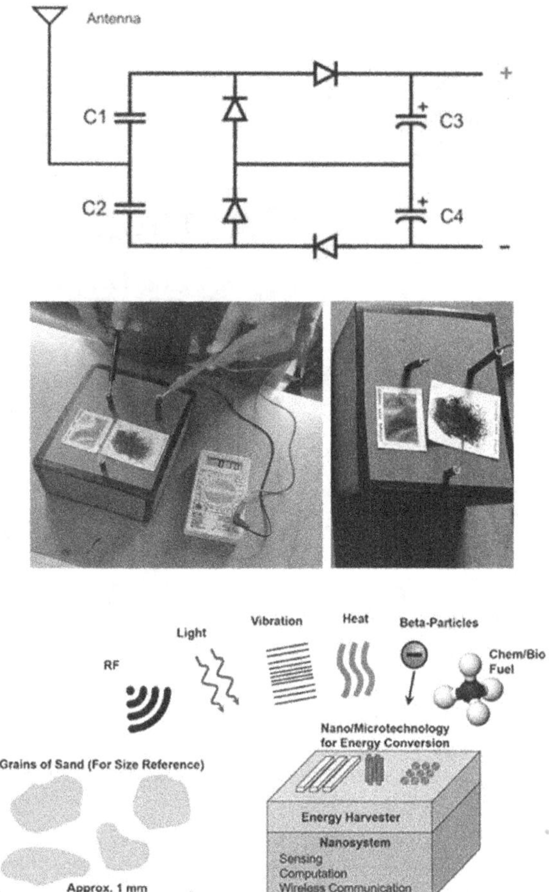

Fig. 30.1 Design of a nano antenna

at higher frequencies, making them promising candidates for nano antenna fabrication. In this paper, we delve into the details of nano antennas based on CNT and graphene, exploring their potential advantages and applications.

The creation of a nano antenna for energy harvesting in telecommunications can be enhanced by using artificial intelligence (AI) and machine learning (ML). These technologies can be applied in designing nano antennas using chemical decomposition methods within IoT-based nanotechnology systems. By applying an AI-ML approach, the designs & developments of an nano based antennas for energies harvestings in IoT-based nanotechnology systems can be enhanced, leading to improved efficiency, cost-effectiveness, and better alignment with the specific requirements of the telecommunications sector (Sambo et.al. 2018). The circuit used for testing is shown in Fig. 30.1. To conclude, the increasing global demand for energy cannot be met solely by non-renewable sources, highlighting the need to maximize energy extraction from renewable sources. Solar energy is a prominent renewable source, but current photovoltaic devices still face challenges in achieving optimal conversion efficiency. In this context, we explore the potential of a new device called the nano antenna, which convert heated energies in to electricel energies & significantly enhance solar cell efficiencies. This technology holds great promise for space applications, with space agencies such as NASA and ISRO utilizing solar based cell as the powered source in different space shuttle. Nano antennas offer a compelling solution due to their efficient fabrication process using Nanoimprinting lithography, requiring minimal material compared to solar cells while delivering higher efficiency. Additionally, nano antennas find application in plasmonic electronic circuits, converting dissipated heated energies from the electricel connections and electronical component in to electricel energies. This functionality contributes to lower operating temperatures for such devices.

The project titled "*Design and Developing an Nano Antennas for the Energy based Harvestings in Internet of Things (IoT) Dependent Nanotechnology Systems Using Chemical Decomposition Methods*" focuses on creating an innovative nano antenna capable of efficiently harvesting ambient energy to power IoT devices in nanotechnology-based systems. The core objective of this research is to design a compact and highly efficient nano antenna that can capture and convert various forms of environmental energy, such as electromagnetic waves, into usable electrical energy. Utilizing chemical decomposition methods, the project aims to develop advanced materials with enhanced conductive and absorptive properties to optimize energy harvesting performance. This approach not only improves energy efficiency but also supports the growing demand for self-sustaining IoT devices in remote or resource-constrained environments. By integrating these nano antennas into IoT-dependent nanotechnology systems, the research work undertaken aimed to give an sustainable & scalable energy solution for future telecommunications, sensor networks, and other IoT applications.

3. CONCLUSIONS

The research project "Designed & Developing an Nano based Antennas for Energies Harvestings for Internet of Things (IoT) Dependent Nanotechnology Systems Using Chemical Decomposition Methods" investigates the creation of a novel nano antenna to efficiently harness ambient energy for powering IoT devices within nanotechnology frameworks. The primary goal of this initiative is to engineer a compact, highly effective nano antenna that can convert environmental electromagnetic energies into electrical power. By employing chemical decomposition techniques, this project seeks to advance material properties to enhance their conductivity and absorption capabilities, thereby boosting the energy harvesting efficacy. This method not only improves the efficiency of energy utilization but also caters to the increasing need for autonomous IoT devices in areas with limited resources. Integrating these nano antennas into IoT-based nanotechnology systems aims to furnish a dependable, scalable energy source for the telecommunications industry, sensor networks, and broader IoT applications, paving the way for more sustainable technology solutions in the future.

REFERENCES

1. Seren H.R. et. al. (2019). "Energy Harvesting With Nanoantennas for Large Area Applications: Current Progress and Future Perspectives," IEEE Journal of Selected Topics in Quantum Electronics, 25(6): 1–11.
2. El Tantawy M.H. et.al. (2020). "Review on Nanostructured Antennas for Energy Harvesting Applications," IEEE Access, 8: 214086–214108.
3. Mohamed K.A. et. al. (2021). "Nanoantenna Design and Fabrication Using Graphene for Energy Harvesting Applications," IEEE Sensors Journal, 21 (9): 10350–10359.
4. Luo Y. and Chan C.T. (2013). "Plasmonic Nanoantennas: From Theoretical Models to Practical Devices," IEEE Journal of Selected Topics in Quantum Electronics, 19(3): 4600614–4600614.
5. Nguyen V.C. and Halvorsen K. (2018). "Rectifying Nanoantennas for Solar Energy Harvesting: A Review," IEEE Journal of Photovoltaics, 8(4): 1135–1142.
6. Sze S.M. and Ng K.K. (2006). Physics of Semiconductor Devices, 3rd ed., Hoboken, NJ, USA: Wiley.
7. Sambo Y.A. et. al. (2018). "Energy Harvesting in Wireless Sensor Networks: A Comprehensive Survey," IEEE Communications Surveys & Tutorials, 20(2): 1292–1331.

Recent Trends in Applied Physics and Material Science – Sudhir Bhardwaj et al. (eds)

A Silkworm Counting Methods with Density Maps Based on Multiscale Feature Fusions in Small Worms Using Physics based Nano-Materials

Manjunath T.C.
Department of Computer Science & Engineering,
Rajarajeswari College of Engineering,
Bengaluru, Karnataka

Pavithra G., Swapnil S. Ninawe
Department of Electronics & Communication Engg.,
Dayananda Sagar College of Engg.,
Bengaluru, Karnataka

Sandeep K.V., Iffath Fawad
Department of Electronics & Telecommunication Engg.,
Dayananda Sagar College of Engg.,
Bengaluru, Karnataka

Abstract: Silkworm rearing is a critical component of agricultural heritage, especially within Han culture, where optimizing silk production relies heavily on accurate population assessment. Traditional counting methods, based on visual inspection and experiential estimation, are often fraught with inaccuracies and inefficiencies. This study introduces the Silkworm Counting Network, a novel convolutional neural network (CNN) designed to enhance the precision of silkworm counting through the generation of density maps. Leveraging a meticulously curated dataset from Guangxi's sericulture areas, which encompasses a wide range of age groups and environmental settings, the network employs the VGG architecture for advanced feature extraction and dilated convolutions to expand the receptive field for accurate density estimation. With rigorous evaluation showing a root means squares errors (RMSEs) of 8.5 & means absolutes error (MAE) of's 5.3, this method outperforms traditional approaches by addressing variability in population distribution and occlusion. The study underscores the potential of deep learning in revolutionizing sericulture by providing a reliable, efficient, and automated counting solution that promises significant advancements in silk production.

Keywords: Silk, Worms, Physics, Nano

1. INTRODUCTION

Silkworm rearing is integral to our agricultural heritage and deeply rooted in Han culture. Accurate counting of silkworms is crucial for optimizing silk production. Traditional counting methods, relying on visual inspection and experiential estimation, are prone to inaccuracies and resource inefficiencies. In response, this study introduces the Silkworm Counting Network, a novel approach aimed at generating density maps and precisely tallying silkworm populations. Using images from the sericulture area of Guangxi, a diverse silkworm dataset was meticulously curated, covering various age groups and backgrounds. Data augmentation techniques were applied to enhance dataset diversity, ensuring robust model performance. The front-end network employs the VGG architecture for feature extraction and integrates multi-scale contextual semantic information extraction modules to enrich silkworm features. The back-end network utilizes dilated convolutions with varying rates to expand the receptive field and generate silkworm density images for accurate counting. Evaluation of the proposed method on the constructed silkworm density dataset demonstrates promising results, with a roots means squares errors (RMSEs) in 8.5 & means absolutes errors (MAEs) in 5.3. By overcoming challenges such as occlusion and varying population distributions, the proposed approach transcends conventional target detection methods. Density map-based counting proves superior in estimating silkworm density per unit area, enabling more accurate quantification than

*Corresponding author: tcmanju@iitbombay.org

DOI: 10.1201/9781003684718-31

simplistic number statistics. A review of related works highlights the scarcity of research in silkworm density statistics, primarily focusing on dense crowd and object detection.

While earlier approaches relied on detection-based algorithms, recent trends favor CNN-based solutions, offering end-to-end regression for crowd counting. Data processing techniques play a pivotal role in preparing the dataset for training. Augmentation methods such as rotation, cropping, and mirroring enrich the dataset, enhancing the detection accuracy and generalization ability of convolutional neural networks. The proposed methodology, integrating VGG-16 for feature extraction and dilated convolutions for receptive field expansion, yields a robust model capable of accurate silkworm counting. The end-to-end training process, optimized using stochastic gradient descent, achieves superior performance compared to state-of-the-art methods. The Silkworm Counting Network emerges as a promising solution for silkworm density estimation, offering unparalleled accuracy and efficiency in silkworm population quantification, with significant implications for the sericulture industry. The research paper introduces the Silkworm Counting Network, employing deep learning to enhance silkworm population assessment. It details the construction of a diverse dataset and the network architecture, integrating VGG-16 and dilated convolutions. Important results include a significant reduction in means absolutes errors (MAEs) to 5.3 & roots means squares errors (RMSEs) to 8.5, outperforming existing methods. The network's ability to accurately count silkworms across diverse backgrounds and age groups demonstrates its potential to revolutionize sericulture, optimizing silk production and enhancing industry productivity.

2. METHODOLOGY

The scope of this research encompasses the development and implementation of the Silkworm Counting Network, a novel deep learning-based approach tailored for accurate and efficient silkworm population assessment. The primary objective is to address the limitations of traditional counting methods and existing automated approaches by leveraging advanced convolutional neural networks (CNNs) to achieve precise and robust population estimates.

Key objectives of the study include:

1. Developing a comprehensive silkworm dataset sourced from the sericulture area of Guangxi, encompassing diverse age groups, backgrounds, and environmental conditions.
2. Constructing a sophisticated network architecture that integrates front-end feature extraction using the VGG-16 architecture with back-end density map generation through dilated convolutions.
3. Enhancing the accuracy and efficiency of silkworm population assessment by automating the counting process and minimizing subjective biases inherent in traditional methods.
4. Evaluating the performance of the Silkworm Counting Network using rigorous metric such-as roots means squares errors (RMSEs) & means absolutes errors (MAEs) to validate its effectiveness in comparison to existing methods.

The importance of this research lies in its potential to revolutionize silkworm population assessment, thereby optimizing silk production and enhancing productivity in the sericulture industry. By automating and improving the accuracy of counting processes, the Silkworm Counting Network promises to streamline operations, reduce labor costs, and minimize errors associated with traditional methods. Furthermore, the integration of deep learning techniques represent an significant advancements in their fields, offerings an scalables & adaptable solutions capable of addressing the complexities of silkworm counting across diverse environments and scenarios. Ultimately, the research holds far-reaching implications for the sustainability and growth of the sericulture industry, positioning it at the forefront of technological innovation and development in agriculture as shown in the Fig. 31.1.

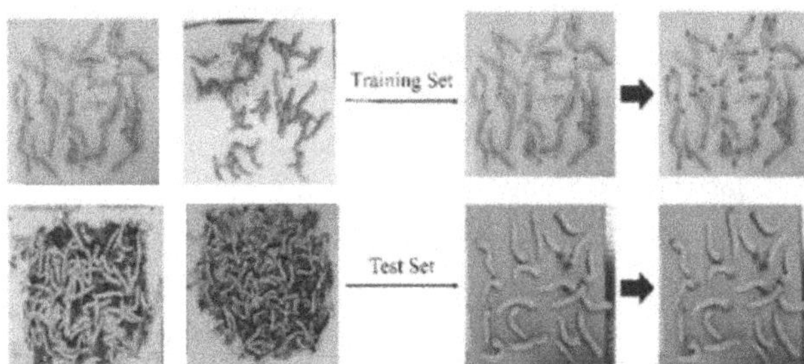

Fig. 31.1 Schematic diagrams of their production and division of the silkworm number dataset

3. CONCLUSIONS

This study introduces the Silkworm Counting Network, an innovative deep learning-based method tailored to enhance the precision and efficiency of silkworm population assessment in the sericulture sector. Traditional counting methods, often reliant on visual inspection and subjective estimation, fall short in accuracy and efficiency, underscoring the need for a more advanced approach. Our network utilizes a meticulously curated dataset from Guangxi's sericulture areas, covering silkworms across various age groups and backgrounds. The adoption of data augmentation techniques bolsters the diversity and robustness of the dataset, significantly benefiting the model's performance.

The Silkworm Counting Network employs the VGG architecture for robust feature extraction and integrates

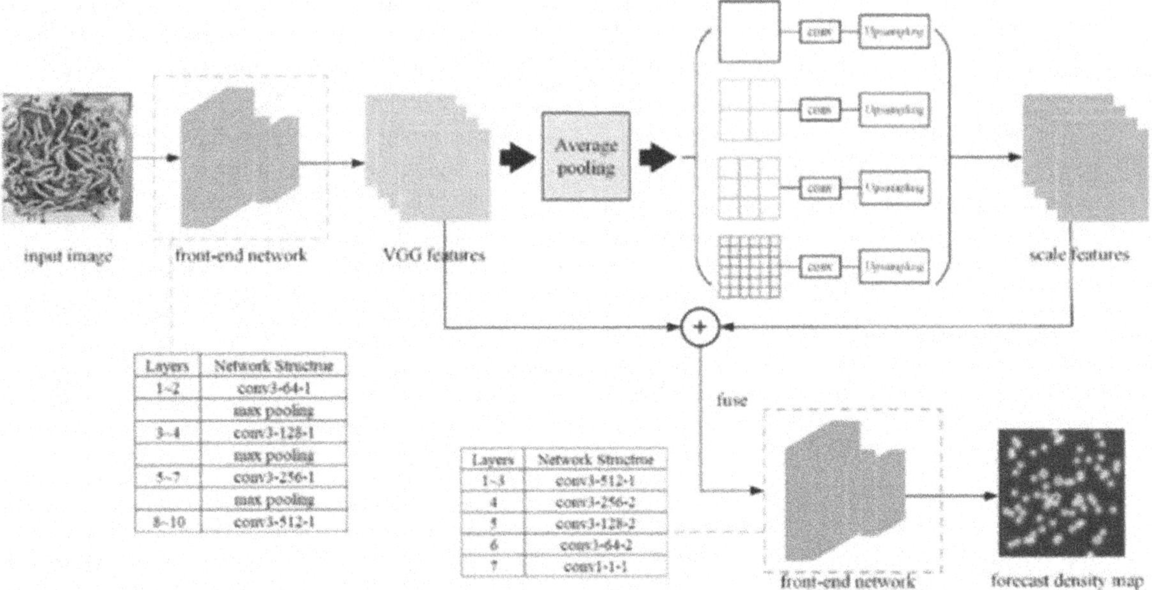

Fig. 31.2 The network structure of silkworm counting net, the parameter in the figure is conv (convolution kernel) – (number of filters) – dilated rate

multi-scale contextual semantic information through dilated convolutions. These elements enable the network to effectively generate density maps, achieving precise silkworm counts. Evaluation metrics such as root mean square error (RMSE) of 8.5 and mean absolute error (MAE) of 5.3 underscore the method's superiority over traditional techniques. This network transcends conventional target detection methods by addressing challenges like occlusion and variable population distributions.

By comparing our approach with existing techniques, we have identified that while earlier methods predominantly focused on detection algorithms, the recent shift towards CNN-based solutions utilizing end-to-end regression for density estimation offers significant improvements in accuracy and generalization capabilities. Our methodology, which integrates cutting-edge techniques such as VGG-16 for feature extraction and dilated convolutions for expanded receptive fields, sets a new standard in the field. The automated, end-to-end training process, optimized via stochastic gradient descent, further validates the efficacy of our approach, positioning the Silkworm Counting Network as a revolutionary tool for the sericulture industry. Ultimately, this research not for their enhancing their process of silkworm population quantification but, also contribute to their broad goals of optimizing silk production. By automating the counting process and improving accuracy, the Silkworm Counting Network offers substantial benefits, including reduced labor costs and minimized errors, thereby supporting sustainable growth and technological advancement in the sericulture industry as shown in the Fig. 31.2.

REFERENCES

1. N. Lakshmi, et.al., "CMOS Implementation of Multipath Fully Differential OTA with Dual Flipped Voltage Follower in 50 nm and 75 nm CMOS Technologies using Cadence Tool," 2024 IEEE ICDCOT, 2024, pp. 1–8. https://doi.org/10.1109/ICDCOT61034.2024.10515482
2. V.K. Suhasini, et.al. "Detection of Skin Cancer using Artificial Intelligence & Machine Learning Concepts," 2022 IEEE 4th Int. Conf. on Cyber., Cogn. & Machine Learning Apps., Goa, 2022, pp. 343–347, https://doi.org/10.1109/ICCCMLA56841.2022.9989146
3. Joshi, et.al., "New Approach of Steganographic Design of Speech Signals & its Application to Voice Recognitions," 2022 IEEE 7th International conference for Convergence in Technology (I2CT), Mumbai, India, 2022, pp. 1–5 https://doi.org/10.1109/I2CT54291.2022.9825419
4. Pritosh Tomar & TC Manjunath, "Numerical Investigation of Thermal Performance Enhancement of Solar Reservoir using Flash Cycle", Scopus Indexed Q3 Journal of Advanced Research in Fluid Mechanics and Thermal Sciences, Volume 123, No. 1, pp. 197–221, ISSN: 22897879, Nov. 2024 https://doi.org/10.37934/arfmts.123.1.197221

Note: All the figures in this chapter were made by the authors.

Recent Trends in Applied Physics and Material Science – Sudhir Bhardwaj et al. (eds)

32 The Persuasive Consequences of Optical, Dielectric and Photoluminescent Properties of Cholesteric Liquid Crystal by the Dispersion of Zinc Oxide (ZnO) Nanoparticles

Manisha Chaudhary*, Divya Ghildyal
Department of Physics, JSS Academy of Technical Education,
Noida, U.P. India

Abstract: A multifaceted system of Zinc oxide (ZnO) nanoparticle dispersed cholesteric liquid crystals revealed the augmented investigations. The significance of dispersion of ZnO nanoparticles at different concentration is very extensive in the cholesteric liquid crystal. Thermodynamical investigation revealed the reduction of transition temperature and enthalpies of different liquid crystal phases. Thermodynamic behaviour of ZnO doped cholesteric liquid crystals shows the transition temperature of different liquid crystal phases. Dielectric investigation documented outcomes of ZnO nano particles doped cholesteric liquid crystal at the biasing. Dielectric studies exhibit the permittivity behaviour of doped cholesteric liquid crystal phases. ZnO nanoparticles doped cholesteric liquid crystal enhanced the efficiency area of doped complex material surface. The increment of Intensity of photoluminescence is responsible due to the functionality of nano particles in cholesteric liquid crystals system.

Keywords: ZnO nanoparticles, Soft mode, Relaxation process, Photoluminescence

1. INTRODUCTION

Liquid crystals are very fascinating material for optical devices. Liquid crystal materials have an anisotropic and fluid properties simultaneously. According to the orientation of molecules liquid crystal show different phases. Cholesteric liquid crystal has spiral structure. (Yang et al.,2013). Various physical properties like viscosity, elasticity, dielectric anisotropic etc are very important for the response time of liquid crystal displays. Cholesteric liquid crystal is very sensitive to light. By applying the biasing, pitch length of chiral liquid crystal has been increased to enhance the reflective band width (Song. et al., 2012). The dispersion of Nano particles in cholesteric liquid crystal enhanced the exciting understanding like dielectric, optical and magnetic properties etc. In the present study, we have investigated thermodynamical behaviour e.g. transition temperatures and enthalpies of different liquid crystalline phases of ZnO nanoparticles dispersed Cholesteryl Olely Carbonate cholesteric liquid crystal. Also reported that photoluminescence properties of ZnO nanoparticles doped CoC cholesteric liquid crystal. (Gan, 2021). Dielectric behaviour explained the total polarisation contribution at the surface of ZnO NPs doped CLC matrix.

2. EXPERIMENT AND METHODS

In the present study, cholesteric liquid crystal (CLC), Cholesteryl Olely Carbonate and ZnO nano-powder

*Corresponding author: dr_manisha@jssaten.ac.in

DOI: 10.1201/9781003684718-32

procured from the sigma Aldrich. We have used pure cholesteric liquid crystal and ZnO nano-powder without any further amendment for sample preparation. ZnO nao-powder and CoC-CLC are used as host and guest elements in the ZnO NPs doped Cholesteryl Olely Carbonate CLC (CoC-CLC) matrix. The particle size was 30 nm in the ZnO nano-powder. We have prepared ZnO NPs doped cholesteric liquid crystals sample. ZnO nanoparticles, 6% by weight and CoC cholesteric liquid crystal were added to toluene. ZnO NPs and CoC cholesteric liquid crystal were mixed consistently by incessant stirring with the method of ultrasonication. For the characterisation of undoped and ZnO doped CoC matrix liquid crystal cells were prepared. Conducting glass substrates have been used for the liquid crystal cells and thickness of the cell was 10 μm. The isotropic phase of the samples filled in the liquid crystal cells with the help of capillary action. We have recorded DSC thermos-grams of samples by using Differential scanning calorimeter (Mettler Toledo), DSC-822e model associated with software STARe, Luminescence properties of undoped CLC and ZnO doped CLC were recorded in the fluorescence mode with fluorescence spectro photo-meter associated with a xenon flash lamp. Agilent Carry Eclipse G9800A is the model's name of fluorescence spectro photo-meter. Using Dielectric measurement documented by the computer-controlled impedance analyser, model name is HP 4192A with the variation of frequency from 100 Hz to 1.0 MHz.

3. RESULT AND DISCUSSION

3.1 Thermodynamic Investigation

DSC thermos-grams of undoped CoC CLC and ZnO doped CoC CLC matrix at the 1.0 °C/min scanning rate, during the heating cycle shown in Fig. 32.1. The mass of samples was 5.0 mg for thermal investigation. There were three peaks in DSC thermograms of undoped and ZnO NPs doped CoC CLC matrix. The first peak of undoped CoC CLC related with crystal (Cr) to smectic phase (SmA*) transition at the 23.0 °C, second peak related to smectic phase (SmA*) to cholesteric phase (N*) at 36.0 °C and cholesteric phase (N*) to isotropic phase (Iso) transition at 37.0 °C. In the DSC thermos-gram of ZnO nanoparticles doped CoC CLC (Fig. 32.1b) peaks at 23.8, 36.5 and 37.7 °C related with crystal (K) to smectic phase (SmA*) and nematic phase. In thermodynamic study, the conversion temperatures of liquid crystal phases increased with the doping of ZnO nanoparticles. This is the significant reason of strong interaction of anisotropic behaviour of liquid crystal with nanoparticles. It is completely agreed with the mean-field theoretical model based on anisotropic nanoparticles interaction (Lu, 2014).

3.2 Photoluminescence

The photoluminescence emission spectra of the undoped and ZnO nano-particles doped CoC-CLC liquid crystals

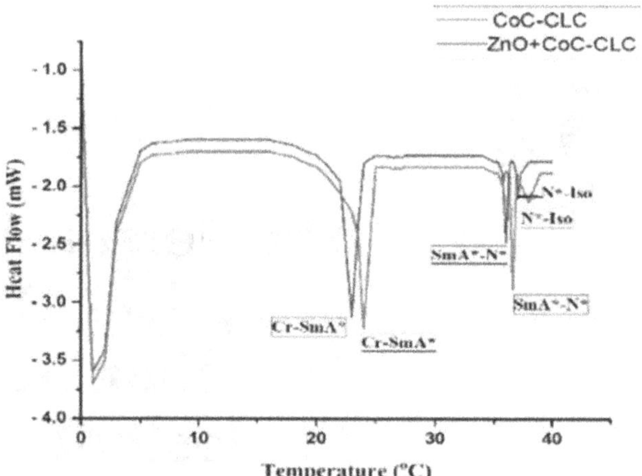

Fig. 32.1 DSC thermo-gram of undoped and ZnO doped CoC-CLC

shown in Fig. 32.2 at the 300K temperature. The 440 nm to 520 nm range of wavelength for the PL spectra performance. The excitement of wavelength of PL measurements is 440 nm. The peaks of PL spectra showed that intensity of photo-luminescence is increased with the doping of ZnO nan-particles. The highest PL intensity achieved at 478 nm. At the surface of nano-particles the movement of electron between the CoC-CLC and ZnO NPs significantly affected by the electromagnetic fields. In the presence of EM radiation, group formation ability increased at the surface of NPs. (Pathinti et al, 2021). Increment of PL intensity of doped CoC-CLC is responsible for the strong interaction NPs phonons and radiation photon.

3.3 Dielectric Studies

Dielectric spectrum for the permittivity with frequency dependence at room temperatures is shown in Fig. 32.3. The dielectric permittivity (real part ε') decreases with the doping of ZnO NPs in CoC-CLC. The decrement of

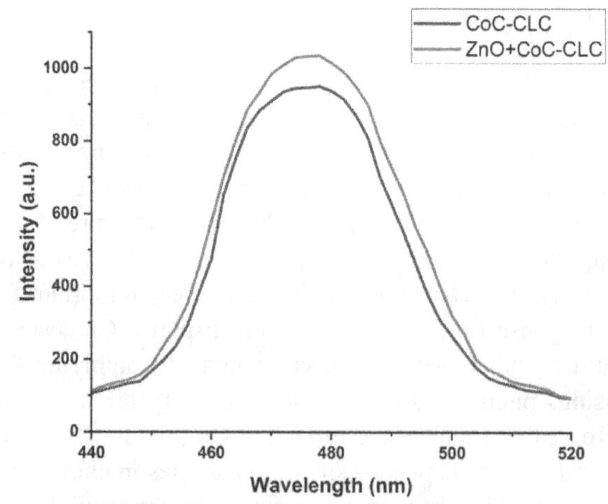

Fig. 32.2 Photoluminescent spectra of undoped and ZnO doped CoC-CLC

permittivity due to the conductivity of ZnO NPs in the frequency range of 10^4 Hz to 10^5 Hz. Dipole moment is completely neglected in ZnP NPs doped CoC-CLC (Pathak 2020).

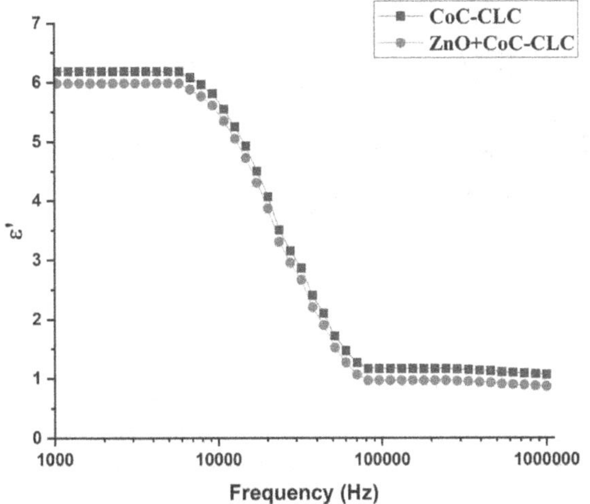

Fig. 32.3 Dielectric permittivity of undoped and ZnO doped CoC-CLC

4. CONCLUSION

Present work investigated the thermodynamic, photoluminescence nature and dielectric behaviour of ZnO NPs doped CoC-CLC matrix. The transition temperature of different liquid crystal phases is increased due to the anisotropic nature of liquid crystals. Interaction between the surface of ZnO NPs and CLC signified signature of the enhancement of photoluminescence intensity. Contribution of total polarization confirmed by the dielectric permittivity.

REFERENCES

1. Gan P et al. (2021) Broadband reflection in polymer-stabilized cholesteric liquid crystal film with zinc oxide nanoparticles film thermal diffusion method. Liquid Crystals. Liquid Crystals. 48: 1959–1968.
2. Lu H, Xu W, Song Z, et al. (2014) Electrically switchable multi-stable cholesteric liquid crystal based on chiral ionic liquid. Opt Lett. 39: 6795–6798.
3. Pathak G, Hegde G and Prasad V. (2020) Octadecylamine-capped CdSe/ZnS quantum dot dispersed cholesteric liquid crystal for potential display application: Investigation on photoluminescence and UV absorbance. Liquid Crystals. 48: 579–587.
4. Pathinti R S, Gollapelli B, Jakka S K, Vallamkondu J. (2021) Green synthesized TiO2 nanoparticles dispersed cholesteric liquid crystal systems for enhanced optical and dielectric properties. Journal of Molecular Liquids. 336: 116877.
5. Song P, Cao H, Wang FF, et al. (2012) Study of polymer-dispersed liquid crystal systems using epoxies/acrylates as hybrid polymer matrix components. Liquid Crystals. 39: 903–909.
6. Yang Y, Zhang Y, Wei Z. (2013) Supramolecular helices: chirality transfer from conjugated molecules to structures. Adv Mater. 25: 6039–6049.

Note: All the figures in this chapter were made by the authors.

Recent Trends in Applied Physics and Material Science – Sudhir Bhardwaj et al. (eds)

Study the Role of Precipitating Agents on Transition Metal Doped ZnO Nanoparticles Synthesized by Green Method

Yojana Sharma, Tanvi Sharma, Nishant Thakur, Vikas Anand
Department of Physics and Astronomical Science, Central University of Himachal Pradesh, Kangra, HP, India

Anup Kumar
Department of Physics, Dr.Y.S.Parmar Government Post Graduate College, Nahan Distt. Sirmaur, HP, India

Pawan Heera*
Department of Physics and Astronomical Science, Central University of Himachal Pradesh, Kangra, HP, India

Abstract: Transition metal doped ZnO nanoparticles (NPs) were prepared by green synthesis method (using Jatropha Plant latex). Further the effect of precipitating agents on the prepared samples has been studied. The morphology, structural and optical properties of the prepared samples has been studied. XRD results confirm a hexagonal wurtzite structure of Cu-doped ZnO particles with good crystallinity. The presence of functional group is confirmed by FTIR analysis for ZnO at 587.81 cm^{-1} and 616.64 cm^{-1} and for Cu doped ZnO at 868.93 cm^{-1} and 875.31 cm^{-1}. The shift in the excitation and emission spectra towards the visible region was observed in the presence of both precipitating agents, as compared to pure ZnO. A direct energy band gap which decreases with the increasing Cu concentration was confirmed by UV-VIS studies. From the SEM analysis the crystallite average size is in range of 13-35 nm. This study highlights how precipitating agents influence the properties of plant-based synthesized nanomaterials.

Keywords: Transition metal, Cu-doped ZnO, Structural properties, Optical band gap, Green synthesis, Precipitating agents

1. INTRODUCTION

Zinc oxide (ZnO) is a versatile material with many applications in materials science and engineering (Raha et al., 2022, Nohavica et al., 2010). Also, ZnO is studied in many fields, including material science, physics, chemistry (Soosen et al., 2009), biochemistry (Nohavica et al., 2010), and solid-state electronics (Kamarulzaman et al., 2015, Joy et al., 2024). ZnO NPs are used in biomedical fields such as in anti-cancer drug delivery and in diabetes treatment (Ansari et al., 2018). Though the ZnO have many characteristics and vast number of applications there is still scope for the further modifications which will enhance their application domain. Significant changes in optical, electrical, and magnetic properties have been noted in reports on the impact of transition metal doping on ZnO (Bresser et al., 2013, Singhal et al., 2012). Various techniques have been used to synthesise NPs of ZnO. Out of these techniques Green synthesis using plant extracts is faster and more stable. Plant extracts contain numerous bioactive molecules that reduces and stabilize the NPs (Karthik et al. ,2022).

In the present work, synthesis of Cu-doped ZnO NPs using latex of Jatropha Curcas has been carried out and the different properties were investigated using XRD, FTIR, and UV-VIS and SEM analysis.

*Corresponding author: pawanheera@hpcu.ac.in

DOI: 10.1201/9781003684718-33

2. EXPERIMENTAL METHODS

Cu-doped ZnO particles with 2.5 wt% and 10 wt% concentration of Cu was synthesized by green method. The latex was collected from Jangal village of Kangra District Himachal Pradesh. The ZnO NPs were prepared by using Zinc Nitrate Hexahydrate(1M) source salt, and 100 ml plant extract as a reducing and capping agent. For doping, 1M solution of ($CuSO_4 \cdot 5H_2O$) is used. 10% aqueous NaOH and NH_4OH each as reducing agent have been used. The resulting precipitates were then collected after washing and filtering them with whatmann filter paper no. 1. The filtrate so obtained was dried for atleast 12 hrs and then calcinated in a muffle furnace at 800°C for 2 hrs resulting into Cu-doped ZnO powder.

3. RESULTS AND DISCUSSION

3.1 Structural Analysis

The X-ray diffraction spectra Fig. 33.1 (a), (b) for Cu doped ZnO and undoped ZnO NPs using NH_4OH and NaOH acting as a precipitating agent. The X-Ray diffractomet-er Rigaku MiniFlex in the range from 10^0 to 80^0 has been used along with monochromatic CuKα (λ = 1.54059 Å) radiation operated at 40 kV and 15 mA. The sharp and intense Peaks confirm the high crystallinity of the prepared samples. The analysis confirms the wurtzite structural phase in all the compositions. The different peaks (110), (002), (101), (102), (110), (103), (200), (112), and (201) were observed. The hexagonal wurtzite structure of ZnO NPs is confirmed by the peak at (101) being the most noticeable (JCPDS card no. 5-0664) (Sajjad and others, 2018). A reflection corresponding to the (111) plane of Cu crystallites, marked as (*) in Fig. 33.1 appears in the XRD pattern (Mukhtar etal., 2012) for Cu (2.5% and 10%). This implied Cu ions have replaced at Zn sites without amending the crystal structure of ZnO as the ionic radii of Cu^{2+} (0.73 Å) and Zn^{2+} (0.74 Å) are similar (Kamaranjan et al., 2022). The (111) peak is faint at 2.5% Cu but more prominent at 10%, reflecting the higher doping concentration. This implies that on increasing the Cu doping, leads to the formation of copper oxide (CuO), as seen by the appearance and growth in intensity of CuO (Khilifi et al., 2022). Scherrer equation was used to find the average crystalline size associated with crystal plane (101) by using the

$$D = \frac{K\lambda}{\beta \cos\theta} \tag{1}$$

Where, D denotes the crystalline size in nm, θ is the Bragg's diffraction angle, k = 0.9, λ = 0.15418 nm wavelength of source of X-ray and, β in radians is the angular peak width at half maximum (FWHM) is Bragg's diffraction angle (Roguai et al., 2020). The crystallite average size for NH_4OH, the size decreased from 34 nm (undoped) to 30 nm and 27 nm with 2.5% and 10% Cu respectively.

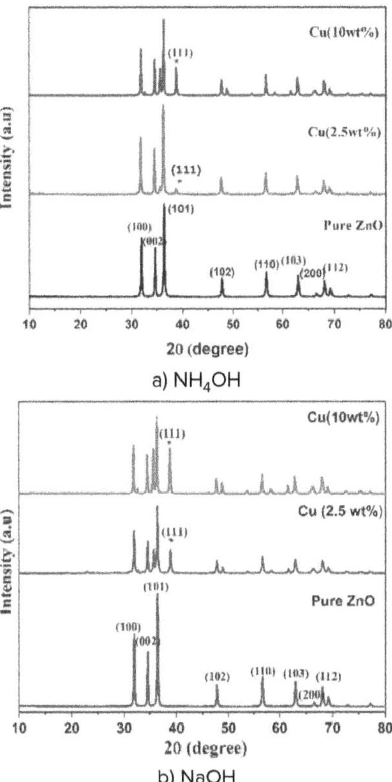

Fig. 33.1 XRD pattern of ZnO and different Cu concentration with precipitating agents a) NH_4OH and b) NaOH

For NaOH, the size dropped from 14.62 nm (undoped) to 14.13 nm and 13.2 nm with 2.5% and 10% Cu. These findings indicate that for both reducing agents, crystallite size decreases as Cu concentration rises.

3.2 FTIR Analysis

The FTIR spectra Fig. 33.2 (a), (b) has been attained using with KBr plates (4000–400) cm^{-1} and Perkin Elmer IR Spectrophotometer

For NH_4OH as the precipitating agent, a Zn-O bond was identified at 587.81 cm^{-1} which is compatible with already reported in literature (Kumar R et al., 2013), the strong bands at 3330.31 cm^{-1}, 2934.41 cm^{-1}, 1410.14 cm^{-1}, and 1144.51 cm^{-1}, indicating OH, CH, C=C, and C-N stretching align with already reported in literature (Prasad et al., 2019). With NaOH, a Zn-O peak appeared at 616.64 cm^{-1}, and O-H groups were observed at 3379.58 cm^{-1}. Additional peaks at 2972.18 cm^{-1} and 2884.04 cm^{-1} indicates C-H stretching, while at 1404.73 cm^{-1} was attributed to carbonyl groups. For Cu doped ZnO NPs, variations in the 550–700 cm^{-1} range confirm the presence of Zn-O peak at 868.93 cm^{-1} and 875.31 cm^{-1}, which is likely due to the presence of Cu dopant. The observed bands correspond with the previously reported in literature (Vishvanath Tiwari et al., 2018).

3.3 SEM Analysis

The SEM images of biosynthesized ZnO NPs are shown in the Fig. 33.3. From the Fig. 33.3 (a), As seen in the

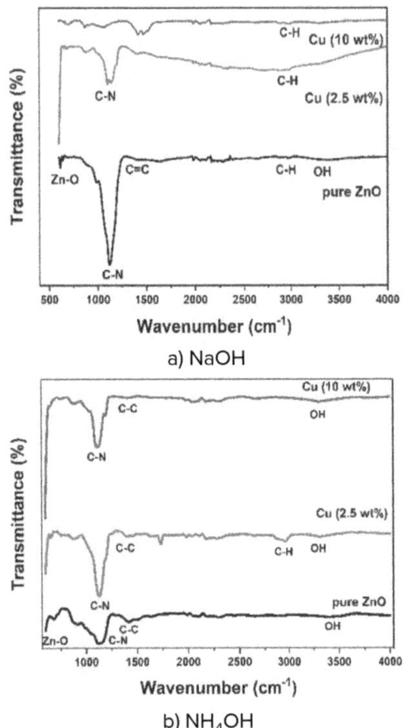

Fig. 33.2 (a), (b) Spectra of pure and different concentration of Cu with precipitating agents NaOH and NH_4OH respectively

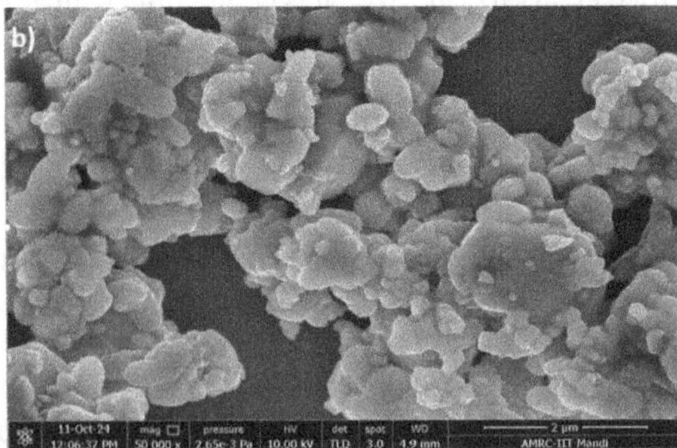

Fig. 33.3 SEM picture of (a) ZnO and (b) Cu doped ZnO NPs

shape of a big cluster for pure ZnO, the NPs' morphology is uniformly spherical. Cu doping has stimulated the growth of ZnO samples and tends to enhance segregation in both situations. This shows that, at higher doping concentrations grain size seems to increase at the same time crystallinity is maintained.

3.4 UV-VIS Analysis

The optical characteristics of both pure and Cu-doped ZnO NPs were analysed using UV-VIS absorption spectroscopy. UV-VIS spectroscopy (200-700 nm) have been obtained in Fig. 33.4 from Shimadzu UV-Vis spectrophotometer. The band gap is calculated by using Tauc's relation (Bhuyan et al., 2015):

$$(\alpha h\nu)^2 = A\,(E_g - h\nu) \tag{2}$$

Where, A is constant, E_g denotes band gap, α the absorption coefficient and h the Planck's constant (4.135667 × 10^{-15} eV s). The calculated band gap from the Tauc's plot is displayed in Fig. 33.5.

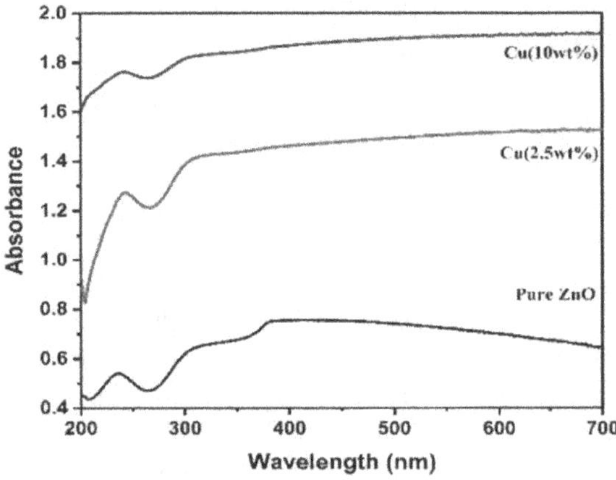

Fig. 33.4 Absorption spectra of ZnO and Cu-doped ZnO NPs

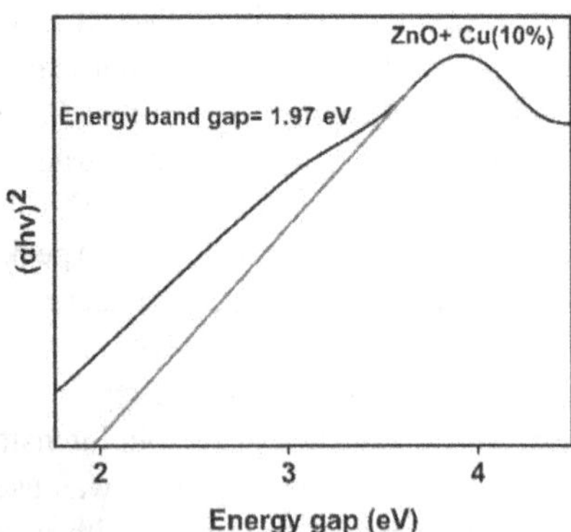

Fig. 33.5 Tauc's plot for Cu-doped ZnO

The largest absorption is in the near band edge (UV) region, between 374–390 nm. For undoped ZnO NPs, the band gap was 3.37 eV. For Cu-doped ZnO NPs with NaOH or NH_4OH the bandgap decreases with increasing concentration from 2.5wt% to 10wt%. This indicates red shift in the absorption edge.

This red shift and decrease in bandgap energy are attributed to the Burstein-Moss band filling effect, as observed in the absorption spectra of Cu-doped ZnO NPs. Our results are in agreement with the already reported in literature (Roguai et al., 2020).

4. CONCLUSION

In conclusion, ZnO and transition metal doped ZnO NPs were successfully synthesized using green and co-precipitation methods. XRD, FTIR, and UV-VIS and SEM analysis confirms the high crystallinity and hexagonal wurtzite structure with a crystallite sizes ranging from 27-34 nm. Cu doping reduces the crystallite size and intensified the CuO (111) peak. FTIR study reveals characteristic bands for OH, CH, C=C, and C-N stretching. A red shift in the band gap from 3.30 eV to 1.97 eV was observed for 10% Cu-doped ZnO. The study emphasizes the role of precipitating agents in determining NPs properties which makes them suitable for electronics and photocatalytic applications.

REFERENCES

1. Ansari, Mohammed T., et al.(2018). "Applications of Zinc nanoparticles in medical and healthcare fields." *Current Nanomedicine (Formerly: Recent Patents on Nanomedicine)* 8(3): 225–233.
2. Bhuyan, Tamanna, et al.(2015). A comparative study of pure and copper (Cu)-doped ZnO nanorods for antibacterial and photocatalytic applications with their mechanism of action. *Journal of Nanoparticle Research* 17: 1–11
3. Bresser, D., Mueller, F., Fiedler, M., Krueger, S., Kloepsch, R., Baither, D., ... &Passerini, S. (2013). Transition-metal-doped zinc oxide nanoparticles as a new lithium-ion anode material. *Chemistry of Materials, 25*(24), 4977–4985.
4. Joy, Amala, et al.(2024). Solar photocatalysts: non-metal (C, N, and S)-doped ZnO synthesized through an industrially sustainable in situ approach for environmental remediation applications. *RSC advances* 14(30): 21655–21667.
5. Kamarajan, D., et al. (2022). Green synthesis of ZnO nanoparticles and their photocatalyst degradation and antibacterial activity. *Journal of Water and Environmental Nanotechnology* 7(2): 180–193.
6. Kamarulzaman, Norlida, Muhd Firdaus Kasim, and RoshidahRusdi.(2015). Band gap narrowing and widening of ZnO nanostructures and doped materials. *Nanoscale research letters* 10: 1–12.
7. Karthik, K. V., et al. (2022). Green synthesis of Cu-doped ZnO nanoparticles and its application for the photocatalytic degradation of hazardous organic pollutants. *Chemosphere* 287: 132081.
8. Khlifi, N., et al. (2022). Non-doped and transition metal-doped CuO nano-powders: structure-physical properties and anti-adhesion activity relationship. *RSC advances* 12(36) 23527–23543.
9. Kumar R, Kumar G, Umar A. ZnO nano-mushrooms for photocatalytic degradation of methyl orange.(2013) Materials Letters. Apr 15(9):100–3.
10. Mukhtar, Mergoramadhayenty, Lusitra Munisa, and Rosari Saleh.(2012).Co-precipitation synthesis and characterization of nanocrystalline zinc oxide particles doped with Cu^{2+} ions.
11. Nohavica, Dušan, and PetarGladkov. ZnO nanoparticles and their applications-new achievements.(2010): *Olomouc, Czech Republic, EU* 10 :12–14.
12. Prasad, Anupama R., et al.(2019). Bio-inspired green synthesis of zinc oxide nanoparticles using Abelmoschusesculentus mucilage and selective degradation of cationic dye pollutants. Journal of Physics and Chemistry of Solids 127: 265–274.
13. Raha, Sauvik, and Md Ahmaruzzaman.(2022).ZnO nanostructured materials and their potential applications: progress, challenges and perspectives. *Nanoscale Advances* 4(8): 1868–1925.
14. Roguai, Sabrina, and Abdelkader Djelloul.(2020). A structural and optical properties of Cu-doped ZnO films prepared by spray pyrolysis. *Applied Physics A* 126(2): 122.
15. Sajjad, Muhammad, et al. "Structural and optical properties of pure and copper doped zinc oxide nanoparticles." *Results in Physics* 9 (2018): 1301–1309.
16. Singhal, Sonal, et al. (2012). Cu-doped ZnO nanoparticles: synthesis, structural and electrical properties. *Physica B: Condensed Matter* 407(8): 1223–1226
17. Soosen Samuel, M., Lekshmi Bose, and K. C. George. (2009). "Optical properties of ZnO nanoparticles. *Academic Review* 16 :57–65.
18. Tiwari, V., Mishra, N., Gadani, K., Solanki, P. S., Shah, N. A., & Tiwari, M. (2018). Mechanism of anti-bacterial activity of zinc oxide nanoparticle against carbapenem-resistant Acinetobacter baumannii. *Frontiers in microbiology, 9*, 1218.

Note: All the figures in this chapter were made by the authors.

34

Transport Properties of n-Type Thallium-Doped Bi_2Se_3 Nanoparticles

Hitesh Chaudhary
Department of Physics, Faculty of Science, The Maharaja Sayajirao University of Baroda, Vadodara, Gujarat, India

Archana Lakhani
UGC-DAE Consortium for Scientific Research, University Campus, Khandwa Road, Indore, Madhya Pradesh, India

P. H. Soni*
Department of Physics, Faculty of Science, The Maharaja Sayajirao University of Baroda, Vadodara, Gujarat, India

Abstract: Bismuth Selenide (Bi_2Se_3) has wide range of applications in thermoelectric and optoelectronic devices. This study is focused on synthesis and characterization of Thallium doped Bi_2Se_3 nanoparticles synthesized via the hydrothermal method. The structural properties were examined using X-ray diffraction (XRD) and Fourier-transform infrared (FTIR) spectroscopy. The optical properties were obtained using Diffuse reflectance spectroscopy (DRS). Additionally, the transport properties were measured using Hall Effect at room temperature and thermoelectric power measurements between room temperature to 160 °C. The developed material is currently being explored for device fabrication for thermoelectric devices.

Keywords: Seebeck effect, Hall effect, Thermoelectric material

1. INTRODUCTION

Thermoelectric (TE) materials are pivotal for solid-state devices, enabling direct conversion of heat into electricity for power generation and refrigeration. Their energy conversion efficiency determined by the dimensionless *'figure of merit'* (ZT) is expressed as $ZT=(S^2\sigma/(K_L+K_e))T$, where the Seebeck coefficient S, electrical conductivity σ, lattice thermal conductivity K_L, electronic thermal conductivity K_e, and absolute temperature T are the key parameters. Enhancing ZT involves increasing the power factor ($S^2\sigma$) and reducing K. and doping have shown promise in improving ZT by leveraging quantum confinement effects and phonon scattering (Shtern et al., 2021). Bismuth selenide (Bi_2Se_3), a narrow bandgap (~0.35 eV) semiconductor with a layered tetradymite structure, is renowned for its thermoelectric properties and optoelectronic applications. Its quasi-layered structure enables nanostructures like nanosheets, which exhibit improved TE performance by reducing lattice thermal conductivity while maintaining electrical conductivity (Heremans et al., 2017). Doping with elements such as Ga or Sn has been reported to influence the electronic and thermal transport properties of Bi_2Se_3; however, the impact of thallium (Tl) doping on Bi_2Se_3 nanoparticles remains unexplored.

This study presents the hydrothermal synthesis of $Bi_{1.94}Tl_{0.06}Se_3$ nanoparticles and provides a comprehensive analysis of their structural, electronic, and thermoelectric properties, offering insights for advanced thermoelectric device applications.

2. EXPERIMENTAL AND CHARACTERIZATIONS

In this material synthesis process (Ota et. al., 2006), a solution was prepared by dissolving 15 mmol of NaOH in 20 ml of deionized water. After that, 9.2 mmol of Se powder was added into the solution while maintaining constant stirring, along with addition of 1.0 ml of hydrazine hydrate. After 1 hour of stirring, 5.9 mmol of $Bi(NO_3)_3 \cdot 5H_2O$ and 0.09 mmol of Tl_2SO_4 titrated in 5 ml of triethanol amine were added to the mixture. After an additional hour of stirring, the mixture was transferred to a 100 ml Teflon-lined stainless-steel autoclave. The

*Corresponding author: ph.soni-phy@msubaroda.ac.in

DOI: 10.1201/9781003684718-34

autoclave was placed into the muffle furnace at 160°C for 24 hours and then allowed to cool naturally to room temperature. The resulting product was collected and thoroughly washed several times with deionized water and ethanol. After that it was allowed to dry at 80°C for 4 hours and subjected to multiple characterization techniques, including X-ray Diffraction (XRD), Field Emission Scanning Electron Microscopy (FE-SEM), Energy-Dispersive X-ray Spectroscopy (EDAX), Fourier Transform Infrared (FTIR) spectroscopy, and UV-Vis Diffuse Reflectance Spectroscopy (DRS), Room-temperature Hall effect measurements, Electrical resistivity was measured using the standard four-probe method and Thermo power measurements were carried out using the differential method.

3. RESULTS AND DISCUSSION

XRD Analysis: X-Ray diffraction pattern of the prepared sample for 24 hrs. at 160°C using hydrothermal synthesis method is shown in figure1. The diffraction pattern shows peaks at (006), (101), (015), (1010), (110), (116), (205), (0210), (1115), (125), and (671). Which are characteristic of the pure hexagonal phase of Bi_2Se_3 (JCPDS card no. 33-0214) and lattice parameters a=b=4.1401Å and c=28.6307Å (space group: R3m) are calculated from Rietveld refinement using the Fullprof suite Software. The information about average crystallite size and strain of nanoparticles are obtained from the Williamson-Hall (W-H) formula, $\beta Cos\theta = \frac{k\lambda}{D} + 4\varepsilon Sin\theta$ where, k is the Scherrer constant (0.94), λ is the wavelength of the X-ray, β is the full width at half maximum (FWHM) of the diffraction peak, ε represents the microstrain, and θ is the Bragg angle determined from the 2θ values corresponding to the peaks in the XRD pattern. According to the Williamson-Hall (W-H) formula, the average crystallite size and strain of the synthesized nanoparticles was calculated to be 10.90 nm and 2.79×10^{-3}, respectively.

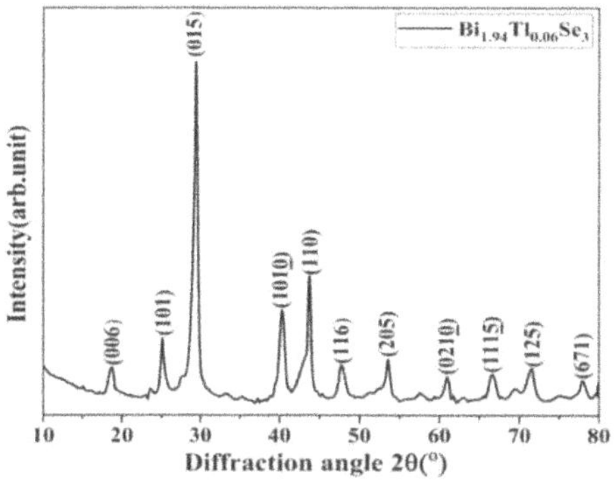

Fig. 34.1 XRD pattern of $Bi_{1.94}Tl_{0.06}Se_3$ nanoparticles

FE-SEM AND EDAX: FE-SEM micrographs of $Bi_{1.94}Tl_{0.06}Se_3$ nanoparticles synthesized via the hydrothermal method, as shown in Fig. 34.3.

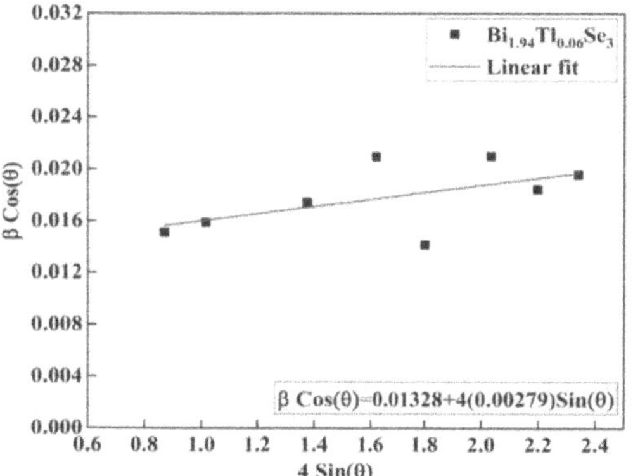

Fig. 34.2 W-H plot of $Bi_{1.94}Tl_{0.06}Se_3$ nanoparticles

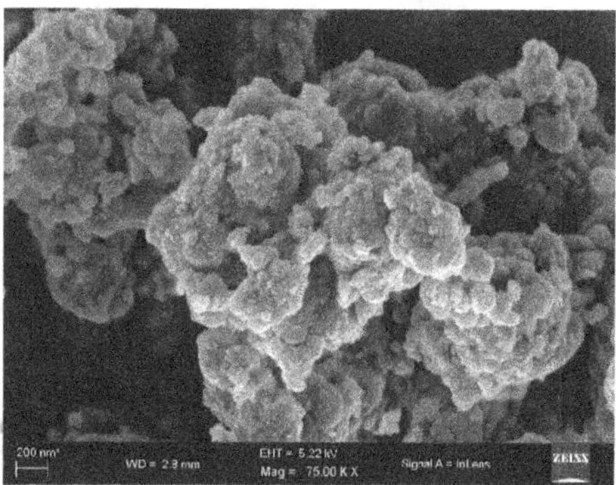

Fig. 34.3 FE-SEM image of $Bi_{1.94}Tl_{0.06}Se_3$ nanoparticles

The surface morphology of $Bi_{1.94}Tl_{0.06}Se_3$ nanoparticles, analyzed at magnifications of 200 nm, revealed a spherical type shape with agglomeration. The micrographs show the agglomeration among the nanoparticles which is likely caused by strong intermolecular interactions such as Van der Waals forces and hydrogen bonding (Endres et. al., 2021). Additionally, the drying stage during sample preparation likely promoted agglomeration due to capillary forces exerted during the removal of the solvent. The elemental composition of the $Bi_{1.94}Tl_{0.06}Se_3$ nanoparticles was evaluated using EDAX, and the results (inset of Fig. 34.4.) confirmed the presence of Bi, Tl, and Se in the expected proportions which confirms the incorporation of Tl was successful in this synthesis process.

Elements	Weight%	Atomic%
Bi M	64.25	41.10
Tl M	0.95	0.62
Se L	34.80	58.28
Total	100	100

0 5 10 15 20
Full Scale 89920 cts Cursor: 0.000 keV

Fig. 34.4 EDAX of $Bi_{1.94}Tl_{0.06}Se_3$ nanoparticles

FTIR: The FTIR spectrum of $Bi_{1.94}Tl_{0.06}Se_3$ nanoparticles shows in the Fig. 34.5. Bi-Se stretching modes are observed at 438 cm^{-1} and 524 cm^{-1} confirms the formation

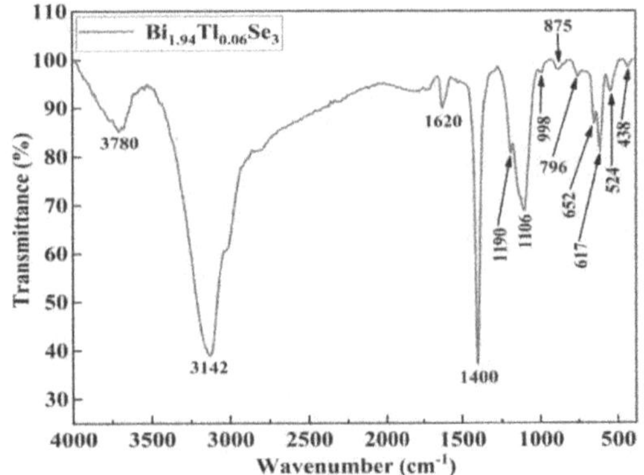

Fig. 34.5 FTIR spectra of $Bi_{1.94}Tl_{0.06}Se_3$ nanoparticles

of the Bi_2Se_3 phase (Abd El-Fattah and Ashoush, 2018). A peak at 617 cm^{-1} corresponds to Bi-O-Bi stretching, indicating minor surface oxidation (Prakash et al., 2024).

A distinct peak at 652 cm^{-1} is attributed to NO_2 wagging, while the peak at 796 cm^{-1} corresponds to Se-O stretching (Vijila et al., 2016). An additional peak at 875 cm^{-1} is due to C-H aromatic stretching, while the peak at 998 cm^{-1} corresponds to CH_2 bending modes, there are two peaks for bending vibrations of C-O at 1110 cm^{-1} and 1190 cm^{-1}, and 1400 cm^{-1} for C-N bending, likely arising from residual organic solvents and reducing agent. The peak at 1620 cm^{-1} is associated with C=O bending, indicating carbonyl group contributions. O-H stretching at 3142 cm^{-1} and C-H stretching at 3780 cm^{-1} indicates surface functional groups.

DRS: The optical properties of thallium-doped Bi_2Se_3 nanoparticles were analyzed using Diffuse Reflectance Spectroscopy (DRS). The reflectance spectra of $Bi_{1.94}Tl_{0.06}Se_3$ nanoparticles shown in inset of Fig. 34.6 were converted into absorbance by applying the Kubelka-Munk function, $[F(R)] = (1-R)^2/2R$ Where, R is the reflectance. The optical band gap of the nanoparticles was estimated using the Tauc plot, which was derived by plotting $[F(R)h\upsilon]^n$ versus $h\upsilon$. Here, n=1/2 for indirect allowed transitions and n=2 for direct allowed transitions, as shown in Fig. 34.6. The band gap of $Bi_{1.94}Tl_{0.06}Se_3$ nanoparticles was determined to be approximately 1.288 eV. The band gap of $Bi_{1.94}Tl_{0.06}Se_3$ nanoparticles is higher than that of bulk Bi_2Se_3 (0.35 eV) due to quantum confinement effects in the nanostructured material and the influence of thallium doping. Incorporation of thallium into the Bi_2Se_3 lattice modifies the electronic structure by altering orbital overlap and introducing defect states (Mustafaeva et al., 2011). These changes shift the conduction and valence bands and improve the optical and electronic properties of the material.

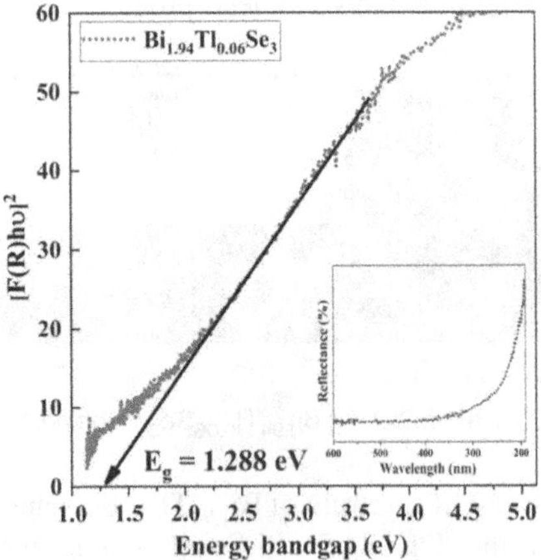

Fig. 34.6 Tauc plot of $Bi_{1.94}Tl_{0.06}Se_3$ nanoparticles

4. TRANSPORT PROPERTIES

(1) Hall effect measurement: Room temperature Hall effect measurements were carried out to analyze the charge carrier type, concentration, and mobility in $Bi_{1.94}Tl_{0.06}Se_3$ nanoparticles. The negative Hall coefficient confirmed n-type conductivity with electrons as the primary charge carriers. The charge carrier concentration was calculated to be 1.86×10^{19} cm^{-3}, and the carrier mobility was approximately 246 $cm^2V^{-1}S^{-1}$. The observed carrier concentration highlights the effect of thallium doping, which reduces the number of free electrons through the replacement of Bi^{3+} with Tl^+ (Kudryashov et al., 2016). This substitution mechanism induces charge compensation by capturing free electrons, thereby decreasing the overall carrier density.

(2) Electrical Conductivity: The electrical conductivity of $Bi_{1.94}Tl_{0.06}Se_3$ nanoparticles was measured over the temperature range of Room temperature to 430 K, as shown in Fig. 34.7. A decreasing trend in electrical conductivity with increasing temperature was observed. This behavior is consistent with pristine Bi_2Se_3. When compared to pristine Bi_2Se_3, the electrical conductivity of $Bi_{1.94}Tl_{0.06}Se_3$ is reduced throughout the measured temperature range. This reduction can be attributed to the substitution of Bi^{3+} with Tl^+, which modifies the electronic structure and reduces the carrier concentration through charge compensation mechanisms that trap free electrons (Kudryashov et al., 2016). Additionally, the altered carrier density increases the material's susceptibility to temperature-dependent scattering processes, such as phonon and defect scattering (Ortiz et al., 2015). These factors collectively lead to the observed reduction in electrical conductivity for $Bi_{1.94}Tl_{0.06}Se_3$ compared to its pristine counterpart.

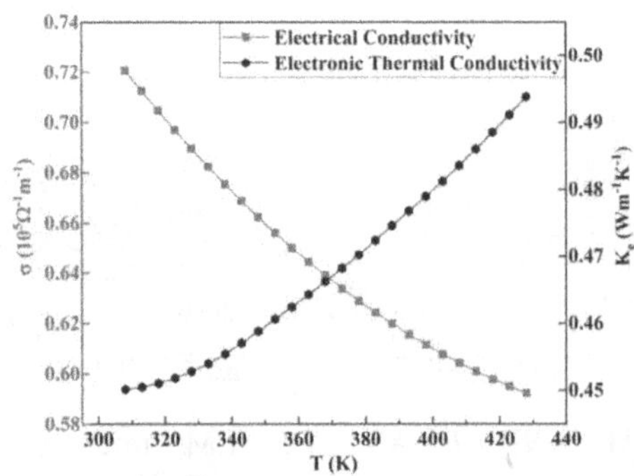

Fig. 34.7 Electrical conductivity and electronic thermal conductivity of $Bi_{1.94}Tl_{0.06}Se_3$ nanoparticles

The total thermal conductivity (K) of a material is the sum of two key components: lattice thermal conductivity (K_L) and electronic thermal conductivity (K_e). The electronic contribution (K_e) is calculated using the Wiedemann–Franz law, which is expressed as $K_e = L_0 \sigma T$, where L_0 is the Lorenz constant, T represents the absolute temperature in Kelvin, and σ represents the electrical conductivity in $\Omega^{-1} m^{-1}$. The Lorenz constant (L_0) can be estimated empirically based on the Seebeck coefficient (S) using an established relation (Putatunda et al., 2019).

$$L_0 = 1.5 + exp\left[-\frac{|S|}{116}\right] \times 10^{-8}\, W\, K^{-2} \quad (1)$$

This relation provides a temperature-dependent estimate of the Lorenz number based on the Seebeck coefficient.

(3) Seebeck coefficient and Power Factor: The Seebeck coefficient (S) and power factor (PF) of $Bi_{1.94}Tl_{0.06}Se_3$ nanoparticles were determined between room temperature and 430 K, as shown in Fig. 34.8**,** respectively, to evaluate their thermoelectric performance. The Seebeck coefficient exhibited an increasing trend with temperature, indicating the enhanced thermoelectric voltage generation due to the incorporation of thallium into the Bi_2Se_3 lattice. At Room temperature, the Seebeck coefficient was approximately -74 μV/K and increased to -93 μV/K at 430 K. This behavior suggests a reduction in carrier concentration and the presence of energy filtering effects, where lower-energy carriers are scattered, leaving higher-energy carriers to contribute to the thermoelectric voltage (Fortulan and Aminorroaya, 2021)

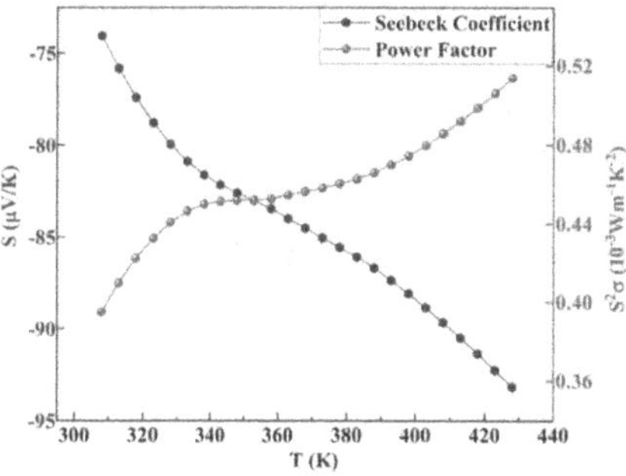

Fig. 34.8 Seebeck coefficient and power factor of $Bi_{1.94}Tl_{0.06}Se_3$ nanoparticles

The power factor (PF), calculated as PF = $S^2\sigma$, exhibited a similar upward trend with temperature, reaching its peak value at 430 K, as shown in Fig. 34.8. The PF reaches a maximum value of 0.51 × 10^{-3} $Wm^{-1}K^{-2}$ at 430 K, compared to 0.40 × 10^{-3} $Wm^{-1}K^{-2}$ at room temperature. The improvement in PF highlights the potential of $Bi_{1.94}Tl_{0.06}Se_3$ nanoparticles for thermoelectric applications within this temperature range.

5. CONCLUSIONS

$Bi_{1.94}Tl_{0.06}Se_3$ nanoparticles were successfully synthesized via the hydrothermal method, and their structural, optical, and thermoelectric properties were systematically investigated. XRD confirmed the hexagonal crystal structure, while FTIR and EDAX verified the presence of Bi-Se stretching and stoichiometric composition. DRS analysis revealed a band gap of 1.288 eV for $Bi_{1.94}Tl_{0.06}Se_3$. Thermoelectric measurements showed an increase in the Seebeck coefficient and power factor with temperature, reaching maxima of -93 μV/K and 0.51 ×10^{-3} $Wm^{-1}K^{-2}$ for $Bi_{1.94}Tl_{0.06}Se_3$ at 430 K. These findings demonstrate the potential of $Bi_{1.94}Tl_{0.06}Se_3$ for thermoelectric applications.

ACKNOWLEDGEMENTS

The authors sincerely express thanks to UGC-DAE Consortium for Scientific Research, Indore for financial supports and other facilities.

REFERENCES

1. Abd El-Fattah, Z. M., and Ashoush, M. A. (2018). Structural characterization of pure and magnetic-doped Bi_2Se_3 nanoparticles. J. Mater. Sci.: Mater. Electron. 29:2593–2599.
2. Endres, S. C., Colombi Ciacchi, L., and Mädler, L. (2021). A review of contact force models between nanoparticles in agglomerates, aggregates, and films. J. Aerosol Sci. 153:105719.
3. Fortulan, R., and Aminorroaya Yamini, S. (2021). Recent progress in multiphase thermoelectric materials. Mater. 14(20):6059.
4. Heremans, J. P., Cava, R. J., and Samarth, N. (2017). Tetradymites as thermoelectrics and topological insulators. Nat. Rev. Mater. 2(10):1–21.
5. Kudryashov, A. A., Kytin, V. G., Lunin, R. A., Kulbachinskii, V. A., and Banerjee, A. (2016). Effect of thallium doping on the mobility of electrons in Bi_2Se_3 and holes in Sb_2Te_3. Semiconductors 50:869–875.
6. Mustafaeva, S. N., Asadov, M. M., and Ismailov, A. A. (2011). Effect of thallium doping on the parameters of localized states in p-GaSe single crystals. Inorg. Mater. 47:941–944.
7. Ortiz, B. R., Peng, H., Lopez, A., Parilla, P. A., Lany, S., and Toberer, E. S. (2015). Effect of extended strain fields on point defect phonon scattering in thermoelectric materials. Phys. Chem. Chem. Phys. 17(29):19410–19423.
8. Ota, J. R., Roy, P., Srivastava, S. K., Popovitz-Biro, R., and Tenne, R. (2006). A simple hydrothermal method for the growth of Bi_2Se_3 nanorods. Nanotechnology 17(6):1700.
9. Prakash, M., Kavitha, H. P., Arulmurugan, S., Vennila, J. P., Abinaya, S., Lohita, D., and Suresh, R. (2024). Ag-doped Bi_2O_3 nanoparticles: synthesis, characterization, antibacterial, larvicidal, and photocatalytic properties. J. Sol-Gel Sci. Technol. 1–12.
10. Putatunda, A., and Singh, D. J. (2019). Lorenz number in relation to estimates based on the Seebeck coefficient. Mater. Today Phys. 8:49–55.
11. Shtern, Y., Sherchenkov, A., Babich, A., Korchagin, E., and Nikulin, D. (2021). Thermoelectric properties of efficient thermoelectric materials on the basis of bismuth and antimony chalcogenides for multisection thermoelements. J. Alloys Compd. 877:160328.
12. Vijila, J. J. J., Mohanraj, K., Henry, J., and Sivakumar, G. (2016). Microwave-assisted Bi_2Se_3 nanoparticles using various organic solvents. Spectrochim. Acta Part A: Mol. Biomol. Spectrosc. 153:457–464.

Note: All the figures in this chapter were made by the authors.

Recent Trends in Applied Physics and Material Science – Sudhir Bhardwaj et al. (eds)

SEM and AFM: Synergistic Approaches for Advanced High-Resolution Surface Characterization

Abhilasha Saini*
Atharva College of Engg., Mumbai, Maharashtra, India

Sudhir Bhardwaj
Bikaner Technical University, Bikaner, Rajasthan, India

P.N. Nemade, Pinky Steffi A.
Atharva College of Engg., Mumbai, Maharashtra, India

Abstract: A variety of analytical techniques are employed for materials characterization, tailored to the specific information required. For high-resolution surface analysis, Atomic Force Microscopy (AFM) and Scanning Electron Microscopy (SEM) are two widely utilized methods, both capable of resolving surface structures at the nanometer scale. Despite their shared resolution capability, the mechanisms of image formation differ significantly, yielding complementary insights into surface morphology. The integration of SEM and AFM in modern analytical laboratories is increasingly prevalent. This article explores the distinct surface measurements provided by each technique, highlighting their complementary nature.

Keywords: AFM, SEM, Surface morphology etc.

1. INTRODUCTION

The inaugural (SEM) had been engineered during the year 1938, who adapted and took the help of the falling electron shaft of a T E Microscope to create what could be considered as (STEM). Since its inception, the SEM has undergone numerous refinements, leading to a remarkable improvement in resolution from 50 nm in 1942 to approximately 0.7 nm in contemporary instruments. In addition to its morphological imaging capabilities, the SEM has evolved to detect a range of signals for example the X-radiations, the electrons scattered backward, cathodoluminescence and some other probes too, that facilitate compositional analysis. The Emergence of Atomic Force Microscopic technique subsequent to the development of the (STM) was spearheaded by Binnig and Rohrer. The microscopy based on atomic force principle recommends and offers a multi-dimensional exterior imaging with a quite small nano level edge ward and sub-angstrom erected resolution, applicable to both insulating and conductive materials. This technological breakthrough paved the way for the development of scanning probe microscopy (SPM), a comprehensive suite of techniques involving the inspection through a razor edged prong across a specimen facet, with uninterrupted surveilling of prong-specimen interactivities to achieve elevated-resolution representation.

2. IMAGING PROCESSES

The operates and uses the principle of involving an appropriate potential difference linking the conductive specimen with the wire-filament, which causes the ejection of electrons from wire-filament towards the chosen specimen. During this interaction a void or vacuum surrounding is maintained. These ejected electrons move to the direction of the chosen specimen via a number of electromagnets put in sequential manner. They fall on the specimen to probe it and to scan and investigate and provide the consequential display. The specimen is judged through the beam current and finalized area of the spot. This is done by the adjustments made with the condensing electromagnets for the minimal fluctuations and irregularities like spherical and chromatic. Electron interactivities with the chosen specimen take place in a range of a few nano to micrometers exterior area. It is also

*Corresponding author: abhilashasaini@atharvacoe.ac.in

DOI: 10.1201/9781003684718-35

a

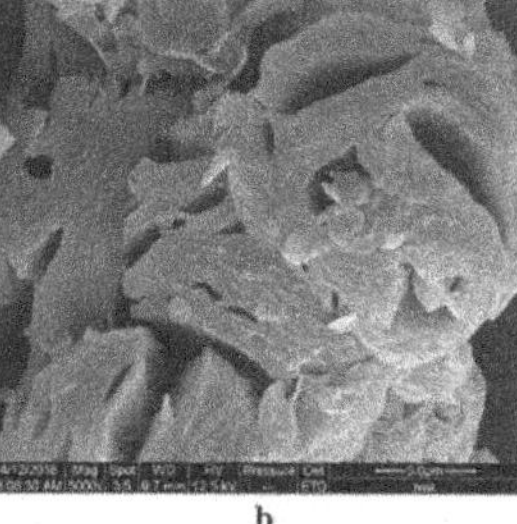
b

Fig. 35.1 Scanning electron micrographs of the samples: (a) SEM image of free HK; (b) SEM image of the HK-PLX (1:4) Solid dispersion

dependent on some electron- shaft parameters as well as the variety of specimens.

The interactivities mostly generate the electrons' emergence through the secondary processes (carry small energy magnitudes of less than fifty electron- volts. These are primarily significant for the surface representation of the sample. The phenomenon behind the emergence of such secondarily generated energy is the energy transfer process of incident electrons to the electrons present in the conduction band, which enables their escape from the surface. Backscattered electrons, scattered high-energy electrons, can also contribute to secondary electron formation. Due to their deeper penetration, backscattered electrons emerge from a larger area of the exterior of chosen specimen. These electrons are precisely gauged with the help of a detector named Everhart-Thornley scintillator-photomultiplier, with the SEM generated representation constructed from the intensity of secondary electron emissions at every rastered spot.

Atomic Force Microscopy sets off via the glancing through a piercing tip, assembled at a flexible cantilever that moves over the exterior of a chosen specimen keeping a sustained but small-scale force. The cantilever is fused with a silicon tip, typically having an end radius of 2–20 nm. Scanning is achieved with the help of a piezoelectricity method which moves the tip of the probe over the chosen specimen in all directions. This interactivities between the probe and the specimen is supervised with the help of a laser beam that suffers reflections and sent to the set of photoconductive diodes. Therefore, it assesses the deflected cantilever positioning and subsequently the oscillation intensity.

AFM is commonly employed in two categories of modes: Contact mode and Tapping Mode. In the first category, the scanning probe assesses the upper exteriors of specimens whereas a response kink regulates the vertical positioning of the scanner for maintaining a continual cantilever deflection, thus ensuring a consistent force (typically 0.1 to 100 nN). The vertical displacement for horizontal and vertical direction points is recorded to form a topographic image. TappingMode involves the cantilever's oscillations on the resonance frequency (~three hundred killoHz) and it gently taps the specimen's exterior at the time of scan. The amplitude of oscillation is maintained via a feedback loop, which records vertical movement to create a topographical map. Tapping Mode's advantage over Contact mode is the reduction of lateral shear forces, making it suitable for imaging soft, fragile, and adhesive surfaces without inducing damage.

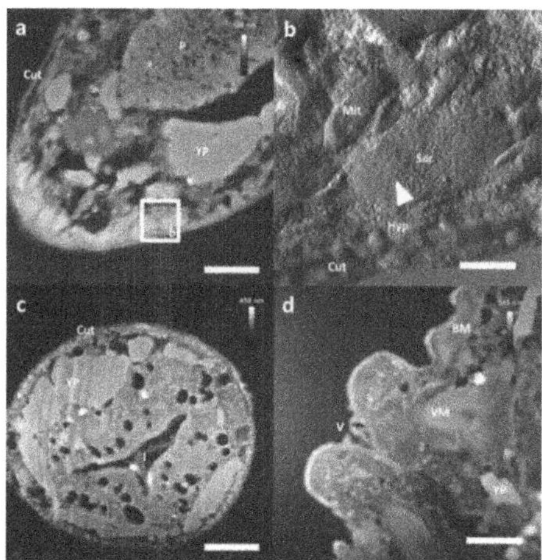

Fig. 35.2 AFM-driven ultra structural examination of caenorhabditis elegans morphology

3. DISCUSSION ON THE MATERIAL CHARACTERIZATION METHODS OF SEM AND AFM

When comparing SEM and AFM, it is important to recognize their underlying similarities despite their apparent differences. In both techniques the scanning tip rasters over the specimen for detecting the interactivities which leads to generating the image representation, with lateral resolutions of comparable scale (although AFM can outperform SEM under certain conditions). Both techniques also exhibit image artifacts that trained operators can identify. However, the mechanism of SEM's

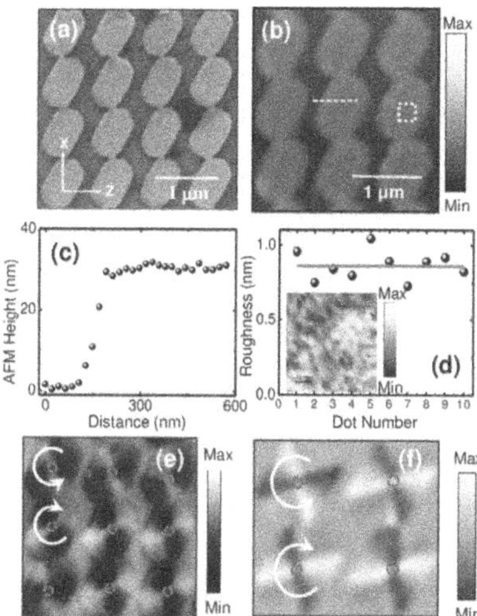

Fig. 35.3 (a) Scanning electron microscopy (SEM) and (b) Atomic force microscopy (AFM) imaging of the sample

working is prolonged to fully grow and arise strategies for artifact recognition and avoidance, AFM's rapid adoption has similarly cultivated a robust understanding of its own artifacts. This article refrains from discussing artifacts unless pertinent to the comparison. Notably, using these complementary techniques allows one to mitigate the artifacts inherent in the other.

However, caution is advised when using integrated AFM/SEM systems, where an AFM is housed within the SEM chamber. A key advantage of AFM makes it better because of its aptness of operation even under normal surroundings i.e. the need of vacuum is not a compulsion. These combined systems often suffer from reduced capabilities, compromising the performance of both instruments.

4. CONCLUSION

SEM and AFM are complementary techniques that, when used together, offer a more comprehensive representation of a surface than either technique alone. Both provide nanometer-scale lateral information, but they differ in capabilities. AFM excels in delivering three-dimensional measurements, including height data with sub-angstrom vertical resolution (<0.5Å), while SEM is adept at imaging rough surfaces because it has a superior deepness of the domain along with a widely-ranged sideward view field. This microscopy is also able to conduct the fundamental investigation via noticing the emitted X-radiation, whereas the atomic force microscopy extends configurationally insights established by the substantial attributes of the specimen.

Distinct operational environments of the two techniques enhance their combined utility. AFM operates without vacuum constraints, allowing it to image samples in fluid or other environments, free from issues like difficult sample preparation or modification. In contrast, SEM's vacuum environment enables techniques such as X-ray analysis. Having both instruments in proximity within an inquisitive establishment broadens the overall scope of capabilities, increasing the facility's flexibility and analytical reach.

REFERENCES

1. Goldstein, J. I., Newbury, D.E., Echlin, P., Joy. D.C., Fiori, C., Lifshin, E.(1981) Scanning Electron Microscopy and X-ray Microanalysis. Plenum Publishing Corp. New York. p. 3.
2. Goldstein, J.(2003). Scanning electron microscopy and x-ray microanalysis, Kluwer Academic/Plenum Publishers, 689.
3. Binning G, Roher H, Gerber C, Weibel E.(1982). "Surface Studies by Scanning Tunneling Microscopy," Phys. Rev. Lett. 49, 57.
4. Reynolds, G.W., Taylor, J.W.(1999). "Correlation of Atomic Force Microscopy Sidewall Roughness Measurements with Scanning Electron Microscopy Line-Edge Roughness Measurements on Chemically Amplified Resists Exposed by X-ray Lithography," J. Vac. Sci. Technol. B 17.2723.
5. Mou, J., Czajkowsky, D.M., Sheng, S., Ho, R., Shao, Z.(1996). "High resolution surface structure of E. coli GroeS Oligomer by Atomic Force Microscopy," FEBS Letters 381.161.

Note: All the figures in this chapter were made by the authors.

Recent Trends in Applied Physics and Material Science – Sudhir Bhardwaj et al. (eds)

Metric *f*(R) Gravity and Black Hole Evolution: A Theoretical Exploration

Suraj Kumar Pati* and Bibekananda Nayak
P.G. Department of Physics, Fakir Mohan University, Balasore, Odisha, India

Abstract: In the present study, Power-law cosmology has been employed to investigate the dynamics of black holes in the metric type *f*(R) gravity obeying Starobinsky's assumption. Specifically, we have examined the variation of black holes size due to accumulation of surrounding energy-matter, named as accretion of black holes. Our finding suggested that a higher accretion rate leads to greater mass as well as longer life span of black holes. Further, it is revealed from our analysis that certain black holes could completely evaporate at a particular value of redshift, regardless of their initial formation mass.

Keywords: Black hole, Starobinsky model, Power-law cosmology, Metric *f*(R) gravity

1. INTRODUCTION

Recent observations on expanding nature of Universe (Riess, 1998), presents a significant challenge to explain theoretically within the scope of standard model of cosmology. One way to address this problem involves modification in gravity such as *f*(R) gravity model (Sotiriou and Faraoni, 2010). This model can describe late time accelerated expansion by considering absence of dark energy component. This theory modifies GTR by introduction of *f*(R) in place Ricci Scalar (R). Metric *f*(R) gravity involving metric formalism, can be often used in cosmological models to describing inflation and late-time cosmic acceleration. On the other hand, a black hole is one of the fascinating objects having strong gravitational field. Generally, black holes are assumed to be formed before the epoch of matter-radiation equality (t_e), approximately 10^{11} seconds after the Big Bang. The maximum mass at formation is estimated as $(M_H)_{t_e} \sim G^{-1}t_e \sim 10^{49}\ gm$. So, some of smaller black holes may have entirely evaporated by now due to Hawking radiation. In this study, we analyse Blackhole dynamics by using Starobinsky-type (Starobinsky, 1982) metric *f*(R) gravity, a modification of Einstein's General Relativity incorporating higher-order curvature corrections. Employing a Lagrangian density $f(R) = R + \frac{R^2}{6M^2}$, where, M^2 is a phenomenological constant, we emphasize the role of the R^2 term in governing large-scale cosmic behaviour, particularly during early inflation. We further explore exact power-law solutions for perfect-fluid-dominated cosmic phases, highlighting their significance as asymptotic or transitional states in Friedmann-Robertson-Walker cosmologies.

2. BASIC SETUP

For a flat FRW universe, Friedmann equation in metric *f*(R) gravity can be expressed as:

$$H^2 = \frac{1}{3\left(\frac{\partial f(R)}{\partial R}\right)} \left[8\pi G\rho + \frac{1}{2}\left(R\frac{\partial f(R)}{\partial R} - f(R) \right) - 3H\dot{R}\frac{\partial^2 f(R)}{\partial R^2} \right], \quad (1)$$

where, $H = \frac{\dot{a}}{a}$ & $R = 12H^2 + 6\dot{H}$ with $\dot{H} = \frac{dH}{dt}$ and $\dot{R} = \frac{dR}{dt}$. Again, cosmological redshift describes the wavelength elongation of light from distant galaxies caused by the universe's expansion. As space stretches, photons

*Corresponding author: surajkumarpati@gmail.com

DOI: 10.1201/9781003684718-36

traveling through it also stretch, resulting in a redshift proportional to the galaxy's distance. This phenomenon is crucial for studying the universe's accelerated expansion, as increasing redshift signifies galaxies receding at an accelerating rate. The connection between redshift and the scale factor (a) through time varies across different cosmic epochs, are given by

$$z = \begin{cases} \left[A\left(\frac{t_e}{t}\right)^{\frac{1}{2}}\right] - 1, (for\ t < t_e) \\ \left[B\left(\frac{t_m}{t}\right)^{\beta_1(t_e)}\right] - 1, (for\ t_e < t < t_m) \\ \left[\left(\frac{t_0}{t}\right)^{\beta_2(t_m)}\right] - 1, (for\ t > t_m) \end{cases} \tag{2}$$

Here, $t_e \approx 10^{11}$ sec, is the radiation-matter equality time, $t_m \approx 3.09996 \times 10^{17}$ sec, is the transition time between decelerated phase and accelerated phase of expansion and $t_0 \approx 4.36 \times 10^{17}$ sec, is the present time of universe, $A = \left(\frac{t_m}{t_e}\right)^{\beta_1(t_e)}\left(\frac{t_0}{t_m}\right)^{\beta_2(t_m)}$ and $B = \left(\frac{t_0}{t_m}\right)^{\beta_2(t_m)}$ are constants. We can derive the values of $\beta_1(t) = \frac{1}{12}\left[\left(11+M^2t^2\right)-\sqrt{\left(5+M^2t^2\right)^2-4M^2t^2}\right]$ and $\beta_2(t) = \frac{1}{12}\left[\left(11+M^2t^2\right)+\sqrt{\left(5+M^2t^2\right)^2-4M^2t^2}\right]$ by using density equation of the fluid filling universe in f(R) model i.e. $\rho = \frac{3\beta^2}{8\pi Gt^2}\left[1-\frac{3(2\beta-1)}{M^2t^2}\right]$ (Pati et al., 2024) and from general power-law (Pati et al., 2020) expression of energy conservation equation $\rho \propto t^{-3\beta(1+\gamma)}$. Where, equation of state parameter γ takes values 0 for matter-dominated and $\frac{1}{3}$ for radiation-dominated eras respectively.

3. BLACK HOLE EVOLUTION IN METRIC *F*(R) MODEL

Black hole mass evolves through two primary mechanisms: accumulation of surrounding energy-matter, via the accretion process and loss of mass-energy due to quantum mechanical effects near the event horizon, termed as Hawking evaporation. These processes are influenced when gravity is modified within the framework of metric formalism.

3.1 Black Hole Accretion

The black hole mass increases during accretion, by obeying the formula:

$$\dot{m}_{acr} = 16\pi G^2 m^2 f\rho \tag{3}$$

Here f is the accretion efficiency used to specify accretion rate of black hole.

The solution of the above differential equation for radiation dominated era can be given by

$$m_{acr} = m_i\left[1+\frac{3f}{2}\left(\left(\frac{z_{acr}+1}{z_i+1}\right)^2-1\right)\right]^{-1}. \tag{4}$$

The variation of black hole accretion mass (m_{acr}) with redshift (z) for various radiation accretion efficiency (f) is given in Fig. 36.1. This figure shows that black hole mas increases with increase in accretion efficiency(f).

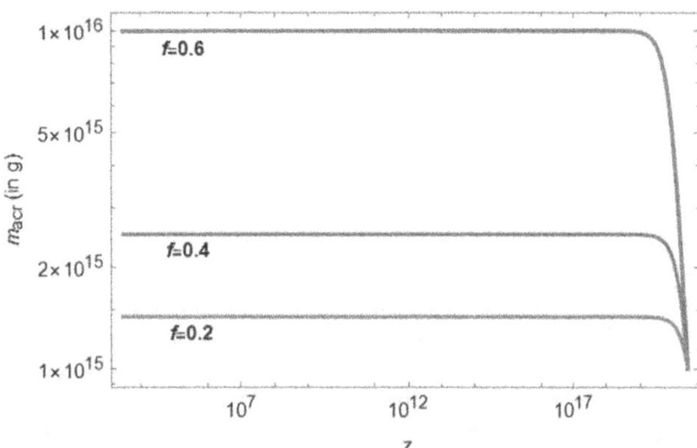

Fig. 36.1 Variation of accreting mass (m_{acr} of m_i = 10^{15} g with redshift (z) for different accretion efficiency (f)

3.2 Black Hole Evaporation

The mass of a black hole decreases via Hawking evaporation, governed by the relation

$$\dot{m}_{evp} = -\frac{a_H}{256\pi^3 G^2 m^2} \tag{5}$$

Here a_H is the dimension less constant related to properties of black hole's horizon.

The solution of the above differential equation for radiation dominated era will be

$$m_{evp} = \left[\frac{1}{m_{acr}} + \frac{a_H}{256\pi^3 G^2}\left(t_0\left(\frac{1}{z_{evp}+1}\right)^{0.72571} - t_e\left(\frac{A}{z_{acr}+1}\right)^2\right)\right]^{-1} \tag{6}$$

The redshift value at the evolution period can be given by the equation below

$$z_{evp} = \left[t_0 \left(\frac{G^2 m_i^3}{3} + \frac{t_e A^2}{(z_i + 1)^2} \right)^{-1} \right]^{1.3779} - 1, \quad (7)$$

We construct Table 36.1 for presenting variation in evaporation redshift for different black holes. We have observed from Table 36.1 that evaporation redshift increases with increase in formation mass.

Table 36.1 Calculation of evaporation redshift (z_{evp}) of black hole for different formation mass (m_i)

z_i	m_i (in g)	z_{evp}
2.817×10^{20}	10^{15}	33.56
1.259×10^{20}	0.5×10^{16}	-0.95541
8.908×10^{19}	10^{16}	-0.99746
3.983×10^{19}	0.5×10^{17}	-0.99999
2.817×10^{19}	10^{17}	-1
8.908×10^{18}	10^{18}	-1

Again, for complete evolution, we have to consider both accretion and evaporation simultaneously and thus the evolution equation can be written as

$$m = m_{acr} + \left[\frac{1}{m_{acr}} + \frac{a_H}{256\pi^3 G^2} \left(t_0 \left(\frac{1}{z_{evp} + 1} \right)^{0.72571} - t_e \left(\frac{A}{z_{acr} + 1} \right)^2 \right) \right]^{-1}.$$

But this equation cannot be solved analytically. Hence, we have kept it for our future study using numerical approximation.

4. CONCLUSION

In this study, we analyse black hole dynamics within the framework of metric-based f(R) model satisfying Starobinsky's assumption. Employing power-law based cosmology to this model, we have predicted the current accelerated expansion of the universe and subsequently investigated black hole evolution. During black holes study we have considered both surrounding energy-matter absorption and Hawking radiation effects. Our analysis also revealed that, black hole mass increases by increase in accretion efficiency during radiation accretion and evaporation redshift increases with increase in formation mass.

REFERENCES

1. Riess, A.G., et al. (1998). Observational evidence from supernovae for an accelerating universe and a cosmological constant. Astron. J. 116:1009–1038.
2. Sotiriou, T.P. and Faraoni, V. (2010). f(R) theories of gravity. Rev. Mod. Phys. 82:451–497.
3. Starobinsky, A.A. (1982). Dynamics of phase transition in the new inflationary universe scenario and generation of perturbations. Phys. Lett. B 117:175–178.
4. Pati, S.K., Swain, S. and Nayak, B. (2024). Cosmological parameters, accelerated expansion of the universe and metric f(R) gravity. Astrophys. Space Sci. 369:72.
5. Pati, S.K., Nayak, B. and Singh, L.P. (2020). Black hole dynamics in power-law based metric f(R) gravity. Gen. Relativ. Gravit. 52:78.

Note: All the figures and tables in this chapter were made by the authors.

Recent Trends in Applied Physics and Material Science – Sudhir Bhardwaj et al. (eds)
© 2026 Taylor & Francis Group, London, ISBN 978-1-041-16452-4

Co-Seismic Ionospheric Effects of Mw 7.5, Noto Peninsula, Japan Earthquake, of 01 January 2024

Sneha Bisht[1], Vishal Chauhan[2], Amit Raj Singh[3]
Department of Physics: Graphic Era Hill University, Dehradun, Uttarakhand, India

Shalini Kumari[4]
Department of Geology: SBAS, SGRR University, Dehradun, Uttarakhand, India

Abstract: This study investigates the co-seismic ionospheric variations observed during the Japan earthquake of Mw7.5 that occurred on 01 January 2024 at 07:10:55 (UTC). For this purpose, the total electron content (TEC) data are computed from three IGS stations, i.e. DAEJ in Korea, MTKA, and USUD in Japan. The TEC variations revealed significant anomalies corresponding to this earthquake. The DAEJ station is 887.4 km from the epicentre, MTKA is 286.7 km, and USUD is the closest station, i.e. 179.15 km from the epicentre. The anomaly was first observed at the USUD station with a change in TEC of about 0.12 TECU after almost 9 minutes of the main event. MTKA Station observed the anomaly after 12 minutes of the main event, with a change in TEC of about 0.065 TECU, and no significant increase in the TEC was observed at station DAEJ. The results are interpreted in terms of the propagation of Acoustic Gravity Waves (AGWs) in the ionosphere during the occurrence of the earthquake. These observations demonstrate the usefulness of TEC monitoring for identifying disturbances in the ionosphere linked to seismic activity. The study reveals that TEC measurements can be a valuable tool in earthquake research, which provides crucial information on the dynamic interactions between seismic events and ionospheric disturbances.

Keywords: AGWs, Total electron content (TEC), Co-seismic ionospheric disturbance (CID), IGS

1. INTRODUCTION

The Ionosphere is the most important layer of the earth's atmosphere, which extends from 60km to 1000km above the earth's surface. It is the only layer of the earth's atmosphere where solar radiation ionizes atmospheric gases, creating a plasma rich in free electrons and ions. The ionosphere is essential for radio communication, GPS navigation, and other satellite-based technologies but is unstable. (Forbes et al., 2000, Rishbeth et al., 2001). Seismic activities, particularly with large earthquakes, can induce significant disturbances in the ionosphere. These perturbations are referred to as co-seismic ionospheric disturbance (CID). These disturbances are triggered by acoustic-gravity waves generated by seismic ruptures and surface displacement (Heki et al., 2011, Liu et al., 2011). There are a variety of propagation characteristics of co-seismic ionospheric disturbances, which includes propagation velocity, period and waveform. (Bolt et al., 1964, Maurya et al., 2016).

Zhang et al. (2021) studied the Alaska earthquake of Mw7.9 in 2018, CID with an amplitude of up to 0.06 TEC units was detected after 10 mins of the main event due to acoustic waves, which were triggered by Rayleigh waves (Zhang et al., 2021). Chen et al. (2014) mentioned in their study the co-seismic ionospheric anomalies on the Lushan earthquake of Mw7.0, which occurred on April 20, 2013, detected using GPS data from 23 reference stations in the region. The disturbances were observed within 15 minutes of the earthquake with an amplitude of 0.24 TEC units caused by acoustic waves propagating upwards. (Chen et al., 2014).

In this study we monitoring co-seismic Ionospheric disturbance caused due to Mw7.5 Japan earthquake of January 01, 2024

[1]snehabisht@gehu.ac.in, [2]vishalchauhan@gehu.ac.in, [3]amitraj.gehu@gmail.com, [4]shalini27888@gmail.com

DOI: 10.1201/9781003684718-37

2. DATA AND METHODOLOGY OF ANALYSIS

The Japan earthquake of Mw7.5 occurred on January 01, 2024, at 07:10:09 (UTC), with the estimated depth of 10km. The epicentre of the earthquake was 37.487°N, 137.271°S located at the northern coast of the Noto Peninsula on the west coast of Honshu, Japan. To study the co-seismic ionospheric disturbance over this area, we used data from three IGS station, i.e. DAEJ at Korea, MTKA and USUD at Japan. We considered these three stations as per the geographical location and the availability of the data. Table 37.1 shows the geographical location and the epicentral distance of the IGS stations: USUD, MTKA, and DAEJ. The GPS data available at these IGS stations in the Rinex format is used to derive the total electron content. To calculate VTEC from the Rinex format, an open software developed by Gopi Seemala is used (seemala.blogspot.in).

In this study, the variation in the VTEC data are observed for PRN- 16, 21 and 27. We applied a bandpass filter of 01 to 10 mHz of 4th order to identify the role of earthquake associated acoustic gravity waves (AGWs) in VTEC data.

Table 37.1 List of IGS stations considered for the study

Sr. No	Station code	Lat (°N)	Lon (°S)	Epicentral Distance
1	USUD	36.13	138.36	179.15
2	MTKA	35.67	139.56	286.71
3	DAEJ	36.39	127.37	887.34

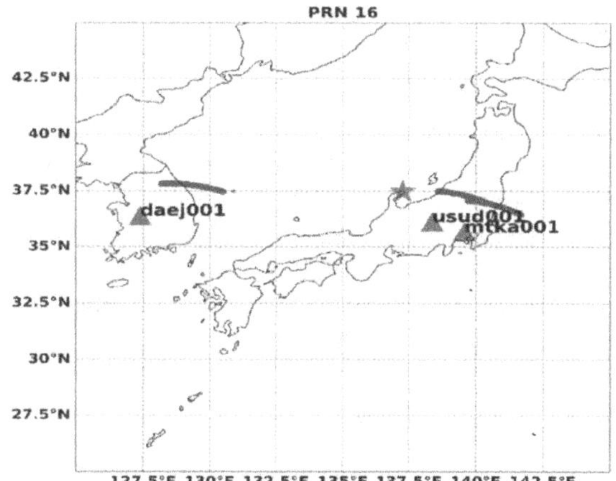

Fig. 37.1 Location of Epicentre and IGS data for PRN-16

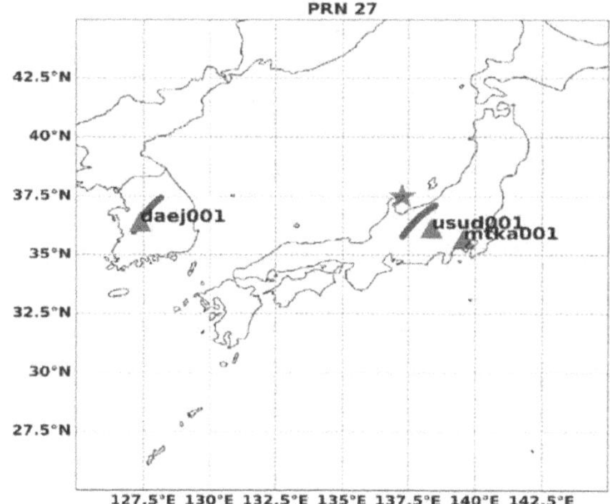

Fig. 37.2 Location of epicentre and IGS data for PRN-27

3. RESULT AND DISCUSSION

The VTEC data corresponding to Mw7.5 Japan earthquake of January 01, 2024, is analysed, which shows the significant variation for PRN – 16, and 27. The time of occurrence of earthquake is 07:10:09 UTC, so we have observed the data between 07:00 UTC TO 08:00 UTC, i.e. one hour data is observed in this study. Figure 37.1 and Fig. 37.2 shows the location of epicentre, IGS stations and IPP locations of PRN-16 and PRN-27 respectively on January 01, 2024. It displays the location of three stations, USUD, MTKA and DAEJ. Here the blue lines represent the IPP locations of PRN-16 and PRN-27. The green triangle indicates the three stations and the three blue diagonal lines that cross the track of PRN-16 and PRN-27. We have only used USUD station in this paper because of the nearness of the station, the disturbance was first reported here.

It can be seen in Fig. 37.3(a) that at USUD station the VTEC values ranges from 8 to 14 TECU and it peaks around 07:19:12 hours UT, approximately 9 minutes after the main event and after that it drops constantly Fig. 37.3(b) represents the filtered VTEC data at station USUD. It is peaking at approximately 0.095 TECU after

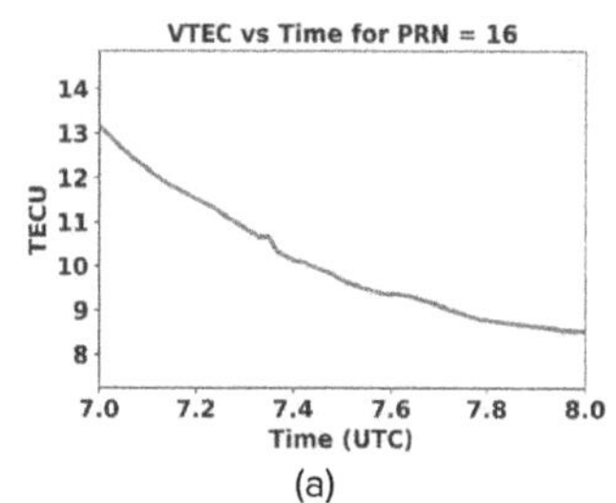

(a)

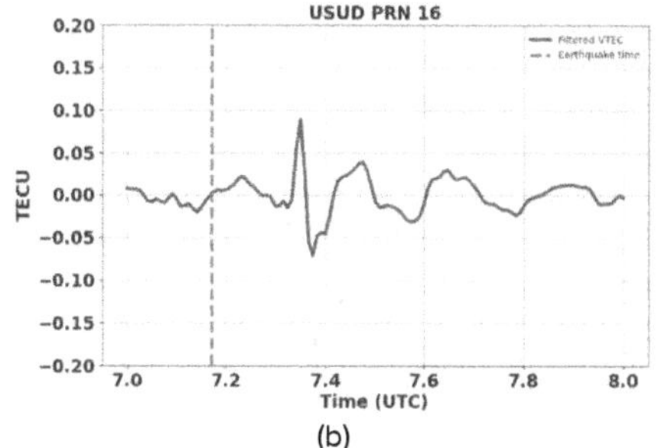

(b)

Fig. 37.3 (a): Variation of VTEC at USUD station for the duration of one hour between 07:00 hrs and 08:00 hrs UT on January 01, 2024, (b): Plot of bandpass filtered VTEC at station USUD for the duration of one hour on January 01, 2024

9 minutes at 07:19:09 hours UT. The values of the filtered VTEC fluctuate between the period -0.2 to 0.2 TECU, showing a prominent peak at USUD.

Figure 37.4(a) shows the VTEC variation at USUD station for PRN-27. The VTEC values range from 6 to 16 TECU and show a peak at 07:19:07 hours UT, approximately 9 minutes after the main shock and after that it keeps on decreasing with time. Figure 37.4(b) shows filtered VTEC at USUD station for PRN-27. It peaks at 07:19:07 hours UT with a value of 0.12 TECU.

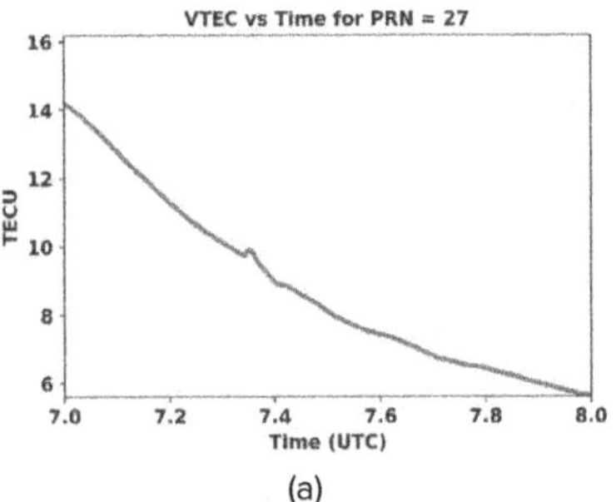

(a)

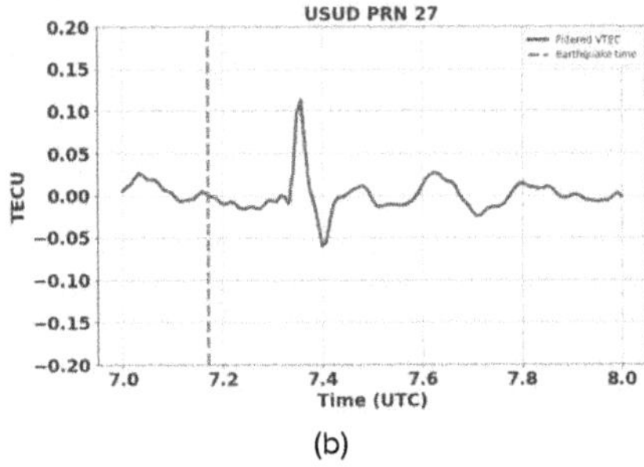

(b)

Fig. 37.4 (a): Variation of VTEC at USUD station for the duration of one hour between 07:00 hrs and 08:00 hrs UT on January 01, 2024 (Remark- This is an original work), (b): Plot of bandpass filtered VTEC at station USUD for the duration of one hour on January 01, 2024 (Remark- This is an original work)

4. CONCLUSION

A strong co-seismic ionospheric disturbance (CIDs) is revealed by the GPS data obtained from USUD and MTKA stations, following the Mw 7.5 earthquake that occurred in Japan, om January 01, 2024. The timing at which the anomaly occurred depends on how far away the stations were from the epicentre. The anomalies were captured 9 minutes within the occurrence of the earthquake. To detect this perturbation, a band pass filter (1mHz-10mHz) was applied to the VTEC. By this study, we demonstrated that a significant earthquake could cause noticeable anomaly in the ionosphere.

REFERENCES

1. Bolt, B. A. (1964). Seismic air waves from the great 1964 Alaskan earthquake. Nature, 202(4937), 1095–1096.
2. Chen, P., Yao, Y., Chen, J., Yao, W., & Zhu, X. (2014). Study of the 2013 Lushan M7. 0 earthquake coseismic ionospheric disturbances. *Advances in Space Research, 54*(11), 2194-2199.
3. Forbes, J. M., Palo, S. E., & Zhang, X. (2000). Variability of the ionosphere. Journal of Atmospheric and Solar-Terrestrial Physics, 62, 685–693.
4. Heki, K. (2011). Ionospheric electron enhancement preceding the 2011 Tohoku-Oki earthquake. Geophysical Research Letters, 38, L17312. https://doi.org/10.1029/2011GL047908
5. Liu, J.-Y., Tsai, H.-F., & Chen, S.-W. (2011). Ionospheric GPS TEC perturbations prior to the 26 December 2004 M9.3 Sumatra–Andaman Earthquake. *Journal of Geophysical Research: Space Physics, 116*(A6).
6. Maurya, A. K., Venkatesham, K., Tiwari, P., Vijaykumar, K., Singh, R., Singh, A. K., & Ramesh, D. S. (2016). The 25 April 2015 Nepal Earthquake: Investigation of precursor in VLFsubionospheric signal, Journalof Geophysical Research: Space Physics, 121, 10,403–10,416. https://doi.org/10.1002/2016JA022721
7. Rishbeth, H., & Mendillo, M. (2001). Pattern of F2-Layer Variability. Journal of Atmospheric and Solar-Terrestrial Physics, 63, 1661-1680. https://doi.org/10.1016/S1364-6826(01)00036-0.
8. Zhang, Y., Liu, X., Guo, J., Shi, K., Zhou, M., & Wang, F. (2021). Co-seismic ionospheric disturbance with Alaska strike-slip Mw7. 9 earthquake on 23 January 2018 monitored by GPS. *Atmosphere, 12*(1), 83.

Note: All the figures and tables in this chapter were made by the authors.

Recent Trends in Applied Physics and Material Science – Sudhir Bhardwaj et al. (eds)

Ionospheric Irregularities Corresponding to the Geomagnetic Storm of 24 April 2023 Using Global Rate of Total Electron Content (TEC) Index (ROTI) Maps

Shristi Singh[1], Vishakha[2], Vishal Chauhan
[a]Department of Physics, Graphic Era Hill University, Dehradun, India

Shalini Kumari[3]
Deparment of Geology SBAS, SGRR University, Dehradun, India

Abstract: In this paper we briefly summarized the work by different workers on ionospheric irregularities using Rate of Total Electron Content Index (ROTI) and scintillation index (S4). We also studied changes in the global ROTI corresponding to a geomagnetic storm of April 24, 2023. The maximum dip in Dst index was observed at 05:00 UTC with a value of -213 nT. Therefore, the global ROTI maps were analyzed from 23 April to 25 April 2023 at 05:10 UTC. It is to be noted here that the global ROTI maps are available at an interval of 10 minutes. We observed that on April 23, there were severe irregularities in the northern region between 60° and 90°N latitudes while in Antarctica, there were mild irregularities. On April 24, ionospheric activities increased in the northern hemisphere while the southern hemisphere showed moderate disturbances. On April 25, the global ionosphere observed a major decline in disturbances, with calm conditions in the northern polar region and mild disturbances near Antarctica. Throughout the event, the mid-latitude and equatorial regions experience the minimum impact. This study shows an understanding of the temporal evolution of ionospheric disturbances associated with geomagnetic storms, shows variation in how total electron content (TEC) behaves in polar and non-polar regions, and emphasizes the significance of ionospheric disturbances effects on satellite-based navigation and communication systems.

Keywords: ROTI, TEC, Scintillation index (S4), Dst

1. INTRODUCTION

The Earth's atmosphere which is ionized by the radiation coming from the sun called ionosphere which in the propagation of radio signals, which are used for communication, navigation, and satellite-based systems. When a GPS signal passes through the ionosphere, the accuracy of measuring the position can be affected due to the variations in electron density. For the ionospheric studies, the most important parameter is the TEC, which measures the free electrons between a satellite and receiver on the Earth. The two bands of enhanced electron density at the magnetic equator is Equatorial Ionization Anomaly (EIA). It plays an important role in ionospheric irregularities and gradients that cause the disturbance in radio signals. These irregularities are often called scintillations. The relationship between the variation in TEC, scintillation index, and EIA is still being investigated for the study of space weather effects and to enhance the reliability of satellite-based systems. Many workers have worked on these parameters are:

1.1 Studies Related to Rate of Change of TEC Index

Samuel et al. (2020) used the RINEX data from 2015 based on ROTI values to measure the ionospheric

[1]shristisingh2298@gmail.com, [2]vishakhachauhan219@gmail.com, [3]shalini27888@gmail.com

DOI: 10.1201/9781003684718-38

abnormalities in Bahir Dar, Ethiopia, from the UNAVCO GNSS station. The high values of ROTI were observed in March, April, and May, while the lowest in June. Their study confirms that the ionospheric irregularities that cause the scintillation at low latitudes. González et al., (2020) analyze the ionospheric irregularities for three geomagnetic storms in Argentina, at EIA. They studied the storms on May 27, 2017, October 12, 2016, and November 7, 2017. The post-midnight disruption was caused by the eastward disturbance dynamo electric fields (DDEF), the May storm was characterized by a protection prompt penetration electric field (PPEF). The October and November months showed high L-band scintillation and spread-F. Kumar et al. (2024) studied that the Equatorial F-region irregularities (EFIs) at Indian region from 2010 to 2023 by using GPS-derived ROTI observations under the different geomagnetic conditions. They found that EFIs are strong depending on solar cycle variations, with more intense occurrences during the maximum solar activity.

1.2 Studies Related to Scintillation Index (S4)

Aggarwal et al. (2013) used GPS data from Rajkot, Gujarat, India, which is close to the EIA crest, to identify low-latitude ionospheric abnormalities caused by a geomagnetic storm that occurred on May 7-8, 2005. They found there is decline in TEC within 12 hours of the storm's beginning and then an increase in TEC during and after the six-hour Sym-H minimum. On 8 May, the values of TEC increased at dusk-time and post-midnight hours. Paul et al. (2015) studied that from the data of September 2011 to April 2012, they calculated TEC and scintillation index data from three sites of India of almost same meridian of 88.5°E. According to the measurements, TEC bites are preceded by periodic oscillations that are limited to a latitude of 19°N to 26°N. Duann et al. (2023) investigated how the GNSS S4 index was affected by the disturbance dynamo (DD) and prompt penetration electric (PPE) fields. The two rather minor geomagnetic storms Commencing on February 3, 2022. We can better understand the ionospheric disruptions thanks to this study. Shucan Ge et al. (2024) compared two Chinese states using three years' worth of ionospheric scintillation data from 2018 to 2020. The statistics show that the occurrence rate (OR) of scintillation events varies with the time of day and season.

1.3 Comparative Studies of ROTI and S4

Abe et al. (2022) used the data of two GPS stations to observe the relation of the S4 index and ROTI at EIA of the West African sector. These studies show that the normalized ROTI and S4 have a substantial correlation, with Cape Verde having higher values than Dakar. Aswathy et al., (2021) examine the relation of EIA and post-sunset ionospheric irregularities using GPS from TEC data from five stations. They studied the base height of the post-sunset F-layer is directly influenced by the amplitude of the EIA crest in the evening. The equatorial spread of F abnormalities is longer when the EIA crests are stronger. Kapil et al. (2021) analyze the relation between the S4 index and ROTI was the main focus of their study on ionospheric disturbances that impact navigation signals. When the correlation is higher than 0.6, ROTI's S4 forecast is still unreliable.

2. DATA METHODOLOGY

For this work, International GNSS Service (IGS) diurnal ROTI maps that use monitor ionospheric irregularities at high latitudes. These maps are available at https://stdb2.isee.nagoya-u.ac.jp/GPS/GPS-TEC/GLOBAL/RMAP/2023/113/05UT.html. The ROTI that have been created using geographic coordinates with 10-minute time intervals.

3. RESULTS AND DISCUSSION

A geomagnetic storm due to interplanetary coronal mass ejection (ICME) caused on April 23–25, 2023, which revealed a complicated interaction between geomagnetic reactions and solar wind dynamics. On April 24, it reached its peak intensity at 5:10 UT with a minimum disturbance storm time (Dstmin) index of -213 nT. During this event, the mid-latitude and equatorial regions experience minimum disturbance in ionospheric conditions. The map represents the ionospheric disturbance due to geomagnetic storm on April 24, 2023, at 05:10 UTC and shows the global distribution of ionospheric irregularities.

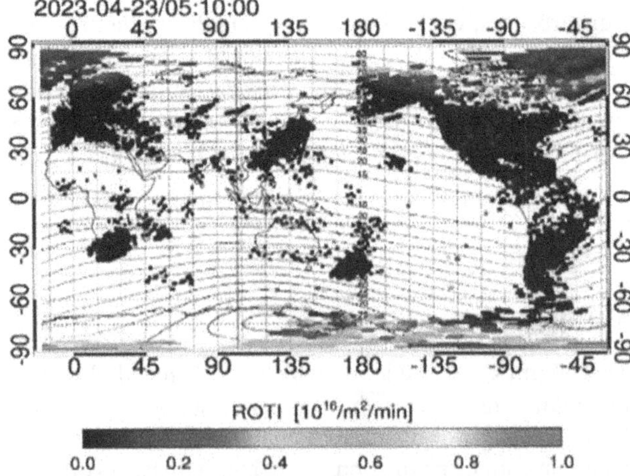

Fig. 38.1 Global ROTI on april 23, 2023, at 05:10 UTC

On April 23, 2023, before geomagnetic storm the northern polar region there is a intense ionospheric irregularities, with a peak in the values of ROTI. The affect the regions are the northern part of Canada, Greenland, and some parts of Europe with latitudes 60° to 90°N and longitudes 90°W to 45°E with mild disturbance near Antarctica with latitudes of 60°S to 90°S.

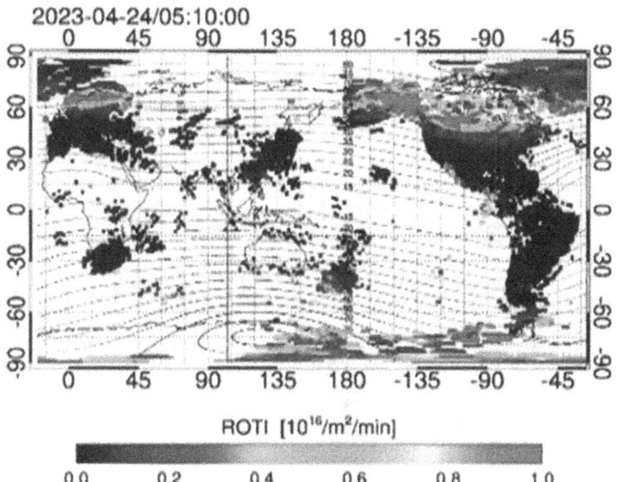

Fig. 38.2 Global ROTI map for April 24, 2023, at 05:10 UTC

On April 24, 2023, in the region around North America and some parts of Europe, there is an increase in ionospheric activities with latitudes of 60° to 90°N, but it moves towards the east a little bit with less disturbance near Antarctica, as shown in blue-colored patches.

The global ionospheric conditions On April 25, 2023, shows a decline in ionospheric disturbances. There were minor irregularities in the region near Antarctica with latitudes of 60°S to 90°S.

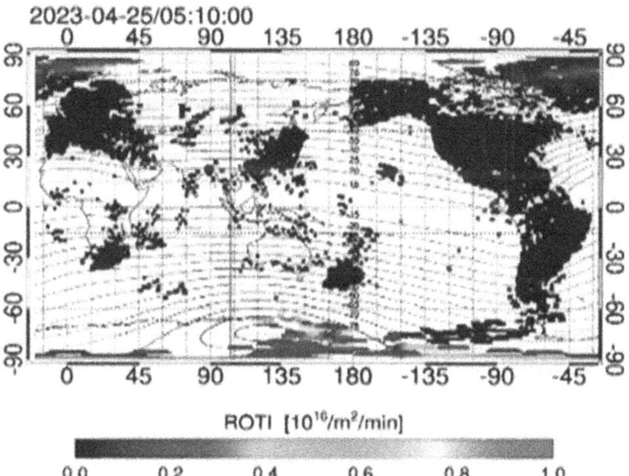

Fig. 38.3 Global ROTI map for april 25, 2023, at 05:10 UTC

4. CONCLUSION

The study shows how geomagnetic storms influence global TEC variations. Here, we analyzed the effect of a geomagnetic storm on global ionospheric disturbance on April 24, 2023, using ROTI for three consecutive days. On April 23-24, the northern polar region experienced the intense ionospheric disturbance. It peaked on April 24 and dropped by April 25, with minimum impact near the southern pole while the mid-latitude and equatorial regions experience the minimum impact. The northern polar region, especially Canada, Greenland, and some parts of Europe, is more prone to ionospheric disruptions during geomagnetic storms, which have consequences for satellite-based communication and navigation systems

REFERENCES

1. Mulugeta, S., & Kassa, T. (2021). Nighttime ionospheric irregularities inferred from rate of total Electron Content Index (ROTI) values over Bahir Dar, Ethiopia. *Advances in space Research, 67*(4), 1261–1266. https://doi.org/10.1016/j.jastp.2020.105309.
2. Abe, O. E., Migoya-Orué, Y. O., & Radicella, S. M. (2023). Correlation analysis of normalized ROTI and S4 as observed during different geomagnetic conditions of the 2013 September equinox over the stations within EIA African sector. *Advances in Space Research, 72*(3), 762–774. https://doi.org/10.1016/j.asr.2021.09.034.
3. Duann, Y., Chang, L. C., & Liu, J. Y. (2023). Impact of the February 3–4, 2022 geomagnetic storm on ionospheric S4 amplitude scintillation index: Observations and implications. *Advances in Space Research, 72*(10), 4379–4391. https://doi.org/10.1029/2022SW003216.
4. De Lourdes González, G. (2021). Spread-F occurrence during geomagnetic storms near the southern crest of the EIA in Argentina. *Advances in Space Research, 67*(3), 1058–1084. https://doi.org/10.1029/2020JA028260.
5. Aswathy, R. P., & Manju, G. (2021). Equatorial ionization anomaly crest magnitude and its implications on the nocturnal equatorial ionospheric plasma irregularity characteristics. *Advances in Space Research, 68*(10), 4129–4136.https://doi.org/10.1016/j.asr.2021.07.019.
6. Kumar, S., & Pathan, B. M. (2024). Equatorial F-region irregularities and geomagnetic storms: A study using ROTI observations along the Indian sector from 2010 to 2023. Journal of Geophysical Research: Space Physics, 129(1), e2023JA031074. https://doi.org/10.1029/2023JA031074.
7. Aggarwal, V. K., & Sharma, A. K. (2013). Low-latitude ionospheric response to the geomagnetic storm of 7–8 May 2005 over India. Annales Geophysical, 31(4), 703–715. https://doi.org/10.5194/angeo-31-703-2013.
8. Ge, S., Li, H., Zhang, S., Zhu, M., Li, J., Xu, B., ... & Wu, J. (2024). On the properties of lower mid-latitudes ionospheric scintillation observed over Chengdu, China. *Advances in Space Research.* https://doi.org/10.1029/2023RS007216.
9. kapil, C., Seemala, G. K., Shetti, D. J., & Acharya, R. (2022). Reckoning ionospheric scintillation S4 from ROTI over Indian region. *Advances in Space Research, 69*(2), 915–925. https://doi.org/10.1016/j.asr.2021.08.006.
10. Paul, A., DasGupta, A., & Ray, S. (2015). Characteristics of periodic formations of total electron content over the Indian equatorial and low-latitude region. Annales Geophysical, 33(8), 1031–1044. https://doi.org/10.5194/angeo-33-1031-2015.

Note: All the figures in this chapter were adapted from (https://stdb2.isee.nagoya-u.ac.jp/GPS/GPS-TEC/GLOBAL/RMAP/2023/113/05UT.html)

Recent Trends in Applied Physics and Material Science – Sudhir Bhardwaj et al. (eds)

Analysis of Major Geomagnetic Storm Events and their Impact on Global Total Electron Content Variations during January to June 2023

Vishakha[1], Vishal Chauhan[2], Amit Raj Singh, Shishupal Singh
Department of Physics, Graphic Era Hill University, Dehradun, India

Rakesh Singh
Department of Petroleum, Graphic Era Deemed to be University, Dehradun, India

Abstract: This study investigates the impact of geomagnetic storms on total electron content (TEC) fluctuations between 01 January 2023 and 30 June 2023. Three significant storms occurred during this period when Dst value dropped to -132 nT on 27 February, -163 nT on 24 March, and -213 nT on 24 April 2023. Significant TEC enhancements were found specifically in the equatorial ionization anomaly (EIA) and low-latitude regions by analysing global TEC (gTEC) maps. During the storm on 27 February 2023, an apparent rise of 20 to 30 TECU was found in TEC values in the Asian region (20° to 30° latitudes and -120° to -90° longitudes). Similarly, on 24 March 2023 and 24 April 202, the TEC values reached to 40 TECU over the latitudes in EIA region. These results are interpreted in terms of fountain effect at EIA region as during geomagnetic storms, the equatorial electrojet and fountain effect are amplified, resulting in higher electron density and enhanced ionization in certain regions. Additionally, our research points to the possibility of travelling ionospheric disturbances (TIDs), which may be a factor in the localized increases in TEC. This study provides significant new understanding about the behaviour of the ionosphere during geomagnetic disturbances and highlights the significance of observing TEC variations in order to understand the impressions of space weather on global communication and navigation systems.

Keywords: TEC, EIA, Ionosphere, CME

1. INTRODUCTION

When the Earth's magnetic field interact with the solar Winds, it produces a large-scale disruption in the Magnetosphere which it termed as geomagnetic storms. In general, fast-moving solar wind streams and CME (coronal mass ejections) are the major source of geomagnetic storms. These solar storms produce various phenomenon, those include compression of the magnetosphere, emission of energetic particles, and formation of electric currents in our atmosphere. The variations in total electron content, further reffered as TEC, is a notable factor to understand the effects of geomagnetic storms. Many technologically based services like Satellite communication, navigation, and positioning services that depend on ionospheric signals for precise operation can also be affected by these variations. Among the various phenomenon in the ionosphere, the EIA is especially vulnerable to these disruptions. It is a distinct phenomenon of the ionosphere, that occurs at two zones located approximately 15° north and south of the magnetic equator with high electron densities. These areas emerge as a result of the EEJ (equatorial fountain effect), which is caused by the interaction of the terrestrial magnetic field with the upward motion of ionospheric plasma. This phenomenon increases the electron density over these latitudes. Mannucci et al. (2005) examined the global behaviour of the ionosphere during geomagnetic storm that

Corresponding author: [1]vishakhachauhan219@gmail.com, [2]vishalchauhan@gehu.ac.in

DOI: 10.1201/9781003684718-39

occurred in October 2003 known as "Halloween Storms". They observed a significant increase in TEC, particularly at mid-latitudes, and complex reactions were detected over the equatorial regions. Trivedi et al. (2011) also studied the effects of magnetic storms occurred in year 2005 and 2007 using vertical TEC values obtained from a GPS station located at Bhopal, India. A significant change in VTEC was observed in the study. A fine TEC variation took place due to modification in the thermospheric composition caused on by Joule heating. It was also observed in the study that there is rise in VTEC values in the course of storm's main phase. VTEC generally increased and showed both positive and negative variations during the recovery phase. Additionally, it was found that Travelling Ionospheric Disturbances (TIDs) were not observed during the storm occurred on August 24, 2005, however TIDs were found during the storm occurred on May 15, 2005. Ranjana et al. (2024) examined the effects of severe magnetic storms occurred in March and April 2023 over the Indian longitude sector. The concluded results of the study showed the existence of Disturbance Dynamo Electric Fields (DDEFs) towards west during the storm's primary and recovery phases, that significantly influenced the TEC variations. Further, a negative storm effect outside the EIA whereas positive over the equatorial region were observed in the course of the recovery of storm. Kumar et al. (2016) noticed the creation of the plasma bubble during geomagnetic storm and suggested that it can be triggered by the Rayleigh–Taylor instability growing due to the storm-induced electric field.

In our study, we focused and examined the effects of geomagnetic storms those occurred from January 2023 to June 2023 over and around the Equatorial region by using gTEC global maps.

2. DATA METHODOLOGY

For this work, the GNSS-derived gTEC maps are utilized to observe spatial TEC fluctuations, with a focus on the Equatorial area. These maps can be found at https://aer-nc-web.nict.go.jp/GPS/DRAWING-TEC/. There are four different types 2-D namely aTEC (absolute TEC) maps, rTEC (ratio of TEC difference) maps, dTEC maps (detrended TEC) and fourth ROTI maps (rate of TEC index) that have been created using geographic coordinates with 10-minute time intervals using RINEX files. Geomagnetic activity is measured using indices including Dst index, Ap index, and the Kp index, which provide details on the duration and intensity of geomagnetic disturbances (https://wdc.kugi.kyoto-u.ac.jp/wdc/Sec3.html).

3. RESULTS AND DISCUSSION

Figure 39.1 depicts the temporal fluctuations of various geomagnetic indices namely mean Dst index, SKp index, and SAp index. The first panel of the figure represent the

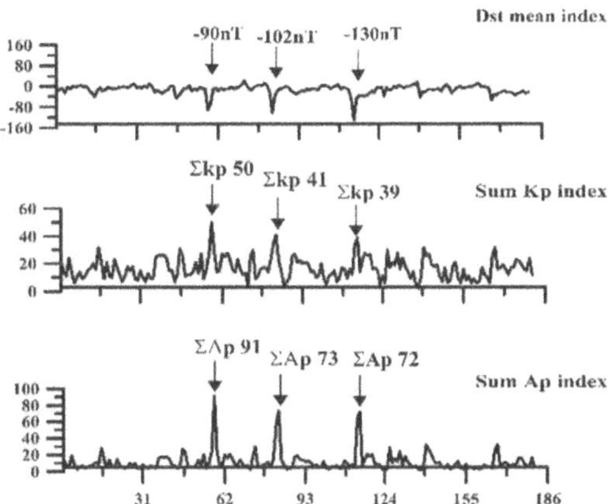

Fig. 39.1 The variation of geomagnetic indices from 01 January to 30 June 2023. (Remark- This is an original work)

fluctuations of mean Dst index. The Dst value dramatically drops to -132 nT at 1300h UTC on February 27, secondly, Dst value takes a drop to -163 nT on 24 March at 0300h UTC and thirdly a dip of -213nT on 24 March at 0600 h UTC. Second panel shows the variation of SKp index which measures the total planetary geomagnetic activity. Third panel shows the variation of SAp index is used to measure the geomagnetic activity and likewise shows peaks during these storm events. For the storm on February 27, the maximum SAp value is 91. This is followed by 73 on March 24, and 72 on April 24. The combination of these indicators indicates the strength and length of geomagnetic activities took place in the first half of year 2023, with the lowest Dst value and the greatest Ap indicating the most severe storm in February.

Event 1: 27 February 2023

Figure 39.2 shows variations in TEC corresponding to the geomagnetic storm that took place on 27 February 2023. In this figure, variation of aTEC is shown for three days, i.e. 26, 27, and 28 February at 1300 h UTC. The chosen grid for aTEC maps is lying between -30° to 30° latitude and 0 to 180° longitude. On February 26, the aTEC map shows moderate TEC levels and stable conditions. But during the storm on February 27, an apparent rise in aTEC of about 20 – 30 TECU is observed in the areas indicated in yellow rectangles. In a particular region the electron density increase can be attributed to the disruption of the ionosphere caused by the geomagnetic storm. Such increases are commonly reported during geomagnetic storms, and the highlighted location is most likely connected with the EIA.

Event 2: 24 March 2023

The Fig. 39.3 shows the aTEC maps for three days from 23 to 25 March at 0300 UTC. On 23 March, TEC levels were moderate however, by 24 March, there had been notable increases in the particular regions marked by yellow rectangle, mostly inside the Equatorial regions, and on

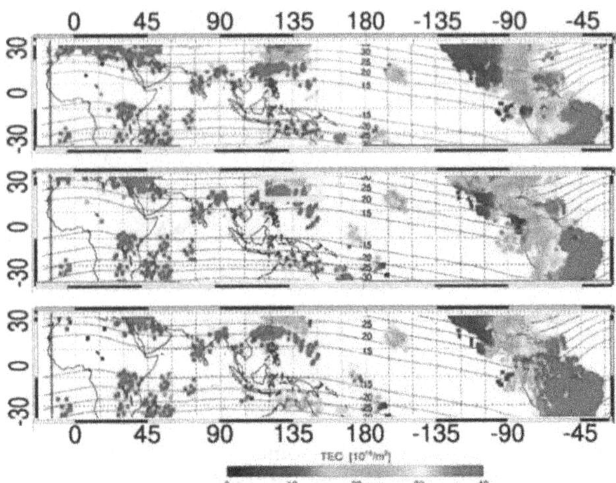

Fig. 39.2 The aTEC maps corresponding to the geomagnetic storm of February 27, 2023 from 26 – 28 February 2023 at 1300 h UTC. The yellow rectangles are highlighting the anomalous aTEC values

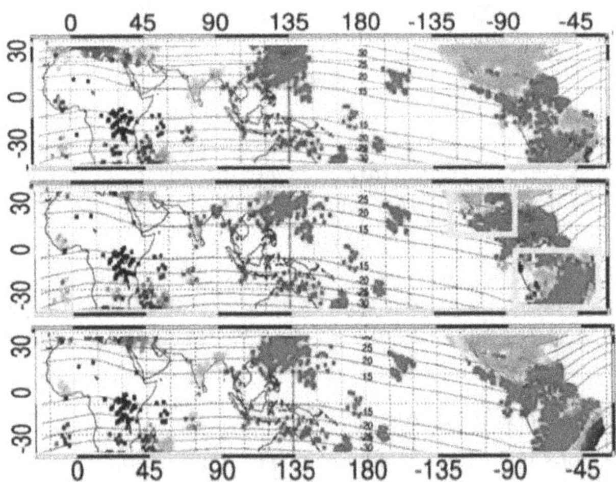

Fig. 39.3 Similar to the Fig. 39.2 but corresponding to the geomagnetic storm of 24 March 2023

25 March, these increases had continued and intensified. The enhanced TEC values of about 35 to 40 TECU are observed in the highlighted regions. The enhanced TEC over the equatorial and low-latitude regions is probably due to increased geomagnetic activity, which enhanced the equatorial electrojet and fountain effect. This led to greater ionisation. The geomagnetic storm's travelling ionospheric disturbances (TIDs), which are especially noticeable in these areas because of the unique dynamics of the EIA during such storms, may also have an impact on this localised TEC increase.

Event 3: 24 April 2023

The Fig. 39.4 shows variation of aTEC during 23 to 25 April at 0600 hrs UTC corresponding to the geomagnetic storm of 24 April 2023. The maps highlight substantial

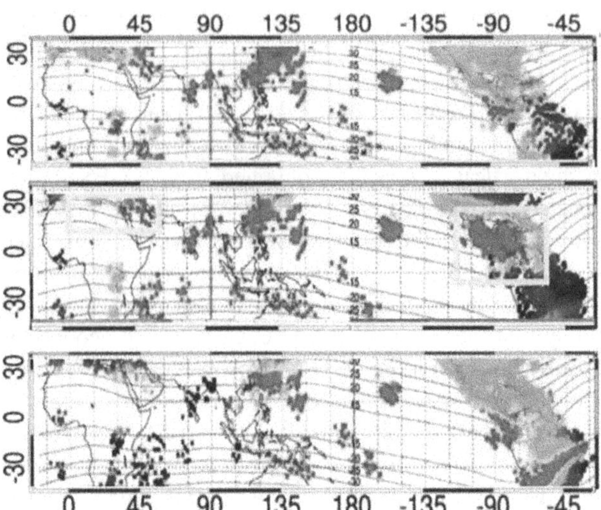

Fig. 39.4 Similar to the Fig. 39.3 but corresponding to the geomagnetic storm of 24 April 2023

TEC enhancements on storm day, particularly over Eastern Hemisphere. During the geomagnetic storm the TEC enhanced at marked region of Africa and America. This area may have been more directly impacted by stronger electric fields and amplified ionospheric currents, like the Equatorial Electrojet. These discrepancies are also influenced by changes in solar radiation and local time during the storm event. This is because locations exposed to daylight receive more solar radiation, which raises TEC values and increases ionization.

4. CONCLUSION

The geomagnetic storms that occurred on 27 February, 24 March and 24 April 2023, caused significant disturbances to the TEC values of the ionosphere, especially over the low-latitudes and equatorial region. We found that the TEC increases significantly during these occurrences demonstrate the strong influence of geomagnetic activity on the Equatorial Ionisation Anomaly (EIA). The area showing large TEC enhancement are indicating of heightened EEJ and fountain effects during storms, leading to higher electron densities. These results demonstrate the crucial role that geomagnetic storms particularly in regions covered by the EIA affecting ionospheric conditions. Abdu et al. (2006) also observed the enhancement in the generation of plasma bubble during geomagnetic storms led to increased TEC fluctuations in the equatorial region. Tsurutani et al. (2004) found that strong IEFs can cause large-scale uplifts in the dayside ionosphere, especially in the Equatorial region, which raises electron densities and TEC values. These results demonstrate the strong connection between ionospheric dynamics and solar wind-magnetosphere interactions.

REFERENCES

1. Abdu, M. A., Batista, I. S., & Reinisch, B. W. (2006). Ionospheric equatorial plasma bubbles: Observations and modeling under quiet and disturbed conditions. *Journal of Atmospheric and Solar-Terrestrial Physics*, 68(7), 766–777.
2. Kumar, S., Chen, W., Liu, Z., & Ji, S. (2016). Effects of solar and geomagnetic activity on the occurrence of equatorial plasma bubbles over Hong Kong. *Journal of Geophysical Research: Space Physics*, *121*(9), 9164–9178.
3. Lissa, D., Srinivasu, V. K. D., Prasad, D. S. V. V. D., & Niranjan, K. (2020). Ionospheric response to the 26 August 2018 geomagnetic storm using GPS-TEC observations along 80 E and 120 E longitudes in the Asian sector. *Advances in Space Research*, *66*(6), 1427–1440.
4. Mannucci, A. J., Tsurutani, B. T., Iijima, B. A., Komjathy, A., Saito, A., Gonzalez, W. D., Guarnieri, F. L., Kozyra, J. U. and Skoug, R. (2005). Dayside global ionospheric response to the major interplanetary events of October 29–30, 2003 "Halloween Storms." *Geophysical Research Letters* , 32(12).
5. Rajana, S. S. K., Panda, S. K., Jade, S. et al. (2024). Impact of two severe geomagnetic storms on the ionosphere over Indian longitude sector during March-April 2023. *Astrophysics and Space*, 369:3.
6. Thiruvarangan, V., Rajavarathan, J., Panda, S. and Swarnalatha Jayakody, J. (2024). Geomagnetic storm effect on equatorial ionosphere over Sri Lanka through total electron content observations from continuously operating reference stations network during Mar–Apr 2022.
7. Trivedi, R., Jain, A., Jain, S. and Gwal, A. K. (2011). Study of TEC changes during geomagnetic storms occurred near the crest of the equatorial ionospheric ionization anomaly in the Indian sector. *Advance in Space Research*, 48(10): 1617–1630.
8. Tsurutani, B., Mannucci, A., Iijima, B., Abdu, M. A., Sobral, J. H. A., Gonzalez, W., ... & Vasyliunas, V. M. (2004). Global dayside ionospheric uplift and enhancement associated with interplanetary electric fields. *Journal of Geophysical Research: Space Physics*, *109*(A8).

Note: All the figures in this chapter were made by the authors.

Recent Trends in Applied Physics and Material Science – Sudhir Bhardwaj et al. (eds)
© 2026 Taylor & Francis Group, London, ISBN 978-1-041-16452-4

Response of GNSS Derived Total Electron Content (TEC) During Geomagnetic Storm of 12 August 2024

Shishupal Singh[1], Vishal Chauhan[2], Vishakha
Department of Physics, Graphic Era Hill University, Dehradun

Rakesh Singh
Department of Petroleum Engineering, Graphic Era (Deemed to be University), Dehradun

Abstract: This study examines the effects of recent geomagnetic storm of 12 August 2024 (Dst = -203 nT) on total electron content (TEC) over and near geographical equator region. To study this event, the TEC data are computed from three IGS stations at different latitudes namely Diego Garcia Island (UK), Bangalore (India) and Vacoas (Mauritius). The stations considered in this study are representing equatorial and low latitudes. The TEC variations on the event day are compared with averaged TEC values of quietest days of the month. The results show that there is no significant change in TEC at equatorial station, however, changes in TEC increases with increase in latitudes. The percentage change of the difference on the event day TEC and averaged quietest days TEC is also calculated and it is found to be varied from 14% to 38%. This change is observed just after the event for both low latitude stations depending on their latitudes but sudden fall is found at the equatorial station. In conclusion, the larger TEC variations were observed at higher latitudes. This suggests the presence of fountain effect in the EIA region. This research work points out the vulnerability of ionospheric TEC arises due to geomagnetic storms and its spatial dependence. This provides insight into the ionospheric character, which could be beneficial for ionospheric corrections required for satellite communication and navigation.

Keywords: GNSS, TEC, Geomagnetic storm

1. INTRODUCTION

Earth is surrounded by a layer of air, called atmosphere, which is further classified into various layers as per their composition and physical changes. Ionosphere is a layer having a complex structure that is affected by various terrestrial and extra-terrestrial events. This layer consists of ions, plasma and electrons generated due to exposition of solar flux. The study and impact of these events on ionosphere is very crucial to understand for humans as navigation and communication signals those may affect and deviate due to these events. These events include geomagnetic storms, solar flare, earthquakes, coronal mass ejections (CMEs) etc. A geomagnetic storm is such an event that occurred due to sudden change in Sun's activity and results in change in magnetic field of Earth. A geomagnetic storm can change the structure of ionosphere that is measured in terms of total electron content (TEC). TEC values can be obtained from ground based GNSS stations located at various places across the globe. However, TEC values depend on various factors such as season, local time, latitude and longitude and above all the geomagnetic disturbances (J. Park et al., 2011, G. Savastano et al., 2019). Many indices are developed by scientific community to understand the intensity of a geomagnetic storm such as Kp, Ap and Dst indices and many more. Paziewski et al. (2022) found in his study that these disturbances are more observed at mid and high

Corresponding author: [1]shishupal_k2002@yahoo.com, [2]vishalchauhan@gehu.ac.in

DOI: 10.1201/9781003684718-40

latitudes. Blagoveshchensky et al. (2018) observed that there is no direct dependency of TEC during geomagnetic storms at different latitudes, in their study. Adebiyi et al. (2014) studied the impacts of geomagnetic storms occurred in year 2000 over southern hemisphere of the earth. They observed significant variation in TEC depending upon the season. The reason behind this behaviour was attributed to thermospheric composition change and circulation in atmosphere. Adimula et al. (2016) had the study about the effect of geomagnetic storms on ionosphere in both northern and southern hemisphere. They observed that during main phase of storm the TEC values increases at all latitudes. The variation occurred during all the storms but the amount of variation was mostly dependent on seasons.

This study deals with the recent strong geomagnetic storm occurred on 12 August 2024 by analysing TEC data considering three stations over and near geographic equator. The purpose of this study is to understand the effects of geomagnetic storms on equatorial locations.

2. DATA AND METHODOLOGY

In this study, we used data obtained from the source site (https://igs.org/data/) for three stations of different latitudes namely Diego Garcia Island (UK), Bangalore (India) and Vacoas (Mauritius). These stations interpret the equatorial and low latitudes. Table 40.1 shows the geographical coordinates of these stations. The considered geomagnetic storm of 2024 falls in the present increasing solar activity cycle. First, the positioning data are obtained for aforementioned stations in RINEX format and then converted into VTEC data by using an open software, designed by Dr. Gopi Seemala. The further information about it can be sourced from seemala.blogspot.in.

Table 40.1 List of GPS stations and their locations considered for study

Sr No	Name of Station	Geographic latitude	Geographic longitude
1	DGAR (Diego Garcia Island, UK)	-7.270	72.370
2	IISC (Bangalore, India)	13.021	77.570
3	VACS (Vacoas, Mauritius)	-20.297	57.497

Further, to examine the effect of geomagnetic storms on TEC, we get the data of Dst index from the web source of World data center, Kyoto (https://wdc.kugi.kyoto-u.ac.jp/qddays/index.html.) and solar event details from https://www.spaceweatherlive.com/en/solar-activity/top-50-solar-flares/year/2024.html. The Dst index data is further analyzed and quietest days were identified from lowest solar activity on the basis of Dst index. To find any sudden variation in VTEC data on event day, we use hourly TEC data. First we calculated the average VTEC values for ten quietest days (as discussed earlier) for the same month August 2024 and compared with actual VTEC values for the event day. We also calculated the percentage change in VTEC values on storm day with respect to averaged VTEC values of quietest days.

3. RESULTS AND DISCUSSION

In this paper, we are presenting the study of a recent geomagnetic storm occurred on 12 August 2024 at three different latitudes near equator. This storm was further followed by another storm on next days but at less intensity. The Dst index for the event day was recorded -203 nT. This storm is considered as one of the strong storm of year 2024 occurred till now on the basis of Dst index.

3.1 Variation of VTEC on 12 August 2024

This is the geomagnetic storm day occurred in 2024 with Dst, Ap and kp index -203 nT, 127 and 8 respectively at event hour. The event started at 0500 hrs and achieved peak at 1700 hrs. To examine any variation in VTEC on event day, we plotted TEC data for 5 days from 10 August to 14 August in Fig. 40.1. The data of event day (12 Aug 2024) is shown at centre whereas other four curves show the TEC variation for two previous days and two later days of the event. The figure depict six panels, in which first and second panels are drawn for DGAR station, third and fourth panels are for IISC and fifth and sixth panels are for VACS station.

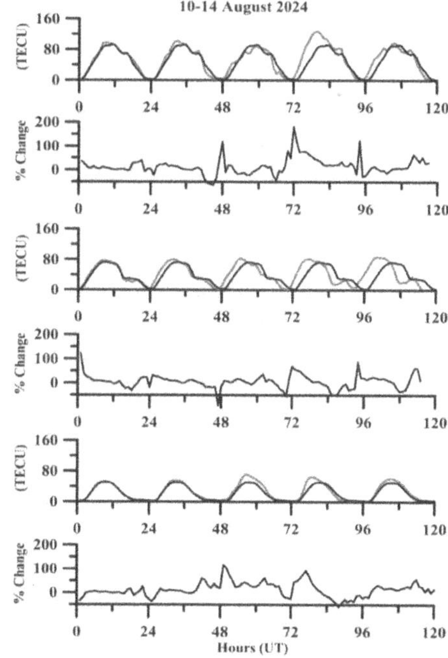

Fig. 40.1 Variations of VTEC from 10 August to 14 August 2024 corresponding to the geomagnetic storm of 12 August 2024. (Remark- This is an original work)

The first two panels are showing results for DGAR station. It can be observed from these panels that the peak value of TEC on the studied disturbed day is observed at 1200 hrs UT equal to 89.32 TECU and it was declined by 1.60 TECU from quietest day average which is equal to 1.76%. During the event the maximum percentage rise in TEC is 14.1% at 1700 hrs and after event the TEC values fell rapidly and shows negative change upto -42.72 %.

The third and fourth panels of the figure are showing the results for IISC station. Here the peak value of TEC on the studied disturbed day is observed at 0800 hrs equal to 83.28 TECU and it was enhanced by 11.73 TECU, equal to 16.40% from quietest day average. During the event the maximum percentage rise in TEC was observed equal to 37.62% at 1600 hrs and after event the TEC values decreases. The fifth and sixth panels show the results for VACS station. It can be seen that the peak value of TEC on the studied disturbed day is observed at 0500 hrs equal to 71.88 TECU and it is increased by 21.92 TECU from quietest day average, that is equal to 43.88% change. During the event, the maximum percentage rise in TEC was observed equal to 36.00 % at 1700 hrs. After event the TEC values and percentage change in TEC decreases. These results are clearly indicating that the increased geomagnetic activity between 1500-1800 hrs (UT) has affected TEC variations dominantly. However, the corresponding changes in TEC are larger at higher latitudes as compared to equatorial station, that shows a rapid decline in TEC values and negative percentage change in TEC values from averaged quietest days TEC observations. The findings of maximum TEC values and change in TEC observed during various instants at all three stations during event is summarized in Table 40.2.

Table 40.2 The maximum TEC values and change in TEC at various stages at different stations during geomagnetic storm day of 12 August 2024

S. No	Stations	Geo Lat.	Max TEC value on event day	Time in hrs (UT)	Change in TEC at peak of TEC values
1	DGAR	-7.27	89.32	1200 hrs	-1.5
2	IISC	13.02	83.28	0800 hrs	11.73
3	VACS	-20.29	71.88	0500 hrs	21.92

The results show that there is no significant change in TEC at equatorial station, however, changes in TEC increases with increase in latitudes. Equatorial latitude station shows a smaller TEC variation and higher latitude show larger TEC variations. The percentage change of the difference in actual TEC values and averaged quietest days TEC is found to be varied from 14% to 38%, during the peak of event on event day. This percentage change in TEC after peak of event falls rapidly at equatorial station just after the event whereas a moderate rate of fall is observed at both the low latitude stations. Both low latitude stations DGAR and IISC show a negative percentage change in TEC after peak of event. Similar kind of results were also obtained by Thiruvarangan et al. (2024) in their study. They observed insignificant irregularities near equatorial locations (Srilanka) during the storm events through rate of TEC index. Wang et al. (2023) also found in their studies that rate of TEC index were small in low and mid latitude and larger at higher latitudes. The results of our study suggest the presence of fountain effect in the EIA region. That explains the fall in TEC values due to EXB drift at equator and rise in TEC at $\pm 20^{\circ}$ latitudes.

4. CONCLUSION

The study of recent geomagnetic storm and its impact on ionospheric TEC gives valuable information about the response of ionosphere during disturbed ionospheric weather conditions. The study is carried out for three different stations situated over and near the geographic equator. A sudden rise was observed at low latitude stations just after the event and the maximum percentage change in VTEC varied between 13% to 36% from averaged quietest days VTEC of the month. However, no significant rise was observed at equatorial station just after the event. The result also concluded that there is a close connection between the disturbed ionospheric weather and its response, depending upon at different latitudes. The effect of storm is observed better at higher latitude except equatorial latitude stations. The fountain effect due to EIA might be the reason behind these results.

REFERENCES

1. Adebiyi, S. J., Adimula, I. A., & Oladipo, O. A. (2014). Seasonal variations of GPS derived TEC at three different latitudes of the southern hemisphere during geomagnetic storms. Advances in Space Research, 53(8), 1246–1254.
2. Adimula, I. A., Oladipo, O. A., & Adebiyi, S. J. (2016). Latitudinal and seasonal investigations of storm-time TEC variation. Pure and Applied Geophysics, 173, 2521–2533.
3. Blagoveshchensky, D. V., Maltseva, O. A., & Sergeeva, M. A. (2018, August). Impact of magnetic storms on the global TEC distribution. In Annales geophysicae (Vol. 36, No. 4, pp. 1057–1071). Copernicus GmbH.
4. Park, J., Von Frese, R. R., Grejner-Brzezinska, D. A., Morton, Y., & Gaya-Pique, L. R. (2011). Ionospheric detection of the 25 May 2009 North Korean underground nuclear test. Geophysical Research Letters, 38(22).
5. Paziewski, J., Høeg, P., Sieradzki, R., Jin, Y., Jarmolowski, W., Hoque, M. M., ... & Orús-Pérez, R. (2022). The implications of ionospheric disturbances for precise GNSS positioning in Greenland. Journal of Space Weather and Space Climate, 12, 33.
6. Savastano, G., Komjathy, A., Shume, E., Vergados, P., Ravanelli, M., Verkhoglyadova, O., ... & Crespi, M. (2019). Advantages of geostationary satellites for ionospheric anomaly studies: Ionospheric plasma depletion following a rocket launch. Remote Sensing, 11(14), 1734.
7. Thiruvarangan, V., Rajavarathan, J., Panda, S. K., & Swarnalatha Jayakody, J. A. (2024). Geomagnetic storm effect on equatorial ionosphere over Sri Lanka through total electron content observations from continuously operating reference stations network during Mar–Apr 2022. Journal of Applied Geodesy, (0)
8. Trivedi, R., Jain, A., Jain, S., & Gwal, A. K. (2011). Study of TEC changes during geomagnetic storms occurred near the crest of the equatorial ionospheric ionization anomaly in the Indian sector. Advances in space research, 48(10), 1617–1630.
9. Wang, Y., Yuan, Y., Li, M., Zhang, T., Geng, H., Wang, G., & Wen, G. (2023). Effects of strong geomagnetic storms on the ionosphere and degradation of precise point positioning accuracy during the 25th solar cycle rising phase: a case study. Remote Sensing, 15(23), 5512.

Note: All the figures and tables in this chapter were made by the authors.

Recent Trends in Applied Physics and Material Science – Sudhir Bhardwaj et al. (eds)

Majorana Neutrino, Torsion and Nature of Dark Matter

Lipika Mullick Ray[1]
Hiralal Mazumdar Memorial College for Women,Dakshineswar, Kolkata, India

Pradip Kumar Ghosh[2]
Jadavpur University, Jadavpur, Kolkata, West Bengal India

Prasanta Kumar Mahato[3]
Narasinha Dutt College, Howrah, West Bengal, India

Abstract: Almost 90 years back we came to know that there is predominantly excessive amount of exotic matter compared to ordinary matter. This exotic matter is nowadays termed as Dark Matter (DM) and this matter only interacts gravitationally. Its existence is understood only through its gravitational nature. Lack of experimental evidence for non-gravitational interaction of DM may suggest that it is something beyond our current theoretical knowledge. Current research investigates the nature of DM as that of scalar field that may originate from Bose-Einstein condensate of decoupled Majorana neutrinos from the early universe. This scalar field is given proper dimension through the tetrads, together which give de Sitter boost in SO(4,1) group. Here the axial part of torsion plays a dominant role in understanding the link between matter known from the Standard Model and dark matter.

Keywords: Dark Matter, Neutrino, Scalar field etc.

1. INTRODUCTION

Starting in 1933 (Zwicky, 1933), no experimentally verified Dark Matter solution existed (Cirelli et al., 2024) & (Bertone and Hooper, 2018). Observations of neutrino oscillations have revealed that neutrinos have tiny masses beyond the Standard Model of Theoretical Physics. An uncharged neutrino particle can be its anti-particle, Majorana Neutrino (Majorana, 1937). This fact is important to build a theoretical mechanism for neutrino masses. In 2021, Grensinge G. (Grensing, 2021) presented the concept of dark-matter-torsion theory as a viable solution to the DM problem, with supermassive axial torsion serving as a mediating field and superheavy right-handed Majorana fermions as DM constituent. Torsion and two scalar fields are also included in the gravitational degree of freedom of the new gravitational theory with torsion (GWT) by Zhao X. and Ma Y. (Zhao and Ma, 2024). It is based on Riemann-Cartan geometry. The empirical Baryonic Tully-Fischer relation has been validated by the solution that effectively aligns with data gathered experimentally on rotation velocities of stars outside the Milky Way's stellar disk when parameters are accurately specified.

Considering a previous research paper of one of the authors (Mahato, 2007), here we have proposed a gravitational Lagrangian in Riemann-Cartan space together with a scalar field ϕ and fermionic field ψ to investigate apparent DM emergence from the interaction of the spinor field with torsion and scalar field.

2. NEUTRINO OSCILLATION AND TORSION

A neutrino produced with a certain Lepton family number is subsequently measured for having an alternative Lepton family number, a phenomenon referred to as neutrino oscillation in quantum mechanics. When a neutrino traverses through space, its probability of being measured varies amongst three known states. Since neutrino oscillation indicates that neutrino has mass even in the absence of Einstein-Cartan torsion (Sabbata and Gasperini, 1981) & (Chakrabarty and Lahiri, 2019), we demand an alternative to the Standard Model of particle physics. Also, the experimental evidence of a massive neutrino flavor suggests the presence of a superheavy right-handed Majorana neutrino that go beyond the standard model (Barger et al., 2012).

[1]lmullickray@gmail.com, [2]pradipkghosh1957@gmail.com, [3]pmahatoprincipal@gmail.com

DOI: 10.1201/9781003684718-41

It has been demonstrated by Sabbata V. De & Gasperini M. (Sabbata and Gasperini, 1981) that neutrino oscillation can be performed even in the case of massless neutrinos and degenerate (but nonzero) neutrino masses. For the latter case, oscillations may be generated via Einstein-Cartan torsion. Weak neutrino current is projected to deviate from the usual vector-axial vector form, and torsion eigenstates are interpreted as distinct combinations of right-handed and left-handed fields.

3. TORSION, SCALAR FIELD, AND SPINOR FIELD

Structural equations of Cartan for Riemann-Cartan space-time, U_4, are as follows (Cartan, 1922 & 1924):

$$T^a = de^a + \omega^a_b \wedge e^b \quad (1)$$

$$R^a_b = d\omega^a_b + \omega^a_c \wedge e^b \quad (2)$$

here $\omega^a{}_b$ represent spin connection and e^a is local frame.

In U_4 there exist 2 invariants closed 4 forms. Pontryagin density (Chern and Simons, 1971 & 1974) P is more prevalent and Nieh-Yan density (Nieh and Yan, 1982) $\tilde{N}$ less-prevalent as described below

$$P = R^{ab} \wedge \mathrm{R_{ab}} \quad (3)$$

and, $$\tilde{N} = d(e_a \wedge \mathrm{T^a}) = T^a \wedge T_a - R_{ab} \wedge e^a \wedge e^b \quad (4)$$

We investigate Riemann-Cartan geometry where the axial vector component of torsion is only relevant. Thus, from (4), Nieh-Yan density is naturally derived.

$$\tilde{N} = 3 = -R_{ab} \wedge e^a \wedge e^b = 3 * N\eta \quad (5)$$

where $$\eta = \frac{1}{4!}\epsilon_{abcd} e^a \wedge e^b \wedge e^c \wedge e^d \quad (6)$$

$\eta$is the invariant volume element. It determines that $*N$, Hodge dual of N, is the scalar density of the dimension $(length)^{-2}$

Gravitational Lagrangian, in the given approach, is described as:

$$\mathcal{L}_G = -N(\mathcal{R} - \beta\phi^2) +* (b_a \wedge \bar{\nabla}^a_e)(b_a \wedge \nabla^a_e)$$
$$-f(\phi)d\phi \wedge* d\phi - h(\phi)\eta + \frac{1}{2}\sigma\phi^2 T \wedge* T \quad (7)$$

where $*$ is Hodge duality operator, $\mathcal{R}\ \eta = \frac{1}{2}\bar{R}^{ab} \wedge \eta_{ab}$, $\bar{R}^b_a = d\bar{\omega}^b_a + \bar{\omega}^b_c \wedge \bar{\omega}^c_a$, $\bar{\omega}^a_b = \omega^a_b - T^a_b$, $\eta_a = \frac{1}{3!}\epsilon_{abcd}\, e^b \wedge e^c \wedge e^d$, $T \wedge* T = \frac{1}{3!}T_{abc}T^{abc}\eta$ and $\eta_{ab} = *(e_a \wedge e_b)$.

Here two dimensionless coupling constants are β and $\sigma \cdot \bar{\nabla}$ is covariant differentiation concerning connection one form $\bar{\omega}^{ab}$, b_a represents 2-form with 1 internal index and dimension $(length)^{-1}$ and functions $f(\phi)$ and $h(\phi)$ are unspecified functions of ϕ, whose forms needs to be established as per the manifold's geometric structure. It is to be noted that Lagrangian density $\mathcal{L}_G$ is a modified version of that of (Mahato, 2007) by addition of the last term $\frac{1}{2}\sigma\phi^2 \mathrm{T} \wedge* T$

We are now prepared to formulate total gravity Lagrangian incorporating spinorial matter field, expressed as

$$\mathcal{L}_{tot.} = \mathcal{L}_G + \mathcal{L}_{D'}, \quad (8)$$

Here,

$$\mathcal{L}_{D'} = \varphi^2 \begin{bmatrix} \frac{i}{2}\left\{\bar{\psi} * \gamma \wedge D\psi + \overline{D\psi} \wedge* \gamma\psi\right\} \\ -\frac{g}{4}\bar{\psi}\gamma_5\psi \wedge T + c_\psi\sqrt{*dT}\bar{\psi}\psi\eta \end{bmatrix} \quad (9)$$

$$\gamma_\mu := \gamma_a e^a_\mu, * \gamma := \gamma^a\eta_a, D := d + \Gamma \quad (10)$$

$$\Gamma := \frac{1}{4}\gamma^\mu D^{\{\}}\gamma_{\mu=} \frac{1}{4}\gamma^\mu\gamma_{\mu:\nu}dx^\nu$$
$$= -\frac{i}{4}\sigma_{ab}e^{a\mu}e^b_{\mu:\nu}dx^\nu \quad (11)$$

Where, $D^{\{\}}$, *or* ':' in tensorial notation, that is Riemannian torsion-free covariant differentiation acting only on external indices; $\sigma^{ab} = \frac{i}{2}(\gamma^a\gamma^b - \gamma^b\gamma^a)$, $\bar{\psi} = \psi^\dagger\gamma^0$ and dimensionless coupling constants are g, c_ψ, $\psi$and $\bar{\psi}$ have dimension $(length)^{-\frac{1}{2}}$ as well as conformal weight $-\frac{1}{2}$.

Following references (Hehl et al., 1995) and (Mahato, 2007) we independently vary e^a, $\bar{\nabla}e^a$,, $\bar{R}^{ab}$, ϕ, $d\phi$, b^a, ψ, $\bar{\psi}$, $D\psi$, $\overline{D\psi}$, T and dT in $\mathcal{L}_{tot.}$ and get, after some simplification

$$\bar{\nabla}e_a = 0 \quad (12)$$

$$* N = \frac{1}{k} \quad (13)$$

i.e. $\bar{\nabla}$ is torsion-free and $\frac{1}{k}$ is an integration constant s.t. κ has $(length)^2$ dimension.

Employing (13), the mass of field ψ may be described as

$$m_\psi = c_\psi\sqrt{*\, dT} = \frac{c_\psi}{\sqrt{\kappa}}, \quad (14)$$

Equation (14) together with $\Psi = \phi\psi$ as the Dirac field having appropriate dimensions and conformal weight and $m_\Psi = m_\psi$, *the* field equations for the spinor field Ψ simplified to the standard form in a specific class of geometry in U_4 space (Mielke, 2001)

$$i * \gamma \wedge D\Psi - \frac{g}{4}\gamma_5\gamma \wedge T\Psi + m_\Psi\Psi\eta = 0,$$
$$i\overline{D\Psi} \wedge * \gamma - \frac{g}{4}\bar{\Psi}\gamma_5\gamma \wedge T + m_\Psi\bar{\Psi}\eta = 0, \quad (15)$$

$$G_a^b\eta = \kappa \begin{bmatrix} \frac{i}{8}\left\{ \begin{matrix} \bar{\Psi}\left(\gamma^b D_a + \gamma_a D^b\right)\Psi - \\ \left(\overline{D_a\Psi}\gamma^b + \overline{D^b\Psi}\gamma_a\right)\Psi \end{matrix} \right\}\eta \\ -\frac{g}{16}\bar{\Psi}\gamma_5\left(\gamma_a{*}T^b + \gamma^b{*}T_a\right)\Psi\eta \\ + f\partial_a\phi\partial^b\phi\eta + \frac{1}{2}(h)\eta\delta_a^b + \frac{1}{4}\sigma\phi^2 \\ \left(-T_{acd}T^{bcd} + \frac{1}{3}T_{ecd}T^{ecd}\delta_a^b\right)\eta \end{bmatrix} \tag{16}$$

$$0 = \begin{bmatrix} \frac{1}{2}\bar{\nabla}_\nu\bar{\Psi}\left\{\frac{\sigma_a^b}{2}, \gamma^\nu\right\}\Psi \\ +\frac{i}{2}\left\{ \begin{matrix} \bar{\Psi}\left(\gamma^b D_a - \gamma_a D^b\right)\Psi \\ -\left(\overline{D_a\Psi}\gamma^b - D^b\Psi\gamma_a\right)\Psi \end{matrix} \right\}\eta \\ -\frac{g}{4}\bar{\Psi}\gamma_5\left(\gamma_a * T^b - \gamma^{b*}T_a\right)\Psi \end{bmatrix} \tag{17}$$

Again using (13), the equations for the variation of T and ϕ reduce to

$$\kappa d\left[\frac{g}{4} * \left(\bar{\Psi}\gamma_5\gamma\Psi \wedge T\right) - \Sigma - \frac{1}{2}\sigma\phi^2 * (T \wedge * T)\right]$$

$$= -\frac{g}{4}\bar{\Psi}\gamma_5\gamma\Psi - \sigma\phi^2 * T \tag{18}$$

$$-\left\{\frac{2}{\kappa}\beta\phi + h'(\phi)\right\}\eta + f'(\phi)d\phi \wedge * d\phi$$

$$+ 2\text{fd} * d\phi + \sigma\phi T\rangle \;\; * T = 0 \tag{19}$$

where, $\Sigma = f*(d\phi \wedge * d\phi) - 2h + \frac{\beta}{\kappa}\phi^2$ (20)

4. TORSION, FRW COSMOLOGY, AND DARK MATTER

The Eqn. (16) may be re-expressed as,

$$G^b{}_a = \kappa[\bar{E}_a^b + \hat{E}^b{}_a] \tag{21}$$

where

$$\bar{E}_a^b = \frac{i}{8}\left\{ \begin{matrix} \bar{\Psi}\left(\gamma^b D_a + \gamma_a D^b\right)\Psi - \\ \left(\overline{D_a\Psi}\gamma^b + D^b\Psi\gamma_a\right)\Psi \end{matrix} \right\} \tag{22}$$

$$\hat{E}^b{}_a = -\frac{g}{16}\bar{\Psi}\gamma_5\left(\gamma_a T^b + \gamma^b * T_a\right)\Psi + \text{f}\partial_a\phi\partial^b\phi$$

$$+\frac{1}{2}h\delta_a^b - \frac{1}{4}\sigma\phi^2\left(-T_{acd}T^{bcd} + \frac{1}{3}T_{ecd}T^{ecd}\delta_a^b\right) \tag{23}$$

represent energy-momentum tensor densities due to baryonic and non-baryonic (apparent non-luminous) matter.

It has been shown (Mahato, 2009) & (Mahato and Bhattacharya, 2011) that the term Σ, as defined in Eqn. (20), if it takes constant value in the absence of the σ-term in the Lagrangian $\mathcal{L}_G$, then considering Friedmann–Robertson–Walker (FRW) cosmology, standard cosmology can be derived with standard energy density with DM, dark radiation, and cosmological constant, where $\phi^2 \propto \frac{1}{e} \propto a^{-3}$, a is the FRW parameter s.t. Hubble's constant is given by $H = \frac{\dot{a}}{a}$.

Böhmer et al. (2008) address DM problem within context of $f(R)$ modified theories of gravitation, focusing on galactic dynamics by limiting their analysis within the static in addition to spherically symmetric metric represented by

$$ds^2 = -e^{\nu(r)}dt^2 + e^{\lambda(r)}dr^2 + r^2 d\Omega^2 \tag{24}$$

where $d\Omega^2 = d\theta^2 + \sin^2\theta d\phi^2$. It has been determined that in the region of flat velocity curves, metric coefficients are expressed as

$$\nu = 2v_{tg}^2 \ln\left(\frac{r}{r_0}\right) \; and \; e^\lambda \approx 1 + 2v_{tg}^2 \tag{25}$$

There exists v_{tg} constant tangential velocity of stars and gas clouds in circular orbits around the periphery of spiral galaxies. In the limit of large distance r, Newtonian potential $\Phi_N(r) \approx v_{tg}^2 \ln\left(\frac{r}{r_0}\right)$ exhibits logarithmic dependency on radial distance r. Consequently, with a definitive Newtonian limit, metric (24) could effectively characterize the geometry of space and time in regions dominated by DM.

Recently, Zhao and Ma (2024) analyzed the metric from Eqn. (24) within a gravitational theory that includes torsion, on the basis of Riemann–Cartan geometry, and then demonstrated that their solution correlates with the observed measurements of Milky Way's rotation curve beyond stellar disk.

Zhao and Ma (2024) have shown that, in empty space, the contribution of torsion to Einstein's tensor together with a scalar field Λ is given by the field equation

$$\tilde{G}_a^b + \Lambda\delta_a^b + T_{cda}T^{bcd} = 0, \tag{26}$$

Whereas our Eqns. (21,22,23) give us, in empty space,

$$G_a^b + \kappa\left(f\partial_a\phi\partial^b\phi + \frac{1}{2}h\delta_a^b\right) - ,$$

$$\frac{1}{4}\kappa\sigma\phi^2\left(-T_{acd}T^{bcd} + \frac{1}{3}T_{ecd}T^{ecd}\delta_a^b\right) = 0 \tag{27}$$

where contribution from the spinor field Ψ is negligible. The other ϕ dependent term, in the galactic scale where ϕ takes nearly a constant, can be approximated to be

$$\kappa\left(f\partial_a\phi\partial^b\phi + \frac{1}{2}h\delta_a^b \right) \approx \Lambda\delta_a^b, \tag{28}$$

the Λ term of Eqn. (26). The real difference lies in the torsional last terms of both equations. In Eqn. (26) it's given by $T_{cda}T^{bcd}$ whereas in our Eqn. (27) it's a different term

$$E^{Tb}{}_a = \frac{1}{4}\kappa\sigma\phi^2\left(-T_{acd}T^{bcd} + \frac{1}{3}T_{ecd}T^{ecd}\delta^b{}_a \right)$$

$$= -\mathcal{T}_{acd}\mathcal{T}^{bcd} + \frac{1}{3}\mathcal{T}_{ecd}\mathcal{T}^{ecd}\delta^b{}_a \tag{29}$$

where we have defined the three form $\mathcal{T}$, by absorbing the contribution from the scalar field ϕ, as given by

$$\mathcal{T} = \frac{1}{2}\sqrt{\kappa\sigma}\phi T \tag{30}$$

Hence Einstein's field equation in our case reduces to

$$G^b{}_a + \Lambda\delta^b{}_a - E^{Tb}{}_a = 0. \tag{31}$$

Considering the metric structure of Eqn. (24), we can get the following Einstein's tensor components (Liang and Zhou, 2023)

$$G_0^0 = e^{-\lambda}\left(r^{-2} - r^{-1}\lambda'\right) - r^{-2}$$

$$G_1^1 = e^{-\lambda}\left(r^{-2} + r^{-1}v'\right) - r^{-2}$$

$$G_2^2 = -\frac{1}{4}e^{-\lambda}\left[-2v'' + v'\lambda' - v'^2 - 2r^{-1}\left(v' - \lambda'\right)\right]$$

$$= G_3^3 \tag{32}$$

& $G_v^\mu = 0 \; for \; \mu \neq v$

We consider $*\mathcal{T}_a$ Hodge dual of 3-form $\mathcal{T}_{abc}$ and have only time component i.e. $*\mathcal{T}_a \neq 0$ only for $a = 0$, given by (Zhao and Ma, 2024)

$$*\mathcal{T}^a = D(r)\left(\frac{\partial}{\partial t}\right)^a, \tag{33}$$

then using the form of the metric of Eqn. (24), we get

$$\mathcal{T}_{123} = -D(r)e_\mu^a = -D(r)e^{\frac{v}{2}}, a = 0 = \mu \tag{34}$$

and,

$$\mathcal{T}_{abc} = 0, \textit{ either one of a, b, c} = 0 \tag{35}$$

Putting these values in Eqn. (29), we get

$$E_a^{Tb} = 2D^2 e^v, a = b = 0$$

$$= 0, \text{otherwise} \tag{36}$$

Assuming, $D(r) = D$ = constant and using Eqn. (36) in Eqns. (31) for the metric (24), we get (Zhao and Ma, 2024)

$$\frac{d^2\tilde{f}}{dr^2} + \frac{r}{r^2+\alpha^2}\frac{d\tilde{f}}{dr} + \left(\frac{2}{r^2+\alpha^2} - \frac{2}{r^2}\right)\tilde{f}$$

$$= -\frac{2D^{-2}}{r^2\left(r^2+\alpha^2\right)} \tag{37}$$

here a is an integration constant s.t.

$$\tilde{f} = e^v, \tilde{g} = e^\lambda,$$

$$\tilde{f}\tilde{g} = e^{\lambda+v} = \frac{D^{-2}}{r^2+\alpha^2}. \tag{38}$$

The general solution of Eqn. (41) is (Zhao and Ma, 2024)

$$\tilde{f}(x) = \tilde{f}(Dr) = 1 + \beta\frac{\sqrt{x^2+1}}{x}$$

$$+ y\left\{\frac{\sqrt{x^2+1}}{x}\ln\left[x + \sqrt{x^2+1}\right] - 1\right\} \tag{39}$$

where β, γ are two integration constants with proper dimensions and here we have adjusted the other integration constant $a - D^{-1}$. There is no need to repeat calculations by Zhao and Ma (2024). We simply conclude that our approach in Einstein-Cartan geometry together with a scalar field bears the same result obtained by Zhao and Ma (2024), but with a different interpretation. Comparing Eqns. (26) & (31) we see that the torsional contribution to energy-momentum tensor in the current case is different from that of the said paper. In the same referred paper, energy density is zero whereas in our present case, it is not so due to the triviality of the space components of the axial vector $*T$. Alternatively, solution (43) could correspond with observational data of a rotation curve beyond the stellar disk in the Milky Way. Hence, galactic DM might only be the gravitational consequence of a theory incorporating torsion.

5. DISCUSSION

One may consider the dark-matter-torsion theory with superheavy right-handed Majorana fermions constituting dark matter and supermassive axial torsion the mediating field, as a promising candidate to resolve the dark matter problem as declared by Grensing (2021). In another approach, the torsion is proportional to the spin density of the background normal matter so the effect of oscillation is very small. But by breaking chiral symmetry in the coupling of fermions to torsion, and by using the fact that all fermions couple to torsion we find a much larger effect as described by Chakraborty and Lahiri (2019).

In our *torsion* ⊗ *curvature* approach of gravity with the modification of the Lagrangian taken in the paper of Mahato (2007) by the addition of a $(torsion)^2$ term, here we are able to derive standard cosmology with standard energy density together with dark matter, dark radiation and cosmological constant.

It turns out that the solution does admit the asymptotic behaviour of a galactic rotation curve. In particular, by suitably choosing the undetermined parameters, the solution can fit very well the experimental data of the rotation velocity of stars outside the Milky Way's stellar disk, while that of GR cannot. Hence, from the perspective of GR with torsion (GWT), the galactic DM effect on the rotation curve is due to the additional gravitational degrees of freedom, including spacetime torsion.

Therefore, we can assert that our *torsion ⊗ curvature* gravity approach bears the same result as that of GWT proposed by Zhao and Ma, claiming that the so-called DM may just be a gravitational effect.

We have achieved our result in the galactic empty space where the contribution to the energy-momentum tensor $\hat{E}^b{}_a$ from the torsion ⊗ matter interaction term $\left(\bar{\Psi}\gamma_5\gamma\Psi\right)\wedge T$ has been ignored. But that should not be the case in general. So, further investigation is necessary to find out the effect of this term, at least in the early universe or on neutrino oscillation, as done by Sabbata and Gasperini (1976) and Sabbata (1994).

ACKNOWLEDGEMENT

The three authors individually created this research paper. We received immense technical support from Sri Partha Bhattacharya of Bangabasi Morning College, Kolkata, in this endeavor. The authors are also indebted to Prof. Pinak Pani Pal of the Indian Statistical Institute, Kolkata.

REFERENCES

1. Barger, V., Marfatia, D. and Whisnant. K. L. (2012) The Physics of Neutrinos. Princeton University Press, Princeton, New Jersey: 1–240.
2. Bertone, G. and Hooper, D. (2018). History of dark matter. Rev.Mod.Phys. 90:045002-1 – 045002-32.
3. B¨ohmer, C.G., Harko, T. and Lobo, F. S. N. (2008). Astropart. Phys., 29:386–394.
4. Cartan. E. (1922) Sur une généralisation de la notion de courbure de Riemann et les espaces à torsion. C. R. Acad. Sci (Paris) 174 (1922), 593–595.
5. Cartan É. (1924) Sur les variétés à connexion affine et la théorie de la relativité généralisée Part I, Ann. Éc. Norm. 40: 325– 412 and 41 (1923), 1–25; Part II: 42 17–88.
6. Chakrabarty, S. and Lahiri, A. (2019). Neutrino oscillations in the presence of torsion. The European Physical Journal C, 79 (8):697, 1–7.
7. Chern S. and Simons. J. (1971) Some Cohomology Classes in Principal Fiber Bundles and Their Application to Riemannian Geometry. Proc. Natl. Acad. Sci. (USA), 68:791–794.
8. Chern. S. and Simons. J. (1974) Characteristic forms and geometric invariants. Ann. Math., 99:48–69.
9. Cirelli, M., Strumia, A. and Zupan, J. (2024). Dark matter. https://doi.org/10.48550/arXiv.2406.01705: 1-515.
10. Grensing, G. (2021). Dark matter and torsion. General Relativity and Gravitation. 53(49): 1–14.
11. Hehl, F. W., McCrea, J.D., Mielke, E. W. and Ne'eman, Y. (1995) Metric-Affine Gauge Theory of Gravity: Field Equations, Noether Identities, World Spinors, and Breaking of Dilation Invariance Phys. Rep. 258:1–171.
12. Liang, C and Zhou, B. (2023) Differential Geometry and General Relativity, I edn vol 1. Springer, New York.
13. Mahato, P. (2007). Torsion, Dirac Field, Dark Matter and Dark Radiation. Int. J. Mod. Phys. A. 22 (4):835–850.
14. Mahato, P. (2009) Torsion and axial current. ICFAI Univ. Jour. of Phys., 2:92–104.
15. Mahato, P. and Bhattacharya, P. (2011). Torsion, chern-simons term, and diffeomorphism invariance. Mod. Phys. Lett. A, 26(6):415–421.
16. Majorana, E. (1937). Theory of the symmetry of electrons and positrons. Nuovo Cimento. 14:171–184.
17. Mielke, E. W. (2001) Beautiful Gauge Field Equations in Clifforms. Int. J. Theor. Phys., 40:171–189.
18. Nieh. H. T. and Yan. M. L. (1982) An identity in Riemann–Cartan geometry J. Math. Phys., 23:373–374.
19. Sabbata, V. De and Gasperini, M. (1976) Neutrino oscillations in the presence of torsion. Il Nuovo Cimento A. 65 (4):1971–1996.
20. Sabbata, V. De (1994) Spin and torsion in the early universe. Cosmology and Particle Physics, 427:97–128.
21. Zhao, X. and Ma, Y. (2024) A new gravitational theory and dark matter problem. Commun. Theor. Phys. 76(6):065403-1 – 065403-8.
22. Zwicky, F. (1933). The red shift of extragalactic nebulae, (in German). Helvetica Phys. Acta. 6:110–127.

Recent Trends in Applied Physics and Material Science – Sudhir Bhardwaj et al. (eds)
 ISBN 978-1-041-16452-4

BEC of Neutrino—A Dark Matter Constituens

Pradip K. Ghosh[1]
Ex Pro V.C., Jadavpur University, Kolkata

A. Sarkar[2]
Ex Asso. Prof. Phys., B. K. Gils' College, Howrah, India

Abstract: Early prediction of dark matter (DM) by Zwicky in 1930 and then dark energy (DE) are still remain a mystery. Recently Bose-Einstein Condensate (BEC) leads to cosmological scalar fields for DE and DM has considered where BEC is a collective super-atom formed by BEC. It considered BEC of ultralight boson to explain the phenomenology of DM and DE. It has been studied DM as BEC in Minkowski space. A general case was studied on curve space time. In BEC cosmological model there exists few parameters to incorporate quantum correction in general theory of relativity (GTR) of which BEC mass parameter, directly relates critical temperature (T_c) of BEC, which is about $\leq 10^{-26}$ eV and it may be enhanced two order considering curved space formalism. BEC model may be summarized as cold dark matter (ʌCDM) and a left over part as DE. SM and its extension predicted WIMP's like axon, sterile neutrinos (v), graviton could be DM candidate. In this work authors revisited the role of active neutrino (v) those decoupled 0.5 sec after big bang, following their quantum mechanical behaviour as spin 1/2 particle along with some other criterion like v's (same flavour) and also following BCS-BEC cross over may be conditionally accepted as bosonic DM particles. In this work authors attempt to incorporate Casimir cosmology over BEC model in a consistent way. The Casimir force is attractive and of quantum origin in nm scale. Enhancement of BEC mass limit is thus analysed.

Keywords: Bose-einstein condensate, Dark matter, Dark energy, Casimir cosmology, CDM, neutrino

1. INTRODUCTION

We have achieved the revolutionary idea through Standard Model (SM) which tried to explain the elementary constituents of physical world yet there are a number of limitation. The study of cosmology, which gathers momentum from introduction of Cosmological constant by Einstein in 1916 and studies of the Coma Cluster of galaxies, where it was noticed that visible mass as per SM i.e baryonic mass is unable to explain cluster's gravitational effect in 1930 by Zwicky (Zwicky,1933) and then the study of rotation curves and observing its flatness by Rubin and Ford in 1970 allowed to rethink both Newton's Gravitation and the SM. Then observation of expansion of universe by Hubble (Hubble, 1929) and the two independent research on type la Supernovae in 1998 (Perlmutter et.al 1999); (Riess et.al,1998) established that the expansion of universe is accelerating in nature.

Zwicky's observation, is the Astronomical evidence of DM, where it was established that the velocities of individual galaxies in clusters (e.g. Coma cluster) exceed substantially than expected and this lead to requirement of adequate amount of DM. Later it has been analysed that DM density contribute about 25% of critical density. Gravitational lensing observations on large scale structures e.g galaxies and cluster galaxies confirm DM and its mass distribution in universe. The overall mass distribution of universe is given by, Ω_m (matter) =0.25, Ω_b (baryon) =0.045, Ω_{DM} =0.2 and rest are DE. DM necessarily to be non-baryonic (some baryonic matter may be dark), cold, non-relativistic movement with almost non-interacting with ordinary matter (saving from self-interacting nature). Matter distribution of universe is also accounted from CMB data and found to be in good agreement with Astronomical estimate. Many attempts have been made to eliminate DM and to change physics of gravitation by introducing modified Newtonian dynamics (MOND) but results are not satisfactory enough. In Cosmic dynamics "Cosmological Constant Cold Dark Matter" (ΛCDM) model gained amazing success yet it is not the final word

[1]pradipkghosh1957@gmail.com, [2]alokesarkar2004@gmail.com

DOI: 10.1201/9781003684718-42

in cosmology. Matter particles with DM characteristics have been received very high attention by Peskin (Peskin, 2015) and also by Fank (Fank, 2015) with their active research in the framework of SM and BSM (beyond SM) between mass range sub eV to GeV but only active neutrinos qualify the exam. A brief comprehensive review on DM may found in ref. (Bahcall, 2015).

Further the study of neutrino oscillations that indicates neutrinos are massive needs explanation beyond the SM. As one gazes beyond the microscopic world to look into the very large scale of the cosmos it requires the presence of extra non visible nonbaryonic mass having only gravitational interaction in the form of dark matter (DM) and some form of energy called dark energy (DE) . The dark matter and dark energy are now integral parts of the matter-energy mix that make up 27% and 68% of the universe but yet they are not explainable. To understand the particle nature of DM and quantum origin of DE modern cosmology should interplays a unique role in the unity of macroscopic and microscopic world, (Sciama,1994) and also (Kolb 1990).

With this background several researches are going on in finding the nature and constituent of DM and the origin of DE with very little success. Following some recent works (Das and Bhadury,(2015a));(Das and Bhadury, (2018)); (Das,and Bhadury, 2015b) where bosons of BEC are taken as a candidate for DM, here present authors made an attempt to explain the formation of bosons from large neutrino field emerges in the early epoch after decoupling (Schnabel,2023); and also (Schnabel,2021).

2. REVIEW OF EARLIER WORKS

2.1 Bose Einstein Condensate (BEC)

BEC (Ornes, 2017) was first observed in 1995 in the field of cold atom physics. It formed at ultra-cold condition few degree above 0K. Its origin lies with self-organization of identical bosons. According to Das and Bhaduri (Das and Bhaduri, 2018) BEC may be consider as DM, a collapsed form of DE and can exert negative pressure. On formation of BEC the bosons lost their individual identity rather they exhibit a collective super atom - like photons in a laser beam. A BEC may exhibit many strange behaviour, like tickle it with laser, trapped in magnetic field, solid flow that flow through itself. Later it was considered as 5th state of matter.

2.2 Cosmic BEC View of DM/DE

Studies on BEC based cosmological model predicts DE as BEC in the form gas DM halo and DM as its condensed form. They supposed to exhibit negative pressure to ensure the cosmic acceleration. Such BEC, is a result of self-attractive nature of boson field, are localized CDM and do not violate cosmological principle. Outcome of BEC is DE to follow a rapid collapse at some critical condition to form compact object like boson stars and/or black holes. These condensed state considered as seed of galaxies and other light emitting objects. The process supposed to be self-organize such that DE/DM ratio ~ unity. Theoretical analysis of this process may be found in (Morikawa, 2004) with the use of Klien-Gordon equation. There are demonstration by Morikawa that thermodynamics sets mass limit of bosons (which is about neutrino mass) for continuous BEC formation.

In an attempt Das and Sur (Das and Sur, 2023) proposed a unified picture of dark sector from BEC. In the work they considered energy density of BEC along with its quantum potential to throw light on the said unification. Their attempt extended towards determination of BEC mass taking quantum back reaction from baryons into account. They recorded a BEC mass enhancement upto three order of magnitude (over Hubble's value) leaving some deviation from the results of ΛCDM model.

2.3 BCS-BEC Crossover

Around 1950's phenomenon superconductivity was explained following BCS theory, as BEC of Cooper pairs (CP). Detail aspect of it is beyond the scope of this article. Authors unfolded only a brief portion relevant to this article. On increase in pairing interaction between fermion forming CP, it assume a small molecular boson like form and under a favourable condition CP's become Bose gas (Bose, 1924) BEC. This transition or cross-over is not like general phase transition rather a weak coupled BCS system passes to strong coupled BEC system. The mentioned cross-over experimentally achieved in 2004, is essentially a verification of tunability in interaction strength (Ohashi et al. 2020). The said criterion follows a non-phonon mechanism known as Feshbach mechanism (Chin et al, 2009) rather than phonon induced BCS mechanism.

Numerical simulation experiment following ΛCDM Model and observational data indicates DM distribution profile follows $\rho_{DM} \infty 1/r$, a singular distribution forming a cusp (Navarro, et al, 1996). Astrophysical observation (Genina, et al, 2018) predicts DM distribution follows a core with constant density. It is great contradiction with earlier studies and known as core- cusp tension. In fact ΛCDM and other DM models ignores self-interaction in DM that leads to some crisis. BEC theories taken into account the self-interaction contribution in theories and thus gained reasonable success in explaining DM characteristics by a recent work by authors (Caracium and Harko, 2020). Testing BEC DM models with SPARC (Spitzer Photometry & Accurate Rotation Curve) galactic rotation curves data have accounted a fair result is obtained using mentioned consideration.

3. CASIMIR COSMOLOGY

Prediction of this effect follows the early work by Hendrik Casimir in 1948. It is small attraction force between two closely spaced dielectric object in vacuum. It originates from quantum fluctuation in vacuum field. Casimir Effect (CE) is found to play integral role in cosmology. Here only a brief introduction of Casimir Cosmology (CC) related to the present work is furnished. The detail of it may be found in literature (Leonhardt, 2020). In1996 the Casimir force was measured (Lamoreaux, 1996) to 5% proximity of theoretical result. The CC is a probable route to resolve a great crisis in cosmology namely "Hubble tension' (Leonhardt,2020).

In 1998 it has been discovered that universe undergoes an accelerated expansion due to an apparent repulsive gravity in cosmological scale. Analysis conclude that the mentioned phenomenon is due to dark energy, which Einstein introduced as cosmological constant Λ, which shares 70% of universe's total energy density. Non-zero Λ is an observational evidence (Bergstrom and Danielsson, 2000). Λ affects the cosmic dynamics and formation of larger structure of universe. The effect of Λ in galactic scale are small and negligible on rotation curves. It has been analysed that Λ is vacuum energy originating from quantum fluctuation and QFT failed to account its correct order of magnitude. Among numerous attempts by different worker/groups notable successful estimation of vacuum energy comes from careful analysis of empirical early evidence of Casimir effect (CE). The early attempt by Lifshtiz analysis on CE predicts correct order of magnitude Λ in a consistent and analytical way. Following Lifshtiz's work it was Zel'dovich (Zel'dovich, 1968) who pointed out the correct origin of Λ despite disagreement with observed result. His starting point was quantum vacuum the ground state (GS) of all physical fields described by the fundamental interactions. The GS of vacuum is that of harmonic oscillator describes GS that carry energy with fluctuating amplitudes. The Casimir cosmology (Leonhardt, 2020) can take care of Zel'dovich's result on disagreement. The QFT estimated of GS energy density of universe, the sum of all harmonic oscillator GS energies which is divergent. In an early work Pauli suggested that possible coupling of it with gravitation could give rise correct estimate of Λ (Wienberg, 2000). In practice a medium needs to be considered as dielectric as in early observation from Casimir. Lifshitz consider dynamical effect rather than static modes vibration inside dielectric. He started from fluctuation dissipation theorem (Balakrishnan,1979) that results in dynamical Casimir effect. The theorem is applied towards explanation of hydrodynamic features (Ota, 2024) in CMB. Overall analysis is undertaken by Leonhardt (Leonhardt, 2020) lead to non- constancy of Λ with space–time. It can remove some Cosmological crisis like Hubble tension. The Casimir effect and dark energy, also known as the cosmological constant, are both caused by vacuum fluctuations. The main difference between the two is the former obeys local and later obeys global boundary conditions. Physically renormalization process in estimation of Λ may be represented as follows, Cosmological expansion factor 'n' is equivalent to refractive index of the medium, which is uniform in space but not with time. For a broad band dielectric like vacuum limiting value of 'n' is unity, since all materials becomes transparent at incident frequency $\nu \rightarrow \infty$. The nature of this attractive force F_{cas} varies as $1/L^4$, L is plate separation. This small force, is active within distances below a nanometer that becomes the strongest force between two neutral objects analogues to molecular force. It may be considered as a dubbed Dark Energy (DE). It is likely to be this Casimir force which is experimentally relevant, as the Compton wavelength of massive neutrinos is of the order of a micron, much larger than the atomic scale.

4. THEORY

The modified version of the SMC, the ΛCDM Model is essentially a parameterization of the Big Bang cosmological model in which the GTR contains a cosmological constant, Λ, which is associated with dark energy, and the universe contains sufficiently massive dark matter particles, i.e., "cold dark matter." However, both dark energy and dark matter are simply names describing unknown entities.

The quantum mechanical theory underlying DM as BEC has been discussed at length in the wok (Caracium, and Harko,2020). In section 2 of this reference provides a good account of DM distribution using external potential $v_{ex}(\mathbf{r})$ and is given by,

$$v_{ex}(\mathbf{r}) = v_{gr}(\mathbf{r}) + v_b(\mathbf{r}) + v_{GP}(\mathbf{r}) \quad \mathbf{(1)}$$

the first term in (1) is the trapping gravitational potential including self-interaction of DM and second term is the effect of baryonic matter and last term represent the random Gaussian potential. Following hydrodynamic consideration over the solution of corresponding Heisenberg's equation of motion (Griffin el al. 2009) by neglecting fluctuating last term, the theory yields expression for radius R_{halo} of DM halo (outside the core), and expression for mass of DM particle m_{DM} respectively which are expressed as ,

$$R_{halo} = K\, r_0^{1/2} \quad and$$
$$m_{DM} = K' r_0^{1/3} / R_{halo}^{2/3} \quad (2)$$

Here K and K' are constant containing h, Planck constant and G, Newton's constant.

Present work incorporated Casimir Effect over BEC DM theory where an additive term $v_{Cas}(\mathbf{r})$ in equation (1) is considered. The $v_{Cas}(\mathbf{r})$ is potential corresponding to Casimir Force given by $F_{cas} = K\ 1/L^4$, L is separation distance.

4.1 Present Work

In this work authors hypothesized formation of micro-cluster (μC) of DM from neutrinos by BEC mechanism. Since F_{cas} and F_{Λ} are opposite in nature within localized μC's and form DM clump. The analytic, "frozen-in" approximation (Bekenstien(2010).) assumes that the pressure and self-gravity of the DM clump are negligible at this point. Attractive Casimir Effect (CE) and self-interaction in DM particles forming the cluster. But it is balanced by repulsive cosmological force due to Cosmological constant. Present work makes an attempt to explain mystery of DM/DE by exploiting BEC and a further correction over it from CC. The goal of this communication is to incorporate neutrino as major constituent of DM/DE.

5. RESULTS AND DISCUSSIONS

Preliminary result on formation of neutrino μC is found to be satisfactory pending details results. Since neutrino mass is found to be below 0.018eV, they are considered to be non-relativistic just now. The experimental results obtained (for sum over three flavor) so far is Mass, $<$ 0.120 eV ($< 2.14 \times 10^{-37}$ kg), 95% confidence level, .The small mass implies that it has been analyzed that they can contribute only 2% fraction of the DM. since 1998, the reactor experiment like KamLAND has identified neutrino oscillations as the neutrino flavor mechanism transition involved that in solar electron-neutrinos. The work (Carlos, Arguelles et al, 2021) discussed the possible characteristics of fermionic DM halos. Authors (Carlos, Arguelles et al,2021) also throw light on the formation and stability of fermionic dark matter haloes in a cosmological framework, predicted on stability and formation such fermionic DM in a possible core-shell structure where a compact core enclosed by a dilute halo such core shell structure supposed to play important role in structure formation in universe following non-linear process. Authors in [28] claims the mentioned mechanism may be important in the formation of supermassive black hole.

Attempts are also going on to detect DM by direct, indirect and using LHC experiments but confirmative positive results are awaiting.

5.1 Probable Candidate for Boson in BEC

Considering the crucial role of bosons of BEC in the understanding of DM/DE, Several attempt has been made to understand a physical explanation and mathematical origin of such bosons in the condensates. In literatures, constituent bosons as gravitons or ultra-light scalar particles were considered. But they are yet to be detected experimentally. Study was also made on the idea that neutrinos forming a CP under the mechanism mentioned in the work (Ornes, 2017)..

With this concept DE is proposed to be created dynamically from the condensate of a singlet neutrino at a late epoch of the early Universe through a self –interaction by some authors.

In this scenario one may consider with neutrino with the study of behavior of them after decoupling (0.5 sec after big bang). The comparison of the average one dimensional distance between two identical neutrinos as quantum degenerate Fermi gas of neutrinos, it finally allows to take the state of neutrino as pure kinetic at and after decoupling.. With this background Schnabel able to established the formation of bosons in the form of 'Cooper pair' considering neutrino's temperature variation after decoupling because of relativistic increase in position uncertainty and the presence of gravitational force because of neutrino's rest mass.

Theories and simulations on structure formation show the nature DM may be in the form of cold and weakly interacting massive particles (WIMP). The standard model (SM) of particle physics failed to account it however beyond SM like super symmetry (SUSY) can predict existence of WIMPs. In this context a wide range of attempts (Jungman, Kamionkowski and griest,1996) have been made so far. On the Other hand the sum of neutrino mass (sum over flavours) provides some aspects of DM. The concept of DM is important to analyse conflict between two types of estimates of astrophysical objects (i) based on luminous and visible parts (ii) dynamical behaviour of its constituents. Many reviews and articles have been published (Bekenstien and Searches, 2010) on DM recent times. Studies indicate that WIMP's may form Cold Dark Matter (CDM). In practice no such particles has been detected so far. The possible other DM components may be solitons, massive compact objects (halo), primordial black holes, graviton and others. According to analysis (Yong 2017) these components have limitations like WIMPs.

6. CONCLUSION

Preliminary result on formation of neutrino μC is found to be satisfactory pending details results. The DN particle mass is estimated to be below 1eV. Inclusion of CE improved the result over standard BEC about 10%. It is indirect hint toward neutrino. Here DM halo radius considered as parameter and scattering length taken laboratory measured value.

REFERENCES

1. Arguelles, C.R. et al,(2021). On the formation and stability of fermionic dark matter haloes in a cosmological framework. Monthly Notice Royal Astronomical Society, 502: 4227–4246. doi:10.1093/mnras/staa3986 Advance Access publication (2020).

2. Bahcall, N. A. (2015) Dark Matter universe. PNAS 6, vol 112 (40) : 12243 – 12245. Balakrishnan, V. (1979). Fluctuation Dissipation Theorem from the generalized Langevin equation. Pramana, vol12, no 4, :301–315.
3. Bekenstien, J.D. (2010). *Particle Dark Matter: Observations, Models and Searches*. Ed. Bertone, Ch 6, pp95-114,CUP). arXiv:1001.3876.
4. Bergstrom, L. and Danielsson, W. (2000). arXiv: astro-ph10002152v2,16.
5. Bose, S.N. (1924). Plank gesetz und lichtquaenhypotheses. Z. Phys.,26:178.
6. Caracium, M. and Harko,T.(2020) arXiv:2007.12222v2 [gr-qc].
7. Carlos, A. et al,(2021). Monthly Notice Royal Astronomical Society, 502, 4227–4246. doi:10.1093/mnras/staa3986 Advance Access publication (2020).
8. Chin, C. Grim, R. Juleenne, P. and Tiesinga, E. (2010). Feshbach resonance in ultra-cold gases. Review of Modern Physics 82(2):1225. DOI:10.1103/RevModPhys.82.1225
9. Das, S. and Bhadury, R.K, (2018).Bose-Einstein condensate in cosmology. arXiv:1808.10505v3[gr-qc];
10. Das, S. and Bhadury, R.K, (2.015). Dark matter and dark energy from Bose – Einstein condensate Class. Quant. Grav, 32 :105003;arXiv1411.0753.
11. Das, S. and Sur, S.(2023).A Unified Cosmological Dark sector from a BEC. arXiv:2203.16402v2 [gr-pc] 28 Sep 2023]. Class. quant. Grav.32, 105003.
12. Fank, S. (2015). Proc. Nal. Acad. Sci. USA 112: 12264-12271
13. Griffin, A. el al (2009). *Bose –condensed gases at finite temperatures*. Cambridge, CUP
14. Hubble, E. (1929). A relation between distance and radial velocity among extra galactic nebulae. PNAS15, 3: 168173.
15. Jungman G., Kamionkowski M. and griest K. (1996). Supersymmetric dark matter. Phys. Rep. **267** :195
16. Kolb, E.W. (1990). *The Early Universe, CRC Press*
17. Lamoreaux S. (1997).Demonstration of the Casimir force in the 0.6 to 6 micrometer range. Phys. Rev. Lett. **78**: 5.
18. Leonhardt U. (2022). Casimir Cosmology. Int. J. of Mod. Phys. A (open access), vol 37, no.19, p2241006. DOI : 10.1142/S0217751X22410068
19. Morikawa, M. (2004). Structure formation through BEC – Unified view of DM & DE. 22nd Texas Symposium on Relativistic Astrophysics at Standford univ. Dec. 13-17:1-6.
20. Ohashi,Y.Tajima, H. and Wyk, R. van. (2020). BCS-BEC cross-over in cold atom and in nuclear system. Prog. Part. Nuc. Phys. vol 111 :133739.
21. Ornes, S. (2017). Core concept – How BE Condensates keep revealing weird physics. pron. Nat. Acad. Sci. USA,114(23) :5766–5769
22. Ota, A. (20240. arXiv:2402.07623v1 [hep-th] 12 Feb 2024
23. Perlmutter, S. et. al, (1999). Measurement of Ω and Λ from 42 high redshift supernovae. Astrophysical J, 517: 2.
24. Peskin, M. E. (2015). Proc. Nal. Acad. Sci. USA, 112 :12256–12263.
25. Riess et al (1998). Observational evidence from supernovae for an accelerating universe and a cosmological constant. Astronomy.(1998). J 116:1009.
26. Rubin, V and Kent, w. (1970). Rotation of the Andromeda nebula from a spectroscopic survey of emission regions. Astrophysical J. 159:379.
27. Sciama, D.W. (1994). *Modern Cosmology and the Dark Matter Problem,* CU Press .
28. Wienberg, S. (2000). In dark matters. arXiv (astro-phy)-005265.
29. Yong, BTL. (2017). A survey of dark matter and related topics in cosmology. Front. Phys. 12: 121201. ; Pleln T., arXiv; 1705.01987
30. Zel'dovich, Y (1968). The cosmological constant and theory of elementary particles, Sov. Phys.,Uspekhi, 11
31. Zwicky, F. (1937). The red shift of extragalactic nebulae. Astrophisics.J , 86,3: 217–246.

Recent Trends in Applied Physics and Material Science – Sudhir Bhardwaj et al. (eds)

Understanding the Conductivity of Martian Environment by using FFT Model with FDTD Method

M. B. Chaudhari[1]
The Charutar Vidhya Mandal (CVM) University,
Vallabh Vidyanagar, Anand, Gujarat, India

F. M. Joshi[2]
Madhuben & Bhanubhai Patel Institute of Technology,
The Charutar Vidhya Mandal (CVM) University,
Vallabh Vidyanagar, Gujarat, India

Abstract: We examine the conductivity of a dust devil that analyzes charging in terrestrial thunderstorms using a Finite Difference Time Domain and Fast Fourier Transform (FFT) method. Extensive research has been conducted on the transmission of Extremely Low Frequency (ELF) electromagnetic waves and resonance events in the Mars atmosphere, including their impact on ionospheric dynamics, thunderstorms, and lightning. It is possible that the resonant cavity formed by Mars' surface and lower ionosphere captures natural electromagnetic waves generated near the surface by dust devil electrostatic discharges. The lightning discharge plays an important role in the chemistry, energetics, and dynamics of planetary atmospheres. The electric field is the primary cause of lightning on the Martian surface. The FFT method takes the time-domain electric field data from the calculation and converts it into the frequency domain through the electric field component [E_x]. In the lower atmosphere, the predicted ion concentrations are insufficient to influence electrical conductivity. Direct calculations of electrical conductivity are made using electron density and electron-CO_2 collisions.

Keywords: FDTD, Lightning effect, Schumann resonance, FFT, Dust devils

1. INTRODUCTION

Dust devils, which are larger convective vortex structures, can form from minute dust grains. When dust devil granules of different sizes and compositions come into touch, they create and exchange charge using a mechanism known as triboelectric charging (Farrell et al.,2006). It is known that the ionosphere-ground gap acts as a resonator for extremely low frequency (ELF) waves. This worldwide resonance, also called Schumann resonances, is caused by lightning discharges (Hayakawa, Otsuyama,2002). A numerical analytic approach, the Finite-Difference Time-Domain (FDTD) model is based on time-dependent differential Maxwell equations (Oberst Ulrich,2007). It was adapted for spherical coordinates and implemented in very low frequency (VLF) and extremely low frequency (ELF) studies. We use an FDTD approach and an FFT model to solve the Maxwell time-dependent equation and explore electric field distribution (Amin Gasmi,2022). We employ the perfectly matched layer (PML) boundary condition because it produces an accurate electromagnetic wave graph.

2. METHOD

The following are the Maxwell's equations that need to be resolved:

$$\nabla \times E = -\mu_0 \frac{\partial H}{\partial t} \tag{1}$$

$$\nabla \times H = \varepsilon_0 \frac{\partial E}{\partial t} + \sigma_0 E \tag{2}$$

Where, E is the electric field, H is the magnetic field, ε_0 and μ_0 are the dielectric constant and permeability, σ is the electric conductivity of Mars. We express the field

[1]manoutichaudhari22@gmail.com, [2]foram.joshi@cvmu.edu.in

DOI: 10.1201/9781003684718-43

updating equations in Spherical coordinates. The FDTD approach solves Maxwell's equation by discretizing both time and space. When electric and magnetic fields are staggered in both time and space, Yee's algorithm is applied (Jamesina J. Simpson, 2007). After the FDTD time stepping, the time domain electric field at the source position, the Fast Fourier Transform (FFT) is used to convert it into the frequency domain. S

$$E_z^{FFT}(f) = \sum_{n=0}^{N-1} E_z(n) e^{-j2\pi \frac{f_n}{N}}$$

Where, f is the spectrum's frequencies and N is the number of time steps. The flow chart presented in Fig. 43.1 describes the entire process of FDTD using FFT model.

The flowchart for the FDTD method with FFT for simulating electromagnetic wave propagation on Mars includes critical parameters and boundary conditions, giving precise information about the Martian ionosphere. The model captures wave transition and attenuation properties in the Martian ionosphere, providing vital insights into phenomena such as Schumann resonances. This is especially applicable to Mars, where the thin CO_2-dominated atmosphere, combined with its unique electrical properties, necessitate thorough time-domain models (J. P. Pabari, 2015). The FDTD approach models wave propagation effectively by leveraging fine temporal resolution and correct boundary conditions, whilst the FFT provides a powerful tool for obtaining resonance frequencies, which are crucial for comprehending the Martian environment.

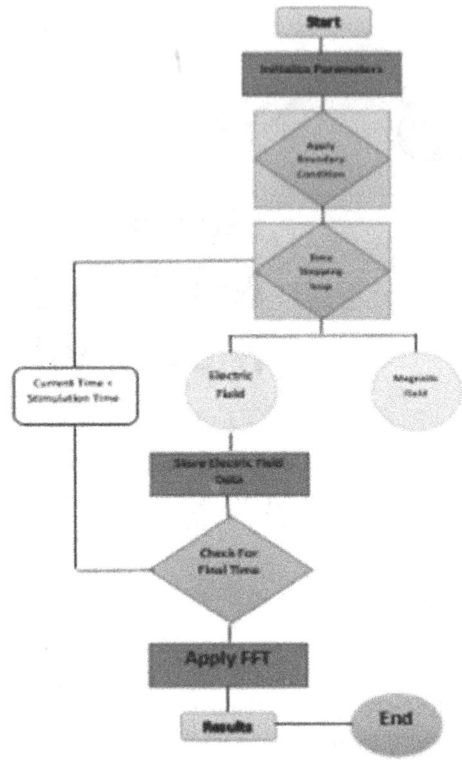

Fig. 43.1 Flow chart for the complete process of the FDTD method with the FFT model

3. DISCUSSION

Figure 43.2: Simulations of the electric field E in the Martian ionosphere at two moments in time, T = 300 and T = 400, using modeling through the finite difference time domain method. Simulation studies involve an interaction of electromagnetic waves with the Martian ionosphere while allowing for a change in permittivity conditions and accounting for a time-dependent electrical conductivity. The left-hand plot shows oscillations of electric field, which proves to be wave propagation in the given low-conductivity medium. However, oscillations are reasonably defined with minimal loss, so energy is retained through this area. The vertical dashed red line is the spot where conductivity is high. Post this point, the amplitudes of waves declined steeply due to losing energy in the ionized area of the Martian atmosphere ionosphere.

More frequently, the oscillations that occurred before the switching for T=400 than for T=300. This means that in effect, the system evolves with time dynamically, possibly because it is time-dependent due to conductivity profile. The dynamics obey the theoretical expectations of wave electromagnetic propagation in conducting and non-conducting media. In regions of low conductivity, waves propagate freely but become absorbed or lose energy once they enter regions of high conductivity. This is an approximate gradual change of waves as they interact with time-dependent permittivity and conductivity of the Martian environment.

4. CONCLUSION

The study neatly presents how to apply the FFT technique together with the FDTD method for the analysis of electromagnetic wave propagation in the Martian atmosphere. We obtained this model, which dynamically describes the behavior of waves on Mars, by applying Maxwell's equations, appropriate boundary conditions, and time-dependent conductivity profiles. The results indicate that the wave propagates with minimum loss in the low conductivity regions but loses most of its energy in regions of high conductivity. This example demonstrates how the ionospheric conductivity dictates the phenomena of Schumann resonance and ELF waves at the Martian surface. It represents new work since it takes a time-domain numerical method to simulate the intricate interactions of the wave in the ionosphere of Mars. The method therefore delivers even more realistic information regarding the electromagnetic dynamics of Mars by expanding the scope of studies on planetary atmospheres. This idea can be expanded by further studies using additional Martian environmental parameters such as seasonal variations in atmospheric conductivity or effects from dust storms. Adding 3D simulation to the model may add further

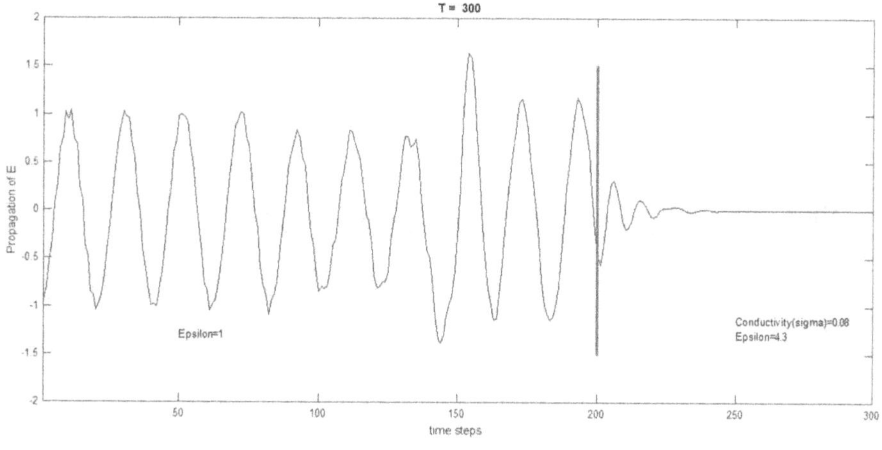

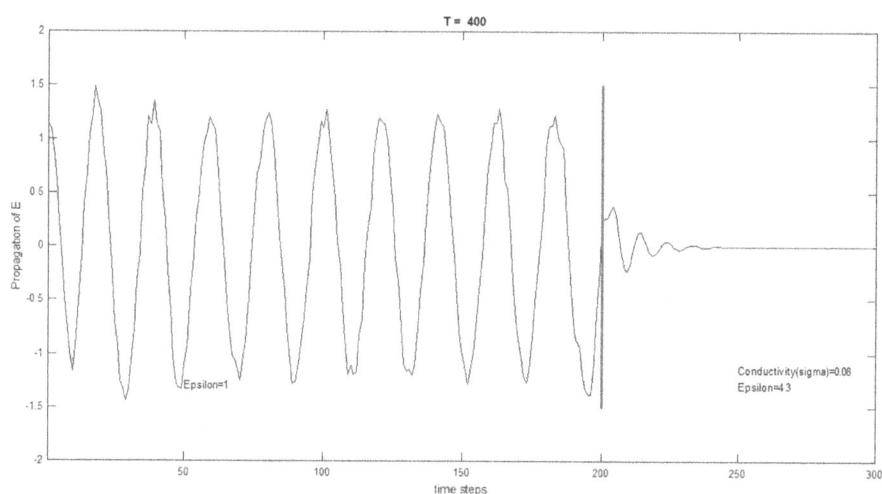

Fig. 43.2 Electric field (E) propagation in the Martian ionosphere for T=300 and T=400-time steps

predictive power and help make the model applicable for mission planning of Martian exploratory missions.

5. ACKNOWLEDGMENTS

I would like to express my heartfelt gratitude to Dr. Jayesh Pabari of the Physical Research Laboratory in Ahmedabad for his valuable thoughts and guidance.

REFERENCES

1. Amin Gasmi, (2022) what is Fast Fourier Transform? 10.13140/RG.2.2.28731.49444.
2. J. P. Pabari, (2015) Dust in Solar System, PLANEX Newsletter (ISSN: 2320-7108), Physical Research Laboratory, Ahmedabad, Vol. 5, Issue 3, pp. 20–25.
3. Jamesina J. Simpson, (2007) A review of progress in FDTD maxwell's equation modelling of Impulsive Subionospheric propagation below 300kHz, IEEE Transactions on Antennas and Propagations, 55 (6), 1582–1590.
4. M. Hayakawa, and T. Otsuyama, (2002) FDTD Analysis of ELF Wave Propagation in Inhomogeneous Subionospheric Waveguide Models, Aces Journal, VOL. 17, NO. 3.
5. Oberst Ulrich, (2007) The Fast Fourier Transform. SIAM J. Control and Optimization. 46. 496-540. 10.1137/060658242.
6. W.M. Farrell et al., (2006) Integration of electrostatic and fluid dynamics within a dust devil, J. Geophys. Res., 111, E01006, 1–10, doi:10.1029/2005JE002527.

Note: All the figures in this chapter were made by the authors.

Recent Trends in Applied Physics and Material Science – Sudhir Bhardwaj et al. (eds)

Radiation Accretion and Rotating Black Hole Longevity in Brans-Dicke Theory

Bijaya Kumar Sahoo
P. G. Department of Physics, Fakir Mohan University, Balasore, Odisha, India

Suryakanta Swain
Dept. of Physics and Astronomical Science, Central University of Himachal Pradesh, Dharamshala, India

Bibekananda Nayak*
P. G. Department of Physics, Fakir Mohan University, Balasore, Odisha, India

Abstract: In the present study, we consider the rotating black holes dynamics in the framework of Brans-Dicke Theory. Through our analysis, we find that the radiation accretion by rotating black holes dominates over the Hawking radiation effects during the early phase of evolution, whereas the evaporation term dominates in later cosmological epochs. Furthermore, both the processes are affected by the time-dependent variation of the gravitational constant. As a result, they may persist for significantly longer durations than their counterparts in standard cosmological models.

Keywords: Brans-dicke theory, Kerr black holes

1. INTRODUCTION

Brans-Dicke Theory (BDT) (Brans and Dicke, 1961) is often examined as a viable replacement of standard model of cosmology involving GTR. In this theory, the gravity and the scalar field, which can be taken as inverse of time-dependent Newton's Gravitational constant, are coupled with a coupling parameter ω. Using this BDT, researchers are able to explain the expanding nature of universe, early and latter stage dynamics of universe and cosmic coincidence problem etc. On the other hand, according to no-hair theorem, a black hole can be explained by the help of mass, angular momentum and charge. Here we have considered a rotating black hole (RBH), which was first given by R. P. Kerr in 1963 (Kerr, 1963), and can be characterized only by other two parameters excluding charge. Further, the evolution mechanism of RBH comprises of accretion of radiation and Hawking evaporation (Hawking, 1975). So, it is expected that consideration of Brans-Dicke formalism could provide an interesting result for studying the evolution of RBH in different cosmic eras. In this backdrop, here we have investigated the evolution of RBH in BDT. Again, by considering BDT, we have analysed the impact of accretion efficiency (f) and rotation through rotating parameter (a^*) on the lifespan of RBHs.

2. MATHEMATICAL FORMALISM

The Friedmann equation for flat FRW universe with a as scale factor and ϕ as BD scalar field, can be written as

$$H^2 + H\left(\frac{\dot{\phi}}{\phi}\right) - \left(\frac{\omega}{6}\right)\left(\frac{\dot{\phi}}{\phi}\right)^2 = \frac{8\pi\rho}{3\phi} \tag{1}$$

Here $H = \frac{\dot{a}}{a}$ and ρ represents Hubble parameter and total energy density respectively. In our assumption, the universe had witnessed radiation domination ($t < t_e$) before it reached present matter dominated era ($t > t_e$). As we are interested about radiation dominated era, we focus our calculation based on some background cosmological relation in that particular era. Now we can write the expression for density and scale factor for ($t < t_e$) as $\rho(a)$ α a^{-4} & $a(t) \propto t^{\frac{1}{2}}$. Further, the solution of $G(t)$ calculated by Barrow and Carr (Barrow and Carr, 1996) for era of radiation domination ($t < t_e$) is $G(t) = G_0\left(\frac{t_0}{t_e}\right)^n$. Here

*Corresponding author: bibekanandafm@gmail.com

DOI: 10.1201/9781003684718-44

G_0, t_0, t_e and n denote the standard value of gravitational constant, the present time, the end time of radiation era and the parameter connected with coupling parameter ω as $n = \frac{2}{4+3\omega}$ respectively. Now the density expression for radiation dominated era becomes

$$\rho_R = \left(\frac{3}{32\pi G_0}\right)\left(\frac{t_e}{t_0}\right)^n\left(\frac{1}{t^2}\right) \quad (2)$$

3. EVOLUTION OF RBHS IN BRANS-DICKE THEORY

3.1 Radiation-Based Accretion

In this segment, we have studied the impact of accretion of radiation on the growth mechanism of RBHs. Now the rate of accretion can be (Nayak et al., 2011) given by $\dot{M}_{acc} = 4\pi f_{rad} R_{RBH}^2 \rho_R$, which further becomes

$$\dot{M}_{acc} = \frac{3fG_0}{8c^3}\left(M + \sqrt{\left(M^2 - \left(a^*\right)^2\right)}\right)^2\left(\frac{t_e}{t_0}\right)^n\left(\frac{1}{t^2}\right) \quad (3)$$

where $R_{RBH} = \frac{G_0}{c^2}\left(M + \sqrt{\left(M^2 - \left(a^*\right)^2\right)}\right)^2$.

To understand the whole mechanism of radiation accretion, we plot Fig. 44.1 and Fig. 44.2. Figure 44.1 explains that the black hole mass increases with increase in accretion efficiency and Fig. 44.2 indicates that when we increase the rotating parameter, the black hole mass also increases. Here it is worth noting that in the early era, the surrounding energy-density is extremely high. So, the energy accumulation by the black holes would be higher and hence the lifetimes of black holes would be more.

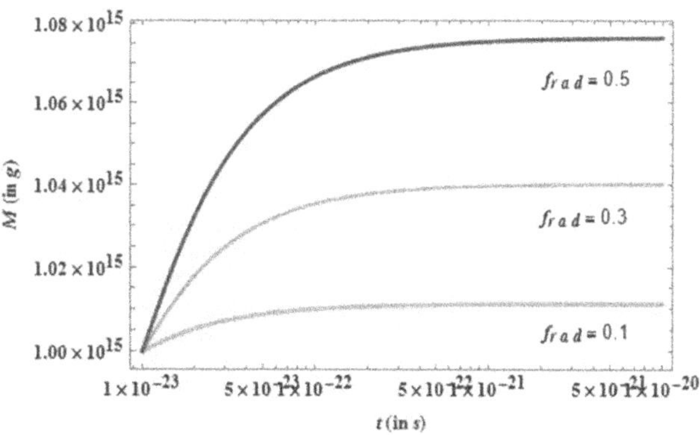

Fig. 44.1 Change in RBH mass with several accretion efficiencies (f) while keeping rotating parameter (a^*) fixed

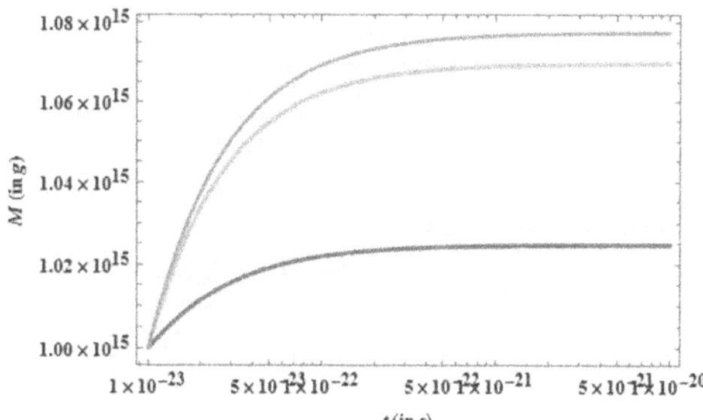

Fig. 44.2 Change in RBH mass by varying rotating parameter ($a^* = M_i$ (Red), $a^* = \frac{M_i}{2}$ (Cyan), $a^* = \frac{M_i}{4}$ (Orange)) at fixed accretion efficiency (f)

3.2 RBHs Evolution under Hawking Evaporation

We can write the expression for decrease of mass of RBH due to Hawking evaporation (Mahapatra and Nayak, 2016) as $\dot{M}_{evp} = -4\pi R_{RBH}^2 a_H T_{BH}^4$, which gives

$$\dot{M}_{evap} = -\left(\frac{\hbar^4 c^6 \sigma_H}{64\pi^3 G_0^2}\right)\left(\frac{\left(M^2 - \left(a^*\right)^2\right)}{M^4\left(M + \sqrt{\left(M^2 - \left(a^*\right)^2\right)}\right)^2}\right)\left(\frac{t_e}{t_0}\right)^{2n} \quad (4)$$

where a_H is the blackbody constant and T_{BH} indicates the Hawking temperature. By considering Hawking evaporation in addition to radiation accretion, the complete evolution of RBH can be described by the equation

$$\dot{M}_{evol} = -\left(\frac{\hbar^4 c^6 \sigma_H}{64\pi^3 G_0^2}\right)\left(\frac{\left(M^2 - \left(a^*\right)^2\right)}{M^4\left(M + \sqrt{\left(M^2 - \left(a^*\right)^2\right)}\right)^2}\right)\left(\frac{t_e}{t_0}\right)^{2n} + \frac{3fG_0}{8c^3}\left(M + \sqrt{\left(M^2 - \left(a^*\right)^2\right)}\right)^2\left(\frac{t_e}{t_0}\right)^n\left(\frac{1}{t^2}\right) \quad (5)$$

The first term of the above equation is dominant in the early formation part and the second term is dominated in the latter part of black hole evolution. Now we make Table 44.1 to present variation of life times of RBHs due to accretion.

Table 44.1 indicates that accretion efficiency extends the lifespan of RBH in BDT than that of in standard model of cosmology. We have further found that for the given accretion efficiencies the evaporation time of a black

Table 44.1 Calculation of evaporation time (t_{evap}) of a particular RBH for different accretion efficiencies keeping a^* constant

$t_i = 10^{-23}$ s, $M_i = 10^{15}$ g, $(a^*)^2 = 10^{-9}\, M_i^2$	
f	t_{evap} (in s)
0	8.84×10^{16}
0.1	9.90×10^{16}
0.2	1.11×10^{17}
0.3	1.26×10^{17}
0.4	1.43×10^{17}

hole in B-D theory is greater than that in the case of loop quantum cosmology (Swain et al., 2024) and standard cosmology.

4. CONCLUSION

From the study of RBH evolution in BDT, here we have observed that the accretion of radiation for the growth mechanism of black hole is very much sensitive to both the accretion efficiency (f) and the rotating parameter (a^*). Again, the RBH accreting mass is found to be larger in rotating case than its non-rotating counterpart. Further, considering Hawking radiation, we have estimated the lifetimes of RBHs. Within this environment, we found that the chances of getting the signature of black hole is more, when we have increased the value of accretion efficiency for a fixed rotation.

REFERENCES

1. Brans, C. and Dicke, R. H. (1961). Mach's Principle and a Relativistic Theory of Gravitation. Phys. Rev. D. 124(3):925–935.
2. Kerr, R. P. (1963). Gravitational field of a spinning mass as an example of algebraically special metrics. Phys. Rev. Lett. 11:237.
3. Hawking, S. W. (1975). Particle creation by black holes. Commun. Math. Phys. 43:199–220.
4. Barrow, J. D. and Carr, B. J. (1996). Formation and evaporation of primordial black holes in scalar-tensor gravity theories. Phys. Rev. D 54:3920.
5. Nayak, B. and Singh, L. P. (2011). Accretion, primordial black holes and standard cosmology. Pramana - J. Phys. 76:173–181.
6. Mohapatra, S. and Nayak, B. (2016). Accretion of radiation and rotating primordial black holes. J. Exp. Theor. Phys. 122:243–247.
7. Swain, S., Sahoo, G. and Nayak, B. (2024). Unveiling the evolution of rotating black holes in loop quantum cosmology. Sci Rep 14:16928.

Note: All the figures and tables in this chapter were made by the authors.

Recent Trends in Applied Physics and Material Science – Sudhir Bhardwaj et al. (eds)
 ISBN 978-1-041-16452-4

Enhancing Early Warning Systems for Extreme Rainfall Events using a Novel Cloud Detection Index: A Case Study of Mumbai, India

Charitarth A. Vyas*, Niket Shastri
Sarvajanik University, Dr. R. K. Desai Marg, Athwalines, Surat, India

Abstract: Extreme weather events, including intense rainfall, flash floods, and cloudburst incidents, have increased significantly in recent years, posing severe threats to human life, the environment, and causing global economic disruptions. Developing effective forecasting and nowcasting tools for these events is essential for safeguarding communities and ensuring efficient disaster management. This study addresses the prediction of extreme rainfall by focusing on cloud formation mechanisms and their role in precipitation. We employed a novel Cloud Detection Index (CDI) that integrates critical atmospheric parameters, namely relative humidity and precipitable water vapor, to identify cloud presence and assess the potential for extreme precipitation. Building on the methodology of Shastri and Pathak (2019) [Advances in Remote Sensing, 8, 30–39], this research utilized CDI to detect cloud patterns associated with extreme rainfall event in Mumbai, India, several days in advance. The results demonstrate a strong correlation between CDI predictions and observed extreme rainfall, effectively identifying low- and mid-level clouds and estimating their thicknesses. The CDI shows higher sensitivity and specificity in forecasting extreme precipitation than conventional cloud detection indices. Our findings validate CDI as a reliable early warning tool, capable of contributing to local extreme weather prediction systems. By integrating these findings with existing meteorological models, we aim to provide a more accurate framework for early warning systems to mitigate the impact of extreme weather events. This approach offers a promising pathway for improving early warning systems, enabling proactive measures for disaster preparedness and risk reduction.

Keywords: Extreme rainfall, Forecasting, Cloud detection index, Early warning systems

1. INTRODUCTION

Clouds play an essential role in Earth's atmospheric and climatic systems. They regulate the hydrological cycle, influence energy balances, and serve as critical indicators of weather patterns. Understanding cloud formation mechanisms is pivotal for predicting extreme weather events like heavy rainfall, which can lead to devastating societal and economic impacts. The increasing frequency of extreme weather events, including heavy rainfall, highlights the importance of understanding atmospheric stability.

On 29th June, 2023, Mumbai experienced an extreme rainfall event characterized by 120 mm of precipitation within 24 h, causing severe disruptions. This case underscores the critical need for accurate early warning systems to mitigate the impacts of such events (Janmani et al., 2006; Patel and Shete, 2015). This study applied the cloud detection index (CDI) (Shastri and Pathak, 2019) to forecast this event. CDI integrates relative humidity and precipitable water vapor, providing a detailed assessment of cloud characteristics associated with extreme precipitation.

2. DATA

This study used radiosonde data from the University of Wyoming (https://weather.uwyo.edu/) for Mumbai station (VABB, 19.11° N, 72.85° E) during June 2023. Data

*Corresponding author: charitarthvyaspro@gmail.com

DOI: 10.1201/9781003684718-45

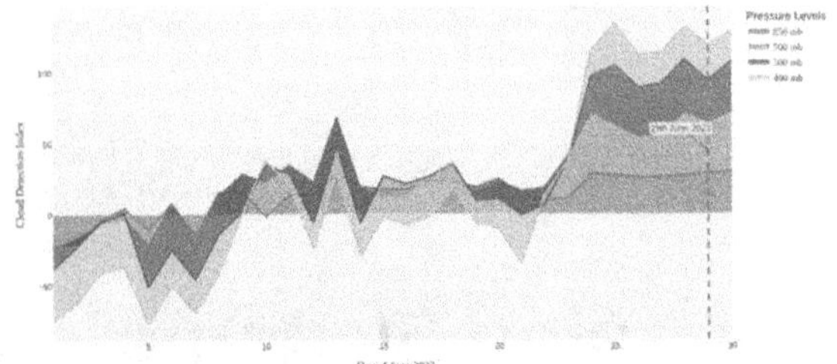

Fig. 45.1 CDI at different pressure levels throughout June 2023. [Original work]

were recorded at 12:00 UT daily. No missing values were found at the selected pressure levels—1000 mb, 850 mb, 700 mb, 500 mb, and 400 mb—used to calculate the CDI. These levels represent the surface and altitudes of 1500 m, 3000 m, 5800 m, and 7500 m, respectively.

3. RESULTS AND DISCUSSION

Analysis of CDI in Fig. 45.1 reveals positive values across all pressure levels on 29th June, coinciding with extreme rainfall in Mumbai. Positive CDI value confirms the presence of cloud at that height (Shastri and Pathak 2019). Leading up to this event, CDI values generally increased, particularly at lower pressure levels (850 mb and 700 mb), indicating increased cloud cover and moisture. The vertical distribution of CDI shows positive values appearing first at lower levels and gradually extending upwards, suggesting the development and growth of rain-producing systems. Given the CDI distribution, we further analyzed cloud activities between 24–29 June 2023.

Figure 45.2 illustrates convective inhibition (CINE) and convective available potential energy (CAPE) values 24 h prior to 24–29 June. CINE values generally decreased from June 23rd to 27th, indicating reduced energy required for convection initiation, while CAPE values peaked on 27th June. However, CDI exhibited a more consistent increase leading up to 29th June, suggesting it may be a more reliable predictor of heavy rainfall potential.

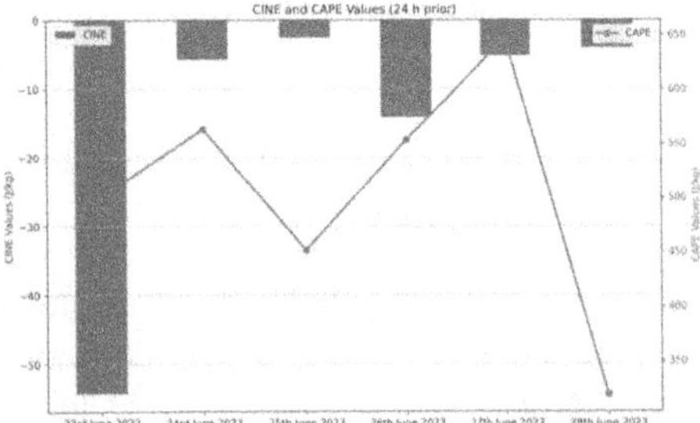

Fig. 45.2 CINE and CAPE values on 23–28 June 2023. [original work]

Incorporating wind profiles and equivalent potential temperature (Fig. 45.3) reveals wind shear (1st panel) and directional shear (2nd panel), both of which can contribute to the development of thunderstorms and heavy rainfall. The shift from easterly to westerly winds with height is evident in the wind direction profiles. A westerly shift indicates the onset or strengthening of a monsoonal or synoptic-scale circulation system, bringing moisture from Arabian Sea and dynamic conditions favorable for extreme

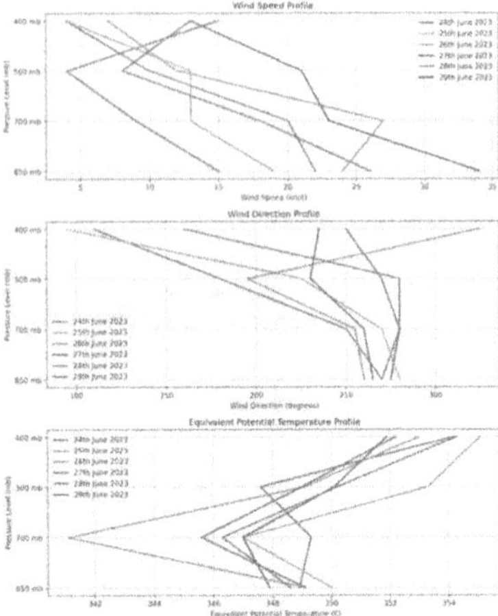

Fig. 45.3 Vertical profiles of wind speed, direction, and equivalent potential temperature for 24–29 June 2023. [original work]

rainfall (Jenamani et al., 2006). Equivalent potential temperature profiles (3rd panel) generally increase with height, indicating potential instability, with steeper lapse rates on 24th and 27th June and suggesting greater convective instability. Wind profiles show strengthening activity at low-to-mid pressure levels, favoring moisture transport inland and organized convection. Furthermore, average thickness of clouds for 24th–29th June derived using CDI was approximately 6000 m. Such thick cloud can retain high humidity, favoring extreme precipitation.

4. CONCLUSION

This study underscores the potential of the CDI as a valuable tool for predicting extreme rainfall events. The ability of CDI to identify cloud patterns associated with heavy precipitation, several days in advance, offers a significant advantage over conventional indices. By integrating CDI with other meteorological parameters, such as wind shear and atmospheric instability, a more comprehensive understanding of the convective environment and its potential to generate extreme rainfall can be achieved. This integrated approach can lead to improved accuracy and lead time in rainfall prediction, crucial for effective disaster preparedness and mitigation strategies. However, we note our work is limited to one case study. Hence, we aim to incorporate a wider range of extreme rainfall events across diverse geographical regions within India, further refining the predictive capabilities of CDI.

REFERENCES

1. Janmani, R. K., Bhan, S. C. and Kalsi, S. R. (2006). Observational/forecasting aspects of the meteorological event that causes a record highest rainfall in Mumbai. Current Science 90(10):1344–1362.
2. Patel, N. R. and Shete D. T. (2015). Analyzing precipitation using concentration indices for north Gujarat agro climatic zone, India. Aquatic Procedia 4:917–924.
3. Shastri, N. and Pathak, K. N. (2009). New cloud detection index (CDI) for forecasting extreme rain events. Advances in Remote Sensing 8(1):30–39.

Note: All the figures in this chapter were made by the authors.

Recent Trends in Applied Physics and Material Science – Sudhir Bhardwaj et al. (eds)

CMBR Constraint on Black Hole Formation in Standard Cosmology

Debasis Sahu*, Bibekananda Nayak
P. G. Department of Physics, Fakir Mohan University, Balasore, Odisha, India

Abstract: Present study aims to examine the black holes on the basis of their abundance in the framework of standard model of cosmology. Mainly two processes are involved in describing the dynamics of black holes: accretion of the surrounding energy-matter, and quantum mechanical Hawking radiation. The emission of Hawking radiation from black holes can significantly influence various astrophysical phenomena during the universe's expansion. A key effect of this radiation is its potential impact on the cosmic microwave background radiation, which is a residue of thermal radiation from early universe, which today has a temperature of approximately 2.725 K. In this context, we have assessed the maximum limits on the initial mass fraction of black holes for several accretion efficiencies, yielding some intriguing results. Furthermore, we have compared our findings with predictions from Brans-Dicke theory, offering additional insights into the interplay between black hole evolution and cosmological models.

Keywords: Black hole, CMBR, Brans-Dicke theory

1. INTRODUCTION

Standard model of cosmology (SMC) is the result of key concepts from general relativity, quantum mechanics, and observational astronomy. According to SMC, the universe emerged about 13.8×10^9 years ago from singularity, a highly hot and dense condition and has been expanding till today. SMC demands the nature of the universe to be homogeneous and isotropic which is described by FRW metric. This model is the elementary platform to study various cosmological and astrophysical processes. Among them black hole (BH) dynamics is a current intriguing subject. The mass evolution of BHs typically depends on two aspects: the Hawking evaporation (Hawking, 1975) that is quantum mechanical in nature, which is responsible for the decay of mass, and the accumulation of surrounding matter and energy contributes to the growth in mass and the process is coined as accretion (Nayak and Singh, 2010). BHs have a great effect on the evolution of the universe in subsequent cosmological epochs, both through their overall contribution to the total matter density and through the emission of Hawking radiation. The presence and characteristics of BHs at different stages of cosmic history are subject to a variety of observational constraints, which include current measurements of the universe's matter density, the spectrum of photons, the cosmic microwave background radiation (CMBR), and the population of light elements (like Helium and Deuterium). These observational data, as discussed in literature (Nayak et al., 2010), place important limits on the mass fraction of BHs in the early universe, and these constraints can be translated into bounds on the initial distribution of BH masses. Moreover, by considering the accretion can lead significant change in their constraints. In this work, In the light of SMC, we have gone through astrophysical constraint on the initial mass fraction of BHs due to distortion of the CMB spectrum. We are sensitive to influence of accretion of radiation during that period.

2. MATHEMATICAL FORMULATION

The evolution equation of BH can be expressed as (Babichev, 2004; Nayak and Singh, 2011)

$$\dot{M}_{\text{evo}} = \dot{M}_{\text{evap}} + \dot{M}_{\text{acc}}$$

Or,
$$\dot{M}_{\text{evo}} = -4\pi R_{\text{BH}}^2 a_H T_{\text{BH}}^4 + 4\pi f R_{\text{BH}}^2 \rho \qquad (1)$$

where, R_{BH} (= $2GM$), a_H, T_{BH} (= $1/8\pi GM$), f and $\rho(=3t^{-2}/32\pi G)$ stands for Schwarzschild radius of BH, Stephan-Boltzmann constant, Hawking temperature,

*Corresponding author: debasissahu777@gmail.com

DOI: 10.1201/9781003684718-46

accretion efficiency and density of the surrounding energy-matter in radiation dominated era respectively. The final mass fraction of BH (α_{evap}) in terms of the initial one (α_i) is (Nayak et al., 2010)

$$\alpha_{\text{evap}} = \alpha_i \frac{M(t_c)}{M(t_i)} \times \frac{a(t_{\text{evap}})}{a(t_i)} \tag{2}$$

where $M(t_c)$ and a are the peak accreting mass of BH and scale factor respectively.

Hawking radiation emitted at $t \geq 4 \times 10^{-10} t_0$ cannot be completely thermalized and hence CMB spectrum would be affected by it. This also modified the plank spectrum which can be easily explained through use of chemical potential (μ) (Sunyaev and Zeldovich, 1970). In fact the relative energy density can be expressed as

$$\frac{\rho_{\text{evap}}}{\rho_{\text{rad}}(t)} = 0.71\mu \tag{3}$$

where ρ_{evap} is associated with evaporating BHs. An upper limit on μ can be set by the observational data (Fixsen et al., 1996) as

$$\mu < 9 \times 10^{-5}. \tag{4}$$

Considering that CMB spectrum can be disturbed by the particles bearing half of the emitted energy, we get

$$\frac{1}{2}\alpha_{\text{evap}} = 0.71\mu. \tag{5}$$

Now use of equation (4) can give

$$\alpha_{\text{evap}} < 1.28 \times 10^{-4}. \tag{6}$$

Again, from equations (2) and (6), α_i of BH can be calculated as

$$\alpha_i < 1.28 \times 10^{-4} \times \frac{M(t_i)}{M(t_c)} \times \frac{a(t_i)}{a(t_{\text{evap}})} \tag{7}$$

Or, $$\alpha_i < 1.28 \times 10^{-4} \times \frac{M(t_i)}{M(t_c)} \times \left(\frac{t_i}{t_{\text{evap}}}\right)^{\frac{1}{2}}. \tag{8}$$

3. RESULTS AND DISCUSSION

For the calculation we consider BHs of Schwarzschild type. These BHs have no charge or rotation. Their mass is the central parameter for their evolution. We here estimate the initial mass fraction of BHs (α_i) with different accretion efficiencies (f) for distortion in CMB spectrum. We state our results in Fig. 46.1 and Table 46.1, where $f = 0$ indicates lack of accretion.

Our analysis indicates that the inclusion of accretion process leads to a stronger limit on the mass fraction of BH to be formed, as inferred from the distortion of the CMB spectrum, when compared to the predictions made within the Brans-Dicke theory (Nayak et al., 2010). Additionally, we find that increasing the accretion efficiency results in more stringent upper limits on the initial BH mass fraction.

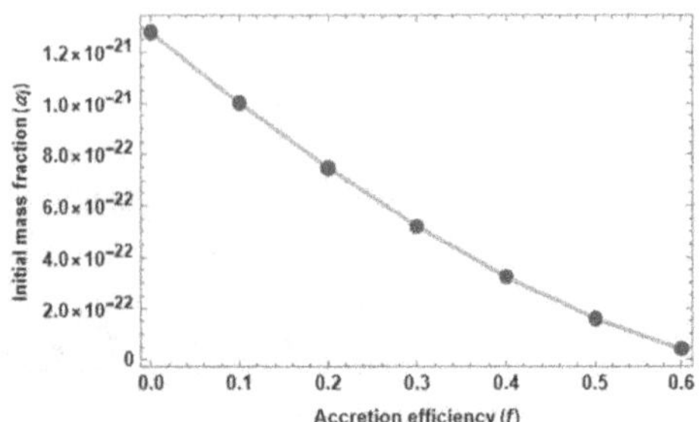

Fig. 46.1 Variation of initial mass fraction of BHs (α_i) with different accretion efficiencies (f) for distortion in CMB spectrum

Table 46.1 Estimation of initial mass fraction of BHs (α_i) for various accretion efficiencies (f)

Distortion of CMB Spectrum	
Accretion efficiency (f)	**Initial mass fraction (α_i) <**
0	1.2771×10^{-21}
0.1	1.0008×10^{-21}
0.2	0.7479×10^{-21}
0.3	0.5209×10^{-21}
0.4	0.3231×10^{-21}
0.5	0.1596×10^{-21}
0.6	0.0404×10^{-21}

4. CONCLUSION

It is a well-known fact that out of several astrophysical processes CMB spectrum distortion can be regarded as one of the vital foot print for estimation of initial mass fraction of black holes. In this analysis we have tried to project a detailed picture cosmological impact of CMBR distortion on formation of BH within standard model of cosmology. Particularly, we have analysed how formation BH density is constrained by the observational CMBR data. Again inclusion of accretion of radiation into the calculation enhances the life span of BHs to a large extent. As a result more number of BHs is available due to accretion and hence the bounds on formation mass of BHs become tightened when we raise the accretion efficiency.

REFERENCES

1. Hawking, S.W. (1975). Particle creation by black holes. Commun. Math. Phys. 43:199–220.
2. Nayak, B., Singh, L.P., (2010). Note on nonstationarity and accretion by primordial black holes in Brans-Dicke theory. Phys. Rev. D 82: 127301.
3. Nayak, B., Majumdar, A. and Singh, L.P. (2010). Astrophysical constraints on primordial black holes in Brans-Dicke theory. J. Cosmol. Astropart. Phys. 2010:039.
4. Babichev, E., Dokuchaev, V. and Eroshenko, Yu. (2004). Black hole mass decreasing due to phantom energy accretion. Phys. Rev. Lett. 93:021102.
5. Nayak, B. and Singh L.P. (2011). Accretion, primordial black holes and standard cosmology. Pramana - J Phys. 76:173–181.
6. Sunyaev, R.A. and Zeldovich, Y.B. (1970). The Interaction of matter and radiation in the hot model of the universe, II. Astrophys. Space Sci. 7:20–30.
7. Fixsen D.J. et al. (1996). The Cosmic Microwave Background Spectrum from the Full COBE/FIRAS Data Set. Astrophys. J. 473:576–587.

Note: All the figures and tables in this chapter were made by the authors.

Recent Trends in Applied Physics and Material Science – Sudhir Bhardwaj et al. (eds)

47 New Structures in Radio Galaxies with RAD@home Citizen Science, GMRT and LOFAR Radio Telescopes

Prasun Machado*
Department of Physics, SIES College of Arts, Science and Commerce (Empowered Autonomous), Sion (West), Mumbai, India
RAD@home Astronomy Collaboratory, Kharghar, Navi Mumbai, Mumbai, India

Ananda Hota
RAD@home Astronomy Collaboratory, Kharghar, Navi Mumbai, Mumbai, India
UM-DAE Centre for Excellence in Basic Sciences, University of Mumbai, Kalina, Mumbai, India

Binayak Ashis Pati, Aditya Sahasranshu, Prakash Apoorva
RAD@home Astronomy Collaboratory, Kharghar, Navi Mumbai, Mumbai, India

Aarti Muley
Department of Physics, SIES College of Arts, Science and Commerce (Empowered Autonomous), Sion (West), Mumbai, India

Arundhati Purohit
RAD@home Astronomy Collaboratory, Kharghar, Navi Mumbai, Mumbai, India

Abstract: Typical radio galaxies with FR I or FR II like linear structures now seem to be "tip of the iceberg" and modern sensitive radio telescopes are revealing new structures in them. Following RAD@home citizen science approach and exploring new data releases we have discovered rare new structures in radio galaxies. We have spotted them, primarily using the LoTSS survey (144 MHz using LOFAR radio telescope). These include the following: (1) a rare Collimated Synchrotron Thread (CST) of a possibly episodic radio galaxy with a size of 180 kpc. This is only the third CST discovered, till date. (2) Radio "burl" of roughly 50-75 kpc size, which are located on the radio jets as compact features other than core, hotspot and kinks. (3) A new 100-400 kpc size relic lobe on one side of standard double lobed radio galaxies but the relic emission is seen on only one side of the host galaxy. This is unlike the well-known double-double radio galaxies with relic lobes on either side of the young lobes. We aim to study these newly found structures in more detail by observing them using GMRT for spectral curvature imaging and ageing studies. This will help us understand their role in merger and feedback driven galaxy evolution.

Keywords: Active galactic nuclei (AGN), Radio galaxies, Citizen science

1. INTRODUCTION

Since 1953 the structure of radio galaxies are well-known where FR I type has core-plume edge-darkened structure and FR II has edge-brightened structure with core-jet-hotspot and lobes with backflow. Recently, due to high dynamic range imaging, new features like wings in X-shaped radio galaxies and faint radio filaments called radio Collimated Synchrotron Thread (CST) and puzzling Odd Radio Circles (ORC) have been discovered (see review by Hardcastle & Croston 2020). In this paper we briefly report a rare new CST, a rare giant relic radio lobe and a possible new structure we named "burl" in the radio jet discovered through the first Indian, nationwide, Inter-University, citizen science research platform RAD@home (Hota et al. 2024).

2. DISCOVERY METHODS

Undergraduate students of India were trained to interpret UV-Optical-IR-Radio RGB-contour images through

*Corresponding author: prasunmachado@gmail.com

DOI: 10.1201/9781003684718-47

image-share and text-comment discussion through a large (4700 member) Facebook group discussion. Selected members were further trained in RAD@home workshops hosted by various research and educational institutes all over India. Training was focused to make possible discovery from the FITS image files of TGSS taken with the GMRT telescope at 150 MHz with 25" resolution and noise of 5 mJy/beam rms noise. When available, these trained citizen scientists (e-/i-astronomers) then incorporate LoTSS DR 1 imaging data from LOFAR radio telescope with better sensitivity in 144MHz and PANSTARRS optical images to understand the host galaxies of those unusual radio sources discovered. These preliminary discoveries by the i-astronomers were further discussed online through Google Meet screen-shares with the professional astronomers for follow up observation with the GMRT (through the ongoing GOOD-RAC project) or direct publications.

3. RESULTS

3.1 RAD-Collimated Synchrotron Thread (CST)

Since the realisation of CSTs as a separate unique structure in the MeerKAT radio image of the galaxy ESO 137-006 (Ramatsoku et al. 2020) astronomers have found a few more CSTs in other galaxies but no consensus on its nature and origin. Some CSTs are seen, in projection, diverging out of the radio core/lobe becoming faint and undetectable at the tail ends, and some connect both the radio lobes without any structure, in brightness or spectral index, along the length. Figure 47.1 (left panel) shows 144 MHz LoTSS (LOFAR) image of RAD-CST. The main radio lobes, displaying a double lobed FR II morphology clearly show the jet and the optical host galaxy (RA: 08:31:10 Dec: +41:47:39). A radio filament (CST) starts from the eastern lobe and gets merged with the western lobe at 230 kpc (labelled in the image). There is no prominent brightness gradient along the length of the CST making it difficult to suggest if lobes supply relativistic plasma originating in the AGN host galaxy.

Further puzzling fact is the CST is seen brighter than the radio jet which possibly show relativistic boosting on the eastern side of the radio core. There is a fainter western extension to the western lobe, suggesting either episodic nature or west-to-east motion of the host galaxy making distortions along the length. Taking clues from this scenario, we speculate that CST is probably a paleo-magnetic channel created by passage of a neighbouring satellite galaxy which has got energised by the recent radio lobe of the AGN-host galaxy.

RAD-Burl: The basic anatomy of a radio galaxy is core-jet-hotspot and lobes with backflow. There are rare U- or C-shaped "kink" structures in otherwise straight radio jets (Dabhade et al. 2022). Kinks could be due to magnetohydrodynamic instabilities, magnetic reconnection or precession of the jet axis. However, we have found a brightened clump in the jet, named "burl" in a radio galaxy (R.A.:14:33:39 Dec.:+63:58:43) which is unlike such kinks. Figure 47.1 (middle panel) shows 144 MHz LoTSS image (in red) from LOFAR along with other radio and optical images. The burl region (labelled in the image) fails to show any background galaxy as an independent radio source. Burl is brighter than the lobe and has no eastern counterpart. As per the VLA FIRST/VLASS data, the size of the burl is as big as 60 kpc and has a flat spectral index. Follow up observations will help reveal the nature and origin of this unique feature.

RAD-One sided relic lobe: As the primary aim of RAD@home collaboratory, i-astronomers would find faint-fuzzy diffuse emission around standard FR II radio galaxies. Similar to the southern relic lobe of the exotic radio galaxy Speca (Hota et al. 2011) a giant 425 kpc size relic lobe was spotted to the north of a small double lobed radio galaxy (R.A.:12:19:12 Dec.: +61:08:17). Its southern counterpart, if any, cannot be seen clearly (Figure 47.1 (right panel)). Since this relic lobe is larger than standard radio galaxies (200-300 kpc) it is unlikely due to a jet-galaxy interaction as we found in RAD12 (Hota et al. 2022), but may be due to a larger obstruction like a group or cluster of galaxies

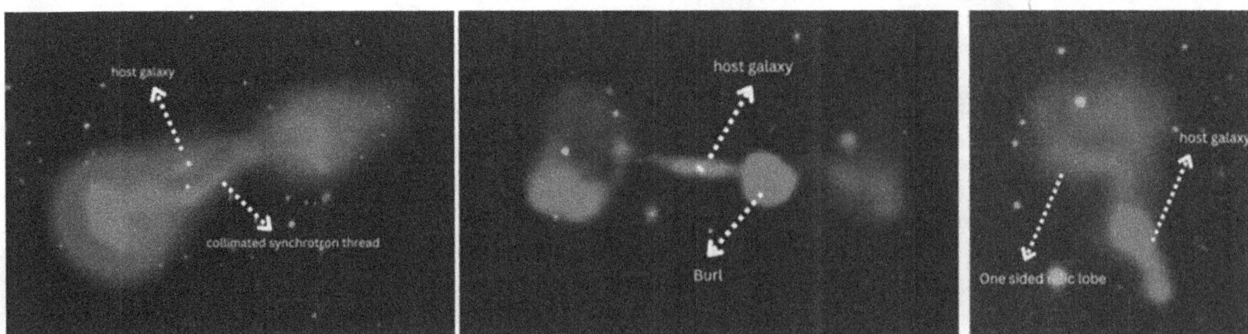

Fig. 47.1 Left Panel: Tricolour image (original) of RAD-CST showing 144 MHz radio image (LoTSS b=6") in red, Optical image (Legacy survey) in green, and 1.4 GHz radio image (NVSS, b=45") in blue. Middle Panel: Similar tricolour image of RAD-Burl showing radio (LoTSS) in red, Optical (SDSS) in green, and 3 GHz radio image (VLASS (JVLA) b=2.5") in blue. Right Panel: Similar tricolour (original) image of RAD-One sided relic lobe showing 144 MHz radio (LoTSS) in red, Optical (SDSS) in green, and 3 GHz radio (VLASS) in blue

on the north. Further investigation would reveal the mechanism of confinement or proliferation of magnetic fields in the large-scale structure of our Universe.

4. CONCLUSION

Following RAD@home citizen science approach, we report here discovery of three new structures from newly released LoTSS 144 MHz radio images. A new 180 kpc CST, a 60 kpc jet-burl and a giant 425 kpc one-sided relic radio lobe is reported. Future follow up through GOOD-RAC will help understand their nature and origin.

REFERENCES

1. Dabhade, P., Shimwell, T. W., Bagchi, J. et.al. (2022),"Barbell-shaped giant radio galaxy with 100 kpc kink in the jet," *Astronomy & Astrophysics*, 668: A64
2. Hardcastle, M and Croston, J., (2020), "Radio galaxies and feedback from AGN jets," *New Astronomy Reviews*.88:101539.
3. Hota,A., Sirothia, S.K., Ohyama, Y., etal. (2011)."Discovery of a spiral-host episodic radio galaxy," *Monthly Notices of the Royal Astronomical Society:Letters*,417(1):L36–L40
4. Hota, A., Dabhade, P., Vaddi, S., et. al. (2022)."RAD@home citizen science discovery of an active galactic nucleus spewing a large unipolar radio bubble onto its merging companion galaxy," . *Monthly Notices of the Royal Astronomical Society: Letters*, 517(1):L86–L91
5. Hota, A., Dabhade, P., Machado, P., et. al. (2024). "Ten years of searching for relics of AGN jet feedback through RAD@home citizen science," [in press] https://arxiv.org/abs/2410.10294
6. Ramatsoku, M., Murgia, M., Vacca, V., et. al.(2020). Collimated synchrotron threads linking the radio lobes of ESO 137-006, *Astronomy & Astrophysics*, 636: L1

Note: The figure in this chapter were made by the authors.

Recent Trends in Applied Physics and Material Science – Sudhir Bhardwaj et al. (eds)

Design a Hemoglobin Concentration based Biosensor using 1D Binary Photonic Crystals

Sanjeev Sharma*
Department of Applied Science & Humanities, IMS Engineering College, Ghaziabad, India

Sri Krishana Singh
Department of Physics, Government College Lamta, Balaghat, Madhya Pradesh, India

Vipin Kumar
Department of Physics, Janta Vedic College Baraut, Uttar Pradesh, India

Kuldeep Singh
Janta Vedic College Baraut, Uttar Pradesh, India

Abstract: A hemoglobin concentration-based biosensor has been examined by using one-dimensional binary photonic crystals. In this device a nanocomposite defect layer (blood) is inserted between two other nanocomposite layers of Si/BaF_2 materials. The concentration of hemoglobin in RBC is a function of the refractive index. The sensing performance of a one-dimensional binary photonic crystal is based on slightly change in the hemoglobin refractive index/ concentrations. The transfer matrix method is employed to analyze the various properties like transmission efficiency, quality factor, sensitivity, and detection limit of a biosensor. The presented structure shows excellent result with small change in the hemoglobin concentration and it shows high sensitivity of about 464.01 nm/RIU. The quality factor and detection limit of the proposed biosensor is recorded as 1419, and 1.40×10^{-3} RIU, respectively.

Keywords: Biosensor, Sensitivity, Quality factor, TMM, Blood sample

1. INTRODUCTION

Generally, Hemoglobin concentration refers to the amount of hemoglobin present in a given volume of blood. Hemoglobin is basically a protein in red blood cells (RBC) which works for transporting oxygen from the lungs to the rest parts of the body. Normal hemoglobin concentrations can vary based on age, sex, and health conditions. For adult men the ranges are approximately 13.8 to 17.2 grams per deciliter (g/dL), for adult ladies it is approximately 12.1 to 15.1 g/dL and for children it can vary, but typically range from about 11 to 16 g/dL, depending on age (Yandamuri at al. 2013).

In the context of photonic crystals, "hemoglobin concentration" typically relates to biomedical applications, specifically in optical sensing and imaging. Generally, photonic crystals are used extensively in biosensors, ODR mirrors, DWDM, optical filters, etc. (Ankita et al. 2020, 2022, 2022A, Sharma et al. 2015, Zhang and Ge 2020, Kumar et al. 2013, 2014, Suthar et al. 2012, 2010, 2009, 2012A). Photonic crystals have wide application in optical fiber communication at mainly second and third transmission windows (Suthar et al. 2022, Ankita et al. 2022, Kumar et al. 2023). S. Sharma et al design various types of devices including DWDM, optical filters etc. using these photonic crystals (Sharma et al. 2015). For instance, in optical biosensors or imaging systems that utilize photonic crystals, the hemoglobin concentration in blood or tissue can be measured to monitor various health parameters (Ankita et al. 2020, 2024). The photonic crystal can be designed to detect changes in the optical properties of the light interacting with the hemoglobin molecules in the blood (Ankita et al. 2020, 2024). In such applications, the sensitivity and accuracy of the measurement depend on the specific design and properties of the photonic crystal, including its periodicity, material composition, and the wavelength of light used. The photonic crystal sensors are designed to detect these concentrations with high precision, often at very low concentrations, which can be crucial for applications like early disease detection

*Corresponding author: sanjeevsharma145@gmail.com

DOI: 10.1201/9781003684718-48

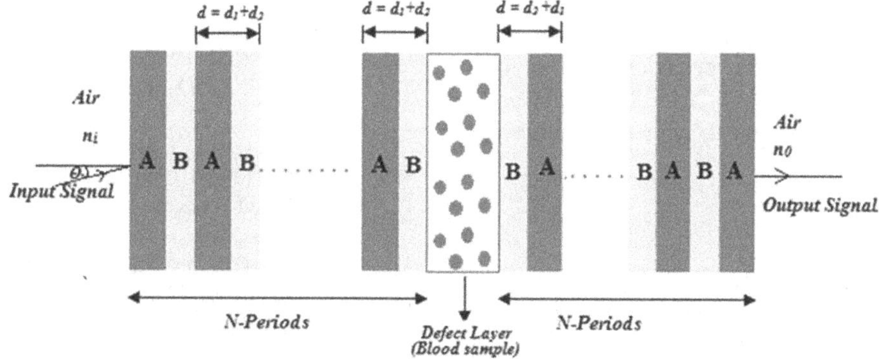

Fig. 48.1 Schematic diagram of 1D PCs with a defect

or monitoring (Suthar and Bhargava 2023, Ankita et al. 2020).

2. THEORETICAL METHOD

The schematic diagram of the proposed device is depicted in Fig. 48.1.

To investigate the proposed biosensor, the refractive indices of silicon and beryllium fluoride materials have been taken at the selected wavelengths as per our choice. The refractive indices of both materials with wavelength λ can be related as (Li 1976):

$$n^2(\lambda) = 1 + \frac{\lambda^2}{0.09385\lambda^2 - 0.00866} \tag{1}$$

$$\&n^2(\lambda) = 1.33973 + \frac{0.81070\lambda^2}{\lambda^2 - (0.10065)^2} + \frac{0.19652\lambda^2}{\lambda^2 - (29.87)^2} + \frac{4.52469\lambda^2}{\lambda^2 - (53.82)^2} \tag{2}$$

To study the transmission properties TMM has been used (Yeh 1988, Suthar et al. 2012). The refractive index of hemoglobin in blood for various concentrations can be related as:

$$n_{Hb} = n_0 + \gamma C_{Hb} \tag{3}$$

Here, $n_0 = 1.3261$ and $\gamma = 0.1834$L/g.

3. RESULTS AND SIMULATIONS

In the schematic diagram $[(Si/BaF_2)^5D(BaF_2/Si)^5]$, D denotes the defect layer of the blood sample with thickness 2000nm which is sandwiched between these photonic crystals. The thickness for Si and BaF_2 material is also taken as d_1 = 230nm and d_2 = 45nm respectively. In the proposed work, a biosensor has been designed by using varying hemoglobin concentration in RBC.

The 3-D view structure of defect layers for the various concentration of hemoglobin in RBC is shown in Fig. 48.2. It`s observed that the defect mode transmission peak shifted towards the lower wavelength range with concentration in the spectrum. At low concentration of hemoglobin, the defect mode shifted towards the shorter wavelength region which responsible the sensitivity. In the spectrum, the transmission peak of defect mode at wavelength 934.5nm indicates the normal range of concentration of hemoglobin (140g/L) in RBC for a healthy person. On the other hand, when hemoglobin concentration in RBC decreases from 140g/L to 120g/L and 120g/L to 100g/L the central wavelength of the defect mode shifts from 934.5nm to 932.7nm, and 932.7nm to 931nm respectively. Up to the concentration of 100g/L in RBC shows the normal state of a person but when the central peak of defect mode shifted from 931nm to 929.3nm it shows the moderate stage (100g/L to 80g/L) of a patient suffering from anemia.

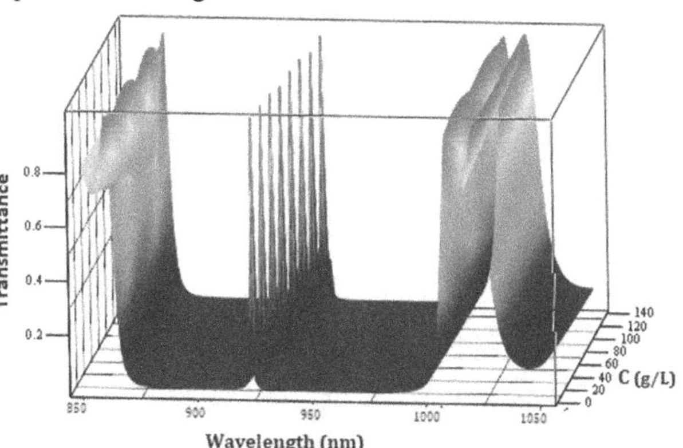

Fig. 48.2 3-D view structure of defect layers for the various concentration of Hemoglobin in RBC

At a concentration of 60g/L of hemoglobin in RBC, the central peak of the defect layer shifted from 929.3 to 927.6nm. Similarly, at concentrations 40g/L, 20g/L, and 0g/L the central peak of the defect layer shifted from 927.6nm to 925.9nm, 925.9nm to 924.2nm, and 924.2nm to 922.5nm respectively. The concentration of hemoglobin below 40g/L in RBC indicates the patient is suffering from vital disease. In this simulation, it`s observed that the transmission peak of defect mode is improved by decreasing hemoglobin concentration. The sensitivity of the proposed structure is recorded as 464.01nm/RIU, which is better than previously published research work [36-38]. In the simulation, the detection limit of this biosensor is recorded as 1.40×10^{-3} RIU. The Q-Factor of this structure is better (1419) than the previously published work (Sharma et al. 2024).

4. CONCLUSION

A hemoglobin concentration-based biosensor has been designed by using 1D binary photonic crystals with a defective layer of blood sample. The proposed device shows high sensitivity for the prediction of different deceases in human. The sensitivity, quality factor, and the detection limit of the device, of the proposed device has been recorded as 464.01nm/RIU, 1419, and 1.40×10^{-3} RIU. The proposed structure can be used as a hemoglobin sensor, glucose, and biomolecules sensors.

REFERENCES

1. Yandamuri at al. (2013). Int. J. Phy. Sci., 5(2), 2086–2089.
2. Ankita et al. (2020) Plasmonics, 16, 59–63.
3. Sharma, S. et al. (2015) Optik, 126 (11–12), 1146–1149.
4. Zhang, Y., Ge, Y. (2020). RCS Advances, 2020, 10, 10972.
5. Kumar et al. (2013). Optik 124 (16), 2527–2530.
6. Kumar et al. (2014). Silicon 6, 73–78.
7. Suthar et al. (2012) Optics Communications 285 (6), 1505–1509
8. Suthar et al. (2010). J. Electron. Sci. Techn. 8(1), 39–42.
9. Suthar et al. (2009). Chalcogenide Letters 6(11), 623–627.
10. Suthar B. and Bhargava A. (2023). Applied Nanoscience 13 (8), 5399–5406.
11. Ankita et al. (2022). Mater. Today: Proc., 62, 5407.
12. Suthar et al. (2012A). Progress in Electromagnetics Research Letters 32, 81–90.
13. Suthar et al. (2022). The European Physical Journal Plus 137 (12), 1301.
14. Ankita et al. (2022A) Macromolecular Symposia 401 (1), 2100319.
15. Kumar et al. (2023). Physica Scripta 98, 065506.
16. Ankita et al. (2024). Optical and Quantum Electronics 56 (7), 1116.
17. Li, H.H. (1976). J. Phys. Chem. Ref. Data, 1976, 5, 329–528.
18. Yeh, P. (1988). Optical Waves in Layered Media. New York: Wiley.
19. Sharma, et al. (2024). Plasmonics, 2024, 19, 1463–1473.

Note: All the figures in this chapter were made by the authors.

Photonic Crystal Fiber-based Refractive Index Sensor for Early Cancer Detection for Terahertz Regime

Shubham Sharma[1], **Ajeet kumar**[2]
Advanced Photonics Simulation Research Lab,
Department of Applied Physics, Delhi Technological University, Delhi, India

Than Singh Saini[3]
Department of Physics,
National Institute of Technology Kurukshetra,
Kurukshetra, HR, India

Abstract: This article reports a novel cancer detection sensor featuring a rectangular core Photonic Crystal Fiber (PCF) designed to identify cancerous cells in the cervix, breast, and skin. The frequency range of 0.9 to 3 THz in the terahertz spectrum has been investigated to achieve higher relative sensitivity with minimal confinement loss. At a frequency of 2.1 THz, significant relative sensitivity responses of 97.75%, 97.40%, and 97.93% have been obtained for cervix, breast, and skin cells, respectively. The confinement losses of x-polarized fundamental mode for cervix, breast, and skin cancer cells have been estimated as low as 2.55×10^{-15}dB/cm, 4.15×10^{-17} dB/cm, and 6.0×10^{-17} dB/cm, respectively.

Keywords: Photonic crystal fiber, Cancer detection, Refractive index, Terahertz, Relative sensitivity

1. INTRODUCTION

Cancer remains a leading global health challenge, accounting for nearly 10 million deaths in 2020, as per WHO data [1]. Early detection is critical to improving survival rates, as cancer arises from the uncontrolled growth of abnormal cells, often driven by environmental factors, lifestyle choices, or carcinogen exposure. Photonic Crystal Fiber (PCF)--based sensors, widely recognized for their high sensitivity, low losses, compact size, and versatility, are increasingly applied in biomedical sensing, including cancer diagnostics. Unlike traditional optical fibers, PCFs offer customizable structural parameters, enabling enhanced optical properties such as improved sensitivity, reduced losses, and optimized mode propagation. In their PCF-based research work, Yadav et al. (2024) introduced a PCF-based sensor for early detection of breast cancer cells in the THz regime, achieving sensitivity values of 65.43% at 1.8 THz. Parvin et al. (2021) developed a spectroscopic optical sensor capable of detecting cancer in multiple tissues, including the adrenal gland, blood, breast, cervix, and skin. Notably, the sensor achieved a high relative sensitivity (RS) of 95.13% for skin cancer detection under x-polarization. As a result, the background material is employed as ZEONEX in the study mentioned above to achieve confinement loss, relative sensitivity, and other optical guiding features (Singh, 2024). Therefore, we have an excellent opportunity to suggest RC-PCF obtain low confinement loss and very relative sensitivity for THz waveguide chemical detection

2. DESIGN AND ANALYSIS

Figure 49.1(a) provides a detailed depiction of the transverse cross-section of the proposed PCF-based cancer cell detector (Singh, 2024). In this configuration, the rectangles that are nearest to the core are designated as R2, with a height of 1050 μm and a width of 150 μm. In the horizontal direction, the rectangles adjacent to the core region, labeled R3, R4, R5, and R6, have the following

[1]shubhamsharma23phdap501@gmail.com, [2]ajeetdph@dtu.ac.in, [3]tsinghdph@gmail.com

DOI: 10.1201/9781003684718-49

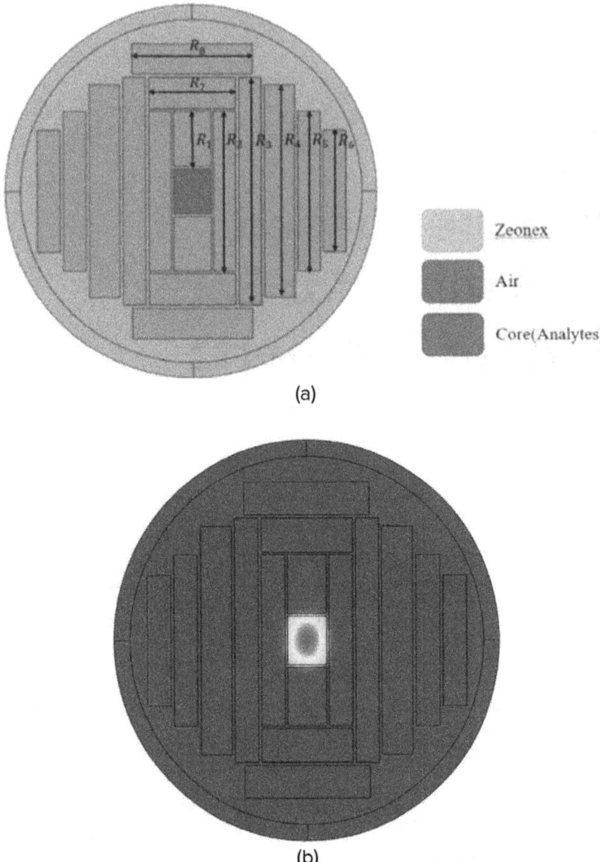

Fig. 49.1 (a) The proposed cross-section of the rectangular core PCF, (b) Distribution profile for cancerous cells

dimensions: R3 measures 1500 μm in height and 150 μm in width, R4 is 1400 μm in height and 200 μm in width, R5 has a height of 200 μm and a width of 1400 μm, and R6 measures 1050 μm in height and 150 μm in width. The rectangles closest to the core are labeled R1, which has dimensions of 250 μm in height and 330 μm in width. In addition, the next nearest rectangles are designated as R7 and R8, with R7 measuring 580 μm in height and 210 μm in width, while R8 has a height of 800 μm and a width of 200 μm. The pitch in both the horizontal and vertical directions is maintained at a constant value of 35 μm. The core dimensions should be maximized to achieve higher core power fraction values.

3. MATHEMATICAL ANALYSES

A crucial parameter for evaluating a sensor's performance is Relative Sensitivity, which can be determined using the following equation.

$$\text{Relative Sensitivity} = \frac{n_a}{n_{eff}} \times PF \tag{1}$$

Confinement loss is another significant optical property of the PCF structure. Notably, a PCF with low confinement loss demonstrates high relative sensitivity. The confinement or leakage loss, L_{conf}, is defined is.

$$L_{conf} = 8.868 \times \left(\frac{2\pi f}{\lambda}\right) \times I_m[n_{eff}]\,\text{dB/m} \tag{2}$$

4. RESULTS AND DISCUSSION

It can be shown that the relative sensitivity increases with frequency. We obtained RS values of 97.75%, 97.40%, and 97.93% for the cervix, breast, and skin in the x-polarized direction at an operating frequency of 2.1 THz.

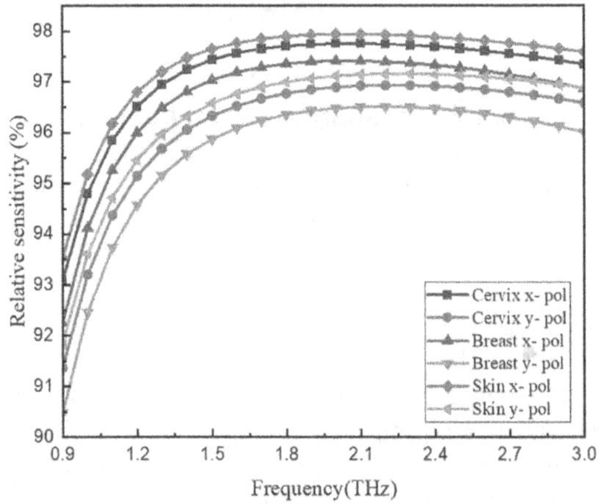

Fig. 49.2 Relative sensitivity variation (x and y pol)

Figure 49.3 presents the graphical representation of confinement loss as a function of frequency. The confinement loss decreases with increasing frequency due to the improved confinement of light at higher frequencies. At a frequency of 2.1 THz, the confinement loss values are 2.55×10^{-15} dB/m for the cervix, 4.15×10^{-17} dB/m for the breast, and 6.0×10^{-17} dB/m for the skin.

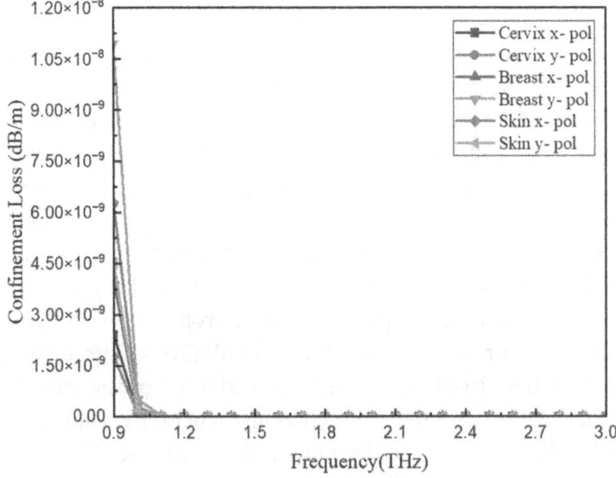

Fig. 49.3 Confinement loss varies with frequency

Table 49.1 Comparison of optical parameters between prior research and the proposed design

Ref. No	Region (THz)	Sensitivity (%)	Confinement Loss(dB/m)	Birefrin-gence
[12]	F= 1.8THz	65.53	1.7 × 10–9	0.0020
[13]	–	94.13	4.3×10^{-9}	–
Proposed Structure	**F = 2.1THz**	**97.93**	**6.0 × 10-17**	**0.00129**

5. CONCLUSION

This paper presents the design and simulation of a rectangular core RI sensor for cancer detection using COMSOL Multiphysics. The optimized sensor design demonstrates high relative sensitivities of 97.75%, 97.40%, and 97.93% for the cervix, breast, and skin at a frequency of 2.1 THz.

REFERENCE

1. Mingomataj, E., Krasniqi, M., Dedushi, K., Sergeevich, K. A., Kust, D., Qadir, A. A., ... & Fatah, G. M. (2024). Cancer Publications in One Year: A Cross-Sectional Study. *Barw Medical Journal*.
2. Yadav, S., Lohia, P., & Dwivedi, D. K. (2024). Quantitative analysis of highly efficient PCF-based sensor for early detection of breast cancer cells in THz regime. Journal of Optics, 53(3): 2642–2655.
3. Singh, J., Khamaru, A., & Kumar, A. (2024). A refractive index-based cancer cells sensor in terahertz spectrum: design and analysis. Journal of Optics, 1–9.

Note: All the figures and tables in this chapter were made by the authors.

Recent Trends in Applied Physics and Material Science – Sudhir Bhardwaj et al. (eds)

50 Photonic Crystal Fibers Infiltrated with Nitrobenzene for Nonlinear Applications

Drishti Singh Tomer[1], Ajeet Kumar[2]

Advanced Photonics Simulation Research Lab,
Department of Applied Physics, Delhi Technological University, Delhi, India

Abstract: This article reports three novel Photonic Crystal Fibers (PCFs) capable of assuring various critical applications such as biomedical diagnostics, explosives detection, enhanced gas sensing capabilities and improvised food quality monitoring systems. COMSOL Multiphysics software based on the finite element method, was used to study and analyze the optical properties of the proposed PCFs at various values of wavelength in order to achieve optimum outcomes. The study has been carried out in near as well as mid-infrared region with a wavelength range of 0.5 μm to 2.5 μm, to achieve relatively flat dispersion profile, minimum effective mode area and low confinement loss. Three designs with rings of air holes embedded in octagonal shape around core , 1st with silica core (#S), 2nd with circular nitrobenzene core (#N), 3rd with elliptical nitrobenzene core with X-polarisation (#NX) & Y-polarisation(#NY) leading to birefringence has been studied . At a pump wavelength of 1.55 μm, the effective mode area of (#S) is 2.75 μm^2, (#N) is 2.19 μm^2, (#NX) is 1.97 μm^2 & (#NY) is 1.90 μm^2. Non-linear coefficient of (#S) is 28.700 $W^{-1}km^{-1}$, (#N) is 31.953 $W^{-1}km^{-1}$, (#NX) is 32.981 $W^{-1}km^{-1}$ & (#NY) is 33.020 $W^{-1}km^{-1}$. The dispersion values are found to be -33.259 ps/nm/km, 9.689 ps/nm/km, 37.940 ps/nm/km & 33.541 ps/nm/km for (#S), (#N), (#NX) & (#NY) respectively. With these improved outcomes, the proposed fiber design can be effectively employed in security applications and sensing technologies.

Keywords: Photonic integrated circuits, Nonlinear optics, Photonic crystal fibers

1. INTRODUCTION

Photonic crystal fibers (PCFs) are an independent technology that was initiated by Phillip Russell (Birks et al. 2003). Traditionally, silica has been used as a fiber material for PCF but it can be used for visible to NIR region but not for MIR due to its limited transparency up to 2.5 μm (Agrawal et al. 2013).It has low nonlinear refractive index (n_2), 2.89×10^{-20} m^2/w at 1550 nm. Recent studies on SCG have focused chalcogenide glasses but they also have compatibility issues with silica for fusion splicing. To overcome these shortcomings recently liquid infiltrated PCFs (using organic liquids) are designed which have high nonlinear refractive index (n_2). They are better in improving the flexibility and modal properties of the fiber. Using liquids in the air holes has proven to be advantageous as it enables tunable properties and high nonlinearity (Tomer et al. 2024). In terms of techniques for infiltration, multiple methods have been used depending upon the requirement like photo-polymerization, thermal collapse, splicer process or UV adhesive process, micro channel milling.

2. PCF GEOMETRY

We have proposed 3 designs with rings of air holes embedded in octagonal shape around core.1st ring being circular and 3 rings are octagonal in each design.1st design has silica with air holes embedded in it, termed as (#S). 2nd design has a circular core filled with nitrobenzene, termed as (#N). 3rd design has been filed with nitrobenzene in an elliptical core. Since the core is asymmetric it possess

[1]ajeetdph@dtu.ac.in, [2]drishti.tomer@gmail.com

DOI: 10.1201/9781003684718-50

two states of polarization. These two states are termed as (#NX) for X-polarisation and (#NY) for Y-polarization respectively. For each design the air hole diameter (d) is 1.5 μm and centre to centre distance between air holes referred as pitch (Λ) is 2.0 μm shown in Fig. 50.1(a), 50.1(b) and 50.1(c). For design #N shown in Fig. 50.1(b).

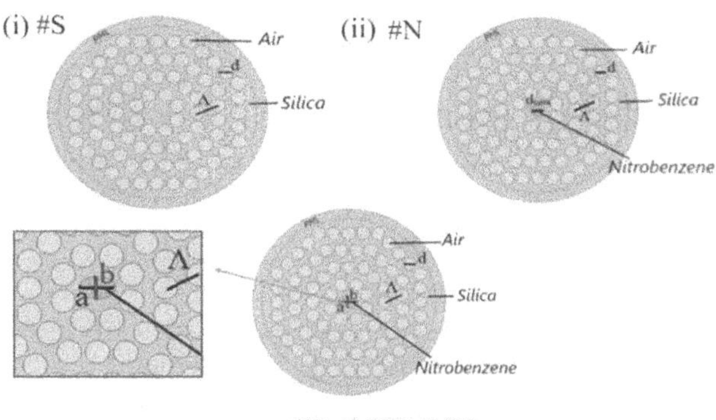

Fig. 50.1 (i) #S, Air hole diameter (d) =1.5 μm, pitch (Λ) =2.0 μm , (ii) #N, Core diameter (d_{core}) = 1.5 μm, air hole diameter (d) = 1.5 μm, pitch (Λ) = 2.0 μm and (iii) #NX & #NY, core dimensions: semi-major axis (a) = 0.75 μm, semi-minor axis (b) = 0.50 μm, Pitch (Λ) =2.0 μm, d=1.5 μm

For design in Fig. 50.1(c) the core is elliptical with dimensions of core as semi-major axis (a) = 0.75 μm, semi-minor axis (b) = 0.50 μm, pitch (Λ) =2.0 μm, and d=1.5 μm.

3. NUMERICAL ANALYSIS

Sellmeier equation: $n^2(\lambda) = 1 + \sum_{k=1}^{N} \frac{B_k \lambda^2}{\lambda^2 - b_k^2}$ .... (1); is used to calculate refractive index of the material. B_1 = 1.30628, B_2 = 0.00502, b_1^2 = 0.02268 μm^2, b_2^2 = 0.18487 μm^2 for nitrobenzene. For silica glass, B_1 = 0.6694226, B_2 = 0.4345839, B_3 = 0.8716947, b_1^2 = 0.0044801 μm^2, b_2^2 = 0.013285 μm^2, b_3^2 = 95.341482 μm^2 with λ as the operating wavelength [4]. The effective mode index has been shown in Fig. 50.2 (i). To calculate the dispersion characteristics of the proposed PCF structures [3]. Dispersion ($D(\lambda)$) $= -\frac{\lambda}{c}\frac{\partial^2 Re(n_{eff})}{\partial \lambda^2}$ ... (2); the dispersion profile has been shown in Fig. (3).

Effective mode area (EMA) has been evaluated in Fig. 50.2(ii). Effective mode area(A_{eff}) $= \frac{\left(\int_{-\infty}^{\infty} |E|^2 \, dx\, dy\right)^2}{\left(\int_{-\infty}^{\infty} |E|^4 \, dx\, dy\right)}$ ...... (3) [3]. Nonlinear coefficient (γ) = $2\pi n_2/\lambda A_{eff}$ ... (4);

n_2 is the nonlinear refractive index 2.0 × 10^{-18} m^2/W for nitrobenzene (Chu et al. 2024) and for silica it is 2.89 × 10^{-20} m^2/W at 1.55 μm (Tomer et al. 2024). The nonlinear coefficient has been evaluated as shown in Fig. 50.2(iii). Confinement loss = 8.686 k × Im (n_{eff}) where, k = 2π/λ shown in Fig. 50.2 (iv). Birefriengence = $|n_x - n_y|$, where n_x & n_y are the x and y components of effective mode index. The value of birefringence at 1.55 mm is 0.3 × 10^{-3}.

4. RESULTS AND DISCUSSION

The dispersion characteristics for all designs has been evaluated in Fig. 50.2(a). It is observed that design #N, #NX & #NY has a flat profile as compared to #S.

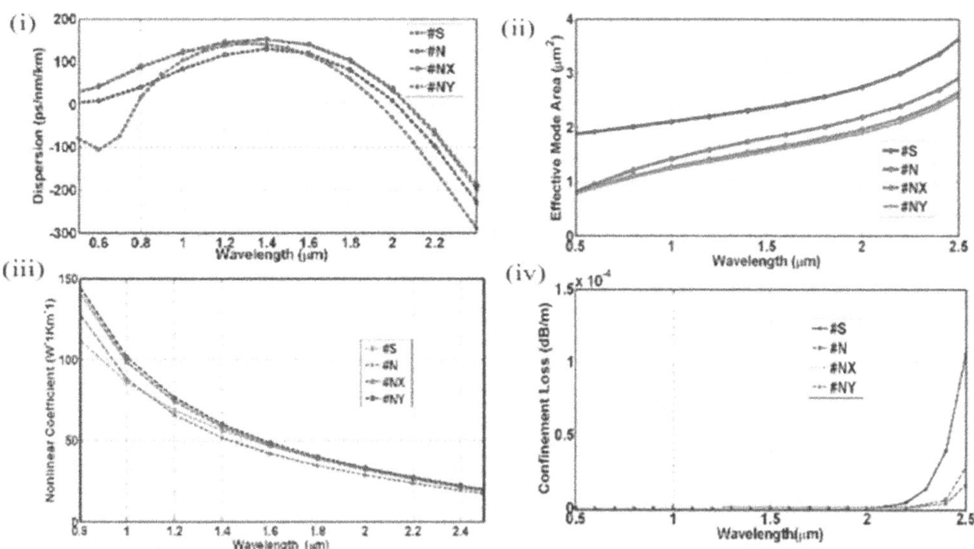

Fig. 50.2 (i) Study of dispersion characteristics of #S, #N, #NX & #NY, (ii) Effective mode area of #S, #N, #NX & #NY, (iii) Nonlinear coefficient of #S, #N, #NX & #NY and (iv) Confinement loss of #S, #N, #NX & #NY

The value of dispersion is large for #NX & #NY as compared to #N and #S at 1.55 μm. Among all designs #N has minimum value of dispersion at 1.55 μm. The dispersion values are found to be -33.259 ps/nm/km, 9.689 ps/nm/km, 37.940 ps/nm/km & 33.541 ps/nm/km for (#S), (#N), (#NX) & (#NY) respectively. Further, EMA has been evaluated in Fig. 50.2(ii) it is observed that its value is minimum for infiltrated elliptical profile and maximum for silica. At a pump wavelength of 1.55 μm, EMA of (#S) is 2.75 μm^2, (#N) is 2.19 μm^2, (#NX) is 1.97 μm^2 & (#NY) is 1.90 μm^2 respectively. Figure 50.2(iii) shows nonlinear coefficient for the proposed designs. Nonlinearity is maximum for elliptical infiltrated core followed by circular profile and least for design (#S). Non-linear coefficient of (#S) is 28.700 $W^{-1}km^{-1}$, (#N) is 31.953 $W^{-1}km^{-1}$, (#NX) is 32.981 $W^{-1}km^{-1}$ & (#NY) is 33.020 $W^{-1}km^{-1}$. Nonlinear coefficient and EMA holds inverse relation if one increases another declines and vice versa. Further, on evaluating confinement loss in Fig. 50.2 (iv), it has been found that the loss has been reduced on infiltration. Also as the core area is increasing (since the area is more in case of elliptical profile as compared to circular) the confinement loss is further decreasing. It is minimum in case of elliptical core.

5. CONCLUSION

These designs can be further employed for various critical applications involving detection of explosives, various gases, cancers, ulcers and enhanced monitoring systems for food quality detection. It is preferable to use low peak power while working with infiltrated profiles.

REFERENCES

1. Birks, T.A., Kakarantzas, G., Russell, P.S.J. and Murphy, D.F. (2003). Photonic crystal fiber devices. In Fiber-based Component Fabrication, Testing, and Connectorization SPIE. 4943:142–151.
2. Agrawal, G.P. (2013). Nonlinear Fiber Optics, 5th ed., Elsevier Academic Press.
3. Tomer, D.S., Kumar, A. (2024). Ethanol-infiltrated circular photonic crystal fiber for low peak power supercontinuum generation from near-infrared to mid-infrared region. J Opt. https://doi.org/10.1007/s12596-024-02152-x
4. Chu Van, L., Nguyen Thi, T., Hoang Trong, D., Le Tran, B.T., Vo Thi Minh, N., Dang Van, T., Le Canh, T., Ho Dinh, Q. and Doan Quoc, K. (2022). Comparison of supercontinuum spectrum generating by hollow core PCFs filled with nitrobenzene with different lattice types. Optical and Quantum Electronics, 54(5):300. https://doi.org/10.1007/s11082-022-03667-y

Note: All the figures in this chapter were made by the authors.

Recent Trends in Applied Physics and Material Science – Sudhir Bhardwaj et al. (eds)

Quantum Efficiency Optimization in Fiber Laser Materials: The Role of Upconversion and Photodarkening

Manish Kapoor*
Department of Physics, Christ Church College,
Kanpur, U.P.

Mohd. Saleem
The Jain International School,
Kanpur, U.P.

Abstract: The drive for higher optical power in commercial laser systems leads to the exploration of suitable materials for high-power fiber lasers. Glass hosts, such as phosphate, silicate, borate, and fluorozirconate doped with trivalent rare-earth ions—such as Neodymium, Erbium, and Thulium—along with optimized nanoparticles, offer broad emission spectra and long radiative lifetimes, making them ideal for high-power lasers. The high quantum efficiency (QE) of these rare-earth ions, a key factor in luminescent transitions, further enhances their performance. Despite extensive research, a comprehensive model for explaining QE remains lacking. This work introduces a novel MKMS model that successfully explains experimental data and identifies the phonon energies that maximize QE.

Keywords: Rare earth ions, Nano particles, Glass host

1. INTRODUCTION

In recent years, the use of rare earth ions doped in glass hosts for high-power lasers has gained significant attention due to their ability to efficiently harness low-power density pump light and convert it into high-power laser output (Chen, Z., et al., 2023). Despite the extensive research, a comprehensive model that links the vast experimental data to theoretical explanations of Quantum Efficiency (QE) remains elusive. The study of QE in rare earth ions is challenging both theoretically and experimentally due to the complexity of their interactions and the need for accurate, model-driven predictions.

To address these challenges, this work adopts a phenomenological approach to provide insights into QE. The MKMS model, developed as part of our ongoing research, integrates several well-established models, including Füchtbauer–Ladenburg, while incorporating new generalizations. It further addresses the mechanisms like multiphonon relaxations, up/down conversions and photodarkening. This model aims to predict the QE for a variety of transitions and ions, focusing specifically on three key trivalent rare earth ions—Neodymium (Nd^{3+}), Erbium (Er^{3+}), and Thulium (Tm^{3+}). These ions are doped into various glass hosts, including Borate, Silicate, Phosphate and ZBLAN, to explore their performance across different materials.

The present work is structured as follows: Section 2 describes the theoretical framework, and Section 3 discusses the results and the future scope.

2. THEORETICAL FRAMEWORK

2.1 Radiative Rate

The radiative rate (w_r) can be expressed using Füchtbauer–Ladenburg equation (Rieder, G. A., 2016) given by,

$$w_r = \upsilon_e^2 \left[\frac{8\pi n^2 \sigma(\lambda_0)}{\lambda_0^2 c_0} \frac{\int I(\lambda_0) d\lambda_0}{I(\lambda_0)} \right] \quad (1)$$

Where, υ_e is the frequency of emitted photon.

Eq. (1) requires two conditions to be met-either all the stark components of the two multiplets must be equally populated or all the transitions must have the same

*Corresponding author: manishcccck@gmail.com

DOI: 10.1201/9781003684718-51

Table 51.1 It gives values of various parameter used in Eq. (4) using the compendium of data [6]. Also, giving the parameters A, B, and H of our MKMS model at room temperature

Rare Earth	Transitions (Type)	Energy Gap ΔE, (In cm^{-1})	$h\upsilon_e$ (in cm^{-1})	Glass Host	C (In s^{-1})	α (In cm)	A (In cm^{-1})	B (In cm^{-2})	H (In cm^{2}s^{-1})
Nd^{3+}	$^4F_{3/2}$ -> $^4I_{11/2}$(4L)	9500	9434	Silicate	1.4*10^{12}	4.7*10^{-3}	21,068	9.65*10^7	3.65*10^{-5}
Nd^{3+}	$^4F_{3/2}$ -> $^4I_{11/2}$(4L)	9500	9434	Phosphate	5.4*10^{12}	4.7*10^{-3}	21268	1.57*10^8	2.21*10^{-5}
Er^{3+}	$^4I_{13/2}$ -> $^4I_{15/2}$(3L)	6500	6515	Borate	2.9*10^{12}	3.8*10^{-3}	15429	4.39*10^8	1.97*10^{-7}
Tm^{3+}	$^3F_4$ -> $^3H_6$(3L)	5400	5405	ZBLAN	1.59*10^{10}	5.19*10^{-3}	11811	7.15*10^6	6.56*10^{-5}

strength. But, because recent studies (Desurvire and E., J. R. Simpson, 1990) have shown that neither of the two conditions satisfy, Eq. (1) leads to significant errors.

To overcome, we present our phenomenological model of radiative rate. This model deals with the emitted frequency of photon in terms of phonon frequency (υ_p) using the known Raman scattering (Weber, W., et al., 1983) described by,

$$w_r = H((h\upsilon_p)^2 + A(h\upsilon_p) + B) \tag{2}$$

Here, H is the arbitrary scale factor in cm^{2}s^{-1}, $h\upsilon_p$ is the phonon energy in cm^{-1} and A, B are the empirical constants in cm^{-1}, cm^{-2} respectively.

2.2 Non-Radiative Rate

A well justified model of non-radiative rate (w_{nr}) already exists (P. D. Dragic, et al., 2018) given by,

$$w_{nr} = Ce^{\Delta E\left(\frac{1}{kT}-\alpha\right)}\left[1-e^{-\left(\frac{h\upsilon_p}{kT}\right)}\right]^{\frac{\Delta E}{h\upsilon_p}} \tag{3}$$

Where, ΔE is the energy gap in cm^{-1}, kT in cm^{-1}, C in s^{-1} and α in cm, are semi-empirical parameters that depend on the host.

2.3 Quantum Efficiency

Employing Eq. (2) and Eq. (3) the expression for QE may be cast as

$$\eta = \frac{H\left((h\upsilon_p)^2 + A(h\upsilon_p) + B\right)}{H\left((h\upsilon_p)^2 + A(h\upsilon_p) + B\right) + Ce^{\Delta E\left(\frac{1}{kT}-\alpha\right)}\left[1-e^{-\left(\frac{h\upsilon_p}{kT}\right)}\right]^{\frac{\Delta E}{h\upsilon_p}}} \tag{4}$$

Inserting the values of the parameters in the above equation for respective transitions leads to the value of phonon energy for which the Quantum efficiency is maximum.

3. RESULTS AND DISCUSSIONS

An excerpt of data has been taken from previous experimental studies (Miniscalco, W., 2001) given in Table 51.1. In the same, the values of parameters A, B, and H corresponding to their respective transitions have been given for various trivalent rare earth ions-Neodymium, Erbium, and Thulium-doped in various glasses.

Employing the parameters of Table 51.1 in Eq. (4), phonon energies are obtained for which quantum efficiency is maximum. These energies are tabulated in Table 51.2 along with their respective experimental values. Data shows high level of correlation between experimental observations and theoretical predictions. It shows that our MKMS model is on the right grounds in explaining QE.

Table 51.2 It provides the experimental phonon energy [6] and calculated phonon energy found through our MKMS model

Rare Earth	Transition	Glass Host	Experimental Phonon energy, $h\upsilon_p$ (in cm^{-1})	Calculated phonon energy $h\upsilon_e$ (in cm^{-1})
Nd^{3+}	$^4F_{3/2}$ -> $^4I_{11/2}$	Silicate	1100	1071.03
Nd^{3+}	$^4F_{3/2}$ -> $^4I_{11/2}$	Phosphate	1200	1196.04
Er^{3+}	$^4I_{13/2}$ -> $^4I_{15/2}$	Borate	1400	1405.7
Tm^{3+}	$^3F_4$ -> $^3H_6$	ZBLAN	500	500.03

In future, the task is to delineate the debatable mechanisms involved in the field and explore other glass family as host doped with vast choices available in experimental observations to predict useful issues of nonlinearity and temperature dependence.

REFERENCES

1. Chen, Z., et al. (2023). Rare earth ion doped luminescent materials: A review of up/down conversion luminescent mechanism, synthesis, and anti-counterfeiting application. *Photonics, 10*(9), 1014.
2. Rieder, G. A. (2016). *Photonics: An Introduction*. Springer, Chapter 6, p. 259.
3. Desurvire, E., J. R. Simpson (1990). Evaluation of 4I15/2 and 4I13/2 Stark-level energies in erbium doped aluminosilicate glass fibers. Opt. Lett. 15:547–549.
4. Weber, W., et al. (1983). Overview of phonon Raman scattering in solids. In M. Cardona (Ed.), *Light Scattering in Solids I* (pp. 1-48). Springer.
5. P. D. Dragic, et al. (2018). Materials for optical fiber lasers: A review. Appl. Phys. Rev. 5, 041301;
6. Miniscalco, W. (2001). Optical and Electronic Properties of Rare Earth Ions in Glasses. In *Optical Properties of Solid Materials* (Chapter 2). CRC Press.

Note: All the tables in this chapter were made by the authors.

Recent Trends in Applied Physics and Material Science – Sudhir Bhardwaj et al. (eds)
© 2026 Taylor & Francis Group, London, ISBN 978-1-041-16452-4

Structural and Cyclic Voltammetry Studies on TiGaN and $TiFe_2O_3$ Synthesized by Sol-Gel Method

Rajani Indrakanti*
Department of Physics,
VNR Vignana Jyothi Institute of Engineering and Technology,
Hyderabad, Telangana, India

Poonam Upadhyay, P. Aman Rajesh, J. Karthik, S. Vamshi Krishna
Department of Electrical and Electronics Engineering,
VNR Vignana Jyothi Institute of Engineering and Technology,
Hyderabad, Telangana, India

N. Suresh Kumar, C. Thirmal
Department of Physics,
VNR Vignana Jyothi Institute of Engineering and Technology,
Hyderabad, Telangana, India

Abstract: This study deals with synthesis of Titanium Gallium Nitride and Titanium Ferrite nanoparticles by Sol-Gel method. The X-ray diffraction (XRD) analysis and cyclic voltammetry (CV) investigations are performed on the synthesized nanomaterials. XRD analysis on TiGaN and $TiFe_2O_3$ has revealed that the material exhibits a simple cubic structure. The crystallite size of TiGaN is in the range of 14-40 nm and $TiFe_2O_3$ 4-40 nm. Analyses are conducted on the electrochemical performance of active material with potassium hydroxide serving as the electrolyte. It has been observed that cyclic voltagrams shape indicates fast electrochemical reactions between the electrode and active material. The Nyquist plots of both the samples shows capacitive behaviour at higher frequency.

Keywords: Sol-Gel method, X-Ray diffraction, Cyclic voltammetry, Nyquist plots

1. INTRODUCTION

In a world that is increasingly driven by advancements in technology, the integration of cutting edge materials into various applications has become pivotal. Nanomaterials, with their size ranging from 1 to 100 nanometers, have garnered significant attention for their remarkable properties that can revolutionize industries. These materials, despite their diminutive size, wield substantial influence, offering unprecedented enhancements in physical, chemical, and electrical properties when applied at the nano scale. These days the necessary and dependable compact source of energy are Batteries, which have been a cornerstone of portable power solutions In this context, our project emerges as a promising exploration into the integration of flexible electrodes as a fundamental component in coin cell batteries, which, in turn, power various portable electronic devices from Mallikarjuna K.et.al (2021). The flexible electrodes at nanoscale have the potential to redefine energy storage solutions, providing versatility and reliability for the diverse range of applications Saritha Yadav.et.al (2020). Further, making the electrode material at Nanoscale will reduce weight, improves efficiency and can withstand for outer environmental conditions. The electrode capacitance range can be from pf-Farads depending upon the type of requirement, can be designed Gebrekidan Gebresilassie Eshetu.et.al(2020). The characteristics of electrodes will be depending upon the type of material used in the making of working electrodes. In this communication, we have synthesized TiGaN and $TiFe_2O_3$ using Titanium butoxide and Gallium Nitride, Ferrite materials by Sol-Gel method.

*Corresponding author: rajini_t@vnrvjiet.in

DOI: 10.1201/9781003684718-52

2. EXPERIMENTAL METHOD

2.1 Materials and Methods

The final products of the study was done by procuring Gallium Nitride (GaN) and Iron (III) Oxide (Fe_2O_3), Titanium butoxide obtained from Sigma Aldrich which is 99.99% pure. Citric acid, Ammonia, and Deionized water were used as ancillary raw materials.

2.2 Synthesis of Samples

The samples TiGaN and $TiFe_2O_3$ were synthesized using the Sol-Gel technique. To begin, GaN and Ti(C4H9O)4 powders were taken 2% by weight, 1% by weight were considered respectively. And fo r the final product of $TiFe_2O_3$ 1.4% by weight of Fe_2O_3 and 2.6% by weight of Ti(C4H9O)4 were taken. An appropriate amount of citric acid was added , thoroughly mixed. The resulting solution was placed within a hot oil bath and heated under constant magnetic stirring at a temperature of 90-95°C with 500 RPM for a duration of 4 hours to ensure even heat distribution throughout the beaker Following this, the solution pH was set to 7 by the gradual addition of Ammonium hydroxide. It's worth noting that this pH value plays a crucial role in determining the Nano Characteristics, with an increase in pH resulting in larger particle sizes and minimal weight loss. After 4 hours, a gel was formed. The Gel was calcinated at 500°C for a period of 4 hours. The resulting powder was collected in a beaker.

3. CHARACTERIZATION

At JNTU,Hyderabad India, researchers used an X-ray diffractometer (XRD) (Philips: PW1830) to determine the phases and grain distribution of sintered materials. Using CH6112E, the produced nanomaterial's electrochemical composition was examined.

4. RESULTS AND DISCUSSIONS

The crystallographic data for the studied samples, including crystallite size, lattice constant (a), and X-ray density (Dx), were determined through the analysis of indexed XRD patterns. X-ray diffraction measurements were conducted at room temperature, employing CuKα radiation with a wavelength (λ) of 1.5406 Å and 2-theta values spanning from 20° to 80°. The X-ray diffraction analysis of Titanium Gallium Nitride revealed well-defined characteristic peaks, as illustrated in Fig. 52.1. These peaks were successfully indexed as (111), (211), (311), (322) and (330), indicative of a simple cubic crystal structure Rajani. I. et. al (2014). And the peaks of $TiFe_2O_3$ are indexed as [111], [200], [211], [220], [310] as shown in Fig. 52.2. The crystallite size (D) was determined using the Debye-Scherrer formula, which is expressed as:

$$D = 0.89\lambda / \beta\cos\theta \tag{1}$$

In this formula, D represents the crystallite size, λ denotes the X-radiation wavelength, and θ represents the angular width of the XRD peak. Additionally, the parameters of XRD are calculated and tabulated in Table 52.1 below.

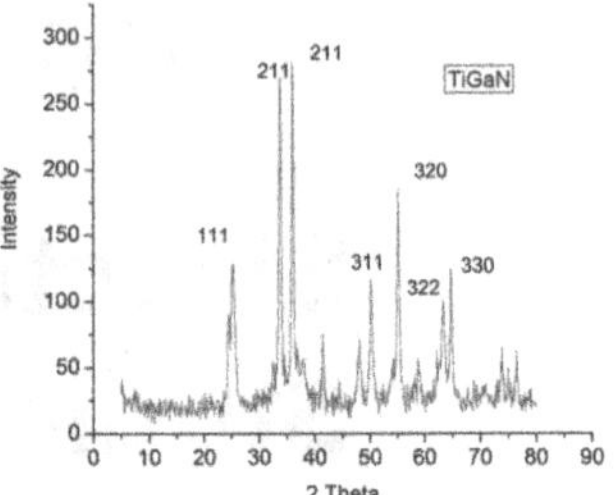

Fig. 52.1 XRD of TiGaN

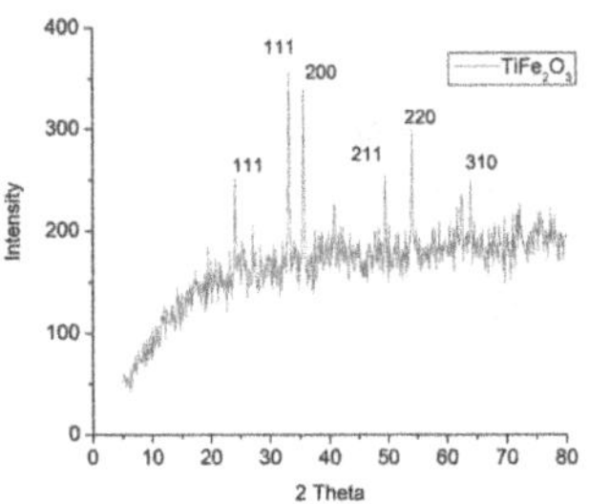

Fig. 52.2 XRD of $TiFe_2O_3$

The interplanar spacing increases for TiFe2O3 when compared to TiGaN due to defects, void Spaces when compared to parental structures of $Fe_2O^3$and GaN are different.

Table 52.1 XRD parameters of TiGaN and $TiFe_2O_3$

Name of Parameter	Parameter Values of TiGaN	Parameter Values of TiFe2O3
Crystallite Size(D)	14-40 nm	4-24 nm
Inter Atomic Spacing(a)	6.0689 A^0	4.721 A^0
Inter Planar Spacing(d)	1.4212 A^0	2.7136 A^0

5. ELECTROCHEMICAL STUDIES

5.1 Electrode Preparation

The separate electrodes were prepared by using 80% of active material. In this 10% of carbon black, 10% of binder (polyvinylidene fluoride (PVDF) were added to 0.08 grams of TiGaN and $TiFe_2O_3$. A few drops of N-methyl pyrrolydine (NMP) was added to the above materials to get the solution. The combined solution was kept in the sonicator and the liquid gel is obtained. This gel were spread on steel plates. These steel plates in a glass jar kept in a furnace at 80 degrees for 5 hours. After 5 hrs the steel plates will get coating and this is given for cyclic voltammetry studies.

5.2 Cyclic Voltammetry Studies

Cyclic voltammetry (CV) investigations were conducted with scan rates of 100 mV/s, covering an applied potential range of -0.2 to 0.4 V, as depicted in Fig. 52.3 and Fig. 52.4 for TiGaN and $TiFe_2O_3$. The above CV voltagramms show two weak reduction peaks. This can be due to the flux of electro active species of ions or molecules reaching the electrode surface is considerably smaller Shireesha, K. et. al (2021)

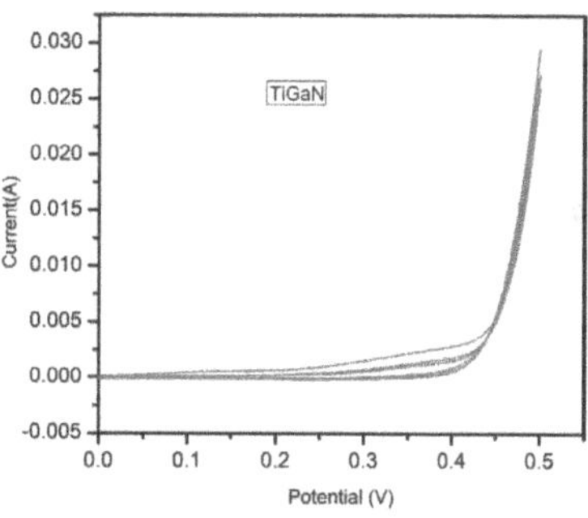

Fig. 52.3 CV analysis of TiGaN at 100 mV/s

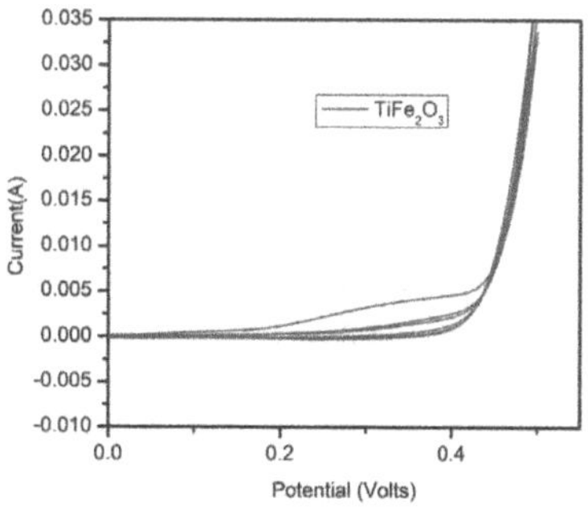

Fig. 52.4 CV analysis of $TiFe_2O_3$ at 100 mV/s

It results from the redox reaction governed by the kinetics of charge transfer. Due to insufficient ion intercalation into the dense centre of the nanostructure, anodic peaks shift to higher potential and cathodic peaks to lower potential.

5.3 Nyquist Plots

The Nyquist plots for synthesized samples are shown in below Fig. 52.5 and Fig. 52.6. The low impedance curves all are near to straight lines which represent capacitive behaviour and at higher frequency there exists no semicircle which is an indication of low internal resistance. This can be due to the decrease in diffusion coefficient there by active species that are supplied rapidly to the electrode surface w.r.t change in concentration also decreases. Thus Warburg impedance Z_w increases with decreasing D. Also the arc and line begin to overlap as the value of *D* decreases. This notifies the diffusive resistance of electrolyte ions.

6. CONCLUSION

In this study TiGaN and $TiFe_2O_3$ were prepared by Sol_Gel method. The effect of combination of Titanium dioxide confirms the Simple cubic structure. The crystallite sizes for TiGaN and $TiFe_2O_3$ was found to be in the range of 14-40 nm and 4-24nm respectively. The Cyclic voltammetry studies results from the redox reaction governed by the kinetics of charge transfer. From Nyquist plots , Warburg impedance Z_w increases with decreasing D . The arc and line begin to overlap as the value of *D* decreases which indicates the diffusive resistance of electrolyte ions.

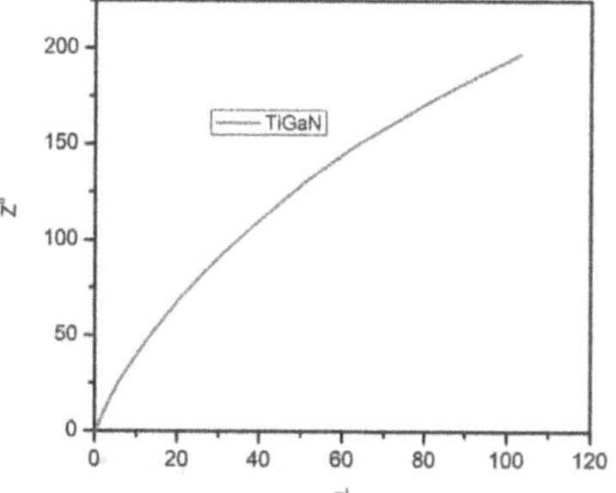

Fig. 52.5 Nyquist plot of TiGaN

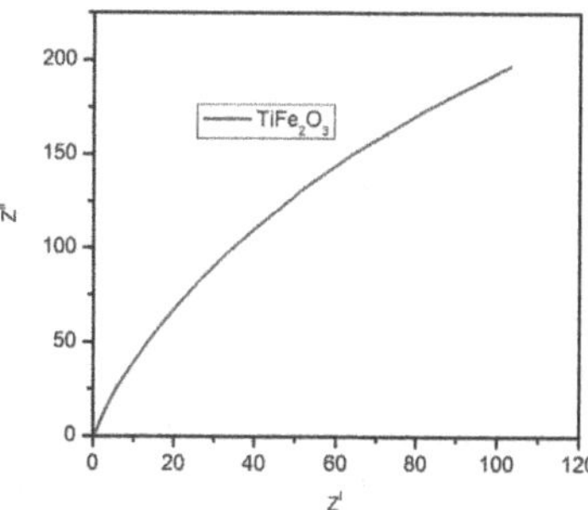

Fig. 52.6 Nyquist plot of $TiFe_2O_3$

REFERENCES

1. Mallikarjuna K, Nasif O, Ali Alharbi S, Chinni SV, Reddy LV, Reddy MRV, Sreeramanan S. . (2021) Phytogenic Synthesis of Pd-Ag/rGO Nanostructures Using Stevia Leaf Extract for Photocatalytic H_2 Production and Antibacterial Studies. *Biomolecules*; 11(2):190. https://doi.org/10.3390/biom110201903.
2. Sarita Yadav, Ambika Devi (2020) Recent advancements of metal oxides/Nitrogen-doped graphene nanocomposites for supercapacitor electrode materials, J. Energy Storage. 30: 101486,
3. Gebrekidan Gebresilassie Eshetu, Giuseppe Antonio Elia, Michel Armand, Maria Forsyth, Shinichi Komabaand Teofilo Rojo .et.al (2020) Electrolytes and Interphases in Sodium-Based Rechargeable Batteries: Recent Advances and Perspectives Adv. Energy Mater **10** 2000093.
4. Rajani I, Udaya Kiran C, Brahmaji Rao V (2014) Structure, Morphology & Infrared Spectroscopic Characterization of Ga(2x+2) N Fe2(49-x) O3 Ferrite synthesized using Sol Gel technique IOSR: J Appl Phys **7** :45
5. R. Indrakanti, V.B. Rao and C.U. Kiran (2015) Studies on conducting nanocomposite with gallium nitride–doped ferrite, part–I *J Nanoeng. Nanosyst* **321:** 43.
6. Rajani Indrakanti, Poonam Upadhyay, Ramasani Sathwick Naidu, Vengaldas Abhilash, Gaddam Rohith Reddy, Shivannagari Vinitha;(2024) Synthesis and electrochemical study of Ti_2GaN electrode material. *AIP Conf. Proc.* 19 3149 (1): 020036. https://doi.org/10.1063/5.0225052
7. Shireesha, K.; Kumar, T.R.; Rajani, T.; Chakra, C.S.; Kumari, M.M.; Divya, V.; Raghava Reddy, K. (2021) Novel NiMgOH-rGO-Based Nanostructured Hybrids for Electrochemical Energy Storage Supercapacitor Applications: Effect of Reducing Agents. *Crystals 11*: 1144. https://doi.org/10.3390/cryst11091144.

Note: All the figures and tables in this chapter were made by the authors.

Recent Trends in Applied Physics and Material Science – Sudhir Bhardwaj et al. (eds)
© 2026 Taylor & Francis Group, London, ISBN 978-1-041-16452-4

Decay Properties of Charmed B Mesons

S. Behera*
Dept. of Physics, Govt. Science College, Chatrapur, Odisha, India

S. Panda
Dept. of Physics, Berhampur University, Berhampur, Odisha, India

Abstract: To elucidate the confining interaction experienced by the free relativistic independent quarks within the meson, we utilize a phenomenologically derived potential that is not dependent on quark flavour. This potential is expressed as a combination of scalar and vector parts in a square root form. Corrections for the spurious centre of mass motion and other residual interactions are accounted for when calculating the meson mass spectra using a perturbative approach. Due to restored chiral symmetry, such interactions include single-gluon exchange over small distances and quark-pion coupling. Using the same potential parameters, we calculate mesonic decay of B- mesons, rare decays of B^0, $B_S{}^0$ mesons. The estimated results of mesonic and rare leptonic decay widths of charmed B mesons are close to some theoretical models like LQCD and experimental predictions.

Keywords: Mass spectra, Mesonic decay widths, Rare decay widths and Branching fractions of B^0 and B_S Meson

1. INTRODUCTION

Different experimental collaborations, such as BaBar, Belle, CDF, and DØ, have observed various resonance states of mesons in both the charm and bottom sectors. The newly discovered B and Bs meson-states, including B_1 (5720), $B_2{}^*$ (5745), $B_{S2}{}^*$ (5839) B_{S2} (5850), B_0 (5732) creates much interest towards hadron spectroscopy(Hayashigaki & Terasaki, 2004). To better understand these meson states, it is crucial to analyze their decay properties. We reviewed several previously studied relativistic independent interquark potentials in this context, incorporating a mix of scalar and vector potentials. These models include the Linear Potential(Barik & Introduction, 1990), Logarithmic Potential (Jena et al., 2009), Harmonic Potential(Jena et al., 2011), Square Root Potential (Behera & Panda, 2024), and others, offering valuable insights into the mesons' characteristics.

2. THEORETICAL FRAMEWORK

As a starting point, we use the confining part of the interaction to define the basic quark dynamics within the meson, described by the quark Lagrangian density,

$$\mathcal{L}_q{}^0(x) = \bar{\psi}_q(x) \left[\frac{i}{2}\gamma^\mu \overrightarrow{\partial\mu} - U_q(r) - m_q\right]\psi_q(x) \quad (1)$$

We considered the qQ or qQ, within a meson bound independently with an average potential of the form (Behera & Panda, 2024), for the present study

$$U_q(r) = ½(1 + \gamma_0)U(r) \text{ and } U(r) = (a^{3/2}\, r^{0.5} + U_0) \quad (2)$$

Here, a, U_0 are the potential parameter, $a > 0$.

From equation (1), we derive the Dirac equation as

$$\left[\gamma^0 E_q - \vec{\gamma}.\vec{P} - m_q - U_q(r)\right]\psi_q\left(\vec{r}\right) = 0 \quad (3)$$

$$\text{The wave function, } \psi_q\left(\vec{r}\right) = \psi_{nlj}(r) = \begin{pmatrix} \psi_{nlj}^{(+)} \\ \psi_{nlj}^{(-)} \end{pmatrix} \quad (4)$$

The ground state meson mass (Behera & Panda, 2024), by treating the various corrections as though they are of the same order of magnitude, shall be written as:

$$M = E_{M_{Q\bar{q}}}(n_1l_1j_1, n_2l_2j_2) + (\Delta E_M)_{c.m} + (\Delta E_M)_g^e + (\Delta E_M)_g^m + \delta m_M \quad (5)$$

*Corresponding author: santosh29688@gmail.com

DOI: 10.1201/9781003684718-53

Again $(\Delta E_M)_{c.m}$ (Behera & Panda, 2024), is the necessary centre of mass correction to the energy. $(\Delta E_M)_g^e$ *and* $(\Delta E_M)_g^m$, are meson's colour electric and magnetic energy (Behera & Panda, 2024) corrections.

Table 53.1 Input parameters

Quark masses (In GeV)	Potential strength (a) (In GeV)	U_0 (In GeV)	Coupling constant (α_c)
$m_{u/d}$ = 0.10	1.390	-1.519	0.10
m_s = 0.40			
m_c = 1.27			
m_b = 4.50			

Table 53.2 Mass spectra of B Meson (GeV)

Meson	E_M^0	$(\Delta E_M)_{c.m}$	$(\Delta E_M)_g^e$	$(\Delta E_M)_g^m$	Meson Mass Present Expt. work
$B^{\pm*}$ $B^{\pm}$	5.272	-0.313	0.355	0.024 -0.072	5.333 5.324 5.259 5.279
$B_S^{0\pm}$ $B_S^{\pm}$	5.409	-0.3205	0.293	0.028 -0.085	5.402 5.425 5.343 5.378

2.1 Mesonic Decay of B-Mesons

The initial-state B meson decays into some final-state mesons or baryons through a process known as hadronic decay. We believe that through the fundamental process $(b \to q + u + \bar{d}; q \in s, d)$ the Cabibbo-favored hadronic decays followed, and the decay widths are given by (Shah et al., 2016).

$$\Gamma\left(\mathrm{M_B} \to \mathrm{M_D}\pi^+\left(\acute{A}^+\right)\right) = C_f \frac{G_F^2 |U_{cb}|^2 |U_{ud}|^2 f_{\pi(\rho)}^2}{32\pi M_\mathrm{B}^3} \times \left[\lambda\left(M_B^2, M_\mathrm{D}^2, M_{\pi^+,\rho^+}^2\right)^{\frac{3}{2}}\right] \left|f_+^2\left(q^2\right)\right| \quad (5)$$

Again, C_f, is the color factor, the factor $\lambda\left(M_D^2, M_\mathrm{D}^2, M_{\pi(\rho)}^2\right)$ which can be calculated (Shah et al., 2016), and the CKM matrix are ($|U_{cb}|.|U_{ud}|$). The value of $f_{\pi(\rho)}$, is taken from (Behera & Panda, 2024).

Consequently, the final State of Isgur Wise function is connected to the form factors, $f_\pm(q^2)$ (Shah et al., 2016).

We have taken the lifetime of B meson from (Workman, R.L. et al. 2022) for calculation of Branching Fraction

$$\text{Branching fraction} = \Gamma \times \tau \quad (6)$$

The estimated mesonic decay widths and their branching fraction are listed in Table 53.3 with experimental values.

Table 53.3 Mesonic decays of *B* - mesons

Decays	$\Gamma(B)$ (Gev)	Branch Fraction	
	Our results	Our results	Expt.
$B^0 \to D^-\pi^+$	1.54×10^{-15}	3.57×10^{-3}	2.68×10^{-3}
$B^0 \to D^{*-}\pi^+$	1.13×10^{-15}	2.61×10^{-3}	2.76×10^{-3}
$B^0 \to D^-\rho^+$	3.22×10^{-15}	7.43×10^{-3}	7.8×10^{-3}
$B^0 \to D^{*-}\rho^+$	2.39×10^{-15}	5.52×10^{-3}	6.8×10^{-3}
$B_S \to D_S^-\pi^+$	1.42×10^{-15}	3.29×10^{-3}	3.2×10^{-3}
$B_S \to D_S^{*-}\pi^+$	1.32×10^{-15}	3.04×10^{-3}	2.1×10^{-3}
$B_S \to D_S^-\rho^+$	2.55×10^{-15}	5.87×10^{-3}	7.4×10^{-3}
$B_S \to D_S^{*-}\rho^+$	4.69×10^{-15}	1.08×10^{-2}	1.03×10^{-2}

2.2 Rare Decays of B-Mesons

The Standard Model (SM) prediction is relatively small for a particular group of flavour decays known as rare decays. The rare decay widths of B^0 *and* $B_S{}^0$ are expressed as (Shah et al., 2016)

$$\Gamma_{\left(B_q^0 \to l^+l^-\right)} = \frac{G_F^2}{\pi} \frac{\alpha^2 f_{B_q}^2 m_l^2}{\left(4\pi sin^2\Theta_W\right)^2} m_{B_q} \times \sqrt{1 - 4\frac{m_l^2}{m_{B_q}^2}} \left|U_{tb}^* U_{tq}\right|^2 \left|C_{10}\right|^2 \quad (8)$$

Similarly, the branching fraction of $B^0(B_S^0) \to \mu^+\mu^-$ is written as

$$BF = \Gamma_{\left(B_q^0 \to l^+l^-\right)} \times \tau_{B_q} \quad (9)$$

G_F, f_{Bq}, are the meson's Fermi coupling constant and decay constant, respectively. The Wilson coefficient is C_{10} (Buchalla & Buras, 1993).

Table 53.4 Rare decays of B^0 and B_S mesons

Decays	$\Gamma(B \to l^+l^-)$ (Gev)	Branch Fraction	
	Our results	Our results	Expt.
$B^0 \to e^+e^-$	1.10×10^{-27}	2.53×10^{-15}	$< 8.3 \times 10^{-8}$
$B^0 \to \mu^+\mu^-$	4.73×10^{-23}	1.09×10^{-10}	$< 6.3 \times 10^{-10}$
$B^0 \to \tau^+\tau^-$	9.79×10^{-21}	2.26×10^{-8}	$< 4.1 \times 10^{-3}$
$B_S \to e^+e^-$	4.76×10^{-26}	1.09×10^{-13}	$< 2.8 \times 10^{-7}$
$B_S \to \mu^+\mu^-$	2.04×10^{-21}	4.71×10^{-9}	3.7×10^{-9}
$B_S \to \tau^+\tau^-$	4.29×10^{-19}	9.89×10^{-7}	

3. RESULTS AND DISCUSSION

In this work, we have calculated the decay properties of B and B_S mesons using a relativistic independent quark model with square root confinement. The analysis begins with determining key model parameters necessary for numerical computations, including a, U_0, quark masses

(m_q), and quark-binding energies (Eq). These parameters were obtained through hadron spectroscopy by fitting the data for heavy and heavy-light flavoured mesons in their ground states, as shown in Table 53.1. We calculated mass spectra of B-mesons, is listed in the Table 53.2. Additionally, the Cabibbo-favored mesonic branching fractions, such as $B_0 \to D^{*-} \pi^+ (2.619 \times 10^{-3})$, $B^0 \to D^- \rho^+$ (7.4339×10^{-3}), $B_S \to D_S^- \pi^+$ (3.291×10^{-3}), $B_S \to D_S{}^{*-} \rho^+$ (1.080×10^{-2}) show excellent agreement with the PDG (Workman, R.L. et al. 2022) data, as listed in Table 53.3. Furthermore, we have also explored the rare decays of $B_q{}^0 \to l^+ l^-$ of B^0 and $B_S{}^0$ mesons. The calculated rare decay widths and branching fractions of B^0 and $B_S{}^0$ mesons are listed in Table 53.4.

4. CONCLUSION

This study employed a phenomenological potential with scalar and vector components in a square root form to describe the confining interaction for relativistic quarks within mesons. The meson mass spectra and decay properties, including rare leptonic decays of B_0 and B_S mesons, were calculated with corrections for centre-of-mass motion and residual interactions. Our results closely align with theoretical predictions, such as those from lattice QCD, and experimental data, demonstrating the reliability of our model.

REFERENCES

1. Barik, N., & Introduction, I. (1990). *I q l 2 1 1. 41*(5).
2. Behera, S., & Panda, S. (2024). Ground state spectra, decay properties of B and D mesons in a relativistic square root potential. *Modern Physics Letters A, 39*(11). https://doi.org/10.1142/S021773232450038X
3. Buchalla, G., & Buras, A. J. (1993). QCD corrections to rare K- and B-decays for arbitrary top quark mass. *Nuclear Physics, Section B, 400*(1–3), 225–239. https://doi.org/10.1016/0550-3213(93)90405-E
4. Hayashigaki, A., & Terasaki, K. (2004). Charmed-meson spectroscopy in QCD sum rule. *ArXiv E-Prints*, hep-ph/0411285.
5. Jena, S. N., Mishra, R. N., Mohapatra, P. K., & Sahoo, S. (2011). *Quark – pion coupling strength and ground state baryon spectra in a chiral potential model. 1272*, 1261–1272. https://doi.org/10.1139/P11-128
6. Jena, S. N., Muni, H. H., & Mohapatra, P. K. (2009). Light- and Strange-Baryon Spectra in a Relativistic Potential Model. *The Open Nuclear & Particle Physics Journal, 2*(1), 34–46. https://doi.org/10.2174/1874415x00902010034
7. Shah, M., Patel, B., & Vinodkumar, P. C. (2016). Spectroscopy and flavor changing decays of B, Bs mesons in a Dirac formalism. *Physical Review D, 93*(9), 1–17. https://doi.org/10.1103/PhysRevD.93.094028.
8. Workman, R.L. et al. (2022), (Particle Data Group), Prog. Theor.Exp.Phys. 2022, 083C01

Note: All the tables in this chapter were made by the authors.

ZrN Plasmonic Nanoparticles for Enhanced Fluorescence and FRET in Pesticide Detection on Wheat Crops

Soniya Juneja*
Department of Applied Sciences,
Krishna Institute of Engineering and Technology, Ghaziabad, India

Pratima Rajput
Department of Physics,
JSS Academy of Technical Education, Noida, India

Vipin Kumar
Department of Applied Sciences,
Krishna Institute of Engineering and Technology, Ghaziabad, India

Alok Singh
Department of Physics,
JSS Academy of Technical Education, Noida, India

Kapil Kumar Sharma, Sweta Shukla
Department of Applied Sciences,
Krishna Institute of Engineering and Technology, Ghaziabad, India

Abstract: Excessive use of pesticides in agriculture to increase yield production may cause toxic effects on the environment as well as on human health. There is always a need for accurate methods for detecting pesticide residues. For sensing pesticides, plasmonic nanoparticles such as gold and silver are widely used. ZrN is being utilized as an alternative to conventional gold nanoparticles due to its high melting point, superior biocompatibility, and electrical conductivity. The effect of ZrN plasmonic nanospheres and nanoshells is investigated for enhanced electric field, fluorescence, and Förster Resonance Energy Transfer (FRET) to improve the detection of pesticides used in wheat and rice crops in India. CdTe quantum dots are employed for their significant role in molecular fluorescence enhancement and FRET efficiency, further aiding in the sensitive detection of pesticides. Spectral properties of three pesticides, Carbofuran, Deltamethrin, and Thiamethoxam, have been studied.

Keywords: Pesticides, ZrN, Nanosphere, Nanoshell

1. INTRODUCTION

Agriculture, as the cornerstone of food production, continually grapples with the delicate balance between increasing crop yields and safeguarding the environment and human health [C.A Damalas et al., 2011]. In the realm of pesticide detection, recent strides have been made by harnessing the unique properties of plasmonic nanoparticles, specifically gold and silver varieties. These nanoparticles have gained substantial attention for their utility in sensing applications [A. J. Haes et al., 2002]. Nevertheless, an intriguing alternative, Zirconium Nitride (ZrN) nanoparticles, has surfaced as a formidable contender to traditional gold nanoparticles. ZrN stands out due to its extraordinary characteristics, including an exceptionally high melting point, superior biocompatibility, and remarkable electrical conductivity when juxtaposed with its gold counterparts.

2. THEORETICAL MODEL

In the present study ZrN nanosphere of dielectric constant $\varepsilon_m(\omega)$ with radius R and ZrN nanoshell of core radius a (dielectric constant ε_1) and shell radius b (dielectric constant $\varepsilon_m(\omega)$) is considered, The considered plasmonic nanoparticles are assumed to be placed in a uniform external electric field $\vec{E} = E_0 \hat{z}$ along +z-direction. For the calculation of fluorescence and energy transfer interactions CdTe quantum dot is assumed as a point

*Corresponding author: sonia.juneja@kiet.edu

DOI: 10.1201/9781003684718-54

dipole and considered to be placed on diametrically opposite sides of core-shell nanoparticle along the z-axis.

For nanoshell geometry having inner radius R_1 and outer radius R_2 is given by

$$A(\omega)=\left|1+\frac{2R_2^3}{r^3}\frac{\begin{array}{l}[(\varepsilon_m-\varepsilon_d)(\varepsilon_d+2\varepsilon_m)+\\ x^3(\varepsilon_1-\varepsilon_m)(\varepsilon_d+2\varepsilon_m)]\end{array}}{\begin{array}{l}[(2\varepsilon_m+\varepsilon_d)(\varepsilon_1+2\varepsilon_m)+\\ x^3(2\varepsilon_m-2\varepsilon_d)(\varepsilon_1+2\varepsilon_m)]\end{array}}\right|^2$$

Where x=R_1/R_2. The calculation for field enhancement were carried out for ZrN nanosphere (x=0) and nanashell particle in three environment of thee pesticides Carbofuran, Deltamethrin and Thiamethoxam.

The normalized fluorescence yield in the vicinity of plasmonic nano particle is calculated using [Pratima Rajput et al., 2020]

$$\frac{\Gamma_{FL}}{\Gamma_{FL}^0}=\left|1+\frac{2\alpha_1}{r^3}\right|^4\Bigg/\left[1+\frac{3}{2k^3}\sum_l(l+1)^2\frac{\operatorname{Im}\alpha_l}{r^{2l+4}}\right]$$

Where α_l denotes the polarizability of nanoparticle, α_1 is the polarizability of nanoparticle for l=1 mode and r is the distance of molecule from the center of nanoparticle and k is the wave number.

The energy transfer rate enhancement factor (ETREF) η between acceptor-donor pair is calculated using

$$\eta=\left|\frac{J(\omega)}{J_0}\right|^2$$

J_0 and $J(\omega)$ for nanosphere and nanoshell is calculated using [M. S. Shishodia et al, 2016].

3. RESULT AND DISCUSSION

In this section, we will discuss the results of electric field enhancement A(ω), normalized fluorescence yield and energy transfer enhancement factor (ETREF) η in the vicinity of ZrN nanosphere of radius 10 nm and ZrN nanoshell of aspect ratio 0.85 for different pesticides. Optical constant of ZrN is adopted from [M. Kumar et al., 2015]. Core is assumed to be of $SiO_2(\varepsilon_1 = 2.04)$. Dielectric constants of carbofuran, deltamethrin and thiamethoxam is taken as 2.3104, 2.6308 and 2.9756 respectively. CdTe quantum dots are assumed as acceptor and donor having dipole moment pa=pd=5.8233 Debye.

Figure 54.1 shows the spectral variation of electric field enhancement A(ω) for nanosphere and nanoshell particle. A(ω) is in the range of 250 to 425 for nanosphere and ~ 1000 order enhancement is achieved for nanoshell structure in case of three pesticides carbofuran, deltamethrin and thiamethoxam. This is evident that. This can be concluded that an amplification factor ~ 1000 is achieved in the presence of isolated spherical nanoshell particle ingrained in different pesrticides medium. Thiamethexom shows maximum electric field enhancement.

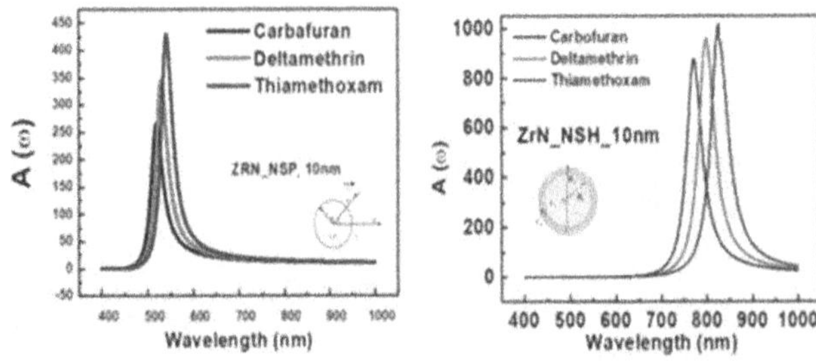

Fig. 54.1 The spectral dependence of h(ω) for nanosphere and nanoshell geometry when ZrN nanoparticle is ingrained in different pesticides mediums (Carbofuran, Delthmethrin, thiamethoxam)

Figure 54.2 shows the spectral variation of normalized fluorescence for nanosphere and nanoshell particle. Fluorescence enhancement of 10^3-10^4 is achieved for nanosphere and ~ 10^5-10^6 enhancement is achieved for nanoshell structure.

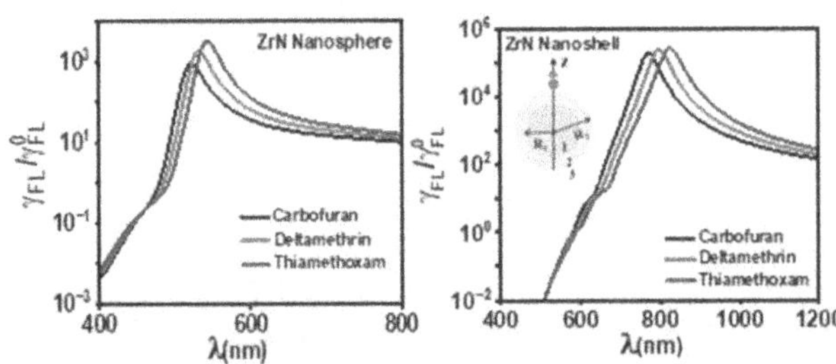

Fig. 54.2 Normalized fluorescence yield in the vicinity of plasmonic nano particles (nanosphere and naoshell) is ingrained in different pesticides

Figure 54.3 shows the spectral variation of ETREF between CdTe quantum dots in the presence of different pesticides order of ~10^4 is achieved for nanosphere and ~ 10^5 enhancement is achieved for nanoshell structure. Only single peak is observed in case of nanosphere in the range of 530 nm to 600 nm. In case of nanoshell 10^5 order enhancement is predicted.

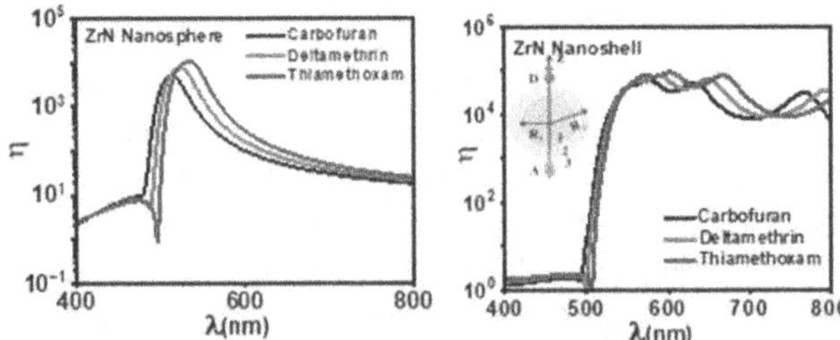

Fig. 54.3 The spectral dependence of η(ω) when ZrN nanoparticle is ingrained in different pesticides mediums (Carbofuran, Delthmethrin, thiamethoxam)

Multiple peaks are observed.. Spectra shows maximum value 425 at resonance wavelength (λ_R) = 530 nm and single resonance peak is observed. In case of ZrN nanoshell approx 1000 order enhancent is predicted.

4. CONCLUSION

The results clearly demonstrate the superior performance of nanoshell structures compared to nanospheres in terms of electric field enhancement, fluorescence enhancement, and energy transfer rate enhancement. The nanoshells achieve amplification factors of ~1000\sim1000~1000 for electric field enhancement, 10^5–10^6 for fluorescence, and 10^5 for ETREF, highlighting their potential for high-sensitivity pesticide detection. These properties, coupled with tunable resonance characteristics, establish nanoshell structures as a powerful platform for sensing pesticides in wheat crops through precise calibration and monitoring.

REFERENCES

1. Damalas C. A. & Eleftherohorinos, I. G. (2011). Pesticide exposure, safety issues, and risk assessment indicators. International Journal of Environmental Research and Public Health, 8: 1402.
2. Haes A. J., & Van Duyne, R. P. (2002). A nanoscale optical biosensor: Sensitivity and selectivity of an approach based on the localized surface plasmon resonance spectroscopy of triangular silver nanoparticles. Journal of the American Chemical Society, 124:10596-.
3. Shishodia M. S. and Juneja S. (2016). Localized surface plasmon mediated energy transfer in the vicinity of core-shell nanoparticle, Journal of Applied Physics 119:201104.
4. Rajput P., Shishodia M. S. (2020). Förster Resonance Energy Transfer and Molecular Fluorescence near Gain Assisted Refractory Nitrides based Plasmonic Core-Shell Nanoparticle. Plasmonics 15:2081.
5. M. Kumar. shii M. S., Umezawa N. and Nagao T. (2015) Band engineering of ternary metal nitride system Ti1-x ZrxN for plasmonic applications, Opt. Mater. Express 6:29.

Note: All the figures in this chapter were made by the authors.

Recent Trends in Applied Physics and Material Science – Sudhir Bhardwaj et al. (eds)
© 2026 Taylor & Francis Group, London, ISBN 978-1-041-16452-4

Investigating the Mass Spectra of the all Strange Tetraquark in Quadratic Quark Confinement

Chetan Lodha* and Ajay Kumar Rai
Department of Physics, Sardar Vallabhbhai National Institute of Technology (SVNIT), Surat, Gujarat, India

Abstract: Over the past two decades, numerous states have been experimentally observed containing more than one strange quark resonances. In this study, the mass spectra for All strange tetraquarks and strangeonium mesons, are computed using semi-relativistic framework. These computations include relativistic mass corrections and employ a Quadratic quark confinement. The study concludes by offering comparisons with other theoretical models and the two-meson threshold.

Keywords: Tetraquark, Meson, Exotic

1. INTRODUCTION

In recent years, experimental efforts at facilities such as Belle, CDF, D∅, CMS, LHCb, BABAR, and BESIII have yielded a wealth of data, leading to the observation of several exotic hadron candidates, as discussed by Singh et al. (2017). The success of the quark model in the 1980s led to the proposal of numerous unconventional or exotic bound states, sparking significant theoretical interest. Over the following decades, these exotic states were explored through various theoretical frameworks, culminating in substantial advancements in our understanding of hadronic structures, as elaborated by Rai (2006) and Rathaud (2021).

Since the first experimental evidence of tetraquarks in 2003, these states have posed an intriguing challenge to particle physicists worldwide. To decode their nature and uncover their mass spectra, researchers have employed diverse methodologies, including Potential Phenomenology, QCD Sum Rules, and Lattice QCD. These approaches have provided valuable insights, yet tetraquarks remain a vibrant area of ongoing research in the quest to deepen our understanding of quantum chromodynamics (QCD). The remarkable discovery of an all-charm tetraquark in 2020 by Aaij (2020) has generated significant interest among researchers, motivating further exploration into the possibilities of all-bottom and all-strange tetraquarks. This finding has opened new avenues in the study of fully heavy and flavor-specific exotic hadrons, enhancing our understanding of the strong interaction and the dynamics of quark confinement.

The primary goal of the present work is to investigate the light-light tetraquark sector. In our previous studies (Lodha, 2023; Lodha, 2024A; Lodha, 2024B; Lodha, 2024C), the well-known Cornell Potential was employed, and the properties of several mesons and tetraquarks were calculated. In this work, we have used a similar approach with modifications to the confinement potential. The acquired masses have been compared with two meson thresholds.

The paper is organized as follows: A brief introduction in Section 1 is followed by Section 2, which describes the diquark-antidiquark formalism and the mass spectra generated by this formalism for all strange tetraquarks. Section 3 summarizes the results and conclusion is drawn in Section 4.

2. THEORETICAL FRAMEWORK

A compact tetraquark state is typically defined as a bound system of four quarks, comprising a diquark and an

*Corresponding author: iamchetanlodha@gmail.com

DOI: 10.1201/9781003684718-55

antidiquark, held together by color forces in a color-neutral arrangement. The exchange of gluons enables interactions between two quarks or two antiquarks, resulting in the creation of bound structures known as diquarks and antidiquarks. In this study, we utilized the Coulomb-plus-quadratic potential $V_{C+Q}(r)$, which incorporates gluonic interactions through a Coulombic term, while quark confinement is described by a quadratic term:

$$V_{C+Q}(r) = \frac{k_s \alpha_s}{r} + br^2 + c$$

Here, αs and ks denote the strong coupling constant and the confinement coefficient, respectively. To incorporate spin-dependent interactions, perturbative corrections are applied to the central potential, following the approach detailed in Lodha (2024A) and Lodha (2024B). The spin-dependent potential $V_{SD}(r)$ is expressed as:

$$V_{SD}(r) = V_{SS}(r) + V_{LS}(r) + V_T(r)$$

The fine structure of the state is governed by the spin-orbit interaction term $V_{LS}(r)$ and the tensor term $V_T(r)$, while the spin-spin interaction term $V_{SS}(r)$ accounts for hyperfine splitting. Within this potential framework, we determined the mass spectra of meson, diquark, antidiquark, and tetraquark states T_{4s}. The masses of the meson, diquark, antidiquark, and tetraquark are given by:

$$M_{s\bar{s}} = m_s + m_{\bar{s}} + E_{[s\bar{s}]} + \langle V^1(r)$$

$$M_{ss} = 2M_s + E_{(ss)} + \langle V^1(r)\rangle$$

$$M_{ss\bar{s}\bar{s}} = m_{ss} + m_{\bar{s}\bar{s}} + E_{[ss][\bar{s}\bar{s}]} + \langle V^1(r)$$

where, m_s and $E_{(ss)}$ are the constituent strange quark mass and the binding energy of the diquark respectively. This framework offers a reliable theoretical method for exploring the mass spectra and structural properties of tetraquark states.

3. RESULTS AND DISCUSSION

In this study, the mass spectra of both S-wave and P-wave all-strange tetraquark states, denoted as T_{4s}, have been calculated and analyzed in comparison to the corresponding two-meson thresholds M_{Th} as well as other theoretical predictions. The findings are summarized in Table 55.1, where the tetraquark states are categorized according to their quantum numbers J^{PC}. The calculated masses are benchmarked against meson thresholds and prior theoretical studies. For this analysis, the parameters were adopted from the latest Particle Data Group (PDG) update, Workman (2022) ensuring consistency with current experimental and theoretical standards.

4. CONCLUSION

In conclusion, all-strange tetraquarks mass spectrum in diquark-antidiquark configurations has been calculated within a non - relativistic framework. The methodology developed in this study will be extended in future research to explore tetraquarks composed of multiple quark flavors. The results from this research could provide valuable insights and serve as a point of comparison for experiments conducted by research facilities like PANDA, J-PARC and others.

Table 55.1 The mass spectra of the all-strange tetraquarks along with the corresponding two-meson threshold (M_{Th}) and other theoretical studies in MeV

State	J^{PC}	Mass	MTh	Lodha (2024A)	Threshold
$^1S_0$	0^{++}	2308	1523	2184	$\eta_s\eta_s$
$^3S_1$	1^{+-}	2366	1781	2248	$\eta_s\phi$
$^5S_2$	2^{++}	2526	2040	2378	$\phi\phi$
$^1P_1$	1^{--}	3278	-	2799	–
$^3P_0$	0^{-+}	2516	2009	2577	$\eta_s f_0$
$^3P_1$	1^{-+}	3258	2219	2800	$\eta_s f_1$
$^3P_2$	2^{-+}	3528	2254	2873	$\eta_s f_2$
$^5P_1$	1^{--}	2466	2198	2574	$\eta_s h_1$
$^5P_2$	2^{--}	3410	2478	2855	ϕf_1
$^5P_3$	3^{--}	3816	2512	2963	ϕf_2

REFERENCES

1. Aaij, R., and LHCb Collaboration. (2020). Observation of structure in the J/ψ-pair mass spectrum *Sci. Bull.*, 65, 1983–99.
2. Lodha, C., Oudichhya, J., Tiwari, R., & Rai, A. K. (2023). Title of the article. *Journal of Condensed Matter*, 1(2), 105–109.
3. Lodha, C., and Rai, A. K. (2024A). Exploring strangeonium meson and all strange tetraquark candidates through mass spectra and decay properties. *Eur. Phys. J. Plus*, 139(7), 663.
4. Lodha, C., and Rai, A. K. (2024B). Investigation of Mass and Decay Characteristics of the All-light Tetraquark *Few Body Syst.*, 65(4), 99.
5. Lodha, C., and Rai, A. K. (2024C). Investigation of Mass and Decay Characteristics of the Light-Strange Tetraquark. *arXiv preprint arXiv:2412.05874.*
6. Rai, A. K., Pandya, J. N., and Vinodkumar, P. C. (2006). Low-lying di-hadronic states in relativistic harmonic model *Indian J. Phys. A*, 80, 387–392.
7. Rai, A. K., Pandya, J. N., and Vinodkumar, P. C. (2007). Multiquark states as di-hadronic molecules. *Nucl. Phys. A*, 782, 406–409.
8. Rathaud, D. P., and Rai, A. K. (2021). Interaction and identification of meson–baryon molecule. *Indian J. Phys.*, 95, 2807–2828.
9. Singh, B., and PANDA Collaboration. (2017). Feasibility study for the measurement of πN transition distribution amplitudes at *PANDA in* $\bar{p}p \to J/\psi\pi^0$ *Phys. Rev. D*, 95(3), 032003.
10. Workman, R. L., and Particle Data Group. (2022). Review of Particle Physics. *Prog. Theor. Exp. Phys.*, 2022, 083C01.

Note: The authors made the table in this chapter.

Recent Trends in Applied Physics and Material Science – Sudhir Bhardwaj et al. (eds)

Exploring the Heavy-Light Pentaquark in Hyper-Central Constituent Quark Model

Gunjan Akbari[1],
Chetan Lodha[2] and Ajay Kumar Rai[3]
Department of Physics, Sardar Vallabhbhai National Institute of Technology (SVNIT),
Surat, Gujarat, India

Abstract: In this study, we examine the mass spectra of pentaquarks using the Hyper-central Constituent Quark Model (hCQM) within the framework of the non-relativistic quark model. For the pentaquarks, we calculate the ground-state masses containing charm quarks for various J^P values. This is accomplished using two different confining potentials, along with an enhanced form of hyperfine interaction. The mass spectra for these pentaquarks are computed across multiple states with corresponding spin-parity assignments. Finally, the calculated pentaquark masses are compared with other theoretical predictions.

Keywords: Pentaquark, Exotic, Hadrons

1. INTRODUCTION

Over the past three decades, extensive theoretical efforts have been dedicated to exploring exotic hadron states, such as tetraquarks, pentaquarks, hybrid mesons, and molecular hadrons, as illustrated in Rai (2006), Rai (2007), Rathaud (2021), Lodha (2024A), Lodha (2024B) and Lodha (2024C). The experimental observation of resonances closely matching theoretical mass predictions has significantly motivated physicists to delve deeper into this dynamic sector of high-energy physics. Among these, the recent discovery of pentaquark candidates by Aaij (2015) has generated considerable interest, highlighting the need to better understand multiquark dynamics within the framework of Quantum Chromodynamics (QCD).

Pentaquarks, consisting of four quarks and one antiquark, present a challenge to the traditional understanding of hadron structures, prompting the development of novel theoretical models to study their unique properties. In this context, the present study investigates the mass spectra of pentaquarks using the Hyper-Central Constituent Quark Model (hCQM) within a non-relativistic framework. Ground-state masses of charm-containing pentaquarks are computed for various spin-parity J^P configurations, employing two types of confining potentials and incorporating enhanced hyperfine interactions. The calculated spectra are rigorously analyzed, validated, and compared against experimental observations and other theoretical predictions, offering insights into the structure and dynamics of these fascinating multiquark states. This work not only advances our understanding of exotic hadrons but also provides a foundation for future experimental and theoretical explorations in this domain.

The rest of the paper is organized as follows: Section 2 outlines the theoretical framework used in the study. Section 3 presents the calculated mass spectra. Section 4 discusses the results in detail, providing insights and interpretations. Finally, Section 5 concludes the paper with a summary of the findings and their implications.

2. THEORETICAL FRAMEWORK

Pentaquark structure is often theorized as either tightly bound diquark-triquark systems or loosely bound meson-baryon molecular states. Experimental evidence

[1]gunjanakbari27@gmail.com, [2]iamchetanlodha@gmail.com, [3]raiajayk@gmail.com

DOI: 10.1201/9781003684718-56

for pentaquarks, such as those observed by the LHCb collaboration, has reignited interest in understanding their formation and properties. Investigating pentaquarks helps uncover new insights into strong interactions and the behavior of quarks in non-traditional configurations.

The Hyper-central Constituent Quark Model (hCQM) is a powerful theoretical framework used to describe the mass spectra of multiquark systems, as demonstrated in Shah (2017), Shah (2023), Menapara (2022) and Menapara (2023). Building upon the traditional quark model, it introduces the hyper-central potential, which relies on the hyper-radius— a collective coordinate that defines the relative positions of all quarks within the system. This approach effectively captures the dynamics of quark interactions in multiquark systems, such as pentaquarks, by incorporating both the confinement potential and hyperfine interactions. Thus, hCQM offers a comprehensive description of the strong force interactions between quarks, making it particularly useful for studying exotic hadrons. For pentaquarks, the dynamics are modeled as a diquark-diquark-antiquark system. In the hCQM framework, the Hamiltonian for the pentaquark system is given by:

$$H = \Sigma_{i=1}^{5}\left(m_i + \frac{p_i^2}{2m_i}\right) - T_{CM} + V_{conf} + V_{hyf}$$

Here, m_i and p_i represent the mass and momentum of the i-th quark, respectively. T_{CM} denotes the center-of-mass correction, V_{conf} corresponds to the confining potential, and V_{hyp} represents the hyperfine interaction, which accounts for spin-dependent forces between quarks. The hyper-central potential, V_{conf} is expressed as a function of hyper radius r:

$$V_{conf(r)} = -\frac{\tau}{r} + \beta r^2$$

Where τ and β represent the model parameters for strength of the Coulomb like term and the confinement respectively. The hyperfine interaction V_{hyp} accounts for spin-spin coupling and is given by:

$$V_{hyp} = i < j \frac{»}{m_i m_j}\left(\vec{S_i} \text{"} \vec{S_j}\right)\delta^3(rij),$$

Where λ is the hyperfine interaction constant, $\vec{S_i}$ and $\vec{S_j}$ are the spins of quarks i and j, and $\delta^3(rij)$ is the delta function representing contact interactions. This methodology allows for the computation of the masses of pentaquarks across various states and spin-parity configurations.

3. RESULT AND DISCUSSION

In this study, the mass spectra of S-wave charm pentaquark states have been calculated using the Hyper-central Constituent Quark Model (hCQM) framework, which effectively captures the dynamics of multiquark systems. The masses of the $\frac{1}{2}^-$, $\frac{3}{2}^-$, and $\frac{5}{2}^-$ states are determined to be 7772 MeV, 7785 MeV, and 7806 MeV, respectively. Yang (2022) estimates masses of the $\frac{1}{2}^-$, $\frac{3}{2}^-$, and $\frac{5}{2}^-$ states to be 8045 MeV, 8095 MeV, and 8137 MeV, respectively. The parameters used in this analysis were adopted from the most recent Particle Data Group (PDG) update by Workman (2022), ensuring consistency with the latest experimental and theoretical standards. This investigation provides valuable insights into the structure and properties of charm pentaquarks and offers comparative data that can be utilized in future experimental searches and theoretical modeling, further advancing our understanding of exotic hadronic states.

4. CONCLUSION

In conclusion, the mass spectra of all-charm pentaquarks have been calculated within a non-relativistic framework, providing valuable insights into their structure and the dynamics governing these exotic states. The methodology developed in this study establishes a robust foundation for extending the analysis to pentaquarks composed of multiple quark flavors in future research. This ongoing investigation will focus on key properties such as masses, decay widths, and magnetic moments, offering a more detailed understanding of their behavior and potential interactions. Additionally, the results presented here could serve as a benchmark for experiments conducted by leading research facilities such as Belle, LHCb, and others, which specialize in studying hadronic resonances. These institutions, dedicated to exploring the complexities of multiquark systems, can use this data to refine their experimental techniques, validate theoretical models, and potentially discover new resonant structures that could open up exciting new areas of research in high-energy physics. Furthermore, the findings could play a significant role in advancing the development of novel theoretical frameworks for multiquark states, driving future exploration in both theoretical and experimental physics.

REFERENCES

1. Aaij, R., and LHCb Collaboration. (2015). Observation of J/ψp Resonances Consistent with Pentaquark States in $\Lambda_b^0 \to J/\psi K^- p$ Decays, *Phys. Rev. Lett. 115 (2015) 072001.*
2. Lodha, C., and Rai, A. K. (2024A). Exploring strangeonium meson and all strange tetraquark candidates through mass spectra and decay properties. *Eur. Phys. J. Plus*, 139(7), 663.
3. Lodha, C., and Rai, A. K. (2024B). Investigation of Mass and Decay Characteristics of the All-light Tetraquark *Few Body Syst.*, 65(4), 99.

4. Lodha, C., and Rai, A. K. (2024C). Investigation of Mass and Decay Characteristics of the Light-Strange Tetraquark. *arXiv preprint arXiv:2412.05874*.
5. Menapara, C., & Rai, A. K. (2023). Mass spectra of singly, doubly, triply light-strange baryons in light of $\mathcal{O}(1m2)$ relativistic correction. *Int. J. Mod. Phys. A,* 38(09n10), 2350053.
6. Menapara, C., & Rai, A. K. (2022). Spectroscopy of light baryons: Δ resonances *Int. J. Mod. Phys. A*, 37(27), 2250177.
7. Rai, A. K., Pandya, J. N., and Vinodkumar, P. C. (2006). Low-lying di-hadronic states in relativistic harmonic model *Indian J. Phys. A*, 80, 387–392.
8. Rai, A. K., Pandya, J. N., and Vinodkumar, P. C. (2007). Multiquark states as di-hadronic molecules. *Nucl. Phys. A*, 782, 406–409.
9. Rathaud, D. P., and Rai, A. K. (2021). Interaction and identification of meson–baryon molecule. *Indian J. Phys.*, 95, 2807–2828.
10. Shah, Z., Kakadiya, A., & Rai, A. K. (2023). Spectra of Triply Heavy Ω_{ccb} and Ω_{bbc} Baryons *Few Body Syst.*, 64(2), 40.
11. Shah, Z., & Rai, A. K. (2017). Masses and Regge trajectories of triply heavy Ω_{ccc} and Ω_{bbb} baryons *Eur. Phys. J. A*, 53(10), 195.
12. Workman, R. L., and Particle Data Group. (2022). Review of Particle Physics. *Prog. Theor. Exp. Phys.*, 2022, 083C01.
13. Yang, G., Ping, J., & Segovia, J. (2022). Fully charm and bottom pentaquarks in a lattice-QCD inspired quark model. *Physical Review D*, 106(1), 014005

Recent Trends in Applied Physics and Material Science – Sudhir Bhardwaj et al. (eds)

Magnetic Moment of Heavy Flavored Baryons

Juhi Oudichhya* and Ajay Kumar Rai
Department of Physics,
Sardar Vallabhbhai National Institute of Technology, Surat, Gujarat, India

Abstract: The magnetic moment is a fundamental intrinsic property of particles that provides deep insights into their internal structure and the dynamics governing transition processes during decay modes. As a result, the investigation of the electromagnetic characteristics of baryons has emerged as a vital domain. In recent decades, substantial advancements have been achieved in the identification of heavy baryons, enhancing our understanding of their properties and behavior. In the present study, we focus on singly heavy-flavour baryons, specifically analysing their electromagnetic properties. We compute the magnetic moments of the ground state baryons with $J^P = \frac{1}{2}^+ \ and \ \frac{3}{2}^+$. Additionally, we provide a comprehensive comparison of our predictions with those obtained through various other theoretical approaches. This comparative analysis highlights the consistency and differences between our results and existing literature, offering a broader perspective on the ongoing investigations in this field.

Keywords: Magnetic moment, Heavy flavored baryons

1. INTRODUCTION

Singly heavy baryons, which consist of one heavy quark and two lighter quarks, hold significant importance within the baryon family of the hadronic spectrum. The presence of charm and bottom quarks as the primary heavy constituents allows these baryons to be classified into charm baryons and bottom baryons. Due to the substantial mass disparity between the heavy quark and the lighter quarks, the internal structure of these baryons can be approximately analyzed. Consequently, singly heavy baryons serve as a valuable framework for advancing the understanding of the nonperturbative aspects of Quantum Chromodynamics (QCD). Recently, substantial progress has been made in identifying heavy baryons (Aaij, 2020). The Particle Data Group (PDG) (Navas, 2024) currently listed 32 charmed baryons with precisely measured masses. However, several observed resonances still require confirmation and quantum number assignment. This extensive experimental data offers theorists a valuable opportunity to evaluate and refine existing theoretical models.

This study is dedicated to examining the magnetic moments of singly charmed baryons, a fundamental property that provides valuable insights into their internal structure and the underlying dynamics governing transitions in their decay processes. Investigating the electromagnetic characteristics of baryons is, consequently, an essential focus of researchers. While substantial progress has been made in studying the Λ_c, Σ_c, Ξ_c, and Ω_c baryons, the precise measurement of their magnetic moments continues to pose a significant challenge. The experimental progress in this area has been limited, emphasizing the importance of further research.

2. MAGNETIC MOMENT

The expression for the magnetic moment is derived by combining the spin-flavor wave function (ϕ_{sf}) with the z-component of the magnetic moment operator (μ_{iz}). The generalized form of the magnetic moment is given as follows (Hazra et. al., 2021 and Kakadiya et. al., 2022),

*Corresponding author: oudichhyajuhi@gmail.com

DOI: 10.1201/9781003684718-57

$$\mu_B = \sum_i \langle \phi_{sf} | \mu_{iz} | \phi_{sf} \rangle. \tag{1}$$

The magnetic moment associated with each individual quark is expressed as follows,

$$\mu_{iz} = \frac{e_i}{2m_i^{eff}} \sigma_{iz}, \tag{2}$$

here e_i indicates the quark charge, σ_{iz} represents the spin component of the respective constituent quark associated with the spin-flavor wave function of the baryonic state, and m_i^{eff} refers to the effective quark mass (Kakadiya et. al., 2022). The magnetic moments of *1S* states of singly charmed baryons can be determined using relations (1) and (2). In our earlier work, we evaluated the mass spectra of all baryon (Oudichhya, 2024, Oudichhya, 2021 and Jakhad, 2023). The mass values utilized in these calculations are taken from our prior study (Oudichhya, 2023).

3. RESULTS AND DISCUSSION

Table 57.1 shows the calculated values of magnetic moments of *1S* states with $J^P = \frac{1}{2}^+$ and $\frac{3}{2}^+$ of Λ_c, Σ_c, Ξ_c, and Ω_c baryons along with the final expression of the magnetic moment. To fully understand these singly charmed baryons' internal quark structure, electromagnetic behavior, and general dynamics, we need to know their magnetic moments. Our calculated results are compared with theoretical predictions from various studies, including those cited in Refs (Gandhi, 2020 and Simonis, 2018).

The close agreement between our calculated values and those predicted by other frameworks not only validates our methodology but also contributes to a deeper understanding of hadronic structures in the context of quantum chromodynamics (QCD). These results are expected to aid future research in refining theoretical models and exploring the electromagnetic characteristics of baryonic systems.

Table 57.1 Magnetic moments (in μ_N) for 1S states of singly charmed baryons

Baryon	J^P	Expression	μ (Ours)	(Gandhi, 2020)	(Simonis, 2018)
Λ_c^+	$\frac{1}{2}^+$	μ_c	0.4210	0.421	0.335
Σ_c^{++}	$\frac{1}{2}^+$	$\frac{4}{3}\mu_u - \frac{1}{3}\mu_c$	1.8312	1.836	2.280
Σ_c^+	$\frac{1}{2}^+$	$\frac{2}{3}\mu_u + \frac{2}{3}\mu_d - \frac{1}{3}\mu_c$	0.3796	0.379	0.487
Σ_c^0	$\frac{1}{2}^+$	$\frac{4}{3}\mu_d - \frac{1}{3}\mu_c$	-1.0903	-1.085	-1.310
Σ_c^{*++}	$\frac{3}{2}^+$	$2\mu_u + \mu_c$	3.2464	3.255	3.980
Σ_c^{*+}	$\frac{3}{2}^+$	$\mu_u + \mu_d + \mu_c$	1.1285	1.127	1.250
Σ_c^{*0}	$\frac{3}{2}^+$	$2\mu_d + \mu_c$	-1.0164	-1.012	-1.490
Ξ_c^+	$\frac{1}{2}^+$	$\frac{2}{3}\mu_u + \frac{2}{3}\mu_s - \frac{1}{3}\mu_c$	0.5591	0.523	0.825
Ξ_c^0	$\frac{1}{2}^+$	$\frac{2}{3}\mu_d + \frac{2}{3}\mu_s - \frac{1}{3}\mu_c$	-1.0115	-1.011	-1.130
Ξ_c^{*+}	$\frac{3}{2}^+$	$\mu_u + \mu_s + \mu_c$	1.3702	1.319	1.470
Ξ_c^{*0}	$\frac{3}{2}^+$	$\mu_d + \mu_s + \mu_c$	-0.8257	-0.825	-1.200
Ω_c	$\frac{1}{2}^+$	$\frac{4}{3}\mu_s - \frac{1}{3}\mu_c$	-0.8395	-0.842	-0.950
Ω_c^*	$\frac{3}{2}^+$	$2\mu_s + \mu_c$	-0.6239	-0.560	-0.936

Note: The authors made the table in this chapter.

ACKNOWLEDGEMENTS

The authors thank the organizers of the Second International Conference on Recent Trends in Applied Physics & Material Science (RAM 2024) for the opportunity to present their work. Ms. Juhi Oudichhya acknowledges financial support from the CSIR under the Direct SRF fellowship scheme (File No. 09/1007(18111)/2024-EMR-I).

REFERENCES

1. Aaij, R. et al., (LHCb Collaboration) (2020). Observation of New Ξ_c^0 Baryons Decaying to $\Lambda_c^+K^-$ Phys. Rev. Lett. 124: 222001.Gandhi, K. and Rai, A. K. (2020). Spectrum of strange singly charmed baryons in the constituent quark model. Eur. Phys. J. Plus 135: 213.
2. Hazra, A., Rakshit, S., Dhir, R. (2021). "Radiative M1 transitions of heavy baryons: Effective quark mass scheme. Phys. Rev. D. 104: 053002."
3. Jakhad, P., Oudichhya, J., Gandhi, K. and Rai, A. K. (2023). "Identification of newly observed singly charmed baryons using relativistic flux tube model. Phys. Rev. D 108: 014011."
4. Kakadiya, A., Shah, Z., Rai, A. K. (2022). Mass spectra and decay properties of singly heavy bottom-strange baryons. Int. J. Mod. Phys. A 37: 2250053.
5. Navas, S., et al. (2024). The review of Particle Physics. Phys. Rev. D. 110: 030001.
6. Oudichhya, J., Gandhi, K. and Rai, A. K. (2021). Mass-spectra of singly, doubly, and triply bottom baryons. Phys. Rev. D 104: 114027.
7. Oudichhya, J. and Rai, A. K. (2023). Spin-parity identification of newly observed singly charmed baryons in Regge phenomenology. Eur. Phys. J. A 59: 123.
8. Oudichhya, J. and Rai, A. K. (2024). Study of singly bottom and doubly heavy baryons within Regge phenomenology. Eur. Phys. J. A 60:125.
9. Simonis, V. (2018). Improved predictions for magnetic moments and M1 decay widths of heavy hadrons. arXiv:1803.01809 [hep-ph].

Recent Trends in Applied Physics and Material Science – Sudhir Bhardwaj et al. (eds)
© 2026 Taylor & Francis Group, London, ISBN 978-1-041-16452-4

Investigating the Mass Spectra of Tetraquarks using Regge Phenomenology

Vandan Patel*, Juhi Oudichhya and Ajay Kumar Rai
Department of Physics, Sardar Vallabhbhai National Institute of Technology, Surat, Gujarat, India

Abstract: This work explores the ground states of fully light tetraquark using Regge Phenomenology, considering tetraquarks as composed of diquark-antidiquark pairs. We employ a quasi-linear Regge trajectory ansatz in the (J, M^2) plane, where J represents the total angular momentum quantum number and M denotes the mass. The model integrates the principles of additivity for both the intercepts and the inverse slopes of the Regge trajectories across different quark flavors. By focusing on light quarks, we try to identify their behaviour in terms of angular momentum and mass. The equations used for the additivity of intercepts and slopes provide a novel approach to constraining the dynamics of the light quark sector, leading to more accurate predictions of the mass spectrum. Our results are expected to shed light on the underlying dynamics governing tetraquarks, offering new insights into their structure and contributing to ongoing discussions on exotic hadrons.

Keywords: Tetraquark, Regge phenomenology

1. INTRODUCTION

The investigation of exotic hadrons like tetraquarks has attracted considerable interest in high-energy physics. Composed of two quarks and two antiquarks, tetraquarks extend the conventional quark model. The identification of the X(3872) state in 2003, the first tetraquark candidate, marked the beginning of a new area in high-energy physics. While the constituent quark model has been highly successful, many nonconforming states have been observed experimentally and predicted theoretically (Yang, Ping, and Segovia., 2020). The most prominent interpretations for these exotic resonances include hadronic molecules, tetraquarks, pentaquarks, hybrids, and others (Rathaud et al., 2021).

Experimental findings show that many tetraquark candidates typically involve at least one heavy quark. However, theoretical research conducted in recent years has increasingly suggested that tetraquarks composed entirely of light quarks may also exist. Notably, the f_0 (500) (previously known as σ), a_0 (980), and f_0 (980) have emerged as prominent candidates for this category (Jaffe, 1977).

Wei et al. (Wei et al., 2008) utilized a quasilinear Regge trajectory framework to establish key mass relations, including quadratic mass equalities, linear mass inequalities, and quadratic mass inequalities for hadrons. Building on this approach, our study employs Regge phenomenology, assuming linear Regge trajectories, to derive connections between the intercept, slope ratios, and tetraquark masses in the (J, M^2) plane. Using these relationships, the ground states of fully light tetraquark $qq\overline{qq}$ have been explored in terms of their masses.

2. THEORETICAL FRAMEWORK

Regge theory is a straightforward and practical phenomenological framework for studying hadron spectroscopy. Several ideas have been introduced to interpret Regge trajectories, with Nambu's approach being the most intuitive. According to his theory, linear Regge trajectories result from the uniform interaction between a quark-antiquark pair connected by a strong flux tube. In Nambu's model, light quarks are envisioned to rotate at the speed of light at the ends of the flux tube, which has a

*Corresponding author: vandankp12998@gmail.com

DOI: 10.1201/9781003684718-58

radius (R). The mass of this flux tube is approximated as (Nambu, 1974; Nambu, 1979).

$$M = 2\int_0^R \frac{\sigma}{\sqrt{1-v^2(r)}} dr = \pi\sigma R \quad (1)$$

Here, σ denotes the mass density per unit length, commonly referred to as the string tension. The angular momentum of the flux tube is determined as follows (Nambu, 1974 and Nambu, 1979):

$$J = 2\int_0^R \frac{\sigma r v(r)}{\sqrt{1-v^2(r)}} dr = \frac{\pi\sigma R^2}{2} + c' \quad (2)$$

So, using the equations (1) and (2) we can get the below expression,

$$J = \frac{M^2}{2\pi\sigma} + c'' \quad (3)$$

Thus, based on the quasilinear Regge theory ansatz, the connection between the total angular quantum number (J) and the mass (M) of a hadron is expressed as:

$$J = \alpha(M) = b(0) + \alpha' M^2 \quad (4)$$

The Regge parameters, including Regge slopes and intercepts, for various quark constituents within a meson multiplet of spin-parity J^P, can be related through the following expressions (Wei et al., 2008; Kaidalov, 1982):

$$b_{i\bar{i}}(0) + b_{j\bar{j}}(0) = 2b_{i\bar{j}}(0) \quad (5)$$

$$\frac{1}{\alpha'_{i\bar{i}}} + \frac{1}{\alpha'_{j\bar{j}}} = \frac{2}{\alpha'_{i\bar{j}}} \quad (6)$$

Where, (i) and (j) denote quark flavours. By equations (5) and (6), we can derive

$$((M_{i\bar{i}} + M_{j\bar{j}})^2 - 4M_{i\bar{j}}^2) = \sqrt{\left(4M_{i\bar{j}}^2 - M_{i\bar{i}}^2 - M_{j\bar{j}}^2\right)^2 - 4M_{i\bar{i}}^2 M_{j\bar{j}}^2} \quad (7)$$

The above equation establishes a relationship between the masses of three mesons, $M_{i\bar{i}}, M_{j\bar{j}}$ *and* $M_{i\bar{j}}$. Consequently, by knowing the masses of any two mesons, the third mass can be predicted using Equation (7).

In the present study, we estimate the ground state masses of all-light tetraquarks by modelling them as bound states of two clusters: a diquark and an antidiquark. The diquarks are treated as pairs of quarks coupled together without internal spatial excitation. Ground state masses of all light tetraquark can be calculated by putting $i = [qq]$, $j = [ss]$ in equation (7),

$$((M_{qq\bar{q}\bar{q}} + M_{ss\bar{s}\bar{s}})^2 - 4M_{qq\bar{s}\bar{s}}^2) = \sqrt{\left(4M_{qq\bar{s}\bar{s}}^2 - M_{qq\bar{q}\bar{q}}^2 - M_{ss\bar{s}\bar{s}}^2\right)^2 - 4M_{qq\bar{q}\bar{q}}^2 M_{ss\bar{s}\bar{s}}^2}. \quad (8)$$

By using above relation, we can calculate ground state masses of all light tetraquark $qq\bar{q}\bar{q}$ for $J^P = 0^+, 1^+$ and 2^+.

3. RESULTS AND DISCUSSION

The ground state masses of fully light tetraquark ($qq\bar{q}\bar{q}$) for $J^P = 0^+, 1^+$ and 2^+ is calculated by taking ground state masses of ($ss\bar{s}\bar{s}$) and ($qq\bar{s}\bar{s}$) as an input from Ref. (Liu et al., 2021) and (Lodha, C et al., 2024) respectively, which are shown in table 1. The calculated values are compared with Ref. (Zhao et al., 2022) and with two meson thresholds also.

The ground state masses of tetraquarks are found to be close to the two-meson threshold. The calculated values are consistent with results from other models and will contribute to advancing future experimental and theoretical investigations of tetraquarks and other exotic hadrons.

Table 58.1 Mass spectra of $ss\bar{s}\bar{s}$ tetraquark (in GeV)

State	J^P	$M_{(calc)}$ Cal. Mass (in GeV)	(Zhao etal., 2022)	Meson Threshold	M_{th} (Thres. mass) (GeV)
1^1S_0	0^+	1.241	1.431	$\eta\eta$	1.096
1^3S_1	1^+	1.529	1.678	$\rho(770)$ $\rho(770)$	1.550
1^5S_2	2^+	2.108	1.978	$\phi(1020)$ $\phi(1020)$	2.038

ACKNOWLEDGEMENT

Vandan Patel acknowledges the financial assistance by University Grant Commission (UGC) under the CSIRUGC Junior Research Fellow (JRF) scheme with Ref No.231610186052.

REFERENCES

1. Jaffe, R. J. (1977). Multiquark hadrons. I. Phenomenology of $Q\bar{Q}^2$ mesons. Phys. Rev. D 15(1): 267–280.
2. Kaidalov, A. B. (1982). Hadronic mass-relations from topological expansion and string model. Z. Phys. C 12:63–66.
3. Liu, F. X., Liu, M. S., Zhong, X. H., and Zhao, Q. (2021). Fully strange tetraquark $ss\bar{s}\bar{s}$ spectrum and possible experimental evidence. Phys. Rev. D 103(1): 016016.
4. Lodha, C et al. (2024). Investigation of Mass and Decay Characteristics of the Light-Strange Tetraquark. arXiv:2412.05874v1.
5. Nambu, Y. (1974). Strings, monopoles, and gauge fields. Phys. Rev. D 10(12):4262–4268.
6. Nambu, Y. (1979). QCD and the string model. Phys. Lett. B 80(4):372–376.
7. Rathaud, D. P., and Rai, A. K. (2021). Interaction and identification of meson–baryon molecule. Indian J. Phys. 95, 2807–2828.
8. Wei, K.-W., Chen, B., and Guo, X.-H. (2008). Some mass relations for mesons and baryons in Regge phenomenology. Phys. Rev. D 78(5):056005.
9. Yang, G., Ping, J., and Segovia., J (2020). Tetra- and Penta-Quark Structures in the Constituent Quark Model. Symmetry, 12(11):1869.
10. Zhao, Z., Xu, K., Kaewsnod, A., Liu, X., Limphirat, A., and Yan, Y. (2022). Study of light tetraquark spectroscopy. Phys. Rev. D 105(3):036001.

Note: The authors made the table in this chapter.

Exploring Ω_c Baryon States within the Relativistic Flux Tube Model

Pooja Jakhad* and Ajay Kumar Rai

Department of Physics, Sardar Vallabhbhai National Institute of Technology, Surat, India

Abstract: Recent advancements in experimental facilities have enabled the direct observation of numerous single-charmed baryons. Motivated by the LHCb's recent detection of the Ω_c^0 baryon states, named $\Omega_c(3000)^0$, $\Omega_c(3050)^0$, $\Omega_c(3065)^0$, $\Omega_c(3090)^0$, and $\Omega_c(3120)^0$, in the $\Xi_c^+ K^-$ channel, we study the mass spectra of Ω_c^0 baryon with its quark diquark structure. The spin average mass is computed using the Regge-like relation which is derived in the relativistic flux tube model. Furthermore, we take into account the effects of spin-dependent interactions. The analysis revealed a strong consistency among the calculated results and the existing experimental masses, which greatly helped predict spin-parity assignments for the states that were observed in the experiments.

Keywords: Ω_c baryon, Relativistic flux tube model

1. INTRODUCTION

Over the last twenty years, there has been a notable surge in the study of strongly interacting heavy hadrons, chiefly attributable to the discovery of several heavy hadron (Navas et al., 2024). For Ω_c baryon, the 1S-states with $J^P = 1/2^-$, $3/2^-$ have been firmly identified. For the Ω_c^0 baryonic family, the LHCb in 2017 made the first observation of many excited states (as indicated in Table 59.1) in the Ξ_c^+ K^- channel, utilizing pp collision data (Aaij et al., 2024). The spin and parity of these states remain undetermined. A number of phenomenological methodologies have been inspired by these observations involving quark model (Ebert et al., 2011; Shah et al., 2016; Gandhi et al., 2020) and Regge phenomenology (Oudichhya et al., 2021).

Table 59.1 The Ω_c baryon resonances reported by LHCb

States	Mass (MeV)	J^P
$\Omega_c(3000)^0$	3000.41 ± 0.22	*Unknown*
$\Omega_c(3050)^0$	3050.19 ± 0.13	*Unknown*
$\Omega_c(3065)^0$	3065.54 ± 0.26	*Unknown*
$\Omega_c(3090)^0$	3090.10 ± 0.50	*Unknown*
$\Omega_c(3120)^0$	3119.10 ± 1.00	*Unknown*

In our prior study (Jakhad, 2023), we investigated singly charmed baryons utilizing a heavy-quark–light-diquark framework within the Relativistic Flux Tube (RFT) model, and we formulated the spin-dependent operators employing a j-j coupling scheme. However, in the case of the Ω_c baryon, the masses of the charm quark and the $\{s, s\}$ diquark are analogous. Therefore, the finite mass effect of the heavy quark may be substantial, necessitating an alternative scheme that is beyond the j-j coupling scheme. Therefore, in this study, we re-examine our calculations utilizing the *Jls* mixing coupling scheme for the masses of the Ω_c baryon to make quark model spectroscopic assignments for the high-mass resonances seen in recent experiments.

2. THEORETICAL FRAMEWORK: RFT MODEL

We investigate the Ω_c baryon in the relativistic flux tube (RFT) model in a configuration in which gluonic field-filled flux tube connects diquark ($\{s, s\}$) and a charm quark (c). The system of a diquark ($\{s, s\}$), a charm quark (c), and a flux tube with string tension T, is rotating with relativistic speed about its centre of mass. Within this context, the Regge relation can be obtained as (Jakhad et al., 2023; Jakhad et al., 2024)

$$\left(\bar{M} - m_c\right)^2 = \sigma L/2 + \left(m_{\{s,s\}} + m_c v_c^2\right), \qquad (1)$$

*Corresponding author: poojajakhad6@gmail.com

DOI: 10.1201/9781003684718-59

where $(\overline{M})$, L, and $m_{\{s,s\}}$ represent the system's mass, angular momentum quantum number, and mass of the diquark, respectively. $\sigma = 2\pi T$. m_c and v_c represents the mass and speed of a charm quark.

The length of the flux tube connecting a charm quark and a diquark is (Chen et al., 2015)

$$r = \left(v_{\{s,s\}} + v_c\right)\sqrt{8L/\sigma}. \quad (2)$$

As the RFT model treats quarks as spinless particles, the spin-dependent interactions must be incorporated from the QCD-inspired quark model (Chen et al., 2022) including the spin-orbit interaction energy

$$H_{so} = \left[\left(\frac{2\alpha}{3r^3} - \frac{b'}{2r}\right)\frac{1}{m_{\{s,s\}}^2} + \frac{4\alpha}{3r^3}\frac{1}{m_{\{s,s\}}m_c}\right]\boldsymbol{L}.\boldsymbol{S}_{\{s,s\}}$$

$$+\left[\left(\frac{2\alpha}{3r^3} - \frac{b'}{2r}\right)\frac{1}{m_c^2} + \frac{4\alpha}{3r^3}\frac{1}{m_{\{s,s\}}m_c}\right]\boldsymbol{L}.\boldsymbol{S}_c \quad (3)$$

the tensor interaction energy

$$H_t = \frac{4\alpha}{3r^3}\frac{1}{m_{\{s,s\}}m_c}\left[\frac{3\left(\boldsymbol{S}_{\{s,s\}}\cdot\boldsymbol{r}\right)\left(\boldsymbol{S}_c\cdot\boldsymbol{r}\right)}{r^2} - \boldsymbol{S}_{\{s,s\}}\cdot\boldsymbol{S}_c\right] \quad (4)$$

and the spin-spin contact hyperfine interaction energy

$$H_{ss} = \frac{32\alpha\sigma_0^3}{9\sqrt{\pi}m_{\{s,s\}}m_c}e^{-\sigma_0^2r^2}\boldsymbol{S}_{\{s,s\}}\cdot\boldsymbol{S}_c \quad (5)$$

Here, $S_{\{s,s\}}$ and S_c stand for the operator of spin of the diquark and the operator of spin of the charm quark, respectively. α represents the coupling constant. b and σ_0 are model parameters. This model's parameters are determined from the measured mass of states of singly charmed baryons. in Jakhad et al. (2023) which are used to determine the spin average mass for the first orbital excitation (1P). Then, we include spin-dependent splitting to find masses of five possible states of the 1P-wave in *Jls* mixing coupling. In the *Jls* mixing coupling, spin multiplet bases diagonalize the H_{so} and H_t interactions, while H_{ss} interaction is treated as a perturbation. The expressions of the spin-dependent splitting in *Jls* mixing coupling are given in a study by Jia et al. (2021)

Table 59.2 Masses of *1P* states of Ω_c baryon in quark-diquark (c{ss}) configuration in MeV

States $\lvert nL, J^P\rangle_{j'}$	This work	(Navas et al., 2024)	(Ebert et al., 2011)	(Oudichhya et al., 2021)
$\lvert 1P, 1/2^-\rangle_{j'=0}$	2999.5	3000.41 ± 0.22	2966	
$\lvert 1P, 1/2^-\rangle_{j'=1}$	3049.2	3050.19 ± 0.13	3055	
$\lvert 1P, 3/2^-\rangle_{j'=1}$	3067.5	3065.54 ± 0.26	3029	3049
$\lvert 1P, 3/2^-\rangle_{j'=2}$	3108.8	3090.10 ± 0.5	3054	
$\lvert 1P, 5/2^-\rangle_{j'=2}$	3128.5	3119.10 ± 1.0	3051	3055

3. RESULTS AND DISCUSSION

The computed masses of the Ω_c baryonic states $\lvert nL, J^P\rangle_{j'}$ associated with the 1P-wave are listed in Table 59.2. The experimentally measured masses of the five narrow excited states of Ω_c baryon as recorded by the Particle Data Group (PDG) (Navas et al., 2024), are presented in the third column. In addition, we compare our findings with those by Ebert et al. (2011) and Oudichhya et al. (2021) in fourth and fifth column of Table 59.2, respectively. We observe that the experimentally determined mass of the $\Omega_c(3000)^0$ and $\Omega_c(3050)^0$ states closely aligns with our prediction for the $\lvert 1P, 1/2^-\rangle_{j'=0}$ and $\lvert 1P, 1/2^-\rangle_{j'=1}$ states, each. Consequently, we designate $J^P = 1/2^-$ for $\Omega_c(3000)^0$ and $\Omega_c(3050)^0$. Further, the LHCb's measured masses for the $\Omega_c(3065)^0$, $\Omega_c(3090)^0$ and $\Omega_c(3120)^0$, is merely 1.96 MeV, 18.7 MeV and 9.4 MeV, different from the $\lvert 1P, 3/2^-\rangle_{j'=1}$, $\lvert 1P, 3/2^-\rangle_{j'=2}$ and $\lvert 1P, 5/2^-\rangle_{j'=2}$ states, respectively. Hence, we assign $J^P = 3/2^-$ to the $\Omega_c(3065)^0$ state, $J^P = 3/2^-$ to the $\Omega_c(3090)^0$ state, and $J^P = 5/2^-$ to the $\Omega_c(3120)^0$ state. As a result, we have described the excited states of the Ω_c baryon as $1P$ states within the RFT model, and the associated J^P value has been accurately assigned.

ACKNOWLEDGEMENTS

The authors thank organizers of Second International Conference on Recent Trends in Applied Physics & Material Science. The primary author expresses gratitude to CSIR for financing her PhD studies.

REFERENCES

1. Aaij, R. et al. (LHCb Collaboration) (2017). Observation of five new narrow Ω_c^0 states decaying to $\Xi_c^+ K^-$. Phys. Rev. Lett. 118(18):182001.
2. Chen, B., Wei, K. W. and Zhang, A. (2015). Investigation of Λ_Q and Ξ_Q baryons in the heavy quark-light diquark picture. J. Eur. Phys. J. A 51(7):82.
3. Chen, B. et al. (2018). *b*-hadron spectroscopy study based on the similarity of double bottom baryon and bottom meson. Phys. Rev. D 105(7): 074014.
4. Ebert, D., Faustov, R. N. and Galkin, V. O. (2011). Spectroscopy and Regge trajectories of heavy baryons in the relativistic quark-diquark picture. Phys. Rev. D 84(1):014025.
5. Gandhi, K. and Rai, A. K. (2020). Spectrum of strange singly charmed baryons in the constituent quark model. Eur. Phys. J. Plus 135(2):213.
6. Jakhad, P. et al. (2023). Identification of newly observed singly charmed baryons using the relativistic flux tube model. Phys. Rev. D 108(1):014011.
7. Jakhad, P. et al. (2024). Interpretation of recently discovered single bottom baryons in the relativistic flux tube model. Phys. Rev. D 110(9): 094005.
8. Jia, D., Pan, J. H. and Pang, C. Q. (2021). A mixing coupling scheme for spectra of singly heavy baryons with spin-1 diquarks in P-waves. Eur. Phys. J. C 81(5):434.
9. Navas, S. et al. (Particle Data Group) (2024). Review of particle physics. To be published in Phys. Rev. D 110(3): 030001.
10. Oudichhya, J., Gandhi, K. and Rai, A. K. (2021). Ground and excited state masses of Ω_c^0, Ω_{cc}^+ and Ω_{ccc}^{++} baryons. Phys. Rev. D 103(11):114030.
11. Shah, Z. et al. (2016). Excited state mass spectra of singly charmed baryons. Eur. Phys. J. A 52(10):313.

Note: All the tables in this chapter were made by the authors.

Study of the Strong Decay of the Λ_c and Ξ_c Baryons

Hardik Rathod*, Pooja Jakhad, and Ajay Kumar Rai
Department of Physics,
Sardar Vallabhbhai National Institute of Technology, Gujarat, India

Abstract: Significant advancement has been made in recent years in our knowledge of the properties of charmed baryons. In this work we study the strong decay of Λ_c and Ξ_c baryons within the framework of Heavy Hadron Chiral Perturbation Theory (HHChPT), which synthesizes a chiral and heavy quark symmetry. We compute the strong decay width of 1P states of Λ_c and Ξ_c baryons. Spin parity quantum numbers of $\Lambda_c(2593)$, $\Lambda_c(2625)$, $X_c(2790)$, and $X_c(2815)$ states have not been measured experimentally yet, but they are expected to be $\frac{1}{2}^-$, $\frac{3}{2}^-$, $\frac{1}{2}^-$, and $\frac{3}{2}^-$, respectively. These estimated spin-parity assignments are supported by our computed decay widths, which correspond closely with the experimentally measured width.

Keywords: Λ_c and Ξ_c baryon, Heavy Hadron Chiral Perturbation Theory etc.

1. INTRODUCTION

Singularly charmed baryons, consisting of one heavy and two light quarks, are used to study the ideas of heavy quark symmetry and the chiral symmetry of light quarks. Numerous states of single charmed baryons have been experimentally detected in the last few years by Belle, BaBar, CLEO, and LHCb (Navas et al., 2024). The excited states of the Λ_C and Ξ_C baryons, identified as $\Lambda_C(2593)$, $\Lambda_C(2625)$, $\Xi_C(2790)$, and $\Xi_C(2815)$, have been found thus far by the Belle and BaBar collaborations; however, their spin and parity are yet unknown (Mizuk et al., 2005; Aubert et al., 2008). Several theoretical approaches have been used in recent years to study the mass spectrum of excited Λ_C and Ξ_C baryons, such as the hyper-central constituent quark model (Shah, et al., 2016; Gandhi, et al., 2020), the QCD-motivated relativistic quark model (Ebert et al., 2011), lattice QCD (Perez-Rubio et al., 2015), QCD sum rules, the Relativistic Flux Tube model (Jakhad at al., 2023), and Reggie phenomenology (Oudichhya and Rai, 2023).

The mass spectra of the baryons $\Lambda_C(2593)$, $\Lambda_C(2625)$, $\Xi_C(2790)$, and $\Xi_C(2815)$ were examined in our earlier study utilizing Regge phenomenology (Oudichhya and Rai, 2023). In this study, we use HHChPT (Heavy Hadron Chiral Perturbation Theory) to analyse the strong decay pattern of the mentioned baryons and give additional evidence for their spin-parity (Cheng and Chua, 2015).

2. METHODOLOGY AND MODEL SPECIFICATIONS

HHChPT provides the best description of the strong decay of singly charmed baryons involving the soft pseudoscalar mesons. The 1P Λ_C baryons decay strongly through the $\Sigma_c\pi$ channels, and the 1P Ξ_C baryons decay strongly through the $\Xi_c'\pi$, and $\Xi_c^*\pi$ channels. The strong decay width for the 1P-wave states of Λ_C and Ξ_C baryons in HHChPT are as follow. (Cheng and Chua, 2007)

$$\Gamma\left[\Lambda_c\left(1^2P_{1/2}\right)\to\Sigma_c\pi\right]=\frac{h_2^2}{2\pi f_\pi^2}\frac{M_{\Sigma_c}}{M_{\Lambda_c\left(1^2P_{1/2}\right)}}E_\pi^2 p_\pi \quad (1)$$

$$\Gamma\left[\Lambda_c\left(1^2P_{3/2}\right)\to\Sigma_c\pi\right]=\frac{2h_8^2}{9\pi f_\pi^2}\frac{M_{\Sigma_c}}{M_{\Lambda_c\left(1^2P_{3/2}\right)}}p_\pi^5 \quad (2)$$

*Corresponding author: hardik13pr@gmail.com

DOI: 10.1201/9781003684718-60

Table 60.1 Predicted mass and decay width of Λ_C and Λ_C baryons (in MeV)

States $N^{2S+1} L_J$	Decay channel	Theoretical M (Oudichhya and Rai, 2023)	Calculated Γ	Experimental Γ (Navas et al., 2024)	Assignment
$\Lambda_c^+(1^2S_{1/2})$	weak	2286.46	–	–	
$\Lambda_c^+(1^2S_{1/2})$	$\Sigma_c\pi$	2592.00	2.13	2.60 ± 0.60	$\Lambda_c^+(2593)$
$\Lambda_c^+(1^2P_{3/2})$	$\Sigma_c\pi$	2630.91	0.117	< 0.97	$\Lambda_c^+(2625)$
$\Xi_c^+(1^2S_{1/2})$	Weak	2467.71	–	–	
$\Xi_c^0(1^2S_{1/2})$	weak	2470.44	–	–	
$\Xi_c^+(1^2P_{1/2})$	$\Xi_c'\pi$	2792.00	8.23	8.90 ± 1.00	$\Xi_c^+(2790)$
$\Xi_c^0(1^2P_{1/2})$	$\Xi_c'\pi$	2792.00	8.23	10.00 ± 1.10	$\Xi_c^0(2790)$
$\Xi_c^+(1^2P_{3/2})$	$\Xi_c'\pi, \Xi_c^*\pi$	2804.99	2.64	2.43 ± 0.26	$\Xi_c^+(2815)$
$\Xi_c^0(1^2P_{3/2})$	$\Xi_c'\pi, \Xi_c^*\pi$	2807.40	2.96	2.54 ± 0.25	$\Xi_c^0(2815)$

Where p_π and E_π are the momentum of pion and energy of pion, respectively, with $f_\pi = 132$. Coupling constant, $h_2 = 0.63$, and $h_8 = 0.0086$. (Piriol et al., 1997; Jakhad at al., 2024)

$$\Gamma\left[\Xi_c^+\left(1^2P_{1/2}\right) \rightarrow \Xi_c'^+\pi^0, \Xi_c'^0\pi^+\right]$$

$$= \frac{h_2^2}{2\pi f_\pi^2}\left(\frac{1}{4}\frac{M_{\Xi_c'^+}}{M_{\Xi_c^+\left(1^2P_{1/2}\right)}}E_{\pi^0}^2 p_{\pi^0} + \frac{1}{2}\frac{M_{\Xi_c'^0}}{M_{\Xi_c^+\left(1^2P_{1/2}\right)}}E_{\pi^+}^2 p_{\pi^+}\right) \quad (3)$$

$$\Gamma\left[\Xi_c^+\left(1^2P_{3/2}\right) \rightarrow \Xi_c'^+\pi^0, \Xi_c'^0\pi^+, \Xi_c^{*+}\pi^0, \Xi_c^{*0}\pi^+\right]$$

$$= \frac{2h_8^2}{9\pi f_\pi^2}\left(\frac{1}{4}\frac{M_{\Xi_c'^+}}{M_{\Xi_c^+\left(1^2P_{3/2}\right)}}p_{\pi^0}^5 + \frac{1}{2}\frac{M_{\Xi_c'^0}}{M_{\Xi_c^+\left(1^2P_{3/2}\right)}}p_{\pi^+}^5\right)$$

$$+ \frac{h_2^2}{2\pi f_\pi^2}\left(\frac{1}{4}\frac{M_{\Xi_c^{*+}}}{M_{\Xi_c^+\left(1^2P_{3/2}\right)}}E_{\pi^0}^2 p_{\pi^0} + \frac{1}{2}\frac{M_{\Xi_c^{*0}}}{M_{\Xi_c^+\left(1^2P_{3/2}\right)}}E_{\pi^+}^2 p_{\pi^+}\right) \quad (4)$$

Similarly, we calculate the decay width of Ξ_c^0.

3. RESULTS AND DISCUSSION

Table 60.1 provides the computed theoretical masses, taken from Ref. (Oudichhya and Rai, 2023), along with the decay widths calculated using HHChPT for the Λ_C and Ξ_C baryons. The experimentally observed masses of $\Lambda_C(2593)$, $\Lambda_C(2625)$, $\Xi_C(2790)$, and $\Xi_C(2815)$ show strong agreement with the theoretical predictions for $\Lambda_c(1^2P_{1/2})$, $\Lambda_c(1^2P_{3/2})$, $\Xi_c(1^2P_{1/2})$, $\Xi_c(1^2P_{3/2})$, respectively. Although the quantum numbers of these states have not yet been determined, the agreement between the computed decay widths and experimentally measured values strongly supports their identification as 1P-wave configurations. The comparison of calculated decay widths with experimental width suggests that the spin-parity assignments for $\Lambda_C(2593)$, $\Lambda_C(2625)$, $\Xi_C(2790)$, and $\Xi_C(2815)$ are most likely $\frac{1}{2}^-$, $\frac{3}{2}^-$, $\frac{1}{2}^-$, and $\frac{3}{2}^-$, respectively.

REFERENCES

1. Aubert, B. et al. (BaBar Collaboration) (2008). Measurements of and and studies of resonances. Phys. Rev. D 78(11):112003.
2. Cheng, H. Y. and Chua, C. K. (2007). Strong decays of charmed baryons in heavy hadron chiral perturbation theory. Phys. Rev. D 75(1):014006.
3. Cheng, H. Y. and Chua, C. K. (2015). Strong decays of charmed baryons in heavy hadron chiral perturbation theory: An update. Phys. Rev. D 92(7):074014.
4. Ebert, D. et al. (2011). Spectroscopy and Regge trajectories of heavy baryons in the relativistic quark-diquark picture. Phys. Rev. D 84(1):014025.
5. Gandhi, K. et al. (2020). Spectrum of Nonstrange Singly Charmed Baryons in the Constituent Quark Model. Int. J. Theor. Phys. 59(4):1129-1156.
6. Jakhad, P. et al. (2023). Identification of newly observed singly charmed baryons using the relativistic flux tube model. Phys. Rev. D 108(1):014011.
7. Jakhad, P. et al. (2023). Spectroscopic exploration of the 1D-wave Σc baryon. DAE Symp. On Nucl. Phys. 67:935.
8. Jakhad, P. et al. (2024). Interpretation of recently discovered single bottom baryons in the relativistic flux tube model. Phys. Rev. D 110(9):094005.
9. Mizuk, R. et al. (Belle Collaboration) (2005). Observation of an Isotriplet of Excited Charmed Baryons Decaying to $\Lambda_c^+\pi$. Phys. Rev. Lett. 94(12):122002.
10. Navas, S. et al. (Particle Data Group), (2024). Review of Particle Physics. Phys. Rev. D 110(3):030001.
11. Oudichhya, J. and Rai, A.K. (2023). Spin-parity identification of newely observed singly charmed baryons in Regge phenomenology. Eur. Phys. J. A 59(6):123.
12. Pirjol, D. et al. (1997). Predictions for s-wave and p-wave heavy baryons from sum rules and the constituent quark model: Strong interactions. Phys. Rev. D 56(3):5483.
13. Perez-Rubio, P. et al. (2015). Charmed baryon spectroscopy and light flavor symmetry from lattice QCD. Phys. Rev. D 92(9):034504.
14. Shah, Z. et al. (2016). Excited state mass spectra of singly charmed baryons. Eup. Phys. J. A 52(10): 313.

Note: The authors made the table in this chapter.

Recent Trends in Applied Physics and Material Science – Sudhir Bhardwaj et al. (eds)

A Comprehensive Study of Mass Spectra of Ξ_{cb} and its Properties Using hCQM

Akram Ansari*
Department of Physics,
Sardar Vallabhbhai National Institute of Technology,
Surat, Gujarat, India

Chandani Menapara
Department of Physics,
Faculty of Science, The Maharaja Sayajirao University of Baroda,
Vadodara, Gujarat, India

Ajay Kumar Rai
Department of Physics,
Sardar Vallabhbhai National Institute of Technology,
Surat, Gujarat, India

Abstract: In the current study, we have explored the properties of the doubly heavy baryon Ξ_{cb} using the "Hyper Central Constituent Quark Model (hCQM)". Our calculations focused on determining the masses of both the ground state as well as excited states of Ξ_{cb}. To enhance accuracy, we incorporated higher-order corrections, including second-order mass corrections in the reliance on spin-related factors. This enabled us to precisely determine the accurate arrangement of spin splitting follows the mass hierarchy. We have successfully determined the spin-parity J^P for both ground and excited states, providing insights into the quantum nature of the Ξ_{cb} baryon. Furthermore, we have done a comprehensive comparison with our predicted resonance masses of Ξ_{cb} and compared it with other theoretical approaches, providing a comprehensive understanding of its properties.

Keywords: Doubly heavy baryon, Mass spectra, Spin-parity

1. INTRODUCTION

Recent data from prominent experimental facilities like LHCb, CMS, and SELEX has sparked increased interest in studying baryons that contain heavy quarks, such as charm or bottom quarks. This influx of data provides a valuable opportunity to enhance our understanding of these intricate particles and gain new insights into the fundamental aspects of particle physics. A variety of baryons with single heavy quarks have been observed in experiments, and their quantum numbers have been accurately determined, offering significant insights into their properties (Zyla et al., 2020). In the framework of the constituent quark model, doubly heavy baryons are composed of two heavier quarks (c or b) paired with a lighter quark (u, d, or s).

These baryons can be categorized into two groups based on their strangeness (S) and isospin (I). The Ξ baryon contains a light strange quark, while the Ω baryon includes an up or down quark. Both types are distinct because they consist of two heavy quarks, which can either be charm (c) or bottom (b). Studying these particles is crucial for advancing our knowledge of hadron spectroscopy and understanding the behaviour of Quantum Chromodynamics (QCD) in the low-energy regime.

The Ξ_{cc}^{++} baryon, characterized by a charm quantum number C=2, was the first doubly heavy baryon detected in experiments (Aaij et al., 2017). At CERN, the LHCb experiment investigated the Ξ_{cb} baryon through the decay channel D^0 pK^- during p-p collisions at a centre-of-mass with 13 TeV energy, but no conclusive evidence of its existence was found. In 2021, the LHCb Collaboration initiated its first study of the baryon and revisited the search for the Ξ_{cb} baryon. This effort targeted a mass range of 6.7 to 7.3 GeV/c2, using data from proton-proton collisions. Recent years have seen substantial research dedicated to understanding the properties of doubly heavy baryons, including their mass spectra, decay mechanisms, and internal structures, employing diverse theoretical

*Corresponding author: akramansari78667@gmail.com

DOI: 10.1201/9781003684718-61

approaches and phenomenological models (Oudichhya et al., 2022; Shah & Rai, 2017).

This article aims to examine the mass spectra of X_{cb} baryons in both their ground and excited states, comparing the results with predictions from various theoretical models.

2. THEORETICAL FRAMEWORK

The "Hypercentral Constituent Quark Model" (hCQM) is for exploring the interactions between constituent quarks within doubly heavy baryons. This model offers a streamlined and effective way to analyse their internal structure and properties. It employs relative coordinates, specifically Jacobian coordinates, to describe the interactions in a three-quark baryonic system, enabling the study of quark-quark correlations. A key feature of the model is its use of a hypercentral potential, which depends only on the hyper-radius (x_1) [. The potential includes a Coulomb term, a confining component, and a spin-dependent interaction (V_{SD}) that accounts for spin-orbit coupling, spin-spin interactions, and tensor effects.

$$V(x_1) = \frac{-\tau_1}{x_1} + \alpha x_1 \quad (1)$$

The running coupling constant α_s expressed as $\tau_1 = \frac{2}{3}\alpha_s$ Initially, a first-order correction, $O\left(\frac{1}{m}\right)$ have been applied to the spin-dependent sector. However, this was insufficient to reproduce the correct spin-splitting pattern. To resolve this discrepancy, a second-order correction, $O\left(\frac{1}{m^2}\right)$ have been introduced in the spin-orbit and spin-tensor terms (Ansari et al., 2023, 2024; Menapara & Rai, 2022, 2023), effectively resolving the mass hierarchy and accurately organizing the spin-splitting structure.

$$V^1(x_1) = -C_F C_A \frac{\alpha_s^2}{4x_1^2} \quad (2)$$

Here, C_F and C_A represent the Casimir elements expressed in terms of fundamental and adjoint representations, respectively.

$$H = \frac{P^2}{2m} + V(x_1) + V_{SD}(x_1) + \frac{1}{m}V^1(x_1) + \frac{1}{m^2}V^2(x_1) \quad (3)$$

3. FINDINGS AND IMPLICATIONS

We have computed the mass spectra of X_{cb} the baryon across a variety of states, including the ground state 1S, excited states up to 3S, and the orbitally excited 1P and 1D states. These findings are summarized in Table 61.1(EAKINS & ROBERTS, 2012; Giannuzzi, 2009; Kakadiya et al., 2023; Shah & Rai, 2017), offering a comprehensive view of the predicted masses for each state.

Our prediction for the ground state mass of the X_{cb} baryon is consistent with other theoretical models, such as those in Ref.(Giannuzzi, 2009), with only a slight discrepancy of around 12 MeV compared to the predictions in Refs. (Kakadiya et al., 2023; Shah & Rai, 2017). We assign the spin-parity to the ground state. For the excited 2S and 3S states, our predictions differ marginally, by approximately 10–15 MeV, from the values suggested by other models. Similarly, the masses of the orbitally excited 1P and 1D examined in our research closely align with other theoretical predictions, showing a small variation of 10 to 25 MeV.

4. CONCLUSION

In this research, we applied the (hCQM) model to examine both the ground states and orbital excitations of the X_{cb} baryon, factoring in higher-order corrections for mass and spin-dependent interactions. These enhanced calculations led to accurate mass spectrum predictions, offering a clear understanding of the excited state structure and spin-parity assignments, which are consistent with predictions from other theoretical models.

Table 61.1 Mass-spectra of Ξ_{cb} baryon and compared with other theoretical approaches

State	J^P	Our	Shah 2017	Kakadiya 2023	Giannuzzi 2009
1S	$\frac{1}{2}^+$	6904	6914	6915	6904
2S	$\frac{1}{2}^+$	7219	7231	7247	7478
3S	$\frac{1}{2}^+$	7469	7492	7481	7904
1P	$\frac{1}{2}^-$	7134	7146	7183	
	$\frac{3}{2}^-$	7145	7135	7179	
1D	$\frac{3}{2}^+$	7329	7303	7268	
	$\frac{5}{2}^+$	7341	7294	7263	

REFERENCES

1. Aaij, R., Adeva, B., Adinolfi, M., Zucchelli, S. (2017). *Physical Review Letters, 119*(11), 112001.
2. Ansari, A., Menapara, C., & Rai, A. K. (2023). Singly charm baryons with higher-order corrections in hCQM: Revisited. *International Journal of Modern Physics A, 38*(21).
3. Ansari, A., Menapara, C., & Rai, A. K. (2024). Mass spectra of all singly bottom baryons with higher-order

correction and their decay properties. *International Journal of Modern Physics A*, *39*(22n23).
4. Eakins, B., & Roberts, W. *International Journal of Modern Physics A*, *27*(08), 1250039.
5. Giannuzzi, F. (2009). *Physical Review D*, *79*(9), 094002.
6. Kakadiya, A., Menapara, C., & Rai, A. K. (2023). Mass spectroscopy and decay properties of Ξcb, Ξbb baryons. *International Journal of Modern Physics A*, *38*(18n19).
7. Menapara, C., & Rai, A. K. (2022). Spectroscopy of light baryons: Δ resonances. *International Journal of Modern Physics A*, *37*(27).
8. Menapara, C., & Rai, A. K. (2023). *International Journal of Modern Physics A*, *38*(09n10).
9. Oudichhya, J., Gandhi, K., & Kumar Rai, A. (2022). Mass spectra of Ξ cc , Ξ bc , Ω cc , and Ω bc baryons in Regge phenomenology. *Physica Scripta*, *97*(5), 054001.
10. Shah, Z., & Rai, A. K. (2017). Excited state mass spectra of doubly heavy Ξ baryons. *The European Physical Journal C*, *77*(2), 129.
11. Zyla, P. A., Barnett, R. M., Beringer, et al. (2020). Review of Particle Physics. *Progress of Theoretical and Experimental Physics*, *2020*(8).

Note: The authors made the table in this chapter.

Recent Trends in Applied Physics and Material Science – Sudhir Bhardwaj et al. (eds)
© 2026 Taylor & Francis Group, London, ISBN 978-1-041-16452-4

Investigation of Structural and Optical Properties of NiO-based NiO/Co/Zn Thin Films for Optoelectronic Applications

Dhara Singh Meena, Arjun Kumawat, M.K. Jangid*
Department of Physics, Vivekananda Global University,
Jaipur, Rajasthan, India

Abstract: The NiO/Co/Zn thin films were fabricated using e-beam evaporation technique, resulting in a uniform thickness of 400 nm. The films were annealed in vacuum at 300°C and 400°C. These are 300°C and 400°C to enhance their crystallinity, as inveterated by X-ray diffraction analysis. The annealed films displayed significant grain growth, indicating improved structural stability. This study indicates that the annealing process, alongside Co and Zn doping, not only improves the crystalline quality and surface characteristics of NiO thin films but also significantly enhances their electrical properties. Photoluminescence (PL) emission peaks detected at approximately 425 nm are likely caused by defect states within the bandgap. Surface morphological images revealed the enhanced surface uniformity in as-deposited as well annealed samples, characterized by smoother and more homogeneous surfaces with fewer defects. These findings suggest that NiO/Co/Zn thin films could be promising candidates for advanced optoelectronic applications.

Keywords: Thin film, XRD, PL, Morphology

1. INTRODUCTION

Recently, there has been a lot of interest in the intriguing characteristics of nickel oxide (NiO) films and devices, notably the broad range of transparency, chemical and thermal stability, conductivity when doped correctly, and flexible production technique (Xu et al., 2019). Transparent conductive oxides, or TCOs, have attracted significant technological interest in recent years (Zrikem et al., 2019). They are extremely useful as transparent electrodes in several kinds of optoelectronic applications owing to their combination optical transparency in the visible region and electrical conductivity (Singh et al., 2021). It is cost-effective, possesses special optical, electrical, and magnetic characteristics, and has outstanding electrochemical stability (Li et al., 2018). Nickel oxide (NiO) is a p-type semiconductor, making it suitable for various electronic and optoelectronic applications. Doping NiO with different elements enhances its electrical conductivity, optical properties, and structural stability. As a result, doped NiO thin films are widely studied for use in solar cells, sensors, transparent electrodes, and energy storage devices. By altering its band structure and defect concentration, doping significantly influences its overall performance. The electrochromic performance of NiO thin films is greatly affected by thickness, porosity, crystallinity, along with specific treatment parameters like annealing. When annealed at the correct temperature, a thick and porous NiO film normally performs a thin and compact film (Tasdemirci et al., 2019). Numerous elements have been added to NiO to enhance its electrochromic qualities (Hou et al., 2018). In this study structural as well as optical characteristics of NiO/Co/Zn have been investigated for optoelectronic applications.

*Corresponding author: mahesh.jangid@vgu.ac.in

DOI: 10.1201/9781003684718-62

2. EXPERIMENTAL

The film was prepared using high-purity components notably NiO (99.99%), Co (99.98%), and Zn (99.98%). In order to ensure consistent mixing, a planetary ball mill (Retsch PM 100 type) was used to combine zinc, cobalt, and nickel oxide at 250 rpm for five hours. The ratio of NiO (75%), Zn (15%), and Co (10%) is used to take the materials. The thin films was prepared unsing e-beam deposition method (Model BC-300 HHV). X-ray diffraction (XRD) analysis was by diffractometer (Panalytical X'Pert Pro diffractometer). The PL characteristics were analyzed using a spectrometer (LS 55, Perkin Elmer). Surface topography of the thin films was analyzed using optical microscope in reflected mode and micrograph images were captured using an input objective lens with a magnification of 10× in microscopic technique.

3. RESULTS AND DISCUSSION

3.1 XRD Analysis

Figure 62.1 demonstrates XRD graph. It illustrates the structural analysis of NiO/Co/Zn thin films. The analysis was conducted under three conditions: (a) as-deposited, (b) annealed at 300°C, and (c) at 400°C. The diffraction peaks correspond to different crystallographic planes, indicating the crystallinity of sample. In as-deposited sample, multiple peaks corresponding to (100), (002), (111), and (200) planes are observed at 2θ =31.84°, 34.42°, 37.06° and 42.88°, indicating the presence of a mixed-phase structure. The peak intensity is relatively lower, suggesting that the crystallinity is not well-developed, likely due to the amorphous or nanocrystalline nature of the deposited film. At 300°C, intensity of the (200) peak significantly increases while additional peaks such as (101) and (220) appear at 2θ=36.52° and 62.59°. This indicates an improvement in crystallinity, as thermal treatment promotes grain growth and the formation of well-defined crystalline phases. Annealing at 400°C consequences in a further increase in the (200) peak intensity along with the appearance of a new peak at 2θ=44.53° corresponding to (222) plane, suggesting continued grain growth and enhanced phase stability. The sharper peaks in this condition confirm better crystallinity and reduced defects, which are expected due to the increased atomic diffusion and reorganization at higher temperatures (Makhlouf et al., 2013 and Chen et al., 2007). XRD analysis indicates that annealing improves the structural quality of the NiO/Co/Zn thin films. (Boukhari et al., 2015 and Naik et al., 2016).

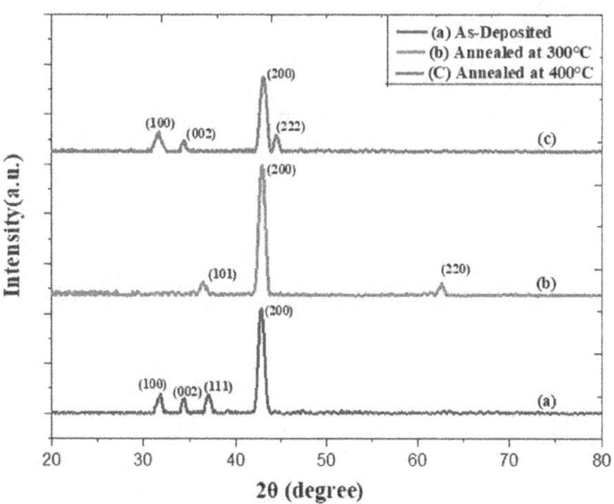

Fig. 62.1 XRD spectra of NiO/Co/Zn thin films

Table 62.1 shows. the values for crystallite size, dislocation. density, and microstrain. The measurements were taken at various annealing temperatures. It was observed that annealing significantly influences structural properties of thin films. The crystallite size decreases from 10.76 nm (as-deposited) to 10.23 nm (at 300°C) and 9.94 nm (at 400°C), indicating structural refinement. Dislocation density increases from 8.64×10^{12} m^{-2} (as-deposited) to 9.56×10^{12} m^{-2}at 300°C and 10.12×10^{12} m^{-2} at 400°C, suggesting defect formation due to dopant incorporation. Strain rises from 3.22 to 3.39 at 300°C and 3.49 at 400°C, indicating lattice distortions caused by dopant penetration and thermal expansion mismatch (Elmassi et al., 2023 and Ponnusamy et al., 2016).

3.2 Photoluminescence Spectra

Figure 62.2 represents PL spectra of NiO/Co/Zn thin films. The photoluminescence (PL) graph illustrates the measure peak of emission intensity observed at 429 nm. The as-deposited sample exhibits the lowest PL intensity, indicating a higher presence of defects or lower radiative recombination efficiency. Upon annealing at 300°C, the PL intensity increases, suggesting improved crystallinity and a reduction in defects, which enhances radiative recombination.

Further the annealing at 400°C results in an even higher PL intensity, implying further structural improvements

Table 62.1 Effect of annealing on intense peak (200), FWHM, Crystallite size, Dislocation density (σ), and strain (ε) of NiO/Co/Zn thin films

Sample	2θ (Deg.)	FWHM	Crystallite size (D) (nm)	Dislocation density (σ) ($\times10^{12}$ m^{-2})	Strain (ε) ($\times10^{-3}$)
As-deposited	42.88	0.79	10.76	8.64	3.22
Annealed at 300°C	42.97	0.83	10.23	9.56	3.39
Annealed at 400°C	43.12	0.85	9.94	10.12	3.49

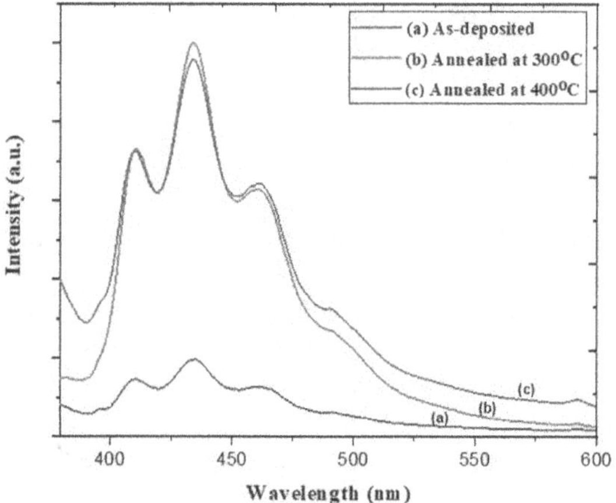

Fig. 62.2 PL spectra of NiO/Co/Zn thin films

and defect passivation. Overall, the annealing process enhances the material's luminescence properties by reducing non-radiative recombination and improving crystallinity (Touati 2023).

3.3 Optical Microscopy

Figure 62.3 shows the optical micrograph of NiO/Co/Zn thin films. The as-deposited sample (a) shows a non-uniform film with small, uneven grains and potential agglomerations, indicating a nanocrystalline structure. Annealing at 300°C (b), the surface becomes more uniform with larger grains, suggesting enhanced atomic mobility and reduced defects. At 400°C (c), the film shows further grain growth, reduced surface roughness, and fewer defects, indicating improved crystallinity and phase formation, consistent with XRD results.

Fig. 62.3 Optical micrograph of NiO/Co/Zn thin films

4. CONCLUSION

This study demonstrates a notable enhancement in structural properties. It also shows an improvement in optical characteristics. These advancements make NiO/Co/Zn thin films suitable for optoelectronic applications. The e-beam evaporation technique used for deposition ensured uniform film thickness, while the annealing at 300°C and 400°C effectively enhanced the crystallinity and structural stability, as confirmed by XRD analysis. The presence of Co and Zn doping further contributed to grain growth and defect reduction. The photoluminescence peaks observed at 425 nm suggest the presence of defect states within the bandgap. Surface morphology analysis revealed smoother, more homogeneous films with fewer defects after annealing. The findings confirm the promising nature of NiO/Co/Zn thin films. They exhibit excellent potential for application in advanced optoelectronic devices due to their improved crystalline quality, surface uniformity, and enhanced electrical properties.

ACKNOWLEDGEMENTS

The authors specially (Dhara Singh Meena) would like to thank CSIR-UGC for providing financial support under the JRF scheme for this work. They also express their gratitude to the Material Research Centre, MNIT Jaipur and Center for non-conventional energy resources, UOR, Jaipur (India) for providing experimental and measurement facilities.

REFERENCES

1. Boukhari, A., Aida, M. S., & Boumaour, M. (2015). Effect of annealing on the structure and optical properties of ZnO and ZnO:Co thin films deposited by sol-gel. *Appl. Surf. Sci., 345*, 46–52.
2. Chen, Y., Qi, D. C., Gao, X. Y., & Wee, A. T. S. (2007). Surface electronic structure evolution of NiO thin films upon thermal annealing. *Phys. Rev. B, 75*(4):045403.
3. Elmassi, S., Beraich, M., El Kissani, A., Alsaad, A., & Outzourhit, A. (2023). Theoretical and experimental investigations of the effect of cobalt doping on the structural, optical, and electrical properties of sputtered NiO films for optoelectronic applications. *Optical Materials, 145*, 114449.
4. Hou, S., et al. (2018). Controllable crystallinity of nickel oxide film with enhanced electrochromic properties. Appl. Surf. Sci. 451:104–111.
5. Li, L., Zhang, J., Lei, J., Jing, X., Shang, B., Liu, L., Li, N., Pan, F. (2018). O-Vacancy-enriched NiO hexagonal platelets fabricated on Ni foam as a self-supported electrode for extraordinary pseudocapacitance. J. Mater. Chem. A. 6(16):7099–7106.
6. Makhlouf, S. A. (2013). Influence of annealing temperature on the structural and optical properties of ZnO thin films. *J. Mater. Sci.: Mater. Electron., 24*(5):1746–1752.
7. Naik, B. S., Sarvaiya, S. N., & Solanki, C. H. (2016). Optical and structural characterization of NiO thin films synthesized by sputtering and thermal treatment. *Thin Solid Films, 603*, 102–108.
8. Ponnusamy, P. M., Agilan, S., Muthukumarasamy, N., Senthil, T. S., & Rajesh, G. (2016). Structural, optical, and magnetic properties of undoped NiO and Fe-doped NiO nanoparticles synthesized by wet-chemical process. *Materials Characterization, 114*, 166–171.
9. Singh, J., Kumar, R., Verma, V., Kumar, R. (2021). Structural and optoelectronic properties of epitaxial Ni-substituted Cr2O3 thin films for p-type TCO applications. Mater. Sci. Semicond. Process. 123:105483.
10. Tasdemirci, T.Ç. (2019). Influence of annealing on properties of SILAR deposited nickel oxide films. Vacuum 167:189–194.
11. Touati, M., Boucherka, T., Barbadj, A., Brihi, N., & Labreche, F. (2023). Low resistivity and transparent Co-doped NiO thin films synthesized
12. Xu, J., Wang, M., Liu, Cui, H. (2019). One-pot solvothermal synthesis of size-controlled NiO nanoparticles. Adv. Powder Technol. 30:861–868.
13. Zrikem, K., Song, G., Rougier, A. (2019). UV treatment for enhanced electrochromic properties of spin coated NiO thin films. Superlattice. Microst. 127:35–42.

Note: All the figures and tables in this chapter were made by the authors.

Recent Trends in Applied Physics and Material Science – Sudhir Bhardwaj et al. (eds)
 ISBN 978-1-041-16452-4

Conduction Mechanism of the PVC-PMMA Polyblend Films

R. Y. Bakale*
Department of Physics, Mahatma Fule Arts, Commerce and Sitaramji Chaudhari Science Mahavidyalaya, Warud, Maharashtra, India

Y. G. Bakale
Department of ESH, Kavikulguru Institute of Technology and Science, Ramtek, Maharashtra, India

Y. S. Tamgadge
Department of Physics, Mahatma Fule Arts, Commerce and Sitaramji Chaudhari Science Mahavidyalaya, Warud, Maharashtra, India

R. P. Ganorkar
Department of Chemistry, Mahatma Fule Arts, Commerce and Sitaramji Chaudhari Science Mahavidyalaya, Warud, Maharashtra, India

S. V. Khangar
Department of Physics, Shri Shivaji Education Society Amravati's, Science College, Congress Nagar, Nagpur, India

P. P. Gedam
Department of Physics, Shri R.L.T Science College, Akola, Maharashtra, India

Abstract: The electrical conduction mechanism of the blend film composed of PVC and PMMA has been investigated across a temperature range of 313K to 353K. The findings are illustrated through I-V characteristics. The analysis has been conducted considering various models, including Poole-Frenkel, Fowler-Nordheim, Schottky, as well as Ln(J) versus T plots, and Richardson and Arrhenius plots. The results indicate that the Schottky-Richardson mechanism predominantly governs the observed conductivity

Keywords: Polyblend film, Conductivity

1. INTRODUCTION

Significant advancements have been made in the field of polymeric materials over the past decade. Polymers are not merely simple covalent crystals as understood in traditional solid-state physics1. Research on both natural and synthetic polymers has been conducted by Japanese scientists2. In the context of organic solids, the conductivity resulting from electrons transitioning from the valence band to the conduction band3,4 is minimal. The complex conduction behavior4,5 is typically elucidated through mechanisms such as electron emission from the cathode, known as the Schottky-Richardson mechanism6 or through the liberation of electrons from traps within the bulk material, referred to as the Poole-Frenkel mechanism7. Additionally, phenomena such as tunneling8 and space charge limited conduction9 have also been explored in the existing literature.

In the current investigation, the direct current conductivity of a blend film composed of PVC and PMMA was assessed to elucidate the electrical conduction mechanisms. The analysis demonstrates how the current-voltage characteristics of the sample can lead to potential conclusions. The findings are elaborated upon through the representation of various mechanisms, including Poole-Frenkel, Fowler-Nordheim, log(J) versus T plots, Schottky plots, Richardson plots, and Arrhenius plots. Notably, the Schottky-Richardson mechanism exhibits a significant temperature dependence of the current, in contrast to the Poole-Frenkel mechanism. Consequently, the examination of the temperature dependence of current density is of considerable significance.

2. EXPERIMENTAL DETAILS

In the present study we prepared a thin polyblend film by using two polymers i.e. PVC and PMMA which we have chosen on electron acceptor and electron donor properties and this polyblend improves the cost-performance ratio. These two polymers are taken in the ratio 3:1 and

*Corresponding author: reenabakale123@gmail.com

DOI: 10.1201/9781003684718-63

these polymers were dissolve in the universal solvent Tetrahydrofuran their solubility parameter value matches with the selected polymer. A mixture solution of these two-polymer heated to dissolve completely then the solution poured on a glass substrate which floated on the pool of mercury and leave it for 12 hours. The polyblend film is prepared by Isothermal evaporation technique. The prepared thin film is cut in small square. The films used for present study having a thickness 80μm. Both the sides of the sample film were coated with quick drying silver paint (supplied by Elteck Pvt. Ltd., Bangalore) to ensure good electrical contacts. The coated sample film was subjected to uniform heating at a temperature of 353K in furnace.

3. RESULT AND DISCUSSION

The voltage characteristics of the sample polyblend film have been analyzed, as illustrated in Fig. 63.1. This figure demonstrates that (i) the current rises with increasing applied voltage at a constant temperature, and (ii) the current also increases with temperature at a constant applied voltage. The underlying mechanisms at play in this scenario are examined in relation to various mechanisms.

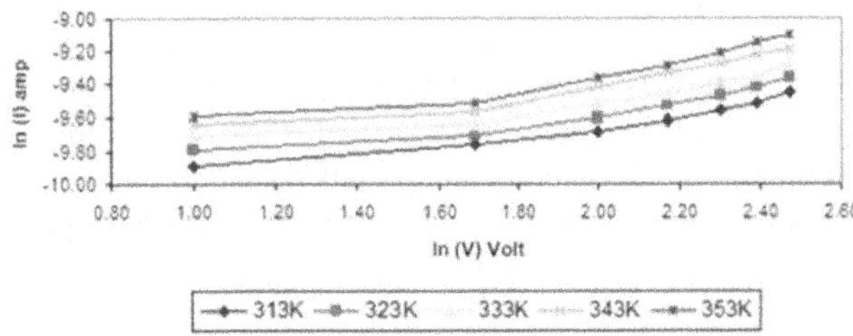

Fig. 63.1 Current voltage characteristics

3.1 Poole-Frenkel Mechanism

$$ln\sigma = ln\sigma_0 + \frac{\beta_{PF}}{2kT} + E^{1/2}$$

The Poole-Frenkel mechanism is defined by the linear relationship observed in ln σ versus E½ plots, where the Poole-Frenkel plots, as predicted by the aforementioned equation, exhibit a linear form with a positive slope. In the current analysis of the PVC and PMMA blend film, the ln σ versus E½ plots are indeed linear; however, they display a negative slope (Fig. 63.2), which suggests that the Poole-Frenkel mechanism is not present.

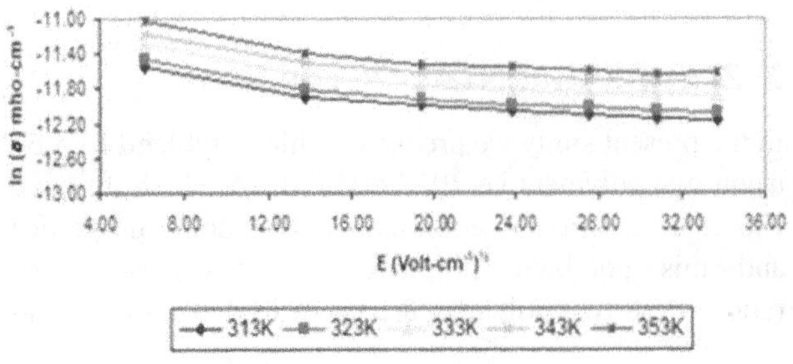

Fig. 63.2 Poole-frenkel plots

3.2 Fowler-Nordheim Mechanism

The Fowler-Nordheim equation for current density is expressed as is is $J = AV^2 e^{-\frac{\varphi}{v}}$ This can be rearranged to yield $ln\frac{J}{V^2} = lnA - \left(\frac{\varphi}{v}\right)$ The plot of ln(J/V²) against 1/V, as illustrated in Fig. 63.3, reveals that, with the exception of a few outlier points, the data points form a nearly linear graph with a positive slope. This observation suggests that tunneling current is absent, in accordance with the implications of the Fowler-Nordheim relation

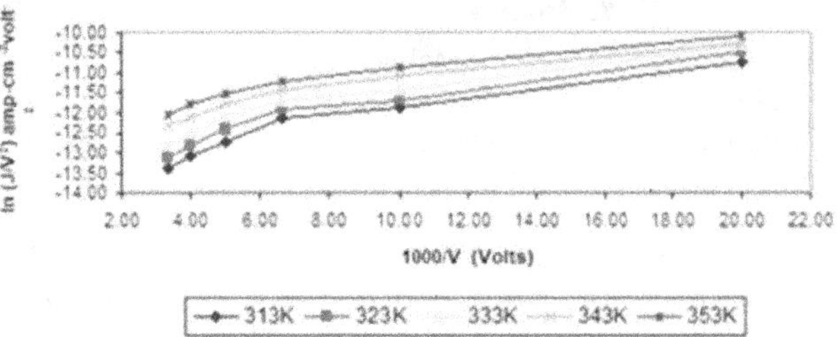

Fig. 63.3 Fowler-nordheim plots

3.3 Schottky Plots

The Schottky-Richardson current-voltage relationship is defined by the equation:

$J = AT^2 e^{\left(\frac{-\varphi_S}{KT}\right) + \beta_{SR}E^{1/2}}$ Consequently, this can be rewritten in logarithmic form as

$lnJ = lnAT^2 - \frac{\varphi_S}{KT} + \beta_{SR}E^{1/2}$ The plot of *ln J* against $E^{1/2}$, as illustrated in Fig. 63.4, reveals a linear relationship with a positive slope, indicating the validity of the mechanism. In the context of the Schottky-Richardson mechanism, the current exhibits a significant dependence on temperature, unlike the Poole-Frenkel mechanism. Therefore, investigating the temperature dependence of current density is of considerable significance.

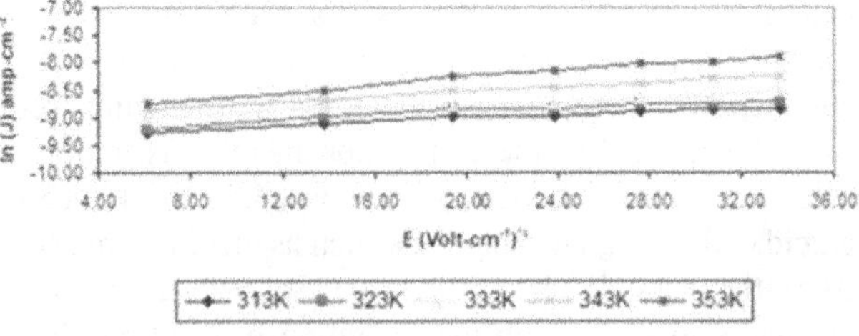

Fig. 63.4 Schottky plots

3.4 Current Density Versus Temperature Plots

The relationship between temperature and current density is illustrated through ln J versus T plots in Fig. 63.5, indicating a non-linear increase of ln J with rising temperature. This pronounced temperature dependence aligns with the Schottky-Richardson mechanism.

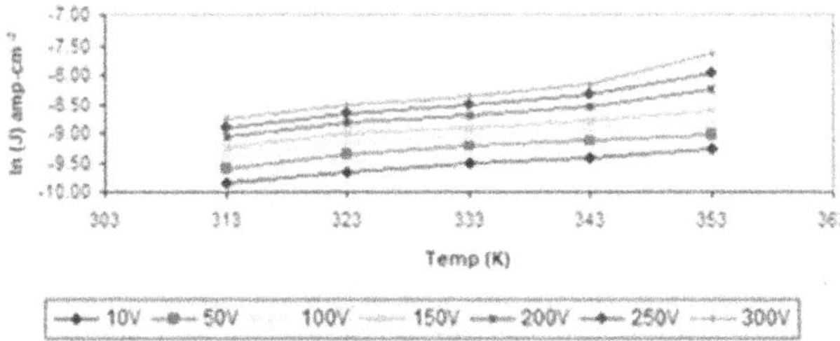

Fig. 63.5 Current density versus temperature plots

3.5 Richardson Mechanism

The Richardson current-voltage relationship is represented by the equation $J = AT^2 e^{\left(\frac{-\varphi_S}{KT}\right)+\beta_{SR}E^{1/2}}$ Consequently, the equation can be rearranged to $ln\frac{J}{T^2} = lnA + \frac{\varphi_S}{KT} + \beta_{SR}E^{1/2}$. According to this relationship, a graph plotting J/T² against 1/kT should yield a straight line with a negative slope. In this study, straight line graphs with a negative slope have been observed, with the exception of one or two data points. The linearity of these plots reinforces the Schottky-Richardson mechanism (see Fig. 63.6)

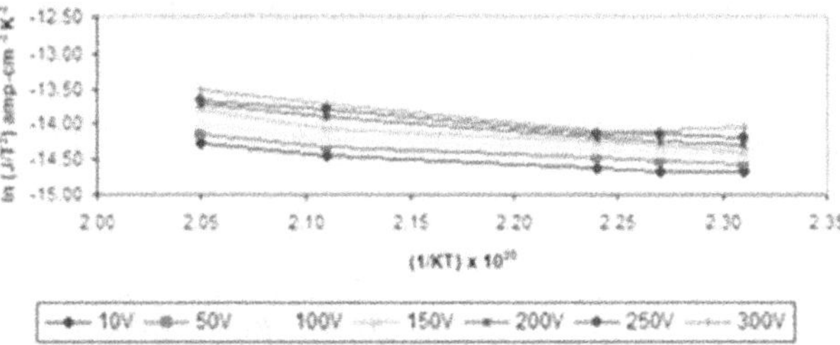

Fig. 63.6 Richardson plots

3.6 Arrhenius Plots

The ln σ versus 1/T graphs (Fig. 63.7) for all applied voltage values exhibit parallel straight lines with a negative slope. The activation energy is determined from the slope of these lines and is approximately 0.13 eV, which aligns well with the previously reported orders of magnitude

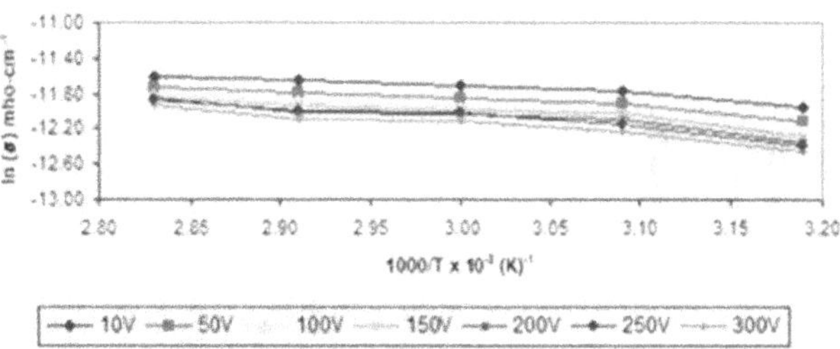

Fig. 63.7 Arrhenius plots

4. CONCLUSION

The S-R conduction mechanism is the dominant process in the sample, overshadowing other mechanisms. The strength of the applied electric field appears inadequate to release electrons from traps, indicating a lack of the PF mechanism. Although the electric field strength and temperature range are insufficient to stimulate contributions from alternative mechanisms, the activation energy can still be accurately determined and is found to be consistent with values observed in similar samples.

REFERENCES

1. Seanor D A, Electrical properties of polymers (D A Seanor Ed) chapter 1 Academic Press New York London 1982
2. Fukada E, Prog poly sci, Jan 2 (1971) 329
3. Keton J E (Ed), Organic semiconducting polymers (Marcel, Dekker, New York) (1968) 267
4. Pavan Khare & Shrivastav A P, Indian J Pure & Appl Phys, 29 (1991) 1410
5. Parak N C & Garg T C, Indian J Pure & Appl Phys, 25 (1987) 110
6. Schottky W Z, Phys, 15 (1994) 872
7. Frenkel J, Phys Rev, 54 (1938) 647
8. Fowler R H & Nordheim L, Proc Roy Soc London, 119 (1928) 173.
9. Rose A, Phys Rev, 97 (1955) 1538.

Note: All the figures in this chapter were made by the authors.

Recent Trends in Applied Physics and Material Science – Sudhir Bhardwaj et al. (eds)
© 2026 Taylor & Francis Group, London, ISBN 978-1-041-16452-4

Synthesis and Characterization of PMMA Polymer Electrospun Fiber

A. B. Patil[1*]
Department of Physics, Mahatma Fule Arts Commerce and Sitaramji Chaudhari Science Mahavidyalaya Warud, Maharashtra, India

R. J. Dhokne (Pathare)[2]
Department of Physics, Shri Shivaji Education Society Amravati's Science College, Congress Nagar, Nagpur, Maharashtra, India

Abstract: PMMA/$SrAlO_7Eu^{2+}Dy^{3+}$ fibres were produced via electrospinning. First, 2 grams of PMMA was liquified in 10 ml of DMF to create 50 wt.% solution stirred for 5 hours at 150-250 rpm. Then, 2 grams of $SrAlO_7Eu^{2+}Dy^{3+}$ was added and stirred for 16 hours to form a unique mixture. This solution was carried to a syringe for electrospinning, with a 15-16 cm distance to the collector, a feed rate of 3 ml/h, and a high voltage of 18-20 kV. The collector rotated at 1200-1400 rpm to form a uniform sheet of fibres. Analyses using SEM, FTIR, and XRD confirmed the morphology of the nanofibers, which contained $SrAlO_7Eu^{2+}Dy^{3+}$ nanoparticles in a monocyclic structure. Photoluminescence studies showed that excitation at 380 nm produced a 530 nm emission, indicating vibrant sea green luminescence. These fibres are suitable for applications in smart textiles, wound dressings, and luminescent fabrics.

Keywords: Polymer, Phosphors, Electrospinning, Nanofibers, etc.

1. INTRODUCTION

Photoluminescent materials in polymers are gaining popularity in various industries, including fashion for nightclub apparel and UV-reactive labels for counterfeit detection and security Mishra (2009), Manjusha (2015). The smart clothing sector, especially in protective textiles, shows significant commercial potential. Functional textiles with antibacterial, superhydrophobic, and flame-retardant properties focus on practical applications rather than aesthetics Khursheed (2020). Divalent Europium (Eu^{2+})-doped strontium aluminate phosphors show extended sea green phosphorescence caused by the 5D to 4F quantum jump in Eu^{2+} ions. Their phosphorescent properties might be intensified by co-doping with rare earth ions like Dy3+ or adding excess AlO_7. These phosphors are used in luminous paints for highways, airports, and buildings, as well as in ceramics, textiles, outdoor displays, glowing clocks, and safety signals. They are ideal for luminous watches, emergency instruments, and alert signals, and they are known for their chemical stability, safety, brightness, and lasting photoluminescence Chang (2019).

Electrospinning is an easy, and money-fruitful, economical method for fabricating fibers, ranging from microscale to nanoscale, by applying an electrostatic field to stretch a polymer solution Feng (2015). Polymer-Phosphor composite electro spun fibres offer benefits like a high shape compactness, large specific surface area, tuneable porosity, flexibility, and customizable fibre composition for enhanced performance Dandekar (2019). This research paper presents a synthesis and characterization of a luminescent phosphor polymer composite, specifically (1:1wt%) PMMA/$SrAlO_7Eu^{2+}Dy^{3+}$ fiber. The fibers produced demonstrate notable brightness and a prolonged afterglow decay. Following a few minutes of exposure to visible light, these fibers are capable of emitting light for over ten hours in the absence of light.

2. EXPERIMENTAL WORK

2.1 Materials

PMMA (Polymethyl methacrylate) average MW120000 by GPC Sigma-Aldrich, Eu^{2+}, Dy^{3+} Doped Strontium Aluminate MW 209.11g/mol Sigma-Aldrich, Dimethyl formamide anhydrous DMF(MW73.09) Sigma-Aldrich. All materials utilized were highly graded and procured from Central Scientific Company, located in Nagpur, Maharashtra.

[1]amrapalisukhadeve03@gmail.com, [2]raginidhokne@gmail.com

DOI: 10.1201/9781003684718-64

2.2 Preparation of PMMA/$SrAlO_7Eu^{2+}Dy^{3+}$ Solution

2gm PMMA was dissolved in 10ml of DMF solvent to give a mass percent composition of 50wt% and magnetically stirred (150 to 200 rpm) for 5hr at room temperature. Then 2gm of $Eu^{2+}Dy^{3+}$ doped Strontium Aluminate was added to PMMA solution and stirred for 16 Hrs. with a magnetic stirrer till homogeneous at room temperature.

2.3 Preparation of PMMA/$SrAlO_7Eu^{2+}Dy^{3+}$ Fiber

The PMMA/$SrAlO_7Eu^{2+}Dy^{3+}$ homogenous blend was laden into a 5 ml syringe and conveyed to the tip at a feed rate of 3 ml/hr using a syringe pump. A high voltage of 20 kV was implicated to the needle, while a stranded electrode was attached to a metallic collector plate covered with aluminum foil, positioned 16 cm away. The fibres were collected on foil and dried in a preheated void furnace for 10 hours to remove residual solvent.

3. CHARACTERIZATION TECHNIQUES

X-ray diffraction (XRD) study was done using a D8-Discover diffractometer (Bruker, USA) with Cu-Kα radiation (1.5405 Å) over a 2theta range of 17° to 90°, at 40 kV and 40 mA. A scanning electron microscope (SEM) model ZU SSX – 550 Super scans from Shimadzu was used to analyze the morphology. FTIR spectral study was done with a Shimadzu FTIR Tracer-100 spectrometer using a Diamond ATR. Photoluminescence (PL) spectra were obtained using a Cary-Eclipse Spectrofluorometer.

4. RESULTS AND DISCUSSION

4.1 SEM

The magnified SEM images with scale bars corresponding to 1 um, and resolutions of x10000 revealed the fibre diameter ranges from 0.68 μm to 1.24 μm. The varying diameters of the electrospun fibres indicate that the choice of polymer affects the viscosity of the PMMA/$SrAlO_7:Eu^{2+},Dy^{3+}$ solution.

This modification is due to the enhanced conductivity of the electrospinning solution from adding $SrAlO_7:Eu^{2+},Dy^{3+}$. The SEM analysis shows that smooth, symmetrical, homogeneous fibres of PMMA/$SrAlO_7:Eu^{2+},Dy^{3+}$ were fabricated. Homogeneous fibres are formed by carefully controlling key electrospinning parameters, including solution viscosity, electric field strength, sample feed rate, etc. Each factor is crucial for producing uniform and symmetrical composite polymers.

4.2 XRD

The XRD analysis of PMMA/$SrAlO_7:Eu^{2+},Dy^{3+}$ strongly correlates with JCPDS PDF card No. 1538528, confirming

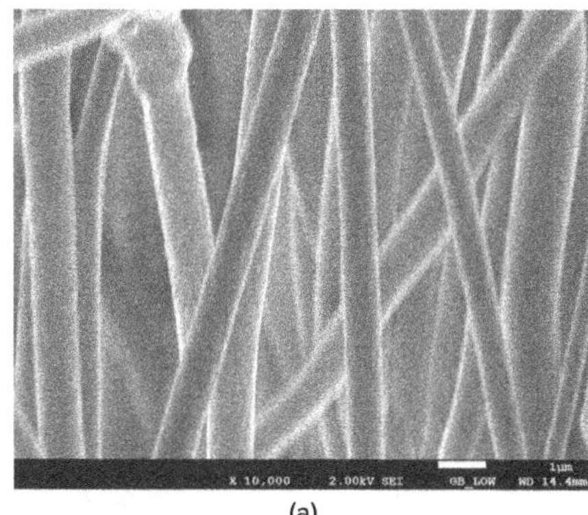

(a)

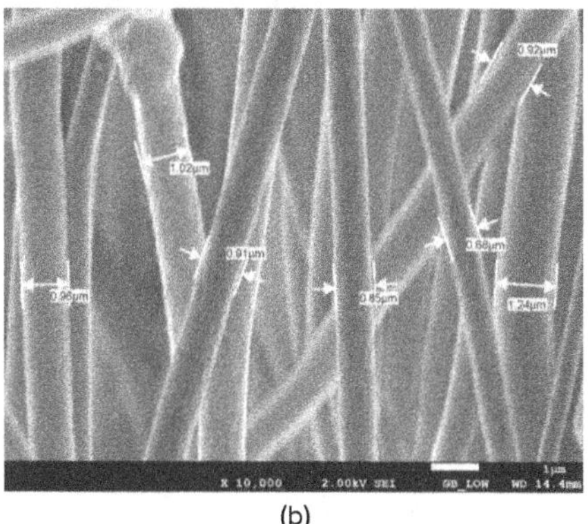

(b)

Fig. 64.1 (a): SEM image of PMMA/SrAl2O4:Eu2+, Dy3+ fiber (b): SEM image of PMMA/SrAl2O4:Eu2+, Dy3+ fibre showing the dimeter of fibre

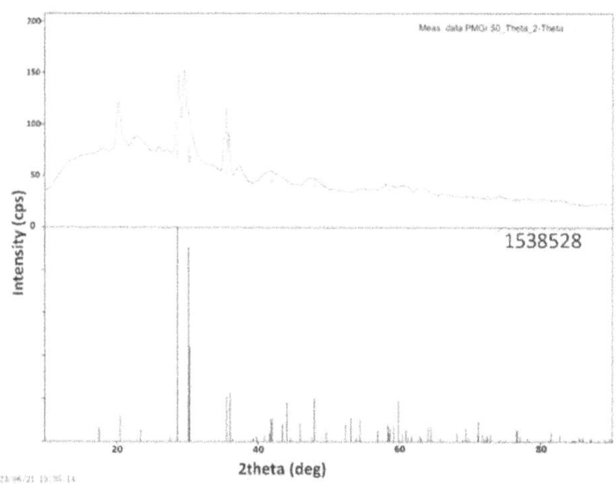

Fig. 64.2 XRD graph image of PMMA/SrAl2O4:Eu2+, Dy3+

a Monoclinic phase with three unequal axes and one oblique angle. The mean particle size, calculated using the Debye-Scherrer equation, ranges from 40 to 45 nm, consistent with pure phosphor characteristics. Lattice parameters obtained from POWD software are a=6.8270 Å, b=8.1180 Å, and c=4.5760 Å. The absence of supplementary peaks in the XRD pattern indicates that PMMA effectively serves as a matrix for the $SrAlO_7:Eu^{2+},Dy^{3+}$ phosphor, confirming the phosphor retains its phase in PMMA.

FTIR Study

FTIR study confirmed the functional groups and chemical bonds in the synthesized PMMA/$SrAlO_7$:$Eu^{2+}Dy^{3+}$ fiber. A peak at 1732 cm^{-1} indicates carbon-oxygen double bond stretching in the carbonyl group, while a peak at 3434 cm^{-1} corresponds to C=C and O-H bond elasticity. The spectral bands from 350 to 1000 cm^{-1} are due to the IR-active vibrations of the $SrAlO_7$:$Eu^{2+}Dy^{3+}$ phosphor. The appropriate dispersion is validated, since the PMMA/$SrAlO_7$:$Eu^{2+}Dy^{3+}$ fibre exhibits all the distinct bands associated with the composite.

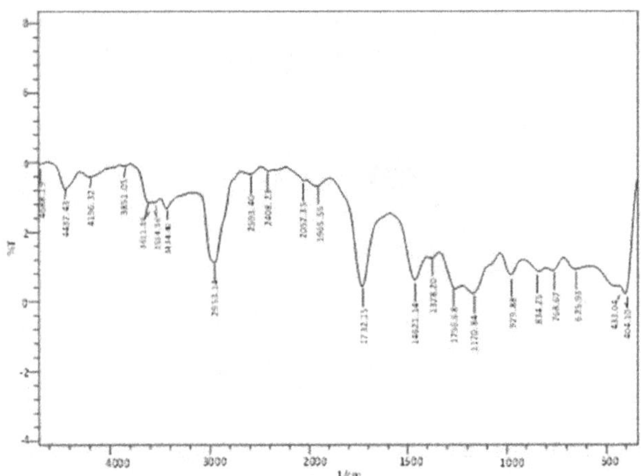

Fig. 64.3 FTIR graph image of PMMA/SrAl2O4:Eu2+, Dy3+

4.3 Photoluminescence Study

The photoluminescence (PL) spectrum of PMMA/$SrAlO_7$:Eu^{2+}, Dy^{3+} fibre revealed a wide emission peak at 530 nm, indicating a sea green hue. The high pointpeak at 530 nm is due to the quantum jump from $4f^6\,5d^1$ to $4f^7$ of the Eu^{2+} ion co-doped with Dy^{3+} ion in the $SrAlO_7$:Eu^{2+}, Dy^{3+} phosphor. Emission spectra were obtained at λex = 380 nm, and excitation spectra were captured at λem = 530 nm. 380nm peak is the characteristic $4f^7 \rightarrow 4f^6\,5d^1$ quantum jump of the divalent europium ion (Eu^{2+}) co-doped with Dy^{3+}in the $SrAlO_7$:Eu^{2+}, Dy^{3+} phosphor.

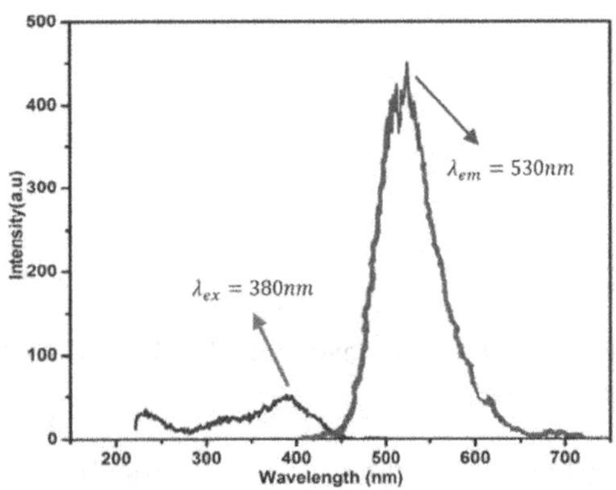

Fig. 64.4 PL spectra image of PMMA/SrAl2O4:Eu2+, Dy3+

5. CONCLUSION

An electrospinning technique has been successfully employed to fabricate PMMA/$SrAlO_7$:Eu^{2+}, Dy^{3+} electrospun fibres, which exhibit an average diameter of 0.94µm. Scanning electron microscopy (SEM) images demonstrate that the Eu^{2+} & Dy^{3+} ions of $SrAlO_7$:Eu^{2+}, Dy^{3+} phosphor are uniformly integrated within the PMMA molecular chain. The resulting fibres significantly influence the photoluminescent (PL) properties of Eu^{2+} ions enhanced by cooping Dy^{3+}ion, attributed to the robust coordination interactions between the Eu^{2+} ions and the polymer matrix. These fibres are suitable for applications in smart textiles, wound dressings, and luminescent fabrics.

REFERENCES

1. S. B. Mishra, A. K. Mishra, N. Revaprasadu, (2009). Strontium Aluminate/Polymer Composites: Morphology, Luminescent Properties, and Durability, Journal of Applied Polymer Science, Vol. 112, 3347–3354
2. Manjusha P. Dandekar. (2015). fabrication of electrospun nanofibers of europium complex Eu(TTA)3phen using electrospinning technique. Procedia Materials Science 10:580–587.
3. Sumara Khursheed. (2020). Phosphor Polymer Nanocomposite: SrAl2O4:Eu2+, Dy3+ Embedded PMMA for Solid-State Applications Materials Today: Proceedings 21 2096–2104
4. Hui-Yi Chang. (2019). Preparation of hydrophobic nanofibers by electrospinning of PMMA dissolved in 2-propanol and water. MATEC Web of Conferences 264, 03004.
5. Feng Ye, Shengjie Dong. (2015). Fabrication and characterization of long-persistent luminescence/ polymer (Ca2MgSi2O7:Eu2+,Dy3+/PLA) composite fibers by electrospinning. Optical Materials.
6. Manjusha Dandekar. (2019). Synthesis and Photoluminescence Study of ElectrospunNanofibers of Eu(TTA)3Phen/PMMA-PVDF Composite For Photoluminescent Fabric Designing. AIP Conference Proceedings 2104, 020011.

Note: All the figures in this chapter were made by the authors.

65

Effect of Salt and Charge Density on the Dynamics of Acrylamide-co-Sodium Acrylate Hydrogel

Susmita Mohanta,
Usharani Mohapatra, and Sidhartha S. Jena
Department of Physics and Astronomy,
National Institute of Technology Rourkela,
Odisha, India

Abstract: We have investigated the diffusion dynamics in the charged gel matrix using dynamic light scattering measurements. Acrylamide-co-sodium acrylate was taken as the sample for polyelectrolyte gel with different charge densities, and the effect of salt on diffusion dynamics was explored. Due to the non-ergodic behavior of the sample, the direct ensemble averaged method was employed to obtain the correlation data. All the correlation curves were fitted with adequate accuracy by single plus stretched exponential functions from which the relaxation times corresponding to fast and slow relaxation modes were acquired. As we increase the charge density of the hydrogel sample without adding salt, an increase in mean relaxation time is noticed. However, when the samples were prepared with the addition of monovalent salt, a reverse trend was detected. Here, the mean relaxation time gradually decreases, implying that the gel network becomes more diffusive. Similar trend is seen for the fast relaxation mode.

Keywords: Diffusion, Charge density, Relaxation time

1. INTRODUCTION

Hydrogels have immense importance in our day to day life due to their biological resemblance and non-toxicity. One can modify the properties of the hydrogels in many ways based on the requirements of various pharmaceutical and industrial applications. The diffusion dynamics in the gel network is governed by these modifications. Here, in this report, the diffusion dynamics of the polyacrylamide-co-sodium acrylate hydrogel is examined. The sodium acrylate monomer contributes the charged part of the samples. For a poly-electrolyte gel, the counter-ion affects significantly in the gel swelling. Generally, due to the osmotic pressure generated by the counter-ions, the swelling of the gel network increases and the dynamics is changed accordingly. Hence it is important to study the charge effect on the dynamics of the polyelectrolyte gels.

2. MATERIALS AND METHOD

2.1 Hydrogel Preparation

Polyacrylamide-co-sodium acrylate hydrogels were prepared by free-radical cross-linking co-polymerization of acrylamide and sodium acrylate with total monomer concentration of 5wt% and the cross-linker concentration of bisacrylamide was fixed 1wt% of the total monomer concentration. To facilitate the polymerization and gelation process, 1.5mg/ml potassium persulfate was added to the solution as the activator and 0.5μl/ml of N,N,N',N'-Tetramethylethylenediamine (TEMED) was added as the initiator.

2.2 Dynamic Light Scattering Measurements

For the dynamic light scattering experiments, a 50mW sapphire LASER from Coherent, USA operating at wavelength of 488nm is used as the light source. The

Corresponding author: sid@nitrkl.ac.in

scattered signal is collected using an avalanche photodiode detector from PerkinElmer, USA. This data is fed to the ALV multi-tau correlator from Germany to give the intensity-intensity correlation function.

3. RESULTS AND DISCUSSIONS

In Fig. 65.1, the field correlation function of the 5wt% polyacrylamide hydrogel is shown in the upper panel with its corresponding residual plots to different fittings are shown in the lower panel.

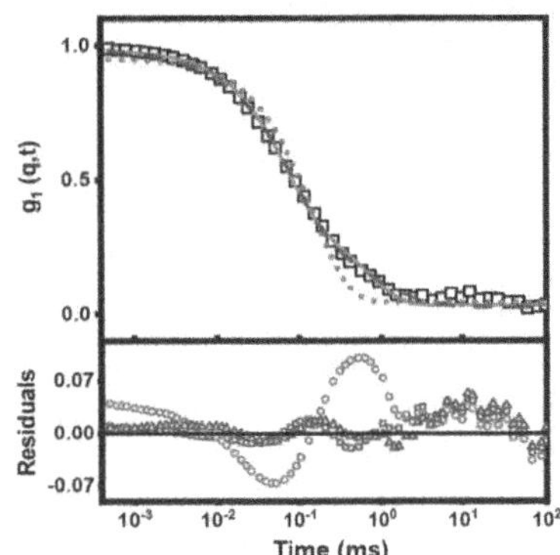

Fig. 65.1 Upper panel: field correlation function of 5wt% polyacrylamide hydrogel at an angle 90° fitted with single exponential(...), single plus stretched exponential(---), double exponential(——) function. Lower panel: residual plots of the corresponding fittings

Here, the field correlation functions for all the samples are fitted well with a single plus stretched exponential equation, given as,

$$g_1(q,t) = a_f \times \exp\left(-\left(t/\tau_f\right)\right) + a_s \times exp\left(-\left(t/\tau_s\right)\right)^{\beta} + g_1(q,\infty) \quad (1)$$

Here, a_s and a_f are the relative scattering amplitudes and τ_s and τ_f are the slow and fast relaxation times respectively. From the fitting of the curves, the above mentioned parameters are obtained. From τ_s, the mean relaxation time τ_m is calculated using the equation,

$$\tau_m = \frac{\tau_s}{\beta}\Gamma\left(\frac{1}{\beta}\right) \quad (2)$$

From Fig. 65.2, it is noticed that, for all the polyelectrolyte gel samples, the correlation functions decays faster at higher angles as compared to the lower ones. When the same functions are re-plotted against q^{2}t (shown in inset), it is seen that the functions do not collapse into one another, Suggesting the restricted Brownian motion in the gel network. Thus, it is concluded that the gel is not a purely diffusive medium.

The relaxation times for both the fast and intermediate mode against the various charge densities of the samples at an angle 90^0 are plotted in the above Fig. 65.3. In case of the samples without salt, it is seen that the mean relaxation time gradually increases with increase in charge

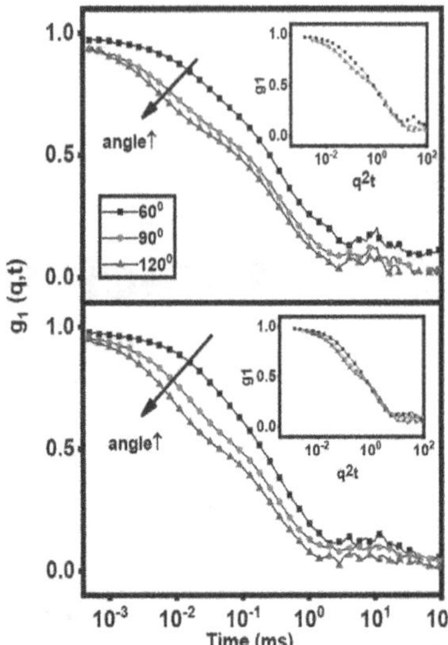

Fig. 65.2 The correlation functions for Aam-co-SA hydrogel with 25% charge density in presence of 0mM (upper) and 100mM (lower) NaCl salt at different angles are shown. Inset: correlation functions of the corresponding samples plotted against q^2t(× 10^{11} sec/m^2)

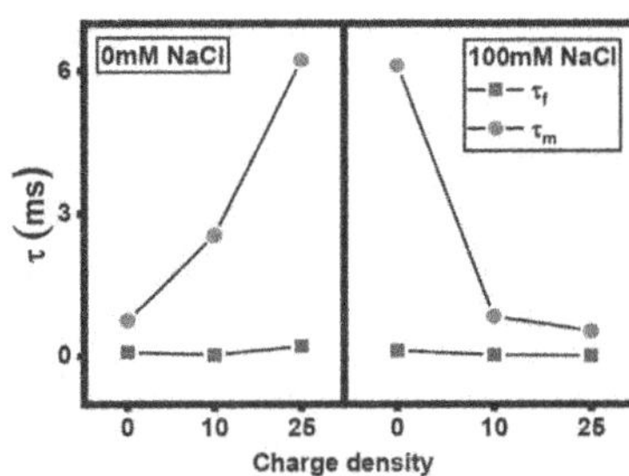

Fig. 65.3 Two characteristic times τ(ms) of the hydrogel samples as a function of charge density with and without salt

density. This mean relaxation time obtained from τ_s refers to the co-operative diffusion of the gel network. So, with increase in charge density, the diffusion of the gel network decreases. However, when the samples were prepared in presence of 100mM NaCl salt, reverse trend is seen. Here, the mean relaxation time gradually decreases and hence the diffusion increases implying more diffusivity of the gel network. For both these kinds of the samples, similar trend is also seen for the fast relaxation time, though the change is not very notable.

REFERENCES

1. Rahalkar, A. and Muthukumar, M. (2017). Diffusion of polyelectrolytes in polyelectrolyte gels. Macromolecules. 50:8158–8168.
2. Joosten, J. G. H. and McCarthy, J. L. (1991). Dynamic light scattering by aqueous polyacrylamide gels. 24:6690–6699.
3. Jena, S. S. and Mithra, K. (2021). Surfactant head group and concentration influence on structure and dynamics of gellan gum hydrogels: crossover from stretched to compressed exponential. 59:1972–1985.

Note: All the figures in this chapter were made by the authors.

Recent Trends in Applied Physics and Material Science – Sudhir Bhardwaj et al. (eds)
 ISBN 978-1-041-16452-4

66

A Comparative Dynamics Analysis of Agar–Polyacrylamide Double Network Hydrogel with its Associated Single Network Hydrogels

Usharani Mohapatra, Susmita Mohanta, Sidhartha S. Jena

Department of Physics and Astronomy,
National Institute of Technology, Rourkela, Odisha, India

Abstract: A mixed agar-polyacrylamide (Ag-PAAM) double network (DN) hydrogel was synthesized using one-pot heating-cooling photo-polymerization method along with its corresponding single network (SN) hydrogels of agar and PAAM. The agar hydrogel was prepared by simply heating followed by cooling it to room temperature, whereas the PAM hydrogel was made with the help of photo-polymerization. Light scattering experiments were performed on the prepared hydrogels to examine and compare their structure and dynamics. Being non-ergodic in nature, direct ensemble average approach was used to analyse the correlation data. Both the DN along with the SN hydrogels shows bimodal relaxation. The magnitude of both the fast and slow mode relaxation time is the maximum for the double network hydrogel of agar-PAAM as compared to the individually prepared single network hydrogel of either agar or PAAM.

Keywords: Double network hydrogel, Hydrogel, Dynamics, Dynamic light scattering

1. INTRODUCTION

Hydrogels are unique soft material having biomimetic and biocompatibility properties as they retain a substantial quantity of water inside their three dimensional porous network structure. So they stand out for their applicability especially in the fields of tissue engineering and drug delivery. Yet, the majority of hydrogels are lack in high mechanical properties which hinders their application requiring toughness. Recent research shows some specially designed double network (DN) hydrogels acquire improved mechanical properties as they have two contrasting polymer networks with strong entanglements. Here, agar -polyacrylamide (Ag-PAAm) DN hydrogel has been prepared using the simple one pot method, due to its effectiveness and controllability. The hydrogen bond linked agar acts as first network while covalently cross linked polyacrylamide serves as the second network for DN network hydrogel.. In this short note we have analysed and compared the structure and dynamics of DN hydrogels with its corresponding SN gels of agar and PAAM by performing dynamic light scattering (DLS) experiments.

2. MATERIAL AND METHODS

2.1 Sample Preparation

Agar and acrylamide are utilized as monomers for the first and second network respectively for DN gel. While N, N'- methylene-bis-acrylamide (MBA) is employed as a cross-linking agent for the second network, Irgacure 2959 is used as photo initiator for the photo-polymerization of acrylamide. In brief, all the mentioned reagents were taken into a test tube along with 1ml of milipore water. The tube was warmed up to 95° C in a hot tub with constant stirring with a magnetic bead. A clear and transparent solution was obtained after 30 minutes. This transparent solution was then transferred into a already clean DLS cuvette and kept at room temperature for cooling. The first network, agar

Corresponding author: sid@nitrkl.ac.in

DOI: 10.1201/9781003684718-66

was formed during this cooling process. Next, the cuvette was exposed to 8 Watt UV light of 365 nm wavelength for one hour for the photo polymerization of acrylamide to form the double network hydrogels. Similarly, the single network gels of agar and PAAM were prepared by heating-cooling and photo-polymerization process respectively.

2.2 Dynamic Light Scattering Method

A hand build DLS set up, consisting of a sapphile LASER from Coherent as the excitation source, an avalanche photo diode from PerkinElmer as the detector and a digital correlator for auto correlation was used for the measurements. The samples being non-ergodic, ensemble averaged intensity correlation functions were accumulated on a rotating sample for 30 minutes and at fixed angle of 90^0. The scattering wave vector, q whose inverse, q^{-1} is known as 'probing length' is given as $q = 4\pi n/\lambda \sin(\Theta/2)$, with n, λ and Θ being the refractive index of the medium, wavelength of the light in space and scattering angle respectively. To study the dynamics of the prepared hydrogels, the direct ensemble averaged intensity-intensity correlation functions (ICF), g_2 (q, t) are acquired using the digital correlator and transformed to field correlation function, g_1 (q, t) using Siegert relation: $g_2(q,t) = 1 + \beta|g1(q,t)|^2$ where β is the instrument coherence component having value ≤ 1.

3. RESULTS AND DISCUSSION

Figure 66.1 upper panel illustrates the field correlation function, g_1 (q, t) for Agar (A), PAAM (B) and Ag-PAAM (C) hydrogels and their corresponding residual plots are displayed in the lower panel. All the correlation curves are best fitted with double exponential function of the form, $g_1(q, t) = a_s \exp(-t/\tau_s) + a_f \exp(-t/\tau_f) + g_1(q, \infty)$ where τ_s and τ_f are slow and fast relaxation time with a_s and a_f as their relative scattering amplitude respectively.

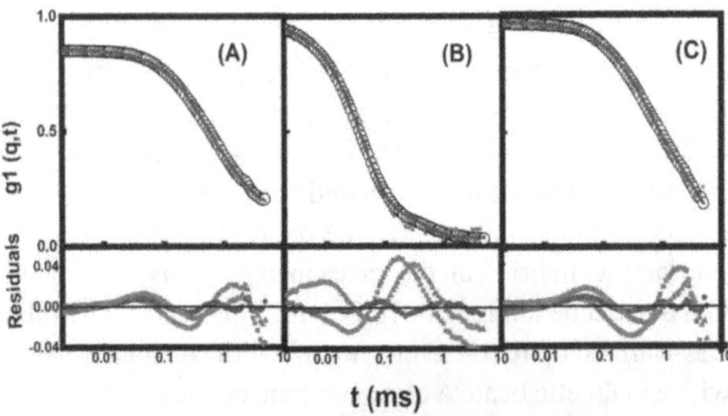

Fig. 66.1 Upper panel: the field correlation curve, $g_1(q, t)$ vs time for (A) Agar SN (B) PAAM SN and (C) Ag-PAAM DN gel and lower panel: their residual plots for single exponential (■), double exponential(●) Single stretched (▲)

All the double and single network hydrogels show bimodal relaxation. The fast relaxation is associated with polymer network relaxation while the slow relaxation mode reflects cluster relaxation. For, all the studied samples, there is at least an order of magnitude difference seen between the fast and slow relaxation mode. The fast relaxation time is found to be the maximum for DN gel of Ag-PAAM. Since τ_f ~ correlation length (ζ), the correlation length of DN is also maximum when compared with corresponding SN gel of Agar and acrylamide.

In the case of purely diffusive gel, the decay rate, Γ has a linear relationship with q^2 and deviates from this linearity for non-diffusive gel. From Fig. 66.2, it is noticed that Γ_f shows a linear reliance with q^2 which is not the case for Γ_s. From the slope of Γ vs q^2 graph, the diffusion coefficient of gels can be estimated. The corresponding diffusion coefficient for PAAM SN, Agar SN and Ag-PAAM DN gels are found to be 1.82×10^{-6} cm^2/s, 1.61×10^{-8} cm^2/s and 1.89×10^{-8} cm^2/s respectively. It is seen from these values that PAAM SN gel has the highest diffusion coefficient value. Agar SN and DN gels have almost nearly equal diffusion coefficient values. This is may be due to the high concentration of agar in DN gel compared to acrylamide. Since the formation of agar network occurred first and has more rigidity than the PAAM network, it can be assumed that the chain density of agar in DN gel is nearly same as the agar SN gel.

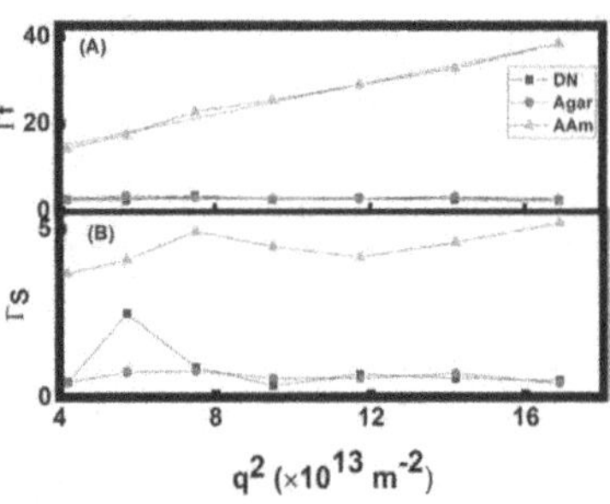

Fig. 66.2 The plot of decay rate for fast Γ_f (A) and slow Γ_s (B) vs q^2.

Figure 66.3 shows the total ensemble averaged scattered intensity, $\langle I_s(q)\rangle_E$ as a function of q for all the three hydrogels of agar, acrylamide and Ag-PAAM. For agar and Ag-PAAM gels the scattered intensity decreases with

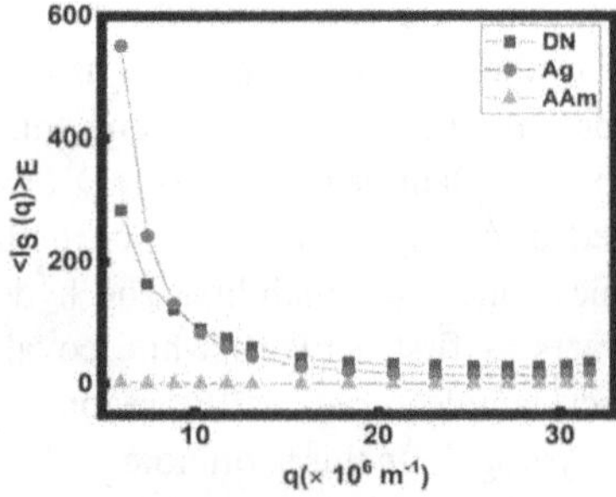

Fig. 66.3 The ensemble averaged total scattered intensity $\langle I_s(q)\rangle_E$ vs scattering vector, q

increase of q e but no noticeable fluctuation in scattered intensity with increasing q is seen for photo-polymerised PAAM gel. This may be due to less inhomogeneity compared to the other two gels.

4. CONCLUSION

In this work, a hybrid Ag-PAAM double network hydrogel was prepared consisting of "non swellable" agar and "highly swellable" PAMM gel network. Single network of Agar and acrylamide gels were also prepared separately to compare their dynamics using DLS. The DN gel was found to be least diffusive and possessed the highest relaxation mode as it contained most number of polymer strand densities arising from inter-network entanglements from Agar and PAAM.

REFERENCES

1. Chen, Flu.(2015). Mechanically strong hybrid double network hydrogel with antifouling properties. J. Mater. Chem. B. (3):5426C.
2. Jia, Muthulumar. (2018). Topologically frustrated dynamics of crowded charged macromolecules in charged hydrogels, USA: University of Massachusetts: 2248.

Note: All the figures in this chapter were made by the authors.

Recent Trends in Applied Physics and Material Science – Sudhir Bhardwaj et al. (eds)
 ISBN 978-1-041-16452-4

Application of Deep Neural Network to Predict GDR Cross-Sections of Zr Isotopes

Manish Kumar Bairwa*
Department of Physics, Indian Institute of Technology Roorkee, Roorkee, Uttarakhand, India

P. Arumugam
Department of Physics, Indian Institute of Technology Roorkee, Roorkee, Uttarakhand, India
Centre for Photonics and Quantum Communication Technology, Indian Institute of Technology Roorkee, Roorkee, Uttarakhand, India

Abstract: The investigation of Giant Dipole Resonance (GDR) parameters plays a key role in enhancing our comprehension of nuclear structure and the behavior of heavy nuclei. The GDR mechanism dominates the photoabsorption cross-section up to 30 MeV, where the cross-section typically exhibits one or more Lorentzian peaks, depending on the deformation of the nucleus. Traditionally, GDR parameters such as peak energy and width are derived from experimental observations or theoretical calculations. This paper presents a methodology for predicting GDR parameters of the ^{90,91,92,94}Zr isotopes using a deep neural network (DNN). We use a comprehensive set of experimental GDR parameters across various nuclei for model training. The neural network is then fine-tuned for highly accurate and reliable predictions. The GDR cross-sections thus obtained using these predicted parameters are compared with experimental observations and show a strong consistency within the bounds of experimental uncertainties. This predictive framework can be extended to other less explored nuclei, providing new insights into nuclear structure and dynamics.

Keywords: Nuclear structure, Giant dipole resonance, Deep neural network

1. INTRODUCTION

The study of GDR cross-sections is important in the interpretation of nuclear structure and reactions, as well as their applications in nuclear physics, astrophysics, and nuclear technology. Traditional models and theoretical frameworks have been used in predicting GDR cross-sections but often fail to capture the intricate details of these phenomena. It has been demonstrated in recent years that data-driven techniques like machine learning (ML) are very helpful in simulating intricate physical processes where significant insights can be obtained from vast amounts of data. The current study proposes the use of an advanced ML technique, namely deep neural network (DNN), to learn and generalize the underlying pattern from experimental as well as theoretical data to predict GDR parameters and hence, cross sections. We feed this model with a dataset containing nuclear properties, excitation energies, etc., and aim to develop a predictive model capable of providing valuable information on GDR cross-sections for a wide range of nuclei.

Our approaches include the following key elements: **Data Integration:** We use a unified dataset from the RIPL-3 library that has experimental measurements and nuclear structure information.

Feature Engineering: The extraction of relevant features from data is the most crucial factor in determining the accuracy of a model.

Validation and Interpretation: We have accomplished rigorous validation processes such as cross-validation

*Corresponding author: mkumarbairwa@ph.iitr.ac.in

DOI: 10.1201/9781003684718-67

and model explainability techniques to judge the model's performance and understanding of the physics behind GDR cross-sections.

The rest of the paper is organized as follows. Section II contains details of the calculation. Section III discusses the findings of the study. Section IV summarises the paper.

2. DETAILS OF THE CALCULATION

The nuclear shapes are connected to the GDR observables using a macroscopic model which involves an anisotropic harmonic oscillator potential coupled to a separable dipole-dipole interaction (Arumugam et al., 2004). The Hamiltonian illustrating GDR excitations is expressed by

$$H = H_{osc} + \eta D^{\dagger} D \quad (1)$$

Here H_{osc}, D and η represent the anisotropic harmonic oscillator Hamiltonian, the dipole operator, and the dipole-dipole interaction strength, respectively. The sum of each of the Lorentzian peaks that correspond to the GDR energies (E_m) provided by the frequencies related to H yields the total GDR cross-section (σ) as follows

$$\sigma\left(E_{\gamma}\right) = \sum_{i} \frac{\sigma_{mi}}{1 + (E_{\gamma}^2 - E_{mi}^2)^2 / E_{\gamma}^2 \Gamma_i^2} \quad (2)$$

where the Lorentz parameters photon energy, full width at half maximum, peak cross-section, and resonance energy are represented by E_{γ}, Γ, σ_m, and E_m, respectively. Here, i stands for the GDR components.

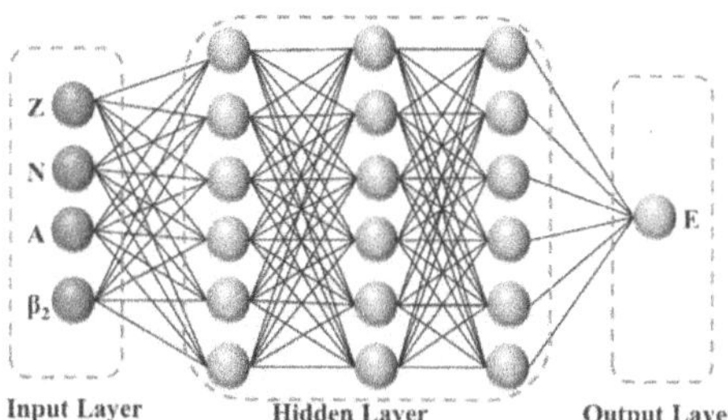

Fig. 67.1 A deep neural network schematic diagram for computing GDR peak energy

There is a strong effect of nuclear deformation (β_2) on the GDR cross-sections. The deformation parameter β_2 is determined from the combination of finite-range liquid-drop (Möller et al., 2016) model (FRDM) results and RIPL-3 data (RIPL-3), when available.

3. RESULTS AND DISCUSSION

The GDR cross-sections can be obtained through the GDR key parameters, peak energy E_m and width Γ_m, as defined through Eq. (2). We use DNN to evaluate the GDR parameters. A schematic diagram for the prediction of peak energy is given in Fig. 67.1. For resonance width, we utilize this energy as an input to the network. We study nuclei not included in the learning set to assess the extrapolation ability of the trained model. Figure 67.2–67.5 displays the resulting cross-sections of ^{90,91,92,94}Zr isotopes derived from these newly obtained parameters. These figures clearly show that the cross-sections obtained from the optimized GDR parameters through DNN reproduce the cross-sections well within the experimental uncertainties.

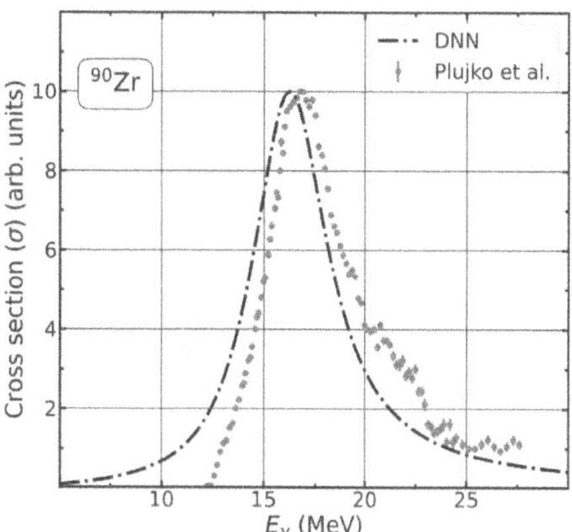

Fig. 67.2 The photoabsorption cross-section of ^{90}Zr. The red-filled circles represent the experimental data (Plujko et al., 2018)

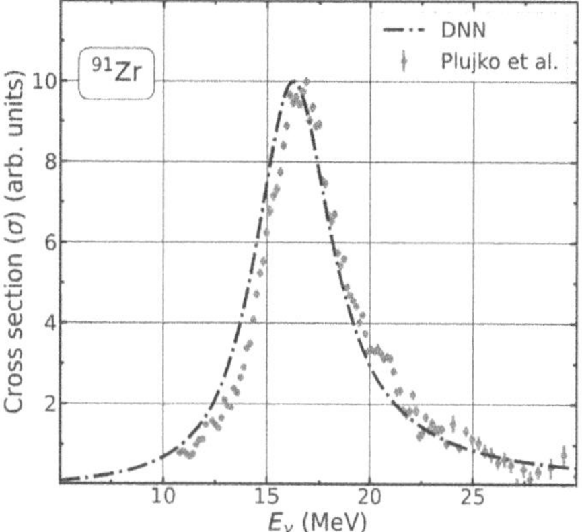

Fig. 67.3 Same as Fig. 67.2 but for ^{91}Zr

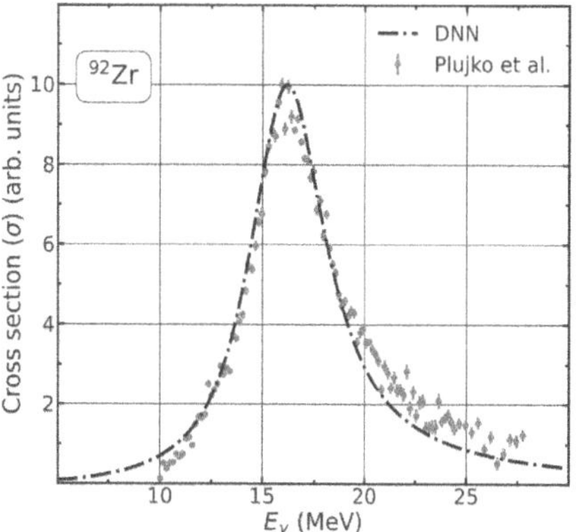

Fig. 67.4 Same as Fig. 67.2 but for ^{92}Zr

4. CONCLUSION

We have trained our models on nuclei between ^{16}O and ^{239}Pu (excluding ^{90,91,92,94}Zr isotopes) using the experimental RIPL-3 database (RIPL-3) to obtain the

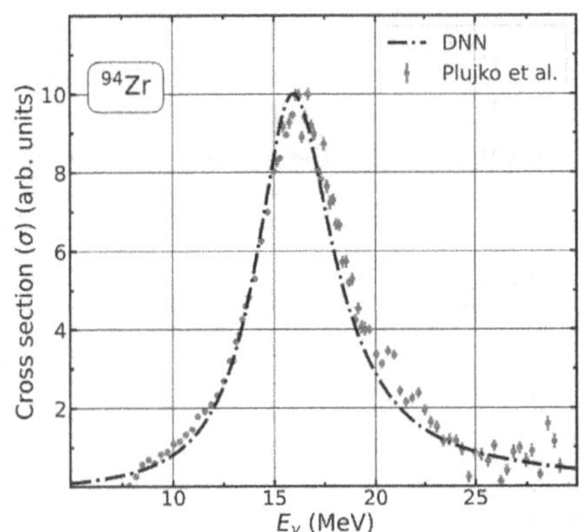

Fig. 67.5 Same as Fig. 67.2 but for ^{94}Zr

GDR parameters. The predicted parameters from the DNN model favor the experimental cross-section studies by Plujko et al. (2018). This justifies the use of the DNN in calculating the GDR cross-sections and aspires to play an important role in studying experimentally unprobed regions.

ACKNOWLEDGEMENTS

We acknowledge a research grant from the SERB, India: CRG/2022/009359.

REFERENCES

1. Arumugam, P., Shanmugam, G. and Patra, S. K. (2004). Giant dipole resonance and Jacobi transition with exact treatment of fluctuations. Phys. Rev. C 69:054313.
2. Möller et al., (2016). Nuclear ground-state masses and deformations: FRDM (2012). At. Data Nucl. Data Tables 109-110:1–204.
3. RIPL-3, https://www-nds.iaea.org/ RIPL-3/.
4. Plujko et al., (2018). Giant dipole resonance parameters of ground-state photoabsorption: Experimental values with uncertainties. At. Data Nucl. Data Tables 123-124:1–85.

Note: All the figures in this chapter were made by the authors.

Recent Trends in Applied Physics and Material Science – Sudhir Bhardwaj et al. (eds)
© 2026 Taylor & Francis Group, London, ISBN 978-1-041-16452-4

Study of ^{23}Mg (p, γ) ^{24}Al Astrophysical Reaction

Pavini Sharma, Sathi Sharma*, Ashok Kumar Mondal
Department of Physics, Manipal University Jaipur, Jaipur, India

Abstract: ^{23}Mg(p,γ)^{24}Al reaction is known to act as possible means for escaping the Ne-Na cycle to form heavier nuclei in stellar hydrogen-burning explosive environments like novae and thermonuclear X-ray bursts. It also plays a key role in linking the Neon-Sodium (Ne-Na) cycle with the Magnesium-Aluminium (Mg-Al) cycle in Oxygen-Neon classical novae. However, for temperatures above 1 GK, the reaction loses its relevance as production of ^{23}Mg is avoided along the proton-rich path. In case of unstable nucleus like ^{24}Al production, the experimental database is scarce. Therefore, theoretical research is equally crucial for improving the estimates of the overall reaction rate. Using Shell Model calculations via NuShellX we determined the energy levels and the γ-ray partial widths of ^{24}Al for significant resonances and identified an essential resonance state at E_x = 2345 keV that contributes considerably to the total reaction rate. This work involves determining the level energies, γ-ray and proton partial widths, resonance strengths, and the reaction rate for the above mentioned first resonance state.

Keywords: Mg-Al cycle, Shell model, Reaction rate

1. INTRODUCTION

In high hydrogen-density and temperature environments, explosive hydrogen burning drives nucleosynthesis through the rapid proton (rp) capture process. This process initiates with the escape from the beta-limited CNO cycle and contributes significantly to the formation of neutron-deficient elements. The ^{23}Mg(p,γ)^{24}Al reaction acts as a crucial link between the Ne-Na cycle and the Mg-Al cycle, influencing the synthesis of heavier nuclei. At typical novae temperatures in the range of 0.2 to 0.4 GK, resonance capture, which results in excited states of ^{24}Al above the proton-separation threshold, dominates the reaction rate with the resonance at E_r = 478 keV, lowest lying, contributing the most to it (Herndl et al., 1998).

Wallace and Woosley (1981) first estimated the reaction rate by considering a single resonance, which was later improved by Wiescher et al. (1986) by including direct-capture processes along with other resonances. Kubono et al. (1995) and Greenfield et al. (1991) used the ^{24}Mg(^{3}He,t)^{24}Al reaction to identify resonance states and refine excitation energies. Later Herndl et al. (1998) used shell-model calculations combined with experimental data to establish that the first resonance state at E_x = 2349(20) keV is dominant in the temperature range of 0.2–1.0 GK. Direct measurement by Erikson et al. (2010) provided energy and resonance strength values for the first state which conflicted with the γ-ray study done by Lotay et al. (2008). Recent work by Vyfers et al. (2024) presented a detailed study on excitation energies and resonance strengths. Excitation energy measured for the lowest resonance, E_x = 2342(4) keV corresponding to E_r = 478(4) keV as well as the calculated resonance strength was found to be in good agreement with Ref. (Erikson et al., 2010) & (Lotay et al., 2008) respectively.

This paper aims to conduct a theoretical analysis of the ^{23}Mg(p,γ)^{24}Al reaction. Using Shell Model as the theoretical framework and NuShellX (Brown & Rae, 2014) as the computational tool, this study aims to identify the key resonance state contributing to the reaction rate at astrophysical temperatures and to determine the γ-ray partial width (Γ_γ), proton partial width (Γ_p), resonance strength (ωγ) along with the resonant reaction rate of the E_r = 478 keV (Vyfers et al., 2024) state.

*Corresponding author: sathisharma1994@gmail.com

DOI: 10.1201/9781003684718-68

2. METHODOLOGY

The energy levels of the excited states of ^{24}Al were computed using NushellX in the sd-model space with the W, USDB and the USDE interactions. These values closely match the previous estimates from Ref. (Herndl et al., 1998), (Vyfers et al., 2024) and (Visser et al., 2007). In this work, the resonance energies (E_r) reported by Vyfers et al. (2024) were used. The spectroscopic factor was also computed using the NuShellx codes, which includes the appropriate Clebsh-Gordan coefficients to give us C^2S.

The γ-ray partial width (Γ_γ), is associated with the lifetime (τ) of a state through the energy uncertainty principle, $\Gamma_\gamma = \frac{\hbar}{\tau}$, where τ can be determined using the transition strengths, computed using NuShellX. The proton partial width (Γ_p) is given by, $\Gamma_p = C^2 S \cdot \Gamma_{sp}$, where, C represents the isospin Clebsch–Gordan coefficient, S stands for spectroscopic factor and Γ_{sp} denotes the partial width for a single-particle.

For a proton capture reaction the resonance strength is expressed as, $\omega\gamma = \frac{2J+1}{2(2j_t+1)} \cdot \frac{\Gamma_p \Gamma_\gamma}{\Gamma_{tot}}$, where j_t and J represent the spin of the target and resonance nucleus, respectively, and $\Gamma_{tot}=\Gamma_p+\Gamma_\gamma$, also known as total width.

For the narrow and isolated resonances of ^{23}Mg(p,γ)^{24}Al reaction, the reaction rate can be determined using (Clayton, 1983),

$$N_A\sigma v = 1.54\times 10^{11}(\mu T_9)^{-3/2}\sum_i(\omega\gamma)_i \times \exp\left(-\frac{11.605E_i}{T_9}\right)\text{cm}^3\,\text{mol}^{-1}\,\text{s}^{-1} \quad (1)$$

where, μ represents reduced mass in amu, T_9 denotes the temperature in GK, $\omega\gamma_i$ is the resonance strength, and E_i is the resonance energy, both in MeV.

3. RESULTS AND DISCUSSIONS

The resonance strength, calculated using the resonance energy (E_r^a) adopted from (Vyfers et al., 2024), is in good agreement with (Herndl et al., 1998) and (Kubono, Kajino, & Kato, 1995). The above parameters were applied to Equation 1 and iterated over T_9 values ranging from 0.2 to 1.9 GK. The dominance shown by the state in this temperature range matches the findings of D.W. Visser et. al. (2007). Refer to Fig. 68.1 which illustrates the dominance of the first excited state in determining the reaction rate within the stated interval.

4. CONCLUSION AND SCOPE OF FUTURE WORK

This study examines the ^{23}Mg(p,γ)^{24}Al reaction, essential for understanding nucleosynthesis in novae. The

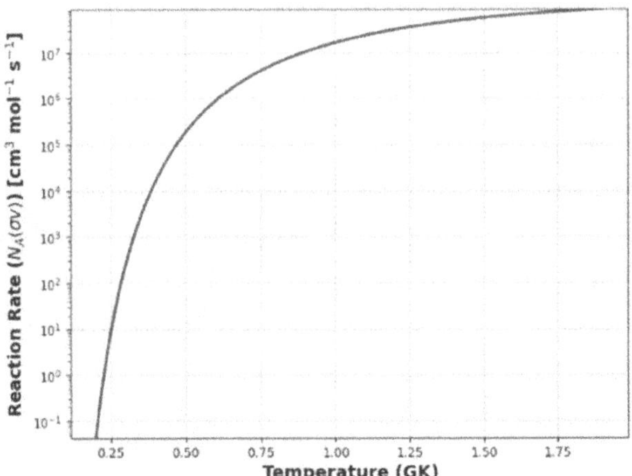

Fig. 68.1 Variation of reaction rate ($N_A\langle\sigma v\rangle$) with temperature ($T_9$)

Table 68.1 Resonant parameters of the first excited state of ^{24}Al

J^π	E_x (kev)	E_r^a (kev)	Γ_γ (meV)	Γ_p (meV)	$\omega\gamma$ (meV)
3^+_3	2348	478(4)	36.6	163.8	26.03

calculated γ-ray partial width and resonance strength for the 3^3_+ state agree well with values reported by (Herndl et al., 1998) and (Kubono, Kajino, & Kato, 1995), and the proton partial width is consistent with (Visser et al., 2007). Future work will include calculating resonant and direct capture reaction rates for other resonances and evaluating their role in synthesis of elements.

REFERENCES

1. Herndl, H., Fantini, M., Iliadis, C., Endt, P. M., & Oberhummer, H. (1998). Thermonuclear reaction rate of ^{23}Mg(p,γ)^{24}Al. *Physical Review C, 58*(3), 1798–1803.
2. Wallace, R. K., & Woosley, S. E. (1981). Explosive hydrogen burning. *The Astrophysical Journal Supplement Series, 45,* 389–420.
3. Wiescher, M., Görres, J., Thielemann, F.-K., & Ritter, H. (1986). Explosive hydrogen burning in novae. *Astronomy & Astrophysics, 160,* 56–72.
4. Kubono, S., Kajino, T., & Kato, S. (1995). Structure of ^{24}Al for the stellar reaction rate of ^{23}Mg(p,γ)^{24}Al. *Nuclear Physics A, 588,* 521–536.
5. Greenfield, M. B., Brandenburg, S., Drentje, A. G., Grasdizk, P., Riezebos, H., Van Der Werf, S. Y., Van Der Woude, A., Harakeh, M. N., Sterrenburg, W. A., & Brown, B. A. (1991). The ^{24}Mg(^{3}He,t)^{24}Al reaction and L=1 giant resonance systematics using an effective ^{3}He–n interaction at 27 MeV/nucleon. *Nuclear Physics A, 524*(2), 228–250.
6. Erikson, L., Ruiz, C., Ames, F., Bricault, P., Buchmann, L., Chen, A. A., Chen, J., Dare, H., Davids, B., Davis, C., Deibel, C. M., Dombsky, M., Foubister, S., Galinski, N., Greife, U., Hager, U., Hussein, A., Hutcheon, D. A., Lassen, J., Martin, L., Ottewell, D. F., Ouellet, C. V., Ruprecht, G., Setoodehnia, K., Shotter, A. C., Teigelhöfer, A., Vockenhuber, C., Wrede, C., ... Wallner, A. (2010).

First direct measurement of the ^{23}Mg(p,γ)^{24}Al reaction. *Physical Review C, 81*(4), 045808.

7. Lotay, G., Woods, P. J., Seweryniak, D., Carpenter, M. P., Hoteling, N., Janssens, R. V. F., Jenkins, D. G., Lauritsen, T., Lister, C. J., Robinson, A., Zhu, S., … & Zhang, Z. (2008). First in-beam γ-ray spectroscopy study of ^{24}Al and its implications for the astrophysical ^{23}Mg(p,γ)^{24}Al reaction rate in ONe novae. *Physical Review C, 77*(4), 042802.
8. Vyfers, E. C., Pesudo, V., Triambak, S., Kamil, M., Adsley, P., Brown, B. A., Jivan, H., Marin-Lambarri, D. J., Neveling, R., Ondze, J. C. Nzobadila, Papka, P., Pellegri, L., Rebeiro, B. M., Singh, B., Smit, F. D., Steyn, G. F., … Smit, F. D. (2024). Proton-unbound states in ^{24}Al relevant for the ^{23}Mg(p,γ) reaction in novae. *Physical Review C, 110*(3), 035803.
9. Visser, D. W., Wrede, C., Caggiano, J. A., Clark, J. A., Deibel, C., Lewis, R., Parikh, A., & Parker, P. D. (2007). Measurement of ^{23}Mg(p,γ)^{24}Al resonance energies. *Physical Review C, 76*(6), 065803.
10. Clayton, D. D. (1983). *Principles of stellar evolution and nucleosynthesis*. The University of Chicago Press.
11. Brown, B. A., & Rae, W. D. M. (2014). The shell-model code NuShellX@MSU. *Nuclear Data Sheets, 120,* 115–118. https://doi.org/10.1016/j.nds.2014.07.022

Note: All the figures and tables in this chapter were made by the authors.

Determination of $E^{r}_{c.m.}$ = 107 keV Resonance Strength ($\omega\gamma$) of ^{25}Mg (p, γ) ^{26}Al Astrophysical Reaction using Shell Model Calculation

Sathi Sharma, Ashok Kumar Mondal*
Department of Physics, Manipal University Jaipur, Jaipur, Rajasthan, India

Abstract: The detection of the 1.809 MeV gamma-ray in 1984 confirmed ongoing ^{26}Al nucleosynthesis in the Galaxy, driven primarily by the ^{25}Mg (p, γ) ^{26}Al reaction. This reaction is dominated by narrow resonances in the ^{26}Al excited states, particularly at $E^{r}_{c.m.}$= 50 - 310 keV, corresponding to astrophysical temperatures of T=0.02 - 2 GK. Despite extensive experimental efforts since 1970, Earth's surface observations are limited due to small cross sections and cosmic-ray background. Using NuShellX code, the present study reproduced a resonance state at $E^{r}_{c.m.}$ = 107 keV. Using the calculated proton width (Γ_p) and spectroscopic factor (C^2S), we have determined the resonance strength of $\omega\gamma = 2.75 \text{ X } 10^{-11}$ eV, consistent with prior studies. Future theoretical work will explore additional resonances and direct capture reactions to inform experimental efforts.

Keywords: Shell model, NuShellX code, Proton width, Gamma width, Resonance strength, Reaction rate

1. INTRODUCTION

Using the satellite HEAO-C, Mahoney et al. (Mahoney, 1984) identified the first 1.809 MeV γ-ray in 1984, indicating that ^{26}Al nucleosynthesis is currently occurring in the Galaxy. The distribution of the Galaxy's 1.809 MeV γ-ray emission was then determined by the COMPTEL telescope on the CGRO satellite (Plüschke, 2001). Its existence in the early solar system is demonstrated by the discovery of excess ^{26}Mg (a decay product of ^{26}Al) in meteorites, which connects the solar system's development to its galactic origin. Understanding the creation and destruction paths of ^{26}Al requires an understanding of reactions involving ^{26}Al, such as ^{25}Mg (p, γ) ^{26}Al (Diehl, 2006). This reaction is significant, particularly at energies between 50–310 keV, which correspond to the astrophysical temperature range of 0.02–2 GK. While numerous studies have been conducted since 1970, direct earth-based observations are limited due to the reaction's low cross section and interference from cosmic ray background.

Additional low-lying resonances at $E^{r}_{c.m.}$ = 37, 57, 92, 107, and 130 keV were likely present based on the known ^{26}Al level structure. As it turns out, these low energy resonances were discovered in indirect investigations using transfer reaction analyses. (Iliadis, 1996, and references therein). Here, we have determined $E^{r}_{c.m.}$ = 107 keV resonance strength ($\omega\gamma$) of ^{25}Mg (p, γ) ^{26}Al astrophysical reaction using Shell Model as the theoretical framework and NuShellX (Brown, 2014) as the computational tool. Finally, the narrow resonance reaction rate in 0.02 – 2 GK temperature range has been computed.

2. THEORETICAL CALCULATION

In this study, we have performed large basis shell model (LBSM) computations using the algorithm NuShellX. The sd model space has been used to calculate energy spectra and wave functions. The ^{16}O core and the valence orbitals $1d_{3/2}$, $1d_{5/2}$, and $2s_{1/2}$ make up the sd model space. The computations have been performed using the USDB

*Corresponding author: mondal.ashok4193@gmail.com

DOI: 10.1201/9781003684718-69

interaction. Without any subshell constraints, the energy spectra have been computed utilizing the complete valence space. Reduced transition probabilities were used to calculate the level lifetime of the mentioned resonant state ^{26}Al nucleus was determined using theoretical transition probabilities, experimental γ-ray energies, and branching ratios as required. Finally, the γ-ray partial width (Γ_γ) was determined using the level lifetime value. The proton partial width (Γ_p) is given by,

$$\Gamma_p = C^2 S \cdot \Gamma_{sp}$$

where, C^2S stands for spectroscopic factor and Γ_{sp} denotes the partial width for a single particle. The spectroscopic factor (C^2S) calculation has been carried out using the NuShellX code.

For a (p, γ) reaction, where a proton is captured by the target nucleus the resonance strength ($\omega\gamma$) is expressed as,

$$\omega\gamma = \frac{2J+1}{2(2j_t+1)} \cdot \frac{\Gamma_p \Gamma_\gamma}{\Gamma_{tot}},$$

where, j_t and J represent the spin of the target and resonance nucleus, respectively, and $\Gamma_{tot} = \Gamma_p + \Gamma_\gamma$ also known as total width.

Finally, the narrow resonance reaction rate has been determined using the formula,

$$N_A\langle\sigma v\rangle = 1.54\times10^{11}(\mu T_9)^{-3/2}\sum_i(\omega\gamma)_i$$
$$\times \exp\left(-\frac{11.605E_i}{T_9}\right) \mathrm{cm^3\,mol^{-1}\,s^{-1}}$$

where, μ represents reduced mass in amu, T_9 denotes the temperature in GK, $\omega\gamma_i$ is the resonance strength, and E_i is the resonance energy, both in MeV.

3. RESULTS AND DISCUSSIONS

Theoretically, the resonance state has been reproduced at E_x = 6361 keV ($E_{expt.}$ = 6414 keV) within only 53 keV energy difference. The calculated spectroscopic factor for this resonance state is, C^2S = 0.0282 which has only l = 2 proton wave capture contribution. The calculated proton width (Γ_p) for this resonant state is 3.3 X 10^{-10} eV. The final resonance strength ($\omega\gamma$) of the $E^r_{c.m.}$ = 107 keV 0^+ resonance state is computed to be 2.75 X 10^{-11} eV. The above-mentioned resonant parameters have been listed in the Table 69.1 below,

Table 69.1 Resonant parameters of the resonance state of ^{26}Al

E_x (keV)	J^π	E_r (keV)	C^2S	Γ_p (eV)	$\omega\gamma$ (eV)
6361	0^+	107	0.0282	3.30 X 10^{-10}	2.75 X 10^{-11}

The above resonance parameters have been utilized to calculate the narrow resonant capture rate in T = 0.02 – 2 GK temperature range. The plot for the reaction rate vs. temperature has been shown in Fig. 69.1.

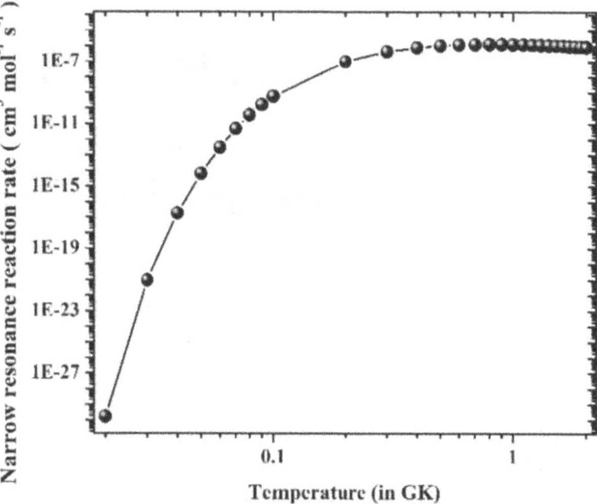

Fig. 69.1 Narrow resonance reaction rate (in cm^{3}mol^{-1} s^{-1}) as a function of temperature (in GK)

4. SUMMARY AND FUTURE PLAN

Here, the estimated resonance strength ($\omega\gamma$) for the $E^r_{c.m.}$ = 107 keV resonance state is equal to 2.75 X 10^{-11} eV, which is in excellent agreement quoted by J. O. Fern´andez Niello et al. (Niello, 2003). This resonance contributes approximately after T = 0.2 GK and then its value almost saturates at higher energies.

Future work will involve calculating reaction rates for other resonant and direct capture processes across various resonances and assessing their contribution to elemental synthesis. The overall outcome serves as a guide for the future ^{25}Mg (p, γ) ^{26}Al direct measurements.

REFERENCES

1. Mahoney, W. A., Ling, J. C., Wheaton, W. A. (1984). HEAO 3 discovery of ^{26}Al in the interstellar medium. Astrophys J, 286: 578–585.
2. Plüschke, S., Diehl, R., Schönfelder, V. (2001). The Comptel 1.809 MeV survey. ESA SP, 459: 55–58.
3. Diehl, R., Halloin, H., Kretschmer, K. (2006). Radioactive ^{26}Al from massive stars in the Galaxy. Nature, 439: 45–47.
4. Iliadis, C., Schange, T., Rolfs, C. (1990). Low-energy resonances in ^{25}Mg (*p*, γ) ^{26}Al, ^{26}Mg (*p*, γ) ^{27}Al and ^{27}Al (*p*, γ) ^{28}Si. Nucl. Phys. A, 1990 512: 509–530.
5. Brown, B. A. and Rae, W. D. M. (2014), The Shell-Model Code NuShellX@MSU. Nucl. Data Sheets 120**:** 115–118.
6. Fern´andez Niello, J. O., Arazi, A., Faestermann, T., Knie, K., Korschinek, G., Richter, E., Rugel, G., and Wallner, A. (2003). An Alternative Method for the Measurement of Stellar Nuclear-Reaction Rates. Brazilian Journal of Physics, 33: 2.

Note: All the figures and tables in this chapter were made by the authors.

Recent Trends in Applied Physics and Material Science – Sudhir Bhardwaj et al. (eds)

The Rate Calculation of the Astrophysical Reaction ^{12}C (α, γ)^{16}O at Stellar Region

Ashok Kumar Mondal, Sathi Sharma*

Department of Physics, Manipal University Jaipur, Jaipur, India

Abstract: In the He-burning stars, the 3α to^{12}C and ^{12}C (α, γ)^{16}O events are the two most significant thermonuclear processes that reduce He abundance. The ^{12}C/^{16}O abundance ratio, subsequence nucleosynthesis in massive stars, and the fate of the stars are all determined by the rate of the 3α to^{12}C and ^{12}C (α, γ)^{16}O reactions near the end of the He-burning phase. The astrophysical S-factors and the Asymptotic Normalisation Coefficient (ANC) values of the dominating states 6.92 (2^+) and 7.12(1^-) MeV of ^{16}O at the stellar area (Temperature T= 0.06-1GK) are used to compute the rate of the astrophysical ^{12}C (α, γ)^{16}O reaction. The rate of this reaction at 0.06 GK is 1.3 x 10^{-25} cm^{3}s^{-1}mol^{-1}. The calculated rates in the temperature range 0.06 GK to 1 GK are good agreement with the literature values.

Keywords: Astrophysical reaction, Asymptotic normalization coefficient, Reaction rate, Stellar region

1. INTRODUCTION

One of the main factors influencing the evolution and related nucleosynthesis of low mass and large stars is the ^{12}C (α, γ)^{16}O reaction (DeBoer, 2017). This includes hydrogen burning in the main sequence, where ^{12}C and ^{16}O produced by the ^{12}C (α, γ)^{16}O reaction in earlier star generations can be crucial. When four hydrogen nuclei fuse to form helium on the main sequence, approximately 25 MeV of energy is released. The human body is composed primarily of hydrogen, with 65% of its mass being oxygen and 18% being carbon. In the Sun and similar main sequence stars, the most prevalent elements heavier than helium are carbon (0.39%) and oxygen (0.85%). Determining the ratio ^{12}C/^{16}O, which is created during helium burning, is a crucial issue in nuclear astrophysics, which is not surprising. The pace of the ^{12}C (α, γ)^{16}O reaction has a significant impact on the ratio of ^{12}C/^{16}O, subsequent nucleosynthesis, and the ultimate destiny of the stars (black hole or neutron star). 10^{-17} barn is the cross-section of the ^{12}C (α, γ)^{16}O reaction at 300 keV (Gamow energy), and it is nearly impossible to measure directly. The ^{12}C (α, γ)^{16}O reaction cross-section at 300 keV can be found using R-matrix extrapolation. The Asymptotic Normalisation Co-efficient (ANC) and resonance characteristics of ^{16}O states (Brune, 1999) determine the extrapolation's outcome. In the ^{12}C (α, γ)^{16}O reaction at 300 keV, two subthreshold states of ^{16}O (6.92 and 7.12 MeV) mainly contributed to the cross-section. The capture cross-section is largely determined by the ANC of these two sub-threshold states. The primary method for determining the ANC has been the ^{12}C (^{6}Li, d) and ^{12}C (^{7}Li, t) α-transfer reactions (Brune, 1999) (Sub-Coulomb measurement) (Mondal, 2017) (above Coulomb measurement). Hence, ^{12}C(^{20}Ne,^{16}O)^{16}O is a suitable substitute for studying the crucial ^{12}C (α, γ) effect. The astrophysical S-factors S_{E1} and S_{E2} values at 300 keV were calculated using R-matrix calculation utilising the ANC of ^{16}O states using a novel reaction ^{12}C (^{20}Ne, ^{16}O)^{16}O for the first time (Mondal, 2021).

The ^{12}C (α, γ)^{16}O reaction rate at the stellar area is determined in this work using the astrophysical S-factors S_{E1} and S_{E2} values. This reaction is occurring at a rate of 1.3 × 10^{-25} cm^{3}s^{-1}mol^{-1} at 0.06 GK.

*Corresponding author: sathisharma1994@gmail.com

DOI: 10.1201/9781003684718-70

2. THEORETICAL CALCULATION

First time, using the transfer reaction $^{12}C(^{20}Ne,\ ^{16}O)^{16}O$ the astrophysical S-factors of the direct capture reaction $^{12}C(\alpha,\gamma)^{16}O$ at stellar energies are calculated using AZURE2 code (Azuma, 2010). At stellar region, using this S-factor the rate of the reaction is calculated by the formula

$$N_A < \sigma v >_{nr} = 7.8327 \times 10^9 \times \frac{Z_p Z_t}{\left(\mu T_9^2\right)^{1/3}} \times S_{eff} \times \exp\left[-4.2487 \times \left(\frac{Z_p^2 Z_t^2 \mu}{T_9}\right)^{\frac{1}{3}}\right] (cm^3 s^{-1} mol^{-1})$$

Where, $S_{eff} \approx S(0)\left[1 + 0.09807 \frac{T_9}{\left(Z_p^2 Z_t^2 \mu\right)^{\frac{1}{3}}}\right]$,

S(0) is the astrophysical S-factor at zero energy, T_9 is the temperature in GK, μ is the reduced mass in atomic units, atomic number of projectile is Z_p and the atomic number of target is Z_t. The value of S(0) which is calculated by R-matrix calculation is (132 ± 51) KeV b (Mondal, 2021).

3. RESULT AND DISCUSSION

At stellar region, using S-factor of the direct capture reaction $^{12}C(\alpha,\gamma)^{16}O$, the rate of the reaction is calculated by the above formula which is shown in the Table 70.1.

Table 70.1 The rate of the direct capture reaction $^{12}C(\alpha,\gamma)^{16}O$ at stellar region

Temperature (GK)	Adopted Rate ($cm^3\ s^{-1}\ mol^{-1}$)	Upper Rate ($cm^3\ s^{-1}\ mol^{-1}$)	Lower Rate ($cm^3\ s^{-1}\ mol^{-1}$)
0.06	1.30671E-25	1.66309E-25	9.50338E-26
0.07	7.18243E-24	9.14128E-24	5.22359E-24
0.08	1.95524E-22	2.48848E-22	1.42199E-22
0.09	3.18833E-21	4.05787E-21	2.31878E-21
0.1	3.52652E-20	4.4883E-20	2.56474E-20
0.11	2.88136E-19	3.66719E-19	2.09554E-19
0.12	1.84815E-18	2.35219E-18	1.34411E-18
0.13	9.73224E-18	1.23865E-17	7.07799E-18
0.14	4.35221E-17	5.53917E-17	3.16524E-17
0.15	1.69655E-16	2.15925E-16	1.23386E-16
0.16	5.88443E-16	7.48927E-16	4.27958E-16
0.18	5.30873E-15	6.75656E-15	3.86089E-15
0.2	3.52547E-14	4.48696E-14	2.56398E-14
0.25	1.55867E-12	1.98376E-12	1.13358E-12
0.3	2.79201E-11	3.55347E-11	2.03055E-11
0.35	2.78769E-10	3.54797E-10	2.02741E-10
0.4	1.85606E-09	2.36226E-09	1.34986E-09
0.45	9.19759E-09	1.1706E-08	6.68916E-09
0.5	3.64472E-08	4.63874E-08	2.65071E-08
0.6	3.51355E-07	4.4718E-07	2.55531E-07
0.7	2.13817E-06	2.72131E-06	1.55503E-06
0.8	9.46003E-06	1.204E-05	6.88002E-06
0.9	3.31803E-05	4.22294E-05	2.41311E-05
1	9.76216E-05	0.000124246	7.09976E-05

The variation of the calculated rate of the direct capture reaction $^{12}C(\alpha,\gamma)^{16}O$ at stellar temperatures using S-factors which are calculated by R-matrix code AZURE2 is shown in Fig. 70.1. The rate of this reaction at 0.06 GK is 1.3 × 10^{-25} cm^3s^{-1}mol-1. The calculated rates in the temperature range 0.06 GK to 1 GK are calculated, and it is good agreement with the literature values.

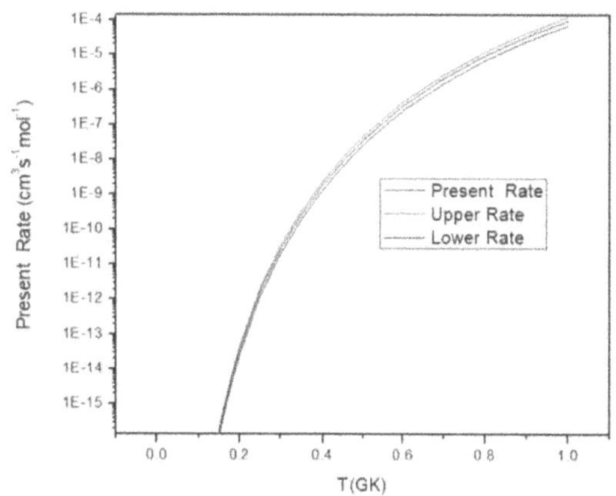

Fig. 70.1 The variation of the calculated present rate of the direct capture reaction $^{12}C(\alpha,\gamma)^{16}O$ at stellar temperatures. Here the black, red and blue solid line are indicating the present adopted rate, upper rate and lower rate respectively

REFERENCES

1. R. J. DeBoer, J. Görres, M. Wiescher et. al. (2017). The 12C(α, γ)16O reaction and its implications for stellar helium burning. REVIEWS OF MODERN PHYSICS. 89: 035007–035073.
2. C. R. Brune, W. H. Geist, R.W. Kavanagh, and K. D. Veal1. (1999). Sub-Coulomb a Transfers on ^{12}C and the $^{12}C(\alpha,\gamma)^{16}O$ *S* Factor. PHYSICAL REVIEW LETTERS. 83: 4025–4028.
3. Ashok Mondal, S. Adhikari, C. Basu. (2017). Effect of compound nuclear reaction mechanism in $^{12}C(^{6}Li,\ d)$ reaction at sub-Coulomb energy. Physics Letters B. 772: 216–218.
4. Ashok Kumar Mondal, C. Basu, S. Adhikari, C. Bhattacharya et. al. (2021). $^{12}C(^{20}Ne,\ ^{16}O)^{16}O$ α-transfer reaction and astrophysical S-factors at 300 keV. International Journal of Modern Physics E. 30: 2150039–2150052.
5. R. E. Azuma, E. Uberseder, E. C. Simpson, C. R. Brune, H. Costantini, R. J. de Boer, J. G¨orres, M. Heil et.al. (2010). AZURE: An R-matrix code for nuclear astrophysics. PHYSICAL REVIEW C. 81:045805–045822.

Note: All the figures and tables in this chapter were made by the authors.

Recent Trends in Applied Physics and Material Science – Sudhir Bhardwaj et al. (eds)
 ISBN 978-1-041-16452-4

Investigating Open Heavy-Flavour Production via Elliptic Flow in Collision Systems with ALICE at the LHC

Abhilasha Saini*
Atharva College of Engg., Mumbai, Maharashtra, India

Sudhir Bhardwaj
Bikaner Technical University, Bikaner, Rajasthan, India

P.N. Nemade, Pinky Steffi A.
Atharva College of Engg., Mumbai, Maharashtra, India

Abstract: The quark–gluon plasmonic state (QGP) established during ultra-relativistic heavy-ion collisions can be extensively studied using heavy quarks as vital probes. The contention centers on the most recent upshots from the ALICE Collaboration dealing with open heavy-flavour creation in proton-proton and lead–lead impacts. Quantifications of the emergence of open heavy flavors in lead–lead impacts are indispensable to comprehend the heavy-quark transport mechanisms and energy loss in the plasmonic state. In-depth apprehension regarding heavy-flavour particles' function within the medium's collective behaviour, the path-length reliance of their in-medium energy drop, and viable rearrangement consequences amid hadronization can be procured by reviewing their elliptical effluence (v_2) coefficient. Open heavy-flavour creation imparts a benchmark for assaying perturbative quantum chromodynamics projections and gauging blazing hot-medium aftermaths in heavy-ion impacts in diminutive hadronic setup such as proton-proton.

Keywords: Heavy-ion collisions, pQCD, LHC etc.,

1. INTRODUCTION

Perturbative Quantum Chromodynamics (pQCD) computations can be deployed to charm and beauty quarks, which are predominantly processed in the preliminary juncture of ultra relativistic level impacts via hard-scattering activities. Quantification of the triggering of such open heavy-flavour hadronic particles during proton-proton encounters, accordingly proffer a consequential means of confirmation of these pQCD expectations. Investigating the hadronizing activity of heavy quarks, with a prominence on heavy-flavour baryons whose hadronizing implementations in pp impacts are still imperfectly acknowledged, is made viable by exploring the producing rates of distinct heavy-flavour hadron classes. A phase changeover connecting the nucleonic substance and the plasmonic phase, a color-deconfined state, is contemplated in the ambience of ultra-relativistic impacts. Since the duration of open heavy-flavour synthesis is shorter than that of plasma emergence, heavy quarks confront the established medium evolvement. These quarks engage in both elastic and inelastic interplays with the medium's integrants as they progress through the Plasmonic phase. On that account, heavy quarks are functional probing to follow up assorted plasmonic attributes, such as the energy dissipation approach, the function of quark recombining process, and the medium's beginning preconditions.

2. ELLIPTIC FLOW AS A PROBE

Deeper insights into the interplays of heavy quarks with plasmonic state are clinched via quantifying the azimuthal anisotropies, which are interpreted through the Fourier coefficients $v_n = \langle \cos f_0\, n(\phi - \Psi_n)\rangle$. The azimuthal inclination of particle's momenta are illustrated by ϕ and

*Corresponding author: abhilashasaini@atharvacoe.ac.in

DOI: 10.1201/9781003684718-71

averaged over all digged out particles. Also, the inclination of the symmetry plane with respect to the nth harmonic is shown by Ψ_n. The domineer harmonic coefficient in semi-central heavy-ion encounters is the elliptic-effuence, depicted as v_2, as evidence in the Fig. 71.1 with dissimilar particle strains.

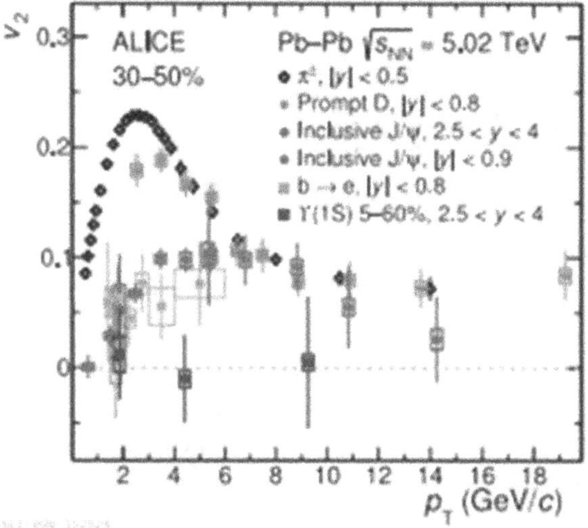

Fig. 71.1 The v_2 for charged pions, non-strange D mesons, inclusive J/ψ, electrons from beauty-hadron decays, and Υ(1S) in 30–50% central Pb–Pb collisions at $\sqrt{s_{NN}}$ = 5.02 TeV

For faint valued transverse momentum (pT), this flow component stipulated the extent of thermalization of heavy quarks. At elevated thermal conditions, it dispenses supplemental details of path-length vulnerability of parton energy drop in the medium. A mass ordering is visible for pT < 3 GeV/c, but v_2 is identical for prompt D mesons and charged particles in the span 3 < pT < 6G eV/c, demonstrating corresponding path-length dependence.

3. CONCLUSION

This handout synopsizes the neoteric arbitrations from the ALICE Collaboration for open heavy-flavour engendering in lead-lead and proton-proton encounters. These estimations, which make use of Run 2 data sets, proffer high-rise statistical fidelity and more contrastive precision than precursory ALICE inquiries. Charm quark hadronizing process suggested a recast even in proton-proton impacts, as evidenced by the substantial nuclear modifications of charm quarks in the quark-gluon plasma (QGP). Strong magnetic fields occur during collisions, and these quarks participate in the system's collective dynamics. Moreover, in order to properly explain charmed-hadron formation in Pb–Pb collisions, theoretical models now need hadronization via recombination.

REFERENCES

1. S. Acharya et al. [ALICE] (2018). JHEP 04 .108 [arXiv:1712.09581 [nucl-ex]].
2. R. Aaij et al. [LHCb](2019). Phys. Rev. D 100 no.3.031102 [arXiv:1902.06794 [hep-ex]].
3. S. Chatrchyan et al.(2012). [CMS], Phys. Lett. B 714 .136-157 [arXiv:1205.0594 [hep-ex]].
4. A. Andronic et al.(2016) Eur. Phys. J. C 76,no.3.107 [arXiv:1506.03981 [nucl-ex]].
5. F. Prino and R. Rapp. (2016),J. Phys. G 43, no.9. 093002 [arXiv:1603.00529 [nucl-ex]].
6. S. Acharya et al. [ALICE]."J/ψ elliptic and triangular flow in Pb-Pb collisions at √ sNN = 5.02 TeV,"arXiv:2005.11131 [nucl-ex].
7. S. Acharya et al. [ALICE].(2020).Elliptic flow of electrons from beauty-hadron decays in Pb–Pb collisions at √sNN = 5.02 TeV arXiv:2005.11130 [nucl-ex].
8. S. Acharya et al. [ALICE].(2020) JHEP 10, 141 [arXiv:2005.14518 [nucl-ex]].

Note: The authors made the figure in this chapter.

Recent Trends in Applied Physics and Material Science – Sudhir Bhardwaj et al. (eds)
 ISBN 978-1-041-16452-4

72

Tribological Aspects and Behaviour of Magnesium Metal Matrix Nanocomposites: An Overview

Gautam Nain[1]
Graphic Era (Deemed to be University), Dehradun, Uttrakhand, India

Harindra Singh[2]
CRA College, Sonipat, Haryana, India

Himani Dahiya[3]
Graphic Era (Deemed to be University), Dehradun, Uttrakhand, India

Abstract: One of the most fundamental objectives of developing metal matrix composites is to enhance the wear resistance of soft and light weight metals. The addition of nano scale reinforcements in metals has been found to improve the wear losses as compared to their pure counterparts. Being light weight and having widespread application in sports, automobile and aviation sectors, the tribological studies of the magnesium based nano components is under scanner recently. Wear performance of machines with sliding/moving parts is of paramount importance for enhancing life of parts and improving efficiency. Thus, Mg based nano composites (MgMNCs) developed by incorporating nano materials viz. Silicon Carbide (SiC) Graphene Nano Platelets (GNP), Carbon Nano Tubes (CNT), Titanium Di-boride (TiB_{2}) and Tungsten Carbide (WC) etc. are being used to address wear issues. The characteristic of such composites depends on the quantity and nature of the additives besides the fabrication process involved in developing these. In present review implications of different parameters viz. load, sliding speed, reinforcement dose & nature and ambient temperature besides fabrication techniques on tribological behaviour and coefficient of friction (COF) of MgMNCs is being examined.

Keywords: Wear resistance, Coefficient of friction, stir casting

1. INTRODUCTION

By making a judicious selection of both matrix phase and reinforcement phase, composite materials can be tailor made to meet desired properties. To meet stringent emission norms and to enhance energy efficiency, advanced materials needs to be searched. Thus strong and lightweight materials are desired and therefore magnesium is preferred as it is about 33% lighter than Al and is abundant. Hence it is continuously substituting conventional materials parts in automotive and aviation sectors where wear behaviour is an important aspect. Fabrication of innovative materials is capable of resolving diverse wear issues. Usually zinc oxide, alumina, Yttrium oxide, silicon carbide, titanium carbide, boron carbide and graphite are added in magnesium and its alloys in order to modify wear behaviour. Particulate dimensions also play a pivotal role in this direction. In case of micron size additives, as large dose is required causing wettability issues due to cluster formation and entrapment of gases. Nano-metric additives resolve these problems and betterment in hardness, tensile strength and improvement in wear resistance is noticed with less than two percent of these in Mg or its alloys. Fabrication of MgMNCs, however is not free from challenges as nano-scale particulates tends to agglomerate and uniform distribution of these in the matrix poses hurdles. Besides this, magnesium being highly reactive metal, so a perfect oxygen free environment is needed otherwise oxidation may happen during composite formation resulting in the formation of magnesium oxide. Thus choosing an appropriate development method and environment is essential for successful fabrication of MgMNCs.

Tribological properties are influenced by parameters such as load, sliding velocity & distance, surfaces in contact,

[1] gautamnain60@gmail.com, [2] dahiya_himani@yahoo.com, [3] hsdahiya14@gmail.com

DOI: 10.1201/9781003684718-72

temperature, extent of lubrication, additives (dose, type and particle size) etc. as shown in Fig. 72.1. Effect of some of these parameters is discussed in this review. Generally, it is noticed that loading pristine metal and alloys with reinforcements improves wear aspects and other properties as depicted in Fig. 72.2. Thus, innumerous researchers have carried out studies on tribological characteristics of Mg-MNCs taking into consideration these parameters. Present review highlights effect of amount, type and dimensions of the reinforcements on wear and other aspects of the MgMNCs. Effect of fabrication technique besides temperature is also under scanner here.

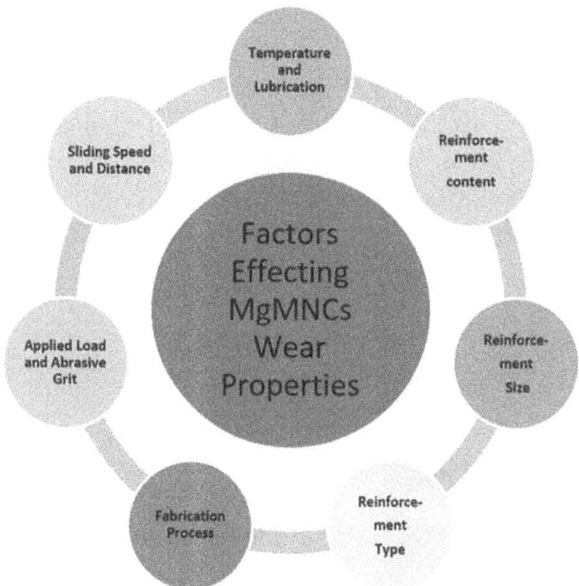

Fig. 72.1 Factors effecting MgMNCs wear properties

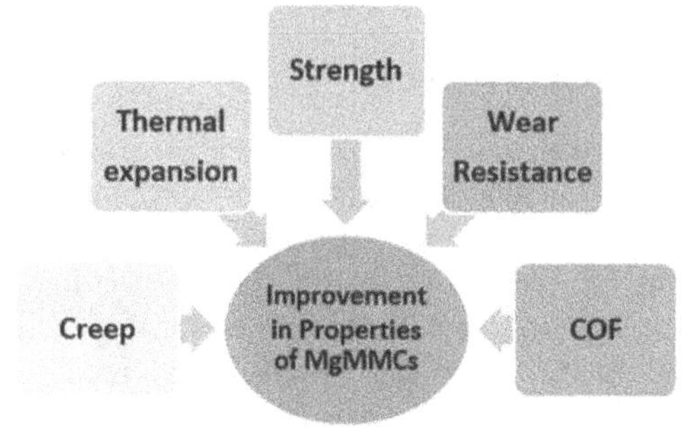

Fig. 72.2 Improvement in properties of MgMMCs on adding reinforcements

2. EFFECT OF AMOUNT OF REINFORCEMENT

Optimization of the amount or dose of the additives is of paramount importance for fabrication of MgMNCs for specific applications. Kumar et.al. (2019) added Graphene nano platelets ranging from 0.1- 0.5 wt.% in pristine magnesium metal as well as magnesium aluminum alloy. The added platelets have width of 5 microns and thickness less than 10nm. Power metallurgy (PM) technique was used for sample fabrication and a pin on disk (POD) tribometer was employed for studying tribological behaviour. It was found that 0.3 wt.% sample gave best results in terms of wear resistance and COF improvement.

By incorporating SiC having 74 μm particle size (0, 10, 15 and 20 vol.%) and graphite of 40 μm (0, 5 vol.%) Mg–SiC–Gr and Mg–Gr composites were fabricated by Al-maamari et.al. (2020) by mechanical alloying technique and wear properties were studied by POD tribometry under 10, 20 and 30N loads with sliding velocities ranging from half to one ms^{-1}. Micro-hardness and UTS were enhanced by 65%–138% and 20%–38% respectively. A subtle change in wear properties is observed on addition of silicon carbide in samples. Mg/5Gr/10SiC sample exhibited lowest value of COF (0.24). By mixing 50nm Al_2O_3 particles in Mg (0.22 -1.11) vol. %, Lim et.al. (2005) prepared composites using disintegrated melt deposition technique (DMD). POD tribometry carried out at 10N load and various sliding velocities ranging from one to ten ms^{-1}, revealed that wear resistance keeps on increasing with increasing alumina content.

3. EFFECT OF TYPE OF REINFORCEMENT

The type of end product desired decides the type of reinforcement to be added. Lim et al. (2003) fabricated hybrid composites by incorporating 14 μm, SiC (8 vol.%) in AZ91 alloy of Mg by powder metallurgy. Tribological investigation was carried out under ten and thirty newton loads at different sliding velocities. Wear resistance kept improving with increasing SiC content for all but 5 ms^{-1} sliding velocity at 10N load. At 30 N load, however, wear increases and composites can be useful only up to 2 ms^{-1} sliding velocity. At 10 N, oxidation seems to be the mechanism governing wear which shifts to abrasion and delamination at higher load.

Hybrid composites comprising of AZ91 alloy of Mg and 40-60 μm $ZrSiO_4$ particles (4 and 8 wt.%) were developed by Mohavanvel et.al. (2021) by stir casting technique. Wear behaviour examined at $2ms^{-1}$ sliding velocity, 20N load and 1000m sliding distance revealed that in the composite both UTS and micro-hardness gets improved which for AZ91-8ZrSiO4 were 183Mpa and 71 HV respectively. Wear resistance was also found better for this particular composition.

AZ31 alloy and graphite nano platelets (GNP) hybrid composite were fabricated by Arab and Marashi (2018) by friction stir processing method and wear studies were carried out on POD tribometer under three load conditions i.e. 5,10 and 20N. Sliding distance was kept fixed at 1Km whereas sliding velocity was varied from 02-1.25 ms^{-1}. A decrease of about 60% in COF was observed on adding GNP and maximum value is exhibited at lowest value of load and sliding velocity. Addition of GNP increased micro-hardness by 41% whereas UTS and Young's Modulus were improved to 278MPa and 217MPa respectively.

4. EFFECT OF REINFORCEMENT SIZE

Size of additives also influences wear and mechanical properties of nano composites as reported by various researchers. Hybrid composites comprising of AZ91D-0.8%Ce and alumina having particle length 300–700 mm & diameter 8–12 μm (8 vol.%) and Graphite of varying particle size (55, 83, 125 and 240 μm) with 15 vol.% were fabricated by Zhang et.al. (2008) employing squeeze-infiltration process. Increasing dimensions of the additives results in an increase in wear resistance and highest value is detected for 240 μm sample. For low load values, oxidation and abrasion wear dominates. Delamination, however is the major wear mechanism for lower Gr particle size samples at high load values.

5. EFFECT OF COMPOSITE FABRICATION TECHNIQUES

Fabrication techniques also plays a dominant role in wear behaviour of hybrid composites. Hybrid composites comprising of Mg and 2wt.% of CaB_6 prepared by Seenuvasaterumal et. al. (2017) by two methods viz. stir and squeeze casting, were used to examine the influence of fabrication technique on hardness, UTS etc. and wear behaviour under loads of 10,20 and 30N at three different sliding velocities viz. 0.4, 0.6, and 0.8 ms^{-1} for fixed 2Km distance. Micro-hardness of stir casted composite was found to be less than that of squeeze casted one. `Enhanced wear resistance was observed in composites as compared to pure Mg samples except under 20N load with sliding velocities beyond 0.8 ms^{-1}. With increasing load and sliding velocity, COF keeps on increasing for both type of samples. A thorough investigation of wear properties of cold as well as hot compressed nanocomposites consisting of Mg/Al_2O_3 was carried out by Rahmani et.al. (2020). Samples prepared with 0,1.5,3 and 5 vol.% of Al_2O_3 having particle size less than 100nm were fabricated and it was found that wear resistance increases with increasing amount of Al_2O_3. Cold compressed samples exhibited poor wear resistance as compared to their hot compressed counterparts. Also, COF was higher with lower doses of Al_2O_3 and it was observed that coefficient of friction of hot compressed samples was lowest ($2.1X10^{-2}$) while that of cold compressed was highest ($2.4X10^{-2}$). As far as micro-hardness is concerned in case of cold compressed Mg-$5Al_2O_3$ composites it increased to 69 HV from 40 HV of Pure Mg whereas in case of hot compressed Mg-$5Al_2O_3$ it increases to 81 HV from 44 HV of Pure Mg.

6. TEMPERATURE EFFECT

Wear resistance vis-a-vis temperature was examined by Banerjee et.al. (2019) for nanocomposites fabricated by adding WC nano particles in AZ31 alloy. It is reported that wear rate kept on increasing with rising temperature whereas, that of nanocomposites remained unaltered upto a threshold limit beyond which wear rate jumped. This threshold temperature for Mg/1.5WC & Mg/2WC was 200^0C while in case of Mg/0.5WC & Mg/1WC, it was reported to be 100^0C. Thus composites exhibited better wear properties at higher operating temperatures as reported by Banerjee et.al. (2019. A minor fall in the COF was observed in all samples with rising temperature.

7. APPLICATIONS

The MgMMCs are used in various areas of manufacturing industry find myriad of applications in different as depicted in Fig. 72.3.

Fig. 72.3 Various applications of MgMMCs

8. CONCLUSIONS

It is found that nanocomposites exhibited better wear behaviour as compared to intrinsic Mg/alloys at all loads, velocities, distance and temperatures. Abrasion is reported at low load & velocity while adhesion starts at high load & velocity due to thermal softening because of increased frictional heating. Nanoparticles helps to reduce wear significantly as a gliding layer is created on the surface which saves these from extreme wear. On addition of nano reinforcements, increased hardness and strength has been reported. Various studies conducted and the results obtained advocates the versatility of Mg nanocomposites in automobile and aviation industry. However widespread commercial applications can only throw light on reliability and cost effectiveness.

REFERENCES

1. Al-maamari, A.E., Iqbal, A. A., and Nuruzzaman, D. M., (2020) Mechanical and tribological

characterization of self- lubricating Mg-SiC-Gr hybrid metal matrix composite (MMC) fabricated via mechanical alloying. J Sci Adv Mater Devices. 5: 535–544.

2. Arab, M. and Marashi. S. P. (2018) Graphene nanoplatelet (GNP)- Incorporated AZ31 magnesium nanocomposite: micro- structural, mechanical and tribological properties. Tribol. Lett. 66: 156.5.
3. Banerjee, S., Poria, S., Sutradhar, G., et al. (2019) Tribological behavior of Mg-WC nano-composites at elevated temperature. Mater. Res. Express.6: 0865c6.
4. Banerjee, S., Sahoo, P. and Davim, J. P. (2021) Tribiological characterization of magnisium matrix nanocomposites: A review. Advances in Mechanical Engg. 13(4):1-39.
5. Kumar, P., Mallick A, Kujur, M.S., et al. (2019). Effects of graphene nano-platelets on the tribological, mechanical, and thermal properties of Mg-3Al alloy nanocomposites. Int. J Mater. Res; 110:534–542.
6. Lim, C. Y., Leo, D. K., Ang, J. J., et al. (2005). Wear of magnesium composites reinforced with nano-sized alumina particulates. Wear. 259: 620–625.
7. Lim, C. Y., Lim, S. C. and Gupta, M. (2003) Wear behaviour of SiCp- reinforced magnesium matrix composites. Wear 255:629–637.
8. Mohanavel, V., Vijay, K., Vigneswaran, A. et al. (2021) Mechanical and tribological behaviour of AZ91/$ZrSiO_4$ composites. Mater Today Proc. 37:1529–1534.
9. Rahmani, K., Sadooghi, A. and Hashemi, S. J. (2020). The effect of cold and hot pressing on mechanical properties and tribological behavior of Mg-Al2O3 nanocomposites. Mater. Res. Express. 7: 085012.
10. Seenuvasaperumal, P., Elayaperumal, A. and Jayavel, R. (2017). Influence of calcium hexaboride reinforced magnesium composite for the mechanical and tribological behaviour. Tribol. Int.111:18–25.
11. Zhang, M. J., Liu, Y. B., Yang, X. H., et al. (2008). Effect of graphite particle size on wear property of graphite and Al_2O_3 rein- forced AZ91D-0.8% Ce composites. Trans. Nonferrous Met. Soc. China. 18:273–277.

Note: All the figures in this chapter were made by the authors.

Recent Trends in Applied Physics and Material Science – Sudhir Bhardwaj et al. (eds)
© 2026 Taylor & Francis Group, London, ISBN 978-1-041-16452-4

Exploring the Potential of Sol-Gel Synthesized Hydroxyapatite for Biomedical Engineering

B. Srimathy
PG & Research Department of Physics,
Seethalakshmi Ramaswami College (Affiliated to Bharathidasan University),
Tiruchirappalli, Tamil Nadu, India

P. Ramesh Babu*, T. Veeramanikandasamy, S. Devendiran
Department of Electronics and Communication System,
Sri Krishna Arts & Science College,
Coimbatore, Tamil Nadu, India

Abstract: Hydroxyapatite is highly biocompatible, bioactive and osteoconductive, making it ideal for biomedical applications like bone grafts, dental implants and tissue engineering. Current work focuses on the synthesis of hydroxyapatite nanoparticles using two different precursors, calcium nitrate tetrahydrate and calcium chloride. The synthesized nanoparticles exhibited hexagonal phase and characteristic absorption peaks of functional groups in the FTIR spectra. Formation of rod like structures and round shaped particles were evident from scanning electron micrographs.

Keywords: Hydroxyapatite, sol-gel synthesis, XRD analysis, FTIR spectra

1. INTRODUCTION

Hydroxyapatite (HAp), renowned for its exceptional biocompatibility and bioactivity plays a vital role in biomedical applications such as bone regeneration and dental repairs (Hendi, 2017). Its versatility extends to environmental, industrial and cosmetic uses driven by its ability to interact with biological and chemical systems effectively. Nanoparticles of HAp play a more significant role and the synthesis process of HAp plays a crucial role in determining its properties and performance (Babu et al., 2021). Several methods have been employed to prepare HAp nanoparticles. However, in the current study, sol-gel method is utilized due to its distinct advantages such as precise control over the composition and the ability to achieve nanostructured particles with excellent bioactivity. Also, the precursor materials used in the synthesis significantly influence the characteristics and properties of HAp. In the present study, calcium nitrate tetrahydrate and calcium chloride were chosen as precursors as they are highly soluble in water ensuring a uniform and controlled release of calcium ions during synthesis and they dissolve completely, minimizing unwanted by-products. Thus the present study focussed on the preparation of HAp nanoparticles using above materials as precursors and their potential characteristics were investigated.

2. EXPERIMENT

Nanoparticle synthesis: HAp nanopowders were prepared by sol-gel method using two different precursors. First, HAp was synthesized using calcium nitrate tetrahydrate and diammonium hydrogen phosphate. Initially, dropwise addition of diammonium hydrogen orthophosphate to calcium nitrate tetrahydrate solution was done for over 3 h. The pH of the mixture was adjusted to 9 by adding ammonia solution. After continuous stirring for 3 h at 70°C, a white precipitate was formed at the bottom of the vessel. The precipitate was washed 7-8 times with double distilled deionised water and dried in a hot oven at 80°C for 24 h. The obtained HAp nanopowder is named as HAp-CNT. A similar method was followed to obtain HAp using calcium chloride and diammonium hydrogen phosphate and HAp nanoparticles thus obtained is named as HAp-CC.

*Corresponding author: prameshbabu8687@gmail.com

DOI: 10.1201/9781003684718-73

Characterization: The structure of the synthesized HAp was elucidated using Bruker D8-Advance X-ray diffractometer. XRD patterns were recorded from 20° to 80° with a step size of 0.2°. Perkin Elmer FTIR spectrophotometer was used to identify functional groups in the range of 400-4000 cm^{-1}. Morphology was examined using HITACHI S2600N scanning electron microscope (SEM) operated at 10 kV in vacuum.

3. RESULTS AND DISCUSSION

3.1 XRD Analysis

Figures 73.1(a) and (b) show the variation in X-ray intensity as a function of 2θ for HAp-CNT and HAp-CC respectively. The XRD pattern was analyzed using the JCPDS standard [09-0432]. The presence of characteristic HAp peaks confirmed its hexagonal structure for both samples. The particle size was determined using Debye's Scherer formula:

$$D = \frac{0.9\lambda}{\beta \cos \theta}$$

The particle size of HAp-CNT was ~54 nm. However, the particle size of HAp-CC was less (~19 nm) compared to HAp-CNT which is due to the higher solubility and ionic mobility of Ca^{2+} ions in solution.

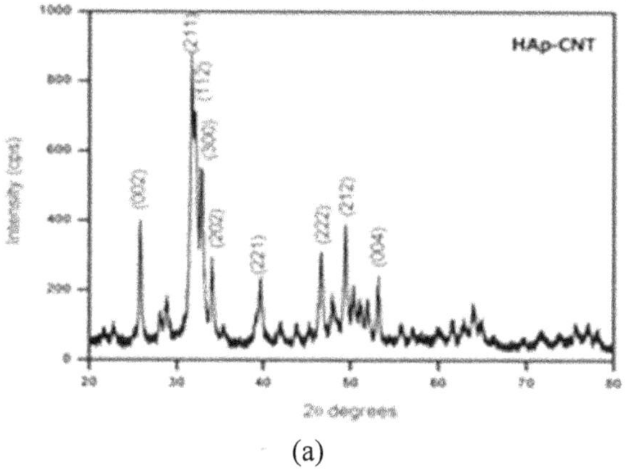

(a)

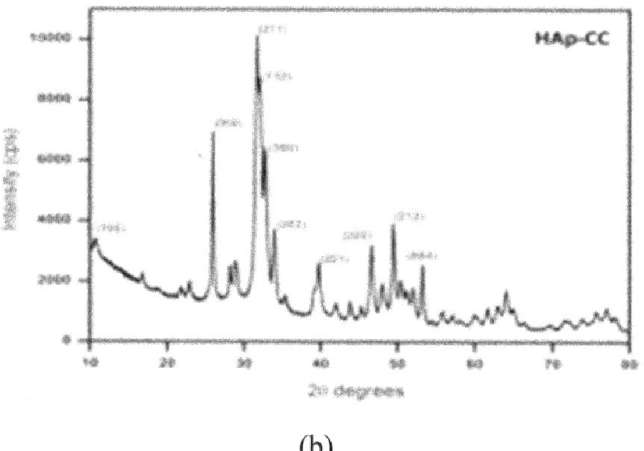

(b)

Fig. 73.1 (a) XRD pattern of HAp-CNT (b) XRD pattern of HAp-CC

3.2 FTIR Spectra

The FTIR spectrum of HAp-CNT and HAp-CC are depicted in Fig. 73.2(a) and (b). HAp-CNT showed a broad absorption peak around 3607 cm^{-1} due to O–H stretching vibration of hydroxyl groups. On the other hand, partial substitution of OH– group by chloride (Cl^{2-}) and carbonate (CO_3^{2-}) ions weakens this peak in HAp-CC. The peak at 1641 cm^{-1} in HAp-CNT corresponds to the H–O–H bending vibration of water molecules. These peaks arise due to the hygroscopic nature of HAp. The in-plane bending vibration (v_4) of the CO_3^{2-} ions gives rise to a peak at 1370 cm^{-1} (Shaltout et al, 2011). Characteristic peaks at 1022cm^{-1} in HAp-CNT and 1016 cm^{-1} in HAp-CC are associated with stretching modes of P–O bonds in PO_4^{3-}. Peak at 630 cm^{-1} is due to the libration (out-of-plane vibration) of OH^- groups and this confirms the presence of hydroxyl groups in both samples. Peaks at 599 cm^{-1} and 559 cm^{-1} arise due to bending modes of P–O bonds in phosphate groups.

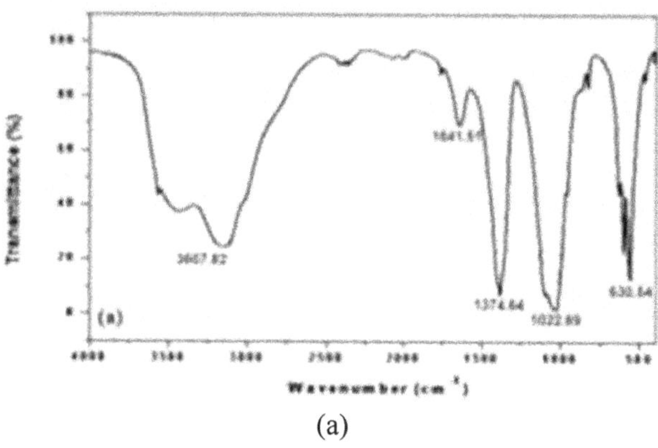

(a)

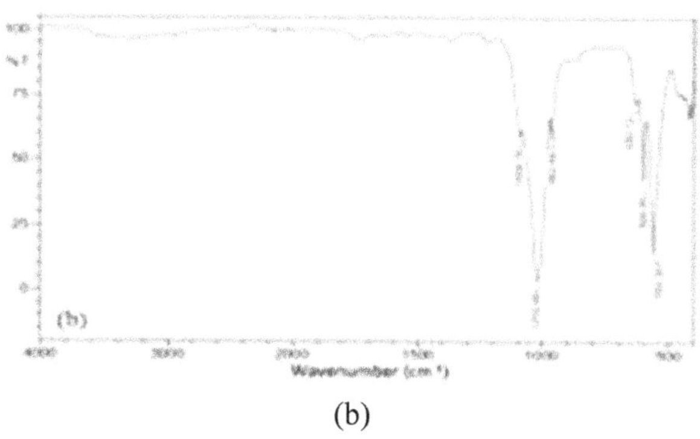

(b)

Fig. 73.2 (a) FTIR spectrum of HAp-CNT, (b) FTIR spectrum of HAp-CC

3.3 Scanning Electron Microscopy

Fig. 73.3(a) and (b) shows the morphology of HAp-CNT and HAp-CC nanoparticles respectively as observed through scanning electron microscopy. HAp-CNT exhibited elongated rod like structures whereas HAp-CC was found to exhibit irregular round shaped particles. Also

HAp-CNT contained few aggregated particles whereas HAp-CC exhibited higher particle agglomeration which is evident from Fig. 73.3.

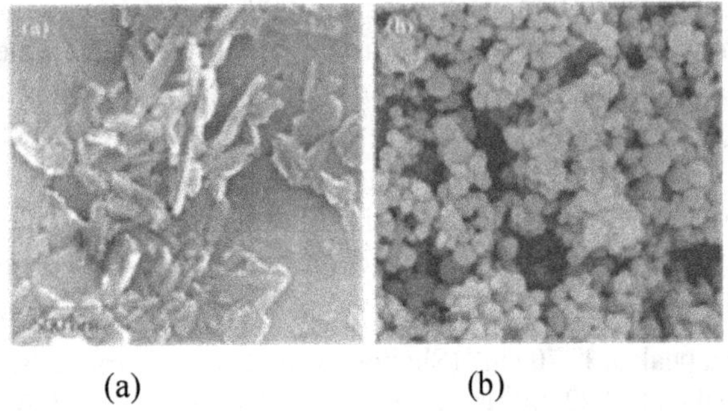

(a) (b)

Fig. 73.3 (a) SEM images of HAp-CNT, (b) HAp-CC

4. CONCLUSION

Hydroxyapatite nanoparticles were prepared by sol-gel technique. XRD spectra demonstrated the formation of hexagonal phase and the incorporation of carbonate and chloride ions are more obvious in HAp-CC than in HAp-CNT. Scanning electron micrographs displayed rod shaped nanoparticles and agglomerated round particles for carbon nitrate tetra hydrate and calcium chloride respectively. From the above observations, it is concluded that calcium nitrate hexahydrate serves as a better precursor to obtain pure hydroxyapatite for biomedical engineering.

REFERENCES

1. Hendi, A.A. (2017). HAp based nanocomposite ceramics. J. Alloys Compd. 712: 147-151.
2. Babu, P.R. Srimathy, B. Shakila, K. Lakshmipriya, M. and Arivuselvi, R. (2021). Investigation on photocatalytic properties of RE-doped titanium dioxide nanoparticles through sol - gel technique. *AIP Conf. Proc.* 2369, 02008.
3. Shaltout, A.A. Allam, M.A. and Moharram, M.A. (2011). FTIR spectroscopic, thermal and XRD characterization of hydroxyapatite from new natural sources. Spectrochim Acta A Mol Biomol Spectrosc. 83(1): 56-60.

Note: All the figures in this chapter were made by the authors.

Recent Trends in Applied Physics and Material Science – Sudhir Bhardwaj et al. (eds)
 ISBN 978-1-041-16452-4

Spectroscopic and Computational Insights into Hydrogen Bonding in (S)-(+)-4-Amino-3-hydroxybutyric Acid Dimer

Mithil Kotyagol, J. Tonannavar, Jayashree Tonannavar*

Vibrational Spectroscopy Group/Molecular Modelling Laboratory,
Department of Physics, Karnatak University, Dharwad, India

Abstract: (S)-(+)-4-amino-3-hydroxybutyric acid is a known neuromodulator in antiepileptic and hypotensive medicines in which its potential role as a H-bond former is of interest. It forms inter-molecular O–H···O/N–H···O bonds zwitterion, particularly showing bifurcated H-bonding between NH_3^+···COO⁻ and OH···COO⁻ groups as confirmed the IR absorption spectroscopy. We computed the bifurcated H-bonded dimer and trimer species in a water medium using MD simulations at production run for 100 ns. For the most stable zwitterion dimer structure, the computed radial distribution functions show well-defined structure corresponding to the inter-molecular O–H···O/N–H···O bonds and satisfy the criterion that the H···O distances for both types of H-bonds are smaller than the sum of the van der Waals radii of the H and O atoms, 2.72 Å. Using the implicit SCRF water solvation model at the B3LYP/6-311++G(d,p) level, the zwitterion dimer exhibits bifurcated inter- O–H···O/N–H···O and inter- N–H···O bonds in agreement with vibrational IR and Raman modes. The stabilization energies for these H-bonds lie in the range of 8 -15 kcal/mol suggesting the identified H-bonds being strong. The results are further supported by AIM and NCI analysis.

Keywords: (S)-(+)-4-amino-3-hydroxybutyric acid, bifurcated N—H···O/O—H···O bonds, Molecular dynamics, DFT

1. INTRODUCTION

(S)-(+)-4-amino-3-hydroxybutyric acid (for short 4A3HBA) is known for its use in treating epilepsy and as a hypotensive agent (Lohray et al., 1996). This molecule contains functional groups such as NH_3^+, COO^-, and OH, which are crucial for forming hydrogen bonds, specifically α and β bonds (in this paper we denoted, α = N—H···O and β = O—H···O) (Yang Bae Kim, 1985). These groups facilitate strong intermolecular interactions, significantly contributing to the molecule's stability and solubility in aqueous environments. The intricate network of hydrogen bonding enhances the compound's overall reactivity and its ability to engage in various biochemical processes. Therefore, the hydrogen bonding capabilities of these functional groups are integral to the molecule's behaviour and functionality. 4A3HBA potential role as a hydrogen bond, former is of particular interest. Therefore, we are interested in investigating the hydrogen bonding properties of 4A3HBA.

This research was prompted by the detection of an asymmetrical broad band structure around 3500 - 2500 (in this paper all wavenumbers are in cm^{-1}) in infrared absorption spectrum, alongside presence of distinct hydrogen-bonded dimeric and trimeric species of 4A3HBA observed in the molecular dynamics (MD) simulation. We determined that the dimer species with the highest stability, at δ (in this paper we denoted, δ = B3LYP/6-311++G (d,p) level), matches IR and Raman band structures observed in the reported data. The stabilization energies linked to α and β interactions, determined through NBO analysis, along with the topological parameters from AIM and NCI analysis, corroborate the previous properties.

*Corresponding author: jjtonannavar@kud.ac.in

DOI: 10.1201/9781003684718-74

2. MATERIALS AND METHODS

2.1 Experimental Techniques

4A3HBA ($C_4H_9NO_3$) with a purity of ≥97%, a molecular weight of 119.12 g/mol, and a melting point ranging from 207 to 212 °C. We employed the same technique and methodology as used by J. Bhovi et al. (J. Bhovi et al., 2024).

2.2 Computational Details

Monomer Structure Calculations

Gaussian 09W and *Gauss View 5* were employed for Density Functional Theory (DFT) calculations. A neutral molecular configuration in the gas phase was examined using B3LYP/3-21G level of theory. PES scans were performed for all torsional angles τ_1: C8-C1-C4-O6, τ_2: C1-C4-O6-H7, τ_3: C8-C9-N12-H14, τ_4: C1-C8-O16-H17) that influence the non-rigid characteristics of the 4A3HBA molecular structure. This analysis yielded a single stable conformer, designated as M1, as depicted in Fig. 74.1(a). Since amino acids exist in a zwitterionic (ZW) state analogous to a solid-phase context XRD report, we optimized 4A3HBA in a solvent medium using a method similar to that employed by S. Yalagi et al. (S. Yalagi et al., 2019). Geometry optimization was performed at δ level using the SCRF-SMD solvation model, resulting in ZW structure ***M*** as depicted in Fig. 74.1(b). This ***M*** structure was used as the foundation for subsequent MD simulation.

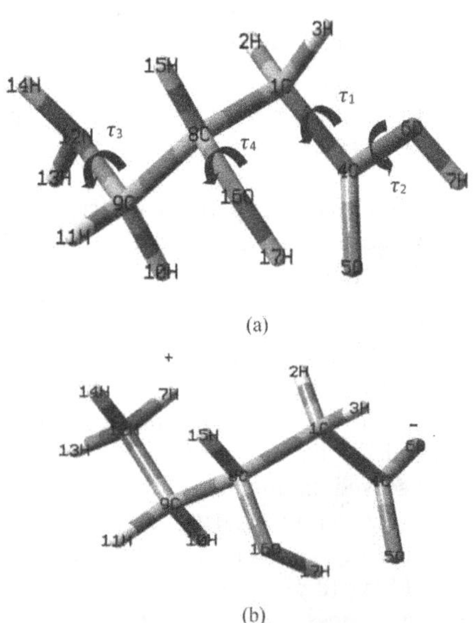

Fig. 74.1 The neutral monomer M1 (a) (τ-Torsional angle), and zwitterionic monomer ***M*** (b) structure of 4A3HBA

Molecular Dynamical Simulation

Gromacs 5.1.1 version was utilized to conduct classical MD simulation. We applied the same MD protocols employed by J. Bhovi et al. (J. Bhovi et al., 2024). The results were analyzed using the Visual Molecular Dynamics (VMD) suite.

3. RESULTS AND DISCUSSION

3.1 MD Analysis

During simulation, we observed α and β bonded dimer and trimer species. Figure 74.2 illustrates the snapshot from the simulations involving the 16 ZW monomer species and 1000 water molecules. Both α and β bonds were characterized using radial distribution functions (see Fig. 74.3) first peak r_{min} 2.14 and 2.25 for α and β respectively, are smaller than the combined van der Waals radii of the O and H atoms, which is 2.72 Å (G.R. Desiraju, et al., 2001). The well-defined RDF peaks represents well-defined structures of 4A3HBA forming α and β bonding.

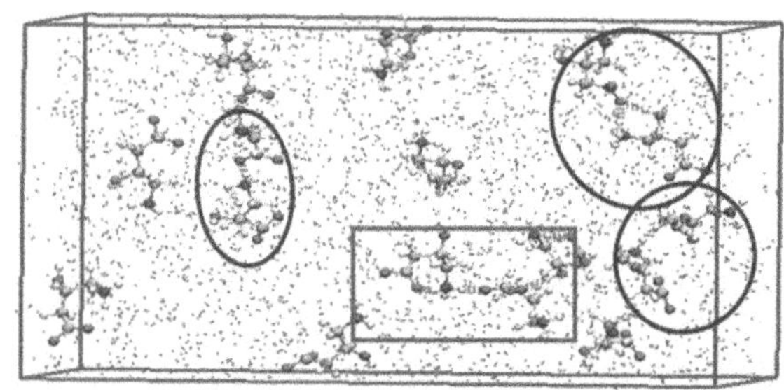

Fig. 74.2 Snapshot of the equilibrated simulation box at 150 ps. 4A3HBA molecules are represented by ball-and-stick style, water molecules are represented as blue dots, and hydrogen bonds are represented by red dotted lines (circle-dimer and rectangle-Trimer)

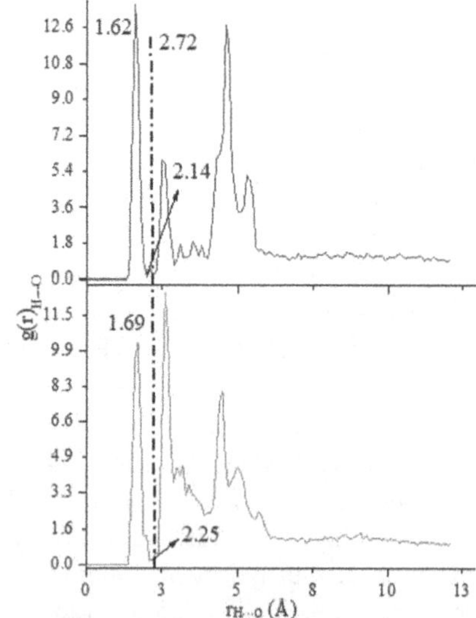

Fig. 74.3 RDF plot of α (up) and β (down) bonds of 4A3HBA. The dotted line indicates the sum of van der Waals radii of the O and H atoms i.e., 2.72 Å

3.2 Structure Analysis

To determine the most stable ZW dimer, we performed single-point energy calculations for all dimer species at δ. We then assessed the Boltzmann population based on their relative single-point energies. Dimer D1 exhibited a Boltzmann population of 94%, while the other dimers showed negligible populations. Consequently, we selected dimer D1 for further optimization and frequency analysis at δ, the resultant structure is shown in Fig. 74.4. The optimized ZW dimer D-I features bifurcated α and β bonds [NH_3^+···COO^- (***a***) and OH···COO^- (***b***)] as well as inter α [NH_3^+···OH (***c***)] bond.

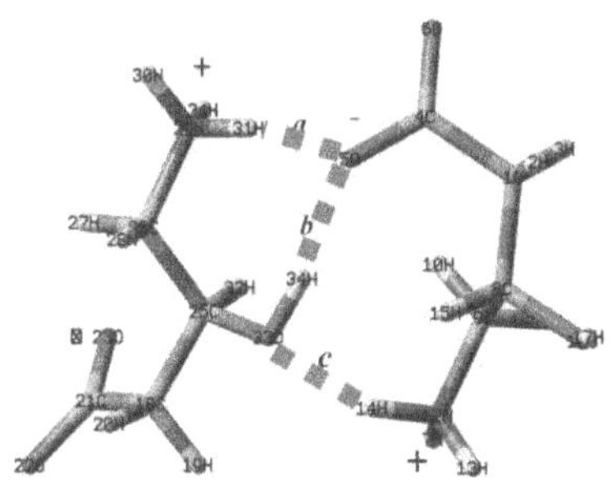

Fig. 74.4 The optimized ZW dimer D-I structure of 4A3HBA

The computed harmonic vibrational frequencies revealed no imaginary values, confirming that this dimer D-I represents the true lowest energy configuration. Table 74.1 provides a comprehensive overview of the optimized parameters for the hydrogen bonds. Global minimum energies for the 4A3HBA monomer and dimer D-I, as determined by the DFT method, are -275135.7523 a.u. and -550279.0934 a.u., respectively. This interaction is characterized by stable hydrogen-bonded contacts: α and β, which contribute to enhanced stability. The supermolecule approach was employed to assess the binding energy (BE) associated with the formation of the dimer through hydrogen bonding. The BE for the 4A3HBA dimer formation (BE = E_{dimer} - 2 $E_{monomer}$) is calculated to be -7.59 kcal/mol (J. Bhovi et al., 2024).

Table 74.1 Dimer D-I optimized geometrical H-bond parameters of 4A3HBA

Parameter	*a*	*b*	*c*
D-H···A	α	β	α
H···O (Å)	1.721	1.812	1.927
D···A(Å)	2.753	2.786	2.904
∠D-H···A	165.4	171.2	155.5

Note: D- Donar, A-Acceptor, a - NH3+···COO–, b -OH···COO–, c -NH3+···OH, α - N–H···O, β = O–H···O,

3.3 Vibrational Analysis

Fig. 74.5 presents the DFT-simulated spectra of the dimer D-I for 4A3HBA, displayed next to the experimental IR and Raman spectra for comparison. The 4A3HBA phase is characterized by intermolecular interactions, with hydrogen bonding being particularly significant. The experimental IR and Raman spectral characteristics suggest α and β interactions.

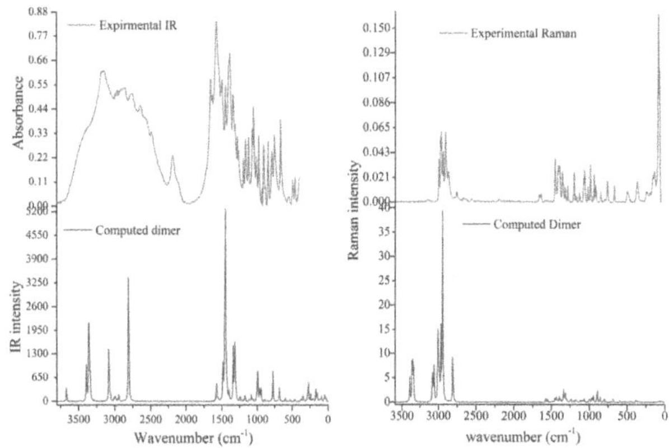

Fig. 74.5 Experimental IR (left) and Raman (right) spectrum (upper) of 4A3HBA is compared with the corresponding simulated dimer D-I spectra (lower)

These dimer species, associated with α and β bonding, are identified by three prominent spectral features: a significant, wide band down-shifted to around 3500–2500, corresponding to the stretching vibrations of N–H/O–H bonds; (ii) a series of faint bands located around 2185, identified as combination bands [1635 (NH_3^+) and 544 (COO^-)]; and (iii) a bonded carbonyl stretching vibration around the 1750–1650 range, which we observed at 1658. In dimer D-I, ***a*** bond showed 17.23% frequency shift between the free (3389) and bonded (2805) N–H stretching frequencies. The ***b*** bond exhibits an 8.49% frequency shift between the free (3664) and bonded (3353) O–H stretching bands. The ***c*** bond exhibits a 9.18% frequency shift between the free (3386) and bonded (3075) N–H stretching bands (N.B. Colthup et al.,1964). The bond lengths for N–H/O–H increase by 0.0286 Å, 0.01446 Å, and 0.0149 Å for bonds ***a***, ***b***, and ***c***, respectively. The Raman spectrum is more sensitive to low wavenumbers below 400, which are attributed to torsional vibration modes. We observed a strong band at 84 (P Ramanna et al., (2021).

3.4 NBO Analysis

Electronic structures arise from the interactions of orbitals between a *D* (*D*=donor) and *A* (*A*=acceptor) in intermolecular α and β bonding within D-I of 4A3HBA, complementing the previously discussed geometrical and vibrational characteristics. We employed the same interpretation method as Prabhu M. D et al. and observed similar orbital overlapping as depicted in Fig. 74.6 (Prabhu M. D et al., 2020). The calculated stabilization energy for dimer D-I of 4A3HBA bonds ***a***, ***b***, and ***c*** were found to be

15.74, 8.69, and 10.62 kcal/mol, respectively (F. Weinhold et al. 2005).

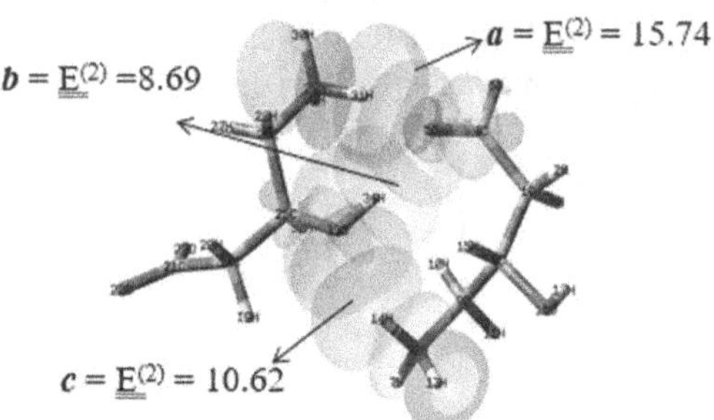

Fig. 74.6 Electron density maps showing orbital overlaps in α and β bonding within D-I of 4A3HBA

3.5 AIM and NCI Analysis

Following the methodology of Pallavi L. et al., we conducted AIM and NCI analyses and observed results consistent with their findings (Pallavi L et al., 2022). Table 74.2 provides the calculated E_{HB} for ***a***, ***b***, and ***c*** bonds in the dimer D-I of 4A3HBA. Here negative sign indicates the attraction. We obtained -12.44, -7.87 and -8.80 E_{HB} values for ***a***, ***b***, and ***c*** bonds respectively (R.F.W. Bader, 1991). Fig. 74.7a and Fig. 74.7b shows a two-dimensional RDG scatter plot that highlights various locations through spikes and their corresponding isosurfaces. The sign of the λ_2 values indicates different types of interactions, with negative λ_2 values and blue isosurfaces corresponds to α and β bonding interactions (Johnson, et al., 2010).

Table 74.2 Topological parameters calculated at BCP's for inter- α and β bond in ZW D-I of 4A3HBA

D-H···A	BCP	ρ(r) in a.u	V(r) in a.u	E_{HB}
α	*a*	0.044	−0.039	−12.44
β	*b*	0.028	−0.021	−7.87
α	*c*	0.033	−0.028	−8.80

Note: ***a*** - NH_3^+···COO−, ***b*** -OH···COO−, ***c*** -NH_3^+···OH, α - N—H···O, β = O—H···O, ρ(r) - electronic density, ρ(r), V(r) -potential energy density, EHB – H-bond interaction energy

4. CONCLUSIONS

In this study, we characterized the bifurcated intermolecular H-bonded dimer structure by α and β bonds analyzed using δ and MD simulations for 4A3HBA. The observed IR spectrum in the region of 3400-2500 provides the evidence for the proposed intermolecular α and β bonding. We obtained various intermolecular α and β bonded dimer and trimer species from MD simulations. The BE between the monomer and dimer is -7.59, indicating the presence of strong hydrogen bonding. The observed frequency shifts of 17.23%, 8.49%, and 9.18% for bonds ***a***, ***b***, and ***c***, respectively, between free and bonded stretching frequencies indicate strong hydrogen

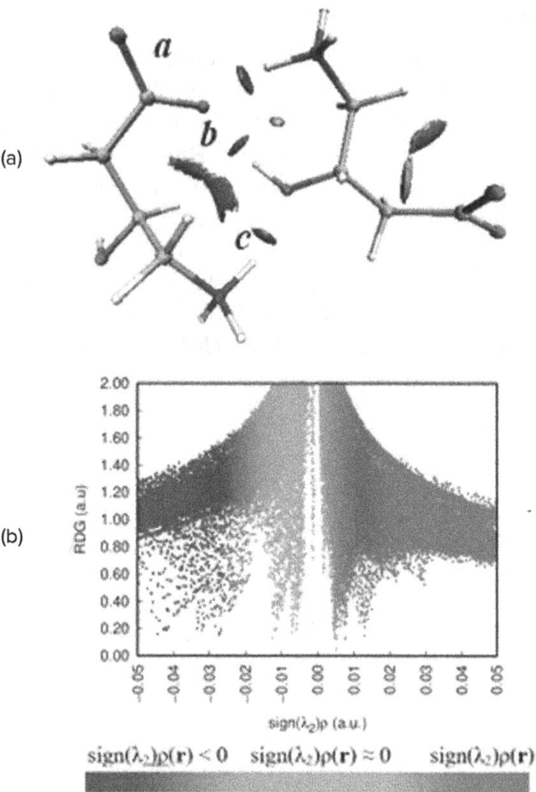

Fig. 74.7 Isosurfaces of dimer D-I (a) and corresponding 2D scatter plots of the sign(λ_2)ρ(r) versus RDG (b) of 4A3HBA. The colour gradient corresponds to the three interaction types shown on it

bonding, with corresponding increases in bond lengths of 0.0286 Å, 0.01446 Å, and 0.0149 Å for bonds ***a***, ***b***, and ***c***, respectively. The predicted vibrational modal properties of the dimer D-I model show a strong correlation with the experimental spectra. NBO analysis of dimer D-I reveals that the α and β interactions arise from the orbital overlapping between *D* and *A*. Stabilization energies of 15.74, 8.69, and 10.62 for bonds ***a***, ***b***, and ***c*** are further evidence of strong hydrogen bonding in the dimer D-I of 4A3HBA. Additionally, the AIM and NCI analysis, with negative λ_2 values and blue isosurfaces for bonds ***a***, ***b***, and ***c*** confirms α and β bonding interactions in the dimer D-I of 4A3HBA. From all analysis, dimer D-I is stabilized by the intermolecular α and β bonds.

5. ACKNOWLEDGEMENTS

We express our gratitude to the Director of USIC at Karnatak University for facilitating the FT-IR and FT-Raman measurements. Mithil Kotyagol sincerely thanks the Ministry of Tribal Affairs, Government of India for awarding National Fellowship for ST students. Additionally, we thank Prof. S. Umapathy for a visit by

Prof. Jayashree Tonannavar to his MD computational lab at the Department of IPC, Indian Institute of Science, Bengaluru.

REFERENCES

1. B. B. Lohray, A. Sekm Reddy and Vidya Bhushan (1996). An Efficient Synthesis of (3R)-4-Amino-3-Hydroxy Butyric Acid (GABOB) v/a Cyclic Sulfite Methodology, Tetrahtdron, 2411-2416.
2. Yang Bae Kim (1985). The Crystal and Molecular Structure of 3-Hydroxy-fl-amino butyric Acid, Arch. Pharm, 1-6.
3. J. Bhovi, J. Tonannavar, Jayashree J. Tonannavar, (2024). IR, Raman spectroscopic, DFT, AIM and NCI characterization of O—H···O/π···π bonds in dimer and trimer species as computed from MD simulations in water for Protocatechuic acid, J Mol Struct. 1299.
4. S. Yalagi, J. Tonannavar, J. Tonannavar (2019), DL-3-Aminoisobutyric acid: Vibrational, NBO and AIM analysis of N–H···O bonded-zwitterionic dimer model, Heliyon 5.
5. G.R. Desiraju, T. Steiner (2001), The Weak Hydrogen Hond, Structural Chemistry and Biology, Oxford University Press.
6. N.B. Colthup, L.H. Daly, S.E. Wiberley (1964). Introduction to Infrared and Raman Spectroscopy, Academic Press Inc, New York and London.
7. P Ramanna, Jayashree Tonannavar, J. Tonannavar (2021). Study of H-bonded cyclic dimer of organic linker 5-Bromoisophthalic acid by DFT and vibrational spectroscopy, J. Mol. Struc. 1241.
8. M.D. Prabhu, Jayashree Tonannavar, J. Tonannavar (2021). Multiple-H-bonded-zwitterionic tetramer structure for L-(+)-2-chlorophenylglycine, as investigated by UV, IR and Raman spectroscopy and electronic structure calculations, J Mol Struct. 1246.
9. F. Weinhold, C.R. Landis (2005). Valency and Bonding: A Natural Bond Orbital Donor-Acceptor Perspective, Cambridge University Press, Cambridge.
10. L. Pallavi, J. Tonannavar, Jayashree Tonannavar (2022). Molecular dynamics simulation, DFT calculations and vibrational spectroscopic study of N—H···O bound dimer models for DL-β-phenylalanine and 3-amino-3-(4-chlorophenyl) propionic acid, J Mol Liq. 352.
11. R.F.W. Bader (1991). A quantum theory of molecular structure and its applications, Chem. Rev. 91:893–928.
12. E.R. Johnson, S. Keinan, P. Mori-Sanchez, J. Contreras-García, A.J. Cohen, W. Yang (2010). Revealing noncovalent interactions, J. Am. Chem. Soc. 132: 6498–6506.

Note: All the figures and tables in this chapter were made by the authors.

Recent Trends in Applied Physics and Material Science – Sudhir Bhardwaj et al. (eds)
© 2026 Taylor & Francis Group, London, ISBN 978-1-041-16452-4

Betamethasone: Low Temperature FTIR Measurement

Archna Sharma*, Vivek K. Gupta
Department of Physics,
University of Jammu, Jammu,
J&K (UT), India

Himal Bhatt
HomiBhabha National Institute,
Mumbai, India
High Pressure & Synchrotron Radiation
Physics Division, BARC, Mumbai, India

ParamJeet Singh
Atomic & Molecular Physics Division,
BARC, Mumbai, India
HomiBhabha National Institute,
Mumbai, India

Abstract: Betamethasone (BM) is characterized by FTIR spectroscopy in KBr pellet at temperatures ranging between RT (300 K) and 103 K, for the first time. Very clear spectral changes are observed in the characteristic stretching regions 2700-3700 cm^{-1} and 1550-1750 cm^{-1}, upon temperature variation. Observed data is compared with the computed data obtained at DFT(B3LYP)/6-311++G(d,p) level of theory. Tentative assignments are provided for some selected observed peaks based on the comparison between experiments and the theory. A possible phase transition is suggested to occur at low temperatures below 280 K.

Keywords: FTIR, Low temperature, Betamethasone, DFT

1. INTRODUCTION

Betamethasone (BM), a widely used glucocorticoid known for its medicinal and commercial applications, serves as an effective anti-inflammatory drug (Cohen *et al*, 2015) (Oliveira *et al*, 2020). Numerous studies have explored its clinical applications (Arthur *et al*, 2004) (Ali *et al*, 2016) (Manassra *et al*, 2010), yet a comprehensive investigation into its detailed spectra-structure analysis remains lacking. This work presents the first-ever report on the FTIR spectra of BM (in KBr) recorded across a temperature range from 293 K to 103 K, addressing this research gap.

2. EXPERIMENTAL DETAILS

Betamethasone (BM) sample with purity ≥98% was purchased from Sigma-Aldrich and was used without further purification. The spectral resolution used is 2cm^{-1}. A total of 256 scans were coadded to obtain each final spectrum. The FTIR of sample dispersed in KBr pellet were measured in the spectral range 600 – 4000cm^{-1} in transmission mode using Bruker Vertex 80V FTIR spectrometer equipped with a liquid nitrogen-cooled HgCdTe detector and KBrbeamsplitter, installed at the experimental station of IR beamline at Indus-1 (Bhawalkar*et al*, 1998). For *in-situ* low temperature studies, sample was mounted inside a chamber made of oxygen free copper. The sample was cooled using liquid nitrogen and temperature control of ~3 K was achieved. The temperature was monitored using Pt100 based temperature sensor. Sufficient time was given for temperature stabilization before the measurements. Additionally, a high concentration of BM was used to prepare KBr pellet to observe temperature variations and low-frequency peaks. This pellet was subjected to a stream of liquid nitrogen, and FTIR spectra were collected at temperatures ranging from 293 K to 103 K.

3. RESULTS AND DISCUSSION

With noticeable changes observed in the FTIR spectra (Fig. 75.1) it was easy to demonstrate that the chemical specie, in spite of having a rigid molecular geometry, is sensitive to subtle changes in temperature. Fig. 75.1 highlights changes in the O-H, C-H and C=O regions of the FTIR spectra of BM. From the figure it is obvious that some shoulders at room temperature emerge as peaks as the temperature is lowered. Especially in the O-H region such

*Corresponding author: dr.archnasharma@gmail.com

DOI: 10.1201/9781003684718-75

variations are very evident (Fig. 75.1a). In the carbonyl region (Fig. 75.1b), an intensity shift between two peaks of a doublet occurs upon decrease in temperature. One peak loses intensity while the other gains it, with a clear change in the peak position. These changes are shown by the vertical dotted lines in both the ranges in Fig. 75.1. The observed FTIR spectrum of BM was compared with the computed wavenumbers and intensities. The comparison showed a reasonably good agreement between the experiment and the theory, validating the theoretical model used inthis study. Most of the observed peaks were well reproduced by the theoretical model used in the present case. In a study by Joe and Soumya (Joe *et al*, 2021) for Dexamethasone, a stereo isomer of BM, a weak absorption attributed to O-H stretching vibration was observed at 3476 cm^{-1}. This absorption appeared as a weak and narrow feature in their study. In the present study, the strong and broad feature around 3500 cm^{-1}, in the FTIR spectrum recorded at room temperature, is likely corresponds to the O-H stretching absorption frequency (Joe *et al*, 2021) (Liu *et al*, 2021). The broadness of the absorption peak was seen to have enhanced in the high concentration spectrum at the same temperature (Fig. 75.1a); it's shoulder appears more visible (Fig. 75.1a) as compared to the low concentration data. As the temperature is lowered, the shoulder gains intensity and appears as a peak as the temperature reaches around 100 K (Fig. 75.1a). The optimized structure of BM is seen to have an intramolecular hydrogen bond of the type O-H···O=C between hydroxyl group of α-carbon (D-ring) and oxygen of the carbonyl moiety of the side chain (D-ring). If we compare shape of the peaks due to O-H stretching absorption frequency in the low concentration and the high concentration FTIR spectra of powder BM recorded at RT, we can clearly see the difference. In the high concentration spectra (Fig. 75.1a), this peak is much broader which may have resulted from presence of higher number of H-bonded networks in the sample (Dai *et al*, 2023). Almost similar changes are observed in the other regions of the spectra as well (Fig. 75.1b). In general FTIR absorption peaks due to C=O and C=C stretching vibrations occur in the broad spectral range of 1750 – 1550 cm^{-1}(Smith B, 1998). Absorptions due to these stretching vibrations in some unbridged diketosteroids are reported to occur in the 1668 – 1660 cm^{-1} region in $CHCl_3$(Jones *et al*, 1965). Occurrence of these peaks at lower wavenumbers obviously indicates towards their involvement in some kind of inter/intra-molecular bonding. In the present study, a total of four peaks are observed consecutively at 1606 cm^{-1}, 1617 cm^{-1}, 1660 cm^{-1} and 1708 cm^{-1}. Out of these, the medium strong intensity peak at 1606 cm^{-1} was assigned to C=C stretching vibration, the 1617 cm^{-1} (ms) and the very strong peak at 1660 cm^{-1} was assigned to the two C=O stretching vibrations in BM. A significant difference in the observed and the calculated (scaled) peaks is seen. Thus, for C=C absorption, a difference of +22 cm^{-1} exists while for the two C=O absorptions that of +62 and +29 cm^{-1} exists. The medium intensity band at 1708 cm^{-1} did not have a theoretical counterpart in the present computed spectrum and thus could not be assigned. Note that present calculations have been performed for the gas phase that excludes any type of inter/intra-molecular interactions. Detailed assignment of peaks in this region was beyond the scope of this manuscript.

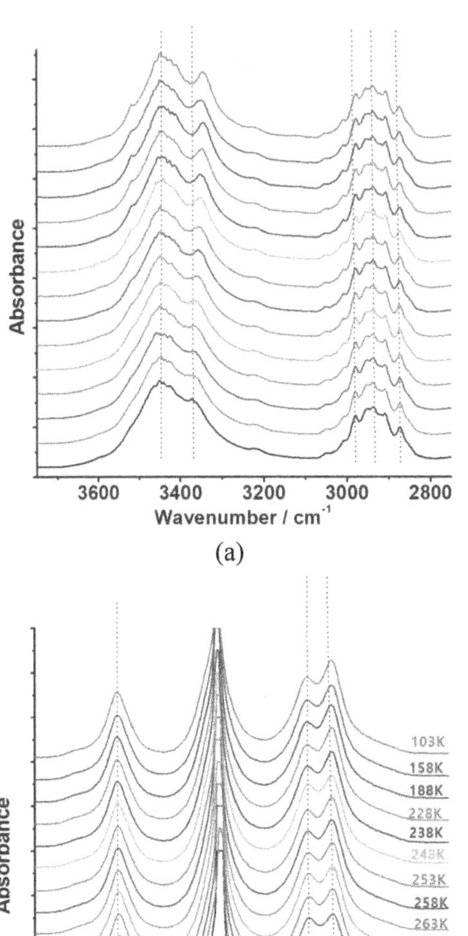

Fig. 75.1 FTIR spectrum (high concentration) of BM recorded in KBr pellet at different temperatures in (a) 3700 – 2700 cm^{-1} and (b) 1750 – 1550 cm^{-1} spectral ranges

4. CONCLUSION

This work marks the first recording of FTIR spectra for powdered BM at low temperatures. The visible changes observed in the spectra are attributed to molecular rearrangements occurring with temperature variations, possibly leading to an increased number ofH-bond networks. Assignments for observed peaks are provided based on comparisons with computed data. The potential onset of a phase transition can not be dismissed.

5. ACKNOWLEDGMENT

The financial support provided by the Department of Science and Technology (DST), New Delhi, INDIA against a financial grant No. SR/WOS-A/PM-3/2018 is gratefully acknowledged by Archna Sharma.

REFERENCES

1. G Cohen, HNatsheh, Y Sunny, C R Bawiec, E Touitou, M A Lerman, P Lazarovici, and P A Lewin *Ultrasound Med Biol.*,**41(9)**, 2449 (2015).
2. I M Oliveira, C Gonçalves, M E Shin, S Lee, Rui L Reis,GKhang, and J M Oliveira, *Biomolecules*, **10(10)**, 1456 (2020)
3. K E Arthur, J-C Wolff and D J Carrier *Rapid Commun. Mass Spectrom.*,**18**, 678 (2004).
4. M B Ali, M Attia, N Bellili and S Fattouch, *Indian J Pharm Sci***78(3)**,402 (2016).
5. A Manassra, M Khamis, M el-Dakiky, Z Abdel-Qader and F Al-Rimawi*Pharm Anal Acta.*, **1(2)**, 113 (2010).
6. D DBhawalkar, G Singh, and R V Nandedkar, Synchrotron radiation sources INDUS-1 and INDUS-2. *Pramana - J Phys.,* **50**, 467 (1998).
7. S Soumya, I H Joe, *J. Mol. Struct.*, **1245**, 130999 (2021).
8. X Liu, *Organic Chemistry I,* Chap 6.3 pp-197 Copyright © 2021.
9. F Dai, Q Zhuang, G Huang, H Deng, and X Zhang, *ACS Omega***8(19)**, 17064 (2023).
10. B Smith, *Infrared spectral interpretation: a systematic approach*, 1998.
11. R N Jones and J B Digiorgi, *Canadian J. Chem.*, **43** 182 (1965).

Note: The authors made the figure in this chapter.

Recent Trends in Applied Physics and Material Science – Sudhir Bhardwaj et al. (eds)

All-optical Logic Half Adder Operation based on Highly Nonlinear CS_2 Filled Triangular Triple Core Photonic Crystal Fiber

T. Uthayakumar*, Namrata Yaduvanshi

Department of Physics, Dayananda Sagar College of Engineering, Kumaraswamy Layout, Bangalore, Karnataka, India

Abstract: This study investigates a triple-core photonic crystal fiber (PCF) structure to achieve all-optical logic half-adder functionality, employing the highly nonlinear liquid CS_2 filled light guiding cores. Through apt modeling of the triple-core PCF, essential optical properties were determined through the finite element analysis, required for numerical computation of transmission characteristics. The propagation dynamics were explored using coupled nonlinear Schrödinger equations following split step Fourier algorithm, focusing on determining the optimal control signal power required for precise extinction ratio calculations. Through these calculations, the demonstration of logic operations necessary for half-adder operation using specific input signals and control signal phases is realized. The results confirm the successful realization of an all-optical half-adder with low input power requirements and a high figure of merit, emphasizing the design's effectiveness in advancing optical computing components for efficient logic operations in integrated photonic circuits.

Keywords: All-optical switches, finite element method, photonic crystal fiber, half adder, NLSE, split step algorithm

1. INTRODUCTION

All-optical couplers are vital in fiber communication systems serving as couplers, switches, logic gates and signal processors (Agrawal, 2020). Especially couplers of triple-core and multi-core stands out for their diverse output states and wide power selectivity. In recent years, photonic crystal fiber (PCF) based optical couplers have gained attention for their design flexibility and superior optical properties, such as high nonlinearity, customizable dispersion, larger mode areas, single-mode operation, and low bending losses (Russell and Knight, 2023).

Studies have highlighted the adaptability of PCF based couplers for various optical functions. Li et al. explored linear and nonlinear pulse propagation in triple core photonic crystal fiber couplers (TPCF) using coupled mode theory (Li, 2010). Followed by the successful demonstration of all-optical logic operations through symmetric and asymmetric (TPCF). Recent reports have demonstrated the feasibility of developing various logic gates based computing devices, such as shift registers, binary counters and half adders (Agrawal, 2020). Among these, half adder is particularly significant for digital processors, as it combines logical XOR and AND gates to perform addition. Menezes et al. implemented a half adder using a TPCF with an appropriate control signal (CS) (Menezes, 2010).

The use of highly nonlinear liquid chloroform filled TPCFs has demonstrated the potential for low power half adder operation (Uthayakumar, 2018). Such nonlinear liquids are crucial for achieving effective phase shift control at minimal control power. This study focuses on leveraging highly nonlinear CS_2-filled TPCF with nonlinearity one

*Corresponding author: uthayapu@gmail.comm

DOI: 10.1201/9781003684718-76

order greater than that of liquid chloroform to implement an all-optical logic half adder. The paper is organized as follows: Sections 2 and 3 detail the TPCF structure and theoretical modelling, respectively. Section 4 examines the transmission characteristics, while section 5 outlines the half adder implementation, Finally, section 6 concludes the study.

2. MODELING OF TPCF

The schematic of CS_2-filled TPCF is presented in Fig. 76.1, with the CS_2-filled cores labelled as 1, 2 and 3. The structural parameters include air hole diameter (d) to the pitch ratio (Λ) d/Λ of 0.8, where the pitch (Λ) is 2 mm and the inter-core separation (C) is $\sqrt{3}$ Λ. The core diameter (dc) is equal to the air hole diameter (d). Using the finite element method (FEM), the effective indices are calculated to derive key optical parameters, including the coupling length (L_c), dispersion (β), nonlinearity (γ), and extinction coefficient (α) are evaluated at an operating wavelength of 1.55 μm.

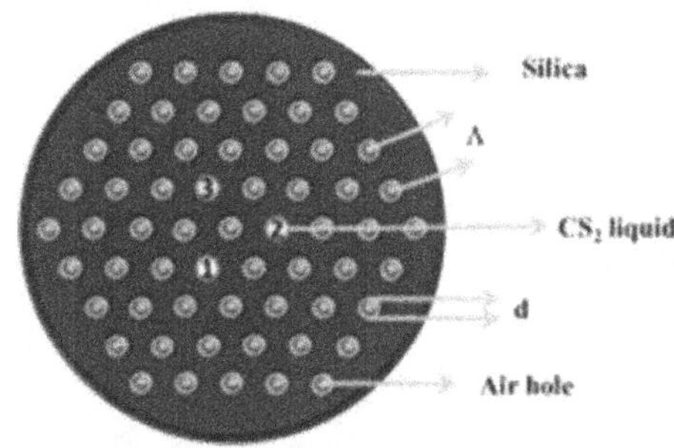

Fig. 76.1 Schematic of liquid CS_2 filled TPCF

3. THEORETICAL MODEL

$$i\frac{\partial A_1}{\partial z}-\frac{\beta_2}{2}\frac{\partial^2 A_1}{\partial t^2}+\gamma|A_1|^2 A_1+\kappa(A_2+A_3)-\frac{i}{2}\alpha A_1=0 \quad (1)$$

$$i\frac{\partial A_2}{\partial z}-\frac{\beta_2}{2}\frac{\partial^2 A_2}{\partial t^2}+\gamma|A_2|^2 A_2+\kappa(A_1+A_3)-\frac{i}{2}\alpha A_2=0 \quad (2)$$

$$i\frac{\partial A_3}{\partial z}-\frac{\beta_2}{2}\frac{\partial^2 A_3}{\partial t^2}+\gamma|A_3|^2 A_3+\kappa(A_1+A_2)-\frac{i}{2}\alpha A_3=0 \quad (3)$$

The propagation of optical pulses through the TPCF is described by a set of coupled nonlinear Schrödinger equations (CNLSE) as outlined in (Uthayakumar, 2018)

Here A_1, A_2, and A_3 represents the optical pulses propagating through cores 1, 2, and 3, respectively. The parameters b_2, γ, κ, and α denotes the group velocity dispersion, nonlinearity, coupling coefficient, and extinction, respectively.

4. TRANSMISSION CHARACTERISTICS

Transmission of optical power using SSFM through TPCF is illustrated in Fig. 76.2 (a). An initial Gaussian pulse, $A(0, t) = A_0 \exp(-t^2/W_0)$, is introduced at z = 0 through the core 1, where, A_0 is the initial amplitude, and W_0 the pulse width. For the remaining cores, only 0.001% of input field from core 1 is considered as input. The parameters used include Lc = 0.395 mm, β_2 = -0.0169 ps^2/m, γ = 11.98 W^{-1} km^{-1}, and α = 3.36 × 10^{-16} m^{-1} at the operating wavelength λ_0 = 1.55 μm. The transmission curves demonstrate excellent switching contrast, even at very low power level, 68% of input power confined to core 1. The remaining input power is evenly distributed among the other cores. This confirms the significant nonlinear contribution from CS_2, leading to strong decoupling via nonlinearity-induced phase shifts. At higher input powers, minor fluctuations in power distribution among the cores are observed, while majority of the power remains confined to the core 1.

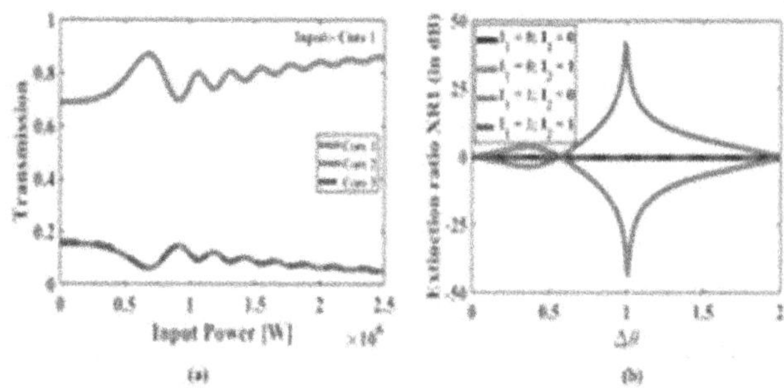

Fig. 76.2 (a) Transmission characteristics and (b) extinction ratio

5. EXTINCTION RATIOS AND HALF ADDER OPERATION

For logic operations, the control signal (CS) is applied through core 1, while inputs I_1 and I_2 are introduced via cores 2 and core 3, respectively. The CS can be either 0 or 1 with a phase difference of ΔΦ = Dqπ between I_1 and I_2, where Dθ ranges from 0 to 2 for desired logic operation. The outputs from core 1 and core 2 are determined using extinction ratios (XR1 & XR2) based on input combinations [I_1; I_2] = [(0,0), (0,1), (1,0), (1,1)], shown in Fig. 76.2(b). Corresponding output combinations [O_1; O_2] are used to perform logic gates with right initial phase [5,6]. The logic half adder operation is observed for the CS phase values varying from 0.1, to 0.5, with optimal figure of merit (FOM) at 0.3, tabulated in Table 76.1.

Table 76.1 Half adder operation for best FOM

I_1	I_2	CS (Δϕ = 0.3 π)	O_1 (XR1) Carry	O_2 (XR2) Sum
0	0	0	**0**	**0**
0	1	0	-0.36 dB **0**	0.36 dB **1**
1	0	1	-3.28 dB **0**	3.28 dB **1**
1	1	1	0.08 dB **1**	-0.08 dB **0**
FOM			3.72 dB	3.72 dB

6. CONCLUSION

We investigated a novel CS_2-filled TPCF for implementing all-optical logic half adder. Using SSFM, we derived transmission curves to calculate control signal power. By applying appropriate control and input power through the cores, the necessary extinction ratios were achieved, enabling half adder functionality with low power consumption and a high figure of merit.

7. ACKNOWLEDGEMENT

TU wishes to thank ANRF SERB for financial support through core research grant, CRG/2023/001074.

REFERENCES

1. Agrawal, G. P. (2020). Applications of Nonlinear Fiber Optics. Academic Press.
2. Russell, P. (2003). Photonic crystal fibers. Science 299 (5605):358-362.
3. Knight, J. C. (2003). Photonic crystal fibres. Nature 424:847-851.
4. Li, P., Zhao, J. and Zhang, X. (2010). Nonlinear coupling in triangular triple-core photonic crystal fibers. Opt. Exp. 18(26):26828-26833.
5. Menezes, J. W. M., Fraga, W. B., Ferreira, A. C., Guimarães, G. F., Filho, A. F. G. F., Sobrinho, C. S. and Sombra, A. S. B. (2010). All-Optical Half-Adder Using All-Optical XOR and AND Gates for Optical Generation of "Sum" and "Carry". Fiber and Integrated Optics 29(4): 254–271.
6. Uthayakumar, T. and Vasantha Jayakantha Raja, R. (2018). Logic gates based all-optical binary half adder using triple core photonic crystal fiber. J. Opt. 20(6): 065503.

Note: All the figures and tables in this chapter were made by the authors.

Recent Trends in Applied Physics and Material Science – Sudhir Bhardwaj et al. (eds)
© 2026 Taylor & Francis Group, London, ISBN 978-1-041-16452-4

Construction of an Artificial Molecule Using Molecular Modelling Software Aided by Atomic and Molecular Physics in Math Model

Manjunath T.C
Department of Computer Science & Engineering,
Rajarajeswari College of Engineering,
Bengaluru, Karnataka

Pavithra G., Swapnil S. Ninawe
Department of Electronics & Communication Engineering,
Dayananda Sagar College of Engineering, Bengaluru, Karnataka

Sandeep K.V., Iffath Fawad
Department of Electronics & Telecommunication Engineering,
Dayananda Sagar College of Engineering, Bengaluru, Karnataka

Abstract: The project titled "*Construction of an Artificial Molecule Using Molecular Modelling Software Aided by Atomic & Molecular Physics*" focuses on the design and simulation of an artificial molecule through advanced computational techniques. Leveraging molecular modeling software, the project integrates principles from atomic and molecular physics to construct and analyze a molecule with custom properties. The modeling process involves the precise calculation of atomic interactions, bond formations, and electron distributions to achieve desired molecular structures and functionalities. By simulating quantum mechanical behaviors and optimizing molecular geometry, this research aims to create a highly stable and functional artificial molecule. The use of atomic and molecular physics ensures that the interactions at the quantum level are accurately represented, facilitating the exploration of novel molecular configurations that may have potential applications in materials science, drug design, and nanotechnology. This project demonstrates the power of molecular modeling software in advancing the field of molecular engineering, providing a framework for designing artificial molecules with specific, targeted properties. Molecular modeling encompasses a broad spectrum of theoreticals & computationals techniques using to depict the structures of molecule, ion, & particle. These methods can be categorized based on the scale of time and length—from electronic to continuous levels—utilized in the modeling process. In the context of solute/solvent systems, molecular modeling is primarily divided into three categories: implicits method, integrals equation with their classicals densities functionals theories, & explicittes method. This paper explores three significant methodologies within the chemical engineering domain: the Poisson'-Boltzmann's equations, classical densities functionals theories & moleculars dynamical simulations. Additionally, the concept of scale integration is discussed as an effective approach for enhancing the accuracy and efficiency of modeling and simulation processes.

Keywords: Molecular modeling, Artificial molecule, simulation

1. INTRODUCTION

Molecular modeling (MM) employs computer-based techniques to simulate and manipulate molecular structures, reactions, and properties that hinge on three-dimensional configurations. This approach spans several disciplines including computational chemistry, drug development, computational biology, materials science, and nanostructures. MM is instrumental in elucidating the physical and chemical interactions that are often challenging to measure experimentally, thereby aiding in the development of new theories, methodologies, products, and models. Key simulation methods in MM include molecular dynamics, Monte Carlo simulations, and

*Corresponding author: tcmanju@iitbombay.org

DOI: 10.1201/9781003684718-77

geometry optimization. Unlike other simulation techniques in MM, Monte Carlo simulations uniquely handle model parameters as random variables instead of fixed ones, with RiskAMP being a notable Monte Carlo simulation engine for Microsoft Excel. MM enables polymer scientists for directly generate & analyze molecular data such as geometrie, energy, electronical property, spectroscopical characteristics, and bulk property. Given the widespread use of synthetic polymers in everyday life and their properties' critical dependence on molecular composition and structure, molecular modeling is vital. It provides a unique window into understanding materials at the molecular level and can significantly enhance traditional experimental approaches when both are used in tandem.

2. AIMS AND OBJECTIVES

After having provided an explanation on the functioning of molecular modelling software and its application in different industries, the aim of the report is to simulate the creation of a molecular model of a 2D nanomaterial, a polypeptide called Aldoxorubicin also called DOXO-EMCH, a compound that sees possible applications in drug delivery systems for cancer treatment (Seminario 2016) as in Fig. 77.1.

Molecular Formula: $C_{37}H_{42}N_4O_{13}$ ...... 2D Structure:

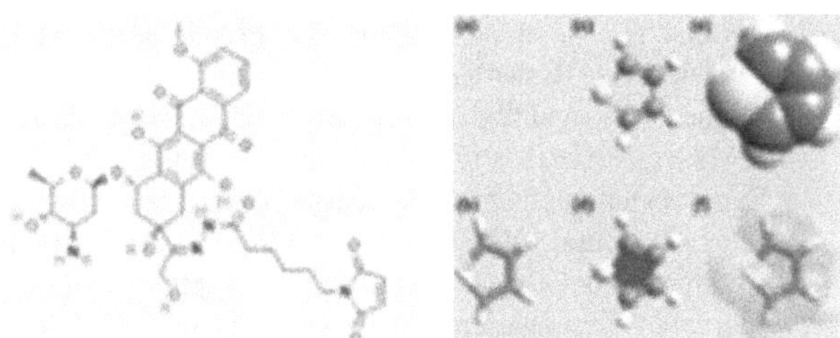

Fig. 77.1 Molecular structure

The (6-maleimidocaproyl) hydrazone derivative of doxorubicin, known as DOXO-EMCH, is a prodrug that binds to albumin and shows promise in cancer treatment due to its acid-sensitive properties. It has exhibited enhanced antitumor effectiveness in various murine tumor models and maintains a favorable safety profile across mice, rats, and dogs, notably featuring significantly reduced cardiotoxic effects. Administered intravenously, DOXO-EMCH quickly attaches to the Cys-34 site on circulating albumin and preferentially accumulates in solid tumors through passive targeting mechanisms. This prodrug has demonstrated its capability to cause regression in tumors that are responsive to anthracyclines, including types like breast cancer, small cell lung cancer, and sarcoma (MacKerell Jr. 2004).

3. METHODOLOGY

The molecule was constructed using Avogadro, a free Molecular Editing software. This tool is a sophisticated molecule editor and visualizer that works across different platforms and is useful in fields like computationals chemistrys, moleculars modelings, bioinformatic, and material sciences. It is known for its flexibles, higher-qualities visual representations & an strong plugind system. Common use includes creating molecule structure, preparing inputted file, & analyzed outputs from various computational chemistry programs. The software includes several tools (Jorgensen et al. 1996):

- Navigation tool: Allows basic scene maneuvers like rotating, panning, tilting, and zooming.
- Draw tool: Enables free-hand drawing of molecules with support for keyboards shortcut, combos box, & an periodical tabular views for selecting element (Frisch et al. 2016).

Additionally, there's an interactive Auto-optimization Tools which allow the user for sculpt their molecule. Users can start optimizing the geometry continuously, and while it's running, they can still use the drawing or adjusting tools to modify the molecule's shape and structure. Changes are visible in real-time as the structure optimizes. This tool can be used alongside the measurement tool for watch how bonded length & angle change as their molecule updates and its geometries is minimizes (Wang et al. 2006).

4. MATHEMATICAL MODEL

Their mathematical model for the construction of an artificial molecule using molecular modeling software, aided by atomic and molecular physics, involves a combination of quantum mechanical and classical methods to describe the behavior of atoms and their interactions. The key aspects of this model include the Schrödinger equation for quantum mechanical treatment, classical force field equations for molecular mechanics, and energy minimization algorithms for optimizing molecular structures. The Schrödinger equation is central to describing the quantum mechanical behavior of electrons within the molecule. For an *N*-electron system, the time-independent, non-relativistic Schrödinger equation for the electronic structure is given as (Shaw et al. 2007)

$$\hat{H}\Psi(r_1, r_2, \ldots, r_N) = E\Psi(r_1, r_2, \ldots, r_N)$$

where, $H^\wedge$ is the Hamiltonian operator, representing the total energy (kinetic + potential) of the system, $\Psi(r_1,r_2,\ldots,r_N)$ is the electronic wavefunction that describes the probability distribution of the electrons in their molecule, 'E' is their totals energies from electrons in the systems, where ri is the position of each electron (Berman et al. 2000). There are various ways to represent the molecular structure of thiophene. These include: (a) a simple wireframe model, (b) a stick or liquorice model, (c) a ball & a sticked model (d') a balls & sticks models thats include a ring, (e) a model showing Van der Waals radii (also known as CPK), and (f) a transparent version of the Van der Waals model with sticks.

5. RESULTS

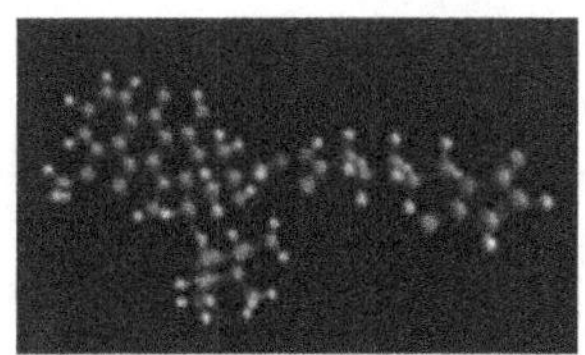

Fig. 77.2 Aldoxorubicin (Doxo-emch)– Single unit structure & Aldoxorubicin – Eight unit structure

6. USES & APPLICATIONS

Computer-Aided Drug Design involves using computer programs to help create and develop new medications. This method includes several techniques, like designing molecules, intelligently creating drugs based on their structure, building new compounds from scratch, modeling how drugs fit into their target receptors, and docking, which predicts how well a drug will bind to its target, along with molecular-dynamics (MD) simulations used for drug discovery. Manuscripts that introduce new methodologies or techniques and their validation are highly encouragads (Genheden et al. 2015). However, research that only uses commercial tools or online services on datasets that anyone can access is not recommended. For docking studies, researchers should use at least three different shapes of the target (Morris et al. 2009). These shapes can come from several crystal structures or be created from a molecular dynamics (MD) simulation that runs for at least 20 nanoseconds (ns). The docking results, or poses, must be shown to be stable by conducting additional MD simulations for at least 20 ns, especially if there's no experimental evidence to support them.

7. CONCLUSION

Molecular modelling is a versatile and powerful computational technique that allows scientists to study and predict the behaviour of molecules and materials at the atomic and molecular level. By using sophisticated computer software and algorithms, molecular modelling enables researchers to gain a deep understanding of the properties and behaviour of molecules, which can facilitate the design of new compounds, materials, and drugs, with further innovations right around the corner.

REFERENCES

1. Seminario J. M. (2016). Molecular modeling: Basic principles and applications, Elsevier.
2. MacKerell Jr. A. D. (2004). Empirical force fields for biological macromolecules: Overview and issues," Journal of Computational Chemistry, 25(13): 1584-1604.
3. Jorgensen W.L., Maxwell D.S. and Tirado-Rives J. (1996). Development and testing of the OPLS all-atom force field on conformational energetics and properties of organic liquids, Journal of the American Chemical Society. 118(45): 11225-11236.
4. Frisch M. J. et al. (2016). Gaussian 16, Revision C.01, Gaussian, Inc., Wallingford CT.
5. Wang J., Wang W., Kollman P.A. and Case D.A. (2006). Automatic atom type and bond type perception in molecular mechanical calculations, Journal of Molecular Graphics and Modelling. 25(2): 247-260.
6. Shaw D.E. et al. (2007). Atomic-level characterization of the structural dynamics of proteins, Science. 330(6002): 341-346.
7. Berman H. M. et al. (2000). The Protein Data Bank, Nucleic Acids Research. 28(1): 235-242.
8. Genheden S. and Ryde U. (2015). The MM/PBSA and MM/GBSA methods to estimate ligand-binding affinities, Expert Opinion on Drug Discovery. 10(5): 449-461.
9. Morris G.M. et al. (2009). AutoDock4 and AutoDockTools4: Automated docking with selective receptor flexibility, Journal of Computational Chemistry. 30(16): 2785-2791.

Note: All the figures in this chapter were made by the authors.

Recent Trends in Applied Physics and Material Science – Sudhir Bhardwaj et al. (eds)
© 2026 Taylor & Francis Group, London, ISBN 978-1-041-16452-4

Exploring the Interaction of the Human Telomeric DNA G-quadruplex with Caffeine Derivative Inhibitors through Molecular Docking

Jwala Ji Prajapati*
Department of physics, Deen Dayal Upadhyaya Gorakhpur University, Gorakhpur, U.P., India

Ramesh Kumar Yadav
Department of physics, B.R.D.P.G. College, Deoria, U.P., India

Umesh Yadava*
Department of physics, Deen Dayal Upadhyaya Gorakhpur University, Gorakhpur, U.P., India

Abstract: G-quadruplexes are formed near the telomeric region of chromosomes for the sequences having multiple of Guanine nucleotides. Plenty of guanines are associated together to build quadruplex secondary structure based on non-Watson-Crick base pair model. We report here the docking of quadruplex structure and its binding affinity with caffeine derivatives. The co-ordinates of the human telomeric DNA G-quadruplex (AGGGAGGGTTAGGGTTAGGG) were retrieved from the RCSB database with PDB ID 6XCL. The structures of caffeine derivatives were downloaded from CHEMBL database and screened through high throughput virtual screening. Ten caffeine derivative ligands complexed with quadruplex were extracted from XP docking result on the basis of docking scores, non-covalent interactions and docking energies. Our study demonstrates that the compound CHEMBL194732 [(4S,5S,8R)-8-((E)-2-(5-chloro-6-methylcyclohexa-2,4-dien-1-ylidene)ethyl)-1,3,9-trimethylhexahydro-1H-purine-2,6-dione] & CHEMBL1927677 [1,3,7-trimethyl-8-(phenethyllamino)-1H-purine-2,6(3H,7H)-dione] exhibit superior binding affinities with glide scores -10.555 and -8.847 kcal/mol respectively than the others.

Keywords: DNA G-quadruplex, G-tetrad, Caffeine derivatives, Glide docking, Human telomers

1. INTRODUCTION

DNA plays a crucial role in cells as data storage and transfer of genetic information for living organisms. DNA exists as a double helical structure which is stabilized by hydrogen bond interactions following Watson-Crick base pairs (Haider S., 2018). In molecular biology, sequences with majority of Guanine residues produces an alternate type of DNA quadruplex nucleic acid secondary structures which lack Watson crick base pairing and differs from the canonical form of DNA referred as G-quadruplexes(G4) (Spiegel J. et al.2020). Most telomeric sequences are found to have higher-order non-canonical DNA secondary structures folded into one, two, or four helical-shaped strands at the ends of telomers in chromosomes (Huppert J. L. et al.2005 and Bryan T.M. et al.2020). Due to interactions between strands within the molecules, it can take on numerous topologies and directions. It forms a G-quadruplex, which is made up of a parallel square planar structure termed as G tetrad. G4 connection depends partially on hydrogen bonding patterns and somewhat on base pairing. The basic unit of a G-quadruplex consisting of a planar adjustment of four adjacent self- guanine bases stabilized by Hoogsteen type hydrogen bonding called as tetrad (Figure 78.1). Four guanines, the donor and acceptor of hydrogen bonds, make up G-tetrad. Two or more guanine tetrads then assembled via various stacking interactions to form a G-quadruplex motif, which is further stabilized by introducing monovalent metal cations (usually $M^+ = K^+$, Na^+, Ca^+, Mg^+....) in the central ionic core (Largy E. et al.2015). K^+ ion is reported

*Corresponding author: jwalajisjlt@gmail.com, u_yadava@yahoo.com

DOI: 10.1201/9781003684718-78

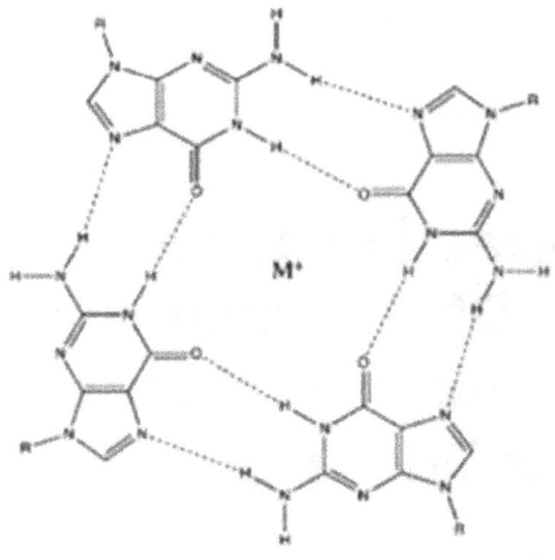

Fig. 78.1. G4 quadruplex structure with centrally coordinated metal ion

to stabilise the structure under study (PDB ID 6XCL). The positions of putative and stabilized G4 structures are not randomly oriented in the nuclear genome but found in specific areas such as telomeric region with G-quadruplex-forming sequences. These promoter G-quadruplexes have been proposed to be involved with the regulation of transcription and have attracted attention as therapeutic targets. G4s are formed not only *in vitro* but also *in vivo* and are attracting considerable interest owing to their potential involvement in biological processes including replication, transcription, mRNA splicing, translation, telomeres maintenance and epigenetic regulation of the genome. G4s can proved to be an important tool to control the unusual pathogenic disease spreading worldwide and challenging its curing. In addition, the orientation of strands can generate G4-structures with different topologies by adopting various conformations such as parallel, antiparallel, hybrid, and looped. Also, putative G4-forming sequences are detected in organelles, such as mitochondria which is also called as the power house of the cell, and to have a functional role during mitochondrial transcription. G4s can associate in polymorphic intra- or intermolecular structures whose space orientations can be affected by selecting different strand polarity and configurational parameters as well as by the nature of loops located in the sequence. The stability of G-quadruplex can be altered by affecting various factors depending on the number of tetrads, loop length, topology, manifold and the sequential arrangement itself within G4 captured regions. The pioneering study of G-quadruplex deciding the existence of possible interacting ligands can withheld within it. To achieve this goal docking program can be used to fit a foreign agent in to the conformation of G4 structure. G-quadruplex can be used for the development of novel inhibitors which will improve the drug molecule against it (Burge S. et al.2006). A recent study has opened doors through the resolution of these non-canonical higher-order structures of nucleic acids and enquiring connections between genome functions with numerous connections to hazardous diseases. Here, we critically investigate the binding of ligands with G4 DNA which may be helpful in highlighting opportunities for drug discovery (Ma Y. et al.2020 and Batool M. et al.2019). The topology and model adopted by the Guanine rich complex enhance the probability of interaction with ligands, as a result conformational changes occur at a large scale for different caffeine derivatives. We address the different configuration and orientations parametric analysis of G-quadruplex to understand the perturbation in it. These findings enhance the growth of G4-targeting approaches for therapeutic discoveries by the means of molecular docking protocols (Sun Z. Y. et al.2019).

2. MATERIALS AND METHODOLOGY

The co-ordinates of G4- ligand complex structure was retrieved from the RCSB database with PDB id 6XCL which contains platinum-based ligand introduced with human telomeric G-quadruplex (Miron, C.E. et al.2021). In our study, we have downloaded thousands of caffeine derivatives from CHEMBL database. Caffeine derivatives having planar moieties can be the better inhibitors that can bind and fitted to the G4 model.

2.1 Molecular Docking

Molecular docking is a computational technique that provide the information and prediction of fitting ligand within binding site of a target molecule. Molecular docking utilizes scoring functions to provide a fast and crude estimation of the binding affinity. This may pave the way in understanding various biological and chemical processes. We have performed the ligand and G4 target preparations and molecular docking with the help of Grid-based Ligand Docking with Energetic (GLIDE) program, as implemented within Schrodinger suite. Glide utilizes exhaustive search-based docking algorithm and predict the binding modes and ranking of ligands. Based on algorithm and scoring methods Glide has it two different protocols SP and XP (Farag M. et al.2023).

2.2 XP Glide Docking

The goal of XP glide methodology is to rank the ability of ligand to bind in particular region of conformation of receptor. All the ligands that we have considered for the study cannot be expected to bind correctly and attain good docking scores. The analysis revealed that the flexibility of the loops was reduced by the ligand, leading to more stable quadruplex complex. The XP Glide score consists of various energy terms as described below (Repasky M. P.et al.2012).

$$XP\ (\text{Glide Score}) = E_{coul} + E_{vdW} + E_{bind} + E_{penalty},$$

$$E_{bind} = E_{hyd\_enclosure} + E_{hb\_nn\_motif} + E_{hb\_cc\_motif} + E_{PI} + E_{hb\_pair} + E_{phobic\_pair}$$

$$E_{penalty} = E_{desolv} + E_{ligand\_strain}$$

2.3 Caffeine Derivatives

Caffeine is an alkaloid widely known as a weak base that reacts with acids to protonate it. It is chemically similar to the adenine and guanine bases of nucleic acids (Figure 78.2). It can act as structurally rigid scaffold which provides enormous possibility for molecular diversity in drug development process and show strong inhibitory activities, therefore, it causes oxidative damage to DNA and amino acids, and biologically active agents. Caffeine derivatives may be capable of interacting with guanine base residues to bind and block its receptors. These new complex systems stand up instinctively as an outcome of intermolecular non-covalent interactions involving hydrogen bonds, π···π, or van der Waals forces. In particulars the caffeine derivatives are aromatic in nature and have a planar moiety of π-surface as G- tetrads have. These derivatives may easily be fitted into the grooves

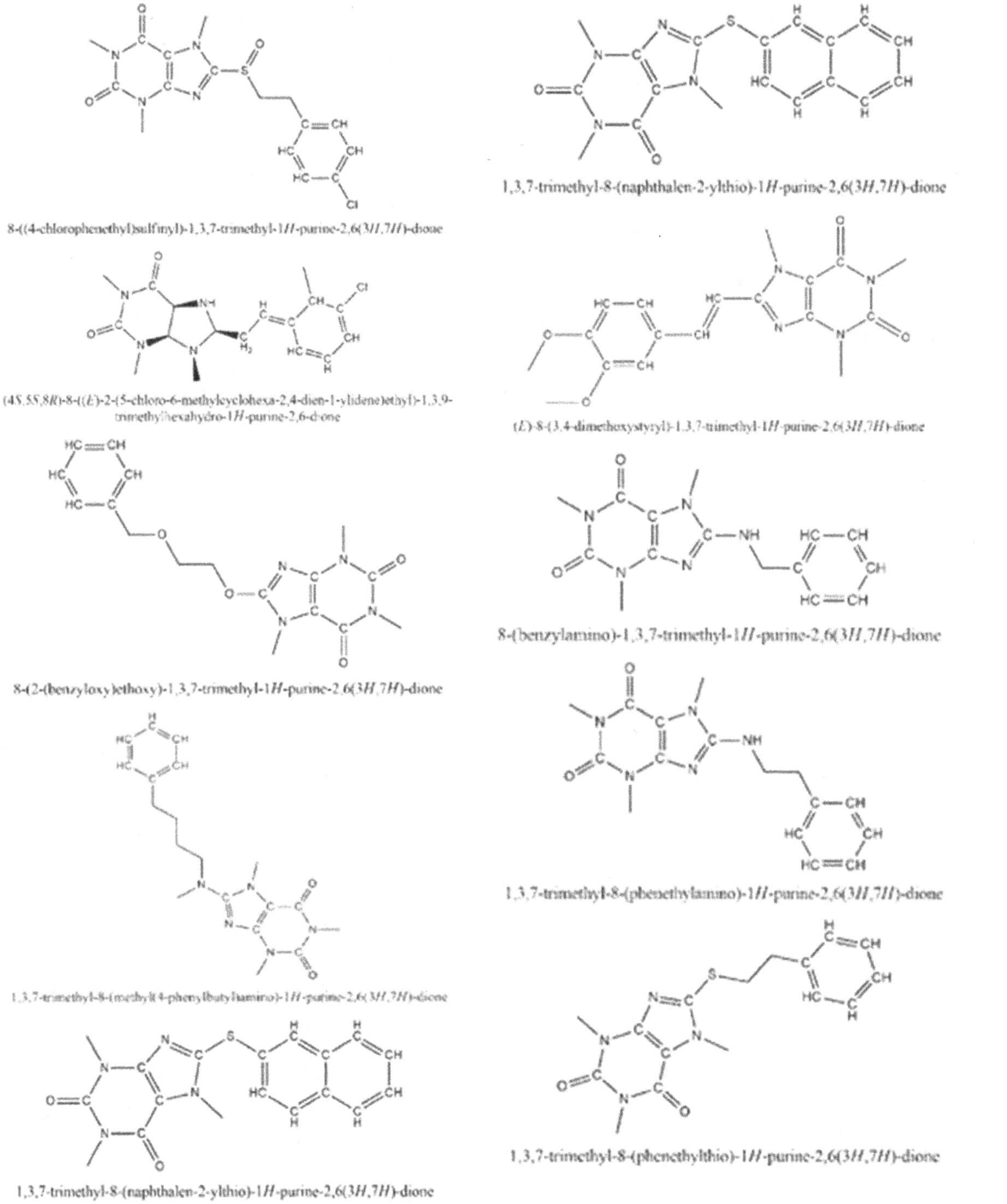

8-((4-chlorophenethyl)sulfinyl)-1,3,7-trimethyl-1*H*-purine-2,6(3*H*,7*H*)-dione

1,3,7-trimethyl-8-(naphthalen-2-ylthio)-1*H*-purine-2,6(3*H*,7*H*)-dione

(4*S*,5*S*,8*R*)-8-((*E*)-2-(5-chloro-6-methylcyclohexa-2,4-dien-1-ylidene)ethyl)-1,3,9-trimethylhexahydro-1*H*-purine-2,6-dione

(*E*)-8-(3,4-dimethoxystyryl)-1,3,7-trimethyl-1*H*-purine-2,6(3*H*,7*H*)-dione

8-(2-(benzyloxy)ethoxy)-1,3,7-trimethyl-1*H*-purine-2,6(3*H*,7*H*)-dione

8-(benzylamino)-1,3,7-trimethyl-1*H*-purine-2,6(3*H*,7*H*)-dione

1,3,7-trimethyl-8-(methyl(4-phenylbutyl)amino)-1*H*-purine-2,6(3*H*,7*H*)-dione

1,3,7-trimethyl-8-(phenethylamino)-1*H*-purine-2,6(3*H*,7*H*)-dione

1,3,7-trimethyl-8-(naphthalen-2-ylthio)-1*H*-purine-2,6(3*H*,7*H*)-dione

1,3,7-trimethyl-8-(phenethylthio)-1*H*-purine-2,6(3*H*,7*H*)-dione

Fig. 78.2 Chemical structure of some caffeine derivatives

which can modulate the function of the quadruplex13 (Ameen A. et al.2024).

3. RESULTS AND DISCUSSIONS

3.1 Extra precision (XP) docking

The prepared caffeine derivatives along with their conformers have been docked within the binding site of the G4 target using extra precision protocol of the Glide. For each ligand, docking scores are different which shows the binding capabilities of the different ligands. The various interactions exhibited by the compounds in their best docking complexes have been shown in Figure 3. The docking scores and various energy values of the top ten compounds (Figure 78.2) as obtained through GLIDE XP docking are shown in Table 78.1.

Docking results show that all the ligands bind to human telomeric G-quadruplex in the grooves (Figure 78.3). Basically, the molecular recognition is done in majority of π and van der Waal's interaction. The ligands cover majority of the interactions in to the middle of the G-tetrad. Most of the ligands show signatures of van der Waal interaction when bound to quadruplex which indicates the diversity in conformations. The scores that are produced for every pose by the scoring procedure are utilized to rank the various ligands and poses. The lower the value is, the better the binding is. Out of all ligands, CHEMBL194732 has the best glide score of -10.55 kcal/mol with the best docking energy and Ligand CHEMBL1451750 show little bit poor tendency of inhibition. Initially best pose with the best Glide scores were selected. It has been observed that Glide Van der Waal energy term is more prominent than the other terms. Figure 78.3 show that all the top ten ligands are interacted with G4 structure through noncovalent interactions.

3.2 G4 Ligand Interaction

The docking results demonstrate hydrogen bonding and π-π stacking interactions as exhibited by all the ligands.

Additionally, ligands CHEMBL1806764 and CHEMBL1927684 show extra π-π T-shaped interactions. π-cation interactions are also demonstrated by all the ligands except CHEMBL2206096. Unfavourable acceptor-acceptor type interaction is shown by the ligand CHEMBL194732 which has more docking score value while unfavourable donor-donor type interaction is shown by the ligand CHEMBL1927677. Out of all the ligands, only two compounds CHEMBL1927684, and CHEMBL2206096 show the π-σ interaction. Carbon hydrogen bond type interaction is shown by all the ligands. Conventional hydrogen bond interaction is shown only by the ligands CHEMBL194732, CHEMBL1451750, CHEMBL1806764, and CHEMBL1927677. Attractive charge interaction is shown by the ligand CHEMBL194732. The ligands CHEMBL194732, CHEMBL1806764, CHEMBL1927677, and CHEMBL1927684 show the maximum number of interactions. The ligand CHEMBL194732 seems to be the superior inhibitor compared to the other ligands under study having involvement in many interactions and appreciable binding energy and Glide score that confirms the stability integrity of quadruplex structure

Table 78.1 Docking score, Glide ligand efficiency(sa,ln), XP Gscores, Glide electrostatic (Ecoul), van der Waals (Evdw), Glide energy, Glide Einternal and Glide Emodel in kcal/mol of the top ten compounds as obtained through GLIDE (XP) docking

Ligand complex	Docking score	Glide ligand efficiency	Glide ligand efficiency (sa)	Glide ligand efficiency (ln)	XP G_{score}	Glide E_{vdw}	Glide $E_{coulomb}$	Glide E_{energy}	Glide $E_{internal}$	Glide E_{model}
CHEMBL194732	-9.564	-0.416	-1.183	-2.313	-10.555	-39.084	-13.022	-52.106	2.956	-72.112
CHEMBL1927677	-8.847	-0.385	-1.094	-2.139	-8.847	-55.751	-2.432	-58.183	5.326	-88.093
CHEMBL280982	-8.799	-0.338	-1.003	-2.066	-8.799	-62.242	-1.166	-63.408	7.025	-92.829
CHEMBL1927675 -8.68		-0.347	-1.015	-2.057	-8.68	-62.202	0.228	-61.974	2.932	-93.885
CHEMBL2206098	-8.47	-0.339	-0.991	-2.008	-8.47	-62.582	-2.081	-64.663	3.838	-95.771
CHEMBL2206096	-8.439	-0.352	-1.014	-2.02	-8.439	-58.199	-2.041	-60.24	3.112	-88.76
CHEMBL2313284	-8.414	-0.366	-1.04	-2.035	-8.414	-55.788	-2.379	-58.167	0.158	-86.667
CHEMBL1806764	-8.404	-0.336	-0.983	-1.992	-8.407	-54.68	-4.622	-59.302	0.502	-87.065
CHEMBL1927684	-8.369	-0.322	-0.954	-1.965	-8.371	-59.208	-2.381	-61.588	6.183	-90.766
CHEMBL1451750	-8.293	-0.377	-1.056	-2.027	-8.293	-52.076	-0.713	-52.789	4.436	-78.031

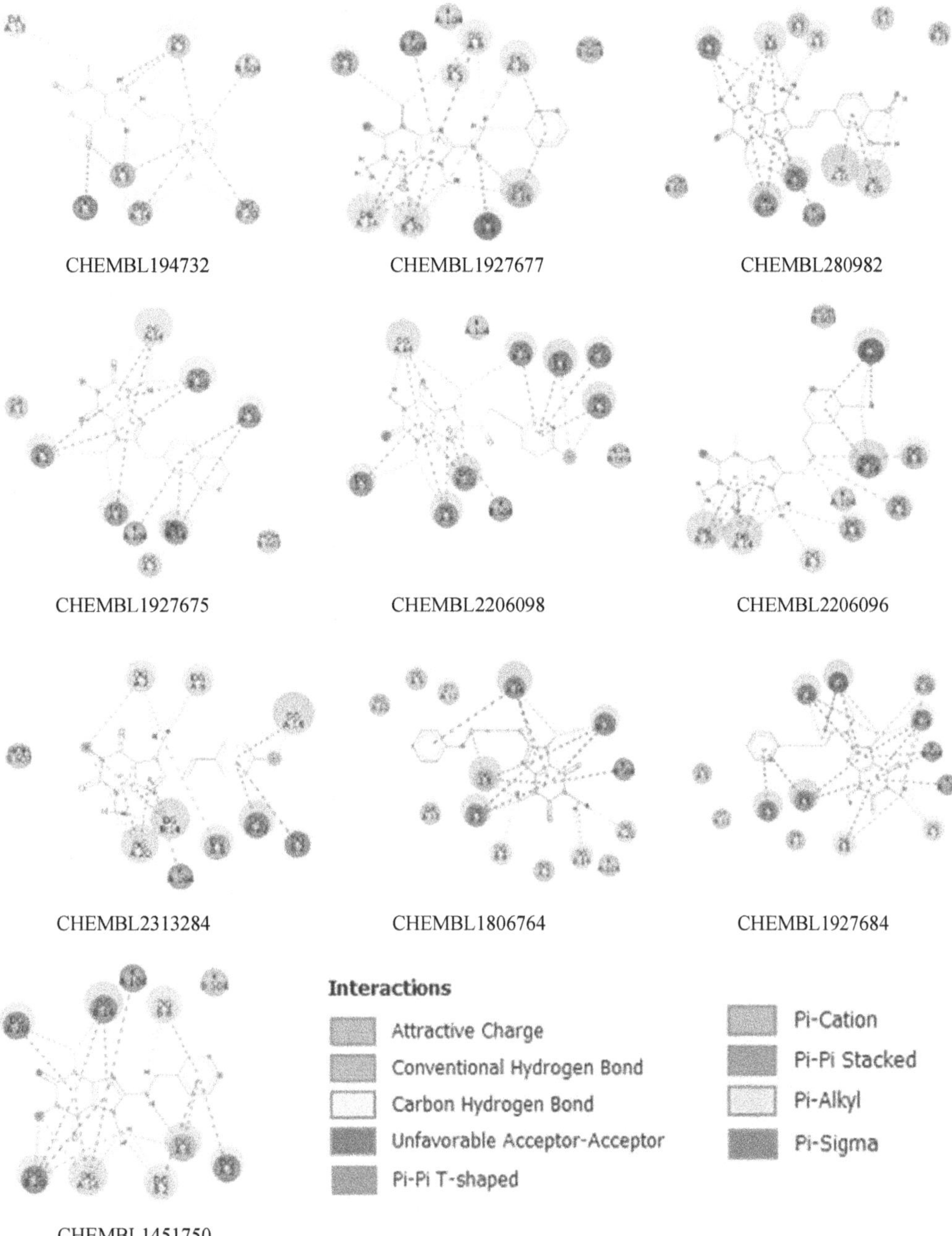

Fig. 78.3 2D molecular interactions (XP Glide docking) Between the functional groups and specific nucleic acid bases of G-quadruplex

4. CONCLUSIONS

In this paper, we have presented the docking and scoring values of diverse caffeine derivatives within the binding site of G4 quadruplex and the prediction of binding energies under the operation of XP Glide docking that can be meaningful in developing and designing novel inhibitors. In spite of having better docking scores and maximum number of interactions, the CHEMBL194732 ligand show some deviation from the centre of groove, while other ligands fitted centrally in the groove of the strands and interact equally with nucleic acid base pairs. As has been observed from the extra precision docking, the carbon hydrogen bond type interaction and van der Waal type interactions play a major role in constituting the docking energy. It has been concluded that the ligand CHEMBL194732 might be the ligand with the best binding capability followed by the ligands CHEMBL1927677 and CHEMBL280982 respectively. The binding affinity can be further studied or evaluated by MD simulations as well as using alchemical binding free energy estimations (Šponer J. et al.2020). In conclusion, we recommend considering and performing docking of small molecules to G4 DNA with caution. Chemical reasoning might allow one to better identify the experimental pose among the best scored poses; however, virtual screening of a large library might not be feasible for G4 DNAs. On the behalf of these calculations, the caffeine derivatives have been proved to be a better and suitable inhibitors, and thus explore the G4 structures to have potential for therapeutic targets.

5. ACKNOWLEDGEMENT

JJP is thankful to joint CSIR-UGC for providing the NET-JRF research fellowship. UY is thankful to CST, UP for the financial support through the project no. CST/ID-1526.

REFERENCES

1. Haider S (2018), Computational Methods to Study G-Quadruplex–Ligand Complexes. *J Indian Inst Sci* 98, 325–339, https://doi.org/10.1007/s41745-018-0083-3
2. Spiegel J, Adhikari S, Balasubramanian S (2020), The Structure and Function of DNA G-Quadruplexes. *Trends Chem.* 2020; 2:123–136.doi: 10.1016/j.trechm.2019.07.002.
3. Huppert J L, Balasubramanian S. (2005), Prevalence of quadruplexes in the human genome. *Nucleic Acids Res.*; 33:2908–2916. doi: 10.1093/nar/gki609.
4. Ma Y., Iida K, Nagasawa K. (2019), Topologies of G-quadruplex: Biological functions and regulation by ligands. *Biochem. Biophys. Res. Commun.* 2020; 531:3–17. doi: 10.1016/j.bbrc.2019.12.103.
5. Largy E, Mergny JL, Gabelica V. (2016), *Metal Ions in Life Sciences.* Volume 16. Springer; Berlin, Germany: Role of Alkali Metal Ions in G-Quadruplex Nucleic Acid Structure and Stability 203–258.
6. Burge S, Parkinson GN, Hazel P, Todd AK, Neidle S. (2006), Quadruplex DNA: Sequence, topology and structure. *Nucleic Acids Res.* 34:5402–5415. doi: 10.1093/nar/gkl655.
7. Bryan TM G-quadruplexes at telomeres: Friend or foe? *Molecules.*2020;25:3686. doi: 10.3390/molecules25163686.
8. Farag M, Messaoudi C, Mouawad I. (2023). An algorithm to calculate advanced structural characteristics of g-quadruplexes, nucleic acids research, 51(50): 2087–2107. https://doi.org/10.1093/nar/gkad060.
9. Miron, C.E., van Staalduinen, L., Rangaswamy, A.M., Chen, M., Liang, Y., Jia, Z., Mergny, J.L., Petitjean, A., (2021) Going Platinum to the Tune of a Remarkable Guanine Quadruplex Binder: Solution- and Solid-State Investigations, Angew Chem Int Ed Engl 60: 2500-2507, https://doi.org/10.1002/anie.202012520
10. Šponer J, Islam B, Stadlbauer P, Haider S. (2020), Molecular dynamics simulations of G-quadruplexes: The basic principles and their application to folding and ligand binding, https://doi.org/10.1016/bs.armc.2020 .04.002
11. Batool M, Ahmad B and Choi S. (2019). A Structure-Based Drug Discovery Paradigm, Int. J. Mol. Sci., 20, 2783; doi:10.3390/ijms20112783.
12. Sun Z Y, Wang X N, Cheng S Q, Su X X, and Ou T M (2019), Developing Novel G-Quadruplex Ligands: From Interaction with Nucleic Acids to Interfering with Nucleic Acid–Protein Interaction, Molecules. Feb; 24(3): 396.doi: 10.3390/molecules24030396, PMCID: PMC6384609, PMID: 30678288.
13. Ameen A. Abu-Hashem, Othman Hakami, Mohamed El-Shazly, Heba A. S. El-Nashar, Mahmoud N. M. Yousif (2024), Caffeine and Purine Derivatives: A Comprehensive Review on the Chemistry, Biosynthetic Pathways, Synthesis-Related Reactions, Biomedical Prospectives and Clinical Applications, https://doi.org/10.1002/cbdv.202400050.
14. Repasky M. P.; Murphy, R. B.; Banks, J. L.; Greenwood, J. R.; Tubert-Brohman, I.; Bhat, S.; Friesner, R. A. (2012) "Docking performance of the Glide program as evaluated on the Astex and DUD datasets: a complete set of Glide SP results and selected results for a new scoring function integrating WaterMap and Glide",J. Comput. Aided Mol. Des., 26: 787-99.

Note: All the figures and tables in this chapter were made by the authors.

Recent Trends in Applied Physics and Material Science – Sudhir Bhardwaj et al. (eds)

Surfactant based Photogalvanic Cells for Solar Energy Conversion and Storage

Lal Mohan*, Gangotri K.M.
Department of Chemistry,
Jai Narain Vyas University,
Jodhpur, Rajasthan, India

Lal Chhagan
Department of Chemistry,
Swami Shraddhanand College,
University of Delhi, Delhi, India

Abstract: Photogalvanic (PG) cells are used to convert solar energy into electrical power and store it. The foundation of PG cells is the photochemical reaction that, upon activation by photon, and yields good electrical results. These energy-dense cells lose electrochemical energy, which causes the production of electricity. For better outcomes, the specially constructed H shaped PG cells were studied. A spectrometer, digital pH meter, microammeter, carbon pot, and resistance key were used in order to measurements. The different solar parameters in a PG cell device were examined. The higher effect caused by solar energy has been investigated by varying the PG cell's different properties. According to observation, the PC and PP were 430 μA and 710 mV, respectively. PG cell outcomes as performance and conversion efficiency were determined 141.00 minutes and 2.1213%, respectively. According to obtained results, PG cells are the most effective and alternative energy source for assisting in the fulfilment of global energy demands. A photochemical mechanism was proposed for the conversion of sunlight into energy in the present generation of PG cells. Increasing significant electrical production was the primary objective of the study. Cell photogeneration is facilitated through a photochemical mechanism.

Keywords: Solar Energy, Methylene blue, Surfactant, PG cells, Photopotential, Photocurrent, Fill factor

1. INTRODUCTION

A key requirement for the growth of a developing country is a substantial and reasonably priced energy source. Energy has become recognized as essential to the modern era's progress. Energy is essential to a nation's economic growth. Without universal access to electricity, no country can claim to be contemporary and advanced. Over half of the electricity produced by power plants is derived from fossil fuels like coal and natural gas, which are known to exacerbate climate change and global warming. Solar energy is abundant, renewable, and free of pollution and carbon emissions.

PG cells are based on solar energy which converts chemical energy into electrical energy in the presence of sunlight. First of all the PG concept was reported in the action of sunlight (Rideal and Williams, 1925) and followed by Thionine-Iron System (Rabinowitch, 1940). Later on, various PG systems were studied by numerous researchers i.e., Use of miscelles in PG cells (Gangotri et al., 1999), role of surfactants in PG cells (Gangotri and Meena, 2006), use of fluoroscein system in PG cell (Madhwani, et al., 2007) and studies in the PG effect in mixed reductants on time to time (Gangotri and Indora, 2010). Energy conversion and storage of PG cell by mixed dyes system with ethylene diaminetetraacetic acid work was reported for current flow (Lal and Gangotri, 2011).

In the way to better finding, PG systems were developed, i.e., A comparative study on the performance of PG cells with mixed surfactant (Lal and Gangotri, 2013), comparative study of mixed photosensitizer system (Gangotri and Mahawar, 2012), and energy efficiency (Genwa and Sagar, 2013). Solar energy research field

*Corresponding author: mohanlalsolarenergylab@gmail.com

DOI: 10.1201/9781003684718-79

provides the evidence on the better role of PG cell and a comparison of conversion efficiencies of various sugars as reducing agents for the photosensitizer (Bhimwal et at., 2013) and study of PG effect in PG cell (Gangotri and Mohan, 2013).

2. LITERATURE REVIEW

Numerous photosensitizers, reductants, and surfactants have been used in PG cells, per a comprehensive study of the literature. Photopotential and photocurrent-based research work plays an important role in electrical output dye reductant surfactant systems. Mall and Solanki (2018) reported spectrophotometric and conductometric work for PG cell. Jayshree et al., (2022) studied on PG cell for electrical output in solar energy conversion and storage. Koli et al., (2022) discussed on Tropaeline O-oxalic acid-benzalkonium chloride photogalvanic cells for solar energy conversion and storage as another research on photosensitizer. Further comparatively better electrical results were reports as innovation for prospective energy by Lal and Gangotri, (2022). Beside these, another category of PG cell also discussed for study of electrical output in PG cell for solar energy by Rathore et al., (2022). Lal and Gangotri (2022) studied on innovation in progressive study for through PG system for various mixed surfactant and they found that significant electrical results to improve the PG cell outcomes. Another study was reported by Lal and Gangotri, (2023) as innovative study on sources of renewable energy using mixed surfactant systems for a more sustainable atmosphere. A thorough study of the literature indicates that it has lately worked on progressive research for possible energy sources utilizing PG systems. Recently, very good results are reported for PG system as comparative study on sunlight induced surfactants system in photogalvanics for solar energy conversion and storage (Lal and Gangotri, 2025).

3. MATERIALS AND METHOD

This system used a mixed surfactant (NaLS+CTAB), xylose (reductant), and MB (photosensitiser). To protect them from sunlight, all solutions were kept in amber-coloured jars. PG solutions were prepared in doubly distilled water and an H-shaped tube of glass that has been blackened with white and black carbon paper was filled with a mixture of MB, xylose, NaLS+CTAB, and NaOH solutions (Fig.79.1. Experimental setup). One side of this cell was connected with a glossy platinum (Pt) wire, while the other arm was filled with a SCE. Before exposing the arm with the Pt electrode to a 200 W light (Philips) source, the entire system was expose. Thermal radiation was blocked using a water filter. The PP and PC produced during experiments, were observed using a digital pH meter and a micro ammeter, respectively, electrical circuit was used of apply an external load in order to study the current voltage characteristics.

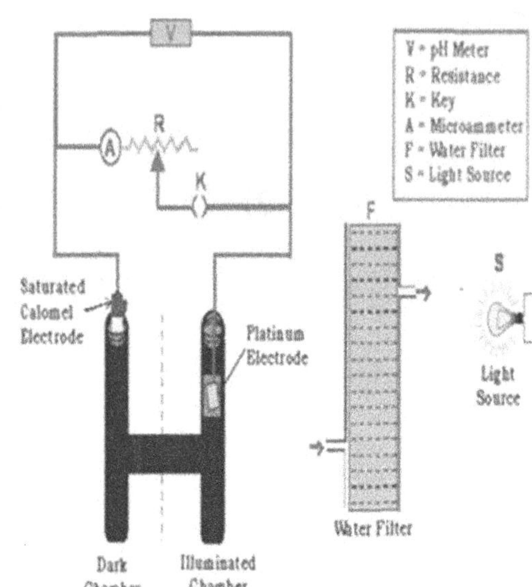

Fig. 79.1 Experimental set up

4. RESULTS AND DISCUSSION

4.1 Effects on the System when Varying the Concentration of MB

It was investigated how the concentration of MB affected the PP and PC. When fewer dye-MB (photosensitiser) to excite and then availed electrons to the platinum electrode, it was discovered that PP and PC decreased with decreasing photosensitiser concentrations. Since the molecules of dye in the pathway absorbed a large number of the light, a higher amount of dye once more led to a decrease in the electrical current as the extent of the light that hit the dye molecules close to the electrode decreased (Table 79.1).

Table 79.1 Impact of MB, Xylose, NaLS, CTAB, and pH

Scale	PP (mV)	PC (μA)
	MB $\times 10^{-5}$ M	
4.05	546.0	335.0
4.10	710.0	430.0
4.15	667.0	234.0
	Xylose $\times 10^{-3}$ M	
2.05	529.0	3420
2.10	710.0	430.0
2.15	543.0	321.0
	NaLS $\times 10^{-3}$ M	
6.40	542.0	332.0
6.45	710.0	430.0
6.50	543.0	231.0
	CTAB $\times 10^{-4}$ M	

7.18	543.0	231.0
7.20	710.0	430.0
7.22	541.0	214.0
pH		
13.19	542.0	243.0
13.20	710.0	430.0
13.21	554.0	232.0
Light Intensity (LT) = 10.4 $mWcm^{-2}$, Temp. = 303 K		

4.2 Variations in the Concentration of the xylose and their Effects

Variations in the system's xylose concentration had an impact on the PG cell's electrical output. Because there were fewer reducing agent particles available to transmit electrons towards the photosensitiser (dye-MB) component, a lower concentration of the reducing agent led to a decrease in electrical output (Table 79.1). At high reducing agent concentrations, the electrical output decreased because the dye molecules could not reach the electrode in the allotted time due to the enormous quantity of reducing agent molecules.

4.3 Effects on the System when Varying the Concentration of a Mixed Surfactant (NaLS+CTAB)

When the concentration of the combined surfactants, NaLS and CTAB, varied, so did the PP and PC of a PG cell contain an MB–xylose-NaLS+CTAB system. In one instance, the amount of CTAB changed but the amount of NaLS remained constant. In the other scenario, the NaLS concentration fluctuated but the CTAB concentration remained fixed. In both situations, a maximum was discovered for a specific CTAB and NaLS saturated value, beyond which the PG cell's electrical output decreased (Table 79.1).

4.4 Effects of pH Fluctuation

The pH changes in the system had an impact on the PG cell's electrical output. The highest was attained at pH 13.20. The PP and PC decreased as the pH increased further. Thus, it was discovered that PG cells employing the method were extremely sensitive to the pH of the mixtures. The optimal pH was found to be higher than the pKa value (pH > pKa), indicating a correlation between the reductant's pKa and pH. The reductant's easy availability in its anionic form which is particularly a better donor form could be one factor.

4.5 Effects of Diffusion Length

H-shaped cells of various sizes have been used to investigate how changing the diffusion length—the distance between the two electrodes—affects the cell's power parameter (imax). Research has demonstrated that during the initial minutes of illumination, the PC rises significantly. Consequently, when the diffusion length grows, the maximum PC (imax) also increases; however, tests do not demonstrate this (Table 79.2). Therefore, the main electroactive species can be regarded as the leuco or semi-leuco version of the dye (photosensitiser: MB) and the colorants in the dark and illuminated chambers, respectively. The reducing agent and its oxidation byproduct only act as electron carriers along the route.

Table 79.2 Diffusion length's impact

DL (mm)	Max PC i_{max} (μA)	Equi PC i_{eq} (μA)	Rate of initial generation of PC (μA min^{-1})
40.0	321.0	198.0	6.01
45.0	430.0	221.0	6.12
50.0	319.0	199.0	6.10
MB = 4.0 × 10^{-5} M, CTAB = 7.2 × 10^{-4} M, Temperature = 303 K, pH = 13.20, NaLS = 6.40 × 10^{-3} M, and xylose = 2.0 × 10^{-3} M			

4.6 Current–voltage (i-V) Characteristics

The FF was calculated using the formula that follows to determine the power point (pp), a point on the i-V curve where the combined value of current and voltage is at its maximum:

$$FF(\eta) = \frac{V_{pp} \times i_{pp}}{V_{oc} \times i_{sc}} \quad ...(1)$$

The fill factor (η) was found to be 0.2789, and the system's cell power point (pp) was found to be 67.56 μW. The typical rectangular topologies of the current-voltage (i-V) trend were observed (Fig. 79.2).

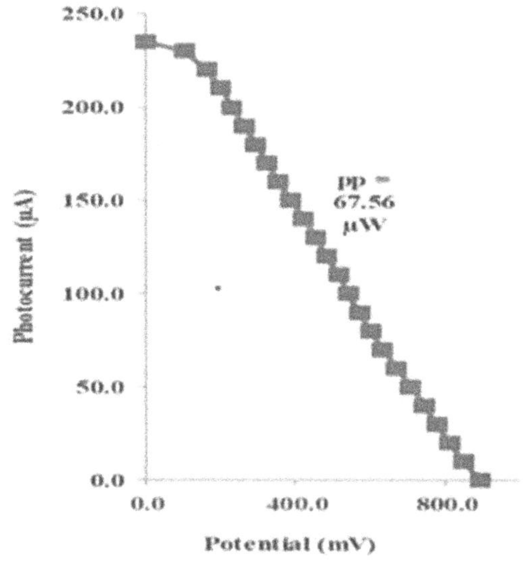

Fig. 79.2 Current voltage (i-v) curve of cell

4.6 Cell Performance and CF

By providing an external load which is required to maintain current at power point and shutting off the illumination as soon as it may have stabilised, researchers assessed the photogalvanic cell's performance. $T_{0.5}$, or the time required for the final result (power) to drop to half at the peak of power in the dark, was used to compute the performance (Fig. 79. 3). The cell's ability to function in the dark for 92 minutes. The cell's CF, calculated using the formula below, was 2.1213%:

$$CE = \frac{V_{pp} \times i_{pp}}{A10.4mWcm^{-2}} \times 100\% \quad ...(2)$$

where A, i_{pp}, and V_{pp} stand for electrode area, PP at power point, and PC at power point, respectively.

4.7 Effect of Electrode Area

It has also been investigated how the electrode area (EA) affects the cell's current characteristics. It was revealed that the value of maximum PC (i_{max}) increased as the electrode area increased up to optimum condition followed by decreased, as reported (Table 79.3).

Table 79.3 The impact of electrode area

PG cell system					
EA	0.70	0.85	1.00	1.15	1.30
i_{max}	228.0	321.0	440.0	320.0	221.0
i_{eq}	197.0	212.0	330.0	231.0	198.0
[Xylose] = 2.0×10^{-3}M, MB = 4.0×10^{-5} M, and LT = 10.4 mW cm^{-2}, Temp = 303 K, pH = 13.20, CTAB = 7.2×10^{-4} M, NaLS = 6.40×10^{-3} M					

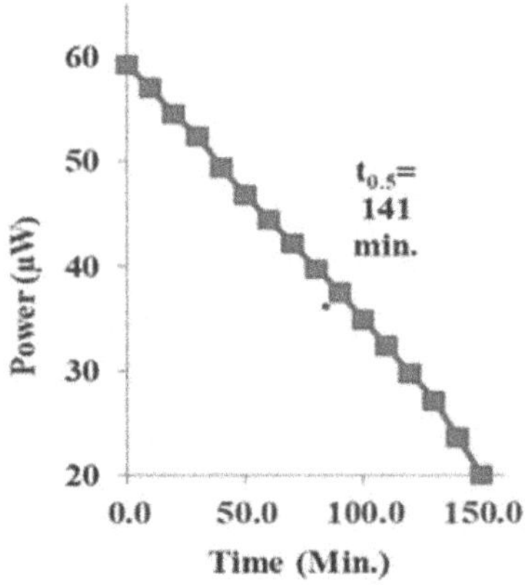

Fig. 79.3 Performance of the PG cell

5. REACTION MECHANISM

The following mechanism for the PC production in the PG cell may be suggested based on the aforementioned investigations:

Illuminate Chamber

$$Dye \xrightarrow{hv} Dye^* \quad ...(3)$$

$$Dye^* + R \longrightarrow Dye^- + R^+ \quad ...(4)$$

$$Dye^- \longrightarrow Dye + e^- \quad ...(5)$$

$$Dye + e^- \longrightarrow Dye \quad ...(6)$$

$$Dye^- + R^+ \longrightarrow Dye + R \quad ...(7)$$

In this case, the dye (MB) is represented by the variable's dye, dye$^-$, respectively. R represents the reductant, while R^+ is the oxidised version.

6. CONCLUSION

Scientists are now interested in PG cell because of the good results of conversion of solar energy. The cell charges only when there is an illuminating source present. The cell only discharges when an external circuit is used for electron transport. If there is no external circuit, the cell will keep storing light energy. In contrast to solar-PV cells, which require additional hardware such as batteries for energy storage, PG cells have built-in storage capacity and can use stored energy even in the absence of light. For surfactant-based PG cells, obtained electrical results as conversion efficiency, cell performance and fill factor are 2.1213%, 141.0 min, and 0.2789, respectively. Conclusively, surfactant-based PG cells are key role for solar energy conversion and storage.

6.1 Photogalvanic Abbreviations

PC stands for photogalvanic, PV for photovoltaic, MB for methylene blue, R for reducer, S for surfactant, ieq for PC at equilibrium, imax for maximum PC, PP for photopotential, PC for photocurrent, ipp for PC at power point, ml for millilitre, FF(for fill factor, t0.5 for cell performance, Voc for open circuit voltage, Vpp for PP at power point, VOC for open circuit voltage, CE for conversion efficiency, SPP for solar PP, DL for diffusion length, SPC for solar PC, SP for solar power, SFF for solar fill factors, and SCE for solar conversion efficiency.

6.2 Acknowledgment

Authors are thankful to Prof. V. Gupta, Head, Department of Chemistry, J.N.V. University, Jodhpur for necessary research facilities. One of the authors, Dr Mohan, is especially grateful to Prof. PP Solanki, BHU, Varanasi for critical and scientific analysis during the research work.

REFERENCES

1. Bhimwal, M. K., Gangotri, K. M. and Bhimwal, M. K. (2013). A comparison of conversion efficiencies of various sugars as reducing agents for the photosensitizer eosin in the photogalvanic cell. Int. J. Energy Res. 37(3):250-258.

2. Gangotri, K. M. and Indora, V. (2010). Studies in the photogalvanic effect in mixed reductants system for solar energy conversion and storage: Dextrose and ethylenediaminetetraacetic acid–Azur A system. Sol Energy. 84(2):271-276.
3. Gangotri, K.M. and Mahawar, A.K. (2012). Comparative study on effect of mixed photosensitizer system for solar energy conversion and storage: Brilliant cresyl blue + toluidine blue–ethylene glycol–NaLS system. Environ. Prog. Sustain. Energy. 31:474-480.
4. Gangotri, K.M., Meena, R.C. and Meena, R. (1999). Use of miscelles in photogalvanic cells for solar energy conversion and storage: Cetyl trimethyl ammonium bromide glucose-toluidine blue system. J of Photochem. and Photobio. A: Chem. 1-3:93-97.
5. Gangotri, K. M., and Meena, J. (2006). Role of surfactants in photogalvanic cells for solar energy conversion and storage. Energy sources, 28(8):771-777.
6. Gangotri, K.M. and Mohan, L. (2013). Study of photogalvanic effect in photogalvanic cell containing mixed surfactant (NaLS+CTAB). Res. J. Chem. Sci. 3(3):20-25.
7. Genwa, K.R. and Sagar, C. P. (2013). Energy efficiency, solar energy conversion and storage in photogalvanic cell. Energy Convers. Manage. 66:121-126.
8. Jayshree, R., Kumar, RK., Pratibha, S. and Mohan, L. (2022). Study of photogalvanic cell for electrical output in solar energy conversion and storage: single surfactant as lauryl glucoside, Tartrazine as a photosensitizer and D-fructose as reductant. Res. J. Chem. Environ. 26(06): 24-29.
9. Koli, P., Kumar, R., Dayma, Y., Pareek, R. K., Meena, A. and Jonwal, M. (2022). Tropaeline O-oxalic acid-benzalkonium chloride photogalvanic cells for solar energy conversion and storage. Battery Energy. 1(4): 20220011.
10. Lal, C. and Gangotri, K.M. (2011). Energy conversion and storage potential of photogalvanic cell based on mixed dyes system: Ethylene diaminetetraacetic acid-Toluidine Blue-Thionine. Environ. Prog. Sustain. Energy. 30:754-761.
11. Lal, M. and Gangotri, K.M. (2013). A Comparative study on the performance of photogalvanic cells with mixed surfactant for solar energy conversion and storage: D-Xylose-Methylene Blue systems. Res. J. Recent Sci. 2(12):19-27.
12. Lal, M. and Gangotri, K.M. (2022). Innovation for prospective energy source through solar cell. J Solar Energy Res. 7(3):1095-1103.
13. Lal, M. and Gangotri, K.M. (2022). Innovation in progressive study for prospective energy source through photo-galvanic-system: D-Xylose+MB+Brij-35+NaLS. Int. J. Energy Res. 46(14):19538-19547.
14. Lal, M. and Gangotri, K.M. (2023). Innovative study in renewable energy source through mixed surfactant system for eco-friendly environment. Environ. Sci. Pollut. Res. 30:98805-98813.
15. Lal, M. and Gangotri, K.M, (2025). Comparative study on sunlight induced surfactants system in photogalvanics for solar energy conversion and storage. Next Sustainability. 6:100101.
16. Madhwani, S., Ameta, R., Vardia, J., Punjabi, P.B. and Sharma, V.K. (2007). Use of Fluoroscein–EDTA system in photogalvanic cell for solar energy conversion. Energy Sources Part A Recov Utiliz Environ Effects. 29:721-729.
17. Mall, C. and Solanki, P.P. (2018). Spectrophotometric and conductometric studies of molecular interaction of brilliant cresyl blue with cationic, anionic and non-ionic surfactant in aqueous medium for application in photogalvanic cells for solar energy conversion and storage. Energy Rep. 4:23-30.
18. Rabinowitch, E. (1940). The Photogalvanic Effect I: The Photochemical Properties of the Thionine-Iron System. J. Chem. Phys. 8(7):551-559.
19. Rathore, J., Arya, RK., Sharma, P. and Lal, M. (2022). Study of electrical output in photogalvanic cell for solar energy conversion and storage: Lauryl glucoside-tartrazine-D-fructose system. Ind. J. Sci. Tech. 15(23):1159-1165.
20. Rideal, E.K. and Williams, E.G. (1925). The action of light on the ferrous iodine iodide equilibrium. J. Chem. Soc. Trans. 127:258-269.

Note: All the figures and tables in this chapter were made by the authors.

Recent Trends in Applied Physics and Material Science – Sudhir Bhardwaj et al. (eds)
© 2026 Taylor & Francis Group, London, ISBN 978-1-041-16452-4

Role of Interaction on the Performance of a Coupled Diffusion Brownian Motor

Ronald Benjamin*
Cochin University of Science and Technology, Cochin, Kerala, India

Abstract: Brownian motors rectify thermal fluctuations in the presence of broken symmetry under non-equilibrium conditions. In this work, we study the effect of coupling between multiple Brownian particles on the performance of Brownian motors subjected to the periodically oscillating temperature of the thermal bath. We observe that coupling enhances the motor's transport coherence while degrading its thermodynamic efficiency.

Keywords: Brownian Motor, Langevin Equation, Thermal fluctuations, Stochastic Energetics

1. INTRODUCTION

Thermal fluctuations are generally considered a nuisance to be avoided. However, research in the last three decades has shown that noise can also play a beneficial role, as in the case of Brownian motors Reimann (2002). These tiny motors move unidirectionally when driven out of equilibrium, provided the symmetry of system is broken.

There are several types of Brownian motors or ratchets differing in terms of the sources of external energy that drive the system out of equilibrium. Among them are the rocking and flashing ratchets and autonomous ratchets such as Feynman and Büttiker-Landauer ratchets Benjamin (2008, 2024). Another type of Brownian motor is the diffusion ratchet in which the temperature of the heat bath changes with time either in a deterministic and time-periodic manner or in a random manner Bao (2000), Reimann (1996).

In studies on Brownian motors, a lot of effort is being expended to obtain criteria for their effective performance as characterized by how regular or coherent the motion of the motor is as well as by the thermodynamic efficiency of the motor i.e. delivering maximum power output with minimal cost of energy input.

While several studies have investigated various transport and thermodynamic characteristics of single-particle diffusion Brownian motor the effect of coupling of multiple Brownian particles on the performance of the diffusion Brownian motor has not been investigated in detail. In the case of some types of motors such as rocking and flashing ratchets, coupling among particles can enhance the current and the motor's coherence and thermodynamic efficiency Wang (2007, 2005), Igarashi (2001), Linke (2005). The coupling of multiple Brownian particles has different effects in different types of Brownian motors and does not enhance or diminish the performance uniformly across the various types of motors.

In this work, we have studied how coupling affects the transport and thermodynamic performance of a coupled diffusion Brownian motor by numerically solving the overdamped Langevin equation. We observe that there are optimal values of the coupling strength at which the transport coherence of the motor reaches a peak. At low temperatures, it is also seen that coherence saturates to a maximum value when the number of motors become large. However, collective effects induced by the coupling of multiple Brownian particles degrade the thermodynamic efficiency of the motor.

The next section describes the model and in the subsequent section we present our results before ending with a conclusion.

*Corresponding author: rbenjamin.phys@gmail.com

DOI: 10.1201/9781003684718-80

2. MODEL

We consider N_m linearly coupled overdamped Brownian particles moving in a sawtooth potential U(x). The dynamics of the system evolves according to the overdamped Langevin equation given by,

$$\gamma\dot{x}_i = -V'(x_{ij}) - U'(x_i) - F_L + \sqrt{2k_B\gamma T(t)}\xi(t) \quad (1)$$

where $V(x_{ij}) = (1/2)k(x_i - x_j - a)^2$ is the harmonic potential through which only the nearest-neighbour particles i and j interact, k being the coupling constant and a is the equilibrium distance between the particles. F_L is the load force acting on every Brownian particle. The potential U(x) is a sawtooth potential given by,

$$\begin{aligned} U(x) &= x/\alpha & 0 \le x < \alpha \\ U(x) &= (L-x)/(L-\alpha) & \alpha \le x < L \end{aligned} \quad (2)$$

The spatial period of U(x), L=1, and α is the asymmetry parameter which has been taken to be 0.1 in all our calculations. We have taken the friction coefficient $\gamma=1$ and the Boltzmann constant $k_B=1$. $\xi(t)$ is Gaussian white noise whose mean is zero and variance is one. The temperature T(t) is time-dependent and switches between T_H and T_C ($T_H > T_C$) in a time-periodic manner with period τ as per the following protocol:

$$\begin{aligned} T(t) &= T_H & n\tau \le t < n\tau + \tau_1 \\ T(t) &= T_C & n\tau + \tau_1 \le t < (n+1)\tau \end{aligned} \quad (3)$$

The source of energy of the motor is when the heat bath is at the larger temperature T_H. We calculate the current <v> and effective diffusion coefficient D_{eff} to determine the peclet number, Pe, as well as the energy input and power output to obtain the thermodynamic efficiency. These quantities are calculated numerically from the following equations:

$$\langle v\rangle = \frac{1}{N_m}\sum_{i=1}^{N_m} \lim \frac{\langle x_i(t)\rangle - \langle x_i(t_s)\rangle}{t - t_s} \quad (4)$$

$$D_{eff} = \frac{1}{N_m}\sum_{i=1}^{N_m} \lim \frac{\langle (x_i(t))\rangle^2 - \langle (x_i(t))\rangle^2}{2t} \quad (5)$$

with the Peclet number given by, Pe=|<v>| L/D_{eff}. When Pe>2, transport is said to be coherent.

The thermodynamic efficiency $\eta = \frac{\dot{W}}{\dot{Q}_H} = \frac{F_L\langle v\rangle}{\dot{Q}_H}$, where

$\dot{W}$ is the work done by the motor which is the power output and $\dot{Q}_H$ is the energy input from the bath at T_H and is given by,

$$\dot{Q}_H = \frac{1}{N_m}\frac{\sum_{i,j=1,j\neq i}^{N_m}\int_{n\tau}^{n\tau+\tau_1}\left\langle \left(U'(x_i) + V'(x_{ij}) + F_L\right).dx_i(t)\right\rangle}{\tau} \quad (6)$$

We solve Eq. (1) numerically using the stochastic Euler algorithm. We use a time-step of 10^{-3} and the number of realizations has been taken to be 1000.

3. RESULTS

In Fig. 80.1, we plot <v>, D_{eff} and Pe versus the coupling constant k at different values of N_m. While the current decreases gradually as the coupling strength increases D_{eff} sharply decreases significantly as soon as the particles get coupled with each other as they begin to move in tandem leading to less dispersion. Upon further increase of k, D_{eff} decreases slowly. It is clear that the coherence reaches a peak (Pe>2) at an optimal value of k. It is also seen that the Peclet number monotonically increases with an increase in N_m at all values of the coupling constant k.

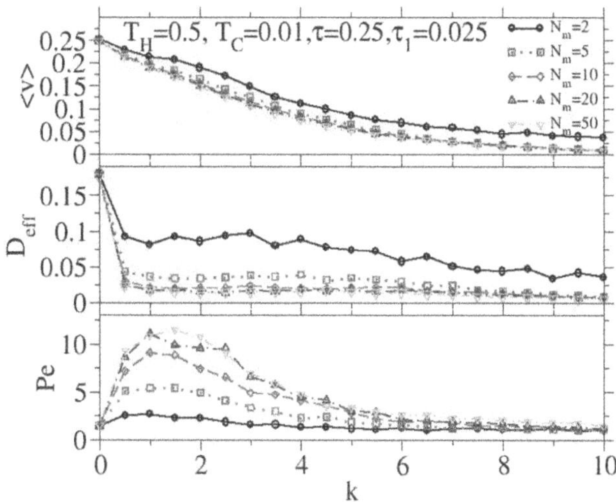

Fig. 80.1 <v>, D_{eff} and Pe as a function of k. Here, a = 0.025

Figure 80.2 shows $\dot{W}$, $\dot{Q}_H$, and η as a function of k. The thermodynamic efficiency monotonically decreases with both k and N_m. As coupling increases after an initial dip the

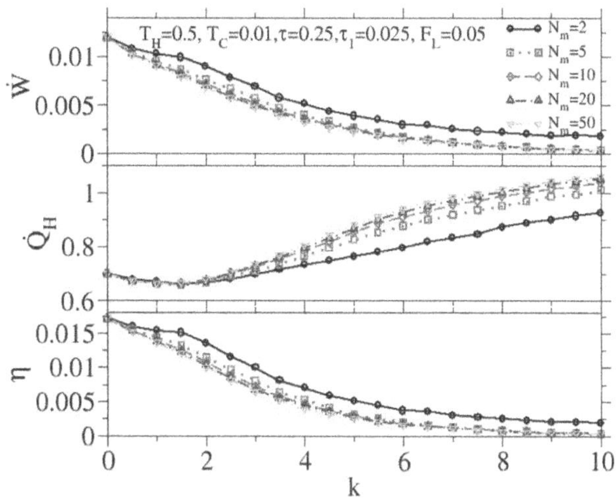

Fig. 80.2 Power output, energy input, and the efficiency as a function of the coupling constant. Here, a = 0.025

Brownian particles consume greater energy from the hot bath while the work output decreases. We have considered other parameter values but have not found a regime where the efficiency is enhanced.

4. CONCLUSION

In this work, we investigated the performance of a Brownian motor driven by the time-dependent temperature of the thermal bath. We observe that in certain parameter regimes, while the motor becomes more coherent as a function of coupling, the interaction between Brownian particles degrades its efficiency. There are several directions along which this study can be extended such as considering non-linear instead of linear coupling, a continuous change of temperature with time rather than oscillating between two values. The underdamped regime may also yield some interesting behaviour. Efforts along these directions are currently underway.

REFERENCES

1. Reimann P (2002). Brownian Motors: Noisy Transport far from Equilibrium, Physics Reports 361 (2-4): 57-265
2. Benjamin R. and R, Kawai (2008). Inertial Effects in the Büttiker-Landauer Motor and Refrigerator in the overdamped limit, Physical Review E 77 (5): 051132
3. Benjamin R. (2024). Noise-induced transport in a periodic square-well potential. Physica Scripta 99 (9): 095257
4. Bao, Jing-Dong (2000). Efficiency of energy transformation in an underdamped diffusion ratchet, Physics Letters A 267 (2-3): 122-126
5. Reimann *P. et al.* (1996). Brownian motors driven by temperature oscillations, Physics Letters A 215 (1-2): 26-31
6. Wang, Hai-Yan and Jing-Dong Bao (2007). Transport coherence in coupled Brownian ratchet, Physica A 374 (1): 33-40
7. Wang Hai-Yan and Jing-Dong Bao (2005). Cooperation behavior of transport process in coupled Brownian motors. Physica A 357 (3-4): 373-382
8. Igarashi, A., Tsukamoto, S. and Goko, H. (2001). Transport properties and efficiency of elastically coupled Brownian motors (2001). Physical Review E 64 (5): 051908
9. Linke, H., Downton, M. and Zuckermann, M. (2005). Performance characteristics of Brownian Motors Chaos 15 (2): 02611.

Note: All the figures in this chapter were made by the authors.

Recent Trends in Applied Physics and Material Science – Sudhir Bhardwaj et al. (eds)

The Fabrication and Characterization of Zinc Phosphate

Dipa Patel[1], Sudheer Lingayat[2], Brahmdev Singh[3]
Department of Chemistry, Khalsa College, University of Mumbai, India

Abstract: Zinc acetate was treated with phosphoric acid and by using the chemical precipitation method, zinc phosphate particles were obtained. Without using chemical hydrazine, the approach aims to synthesize zinc phosphate in the simplest possible way in terms of particle dimensions. To identify the synthesized substance and reveal the information or materials for which it is designed, several characterization procedures were employed. X-ray Diffraction, Ultraviolet-Visible, Scanning Electron Microscopy, Raman Spectroscopy and Fourier Transform Infrared were utilized to analyze the synthesized particle. The FTIR spectra are used to analyze the stretching and bending frequencies of the molecular functional groups in the material. When using XRD, the morphology of the zinc phosphate particles was identified. The obtained product Zinc (II) Phosphate has well-crystallized particles, as seen in the SEM image. The UV spectrum notifies the different electronic excitation states of Zinc (II) Phosphate. The Raman spectra's peaks reveal the existence of zinc phosphate, which was synthesized by combining zinc acetate with phosphoric acid.

Keywords: Zinc Phosphate, XRD, IR, SEM

1. INTRODUCTION

One of the most significant multifunctional metal phosphates due to its intriguing characteristics and numerous uses in a variety of industries is zinc phosphate (ZP), which is not lethal. ZP is frequently used as an anticorrosive pigment, to block anodic and cathodic corrosion reactions on metal. When ZP is present, the drying qualities of the paint layer and its adhesion to the metal substrate can be increased. ZP is expected to be beneficial as white pigments in cosmetics because of its well-known receptivity for living things or organisms. Furthermore, the catalytic qualities of zinc phosphate have been examined in the procedure of hydrocarbon conversion, including the conversion of methanol.

2. EXPERIMENTAL DETAILS

S. Ramesh and *et al*, developed a method of preparation of ZP using precipitation method by taking stoichiometric amounts of zinc acetate and phosphoric acid, followed by addition of hydrazine hydrate. The chemicals were heated at 105°C, till precipitation is complete for an hour. The obtained precipitate was washed with water to achieve pH=3.The precipitate undergoes filtering and then dried in oven at 150 °C to obtain the desired product ZP.

3. CHARACTERIZATION METHODS CONDUCTED

EMPYREAN X-ray Diffractometer system (anode material-Cu, Ka wavelength 1.540598 nm) was used. The morphology of the sample was performed on SEM Carl Zeiss Model Supra 55 Germany instrument. The absorbance mode was practiced to obtain UV spectra in the range of wavelengths of 190 to 900 nm. A Perkin Elmer FTIR spectrophotometer was used to detect an infrared spectrum (FTIR) between 4000 and 400 cm^{-1} .Raman spectrum was recorded using Horiba Japan Xplora Plus for chemical and structural characterisation.

3.1 XRD

Figure 81.1 shows a XRD of ZnP. According to XRD investigations, ZnP is found to be crystalline and nanosized.

[1] gala.dipa000@gmail.com, [2] sklingayat@gmail.com, [3] singhbrahmdev1994@gmail.com

DOI: 10.1201/9781003684718-81

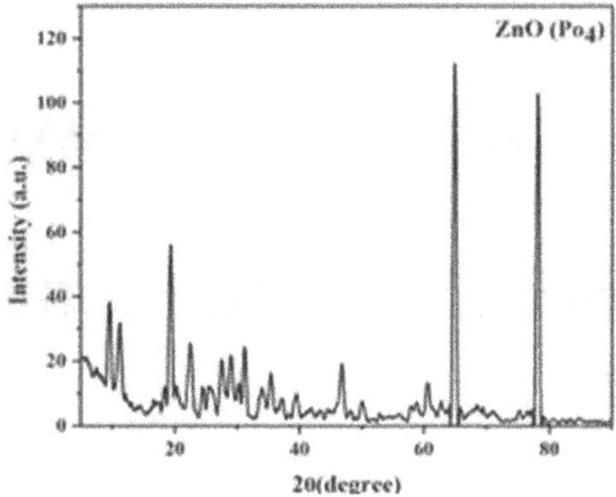

Fig. 81.1 XRD pattern of ZP

3.2 SEM

Figure 81.3 shows a SEM a microscope image of the developed Zinc Phosphate, indicating that the particles have an irregular shape and are agglomerated in nature. The original picture is superimposed on each of the maps for Zn, P, and O shown in Figure 81.3. The focal elements in the examined field are Zn, P, and O, as seen in the EDX spectrum (Fig. 81.2). SEM images of zinc(II) phosphate nanoparticles show well-crystalline pieces with layered crystallized morphology.

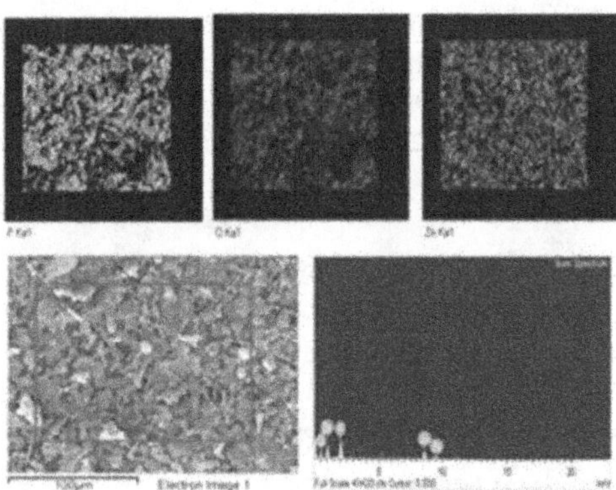

Fig. 81.2 Elemental maps of P, O, Zn, electron image and EDX analysis

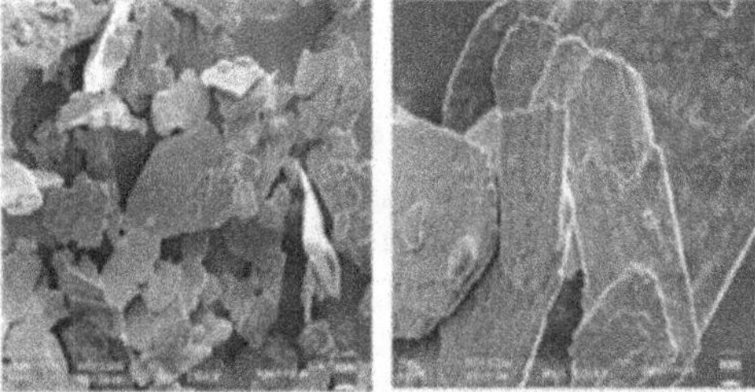

Fig. 81.3 SEM image at 2.00 kx and 10.00 kx

3.3 UV-Vis

Figure 81.4 illustrates the UV-vis reflectance spectra of zinc phosphate. Peaks encountered between 200 and 250 nm (UV area) are the major characteristics and are associated with phosphate-related electrical transitions.

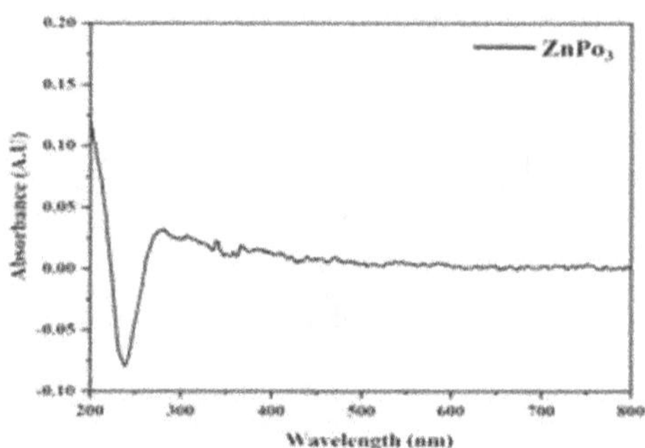

Fig. 81.4 UV spectra of ZP

3.4 FTIR

Figure 81.5 depicts ZP's FTIR Spectra. The asymmetric stretching vibrations of the P-O bonds in the phosphate group are the cause of the peak that occurs between 1050 and 1100 cm^{-1}. It is implied that the phosphate group has symmetric stretching vibrations by the peaks at 924 cm^{-1} and 1002.7 cm^{-1}. Because of the asymmetric stretching of the phosphate group, the absorption band is located between 1056.7 and 1082.8 cm^{-1}. There are peaks at 550–650 cm^{-1} that are caused by the phosphate group bending in-plane. The peak at 3363 cm^{-1} indicates that the OH group is present.Thus FTIR confirms successful synthesis of ZP.

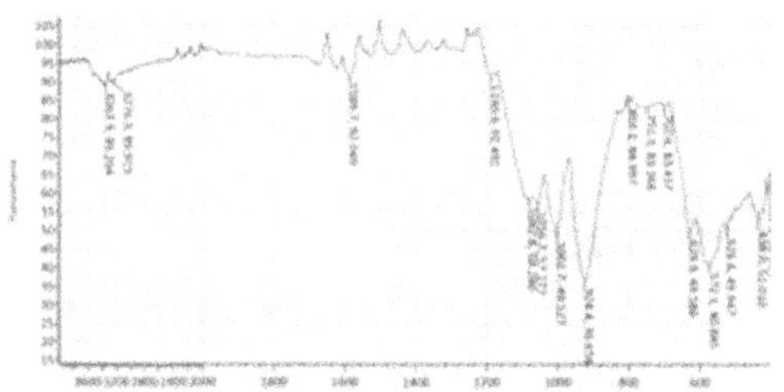

Fig. 81.5 FTIR spectra of ZP

3.5 Raman

Peaks in the 400–500 cm^{-1} range which are displayed in Figure 81.6 of Raman spectra of ZnP correspond to the

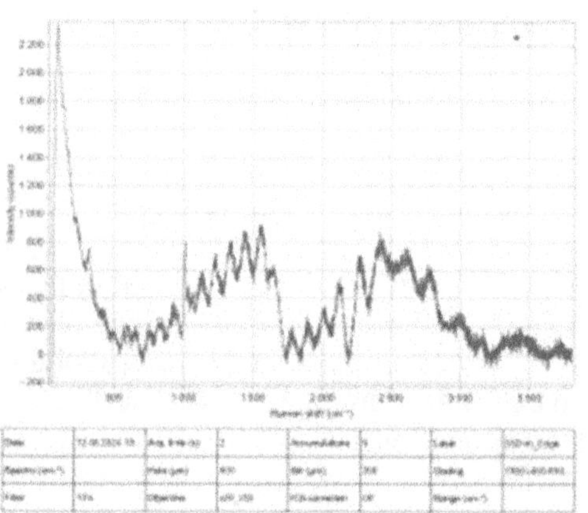

Fig. 81.6 Raman spectra of ZP

PO_4^{3-} group's bending modes (v_2 and v_4).These lower wavenumber peaks between 100 and 200 cm^{-1} might be caused by zinc-related metal-oxygen bond strains or lattice vibrations. The presence of phosphate ions is shown by the peak at 1007 cm–1. Peaks between 1500 and 1650 cm^{-1} show a wide O-H stretching peak.

4. CONCLUSION

XRD, SEM, FT-IR, and UV-Vis, Raman analysis methods were used to evaluate the morphology and optical properties of zinc phosphate nanoparticles generated using precipitation approach. Results indicate that the synthesized zinc phosphate is successfully synthesized without the use of hydrazine hydrate.

REFERENCES

1. Jadhav AJ, Holkar CR, Pandit AB and Pinjari DV (2016). Intensification of Synthesis of Crystalline Zinc Phosphate (Zn3(Po4)2) Nanopowder: Advantage of Sonochemical Method Over Conventional Method. Austin Chem Eng. 3(2): 1028.
2. S. Ramesh, V. Narayanan (2014). Synthesis and Characterisation of Zinc Phosphate Nanoparticles by Precipitation Method. (IJIRSE).
3. Jian Dong Wang, Da Li, Jin Ku Liu, Xiao Hong Yang, Jia Luo He, Yi L. (2011). One- Step Preparation and characterization of Zinc Phosphate Nanocrystals with Modified Surface. Soft Nanoscience Letters. 1:81-85.
4. Tbib Bouazza, Eddya M., Abdelali L. and El Moutarajji A. (2018). Studies on obtaining Zinc phosphate powders. IJREI. 2(1):29-32
5. Tonmoye S., Md. Abdur R. Hasan A.,and Md. Rabiul K. (2022).Recent Progress in Synthesis and Applications of Zinc Phosphate Nanoparticles: A Review. Journal of Nano Research, 73:59-88.

Note: All the figures in this chapter were made by the authors.

Recent Trends in Applied Physics and Material Science – Sudhir Bhardwaj et al. (eds)
© 2026 Taylor & Francis Group, London, ISBN 978-1-041-16452-4

Investigation of $RAlO_3$ (R = Gd, Nd, Sm, and Pr) Perovskites as Substrates for $YBa_2CuO_{7-\delta}$ Superconducting Films

Saji S. K.*
Government Polytechnic College, Attingal,
Thiruvananthapuram, Kerala, India

Abstract: The development of high-temperature superconducting films requires suitable substrate materials that ensure chemical stability and structural compatibility. This study investigates the potential of $RAlO_3$ (R = Gd, Nd, Sm, and Pr) perovskite oxides as substrates for $YBa_2CuO_{7-\delta}$ (YBCO) superconducting films. We evaluate the chemical reactivity between $RAlO_3$ and YBCO through X-ray diffraction measurements and magnetic moment analysis. Our results demonstrate that $RAlO_3$ perovskites exhibit good chemical stability with YBCO, making them promising candidates as substrates for YBCO superconducting films.

Keywords: Perovskites, Superconducting substrate materials, Rare-earth alumiantes

1. INTRODUCTION

High-temperature superconductors (HTSCs) have garnered significant attention due to their potential applications in various fields. However, the development of HTSC films requires suitable substrate materials that ensure chemical stability and structural compatibility. $YBa_2CuO_{7-\delta}$ (YBCO) is a prominent HTSC material, but its chemical reactivity with conventional substrate materials limits its applications. In this study, we investigate the potential of $RAlO_3$ (R = Gd, Nd, Sm, and Pr) perovskite oxides as substrates for YBCO superconducting films.

2. EXPERIMENTAL METHODS

YBCO superconductor powder (99.9% purity) was purchased from Sigma Aldrich Co. The X-ray diffraction (XRD) pattern of the nanocrystalline single-phase YBCO was recorded. The superconducting transition temperature of YBCO was determined by measuring the magnetic moment over a temperature range of 4 K to 300 K in a magnetic field of 20 Oe using a Physical Property Measurement System (PPMS - Quantum Design).

To evaluate the chemical reactivity between $RAlO_3$ and YBCO, equal volumes of YBCO powder and $RAlO_3$ were mixed and pressed into circular discs. The discs underwent heat treatment at 950°C in air for 15 hours. The XRD patterns of the YBCO-$RAlO_3$ composites were recorded and compared with those of pure $RAlO_3$ and YBCO compounds.

3. RESULTS AND DISCUSSION

The XRD patterns of the YBCO-$RAlO_3$ composites show no additional peaks, indicating no detectable chemical reaction between YBCO and $RAlO_3$ at the processing temperature. This suggests that $RAlO_3$ perovskites are chemically stable when in contact with YBCO superconductors.

The magnetic moment measurements confirm the superconducting nature of YBCO, with a transition temperature of 63 K. The absence of any significant changes in the XRD patterns and magnetic moment measurements indicates that $RAlO_3$ perovskites do not adversely affect the superconducting properties of YBCO.

*Corresponding author: sajisk@gmail.com

DOI: 10.1201/9781003684718-82

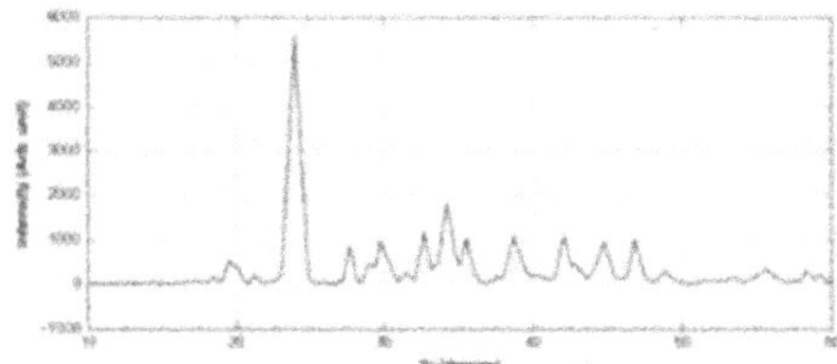

Fig. 82.1 XRD pattern of superconducting YBCO superconductor

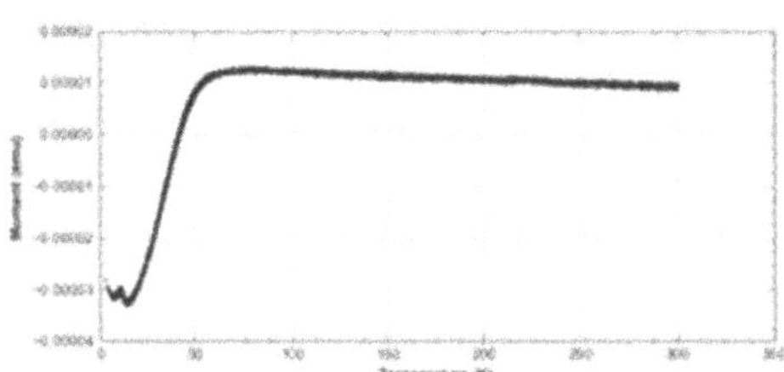

Fig. 82.2 Magnetic moment vs temperature for YBCO at H = 20 Oe

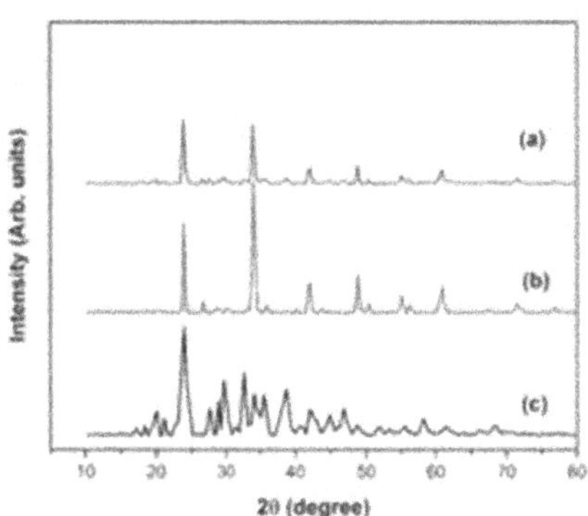

Fig. 82.3 XRD patterns of (a) GdAlO3 and YBCO composite heat treated at 9500C for12 h, (b) GdAlO3 and (c) YBCO

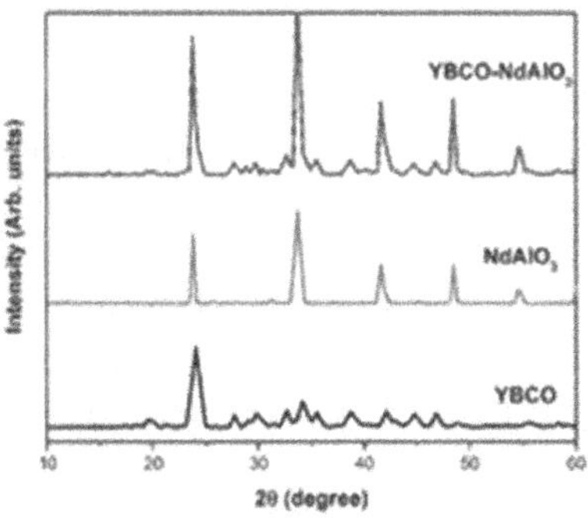

Fig. 82.4 XRD pattern of (a) NdAlO3 and YBCO composite heat treated at 9500C for 12 h, (b) NdAlO3 and (c) YBCO

4. CONCLUSION

In conclusion, our study demonstrates that $RAlO_3$ perovskites are chemically stable with YBCO

Superconductors, making them promising candidates as substrates for YBCO superconducting films. These findings

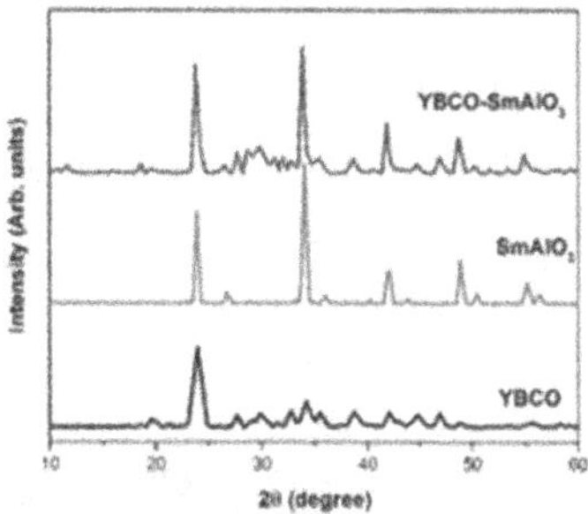

Fig. 82.5 XRD pattern of (a) SmAlO3 and YBCO composite heat treated at 9500C for 12 h, (b) SmAlO3 and (c) YBCO

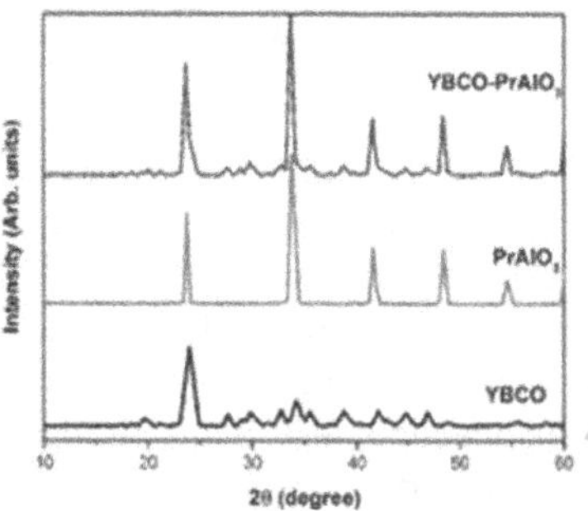

Fig. 82.6 XRD pattern of (a) PrAlO3 and YBCO composite heat treated at 9500C for 12 h, (b) PrAlO3 and (c) YBCO

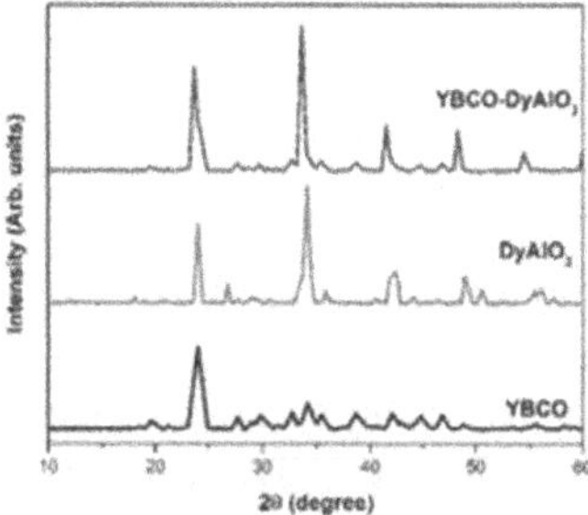

Fig. 82.7 XRD pattern of (a) DyAlO3 and YBCO composite heat treated at 9500C for 12 h, (b) DyAlO3 and (c) YBCO

have significant implications for the development of high-temperature superconducting films and devices. Further research is necessary to explore the potential of $RAlO_3$ perovskites as substrates for YBCO superconducting films and to optimize their properties for practical applications.

REFERENCES

1. Phillips, J.M., 1996. Substrate selection for high-temperature superconducting thin films. Journal of Applied Physics, 79(4), pp.1829-1848.
2. Paranthaman, M.P. and Izumi, T., 2004. High-performance YBCO-coated superconductor wires. *MRS bulletin*, *29*(8), pp.533-541.
3. Scanlan, R.M., Malozemoff, A.P. and Larbalestier, D.C., 2004. Superconducting materials for large scale applications. *Proceedings of the IEEE*, *92*(10), pp.1639-1654.

Note: All the figures in this chapter were made by the authors.

Recent Trends in Applied Physics and Material Science – Sudhir Bhardwaj et al. (eds)

Fractal Geometry, the Da Vinci of Mathematical Science

Bhakti Rajvaidya*
G H Raisoni College of Engineering, Nagpur, Maharashtra, India

Abstract: This paper presents a unique collection of fractal designs that simulate a diverse range of natural and man-made phenomena, from African and Indian masks to human body, face, and even bacterium structures. Paper demonstrates how these fractal designs can be used as a modern art form, transforming spaces such as homes, offices, and hospitals into calming environments. The intricate patterns and self-similarities of fractals have been shown to have a therapeutic effect, reducing anxiety, depression, and mental fatigue. Some fractal art pieces, inspired by Kundalini and Mandala arts, offer a unique blend of mathematical beauty and cultural significance. By integrating these fractal designs into modern spaces, author provides an innovative solution for promoting mental well-being and serenity. This interdisciplinary work bridges the gap between mathematics, art, and therapy, opening new avenues for the application of fractal geometry in enhancing human experience.

Keywords: Self-similarities, Fractal Design, Fractal Geometry

1. INTRODUCTION

Fractals are most fascinating version of mathematics. It can be called as a good-looking Mathematics. Since the first discussion of fractal, many mathematicians and computer designers try to mimic naturally occurring fractals through software. Tree Fractal is the best example of it. Nowadays even basic fractal generating apps are available to create varied range of fractals.

2. LITERATURE REVIEW

It is been more than four decades now since Banoit Mandelbrot (1967) suggested ingenious ways of measuring the coastline of Britain. Since then many different natural events are identifies as fractals and many computerized iterative mathematics have crested man made fractals. Fractals are the repetitive patterns over the increasing or decreasing scale, which makes a miraculously catchy impressions on human mind. Beside the varied uses of fractals as in the fields of medicines, biology, soil mechanics, physics, graphics, seismology, music, computer etc., it has also shown potential uses in textile industries which hiked after Mandelbrot set Julia sets got popular to be printed on the casual wears as studied and published by Weijie Wang et al. (2019), computer wallpapers, screen-savers, product packaging, product designs or even 3D printed porous structure which can be used further in water purifiers or soil management structure stated by A. S. Ullah et al. (2021). A charming use of Fractal art in ceramic pottery is also seen by Xingrui Lin and Wenming Liu (2019). Fractals have been increasingly used in art to create visually striking and thought-provoking pieces. Artists have leveraged the self-similar patterns and intricate details of fractals to explore themes of nature, complexity, and infinity by Boden (2010). Fractal art has also been used to represent the beauty and complexity of the human body, as seen in the works of artists like Scott Olsen (2013). Additionally, fractals have been used in architecture and design to create innovative and sustainable structures as stated by M. J. Winkler (1993). The use of fractals in art has also been shown to have a therapeutic effect, reducing anxiety and promoting relaxation which is depicted in the work of R. Sarhangi (2008). Overall, the artistic use of fractals has opened up new possibilities for creative expression and visual exploration.

*Corresponding author: bhakti.patankar@raisoni.net

DOI: 10.1201/9781003684718-83

3. METHODOLOGY AND MODEL SPECIFICATIONS

Here few such fractals are designed mimicking some more interesting real life phenomenon. The fractals are created using iterative program which iterates for particular number of times till the required image is produced. The image on the left side is the starting image and the image on the right is final image after few iterations. All these fractals can be classified as fractals which end up on a proper patterned fractal which even after thousands of iterations will generate same pattern shown in next figures in Fig. 83.1.

From that examples, it is clear that these patterns are fractals which but can be used as space filling iterative patterns, but are mesh or net kind of structures. Below is the discussion on few fractals which are created randomly but found to mimic certain physical phenomenon which never have been cited ever. One such fractal is mask fractals. Masks are famous artefacts used as a home decor items, on similar grounds such fractal image can be used in the form of painting to replace real costly artefacts. Below shown the starting image and the final image after some finite iterations and the real image it is mimicking, here it is important to note that more iteration may lead to blacking out of the complete pattern and hence these patterns can be used as space fillers.

On the similar ground animal mask from African Artefacts can also be mimicked using fractal geometry. Not only the Artefacts but also some of the well-known paintings can also be mimicked using fractal iterative geometry. It is proven earlier, that watching fractals can reduce anxieties or stress, these fractal art paintings can also be used. As per Ancient Indian Knowledge systems, Meditation and Kundalini awakening is very helpful to maintain a healthy and peaceful lifestyle. Many Yoga and meditation centres use such paintings to create an aesthetic according to the purpose. Fractal iterations can produce image mimicking such human body and aura depicting images, one such is shown in Fig. 83.2.

Not only the Artefacts but also some of the well-known paintings can also be mimicked using fractal iterative geometry. It is proven earlier, that watching fractals can reduce anxieties or stress, these fractal art paintings can also be used. As per Ancient Indian Knowledge systems, Meditation and Kundalini awakening is very helpful to maintain a healthy and peaceful lifestyle. Many Yoga and meditation centres use such paintings to create an aesthetic according to the purpose. Fractal iterations can produce image mimicking such human body and aura depicting images, one such is shown in (Fig 83.3) plots (g) to (i). Fractals are also capable of mimicking many other day to day phenomenon. Below shown are few fractals which can mimic the structures of some random physical phenomenon.

A simple variation in the starting image (a) can lead to some different concluding fractal which resembles human moustache and A slight angular variation in the above starting image will lead to crown shaped fractal structure after few iterations as shown in Fig. 83.4.

4. CONCLUSION

Fractals have a remarkable ability to mimic a wide range of natural objects and structures, from the intricate branching of trees and river networks to the complex patterns of coastlines and mountain ranges. The self-similar and scaling properties of fractals allow them to capture the essence of natural forms, such as the spiral arrangements of seeds in a sunflower, the flow of water in a river delta, or the structure of snowflakes and crystals. Fractals can also replicate the patterns found in biological systems, like the branching of blood vessels, the growth of coral reefs, or the arrangement of leaves on a stem. Addition-ally, fractals can even model the behaviour of complex systems, such as weather patterns, population growth, and financial markets, making them a powerful tool for understanding and simulating the natural world. This versatility has led to fractals being used in various fields, including art, architecture, engineering, and science, to generate realistic models and simulations of natural phenomena.

5. RESULT

The ability of fractals to mimic various natural objects and structures has far-reaching potential for future applications.

By simulating the intricate patterns and behaviours of nature, fractals can help us better understand and predict complex phenomena, leading to breakthroughs in fields like environmental science, materials engineering, and

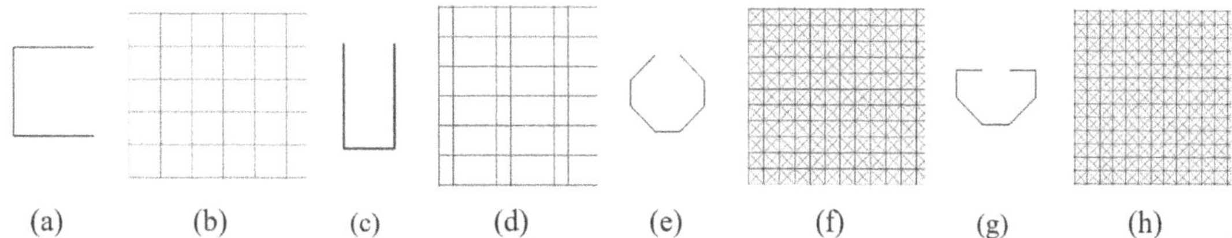

Fig. 83.1 Each figure on the left (a), (c), (e), (g) are the staring images of the iterations while each on the right (b), (d), (f), (h) figures on the right are the final images generated after several iterations (like infinite)

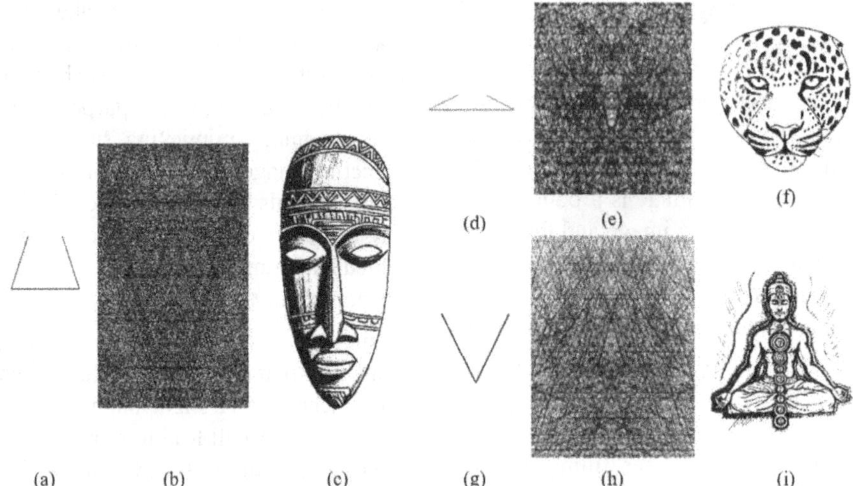

Fig. 83.2 (a) Figure on the left most is the starting images of the iterations while (b) Fractal after few finite iterations which depicts mask like structure in the middle (c) actual mask artefact (d) Figure on the left most is the starting images of the iterations while (e) Fractal after few finite iterations showing animal like image in the middle (f) actual mask artefact of animal, (g) Figure on the left most is the starting images of the iterations while (h) Fractal after few finite iterations showing human body and aura like image in the middle (i) actual painting of kundalini meditation

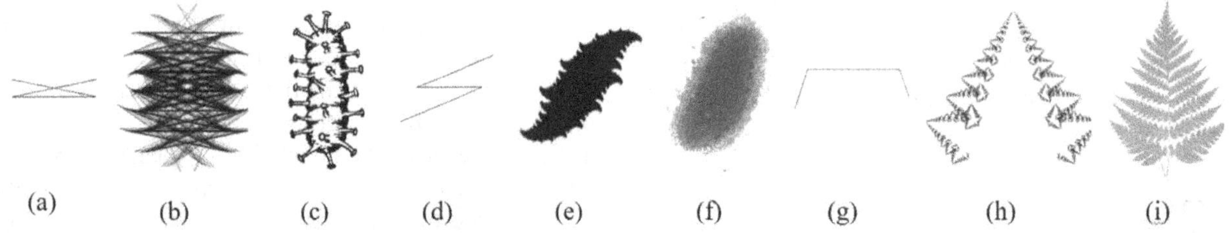

Fig. 83.3 (a) Figure on the left most is the starting images of the iterations while (b) Fractal after few finite iterations showing tentacular structure mimicking a bacterium in the middle (c) actual Salmonella bacterium, (d) Figure on the left most is the starting images of the iterations while (e) Fractal nearly infinite iterations showing another bacterium shape (f) actual Vibrio Vulnificus bacterium, (g) Figure on the left most is the starting images of the iterations while (h) Fractal after few finite iterations showing fern like structure (i) Fern leaf

Fig. 83.4 (a) Figure on the left most is the starting images of the iterations while (b) Fractal after nearly infinite iterations showing human mustache-like structure, (c) Figure on the left most is the starting images of the iterations while (d) Fractal after nearly infinite iterations crown structure formation

medicine. For instance, fractal models of river networks can inform more efficient irrigation systems, while fractal-inspired materials can lead to stronger, lighter, and more sustainable construction materials.

Additionally, fractal analysis of medical images can help diagnose diseases earlier and more accurately. Furthermore, fractals can also inspire new designs for sustainable cities, optimized traffic flow, and even more efficient communication networks. As our understanding of fractals and their natural counterparts continues to grow, we can expect to see innovative solutions emerge across various disciplines, leading to a more harmonious and sustainable relationship between human innovation and the natural world.

The detailed study of fractal classification was made by B. P. Rajvaidya in the previous paper (2024) and in M Kurhekar and B. P. Rajvaidya (2024) discussed other engineering applications of fractals in details.

Fractal designs that mimic real-world patterns and structures have the potential to contribute significantly to sustainable development.

By emulating nature's efficient and adaptive systems, fractal-inspired solutions can optimize resource use, reduce waste, and promote resilience as per J. Buhl (2006). For instance, an old study by Rodriguez-Iturbe (1997) showed fractal-based water management systems can improve irrigation efficiency and reduce evaporation losses. Similarly, fractal-inspired building designs can enhance energy efficiency, natural ventilation, and day lighting quote by F, Ming (2015).

Moreover, fractal-based materials and structures can exhibit improved strength-to-weight ratios, reducing material consumption and environmental impact which is stated in the paper by Y Chen (2015). Thus, By embracing fractal geometry and bio-mimicry, we can develop more sustainable technologies and systems that work in harmony with the natural world.

REFERENCES

1. Boden, M. A. (2010). Fractals in art: A study of self-similarity in visual composition. Leonardo, 43(3), 241–248. https://doi.org/10.1162/leon.2010.43.3.241
2. Buhl, J., et al. (2006). Fractals and the environment. Journal of Environmental Science and Health, Part C, 24(1), 1–15. https://doi.org/10.1080/10590500600614389
3. Chen, Y., et al. (2015). Fractal-based materials and structures for sustainable development. Materials Science and Engineering: C, 46, 33–41. https://doi.org/10.1016/j.msec.2014.10.036
4. Kurhekar, M., et al. (2024). Review on applications of fractals with recent advancements. AIP Conference Proceedings, 3139, 050004-1–050004-7. https://doi.org/10.1063/5.0076527
5. Lin, X., & Liu, W. (2019). Fractal study in material science engineering. IOP Conference Series: Materials Science and Engineering, 573, 012003. https://doi.org/10.1088/1757-899X/573/1/012003
6. Mandelbrot, B. (1967). How long is the coast of Britain? Statistical self-similarity and fractional dimension. Science, 156(3775), 636–638. https://doi.org/10.1126/science.156.3775.636
7. Ming, F., et al. (2015). Fractal-inspired building design for energy efficiency. Building and Environment, 86, 233–241. https://doi.org/10.1016/j.buildenv.2014.12.011
8. Olsen, S. (2013). Fractal art: A new frontier. Art & Design Education, 32(1), 53–63.
9. Rajvaidya, B. P. (2024). Fractal dynamics: Origin, design, and classifications. AIP Conference Proceedings, 3139, 050006-1–050006-9. https://doi.org/10.1063/5.0067528
10. Rodriguez-Iturbe, I., et al. (1997). Fractal river networks and their implications for water resources management. Water Resources Research, 33(10), 2123–2135. https://doi.org/10.1029/97WR01646
11. Sarhangi, R. (2008). Fractal design in Islamic art. Journal of Mathematics and the Arts, 2(1), 23–36. https://doi.org/10.1080/17513470801903723
12. Ullah, A. S., D'Addona, D. M., Seto, Y., Yonehara, S., & Kubo, A. (2021). Utilizing fractals for modeling and 3D printing of porous structures. Fractal and Fractional, 5(40). https://doi.org/10.3390/fractalfract5010040
13. Wang, W., Zhang, G., Yang, L., & Wang, W. (2019). Research on garment pattern design based on fractal graphics. EURASIP Journal on Image and Video Processing, 2019(1), Article 43. https://doi.org/10.1186/s13640-019-0431
14. Winkler, M. J. (1993). Fractal geometry in art and architecture. Journal of Mathematics and the Arts, 7(2), 67–76. https://doi.org/10.1080/17513472.1993.10820712

Note: All the figures in this chapter were made by the authors.

On Generalized Fractional Kinetic Equations Involving Generalized (p,q)-extended Hypergeometric Function

N. S. Solanki[1]
Department of HEAS, Bikaner Technical University, Bikaner, Rajasthan, India

S. Syed Rifayudeen[2]
Department of Mathematics, Pondicherry University, Puducherry, India

Naveen Jha[3]
Govt. Engineering College, Bharatpur-321001, Rajasthan, India

Abstract: Fractional kinetic equations play an important role in certain astrophysical problems. In this paper, we have established further generalization of fractional kinetic equations involving generalized (p,q)-extended Gauss' hypergeometric function. Solutions of these generalized fractional kinetic equations were obtained in terms of Mittag-Leffler function using Laplace transform. Some special cases also contain the extension of extended Gauss' hypergeometric function, extended Gauss' hypergeometric function and Gauss' hypergeometric function.

Keywords: Fractional Calculus; Hypergeometric function; Fractional kinetic equation; Laplace transforms; Mittag-Leffler function.

1. INTRODUCTION

Fractional kinetic equations gained remarkable interest due to their applications in astrophysics and mathematical physics. The extension and generalization of fractional kinetic equations involving many fractional operators were found (Kumar, Purohit, Secer and Atangana 2015; Kachhia and Prajapati 2016; Haubold and Mathai 2000; Choi and Kumar 2015; Luo and Raina 2014; Saxena et al. 2002, 2004 and 2008; Chaurasia and Pandey 2008 and 2010).

If an arbitrary reaction is characterized by a time dependent N = N (t) then it is possible to calculate the rate of change of $\frac{dN}{dt}$ by mathematical equation

$$\frac{dN}{dt} = -d + \rho, \tag{1}$$

where d is the destruction rate and ρ is the production rate of N. Haubold and Mathai (2000) established a functional differential equation between the rate of change of reaction, the destruction rate and the production rate as follows:

$$\frac{dN}{dt} = -d\left(N_t\right) + \rho\left(N_t\right), \tag{2}$$

where N = N (t) is the rate of reaction, $d(N(t))$ is the rate of destruction, $\rho(N_t)$ is the rate of production and N_t denotes the function defined by $N_t(t^*) = N(t) - t^*$, $t^* > 0$.

A special case of (2), when spatial fluctuations or homogeneities in the quantity N (t) are neglected, is given by the following differential equation [Haubold and Mathai (2000) and Kourganoff (1973)]:

$$\frac{dN_i}{dt} = -c_i N_i\left(t\right), \tag{3}$$

where initial condition $N_i(t = 0) = N_0$ is the number of density of species i at time $t = 0$, $c_i > 0$. Solution of standard kinetic Eq. (3) is given by Kourganoff (1973) as

$$N_i(t) = N_0 e^{-c_i t}. \tag{4}$$

[1]solanki_dada@rediffmail.com, [2]syedrifayudeen123@gmail.com, [3]naveenjha2410@gmail.com

DOI: 10.1201/9781003684718-84

If we decline the index i and integrate standard kinetic Eq.(3), we have

$$N(t) - N_0 = -c_0\, {}_0D_t^{-1} N(t), \tag{5}$$

where ${}_0D_t^{-1}$ is standard fractional integral operator.

Haubold and Mathai (2000) obtained the fractional generalization of the standard kinetic Eq.(3) as

$$N(t) - N_0 = -c_0^{\nu}\, {}_0D_t^{-\nu} N(t), \tag{6}$$

Where ${}_0D_t^{-\nu}$ is Riemann-Liouville fractional integral operator defined for $t > 0$, $\Re(\nu) > 0$ as follows [Samko and Kilbas, Marichev (1990)]:

$${}_0D_t^{-\nu} f(t) = \frac{1}{\Gamma(\nu)} \int_0^t (t-u)^{\nu-1} f(u)\,du \tag{7}$$

Solution of Eq. (6) is given by [Haubold and Mathai (2000)]

$$N(t) = N_0 \sum_{n=0}^{\infty} \frac{(-1)^n}{\Gamma(\nu n+1)} (c_0 t)^{\nu n}. \tag{8}$$

The exponential solution (4) of the standard kinetic Eq.(3) can be obtained by taking $\nu = 1$ in (8).

Two parameter Mittag-Leffler function [Choi et al. 2020] is defined as

$$E_{\alpha,\beta}(z) = \sum_{n=0}^{\infty} \frac{z^n}{\Gamma(\alpha n+\beta)},$$
$$\left(\alpha,\beta \in C; \Re(\alpha) > 0, \Re(\beta) > 0\right). \tag{9}$$

Laplace transform [Sneddon (1979)] is defined as

$$L\{f(t);s\} = \int_0^{\infty} e^{-st} f(t)\,dt. \tag{10}$$

Laplace transform of the Riemann-Liouville fractional integral operator is given by [Srivastava and Saxena (2001)]

$$L\{{}_0D_t^{-\rho} f(t);s\} = s^{-\rho} L\{f(t);s\}. \tag{11}$$

This paper is structured as follows: Section-2 we established further generalization of fractional kinetic equations involving generalized (p,q)-extended Gauss' hypergeometric function. Solutions of these generalized fractional kinetic equations were obtained in terms of Mittag-Leffler function using Laplace transform. Some special cases also contain the extension of extended Gauss' hypergeometric function, extended Gauss' hypergeometric function and Gauss' hypergeometric function.

2. GENERALIZED (p, q)-EXTENDED GAUSS' HYPERGEOMETRIC FUNCTION AND ITS SPECIAL CASES

The generalized (p, q)-extended Gauss' hypergeometric function [Ali et al. (2024)] is defined as

$$F_{p,q}^{\gamma,\delta}(a,b;c;z) = \sum_{n=0}^{\infty} \frac{B_{\gamma,\delta}(b+n,c-b;p;q)}{B(b,c-b)} (a)_n \frac{z^n}{n!}$$
$$\left(p,q \ge 0, |z| < 1; \gamma,\delta \in \{z \in C : \Re(z) \succ 1\}\right), \tag{12}$$

where $B_{\gamma,\delta}(x, y; p; q)$ is the generalized (p, q)-extended Beta function introduced by Ali et al. (2024)

$$B_{\gamma,\delta}(x,y) = \int_0^1 t^{x-1}(1-t)^{y-1} E_{\gamma}\left(\frac{-p}{t}\right) S_{\delta}\left(\frac{-q}{1-t}\right) dt$$
$$\left(min\{\Re(x),\Re(y)\} > 0; min\{\Re(p),\Re(q)\} > 0\right), \tag{13}$$

where, $E_{\gamma}(z)$ is Mittag-Leffler function [Choi et al. 2020] defined by

$$E_{\gamma}(z) = \sum_{n=0}^{\infty} \frac{z^n}{\Gamma(\gamma n+1)}, \tag{14}$$

and $S_{\delta}(z)$ is Bessel-Struve Kernel function [Parmar and Choi (2018)] given as follows

$$S_{\delta}(z) = \frac{\Gamma(\delta+1)}{\sqrt{\pi}} \sum_{m=0}^{\infty} \frac{\Gamma\left(\frac{m+1}{2}\right)}{\Gamma(m/2+\delta+1)} \frac{z^m}{m!}. \tag{15}$$

If we take $\gamma = 1$ and $\delta = \frac{-1}{2}$ in (12), then we get extension of extended Gauss' hypergeometric function [Choi, Rathie and Parmar (2014)]:

$$F_{p,q}(a,b;c;z) = \sum_{n=0}^{\infty} \frac{B(b+n,c-b;p;q)}{B(b,c-b)} (a)_n \frac{z^n}{n!} \tag{16}$$

$(p, q \ge 0, |z| < 1)$,

where $B(x, y; p; q)$ is the extension of extended Beta function [Choi, Rathie and Parmar (2014)] defined by

$$B(x,y;p;q) = \int_0^1 t^{x-1}(1-t)^{y-1} e^{\left(\frac{-p}{t} - \frac{q}{1-t}\right)} dt \tag{17}$$
$$\left(min\{\Re(x),\Re(y)\} > 0; min\{\Re(p),\Re(q)\} > 0\right),$$

If we take $\gamma = 1, \delta = \frac{-1}{2}$ and $q = 0$ in (12), then we get extended Gauss' hypergeometric function [Chaudhry et al. 2004]:

$$F_p(a,b;c;z) = \sum_{n=0}^{\infty} \frac{B(b+n,c-b;p)}{B(b,c-b)} (a)_n \frac{z^n}{n!} \tag{18}$$
$$\left(p \ge 0, |z| < 1\right),$$

where $B(x, y; p)$ is the extended Beta function [Chaudhry et al. 1997] defined by

$$B(x,y;p) = \int_0^1 t^{x-1}(1-t)^{y-1} e^{\left(\frac{-p}{t(1-t)}\right)} dt \tag{19}$$
$$\left(min\{\Re(x),\Re(y)\} > 0; \Re(p) > 0\right),$$

If we take $\gamma = 1, \delta = \frac{-1}{2}$ and $p = 0 = q$ in (12), then we get Gauss' hypergeometric function [Chaudhry et al. 2004]:

$${}_2F_1(a,b;c;z) = \sum_{n=0}^{\infty} \frac{B(b+n,c-b)}{B(b,c-b)} (a)_n \frac{z^n}{n!}, \tag{20}$$

where $B(x, y)$ is the Beta function defined by

$$B(x,y)=\int_0^1 t^{x-1}(1-t)^{y-1}\,dt. \tag{21}$$

3. GENERALIZED FRACTIONAL KINETIC EQUATIONS

In this section, generalized fractional kinetic equation is established as Theorems and their special cases (Corollaries).

Theorem 3.1. If $d > 0$, $\rho > 0$, $p, q > 0$ and $\left|\frac{d^\rho}{s^\rho}\right| < 1$, then the solution of equation

$$N(t)-N_0F_{p,q}^{\gamma,\delta}(a,b;c;t)=-d^\rho\,{}_0D_t^{-\rho}N(t) \tag{22}$$

is given by

$$N(t)=N_0\sum_{n=0}^{\infty}(a)_n\frac{B_{\gamma,\delta}(b+n,c-b;p;q)}{B(b,c-b)}\Gamma(n+1)\frac{t^n}{n!}.E_{\rho,n+1}\left(-d^\rho t^\rho\right), \tag{23}$$

where $E_{\alpha,\beta}(x)$ is the generalized Mittag-Leffler function given by (9).

Proof. Applying Laplace transform (10) on (22) and using (11), we have

$$L\{N(t);s\}=N_0L\{F_{p,q}^{\gamma,\delta}(a,b;c;t);s\}-d^\rho L\{{}_0D_t^{-\rho}N(t);s\}$$

$$=N_0\left(\int_0^\infty e^{-st}F_{p,q}^{\gamma,\delta}(a,b;c;t)\,dt\right)-d^\rho s^{-\rho}L\{N(t);s\}$$

$$=N_0\left(\int_0^\infty e^{-st}\sum_{n=0}^{\infty}(a)_n\frac{B_{\gamma,\delta}(b+n,c-b;p;q)}{B(b,c-b)}\frac{t^n}{n!}dt\right)-d^\rho s^{-\rho}L\{N(t);s\} \tag{24}$$

therefore,

$$\left(1+d^\rho s^{-\rho}\right)L\{N(t);s\}=N_0\sum_{n=0}^{\infty}\frac{B_{\gamma,\delta}(b+n,c-b;p;q)}{B(b,c-b)}\frac{(a)_n}{n!}\int_0^\infty e^{-st}t^n dt. \tag{25}$$

This gives,

$$L\{N(t);s\}=N_0\sum_{n=0}^{\infty}\frac{B_{\gamma,\delta}(b+n,c-b;p;q)}{B(b,c-b)}\frac{(a)_n}{n!}\times\frac{\Gamma(n+1)}{s^{n+1}}\frac{1}{\left(1+d^\rho s^{-\rho}\right)} \tag{26}$$

After simplification of above equation, we get

$$L\{N(t);s\}=N_0\sum_{n=0}^{\infty}\frac{B_{\gamma,\delta}(b+n,c-b;p;q)}{B(b,c-b)}\frac{(a)_n}{n!}\Gamma(n+1)\left\{\sum_{k=0}^{\infty}(-1)^k d^{\rho k}s^{-(\rho k+n+1)}\right\} \tag{27}$$

Taking inverse Laplace transform of (27) and using $L^{-1}\{s^{-\rho}\};t\}=\frac{t^{\rho-1}}{\Gamma(\rho)}$, we arrived at

$$N(t)=N_0\sum_{n=0}^{\infty}\frac{B_{\gamma,\delta}(b+n,c-b;p;q)}{B(b,c-b)}\frac{(a)_n}{n!}\Gamma(n+1)L^{-1}\left\{\sum_{k=0}^{\infty}(-1)^k d^{\rho k}s^{-(\rho k+n+1)}\right\}$$

$$=N_0\sum_{n=0}^{\infty}\frac{B_{\gamma,\delta}(b+n,c-b;p;q)}{B(b,c-b)}\frac{(a)_n}{n!}\Gamma(n+1)\left\{\sum_{k=0}^{\infty}(-1)^k d^{\rho k}\frac{t^{\rho k+n}}{\Gamma(\rho k+n+1)}\right\}$$

$$N(t)=N_0\sum_{n=0}^{\infty}(a)_n\frac{B_{\gamma,\delta}(b+n,c-b;p;q)}{B(b,c-b)}\Gamma(n+1)\frac{t^n}{n!}\left\{\sum_{k=0}^{\infty}\frac{\left(-d^\rho t^\rho\right)^k}{\Gamma(\rho k+n+1)}\right\} \tag{28}$$

Hence,

$$N(t)=N_0\sum_{n=0}^{\infty}(a)_n\frac{B_{\gamma,\delta}(b+n,c-b;p;q)}{B(b,c-b)}\Gamma(n+1)\frac{t^n}{n!}\times E_{\rho,n+1}\left(-d^\rho t^\rho\right). \tag{29}$$

For $\gamma = 1$ and $\delta = \frac{-1}{2}$ and in light of Eq.(16), Theorem 3.1 leads to the following Corollary.

Corollary 3.1.1. If $d > 0$, $\rho > 0$, $p, q > 0$ and $\left|\frac{d^\rho}{s^\rho}\right| < 1$, then the solution of equation

$$\mathrm{N}(t)-N_0F_{p,q}(a,b;c;t)=-d^\rho\,{}_0D_t^{-\rho}\mathrm{N}(t) \tag{30}$$

is given by

$$N(t)=N_0\sum_{n=0}^{\infty}(a)_n\frac{B(b+n,c-b;p;q)}{B(b,c-b)}\Gamma(n+1)\frac{t^n}{n!}\times E_{\rho,n+1}\left(-d^\rho t^\rho\right). \tag{31}$$

For $\gamma = 1$, $\delta = \frac{-1}{2}$ and $q = 0$ and in light of Eq.(18), Theorem 3.1 leads to the following Corollary.

Corollary 3.1.2. If $d > 0$, $\rho > 0$, $p > 0$ and $\left|\frac{d^\rho}{s^\rho}\right| < 1$, then the solution of equation

$$N(t)-N_0F_p(a,b;c;t)=-d^\rho\,{}_0D_t^{-\rho}N(t) \tag{32}$$

is given by

$$N(t)=N_0\sum_{n=0}^{\infty}(a)_n\frac{B(b+n,c-b;p)}{B(b,c-b)}\Gamma(n+1)\frac{t^n}{n!}\times E_{\rho,n+1}\left(-d^\rho t^\rho\right). \tag{33}$$

It may be noted in passing that the special case of generalized $(p,\ q)$-extended Gauss' hypergeometric function (12) is Gauss' hypergeometric function (20), when $\gamma = 1$, $\delta = \frac{-1}{2}$ and $p = 0 = q$ immediately leads to the Gauss' hypergeometric function ${}_2F_1\ (a,\ b;\ c;\ t)$, which yields following corollary.

Corollary 3.1.3. If $d > 0$, $\rho > 0$ and $\left|\frac{d^\rho}{s^\rho}\right| < 1$, then the solution of equation

$$N(t) - N_{02}F_1(a,b;c;t) = -d^\rho\ {}_0D_t^{-\rho}N(t) \tag{34}$$

is given by

$$N(t) = N_0 \sum_{n=0}^{\infty} (a)_n \frac{B(b+n,c-b)}{B(b,c-b)}$$
$$\Gamma(n+1)\frac{t^n}{n!} \times E_{\rho,n+1}\left(-d^\rho t^\rho\right). \tag{35}$$

Theorem 3.2. If $d > 0$, $\rho > 0$, $p,\ q > 0$ and $\left|\frac{d^\rho}{s^\rho}\right| < 1$, then the solution of equation

$$N(t) - N_0 F_{p,q}^{\gamma,\delta}\left(a,b;c;d^\rho t^\rho\right) = -d^\rho\ {}_0D_t^{-\rho}N(t) \tag{36}$$

is given by

$$N(t) = N_0 \sum_{n=0}^{\infty} (a)_n \frac{B_{\gamma,\delta}(b+n,c-b;p;q)}{B(b,c-b)}$$
$$\Gamma(\rho n+1)\frac{\left(d^\rho t^\rho\right)^n}{n!} \times E_{\rho,\rho n+1}\left(-d^{\acute{A}} t^{\acute{A}}\right), \tag{37}$$

where $E_{\alpha,\beta}(x)$ is the generalized Mittag-Leffler function given by (9).

Proof. Applying Laplace transform (10) on (36) and using (11), we have

$$L\{N(t);s\} = N_0 L\{F_{p,q}^{\gamma,\delta}\left(a,b;c;d^\rho t^\rho\right);s\}$$
$$- d^\rho L\{{}_0D_t^{-\rho}N(t);s\}$$
$$= N_0\left(\int_0^\infty e^{-st} F_{p,q}^{\gamma,\delta}\left(a,b;c;d^\rho t^\rho\right)dt\right)$$
$$- d^\rho s^{-\rho} L\{N(t);s\}$$
$$= N_0\left(\int_0^\infty e^{-st} \sum_{n=0}^{\infty} (a)_n \frac{B_{\gamma,\delta}(b+n,c-b;p;q)}{B(b,c-b)} \frac{\left(d^\rho t^\rho\right)^n}{n!} dt\right)$$
$$- d^\rho s^{-\rho} L\{N(t);s\} \tag{38}$$

therefore,

$$\left(1+d^\rho s^{-\rho}\right) L\{N(t);s\}$$
$$= N_0 \sum_{n=0}^{\infty} \frac{B_{\gamma,\delta}(b+n,c-b;p;q)}{B(b,c-b)} \frac{(a)_n}{n!} d^{\rho n} \int_0^\infty e^{-st} t^{\rho n} dt. \tag{39}$$

This gives,

$$L\{N(t);s\} = N_0 \sum_{n=0}^{\infty} \frac{B_{\gamma,\delta}(b+n,c-b;p;q)}{B(b,c-b)} \frac{(a)_n}{n!} d^{\rho n}$$
$$\times \frac{\Gamma(\rho n+1)}{s^{\rho n+1}} \frac{1}{\left(1+d^\rho s^{-\rho}\right)} \tag{40}$$

After simplification of above equation, we get

$$L\{N(t);s\} = N_0 \sum_{n=0}^{\infty} \frac{B_{\gamma,\delta}(b+n,c-b;p;q)}{B(b,c-b)} \frac{(a)_n}{n!}$$
$$\Gamma(\rho n+1)\left\{\sum_{k=0}^{\infty} (-1)^k\ d^{\rho k+\rho n} s^{-(\rho k+\rho n+1)}\right\} \tag{41}$$

Taking inverse Laplace transform of (41) and using $L^{-1}\{s^{-\rho};\ t\} = \frac{t^{\rho-1}}{\Gamma(\rho)}$, we arrived at

$$N(t) = N_0 \sum_{n=0}^{\infty} \frac{B_{\gamma,\delta}(b+n,c-b;p;q)}{B(b,c-b)} \frac{(a)_n}{n!}$$
$$\Gamma(\rho n+1) L^{-1}\left\{\sum_{k=0}^{\infty} (-1)^k\ d^{\rho k+\rho n} s^{-(\rho k+\rho n+1)}\right\}$$
$$= N_0 \sum_{n=0}^{\infty} \frac{B_{\gamma,\delta}(b+n,c-b;p;q)}{B(b,c-b)} \frac{(a)_n}{n!}$$
$$\Gamma(\rho n+1)\left\{\sum_{k=0}^{\infty} (-1)^k\ d^{\rho k+\rho n} \frac{t^{\rho k+\rho n}}{\Gamma(\rho k+\rho n+1)}\right\}$$
$$N(t) = N_0 \sum_{n=0}^{\infty} (a)_n \frac{B_{\gamma,\delta}(b+n,c-b;p;q)}{B(b,c-b)}$$
$$\Gamma(\rho n+1)\frac{\left(d^\rho t^\rho\right)^n}{n!}\left\{\sum_{k=0}^{\infty} \frac{\left(-d^\rho t^\rho\right)^k}{\Gamma(\rho k+\rho n+1)}\right\} \tag{42}$$

hence,

$$N(t) = N_0 \sum_{n=0}^{\infty} (a)_n \frac{B_{\gamma,\delta}(b+n,c-b;p;q)}{B(b,c-b)}$$
$$\Gamma(\rho n+1)\frac{\left(d^\rho t^\rho\right)^n}{n!} \times E_{\rho,\rho n+1}\left(-d^\rho t^\rho\right). \tag{43}$$

Corollary 3.2.1. If $d > 0$, $\rho > 0$, $p,\ q > 0$ and $\left|\frac{d^\rho}{s^\rho}\right| < 1$, then the solution of equation

$$N(t) - N_0 F_{p,q}\left(a,b;c;d^\rho t^\rho\right) = -d_0^\rho D_t^{-\rho} N(t) \tag{44}$$

is given by

$$N(t) = N_0 \sum_{n=0}^{\infty} (a)_n \frac{B(b+n,c-b;p;q)}{B(b,c-b)}$$
$$\Gamma(\rho n+1)\frac{\left(d^\rho t^\rho\right)^n}{n!} \times E_{\rho,\rho n+1}\left(-d^\rho t^\rho\right). \tag{45}$$

Corollary 3.2.2. If $d > 0$, $\rho > 0$, $p > 0$ and $\left|\frac{d^\rho}{s^\rho}\right| < 1$, then the solution of equation

$$N(t) - N_0 F_p\left(a,b;c;d^\rho t^\rho\right) = -d^\rho\, {}_0D_t^{-\rho} N(t) \quad (46)$$

is given by

$$N(t) = N_0 \sum_{n=0}^{\infty} (a)_n \frac{B(b+n,c-b;p)}{B(b,c-b)} \Gamma(\rho n+1) \frac{\left(d^\rho t^\rho\right)^n}{n!} \times E_{\rho,\rho n+1}\left(-d^\rho t^\rho\right). \quad (47)$$

Corollary 3.2.3. If $d > 0$, $\rho > 0$ and $\left|\frac{d^\rho}{s^\rho}\right| < 1$, then the solution of equation

$$N(t) - N_{0\,2}F_1\left(a,b;c;d^\rho t^\rho\right) = -d^\rho\, {}_0D_t^{-\rho} N(t) \quad (48)$$

is given by

$$N(t) = N_0 \sum_{n=0}^{\infty} (a)_n \frac{B(b+n,c-b)}{B(b,c-b)} \Gamma(\rho n+1) \frac{\left(d^\rho t^\rho\right)^n}{n!} \times E_{\rho,\rho n+1}\left(-d^\rho t^\rho\right). \quad (49)$$

Theorem 3.3.

If $d > 0$, $\eta > 0$, $\rho > 0$, p, $q > 0$ and $\left|\frac{d^\rho}{s^\rho}\right| < 1$, then the solution of equation

$$N(t) - N_0 F_{p,q}^{\gamma,\delta}\left(a,b;c;d^\rho t^\rho\right) = -\eta^\rho\, {}_0D_t^{-\rho} N(t) \quad (50)$$

is given by

$$N(t) = N_0 \sum_{n=0}^{\infty} (a)_n \frac{B_{\gamma,\delta}(b+n,c-b;p;q)}{B(b,c-b)} \Gamma(\rho n+1) \frac{\left(d^\rho t^\rho\right)^n}{n!} \times E_{\rho,\rho n+1}\left(-\eta^\rho t^\rho\right), \quad (51)$$

where $E_{\alpha,\beta}(x)$ is the generalized Mittag-Leffler function given by (9).

Proof. Applying Laplace transform (10) on (50) and using (11), we have

$$L\{N(t);s\} = N_0 L\{F_{p,q}^{\gamma,\delta}\left(a,b;c;d^\rho t^\rho\right);s\} - \eta^\rho L\{{}_0D_t^{-\rho} N(t);s\}$$

$$= N_0 \left(\int_0^\infty e^{-st} F_{p,q}^{\gamma,\delta}\left(a,b;c;d^\rho t^\rho\right) dt \right) - \eta^\rho s^{-\rho} L\{N(t);s\}$$

$$= N_0 \left(\int_0^\infty e^{-st} \sum_{n=0}^{\infty} (a)_n \frac{B_{\gamma,\delta}(b+n,c-b;p;q)}{B(b,c-b)} \frac{\left(d^\rho t^\rho\right)^n}{n!} dt \right) - \eta^\rho s^{-\rho} L\{N(t);s\} \quad (52)$$

therefore,

$$\left(1+\eta^\rho s^{-\rho}\right) L\{N(t);s\} = N_0 \sum_{n=0}^{\infty} \frac{B_{\gamma,\delta}(b+n,c-b;p;q)}{B(b,c-b)} \frac{(a)_n}{n!} d^{\rho n} \int_0^\infty e^{-st} t^{\rho n} dt. \quad (53)$$

This gives,

$$L\{N(t);s\} = N_0 \sum_{n=0}^{\infty} \frac{B_{\gamma,\delta}(b+n,c-b;p;q)}{B(b,c-b)} \frac{(a)_n}{n!} d^{\rho n} \times \frac{\Gamma(\rho n+1)}{s^{\rho n+1}} \frac{1}{\left(1+\eta^\rho s^{-\rho}\right)} \quad (54)$$

After simplification of above equation, we get

$$L\{N(t);s\} = N_0 \sum_{n=0}^{\infty} \frac{B_{\gamma,\delta}(b+n,c-b;p;q)}{B(b,c-b)} \frac{(a)_n}{n!} d^{\rho n} \Gamma(\rho n+1) \left\{ \sum_{k=0}^{\infty} (-1)^k \eta^{\rho k} s^{-(\rho k+\rho n+1)} \right\} \quad (55)$$

Taking inverse Laplace transform of (55) and using $L^{-1}\{s^{-\rho}; t\} = \frac{t^{\rho-1}}{\Gamma(\rho)}$, we arrived at

$$N(t) = N_0 \sum_{n=0}^{\infty} \frac{B_{\gamma,\delta}(b+n,c-b;p;q)}{B(b,c-b)} \frac{(a)_n}{n!} d^{\rho n} \Gamma(\rho n+1) L^{-1} \left\{ \sum_{k=0}^{\infty} (-1)^k \eta^{\rho k} s^{-(\rho k+\rho n+1)} \right\}$$

$$= N_0 \sum_{n=0}^{\infty} \frac{B_{\gamma,\delta}(b+n,c-b;p;q)}{B(b,c-b)} \frac{(a)_n}{n!} d^{\rho n} \Gamma(\rho n+1) \left\{ \sum_{k=0}^{\infty} (-1)^k \eta^{\rho k} \frac{t^{\rho k+\rho n}}{\Gamma(\rho k+\rho n+1)} \right\}$$

$$N(t) = N_0 \sum_{n=0}^{\infty} (a)_n \frac{B_{\gamma,\delta}(b+n,c-b;p;q)}{B(b,c-b)} \Gamma(\rho n+1) \frac{\left(d^\rho t^\rho\right)^n}{n!} \left\{ \sum_{k=0}^{\infty} \frac{\left(-\eta^\rho t^\rho\right)^k}{\Gamma(\rho k+\rho n+1)} \right\} \quad (56)$$

hence,

$$N(t) = N_0 \sum_{n=0}^{\infty} (a)_n \frac{B_{\gamma,\delta}(b+n,c-b;p;q)}{B(b,c-b)} \Gamma(\rho n+1) \frac{\left(d^\rho t^\rho\right)^n}{n!} \times E_{\rho,\rho n+1}\left(-\eta^\rho t^\rho\right). \quad (57)$$

Corollary 3.3.1. If $d > 0$, $\eta > 0$, $\rho > 0$, $p, q > 0$ and $\left|\frac{d^\rho}{s^\rho}\right| < 1$, then the solution of equation

$$N(t) - N_0 F_{p,q}\left(a,b;c;d^\rho t^\rho\right) = -\eta^\rho {}_0D_t^{-\rho} N(t) \quad (58)$$

is given by

$$N(t) = N_0 \sum_{n=0}^{\infty} (a)_n \frac{B(b+n, c-b; p; q)}{B(b, c-b)} \Gamma(\rho n + 1) \frac{\left(d^\rho t^\rho\right)^n}{n!} \times E_{\rho,\rho n+1}\left(-\eta^\rho t^\rho\right). \quad (59)$$

Corollary 3.3.2. If $d > 0$, $\eta > 0$, $\rho > 0$, $p > 0$ and $\left|\frac{d^\rho}{s^\rho}\right| < 1$, then the solution of equation

$$N(t) - N_0 F_p\left(a,b;c;d^\rho t^\rho\right) = -\eta^\rho {}_0D_t^{-\rho} N(t) \quad (60)$$

is given by

$$N(t) = N_0 \sum_{n=0}^{\infty} (a)_n \frac{B(b+n, c-b; p)}{B(b, c-b)} \Gamma(\rho n + 1) \frac{\left(d^\rho t^\rho\right)^n}{n!} \times E_{\rho,\rho n+1}\left(-\eta^\rho t^\rho\right). \quad (61)$$

Corollary 3.3.3. If $d > 0$, $\eta > 0$, $\rho > 0$ and $\left|\frac{d^\rho}{s^\rho}\right| < 1$, then the solution of equation

$$N(t) - N_0\, {}_2F_1\left(a,b;c;d^\rho t^\rho\right) = -\eta^\rho {}_0D_t^{-\rho} N(t) \quad (62)$$

is given by

$$N(t) = N_0 \sum_{n=0}^{\infty} (a)_n \frac{B(b+n, c-b)}{B(b, c-b)} \Gamma(\rho n + 1) \frac{\left(d^\rho t^\rho\right)^n}{n!} \times E_{\rho,\rho n+1}\left(-\eta^\rho t^\rho\right). \quad (63)$$

4. CONCLUSION

Various kinds of generalized fractional kinetic equation constructed and derived their solutions in view of generalized (p, q)-extended Gauss' hypergeometric function, extension of extended Gauss' hypergeometric function, extended Gauss' hypergeometric function, Gauss' hypergeometric function and other special functions.

REFERENCES

1. Chaudhry M.A., Qadir A., Rafique M. and Zubair S.M. (1997). Extension of Euler's Beta function, Comput.Appl. Math.78:19–32.
2. Chaudhry M.A., Qadir A., Srivastava H.M. and Paris R.B. (2014). Extended hypergeometric and confluent hypergeometric functions, Appl. Math. Comput.159:589–602.
3. Choi J., Rathie A.K. and Parmar R.K. (2014). Extension of extended beta, hypergeometric and confluent hypergeometric functions, Honam Math. J.36(2):339–367.
4. Shoukat Ali, Naresh Kumar Regar, and Subrat Parida (2024). On Generalized extended beta and hypergeometric functions, Honam Mathematical J.46:N0.2,pp.
5. Choi J., Parmar R.K., Chopra P. (2020). Extended Mittag-Leffler function and associated fractional calculus operators, Georgian Mathematical Journal,27(2):199-209.
6. Parmar R.K. and Pogan T.K. (2019). On (p,q)–extension of further members of Bessel-Struve functions, Miskolc Mathematical Notes, 20(1):451–463.
7. Choi J. and Parmar R.K. (2018). Fractional calculus of the (p,q)–extended Struve function, Far East Journal of Mathematical Sciences, 103(2):541–559.
8. Choi J., Parmar R.K. and Pogany T.K. (2017). Mathieu-type series built by (p,q)–extended Gaussian hypergeometric function, Bull. Korean Math.Soc.54(3):789–797.
9. Wiman A. (1905). Uber de fundamental theorie der funktionen, Acta math. 29(1):191-201.
10. Kumar D., Purohit S.D., Secer A., Atangana A. (2015). On generalized fractional kinetic equation involving generalized Bessel functions of the first kind,Math. ProblemEng.7:289-387.
11. Kachhia K.B., Prajapati J.C. (2016). On generalized fractional kinetic equations involving gen- eralized Lommel-Wright functions , Alexandria Engineering Journal 55:2953-2957.
12. Haubold H.J., Mathai A.M. (2000). The fractional kinetic equation and thermonuclear functions, Astrophys,SpaceSci.273(1):53-63.
13. Srivastava H.M., Saxena R.K. (2001). Operators of fractional integration and their applications, Appl.Math. Comput.118:1-52.
14. Sneddon I.N. (1979). The Use of Integral Transform, Tata MeGrawHill, New Delhi, India.
15. Choi J., Kumar D. (2015). Solutions of generalized fractional kinetic equations involving Aleph functions,Math.Commun.20:113-123.
16. Luo M.J., Raina R.K. (2014). On certain classes of fractional kinetic equations, Filomat 28(10):2077- 2090.
17. Saxena R.K., Mathai A.M., Haubold H.J. (2002). On fractional kinetic equation, Astrophys. SpaceSci.282:281-287.
18. Saxena R.K., Mathai A.M., Haubold H.J. (2004). On generalized fractional kinetic equation, Physica A 344:657-664.
19. Saxena R.K., Kalla S.L. (2008). On the solution of certain fractional kinetic equations, Appl. Math. Comput.199:504-511.
20. Samko S.G., Kilbas A., Marichev O. (1990). Fractional Integral and Derivatives: Theory and Applications, Gordon and Breach Sci. Publ., New York.
21. Chaurasia V.B.L., Pandey S.C. (2008). On the new computable solution of the generalized fractional kinetic equations involving the generalized function for fractional calculus and related functions,Astrophys. SpaceSci.317:213-219.
22. Chaurasia V.B.L., Pandey S.C. (2010). Computable extensions of generalized fractional kinetic equations in astrophysics, Res. Astron, Astrophys. 10(1):22-32.
23. Kourganoff V. (1973). Introduction to the Physics of Stellar Interiors, D. Reidel Publishing Company, Dor- drecht.

Recent Trends in Applied Physics and Material Science – Sudhir Bhardwaj et al. (eds)
© 2026 Taylor & Francis Group, London, ISBN 978-1-041-16452-4

Effect of Bismuth doping in surface and mechanical properties of Barium Titanate Ceramics

Nabiya Iqbal, Anju Dixit*
Department of Physics, School of Basic Sciences, UIET,
CSJM University, Kanpur, India

Pramod S. Dobal
Department of Physics, VSSD College,
Kanpur, India

Abstract: In this work, $Ba_{1-x}Bi_xTiO_3$ ceramics were synthesized using the sol – gel method for x = 0.00 and 0.05 compositions (samples were labelled as BT and BBT5 respectively), to investigate the effects of bismuth doping on their surface, dielectric and mechanical properties. Incorporation of bismuth in the crystal lattice resulted in enhanced density, grain size, dielectric constant and moderate improvements in mechanical properties like fracture toughness and hardness. Relaxor behavior was also observed in the bismuth doped sample.

Keywords: Barium titanate, Bismuth doping, Sol – gel method

1. INTRODUCTION

Barium Titanate ($BaTiO_3$), a key perovskite material, is valued for its high dielectric constant and low dielectric loss, making it an ideal dielectric material. However, it lags in ferroelectric and piezoelectric performance as compared to lead – based counterparts like $PbTiO_3$. Bismuth (Z = 83) has electronic configuration $[Xe]4f^{14}5d^{10}6s^2 6p^3$. In the periodic table, just prior to bismuth lies Lead (Z = 82), with a configuration similar to bismuth, i.e., $[Xe]4f^{14}5d^{10}6s^2 6p^2$. Lead is present with a +2 oxidation state in $PbTiO_3$ with $6s^2$ lone pair of electrons which are majorly responsible for its superior ferroelectricity and piezoelectricity. Bismuth as Bi^{3+} ion can partially substitute Ba^{2+} ion in $BaTiO_3$ and introduce $6s^2$ lone pair electrons which were previously absent at the A-site of $BaTiO_3$ (Jallouli Necib *et al.* 2022). Thus, enhancing its ferroelectric and piezoelectric properties. Bismuth, also known to reduce the sintering temperature, serves as an eco- friendly alternative to lead based materials.

2. EXPERIMENTAL

Raw powders of the ceramics were synthesized by sol – gel method, further calcined at 900°C, then pressed into circular pellets (diameter: 10 mm, thickness: 2.5 mm to 3 mm) by using a hydraulic press and conventionally sintered at 1100°C. These temperatures had been selected according to the thermal analysis of the materials mentioned in our previous paper (Nabiya Iqbal *et al.* 2023). Surface analysis of pellets was done by FESEM (NOVA NANOSEM 450) and temperature dependent dielectric characterization by LCR meter along with a High Temperature Probostat.

3. RESULTS AND DISCUSSIONS

Synthesized barium titanate was less dense (3.77 gm cm^{-3}) with noticeable porosity while bismuth doping enhanced the density (4.42 gm cm^{-3}) and reduced porosity. The bulk density (ρ) of the samples was calculated by using the formula,

$$(\rho) = m/\pi r^2 h$$

where *m*, *r* and *h* are the mass, radius and thickness of the prepared pellets respectively. Fig. 85.2(a) and 85.2(b) reveal that grain size of barium titanate increased from 0.31 μm to 0.70 μm with bi addition. The grain size was calculated by using the Gaussian Distribution Function. This grain growth was promoted by Bi due to the formation of liquid phase during sintering, enhanced

*Corresponding author: anjudixit@csjmu.ac.in

DOI: 10.1201/9781003684718-85

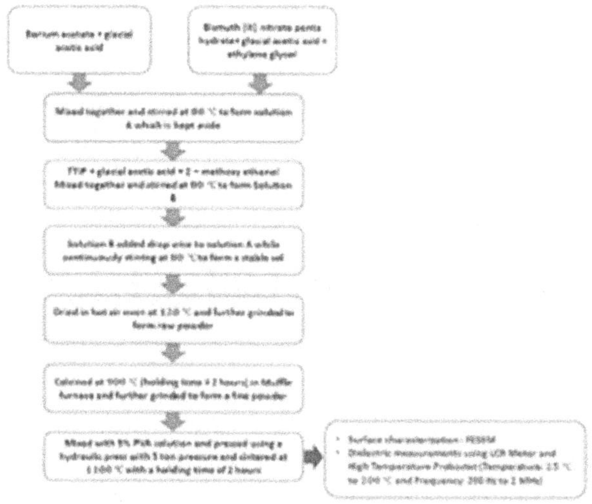

Fig. 85.1 Flowchart representing the synthesis and characterization route followed in this study [Original work]

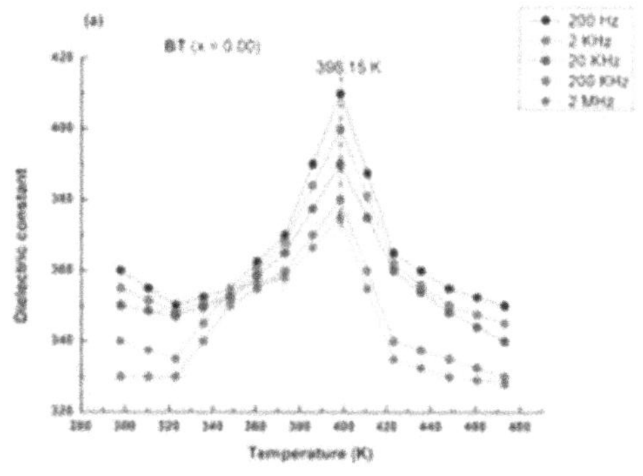

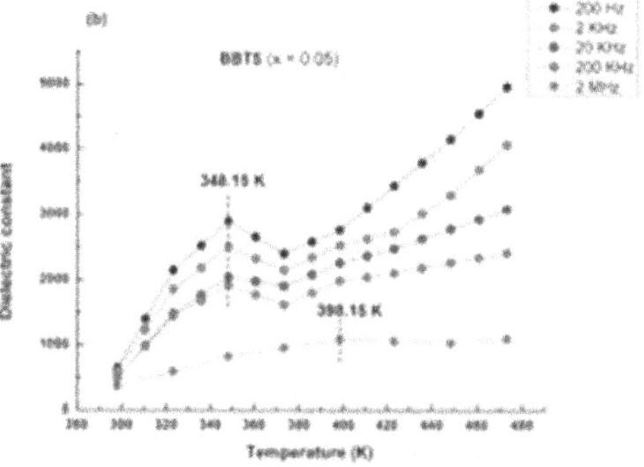

Fig. 85.3 (a) and (b) Dielectric constant vs Temperature graph for BT and BBT5 respectively at frequency 200 Hz to 2 MHz [Original work]

diffusion at the grain boundary and beneficial defect formation in the crystal structure (Suravi Islam *et al.* (2022). The maximum value of the dielectric constant also increased from 410 for BT to 2914 for bismuth doped sample at frequency 200 Hz (Fig. 85.3(a) and 85.3(b)). In low frequency region, the peak of dielectric maxima was obtained at 348.15 K for BBT5 and at 398.15 K for BT. This means that the temperature at which dielectric maxima had been obtained shifted to a lower temperature with bismuth doping. However, this peak shifted to a higher temperature, as the frequency increased in case of BBT5, thus showing a Dielectric Relaxor behavior. The microstructure of the ceramics plays an important role in determining their mechanical properties. Theoretically, fine grained ceramics exhibit better fracture toughness and hardness (Tomasz Trzepiecinski *et al.* 2018). However, in this work, the dense bismuth doped sample along with large grain size are expected to show moderate hardness and fracture toughness provided the grain growth is under control.

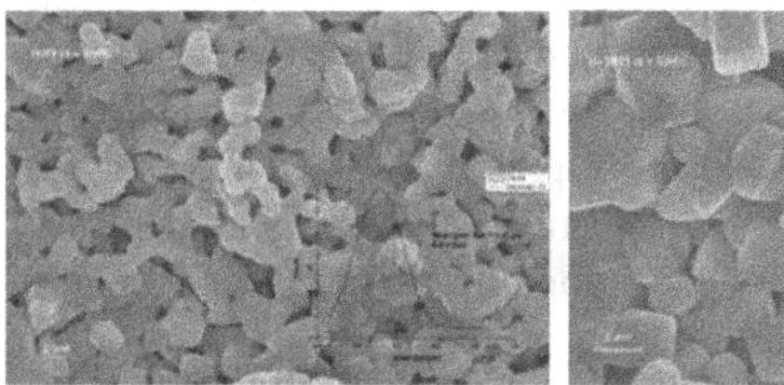

Fig. 85.2 (a) and (b) FESEM micrograph of BT and BBT5 respectively [Original work]

4. CONCLUSIONS

Bismuth doping in barium titanate led to the formation of a dense ceramic with larger grains, enhanced dielectric property with the presence of a relaxor behavior, lowering of the Curie Temperature and is expected to show a moderate fracture toughness and hardness. Thus, this study highlights the potential of using bismuth – doped barium titanate ceramics for applications requiring high dielectric performance and stable mechanical properties.

5. ACKNOWLEDGEMENT

The authors are thankful to Professor Kantesh Balani, IIT Kanpur for his generous cooperation in this work.

REFERENCES

1. Necib, J., López-Sánchez, J., Rubio-Marcos, F., Serrano, A., Navarro, E., Peña, Á., ... & Marín, P. (2022). A feasible pathway to stabilize monoclinic and tetragonal phase coexistence in barium titanate-based ceramics. *Journal of Materials Chemistry C*, *10*(46), 17743-17756.
2. Iqbal, N., Dixit, A., & Dobal, P. S. (2023). The mechanism of formation of bismuth-doped barium titanate ceramics via sol gel. *Ferroelectrics*, *615*(1), 386-395.
3. Islam, S., Khatun, N., Habib, M. S., Farhad, S. F. U., Tanvir, N. I., Shaikh, M. A. A., ... & Siddika, A. (2022). Effects of yttrium doping on structural, electrical and optical properties of barium titanate ceramics. *Heliyon*, *8*(9).
4. Trzepiecinski, T., & Gromada, M. (2018). Characterization of mechanical properties of barium titanate ceramics with different grain sizes. *Mater. Sci*, *36*, 151-156.

Note: All the figures in this chapter were made by the authors.

Recent Trends in Applied Physics and Material Science – Sudhir Bhardwaj et al. (eds)

Electrical Properties of $Li_2Pb_2Dy_2W_2Ti_4V_4O_{30}$ Ferroelectric Ceramics

Aditya K. Sahu*
Center for UG & PG Studies (CUPGS),
Biju Patnaik University of Technology,
Rourkela, Odisha, India

S. Behera
Department of Physics,
Centurion University of Technology
and Management,
Bhubaneswar, Odisha, India

Piyush R. Das
Center for UG & PG Studies (CUPGS),
Biju Patnaik University of Technology,
Rourkela, Odisha, India

Abstract: A polycrystalline ceramic, $Li_2Pb_2Dy_2W_2Ti_4V_4O_{30}$, was synthesized by a traditional mixed-oxide method at a moderate temperature. The material's growth under the specified conditions was previously confirmed using an X-ray diffraction technique. The frequency-dependent change of the dielectric constant might indicate the prevailing polarization within a specific frequency range of the molecule. Analysis of the materials' electrical properties was conducted using the impedance spectroscopy method. Systematic investigation of impedance and its associated factors indicates that the electrical characteristics of the material are significantly influenced by temperature and are well correlated with their microstructure. Electrical phenomena in the material occur as a result of its bulk properties. Accurate determination of the bulk resistance utilizing complex impedance spectra reveals a negative temperature coefficient of resistance. Increasing temperature shifted the peaks in the loss spectrum towards higher frequencies, showing the relaxation process within the system. An analysis of the complex electric modulus indicates that only the grain effect contributes to the total capacitance. The variation of the imaginary part of the electric modulus with frequency is associated with the short-range mobility of charge carriers. There is great interest in studying ferroelectrics of a new class of tungsten-bronze vanadate structure, $Li_2Pb_2Dy_2W_2Ti_4V_4O_{30}$, as it has found application in microwave devices and electric tunable devices.

Keywords: Ferroelectric, Electrical Properties, Impedance, Modulus Spectroscopy

1. INTRODUCTION

The members of the oxygen octahedra family having one or more component oxides play a significant role in ferroelectrics. All the materials of this structure have different crystal structures, electrical and mechanical properties, T_C and polarization (Aiger, 1972). The interest in tungsten bronze (TB) ferroelectrics was reinvigorated in the 1960s because of large optical nonlinearity (Jamieson et al., 1965). TB-structured materials have either tetragonal or orthorhombic symmetry. Many different ferroelectric oxides, including oxygen octahedra, have been the topic of significant research in recent years due to their varied industrial and technological applications. Even though there are several TB-type ferroelectrics available today, there is a growing interest in the electro-optic usage of ferroelectric niobates. Li and Chen (2001) investigated the dielectric and structural properties of a series of TB-structured compounds for microwave applications. They also analyzed the modification of characteristics with non-stoichiometric composition.

A comprehensive literature review indicates that substantial research has been conducted on ferroelectric ceramics belonging to the TB structural family. The physical and device characteristics of ferroelectrics can be readily modified by including appropriate impurities. Another attractive aspect of this family is that it provides

*Corresponding author: adityasahu49@gmail.com

DOI: 10.1201/9781003684718-86

substantial property tailoring by introducing relevant cations with varied ionic radii and valences at various interstitial sites. Based on these, we synthesized (by mixed oxide method) a new group of TB-structure vanadate, $Li_2Pb_2Dy_2W_2Ti_4V_4O_{30}$, and investigated their structural, dielectric, and other electrical properties.

2. EXPERIMENTAL PROCEDURE

The polycrystalline sample $Li_2Pb_2Dy_2W_2Ti_4V_4O_{30}$ was formed using a mixed oxide approach, a solid-state reaction at a moderate temperature. The formation of the chosen compound was verified through experimental structural investigation utilizing the X-ray diffraction (XRD) technique (Parida et al.2013). The sample's dielectric characteristics (ε_r and tanδ) were measured at room temperature with varied frequencies using a Hioki 3532 LCR HiTESTER. The electrical (impedance and modulus) characteristics of the compound were measured using a computer-controlled HIOKI LCR HiTester at an AC signal level of amplitude 1.5 V over a wide temperature range from room temperature to 500^0C.

3. RESULTS AND DISCUSSIONS

3.2 Variation of Relative Dielectric Constant (ε_r) and Loss Tangent (tanδ) with Frequency

The variation of ε_r with frequency can indicate the presence of polarization in the molecule at specific frequencies. Figure 86.1 displays the frequency dependence of the sample's ε_r and tanδ at room temperature. The relative dielectric constant (ε_r) of the sample decreases with increasing frequency, a pattern typical of most dielectrics (Anderson, 1964). When frequency increases from 100 Hz to 1 MHz, ε_r changes from 127 to 90, and tan δ changes from 0.5344 to 0.0105. A decrease in space charge polarization lowers the value of ε_r. The value of ε_r falls when space charge polarization decreases. The tanδ decreases as the frequency increases because of the high packing fraction of the pellet sample.

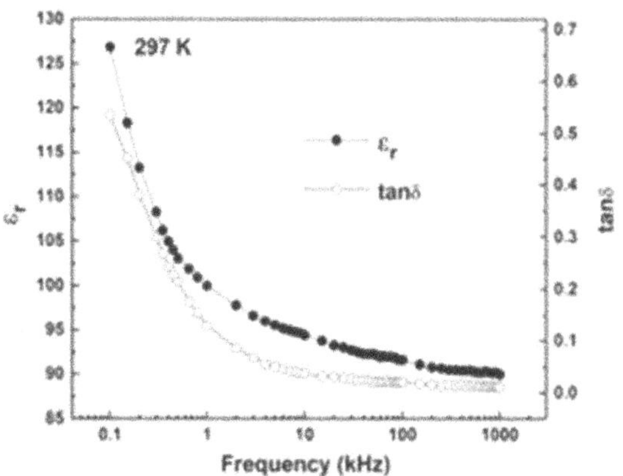

Fig. 86.1 Frequency dependence of dielectric parameters (ε_r & tanδ) at room temperature

3.2 Impedance Analysis

Figure 86.2 illustrates the variations of the real and imaginary components of the sample's impedance (Nyquist plot). At low temperatures, an arc appears, transforming into a semicircle as temperatures rise. Generally, whether a full, partial, or no semicircle is observed depends on the relaxation intensity and the empirically available frequency range (Gerhardt 1994). The intercept of the semicircular arc along the Z' axis (i.e., the x-axis) yields the bulk resistance value, which decreases with increasing temperature, indicating an increase in dc conductivity. The low-frequency semicircle results from the grain boundary effect. The higher frequency semicircle represents the bulk effect. The bulk effect results from the parallel combination of bulk resistance and capacitance.

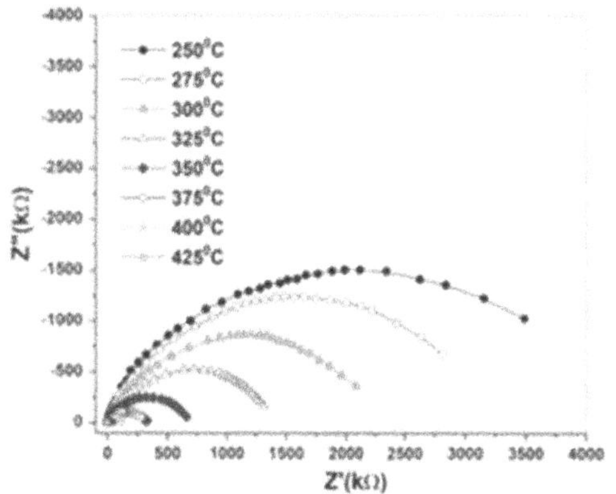

Fig. 86.2 Complex impedance spectra of the compound

3.3 Electric Modulus Study

Figure 86.3 displays the compound's complex electric modulus spectrum (M' vs. M''). The curves did not form the precise semicircles necessary for the ideal Debye model. They have the shape of a distorted semicircle, with the centers located below the x-axis. The non-Debye type of relaxation in the materials is thus concluded based on the relaxation dispersion with various (mean) time constants. The single semicircular arc confirms a single-phase compound's development, indicating the total capacitance contributed by the grain effect only.

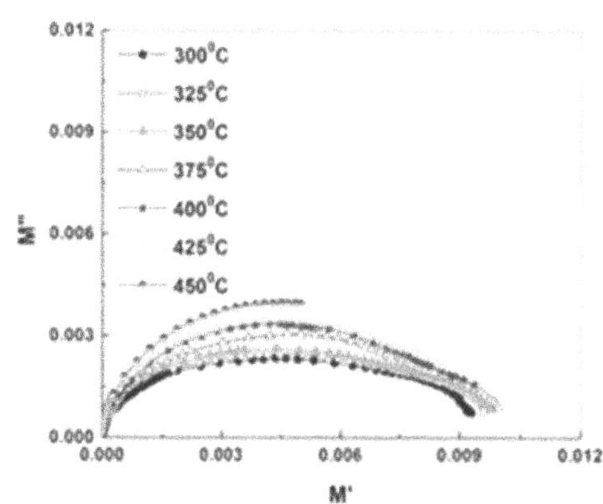

Fig. 86.3 Complex modulus spectrum of the compound

4. CONCLUSION

This study looks into dielectric characteristics (ε_r and tan δ) with frequency variation, complex impedance spectroscopy, and modulus analysis to investigate the electrical properties of new tungsten bronze ferroelectric vanadates, $Li_2Pb_2Dy_2W_2Ti_4V_4O_{30}$. The variation of the dielectric constant with frequency can reveal the predominant polarization in the molecule in a specific range of frequencies. Complex modulus research reveals that the grain boundary causes the effect in the compound. Complex impedance analysis shows that the material's dielectric relaxation is polydispersive and non-Debye. Studying the resistance variation with temperature expedites the material's application as an NTC-type thermistor.

REFERENCES

1. Aiger, F. W. (1972), Modern Oxide Materials-Preparation, properties and device applications, Academic Press. Inc. London.
2. Jamieson, P. B., Abrahams, S. C., Bernstein, J. L., (1965) J. Chem. Phys. 48, 5048.
3. Li, J., Chen, X. M., (2001) J. Eur. Ceram. Soc. 21, 155.
4. B. N. Parida B.N., Das, P. R., Padhee R., Choudhary, R. N. P. (2013) J. Electron. Mater. 42, 2587-2594.
5. Anderson, J. C., (1964): Dielectrics (Chapman & Hall, London).
6. Gerhardt, R. (1994) J. Phys. Chem. Solids 55, 1491.

Note: All the figures in this chapter were made by the authors.

Recent Trends in Applied Physics and Material Science – Sudhir Bhardwaj et al. (eds)
© 2026 Taylor & Francis Group, London, ISBN 978-1-041-16452-4

Physical, Optical and EPR Studies of Alumino Magnesium Borate Glasses Containing Cu^{2+} ions

Mohamad Raheem Ahmed*, Ummey Haani, Sayyad Muskan, Saba Farheen, Syed Afsheen Fatima, Shaik Kareem Ahmmad
Muffakham Jah College of Engineering and Technology, Hyderabad, India

Abstract: In this study, xAl_2O_3-(30-x) MgO-69.5B_2O_3-0.5CuO glasswasproduced using the method melt- quench technique. The role of MgO and Al_2O_3modifiers was studied using physical (density and molar volume), optical and EPR techniques. Sharp peak free X-ray figures of all glass samples in the present work confirmed the amorphous nature i.e., non-crystalline nature. The decreasing density values of the current glasses suggest structural reform as a result of theincrease in Al_2O_3 content. Al_2O_3 interacts with the glass composition, forming AlO_4 and BO_3 units along with non-bridging oxygens (NBOs). The calculated molar volume is based on the observed density measurements. This study employs ESR and optical absorption techniques to explore the behaviour of the dopant CuO and the modifier MgO. The spin-Hamiltonian parameters (A-parallel, perpendicular, g-parallel, perpendicular, α2, β12) were determined from the ESR spectra, revealing that the state of Cu^{2+} is $d_{x^2-y^2}$($^2B_{1g}$ ground state). The Cu^{2+} ion is situated in a tetragonally observable distorted (elongated along c axis) octahedral environment. All MABC glasses containing Cu^{2+} ions as dopant exhibited a broad optical absorption band associated with $2B_{1g} \rightarrow 2B_{2g}$ transition. Observations from optical spectra revealed the shifting of a single broad peak (787 nm), along with changes in the optical band gap ε_g (from 3.49 to 2.98) and Urbach energy values (from 0.720 to 0.672), indicating structural changes of glass due to the variation of modifiers.

Keywords: Alumino borate glasses, Borate glasses, Copper doped glasses, EPR studies

1. INTRODUCTION

In recent years, Alumino borate glasses have significant attention from researchers due its scientific and technological interest. In addition to the doping of transition metal(Cuo)ions which leads to several applications in the fields of solid-state devices, lasers, optical materials and optoelectronic devices (Aboushaswa et al., 2024, Narsimha et al., 2024). Borate oxide-based alumino borate glasses, which incorporate alkaline (MgO, CaO, BaO) oxides as modifiers, offer several benefits, including high refractive index, excellent electrical resistance, low dispersion, low thermal expansion coefficient, and high density. These properties make them suitable for optoelectronic and radiation shielding applications. MR Ahmed et al suggested that in borate glasses, Al_2O_3 exists as AlO_3 and AlO_4 structural elements. In that, tetra Alumino units (AlO_4)function as glass formers, strengthening the glass structure, while the distorted observed octahedral hexa alumino-AlO_6 units behaves as modifiers, introducing bonding defects and reducing the overall glass strength Al^{3+} forms tetrahedral structures only when alkaline earth metals are present, leading to notable changes in the optical (band gap) and electrical properties of the glasses (Mohamad Raheem Ahmed et al., 2022, Krisha Priya et al., 2022, Ahmed at al., 2019, Aljawhara et al., 2024). However, the technical literature lacks adequate information regarding the conditions for glass formation and the role of Al_2O_3 in alkaline borate glasses, which are essential for justifying their production processes and determining the optimal composition.

*Corresponding author: mohdraheem1980@gmail.com

DOI: 10.1201/9781003684718-87

In this study, xAl_2O_3-(30-x) MgO-69.5B_2O_3-0.5CuO (0 to15 with 2.5 mole% regular increment) glasses were examined through physical, optical and EPR analyses. The primary focus of this article mainly examines the impact of Al_2O_3 in the borate glass matrix, particularly in the presence of MgO and transition metal ions.

2. EXPERIMENTAL

Copper-doped magnesium alumino borate glasses named as MABC were synthesized using the melt-quench technique. The glass composition xAl_2O_3-(30-x) MgO-69.5B_2O_3-0.5CuO with x values of 5, 7.5, 10, 12.5 and 15 mole% (denoted as MABC1, MABC2, MABC3, MABC4, MABC5 respectively) was melted at 1000°C in an electric muffle silicon rod furnace. The MABC chemical mixture was prepared through weighing the components according to their molar ratios and scaled to a total of 15 grams. The raw material wassettled in a J-series porcelain crucible and subjected to heat in an electrically controlled furnace bearing upto1100°C, where it took approximately 60 minutes to achieve a homogeneous melt. The molten mixture was then rapidly (within a minute) cooled by spew it onto a preheated stainless-steel plate (150°C) using an electrical filament heater. To alleviate internal stresses, the glasses were continues and study heated at 300°C for 5 hours in separate furnace. The final glasses were transparent and coloured glasses, with a thickness between 0.5 and 1.5 mm.

The prepared glasses were examined using an XRD (Philip Xpt Pro) at a continues scanning rate of 2° (2θ values from 0 to 100) /min to confirm their amorphous nature. UV-Vis spectra of MABC glasses were obtained using a Shimjadzu UV-1800 spectrophotometer (wavelength ranges from 0 to 1000nm) at room (300K) temperature. EPR spectra of the MABC glass samples were annals on an advance level of BRUKER electron paramagnetic resonance spectrometer which was 7operates at an X-band frequency of 9.7 GHz with higher field modulation of 100 kilo Hz.

3. RESULTS AND DISCUSSION

Figure 87.1 displays the X-ray of all MABC diffraction pattern of the glasses studied. The lack of sharp edge Bragg peaks in the diffraction pattern indicates the amorphous nature(not a crystalline material) of the prepared all MABC glass. The calculated physical parameters (depicted in Table 87.1), such as density (ρ), are crucial for observing structural changes in glass. The density (be entitled as ρ) of all MABC glasses was deduced from Archimedis principle (present in the reference 3)with xylene serving as the reference liquid, and the calculated density and molar volume (indicated as V) values, along with the appropriate formulas discussed in our previous article (Mohamad Raheem Ahmed et al., 2022) are provided in Table 87.1. It was found that the density (ρ values) decreases from 3.452 to2.732 while the molar volume (V values) increases from18.52 to25.66 with notable change of Al_2O_3 content. The incorporation of Al_2O_3 (increased from 5 to 15 mol%) into the glass leads to an increase in BO_3 units, which consume some of the oxygen atoms in the MgO. These oxygen atoms are required to convert tri-BO_3 to tetra-BO_4, and the priority shifts to transforming AlO_3 (Trivalent alumino oxygen) to AlO_4(Tetravalent alumino oxygen) This process results in an increase in glass volume, thereby reducing the overall glass density (Mohamad Raheem Ahmed et al., 2022, Ahmed et al., 2019).

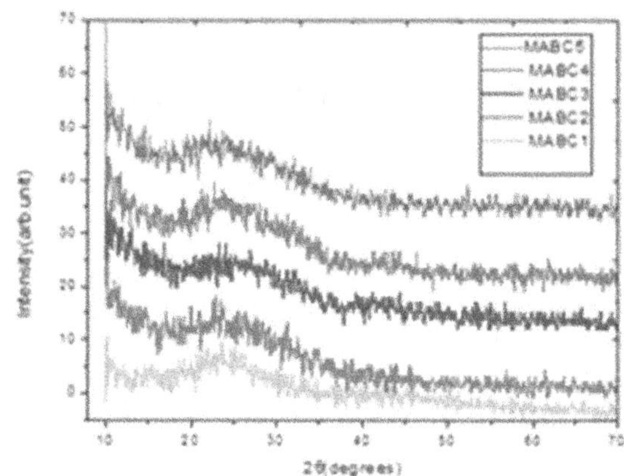

Fig. 87.1 XRD spectra of MABC glass

Figure 87.2 depicts the optical (UV-Visi) 300 to 1000nm absorption spectra of MABC glass, highlighting a single, wide absorption bump i.e., band pivoting around 787 nm. This generalization is related to the$^2B_{1g}$→$^2B_{2g}$energytransition of Cu^{2+} ions observed within a distorted structure along an axial octahedral environment (Ahmed et al., 2019). The intensity of this single band at 787nm is decreases as the Al_2O_3 mole percentage is increases from 5 to 15.

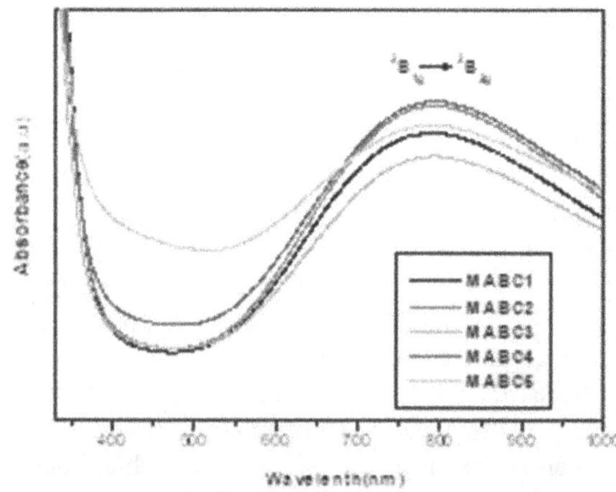

Fig. 87.2 Optical (UV-Visible) absorption of MABC glassspectra

Figure 87.3 illustrates the indirect band gap, with its values determined using an equation. (Aljawhara et al., 2024)

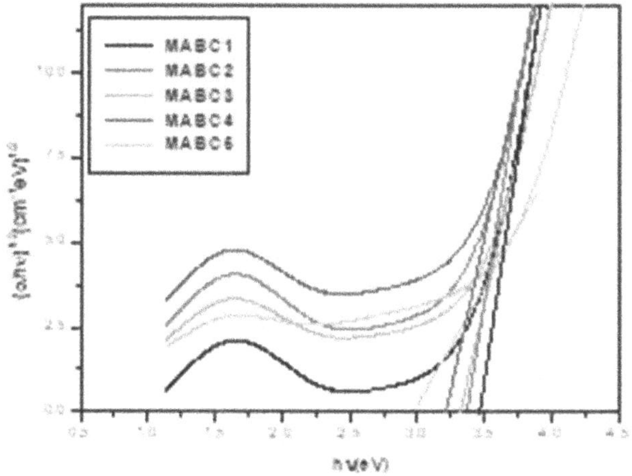

Fig. 87.3 Indirect bands of MABC glass

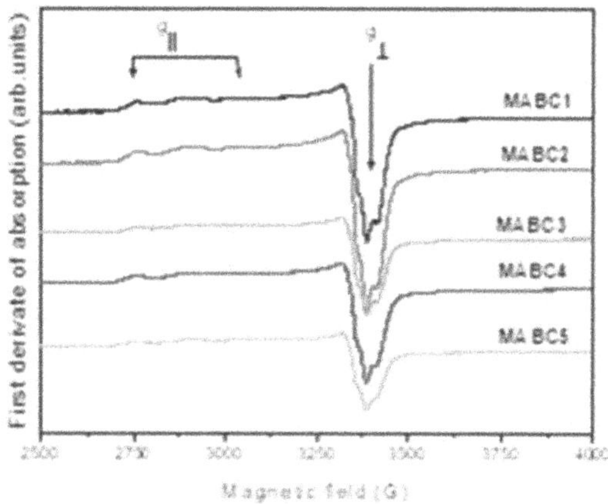

Fig. 87.4 The EPR spectra of MABC glass

$$E_{opt} = h\nu - \left(\frac{\alpha h\nu}{A}\right)^{1/2} \quad (1)$$

Here, α depends on ν, and A is a constant.

The Urbach 0.5 to 4.5 eV energy values of all MABC are determined from plots of lnα(ν) versus hν, employing the subsequent equation (Mohamad Raheem Ahmed et al., 2022)

$$\alpha(\nu) = \text{const}\exp\left(\frac{h\nu}{\Delta E}\right) \quad (2)$$

ΔE denotes the width of the optical absorption band tails in the localized states of all MABC glasses. The optical band gap energies (E_{opt}) and Urbach ΔE energy values of all MABC glasses are summarized as1to5 MABC in Table 87.1. The observed decrease is likely attributed to structural modifications within the glass network, which may result from increased disorder and the addition of the modifier Al_2O_3, which induces the introduction of localized states inside the band gap. Lower band gap values suggest the glass may exhibit enhanced absorption in the visible or ultraviolet regions, potentially altering its optical and electronic properties for specific applications.

From Figure 87.4, it is clearly observed that there are three parallel bumps i.e., components ($g_{||}$) and a fourth component that overlaps with the perpendicular components ($g_\perp$). These four components parallel and perpendicular are consistent with the relation (2I+1), where the nuclear spin I = 3/2 corresponds to 63Cu, 65Cu and spin S=1/2.

The EPR graph was analysed using a Hamiltonian of the following form

$$H = \beta[g_{|}B_zS_z + g_\perp(B_xS_x + B_yS_y)] + A_{|}S_zI_z + A_\perp(S_xI_x + S_yI_y) \quad (3)$$

In the context of the previously mentioned equation, the parameter (z) denotes the symmetry axis associated with each copper ion, which is in the oxidation state of plus 2.β direct to Bohr magneton, $g_{||}$ (parallel values are depicted in Table 87.1) and $g_\perp$ (perpendicular values are calculated using equation present in the reference 5) components of the Lande factor. In EPR spectra $A_{||}$, $A_\perp$ are components of hyperfine coupling tensor A and β is component of magnetic field.

The Spin-Hamiltonian g, A parameters were determined from the EPR x-band spectra and are listed in Table 87.1. From calculated data it is observed that g values and A values shows relation $g_{||} > g_\perp > g_e$ and $A_| > A_\perp$are typical for Cu^{2+} ions surrounded (co-ordinated) by six oxygen ligands, forming generally an octahedron elongated towards the z or c-axis is $d_{x^2-y^2}$($^2B_{1g}$ ground state).

The in-plane sigma σ bonding is denoted by α^2, with a value of approximately 0.77, indicating a moderately covalent character. The π bonding, represented by β_1^2, has a value

Table 87.1 Physical, Optical and EPR Spin Hamiltonian (g, A, α and β) parameters of MABC glasses

Glass Code	Density(ρ) g/cm³	Molar Volume (V)	Optical Band Gap E_{opt} (eV)	Urbaach Energy ΔE(eV)	$g_{\|\|}$	$g_\perp$	$A_{\|\|}$ (10^{-4} cm^{-1})	$A_\perp$ (10^{-4} cm^{-1})	α^2	β_1^2	β^2
MABC1	3.452	18.52	3.49	0.720	2.336	2.073	147	23	0.802	0.794	0.951
MABC2	3.325	19.69	3.35	0.742	2.343	2.071	141	21	0.689	0.940	0.947
MABC3	2.961	22.64	3.30	0.864	2.362	2.066	138	20	0.774	0.883	0.993
MABC4	2.859	23.98	3.24	0.785	2.352	2.064	131	20	0.773	0.858	0.995
MABC5	2.732	25.66	2.98	0.672	2.251	2.063	128	21	0.782	0.838	0.982

of about 0.85, suggesting an ionic nature. Additionally, β^2 corresponds to out-of-plane π bonding, and its value, close to 0.98, also indicates ionic characteristics (Ramadevudu,et al., 2000).

4. CONCLUSION

The lack of distinct Bragg peaks in the diffraction pattern suggests the material's amorphous in nature. As the Al_2O_3 content increases from 5 to 15 mole percent, the density decreases while the molar volume increases. The optical wide band is ascribed to the (vibrational energy $^2B_{1g} \rightarrow {}^2B_{2g}$) energy level transition of Cu^{2+} ions in distorted structure octahedral environments, with its intensity diminishing as Al_2O_3 is incorporated. ESR spectra show that the relationship $g_{\|} > g_{\perp} > g_e$ and $A_{\|} > A_{\perp}$ is characteristic of Cu^{2+} ions coordinated by six ligands in an octahedral structure which is elongated along the z-axis. The ground state is $d_{x^2-y^2}$($^2B_{1g}$state), and binding coefficients indicate that the bonding between copper ions and their ligands is predominantly ionic in nature.

REFERENCES

1. Abouhaswa, A.S.,Abdelghany,A.M., Alfryyan, N. et.al. (2024). The impact of B_2O_3/Al_2O_3 substitution on physical properties and γ-ray shielding competence of aluminum-borate glasses: comparative study. J Mater Sci:Mater Electron 35:845
2. Narsimha,B., Sekhar, K.C., Shareefuddin, M. (2024). Influence of In_2O_3 on Optical, EPR, and Structural Properties of Barium Borate Glasses Containing Cu^{2+} Spin Probe. Braz.J.Phys 54:40.
3. Mohamad Raheem Ahmed, Chandra Sekhar. k, Sheik Ahammed, Vasant Sathe, Alrowaili.Z.A, Mongi Amami, I.O. Olarinoye, M.S. Al-Buriahi, Tonguc B.T, Shareefuddin, M.D,(2022) Synthesis, physical, optical, structural and radiation shielding characterization of borate glasses: A focus on the role of SrO/Al_2O_3 substitution, Cera. Inter. ,48(2):2124-2137.
4. Krishna Priya,G.Yusub, S. Ramesh Babu,A. Sree Ram, N. Aruna, V.(2022), Electricaland spectroscopic characteristics of B_2O_3–Bi_2O_3–Al_2O_3–MgO glasses alloyed with MnO, J. Phy. Chem.Soli.,170:110957.
5. Ahmed, M.R., Shareefuddin, M (2019). EPR, optical, physical and structural studies of strontium alumino-borate glasses containing Cu2+ ions. SN Appl. Sci. 1:209
6. Aljawhara H. Almuqrin, K.A. Mahmoud, U. Rilwan, Sayyed,M.I.(2024), Influence of various metal oxides (PbO, Fe_2O_3, MgO, and Al_2O_3) on the mechanical properties and γ-ray attenuation performance of zinc barium borate glasses, Nucl. Eng Tech., 56(7):1738-5733.
7. Ramadevudu,G.Shareefuddin, Md. Sunitha Bai, N. Lakshmipathi Rao, M. Narasimha Chary, M(2000). Electron paramagnetic resonance and optical absorption studies of Cu^{2+} spin probe in MgO–Na_2O–B_2O_3 ternary glasses, J. Non-Cryst. Soli. ,278:205-212.

Note: All the figures and tables in this chapter were made by the authors.

Recent Trends in Applied Physics and Material Science – Sudhir Bhardwaj et al. (eds)
 ISBN 978-1-041-16452-4

Microwave Dielectric Properties of Combustion Synthesized. $RAlO_3$ (R = Gd, Nd, Sm, Pr and Dy) Ceramics

Saji S. K.*

Government Polytechnic College, Attingal, Kerala, India

Abstract: High-purity $RAlO_3$ dielectric ceramics (R = Gd, Nd, Sm, Pr, Dy) were synthesized using the solution combustion method. X-ray diffraction analysis confirmed the phase purity and determined the crystal symmetry of the synthesized materials. The microwave properties were assessed using the Hakki-Coleman method, which employed a custom-built post-resonator test fixture to identify the TM010 mode within the microwave frequency region. Measurements were taken using a network analyzer. The examined materials exhibit promising permittivity levels, making them well-suited for applications in dielectric resonators.

Keywords: Rare-earth aluminates, Perovskites, Microwave dielectrics

1. INTRODUCTION

Perovskites are an important class of inorganic crystals represented by the structure ABX_3, where A and B stand for cations and X for an anion. Their unique physical properties make them highly valuable for various technological applications, primarily due to their diverse range of pseudo-symmetrically related crystal structures. This study presents the synthesis of rare-earth aluminate systems, a significant class of perovskite-structured materials, using a solution combustion technique. The dielectric properties of these materials were also investigated and measured [1,2].

Rest of the paper is structured as follows. Section-2 reviews the extant literature. Section-3 describes the sample and variables. Section-4 explains the research methodology. Secttion-5 discusses the empirical findings. Section-6 summarises the paper.

2. EXPERIMENTAL DETAILS

Aqueous solutions containing ions of aluminum (Al) and rare earth elements (R), where R can be Gd, Nd, Sm, Pr, or Dy, were prepared by dissolving precisely measured high-purity aluminum nitrate nonahydrate $(Al(NO_3)_3{\cdot}9H_2O$ and rare earth oxides (R_2O_3 or R_3O_6 in a specific ratio. The synthesized powders underwent calcination at 1000°C for 2 hours, resulting in successfully forming the $RAlO_3$ phase. The synthesized powders were sintered to achieve a high density, reaching up to 97% of the theoretical density. X-ray diffraction (XRD) analysis was carried out on a powder diffractometer (Bruker AXS D8 Advanced). The dielectric properties of the sintered specimens were characterized in the X-band frequency range using a network analyzer (Aeroflex IFR6823). The dielectric permittivities were then calculated using the post-resonator method.

3. RESULTS AND DISCUSSION

XRD analysis indicated that $GdAlO_3$, $SmAlO_3$, and $DyAlO_3$ crystallized in an orthorhombic structure, whereas $NdAlO_3$ and $PrAlO_3$ adopted a rhombohedral crystal structure. Figure 88.1 and Figure 88.2 presents the overlaid XRD patterns of the synthesized $RAlO_3$ samples. The XRD peaks of the synthesized $RAlO_3$ systems show excellent agreement with reported data from the International Centre for Diffraction Data (ICDD), confirming the phase purity of the samples.

*Corresponding author: brahmadev.panda@gmail.com

DOI: 10.1201/9781003684718-88

Remarkably, no diffraction peaks corresponding to impurities or secondary phases were detected in the XRD pattern.

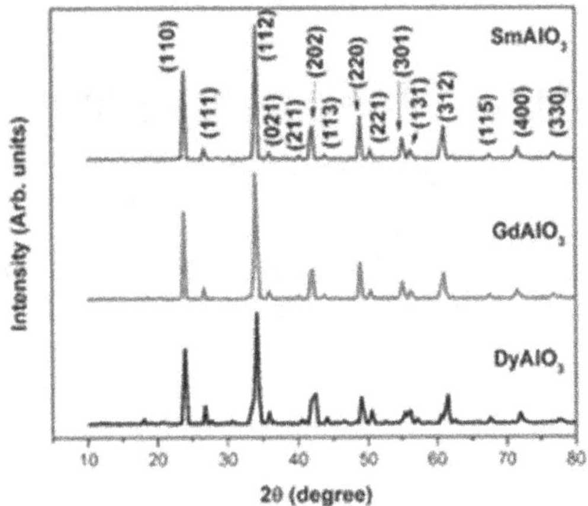

Fig. 88.1 XRD pattern of the orthorhombic system

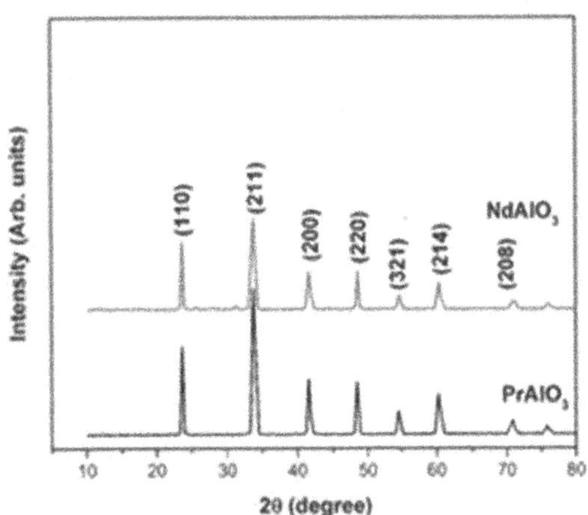

Fig. 88.2 XRD pattern of the rhombohedral system

Dielectric behavior at microwave frequencies of $RAlO_3$ perovskites are presented in Table 88.1. Microwave dielectric Studies indicate that the permittivity of rhombohedral structured materials is generally higher than those of orthorhombic structured materials. The elevated permittivity in rhombohedral-structured rare-earth aluminates is primarily due to the increased polarizability of the incorporated rare-earth ions, which enhances the overall dielectric properties of these compounds.

Ionic polarizability, which contributes significantly to microwave dielectric permittivity, results from the displacement of cations and anions in opposite directions when an electric field is applied [3]. The displacement of ions in ionic compounds is influenced by the interplay between bond strength and bond length[4]. Specifically, stronger bonds and shorter bond lengths between anions and cations result in reduced ionic displacement

Table 88.1 Microwave dielectric properties of $RAlO_3$ specimens

System	Dielectric Constant	Resonant Frequency
$GdAlO_3$	14.70	7.20
$NdAlO_3$	16.92	6.35
$SmAlO_3$	18.56	6.60
$PrAlO_3$	19.08	6.42
$DyAlO_3$	11.99	12.00

4. CONCLUSION

It was found that among the studied $RAlO_3$ perovskites, $NdAlO_3$ and $PrAlO_3$ with rhombohedral crystal symmetry show a higher permittivity than, $DyAlO_3$ and $SmAlO_3$ with orthorhombic crystal symmetry. A comparative investigation of $GdAlO_3$ microwave dielectric properties in Rare-Earth aluminate compounds exhibiting rhombohedral structures consistently displayed higher permittivity values than those with orthorhombic structures. The increased permittivity observed in rhombohedral structures over orthorhombic ones is due to differences in ionic size and coordination numbers. These materials have shown great potential for use in dielectric resonators due to their favorable dielectric characteristics.

REFERENCES

1. Atwood, D.A. and Yearwood, B.C., 2000. The future of aluminum chemistry. Journal of Organometallic Chemistry, 600(1-2), pp.186-197.
2. Khirade, P.P. and Raut, A.V., 2022. Perovskite structured materials: synthesis, structure, physical properties and applications. In Recent advances in multifunctional perovskite materials. IntechOpen.
3. Cho, S.Y., Kim, I.T. and Hong, K.S., 1999. Microwave dielectric properties and applications of rare-earth aluminates. Journal of materials research, 14, pp.114-119.
4. Attfield, J.P., 2001. Structure-property relations in doped perovskite oxides. International Journal of Inorganic Materials, 3(8), pp.1147-1152.

Note: All the figures and tables in this chapter were made by the authors.

Recent Trends in Applied Physics and Material Science – Sudhir Bhardwaj et al. (eds)

Giant Energy-storage Density with Ultrahigh - Efficiency in Pb-free $0.94NaBiTiO_3$–$0.06BaTiO_3$ Bulk Ceramics

T. V. Narmada
Jawaharlal Nehru Technological University Anantapur, Ananthapurum, AP., India
Department of Physics, Seshadrirao Gudlavalleru Engineering College, Gudlavalleru

D. Zarena*
Department of Physics, JNTUA College of Engineering, Ananthapurum, AP., India

Abstract: Among the notable features of the lead-free material $NaNbO_3$ (NNb) are its exceptional (W_{rec}) and high (E_b). However, the considerable energy loss associated with NNb limits its applicability in energy storage, adversely affecting both (W_{rec}) & efficiency (η). This experiment explores a conceive new way to enhance the capacity of $(1\text{-}x)0.94NBT\text{-}0.06BaTiO_3\text{-}xNaNbO_3$ ($x = 0.1\text{-}0.4$) ceramics. The discoloration of NNb disrupts the long-range ordered structure and effectively reduces grain size, resulting in the formation of internal regions , that help titivate energy density loss. As the concentration of NNb increases, significant improvements in both W_{rec} and η are observed. For $x = 0.3$ exhibit the best performance among all variations, achieving density ($W_{rec} = 3.13$ J/cm³), rapid conversion efficiency ($\eta = 85\%$), and excellent breakdown strength. These Pb-free based ceramics present compelling hope for current exertions as well as next-generation electro ceramics.

Keywords: X-ray diffraction, Polarization, Energy storage

1. INTRODUCTION

Dielectric energy storage capacitors are crucial components in essential electrical power systems, Due to their deftness to efficiently store and rapidly deliver high-power electrical energy. Their performance is vital in applications where quick energy release and reliability are paramount ensuring the proper functioning of devices and systems that require instantaneous bursts of power by (H. Palneedietal, 2018). Strong polarization and high breakdown strength are essential for polar and dielectric materials to achieve high energy density. Their enhanced polarization characteristics enable them to store more energy, making them ideal for high-performance energy storage systems by (V. Misra et.al, 2023). This includes relaxors, antiferroelectrics, and normal ferroelectrics, all of which exhibit greater energy density because of the spontaneously polarized electric dipoles that can be reoriented by an electric field, resulting in significantly enhanced electric polarization by (R. Khosla & S. K. Sharma,2021)

The aim of these initiatives is to enhance the capabilities and performance of pulsed power systems across various applications. However, progress in developing bulk dielectric ceramics with comprehensive energy storage capabilities has been limited despite on-going inquiry. To achieve both high W_{rec} and η, materials must possess high dielectric breakdown strength (E_b), high maximum power density (P_{max}), and low remnant polarization (P_r). Achieving these characteristics in dielectric materials remains a significant challenge. To fully utilize bulk dielectric ceramics in energy storage applications, several criteria must be met.

*Corresponding author: zareenajntua@gmail.com

DOI: 10.1201/9781003684718-89

2. EXPERIMENTAL PROCEDURE

Crystalline ceramics with the composition xNaNbO$_3$-(1-x)0.94NBT-0.06BaTiO$_3$-(x=0.1-0.4) were fabricated standard solid-state(reaction) techniques.

3. RESULTS AND DISCUSSION

Analyse the structural characteristics of these ceramics, a dual-phase model is employed, effectively representing both the rhombohedral phase (R3c) and tetragonal phase (P4mm). Due to the non-centrosymmetric and polar nature of the rhombohedral structure, polarities shift parallel to the A-site and B-site cations, along with the directions of the (111) Bragg peaks. Consequently, the introduction of NNb disrupts the long-range polar phase. XRD measurements, complemented by Rietveld refinement using the FULLPROF software package (version 2000), quantify phase fractions and elucidate the role of structural phases in the substitution of present composition. Several dual-phase models yield satisfactory fits with stable parameters, specifically the tetragonal (P4mm) and rhombohedral (R3c) phases. The R3c phase demonstrates a significant contribution compared to the P4mm space group. The structural phase contributions and quantitative phase fractions were confirmed through Rietveld refinement of the XRD data for (1-x) 0.94NBT-0.06BaTiO$_3$- xNaNbO$_3$(x=0.3). Various dual-phase models with stable fitting parameters further validate these findings, reinforcing the substantial role of the R3c phase over the P4mm phase. The improved fitting parameters, essential for assessing the quality of the fit. The refinement graph, along with the Bragg positions and pseudo cubic representations, is depicted in Fig. 89.1.

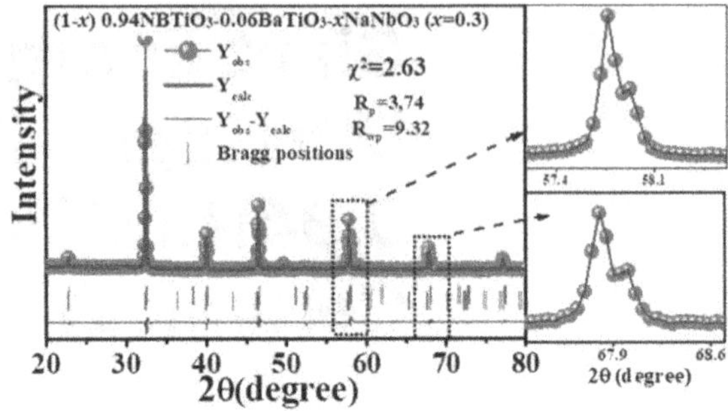

Fig. 89.1 Rietveld refinement (x = 0.3)

Figure 89.2a shows the P-E curves of (1-x) 0.94NBT-0.06BaTiO$_3$- xNaNbO$_3$ (x = 0.1-0.4) ceramics under distiunguish applied(E) - 236 kV/cm, curves are approximately linear, suggesting that the material remains in the antiferroelectric (AFE) phase and does not transition to the ferroelectric (FE) phase below this threshold. Except for (1-x) 0.94NBT-0.06BaTiO$_3$- xNaNbO$_3$ (x = 0), the (E_C),(P_r), and polarization (P_S) values are consistent across all specimens. The material's energy storage capacity, loss capacity, and storage performance can be derived positive

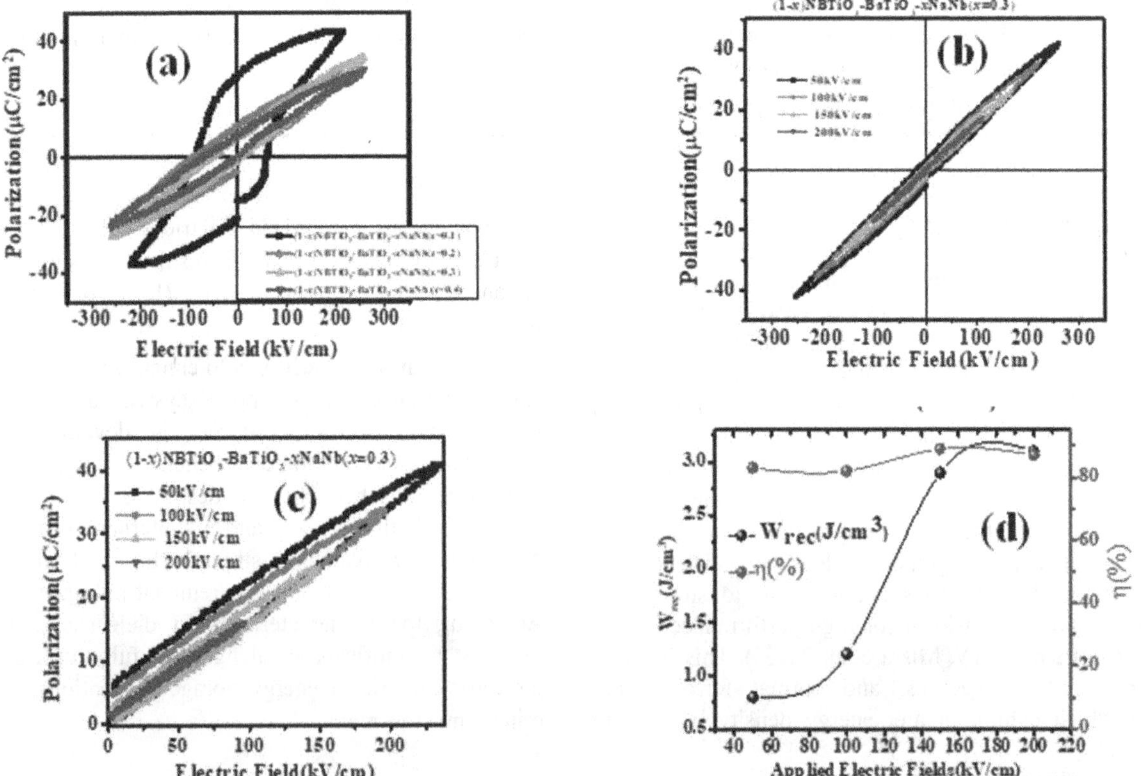

Fig. 89.2 (a) P-E curves of (x = 0.1-0.4) ceramicsj. (b) (x = 0.3) at variant fields. (d) Recoiel energy (x = 0.3)

quadrant of the P-E loop, attributed to the disparity in the P Vs E graph (H. Zhang et.al 2010) Notably, for the room temperature polarization loops.($W_{total\ energy\ capacity}$) and energy storage capacity (recoverable) (W_{rec}) of the NBT-SrT-*x*KNb (*x*=0.1-0.4) ceramics exhibit a roughly straight increase from (50 kV/cm) to (236 kV/cm,) as shown in Fig. 89.2d. Antiferroelectric ceramics that do not undergo an AFE-FE change behave similarly to linear ceramics, and the linear association between W_{rec} and E can be attributed to the inverse polarization displacement. Notably, NBT-SrT-*x*KNb (*x*=0.3) ceramics demonstrate impressive energy storage performance, achieving an efficiency (η) of 76% and a W_{rec} of 3.13 J/cm³. In comparison, KN ceramics, which have a W_{rec} of 2.66 J/cm³ and an η of 85%, show increases of 24% in W_{rec} and 245% in energy storage capacity, respectively.

REFERENCES

1. H.Palneedi, M.Peddigari, G.T. Hwang,D.Y. Jeong & J. Ryu, High-performance dielectric ceramic films for energy storage capacitors: progress and outlook. *Adv. Funct. Mater*, *28*(42), 1803665(2018).
2. V.Misra, S. Khan,M. Yadav, A. Singh &V. Singh, Recent Advances in Dielectric Materials for Energy StorageDevices:A Comprehensive Overview. *Energy Harvesting and Storage Devices*, 211-243.
3. R. Khosla&S. K. Sharma, Integration of ferroelectric materials: an ultimate solution for next-generation computing and storage devices. *ACS Appl. Electron*, *3*(7), 2862-2897(2021).
4. H. Zhang, X. Chen, F. Cao, G.Wang, X.Dong, Z.Hu & T. Du, Charge–discharge properties of an antiferroelectric ceramics capacitor under different electric fields. *JACerS*, *93*(12), 4015-4017(2010).

Note: All the figures in this chapter were made by the authors.

Recent Trends in Applied Physics and Material Science – Sudhir Bhardwaj et al. (eds)

Integrating Remote Sensing and Field Survey for Vegetation Classification Based on Chlorophyll Content Using NDVI and Random Forest Techniques

Vishal Shirsat*
J.E.S. College, Jalna, Maharastra, India

Sanjay Tupe
Kalika Devi Mahavidyalay Shirur Kasar, Beed, Maharastra, India

Shafiyoddin Sayyad
Miliya College, Beed, Maharastra, India

Abstract: Vegetation classification based on chlorophyll content provides critical insights into plant health and ecosystem sustainability. This study combined satellite-based analysis and field survey data to classify vegetation around the region at latitude 19.00445 and longitude 75.376804. Sentinel-2 satellite imagery from the European Space Agency was processed using SNAP software to calculate the Normalized Difference Vegetation Index (NDVI), an established indicator of vegetation health. The NDVI values were used to classify vegetation based on its chlorophyll content. Concurrently, a field survey was conducted to collect crop samples from the same location. Chlorophyll content was measured in the laboratory for the collected samples to validate the NDVI-derived classifications. A Random Forest classification technique was applied to the processed imagery for detailed vegetation mapping. This integrated approach of remote sensing data and ground truth analysis enabled accurate classification of vegetation types based on their chlorophyll content. The study demonstrates the effectiveness of SNAP software and NDVI analysis for monitoring vegetation health and offers a scalable method for assessing large agricultural areas. The results have significant implications for improving farming practices, monitoring ecosystem changes, and optimizing land use management.

Keywords: Sentinel-2, NDVI, SNAP

1. GROUND TRUTHING

Vegetation was categorized according to chlorophyll content using NDVI values. Simultaneously, a field survey was carried out at the same site to gather crop samples for verification. In total, nine crop samples were collected, encompassing a variety of vegetation types: (Dutta, S., and Banerjee, S. (2020)

1. wet wheat
2. parched wheat
3. millet
4. millet
5. sugarcane
6. guava
7. maize
8. maize
9. wet wheat

2. LABORATORY ANALYSIS

The chlorophyll content of the crop samples was analyzed using the methodology outlined by S. Sadasivam and A. Manickam in *Biochemical Methods*. The steps of the procedure were as follows:(Sadasivam, S., & Manickam, A. (1996)

1. The samples were cut into small pieces.
2. To make 100 mL, a solvent mixture of 80% acetone (80 mL) and 20% distilled water (20 mL) was prepared.

*Corresponding author: sanjaytupek@gmail.com

DOI: 10.1201/9781003684718-90

3. A 0.1 g sample was ground with 50 mL of the 80% acetone solution and filtered through muslin cloth. The final volume was adjusted to 10 mL with 80% acetone.
4. The mixture was centrifuged at 5000 rpm for 10 minutes. The supernatant was collected in a 100 mL flask.
5. 5 mL of 80% acetone was added to the residue and centrifuged at 5000 rpm for 5 minutes. This step was repeated until the residue became colorless.
6. All collected supernatants were combined, and the final volume was adjusted to 20 mL.

The optical density (OD) was measured at 645 nm and 663 nm to determine chlorophyll content.

3. RESULTS

The chlorophyll content of the crop samples was calculated based on the OD readings. The results are as follows:

1. Wet wheat: 30.6
2. Parched wheat: 16.7
3. Millet: 23.6
4. Millet: 26.4
5. Sugarcane: 12.4
6. Guava: 15.8
7. Maize: 12.4
8. Maize: 13.3
9. Wet wheat: 9.82

4. RANDOM FOREST TECHNIQUE

Random Forest is a versatile and widely used ensemble learning technique primarily for classification and regression tasks. It builds multiple decision trees during training and merges their outputs to improve accuracy and control overfitting. (Belgiu, M., & Drăguţ, L. (2016)

4.2 Key Concepts in Random Forest

Ensemble Learning

- Combines the predictions of multiple base models to produce a robust final model.
- Random Forest uses **bagging (Bootstrap Aggregating)**, where each tree is trained on a random subset of the data.

Decision Trees

- Random Forest consists of numerous decision trees, each trained independently.
- Trees in the forest are not pruned, ensuring that each is highly specialized to its subset of data.

Random Sampling

Two types of randomness are introduced:

- **Bootstrap Sampling**: Each tree is trained on a random sample (with replacement) of the training data.
- **Feature Subsetting**: During the construction of each tree, only a random subset of features is considered for splitting at each node.

Aggregation

- For classification: Combines the majority vote of individual trees.
- For regression: Averages the predictions of individual tree.

Classified Image

Fig. 90.1 NDVI colours showing vegetation density

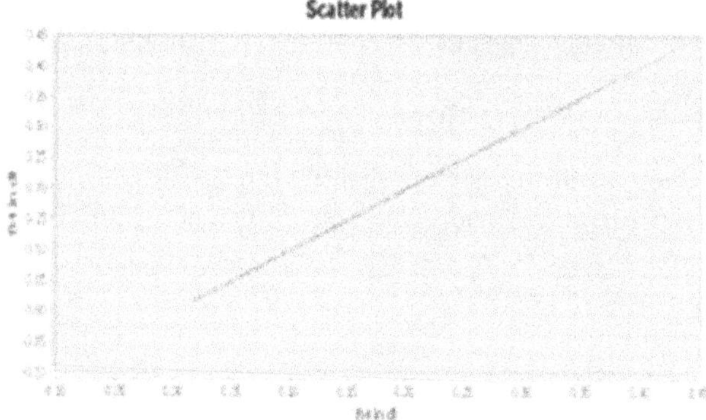

Fig. 90.2 Scalter plot showing NDVI Numbers

5. NDVI CALCULATIONS

The histogram represents the distribution of satellite imagery values for Band 4 (Red). To compute the **Normalized Difference Vegetation Index (NDVI)**, you also need Band 8 (NIR) values. NDVI is calculated as: (Pettorelli, N., Vik, J. O., Mysterud, A., Gaillard, J. M., Tucker, C. J., & Stenseth, N. C. (2005).

$$\text{NDVI} = \text{NIR} - \text{Red} / \text{NIR} + \text{Red}$$

5.1 NDVI Value Range

NDVI ranges from **-1 to +1**:

- Negative values (close to -1): Water, snow, or non-vegetative surfaces.
- Near 0: Bare soil.
- Positive values (close to +1): Dense vegetation

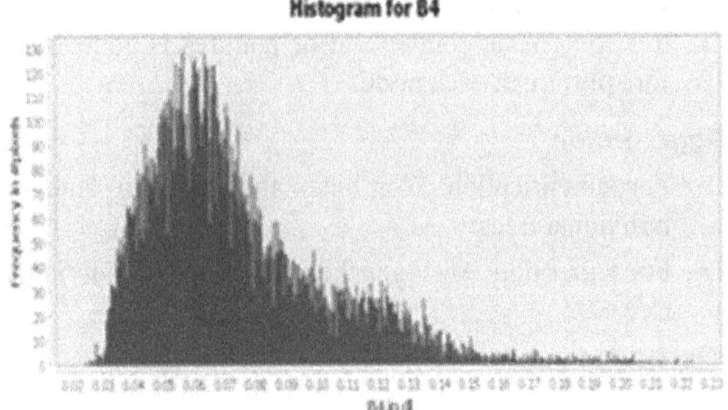

Fig. 90.3 Histograph showing NDVI values

The results obtained from laboratory analysis and the application of the Random Forest classification technique demonstrate a strong correlation between NDVI values and chlorophyll content. The results from both approaches consistently show that elevated NDVI values are associated with increased chlorophyll levels, emphasizing NDVI's strength as a dependable measure of vegetation health and productivity. The convergence of these methods highlights their reliability and efficiency in evaluating vegetation attributes. This study emphasizes the value of combining machine learning techniques, such as Random Forest, with conventional laboratory methods to achieve accurate and thorough vegetation monitoring (Jensen, J. R. (2007).

6. CONCLUSION

The study demonstrates a significant relationship between NDVI values and chlorophyll content, confirmed through laboratory experiments and Random Forest classification. The alignment of outcomes from both methods underscores NDVI's effectiveness as a reliable measure of vegetation health and productivity. Moreover, the concurrence between these approaches emphasizes the advantages of combining machine learning techniques with conventional laboratory practices. This synergy offers a precise and dependable framework for detailed vegetation evaluation, proving to be an essential asset in ecological and environmental monitoring.

REFERENCES

1. Dutta, S., and Banerjee, S. (2020). Applications of ground-truthing in satellite image analysis. Journal of Remote Sensing Applications, 12, 45-60.
2. Sadasivam, S., & Manickam, A. (1996). *Biochemical Methods* (2nd ed.). New Age International Publishers.
3. Belgiu, M., & Drăguţ, L. (2016). Random forest in remote sensing: A review of applications and future directions. *ISPRS Journal of Photogrammetry and Remote Sensing, 114*, 24-31.
4. Pettorelli, N., Vik, J. O., Mysterud, A., Gaillard, J. M., Tucker, C. J., & Stenseth, N. C. (2005). Using the satellite-derived NDVI to assess ecological responses to environmental change. *Trends in Ecology & Evolution, 20*(9), 503-510.
5. Jensen, J. R. (2007). *Remote Sensing of the Environment: An Earth Resource Perspective* (2nd ed.). Pearson Education.

Note: All the figures in this chapter were made by the authors.

Recent Trends in Applied Physics and Material Science – Sudhir Bhardwaj et al. (eds)
© 2026 Taylor & Francis Group, London, ISBN 978-1-041-16452-4

91

Preparation of 4-Hydroxy-3-((4-hydroxy-2-oxo-2H-chromen-3-yl) (2,3-dimethoxyphenyl)methyl)-2H-chromen-2-one Molecule: Structural, Optical and Thermal Studies using DFT

Venugopal N., Suchetha M., Renuka C. G. and Meera B. N.*
Department of Physics, Jnanabharathi, Bangalore University, Bengaluru, Karnataka, India

Abstract: In this report, a facile and efficient protocol has been followed for the synthesis of novel 4-Hydroxy-3-((4-hydroxy-2-oxo-2H-chromen-3-yl)(2,3-dimethoxyphenyl)methyl)-2H-chromen-2-one via a one pot reaction of aryl aldehydes with 4-hydroxycoumarin and in the presence of ethanol at 80 ^{0}C. The synthesized molecule is being further tested for structural confirmation with nuclear magnetic resonance and Fourier transform infrared spectroscopic technique. From the obtained result of FTIR spectroscopy it has been confirmed that the synthesized molecule is stable with functional groups. The optical absorption measurement was carried out using density functional studies in order determine the optical property of the synthesized molecule such as the bandgap and molar extinction coefficient. Further the bond angles and bond lengths of the molecule the theoretical calculated conceptual DFT descriptors of complexes of the titled molecule has been estimated. From over all outcomes the molecule exhibited a considerable result in order to derive the optical device application.

Keywords: Synthesis, Aldehydes, Structure

1. INTRODUCTION

In this current decade the organic materials were receiving very much importance due to their most promising and reliable applicative point of view. In many of the cases the organic molecule that to the highly fluorescent characteristic exhibiting compounds are receiving vast demand due their optoelectronic point of view [1]. Many of the researchers has made an attempt to build such optoelectronic exhibiting materials using different synthesis scheme and they have published many reports on the organometallic materials. Hence, we have choose such a property exhibiting material and we attempted theoretical estimations for the synthesized molecule [2]. In order to expand the structure of the molecule the two or higher coordinating atoms example nitrogen and oxygen used as an organic ligand further it has been confirmed that in order to increase the fluorescent property the aromatic ligand with a specified conjugate structures were utilized. The computational chemistry is a significant area of chemistry that studies the characteristics of compounds. It is very effective instruments enable the identification and provision of data regarding the electrical and structural characteristics of the materials (Pramod et al. 2019). Furthermore, we have to analyze and understand the relationship between the azo dyes structure, computational studies and spectroscopic nature of the molecule is important parameters to study although the photochromatic and the optical properties of 4,3 HDC molecule is monitored. In this present investigation, we have prepared an organic molecule

*Corresponding author: renubub@gmail.com

DOI: 10.1201/9781003684718-91

and it has abbreviated as 4,3 HDC molecule so for the computational investigation has not been carried out for this compound and we have estimated the structural, optical and thermal properties [3] for the molecule and the detailed observations are discussed in the further section.

2. COMPUTATIONAL DETAILS

The DFT calculations were performed and the results of 4,3 HDC molecule has been carried out using Gaussian 09W software. There are many other platforms/software's are available to perform the theoretical studies which were based on the quantum chemical calculations. Initially the molecule has to be optimized, this process of optimizing is done because the molecule will become stable and the further structural changes is not possible. To perform this calculation, the DFT (B3LYP) [4] basis set this basis set is considered to be the stable one because of the results obtained. The different basis set can also be used in order to perform the computational studies but the results aren't in consideration with the B3LYP basis set hence we perform the calculation DFT (B3LYP) basis set (Periyasamy et al. 2022).

3. RESULT AND DISCUSSION

The FTIR spectra of the studied has been represents in Fig. 91.1 from this we can infer that we have observed harmonic vibrational frequencies by using quantum chemical calculation the C-H or C-C vibrations of aromatic composites are examined as a separate ring. Because of substituent group we have observed C-H bending vibrations at 900 to 1100 cm^{-1} also we have observed predominant three C-H vibrations at frequency range 1000 to 1300 cm^{-1} in general the C-C vibrations and C-C stretching vibration are found in the frequency range of 1400 to 1600 cm^{-1}. From all these observations we can confirm the functional group of the molecule.

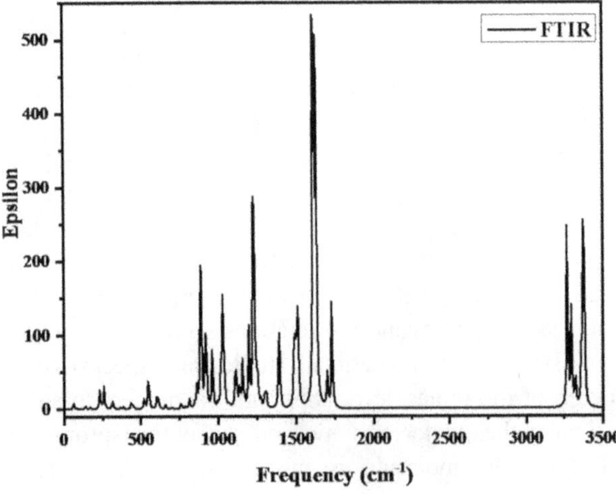

Fig. 91.1 FTIR spectra of 4,3 HDC molecule

Further the optical studies were performed for the synthesized 4,3 HDC molecule and the theoretical spectra of absorption is showed in Fig. 91.2. A plot of wavelength v/s absorption is plotted it has been observed that the molecule exhibited the single peak at a wavelength range 217 nm the transitions took place at 217 nm is attributed to a $\pi \rightarrow \pi^*$ transition [1]. This transition of estimate absorption spectra indicates that the electronic transition from the HOMO to the LUMO with severe involvement is similar to the extreme absorption wavelength. Due to the inclusion of a substituent group, the chromophore at this molecule changed its properties from nitro aniline to tri-nitro aniline. Although the bandgap is also estimated using the HOMO and LUMO studies from that molecular orbital studies we have determined energy bandgap value and is found to be 6.2 eV. The HOMO LUMO plot is presented in Fig. 91.3.

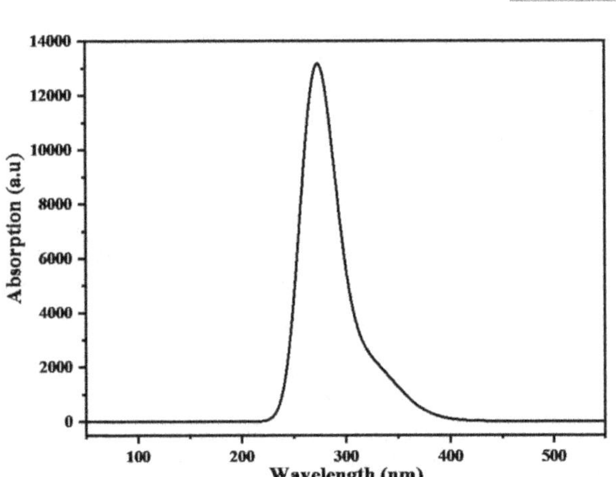

Fig. 91.2 UV-Visible absorption spectra using DFT

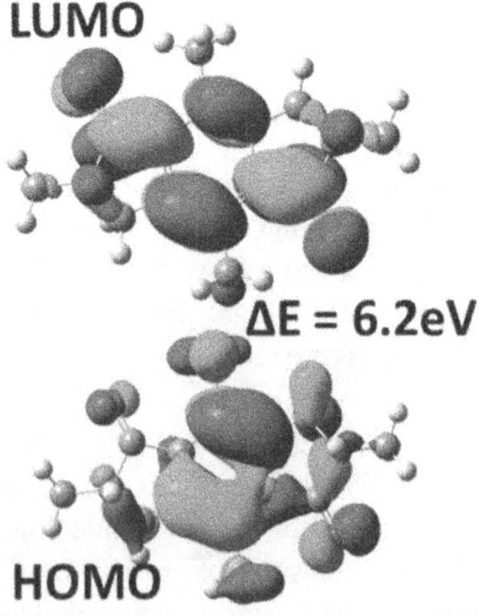

Fig. 91.3 HOMO LUMO molecular orbital spectra 4,3 HDC molecule

Table 91.1 lists the standard thermochemical functions that were computed using the DFTB3LYP/6311++G (dp) basis set: the different parameter which were related to thermal studies are estimated through Gaussian 09W code and the parameter were mentioned in the table. Information for additional research on the molecule is provided by the thermochemical data.

Table 91.1 Thermal simulated parameters of 4,3 HDC molecule

Thermochemical Parameter	Values (DFTB3LYP/6-311++)
Thermal Energy (Hatree)	0.301
Enthalpy	0.321
Gibbs Energy	0.243
Translational Energy	0.989
Rotational Energy	0.889
Vibrational Energy	186.3
Zero point Energy	180.3

It is evident from Table 91.1 that the results of these thermodynamic functions can be used as a descriptor to illustrate how the positive and negative charge centers affect the transport of charge through a molecule. Additional research on the chemical can be conducted using the parameters. To determine the directions of chemical reactions based on the correlations between thermodynamic functions and compute the other thermodynamic energies[4,5].

3. CONCLUSION

Density functional theory (DFT) has been employed in order to study the theoretical properties of 4,3 HDC molecule from the FTIR spectra we found that the functional group molecules used in synthesis has confirmed, from the UV-Visible spectra it has been found that at 217 nm a absorption band has formed and attributed to $\pi \rightarrow \pi^*$ transition energy bandgap is found to be 6.2 eV. From thermal data we infer how the positive and negative charge centers affect the transport of charge through a molecule.

REFERENCE

1. Pramod, A. G., Nadaf, Y. F. & Renuka, C. G. (2019). Synthesis, photophysical, quantum chemical investigation, linear and non-linear optical properties of coumarin derivative: Optoelectronic and optical limiting application. Spectrochim. Acta - Part A Mol. Biomol. Spectrosc. 223, 117288.
2. Periyasamy, K. et al. (2022). Synthesis, photophysical, electrochemical, and DFT examinations of two new organic dye molecules based on phenothiazine and dibenzofuran. J. Mol. Model. 28, 1–14.

Note: All the figures and tables in this chapter were made by the authors.

Recent Trends in Applied Physics and Material Science – Sudhir Bhardwaj et al. (eds)
© 2026 Taylor & Francis Group, London, ISBN 978-1-041-16452-4

Understanding Fluid Behaviour through the Application of Partial Differential Equation in Fluid Physics

Manoj kumar[1], Binny Kakkar[2]
Department of Mathematics, SKD University, Hanumangarh, Rajasthan, india

Abstract: Partial differential equations (PDEs) are essential for analysing and simulating a wide range of fluid dynamics processes. These equations are widely used in fields such as weather forecasting, ocean currents, fire dynamics, and fluid flow systems, governing the behaviour of quantities like velocity, pressure, temperature, and viscosity. PDEs describe complex interactions like precipitation changes, diffusion, and fluid-solid interactions. They are vital for implementing advanced techniques and find use in fields like electrostatics, thermal conduction, transmission lines, quantum physics, and wave theory. There are two main types of PDEs: linear and nonlinear. This paper explores the theoretical aspects of these equations, highlighting their significance in scientific and engineering disciplines, where they model most real-world systems and provide insight into complex behaviours across different domains.

Keywords: Partial differential equation, Fluids mechanic

1. INTRODUCTION

Fluid physics is fundamental in many scientific and engineering disciplines, as it helps understand the behavior of liquids and gases in various environments. Partial differential equations (PDEs) are essential tools in this field, allowing the description of changes in variables like pressure, velocity, temperature, and density over time and space. In fluid mechanics, PDEs, especially the Navier-Stokes equations, are used to model the movement of fluids, including phenomena like flow, turbulence, and heat transfer. Other key PDEs include the heat equation, which models temperature changes in fluids, and the wave equation, which governs the propagation of waves. By solving these equations, researchers can predict fluid behavior in different scenarios, aiding in fields such as oceanography, aerospace engineering, and medical imaging. In summary, PDEs provide a powerful mathematical framework to model complex fluid behaviours and make accurate predictions in various applications.

2. LITERATURE REVIEW

Chorin and Marsden (2014), provides a concise summary of Fluid Mechanics highlighting its focus on mathematical principles in fluid mechanics. It notes that the book is valuable for both professionals and students, covering key topics like Eulerian and Lagrangian flow descriptions, safety rules, and the Navier-Stokes equations. Chow (2014) explores stochastic partial differential equations (SPDEs), which extend traditional partial differential equations by incorporating random variations, making them highly effective for modeling systems affected by uncertainty or random forces. Chow also provides valuable insights into the behavior of complex systems in fluid mechanics and related fields, focusing on the interplay between deterministic. Dobek's (2012) work concentrates on the Navier-Stokes equations, which control the movement of viscous fluids and are fundamental to the study of fluid dynamics. The book guides readers through the derivation and analysis of these equations, addressing both theoretical aspects and practical applications with clear explanations and examples. Dobek explores key concepts such as

[1]amanojrajera@gmail.com, [2]bbinnygupta1519@gmail.com

DOI: 10.1201/9781003684718-92

turbulence, Reynolds number, and boundary layer theory, shedding light on the transition from laminar flow to turbulent fluid behavior. Fritz John's *Partial Differential Equations* (2011), now in its third edition, provides a comprehensive and in-depth exploration of the subject of partial differential equations (PDEs). The book covers essential topics such as existence and uniqueness theories, solution methods, and the classification of PDEs. Robinson and Rodrigo (2009) offers a comprehensive collection of scientific papers focused on the role of partial differential equations (PDEs) in fluid mechanics. With contributions from leading experts in the field, the book covers a wide range of topics including geophysical fluid dynamics, existence and regularity of solutions, turbulence models, and mathematical analysis of the Navier-Stokes equations.

3. PARTIAL DIFFERENTIAL EQUATION DEFINITION AND TYPES

A **Partial Differential Equation (PDE)** is an equation that involves a function of multiple variables and its partial derivatives. Unlike ordinary differential equations (ODEs) that involve derivatives with respect to a single variable, PDEs describe phenomena where multiple independent variables influence a dependent variable. For example, in physics, a PDE might describe how the temperature in a material changes over both space and time, or how the concentration of a chemical evolves in a fluid across space. Representation of PDE

$$F\left(x_1, x_2, x_3, \ldots\ldots\ldots, x_n, \frac{\partial u}{\partial x_1}, \frac{\partial u}{\partial x_2}, \ldots\right) = 0$$

where $u(x_1, x_2, \ldots \ldots x_n)$ is the unknown function.

3.1 Types of Partial Differential Equations

1. By Linearity:

Linear PDEs: The unknown function and its derivatives appear only to the first power, and there's no multiplication between them.

Example: **Heat equation**: $\frac{\partial u}{\partial t} = \alpha \nabla^2 u$

Nonlinear PDEs: Nonlinear PDEs involve the unknown function u or its derivatives raised to powers greater than one or products of the function and its derivatives.

Example: **Burgers' equation** $\frac{\partial u}{\partial t} + u\frac{\partial u}{\partial x} = v\frac{\partial^2 u}{\partial x^2}$

2. By Order: First-Order PDEs:

Only first derivatives are involved.

Example: **Advection equation**: $\frac{\partial u}{\partial t} + c\frac{\partial u}{\partial x} = 0$

Second-Order PDEs: These involve second derivatives of the unknown function.

Example: **Wave equation**: $\frac{\partial^2 u}{\partial t^2} = c^2 \nabla^2 u$

By Classification (Based on Geometry/Physical Interpretation):

PDEs can be classified based on their geometric properties or their solution behaviour, typically into **elliptic**, **parabolic**, and **hyperbolic** types.

Elliptic PDEs: These describe steady-state solutions where there is no time dependency. The solution often represents something like an equilibrium state, such as the potential field in electrostatics.

Example: **Laplace equation**: $\nabla^2 u = 0$

Parabolic PDEs: These describe diffusion-like processes, such as the flow of heat over time. The solutions depend on both space and time, and they typically model time-dependent phenomena that evolve toward a steady state.

Example: **Heat equation**: $\frac{\partial u}{\partial t} = \alpha \nabla^2 u$

Hyperbolic PDEs: These describe wave-like phenomena and are often used to model propagation of sound, light, or other waves.

3. By the Number of Independent Variables:

2-D or 3-D PDEs: These involve two or three spatial variables. These equations typically model real-world phenomena in multiple dimensions, such as fluid flow or heat distribution in a material.

3.2 Fluid Flow Modeling: A Key to Innovation Across Industries

Fluid flow modeling is key in many fields like engineering and biology, using mathematical models and partial differential equations (PDEs) to predict fluid behavior. These models rely on conservation laws (mass, momentum, energy) and the Navier-Stokes equations to simulate complex phenomena. Key applications include automotive and aerospace design, environmental engineering, biomedical simulations, oil and gas extraction, and manufacturing processes. Advancements in computational power and AI are enhancing the accuracy and efficiency of fluid flow models, helping solve global challenges like pollution and resource management. PDEs remain crucial for understanding fluid dynamics and mechanical systems

3.3 Extending the Concept to Fluid Mechanics

In fluid dynamics, PDEs govern the behaviour of fluid flow, which is governed by both the velocity field of the fluid and external forces acting on it. The Navier-Stokes equations are the fundamental PDEs that describe the motion of viscous fluid substances:

$$\rho\left(\frac{\partial v}{\partial t} + v.\nabla v\right) = -\nabla p + \mu \nabla^2 v + f$$

where ρ is the fluid density, v is the velocity vector, p is the pressure, μ is the dynamic viscosity, and f represents external body forces such as gravity. The term $\nabla^2 v$ accounts for the diffusion of momentum, while the first term on the left-hand side represents the inertial forces due to the velocity of the fluid.

3.4 Unifying Mechanical and Fluid Systems: A New Approach

PDEs for a stretched string and Navier-Stokes equations differ in interpretation but can be coupled for fluid-structure interactions (FSI). In such systems, the fluid's movement (Navier-Stokes) and the string's displacement (wave-like PDE) influence each other, creating a feedback loop. An example of such a coupled system might look like:

$$\rho\left(\frac{\partial v}{\partial t}+v.\nabla v\right)=-\nabla p+\mu\nabla^2 v+f\left(u,v\right)$$

$$F\left(x\right)=T\frac{\partial^2 u}{\partial x^2}+\int_0^L C\left(x,v\right)dx$$

Here, C(x, v) represents the force exerted by the fluid on the string, which is a function of both the fluid velocity and the string's displacement. This system allows us to simulate more realistic interactions between solids and fluids, as seen in applications such as underwater cables, flexible pipes in turbulent flows, or sails on boats.

4. FLUID MECHANICS

Fluid mechanics is a branch of physics that studies the behavior of fluids (liquids and gases) under various forces. It plays a crucial role in industries like engineering, hydraulics, aeronautics, meteorology, and biology. Water, the most studied fluid, is analyzed in two subfields: hydrostatics (stationary fluids) and hydrodynamics (moving fluids). Archimedes laid the groundwork for hydrostatics around 250 BC with his principle of buoyancy. Hydrodynamics developed in the 18th century, thanks to mathematicians like Euler and Bernoulli, who built on Newton's theories of motion. In the 19th century, scientists like Stokes and Thomson expanded fluid theory to explain phenomena such as pipe flow and raindrops. A key breakthrough came in 1904 when Ludwig Prandtl introduced the boundary layer concept, essential for understanding high-speed fluid flows, especially in aviation. Geoffrey Taylor, in the 20th century, made significant contributions to turbulence, a chaotic fluid behaviour .Fluid mechanics is a dynamic and evolving field with broad applications, and its complex nature ensures it will remain an active research area for years.

4.1 The Evolution and Significance of Fluid Mechanics

Fluid mechanics is a branch of physics that studies the behavior of fluids—liquids, gases, and other substances that flow—under various forces and conditions. It is a field with vast applications in engineering, natural sciences, medicine, and beyond. The study of fluid mechanics traces its roots back to ancient civilizations, and its development over centuries has led to major advances in technology and scientific understanding. From Archimedes' discoveries in ancient Greece to modern-day breakthroughs in turbulence and computational fluid dynamics, fluid mechanics has evolved into a key field that shapes numerous industries, including aerospace, hydrodynamics, weather forecasting, and medicine.

4.2 Historical Foundations of Fluid Mechanics

Fluid mechanics has its roots in ancient Greece, with Archimedes formulating the buoyancy principle, a foundation of fluid statics. The field grew during the Islamic Golden Age, with scholars like Biruni and Al-Khazini expanding fluid theory. In the Renaissance, Leonardo da Vinci's experiments set the stage for future breakthroughs. The 17th century saw key developments, including Torricelli's barometer, Newton's research on viscosity, and Pascal's work on hydrostatics. A major turning point came in 1739 with Daniel Bernoulli's "Hydrodynamica," introducing the Bernoulli equation and solidifying fluid mechanics as a formal science.

5. THE FUTURE OF FLUID MECHANICS

The future of fluid mechanics is focused on understanding complex fluid behaviors like turbulence, multiphase flow, and fluid-structure interactions. With advanced computational techniques, researchers can now predict fluid behaviour more accurately, driving innovation in aerospace, automotive, and environmental industries. Fluid mechanics is also expanding into fields like biomedical engineering, where it helps in understanding blood flow for better diagnostics and treatments. Microfluidics, the study of fluids at tiny scales, is opening up new possibilities for drug delivery and lab-on-a-chip technologies. Overall, fluid mechanics is evolving and continues to impact scientific discovery and technological innovation across various industries.

6. CONCLUSION AND FUTURE PROSPECTS

Partial differential equations (PDEs) are key to understanding fluid dynamics, allowing precise modeling of fluid flow. These equations help drive advances in aerospace, biomedical engineering, and material science. By exploring different types of fluids and incorporating factors like temperature and deformation, we improve designs in healthcare and engineering. As computational techniques improve, PDEs will continue to unlock new possibilities, especially in coupled systems that simulate both solids and fluids. This will lead to exciting innovations in multiple fields.

REFERENCES

1. Abbott,M.B.,and Basco D.R.,Computational Fluid Dynamics: An Introduction For Engineers, Longman Singapore Publishers (Pte) Ltd. 2015
2. Großmann,C., and Roos,H., Numerics of partial differential equations. Teubner Studienb¨ucher: Mathematik. Stuttgart: B.G. Teubner. 477 p., 2012.
3. Chetverushkin, B. N., Fitzgibbon, W., Kuznetsov, Y. A., Neittaanmäki, P., Periaux, J., &Pironneau, (2019). Contributions to Partial Differential Equations and Applications. Springer International Publishing.
4. Debnath, L., & Debnath, L. (2005). Nonlinear partial differential equations for scientists and engineers [6] Dobek, Steven, 2012. Fluid Dynamics and the Navier-Stokes Equation. CMSC498A: Spring'12 Semester
5. Robinson ,C., & Rodrigo, J. L. (Eds.). (2009). Partial differential equations and fluid mechanics . Cambridge University Press.
6. Fritz John, Partial Differential Equations (3th Edn), Applied Mathematical Sciences 1,SpringerVerlag, Heidelberg-Berlin-New York, 2011.
7. PAUL A., Smith And Eilenberg S. (1949). Partial differential equations of physics. PURE and Applied Mathematics.
8. Herman, R., L., (2015). Introduction to partial differential equations. North Carolina, NC, USA.

Recent Trends in Applied Physics and Material Science – Sudhir Bhardwaj et al. (eds)
 ISBN 978-1-041-16452-4

Effects of Arsenic Doping (As) on Structural and Electronic Properties of Zigzag Germanium Sulfide Nanoribbons: A First-Principles Approach

Banti Yadav[1], Pankaj Srivastava[2]
Department of Engineering Sciences, Nanomaterial Research Laboratory,
ABV- Indian Institute of Information Technology and Management (IIITM),
Gwalior, Madhya Pradesh, India

Varun Sharma[3]
Department of Electronics Engineering,
Madhav Institute of Technology and Science (MITS),
Gwalior, Madhya Pradesh, India

Abstract: Density functional theory (DFT) is employed to study the structural and electronic properties of zigzag germanium sulfide nanoribbons (ZGeSNR) via substitutional doping of Arsenic (As) atom (n-type impurity). We observed the effect of Arsenic (As) doping on the Ge-edge, S-edge, and center of ZGeSNR. Post-substitutional doping structural stability of the doped structures was evaluated via formation energy (EFE) calculation, and we found that all the configurations were thermodynamically stable. Further, we have also calculated the electronic properties on the basis of the E-k diagram and density of state (DOS) profile, which shows that the pristine nanoribbons turn metallic upon doping. This induced metallicity in ZGeSNR warrants their applicability in the field of 1-D nano interconnects.

Keywords: DFT, ZGeSNR nanoribbons, Doping, Structural stability, Electronic properties, Metal interconnects

1. INTRODUCTION

The MX nanostructures (where M = Ge/Sn; X = S/Se) have recently gained new research interest due to their novel electronic and structural properties, like phosphorene[1]. The group-IV monochalcogenides are defined as a set of binary elements as MX[2]. The experimental synthesis of germanium sulfide (GeS) and germanium selenide (GeSe) nanosheets by using a one-pot strategy shows their semiconducting band gap nature [3]. There are other methods also by which we can synthesize GeS monolayers, such as; chemical vapor deposition (CVD) and chemical vapor transfer (CVT)[4]. To fabricate ZGeSNR, we can cut the GeS monolayer in a specific direction. However, theoretically, the GeS monolayer has an indirect band gap (2.34 eV) semiconductor, as reported by Li et al.[5]. The various properties (like electronic, structural, and transport) of nanoribbons can be adjusted effectively by the variation of width, edge functionalization, and doping. Recently, the substitutional effect of doping with N and P atoms has been reported for metal interconnect applications[6]. In our present work, we have explored the effect of Arsenic (As) atom doping (n-type impurity) on the structural and electronic properties of zigzag germanium sulfide nanoribbons (ZGeSNR). Further, we can use their applicability in the areas of nano interconnects like graphene nanoribbons, germanene nanoribbons, and silicene nanoribbons, which are already reported work [7]

2. COMPUTATIONAL DETAILS

Our calculations were performed using the Quantum ATK software package based on the Density Functional Theory (DFT) mechanism. A double zeta basis set was employed

[1]banti@iiitm.ac.in, banti1526sv@gmail.com, [2]pankajs@iiitm.ac.in, [3]varunusha9786@gmail.com

DOI: 10.1201/9781003684718-93

for all atoms. To account for electron-electron interaction, we used local density approximations (LDA) exchange correlation with Perdew-Zunger (PZ). The Brillouin zone was sampled with 1 × 1 × 20 k-point sampling, and for DOS, the k-point was sampled as 1 × 1 × 60. However, the energy mesh cut-off was set at 150 Ry, and for geometry relaxation, the force criteria were set at 0.01eV/Å. To prevent interaction with periodic images in the z-direction, we applied 20 Å vacuum padding in the x and y directions. Formation energy (E_{FE}) calculates the stability of our doped configurations, negating values showing that they are thermodynamically stable structures.

3. RESULT AND DISCUSSION

We considered the four different cases, (i) 10z-GeSNR-top-S-edge-As-doped, (ii) 10z-GeSNR-centre-S-edge-As-doped, (iii) 10z-GeSNR-bottom-Ge-edge-As-doped, and, (iv) 10z-GeSNR-bottom-S-edge-As-doped (as shown in Fig. 93.1 and Fig. 93.2). In the structural stability, the maximum structural stability is calculated by 10z-GeSNR-bottom-Ge-edge-As-doped case with the highest E_{FE} = -5.52 eV among all configurations, as depicted in Table 93.1. Further, we have also calculated the electronic properties, which are calculated by band structure diagram and DOS profile, which shows the metallic nature. By the doping of As-atom at different sites, the result of doping turns the semiconducting nature of pristine ZGeSNR to metallic, as shown in Fig. 93.3 and Fig. 93.4. On the basis of metallic nature, we can use As-atom doped ZGeSNR for nano interconnects applications.

Table 93.1 Calculated formation energy and band gap of As-doped ZGeSNR at different edges/sites

Configurations	Band gap (E_g)	Formation energy (E_{FE})
10z-GeSNR-top-S-edge-As-doped	M	-5.50 eV
10z-GeSNR-centre-S-edge-As-doped	M	-5.49 eV
10z-GeSNR-bottom-Ge-edge-As-doped	M	-5.52 eV
10z-GeSNR-bottom-S-edge-As-doped	M	-5.49 eV

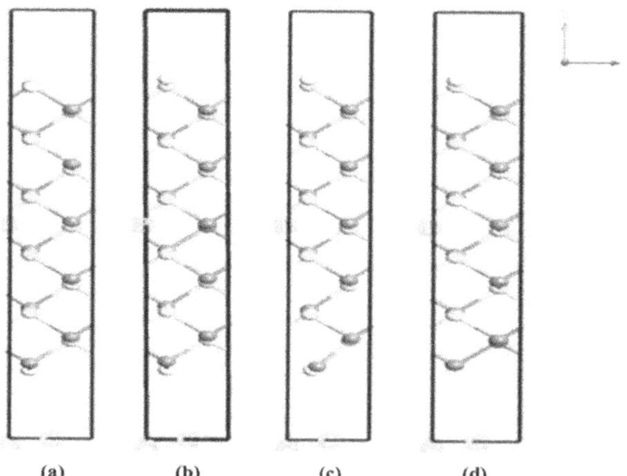

Fig. 93.1 Front view of optimized configuration (a) top-S-edge-As-doped, (b) centre-S-edge-As-doped, (c) bottom-Ge-edge-As-doped, (d) bottom-S-edge-As-doped, of ZGeSNR, respectively. Here, Grey, Yellow, white, and violet colors represent the Germanium atom, Sulfure atom, Hydrogen atom, and Arsenic atom, respectively

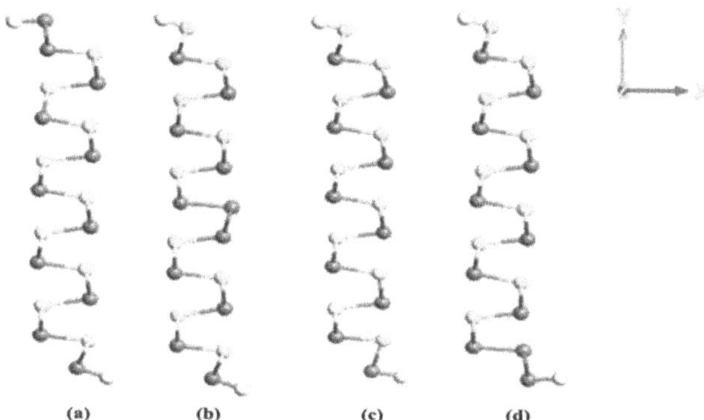

Fig. 93.2 Side view of optimized configuration (a) top-S-edge-As-doped (b) centre-S-edge-As-doped (c) bottom-Ge-edge-As-doped (d) bottom-S-edge-As-doped of ZGeSNR, respectively

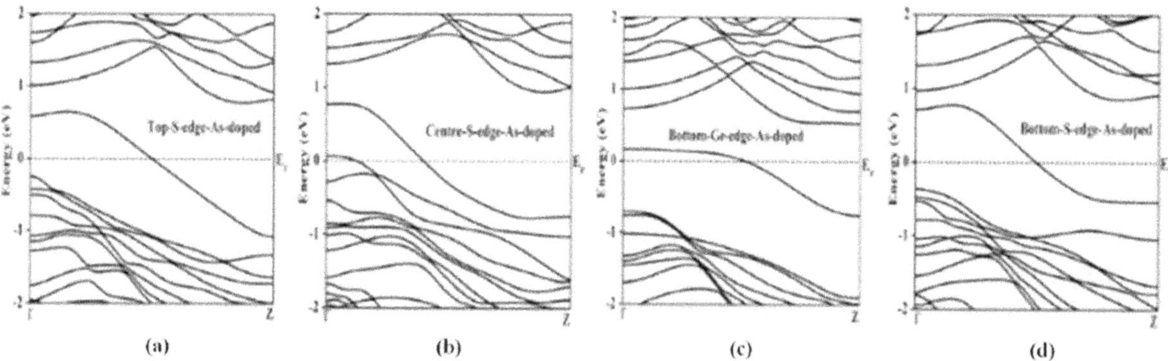

Fig. 93.3 Computed band structure diagrams of (a) top-S-edge-As-doped (b) centre-S-edge-As-doped (c) bottom-Ge-edge-As-doped (d) bottom-S-edge-As-doped of ZGeSNR, respectively

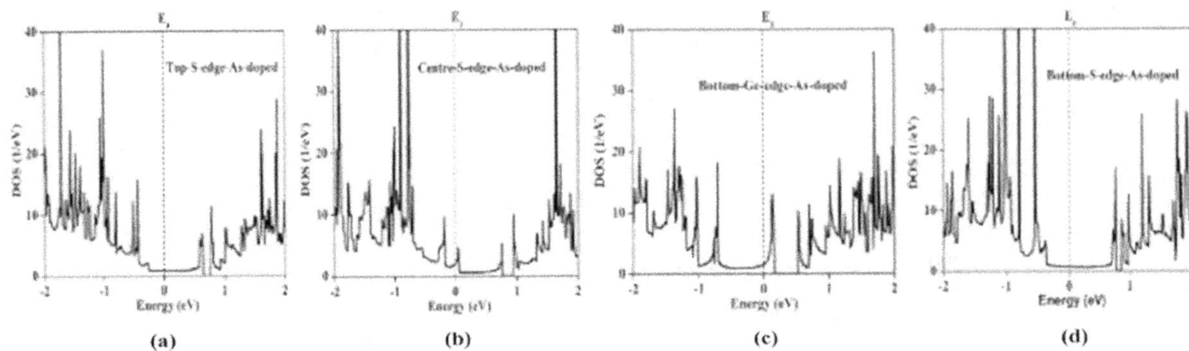

Fig. 93.4 Computed Density of states (DOS) profile of (a) top-S-edge-As-doped (b) centre-S-edge-As-doped (c) bottom-Ge-edge-As-doped (d) bottom-S-edge-As-doped of ZGeSNR, respectively

4. CONCLUSION AND FUTURE PROSPECTIVE

We study the structural and electronic properties (based on DFT) of ZGeSNR via substitutional doping of As-atom. We observed the effect of As-atom doping on the Ge-edge, S-edge, and center of ZGeSNR. Post-substitutional doping structural stability of the doped structures was evaluated via formation energy calculation and we found that all the configurations were thermodynamically stable. Further, we have also calculated the electronic properties on the basis of the *E-k* diagram and density of state (DOS) profile, which shows that the pristine nanoribbons turn metallic upon doping. This induced metallicity in ZGeSNR warrants their applicability in the field of 1-D nano interconnects.

REFERENCES

1. Zhang, M.; An, Y.; Sun, Y.; Wu, D.; Chen, X.; Wang, T.; Xu, G.; Wang, K. The Electronic Transport Properties of Zigzag Phosphorene-like MX (M= Ge/Sn, X= S/Se) Nanostructures. *Physical Chemistry Chemical Physics*, 2017, *19* (26), 17210–17215.
2. Li, R.; Cao, H.; Dong, J. Electronic Properties of Group-IV Monochalcogenide Nanoribbons: Studied from First-Principles Calculations. *Phys Lett A*, 2017, *381* (44), 3747–3753.
3. Vaughn, D. D.; Patel, R. J.; Hickner, M. A.; Schaak, R. E. Single-Crystal Colloidal Nanosheets of GeS and GeSe. *J Am Chem Soc*, 2010, *132* (43), 15170–15172.
4. Tan, D.; Lim, H. E.; Wang, F.; Mohamed, N. B.; Mouri, S.; Zhang, W.; Miyauchi, Y.; Ohfuchi, M.; Matsuda, K. Anisotropic Optical and Electronic Properties of Two-Dimensional Layered Germanium Sulfide. *Nano Res*, 2017, *10*, 546–555.
5. Li, F.; Liu, X.; Wang, Y.; Li, Y. Germanium Monosulfide Monolayer: A Novel Two-Dimensional Semiconductor with a High Carrier Mobility. *J Mater Chem C Mater*, 2016, *4* (11), 2155–2159.
6. Yadav, B.; Srivastava, P.; Sharma, V. Substitutionally Doped Zigzag Germanium Sulfide Nanoribbon for Interconnect Applications: DFT-NEGF Approach. *IEEE Trans Nanotechnol*, 2024, *23*, 809–814. https://doi.org/10.1109/TNANO.2024.3504601.
7. Soldano, C.; Talapatra, S.; Kar, S. Carbon Nanotubes and Graphene Nanoribbons: Potentials for Nanoscale Electrical Interconnects. *Electronics (Basel)*, 2013, *2* (3), 280–314.

Note: All the figures and tables in this chapter were made by the authors.

Recent Trends in Applied Physics and Material Science – Sudhir Bhardwaj et al. (eds)

Compositional Dependence on Glass Transition Temperature in Germanium Telluride Based Glasses on The Network Connectivity

Shiv Ruyal, A. Bhargava*, B. Suthar
Nanophysics Laboratory, Department of Physics, Govt. Dungar College, Bikaner, India

Abstract: An analytical framework is presented describing the viscosity increase due to agglomeration of covalant bonds in the glass network as a result of changes in the structural units. The glass transition temperature as a function of network parameter in binary GeTe system is analyzed. The issue related to network entropy in modified network is also discussed.

Keywords: Glass Transition Temperature, Glasses, Network etc.

1. INTRODUCTION

In the case of low modified network glasses such as metallic chalcogenides, the variation of Tg is mostly controlled by the entropy of the glass network, being related to its connectedness (Feng et al. 1997). The magnitude of the glass-transition temperature in certain systems can be quantitatively understood in terms of combinatorial arrangements of metal–chalcogen bondings, which are again related to the network entropy (Mahadevan et al. 1994).

The glass-transition temperature, Tg, is one of the most important parameters for the characterization of the glassy state of these materials (Elliot 1989). Although its does not seem to play an essential role in the description of the kinetics of glass formation, this quantity remains of considerable interest in applied sciences, as well as for the glass manufacturing (Uhlmann and Yinnon 1983). Although thermal kinetic effects change the absolute value of Tg, besides that there are also large structure-related effects that apparently control the magnitude of Tg in glass-forming alloys (Tanaka 1984, Phillips 1979).

Several authors have proposed to relate Tg with some property of the glass-forming melt, involving either the melting temperature or the Debye temperature of the phonon spectrum (Bhargava and Suthar 2013). In chalcogenide glasses, one can change the concentration of modifier atoms with coordination number r. This produces a change in the network connectivity or in the mean coordination number $< r >$ and a continuous variation of structure-related physical properties (Bhargava and Kalla 2016, Elliot 1989).

Tanaka has derived from viscoelastic considerations the relation:

$$\ln T_{gt} = 1.6\langle r\rangle + 2.3 \tag{1}$$

Additionally, it is noted that in multi-component chalcogenide glasses, T_{gt} obeys a modified Gibbs-DiMarzio formula, which is written as follows:

$$T_{gt} = \frac{T_0}{1 - \beta < r > -2} \tag{2}$$

here T_0 represents the glass-transition temperature's limiting value as concentration of modifier (as Ge in Ge–Te systems) approaches zero. This is a system-specific

*Corresponding Author: anamib6@gmail.com

DOI: 10.1201/9781003684718-94

constant that correlates to a glass with a chain-like structure, such vitreous tellurium, which has an average coordination number of two (Elliot 1989, Bhargava and Jain 1994). In general, glass-transition temperature increases with the network connectivity, and vice versa.

If a graph is plotted as the ratio T_{gt}/T_0 with average coordination number for a binary chalcogenide system as an element of Group IV (r_B = 4) and = 2 + 2x, then almost all systems behave very similarly up to ~2.35. For a higher mean coordination number, the slope changes for the telluride systems are almost constant. For the case of variation of overall mean bond energy of the network, following the covalent bond approach (CBA) (Thorpe 1983), this analysis, concludes that T_g is here also a linear function.

This analysis provides a measure of the stochastic nature of the network, in which the chalcogen rich region metallic atoms randomly cross-link the chalcogen chains (Lebaudy and Saiter 1991). The formation of covalent connections between atomic or molecule species in network glass formers, also known as strong glass formers, is directly linked to the Arrhenius-like increase of viscosity in the supercooled state, which gradually increases the likelihood of molecular motion. The agglomeration process, in which typical local structural configurations with coordination numbers r_i adhere to one another and form new covalent connections, is thought to be the most significant mechanism during glass transition. Structural arrest occurs at glass transition temperature, causing certain designs to become stuck and stop moving. T_{gt} is defined by this criterion.

According to an initial glass transition temperature T_o, glass-transition temperature fluctuation seems to be completely governed by connectedness (coordination numbers r_B and r_A of the relevant atoms A and B). Kinetic or thermal contributions do not contribute and bond energies are absent (Lebaudy and Saiter 1991). As the entropy of the network depends on the number of equivalent ways. Therefore, all conceivable bonds A–B and A–A can be formed by the joining of A and B atoms. An increase in the number of equivalent accessible structural states can also lead to a network being more connected. The average coordination number prediction of T_{gt} for ternary glasses is parameter-free. As a result, the deviation for telluride glasses stays linear and is predictable across the concentration range of interest for all systems examined near = 2.35.

Glass-forming ability (GFA) is a measure of thermal stability of the material. The decreased glass transition temperature ($T_{rg} = T_{gt}/T_m$), which Kauzmann created, is used to evaluate GFA. For a good glass, the value of (T_{rg}) is almost 2/3. This value fits the examined glasses well. The fragility (Fi) characterizes and quantifies the anomalous non-Arrhenius transport behavior of glassy materials near the ergodicity breaking glass transition region (Elliot 1989). Fragile glasses are substances with non-directional interatomic/intermolecular bonds. While glasses with greater values of (Fi) are categorized as fragile, those with lower values are categorized as strong. The fragility index of the studied glasses decreased with the increase in the heating rate, Strong glass-forming liquids are indicated by a low value of Fi (Fi ≈ 16).

REFERENCES

1. Bhargava, A., Jain, I.P. (1994). Journal of Physics D: Applied Physics 27 (4), 830.
2. Bhargava, A., Kalla, (2016). J. International Journal of Materials Science and Engineering 4 (2), 126-132
3. Bhargava, A., Suthar, B. (2013). AIP Conference Proceedings 1536 (1), 15-18.
4. Elliot, S.R. (1989). Physics of amorphous materials, John Wiley & Sons, New York.
5. Feng, X.W., Bresser, W.J., Boolchand, P. (1997), Phys. Rev. Lett. 78, 4422.
6. Lebaudy, P., Saiter, J.M. (1991). J. Grenet, M. Belhadji, C. Vautier, Mater. Sci. Eng. A 132, 273.
7. Mahadevan, S., Giridhar, A., Singh, A.K. (1994) J. Non-Cryst. Solids 169, 133.
8. Phillips, J.C. (1979). Topology of covalent non-crystalline solids. I: Short range order in chalcogenide alloys, J. Non-Cryst. Solids 34, 153.
9. Tanaka, S. (1984). Glass transition of covalent glasses, Solid State Commun. 54, 867
10. Thorpe, M.F. (1983) J. Non Cryst. Solids 57, 355.
11. Uhlmann, D.R., Yinnon, H. (1983) The formation of glasses, Glass Science and Technology, Academic Press, New York.

Recent Trends in Applied Physics and Material Science – Sudhir Bhardwaj et al. (eds)
© 2026 Taylor & Francis Group, London, ISBN 978-1-041-16452-4

Design and Fabrication of Piezo Actuated Miniature Fabry-Pérot Interferometer Filter for Spectroscopic Applications

Shahabas Ahammed N.*
Laboratory for Electro-Optics Systems (LEOS-ISRO), Bengaluru
Cochin University of Science and Technology (CUSAT), Kochi

Mahendra K., Ashwini Jambhalikar, Giridhar M. S., Shivaprasada Karantha, K. V. Sriram
Laboratory for Electro-Optics Systems (LEOS-ISRO), Bengaluru

Abstract: This paper presents design, fabrication, and testing of a MEMS-based Fabry-Pérot interferometer (FPI) using a multilayer piezoelectric actuator for tunable spectral filtering across the visible to SWIR range. The design integrates MEMS processes, including lithography, sputtering, and dicing, with a ring-type piezo actuation mechanism for voltage-controlled tuning. High-reflectivity mirrors with approximately 80% reflectivity are fabricated by sputtering aluminium film onto 1 mm thick Pyrex glass substrates, featuring a 5.5 mm central aperture. A 160 nm aluminium stopper is deposited to non-aperture areas to maintain a minimum gap and protect mirror surfaces during assembly. Mirror parameters, such as surface roughness and tilt, were characterized using an Optical Profiler, yielding an effective finesse of 6.67. FPI's optical performance testing with a 632.8 nm He-Ne laser confirmed resonance peaks at specific cavity lengths, with an experimentally observed finesse of 5. The device tuned by piezoelectric actuation from 0 V to 80 V. Finesse discrepancies are attributed to mirror tilt, suggesting that alignment improvements could enhance performance. This FPI is being developed for spectral analysis using single-point detectors and offers potential for space-based applications such as environmental monitoring.

Keywords: Tunable Fabry-Perot Interferometer, Piezo tuning, Spectrum analyser, MEMS process technology.

1. INTRODUCTION

Spectroscopy is a proven, powerful method for non-invasive compositional analysis across diverse samples, with applications in gas leak detection, wildfire monitoring, and precision agriculture. However, traditional instruments, despite their high resolution and broad wavelength range, are often bulky, heavy, require cooling, and consume significant power, making them costly, and less portable. Recent advancements in micro electro mechanical systems (MEMS) technology have enabled the miniaturization of various sensing devices. Spectroscopic instruments are now following this trend, spurred by the development of MEMS-based tunable Fabry-Pérot (FP) interferometer filters (Martin, et.al. 2016). This research focuses on the development of miniature piezo-actuated Fabry-Pérot tunable filters with a 5.5 mm mirror aperture, utilizing MEMS process technologies. These filters are basic building blocks of miniature spectroscopic instruments. Moreover, these kind of MEMS FPI filters can also be implemented in various other applications like hyperspectral imaging, thanks to its high definition and high-resolution capabilities.

2. THEORY

A Fabry-Pérot interferometer consists of two parallel reflecting surfaces that form a resonant cavity. When light

*Corresponding Author: shahabas1199@gmail.com, ashwini@leos.gov.in

DOI: 10.1201/9781003684718-95

enters this cavity, it undergoes multiple reflections between the surfaces, resulting in the formation of an interference pattern. The conditions for interference depend on the wavelength of the incident light with intensity (I_O) and the separation between the mirrors. The transmission intensity (I_T) is given by Silfvast, (2004),

$$I_T = \frac{I_O(1-R)^2}{(1-R)^2 + 4R\sin^2\delta} = \frac{I_O}{(1+F\sin^2\delta)} \quad (1)$$

Where $F = \frac{4R}{(1-R)^2}$ is the Finesse

In this equation R represents the reflectivity of the mirrors, and the phase shift δ is expressed as $\delta = 2\pi$ nd cos θ/λ, where n is the refractive index of the cavity (here it is air and $n = 1$), d is the distance between the mirrors, and θ is the angle of incidence. The Finesse is the figure of merit which is also influenced by tilt (d_t), roughness (d_r) & bow (d_b) of the mirrors.

Table 95.1 Reflective and imperfection finesses in FPI

Reflective Finesse (F_{Ref})	Component of defect finesse		
	Bow (F_b)	Tilt (F_t)	Roughness (F_r)
$\frac{\pi\sqrt{R}}{1-R}$	$\frac{\lambda}{2d_b}$	$\frac{\lambda}{\sqrt{3}d_t}$	$\frac{\lambda}{\sqrt{22}d_r}$

The effective finesse is given by,

$$\frac{1}{F_{eff}^2} = \frac{1}{F_{Ref}^2} + \frac{1}{F_{Def}^2} \quad (2)$$

Where F_{Def} is the defect finesse attributed due to mentioned imperfection in fabrication and is given by,

$$\frac{1}{F_{Def}^2} = \frac{1}{F_b^2} + \frac{1}{F_t^2} + \frac{1}{F_r^2} \quad (3)$$

3. DESIGN AND FABRICATION

The design integrates MEMS technologies with a multilayer piezoelectric actuator tuning mechanism. Uniform tuning is achieved using a ring-type mechanism via voltage scanning. High-reflectivity mirrors, with a reflectivity of approximately 80% in visible range verified by spectrophotometer, are fabricated by sputtering aluminium onto 1 mm thick Pyrex glass substrates. Each mirror has dimensions of 6 mm x 6 mm, with a 5.5 mm central aperture. The remaining area acts as a stopper during initial assembly conditions, thus protecting the mirror surfaces. The fabrication process is as follows (Fig. 95.1)

After development and removing PR, the mirrors are diced into individual units, as shown in Fig. 95.2(a). These mirrors, are assembled with suitable support and piezo ring positioned between them, completing the Fabry-Pérot interferometer (FPI) structure, as illustrated in Fig. 95.2(b).

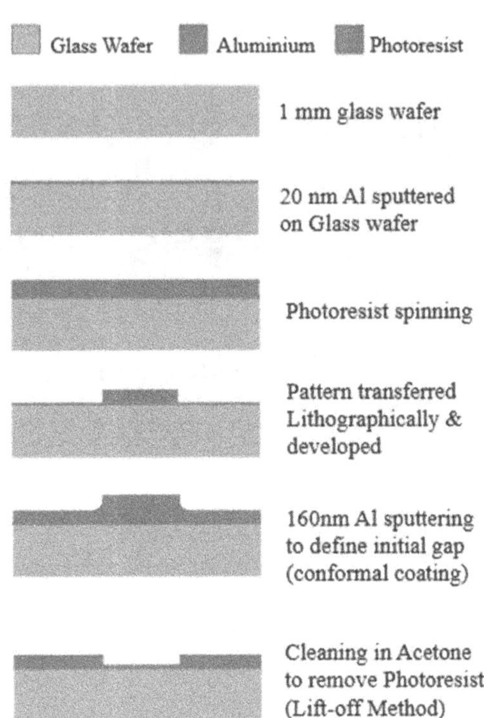

Fig. 95.1 Process flow for mirror fabrication

4. EXPERIMENT AND RESULTS

FPI Mirrors are characterized using a Wyko NT1100 Optical Profiler to assess surface quality and alignment requirement defined by stopper step. Resulted the Surface Roughness *(d_r)* 10 nm, Stopper thickness variation *(d_t)* 40 nm which are mostly contributed to the defect finesse and thus effective finesse estimated by equation (2) is 6.67.

The FPI's optical performance evaluated experimentally, for this a 632.8 nm He-Ne laser (Spectra-physics) served as the coherent light source, directed into the FPI assembly. The cavity length was modulated by applying a scanning voltage ranging from 0 V to 80 V to the piezoelectric actuator. Corresponding transmission intensity is observed (Fig. 95.3) in the oscilloscope coupled via silicon photo detector with inbuilt amplifier (PDA10A2 -Thorlabs).

- Resonance Peaks: Clear and distinct resonance peaks were observed corresponding to specific cavity lengths where constructive interference occurs.
- The experimentally observed finesse was 5.

5. CONCLUSION

The results confirm that the fabricated FPI operates as intended, confirming tunability. The discrepancy between the effective and observed finesse values can be attributed mostly to mirror tilt. Improving tilt parameters by better alignment during assembly, could further increase the

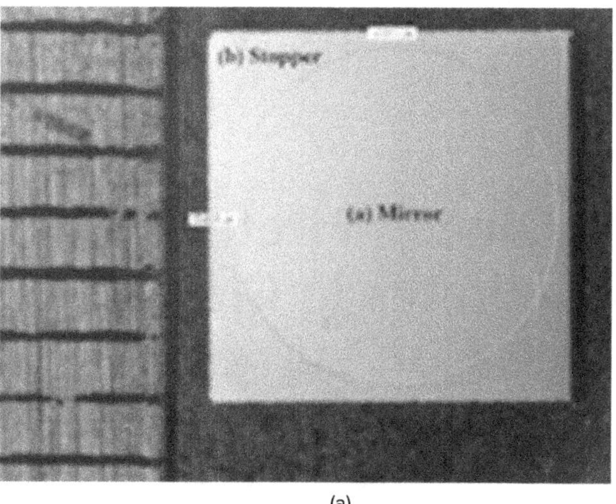

(a)

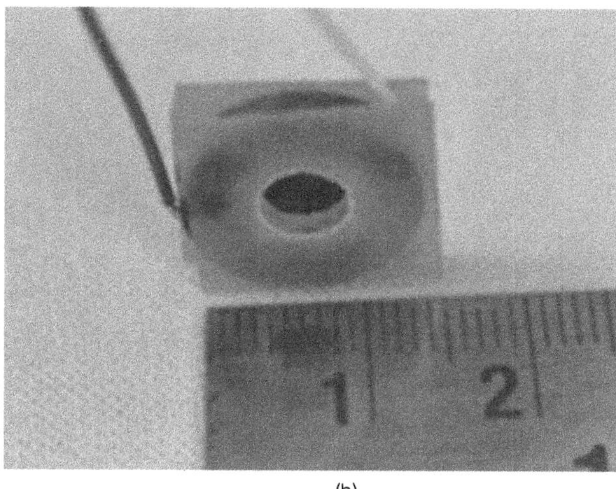

(b)

Fig. 95.2 (a) Diced mirror with stopper step, (b) Fabricated FPI using the diced mirrors and ring piezo with OD 15mm, ID 9mm and travel range 2μm

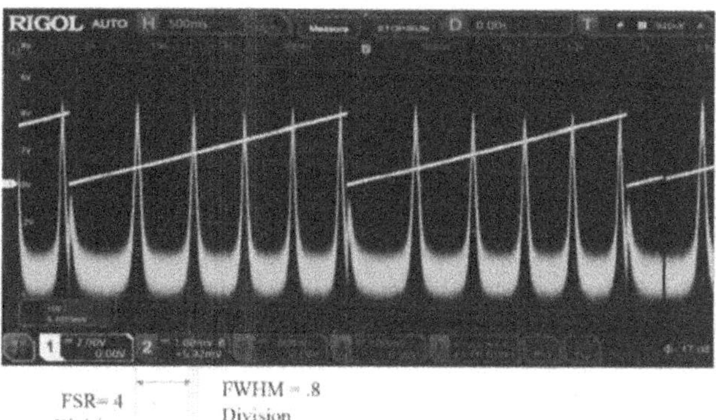

Fig. 95.3 Observed output with scanning piezo actuation

finesse and overall performance of the device. This, can be employed for spectrum analysis of target materials using single-point detector in required wavelength range. The analyser electronics can be implemented to generate final outputs through graphical user interfaces (GUIs).

REFERENCES

1. E. Martin, et.al. (2016) Tunable MEMS Fabry-Pérot filters for infrared micro spectrometers: A review, Proc. of SPIE 9760, 97600H-1-20
2. W. T. Silfvast, (2004) Laser Fundamentals", (Cambridge University Press.

Note: All the figures and tables in this chapter were made by the authors.

Recent Trends in Applied Physics and Material Science – Sudhir Bhardwaj et al. (eds)
© 2026 Taylor & Francis Group, London, ISBN 978-1-041-16452-4

Review on Importance of Rate of cooling of Gas Hydrates in Clay Environments with Non-stirred Reactor

Radhika Ikkurti*
Department of Basic Sciences, G. Narayanamma Institute of Technology & Science for Women,
Shaikpet, Hyderabad, Telangana, India

Abstract: This study experimented with gas hydrate in various clay media. It is discovered that in a clay environment, the rates of methane conversion were evidently smaller for higher Rw ratios than those in other media experimentally. We express that it is because of the distinct water compositions at various pressures in different media that caused the differences in induction time and rate of cooling. In sediments like clay, because of its small particle size, the proportion of bound water is much higher than in sand, and the bound water is more difficult to hydrate conversion. During thawing cycles, our experiments shows that the rate of formation after sub-cooling temperature has rapid growth through the saturation level.

Keywords: Rate of cooling, Gas hydrate, Formation, Sub cooling

1. INTRODUCTION

Gas Hydrate formation mostly takes place at the interface amongst H_2O and gas. In the case of water soothing residues that are compatible with water, hydrate formation is kinetically accelerated over time as a result of an intensification in methane intake for various sediments per varying water ratios. Diffusion controls subsequent conversion. When a convinced amount of liquid adapts to G H, it customs a amply profuse layer over the H_2O phase, serving as a noteworthy weight-transfer barrier to supplementary hydrate evolution (Sloan, 2007, Simon,2010). It is unclear, though, whether the growth of additional hydrates is controlled by the carrying of the guest (water) or the congregation molecule through the mineral film. In order to store methane gas using hydrates, the unqualified gas amount trapped in Ghydrate essential be high the combined weight of the matrix solid or the total artefact ,of the system must be low mean that the permeable medium carrying gas hydrate must be saturated or supersaturated, though water in order to transform the majority to transform into hydrates (L Wang, 2022). Reports have Primarily demonstrated that hydrate yields increases only when the apertures are somewhat saturated with water (Ngoc N,2020).Nano clay in particular exhibits favourable thermodynamic behaviour (Radhika etal,2018). Due to the increased ratio of the matrix to water, there is consequently less gas in the system. This is because there is less gas stock in the system. It has recently been demonstrated that increasing the body of water/matrix ratio and, consequently, the composite yields (out-and-out methane resonant capacity) may be achieved more quickly by using dry water or reactive carbons(Wei,2023)

2. EXPERIMENTAL SETUP

The high pressure vessel (SS-316) of volume 400 ml with stirred head was used. The reactor can sustain pressures up to 20MPa. All the experiments were conducted at constant stirring rate of 500 rotations per minute. Cold fluid (water

*Corresponding Author: daakshayini.radhika@gmail.com

DOI: 10.1201/9781003684718-96

and glycol in 1:1 ratio) was circulated around the vessel to maintain temperature with the help of lab acquaintance (R W-0525G) circulator. Platinum sensor thermometer (Pt 100) was inserted into the container to degree temperature with uncertainties about 0.2 K. WIKA compression transducer (WIKA, type A-10 for pressure range 0 to 16 MPa) was used for pressure measurement inside the vessel.

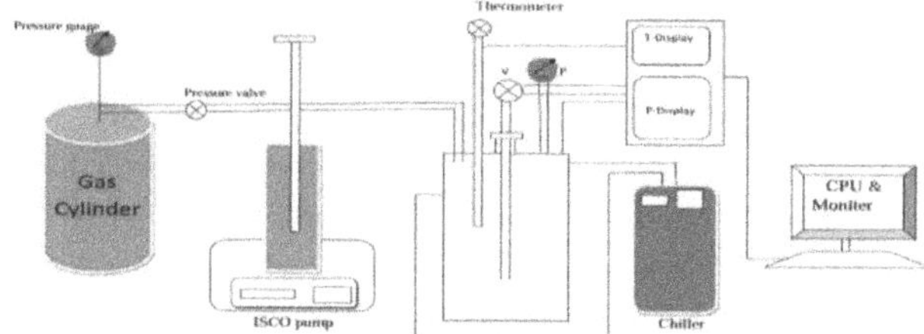

Fig. 96.1 Schematic diagram for hydrate synthesis

Sample solution taken was 30 % of the total available volume of the vessel. Methane gas was filled in the vessel from the cylinder up to 7.5 MPa using TSyringe pump (model 100DX) at room temperature, then valve inline linking the vessel and I S C O chamber was closed. Then temp was decreased slowly to bring the system into CH4hydrate construction region. At certain temperature and pressure conditions GH formation was detected by sharp pressure drop.

P-T phase diagram: T & P inside the vessel were recorded with time scale by the computer software (AMAR). In this mode a pressure temperature hysteresis loop was obtained for each experiment. Thermodynamics of hydrate development and separation were studied with this graph. The molar deliberation of methane gas in the hydrate stage throughout the trial at t, is clear by equation (sloan,2007). Where the Z was referred from Perry's C Engineer's Reference. Force/A (P) and T were registered at a fixed interval intermission (60 seconds) as the hydrate formation headways, for dissociation log rate was 30 seconds (Fig. 96.2).

Effect of clay sediment: Methane H have been successfully formed on several occasions in this study to explore developments matted during CH4 hydrate formation on addition of additives like clay. The effect of clay has influenced the rate of formation to be rapid in the later ratios than the initial water ratios. This could be due to memory effect and in sufficient water absorption in to clay. This leads to cementation of the clay and difficult for the gas to penetrate to form Hydrate. On the other hand, clay by its nature is more permeable when goes with higher water ratios. With the escalation of montmorillite gratified, the upholding effect progressively turns into hindering result. Therefore, the degree of the surround-water content in MC play vital role for research on hydrate construction (Tao,2020).

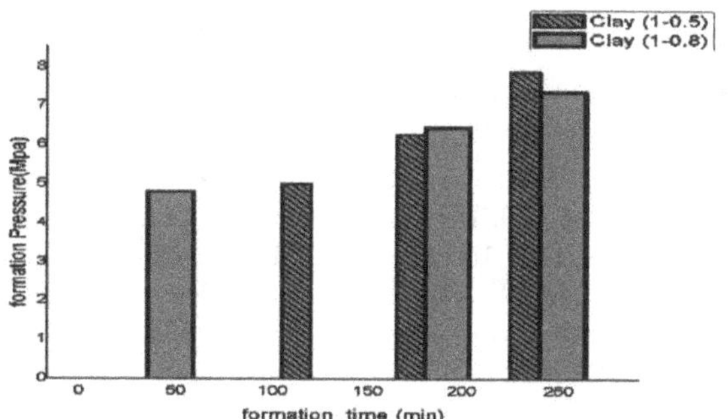

Fig. 96.2 Foramtion time vs formation pressure

3. CONCLUSIONS

The GH inside superior pores (some tens of microns in one hundred μm) disconnected dissolute and those in tiny pores (less than a few microns) underwent two stages of dissociation before finally serving temperature was around 260 K (Prasad etal, 2020). However, we may well not find huge deviation in the isolation temperatures in case of clay (nano powders) for mutually the water ratios with diverse pressures. Nevertheless, since experiments, we could illustrate 30-40% upturn in the return in circumstance of clay as the mechanical chattels like cementation of particles by gas hydrate has great effects on shear modulus but rare effects on the bulk modulus (Lu X.B,2013).The reinforce impact to cut power of clay declines with upsurge in definite surface area of soil reserves. This strength be since of agitating effect and there by H2O to gas superficial area falls at more attentions of clay(Radhika etal,2018). This can be confirmed from the results using BET studies discussed in Table 96.1 and Fig. 96.3.

Table 96.1 BET studies parameters

Property	Clay (44.17nm)
Average pore diameter(Å)	68.850 Å
All pore diameter(Å)	111.039 Å
Specific Surface area	9.007 m^2/g
Total Pore Volume	0.015 cc/g

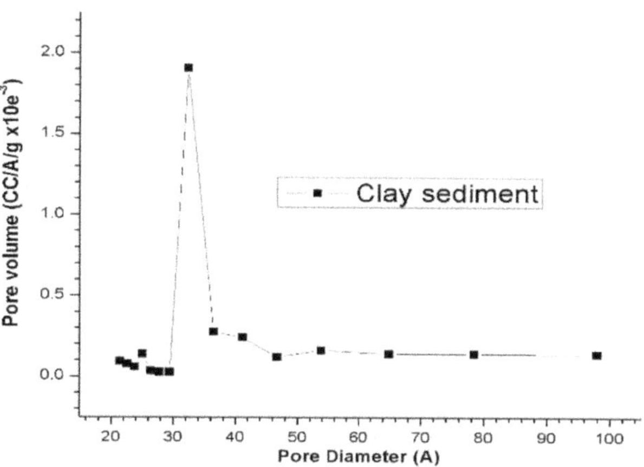

Fig. 96.3 BET studies for clay sediment

REFERENCES

1. B Sloan, Jr, E.D. &Koh, CA Clatharate hydrate of natural gasses, 3rd edition(CRC press 2007).
2. Simon R. Davies, E. Dendy Sloan, Amadeu K. Sum, and Carolyn A. Koh,The Journal of Physical Chemistry C 2010 114 (2), 1173-1180, DOI: 10.1021/jp909416y
3. A Review of the Effect of Porous Media on Gas Hydrate Formation Lanyun Wang, ACS Omega 2022 7 (38), 33666-33679DOI: 10.1021/acsomega.2c03048.
4. Critical Review on Gas Hydrate Formation at Solid Surfaces and in Confined Spaces—Why and How Does Interfacial Regime Matter? Ngoc N. Nguyen, Mirza Galib, and Anh V. Nguyen Energy & Fuels 2020 34 (6), 6751-6760, DOI: 10.1021/acs.energyfuels.0c01291.
5. Radhika,etal, Influence Of Sediment Structural Properties And Their Dynamics In The Formation And Dissociation Of Methane Hydrates. Mater. Today Proc. 2018, 5, 17572–17578.
6. Wei, Y.; Maeda, N. Dry Water as a Promoter for Gas Hydrate Formation: A Review. Molecules 2023, 28, 3731.
7. Tao, Y.; Yan, K.; Li, X.; Chen, Z.; Yu, Y.; Xu, C. Effects of Salinity on Formation Behavior of Methane Hydrate in Montmorillonite. Energies 2020, 13, 231.
8. Pinnelli S. R. Prasad, Burla Sai Kiran and Kandadai Sowjanya RSC Adv., 2020, 10, 17795, DOI: 10.1039/D0RA01754J.
9. Advances in Study of Mechanical Properties of Gas Hydrate-Bearing Sediments Lu X.B* , Zhang X.H and Wang S.YThe Open Ocean Engineering Journal, 2013, 6, 26-40.
10. Radhika ikkurti, S. R. Prasad, P., & Suresh Babu, D. (2018). Experimental investigations of methane hydrate in sediment suspensions. International Journal of Engineering & Technology, 7(4.5), 651-653.

Note: All the figures and tables in this chapter were made by the authors.

Camphene - Fatty Alcohol-based Eutectic Phase Change Materials for Thermal Energy Storage Applications

Rahul Bidiyasar
Department of Physics, University of Rajasthan, Jaipur, Rajasthan, India

Rohitash Kumar
Defence Laboratory, DRDO Jodhpur, Rajasthan, India

Narendra Jakhar*
Department of Physics, University of Rajasthan,Jaipur, Rajasthan, India

Abstract: Despite extensive research on phase change materials (PCMs), most studies rely on experimental trial-and-error methods to test component combinations for eutectic PCMs, which are costly, time-consuming and inefficient for systematically discovering new eutectics when screening large numbers of candidate materials. In this study, the thermophysical properties of novel eutectic PCMs composed of camphene and various fatty alcohols, including 1-dodecanol, 1-tetradecanol, 1-hexadecanol and 1-octadecanol, were theoretically calculated using the thermodynamical principles based on solid-liquid equilibria. The eutectic mixtures have eutectic compositions ranging from 26-71 mol% camphene, with melting temperatures between 17.55 °C to 40.76 °C and latent heats of fusion varying from 211-234 J/g. These theoretical studies offer an efficient and cost-effective alternative to experimental screening. They provide valuable insights into the thermophysical properties of camphene-fatty alcohol-based eutectic PCMs, potentially expanding their applications in thermal energy storage systems, such as building heating and cooling.

Keywords: Phase change material, Eutectic, Theoretical, Melting temperature, Thermal energy storage

1. INTRODUCTION

The rising global demand for sustainable energy highlights the importance of efficient thermal energy storage (TES) technologies. TES helps balance energy supply and demand, improve renewable energy efficiency, and reduce dependence on fossil fuels. Among TES methods, latent heat storage using phase change materials (PCMs) is particularly promising because they can store and release significant energy within narrow temperature ranges, making them ideal for applications like building energy management, electronic cooling and waste heat recovery (Sharma et al. 2009). Binary eutectic PCMs (BEPCMs), formed by combining two components at specific ratios, have a single melting point lower than that of the individual constituents. They offer customizable thermal properties, such as eutectic composition and melting temperature, that can be fine-tuned to meet specific application requirements for precise thermal management (Singh et al. 2021). However, determining the eutectic composition, melting temperature and latent heat of fusion often requires extensive experimental work, which is time-consuming and costly (Alipour et al. 2023).

Camphene, a terpene-based compound, is a promising candidate for eutectic PCM formulations due to its low toxicity, environmental friendliness, and relatively high latent heat of fusion. Similarly, fatty alcohols are widely recognized for their favorable thermal properties, including chemical stability and compatibility with various applications. Combining camphene with fatty alcohols offers a promising pathway for developing efficient and sustainable BEPCMs. Despite their potential, the phase behavior and thermophysical properties of such systems remain underexplored, necessitating a systematic approach to design and optimization.

*Corresponding Author: jakharnaren@yahoo.com

DOI: 10.1201/9781003684718-97

To simplify and speed up PCM development, precise thermodynamic models like the Schroder-van Laar equation are invaluable. This model predicts key thermal properties, reducing experimental effort and enabling a more efficient PCM design process (Kumar et al. 2016). The theoretical values of eutectic composition, melting point and latent enthalpy obtained using the Schroder-van Laar model have been extensively validated through experimental results reported in numerous studies in the existing literature. Kumar et al. developed phase diagrams of 1-dodecanol with various fatty acids to design BEPCMs with melting points of 15.5-19 °C and latent heat of 183-188 J/g. Eutectic compositions were experimentally validated, showing strong agreement with computed values with a mean relative deviation of 6.61% for melting temperature and 2.8% for latent heat (Kumar et al. 2017). Fan et al. developed BEPCMs consisting of various combinations of fatty acids having melting temperatures ranging from 35.10 to 39.29 °C and latent enthalpy ranging from 166.18 to 189.50 J/g. Experimental values were in good agreement with the calculated result obtained from the Schroder Van Laar model, with a mean relative deviation of 2.81% and 4.17 % for melting temperature and latent heat, respectively (Fan et al. 2022).

This study applies the Schroder-van Laar model to camphene and various fatty alcohols like 1-dodecanol (1-DD), 1-tetradecanol (1-TD), 1-hexadecanol (1-HD) and 1-octadecanol (1-OD) systems to predict their eutectic composition, melting point, and latent heat. By providing a theoretical foundation, this work aims to facilitate the development of advanced BEPCMs for TES applications.

2. COMPUTATIONAL METHODS

The eutectic composition, phase transition temperature, and latent enthalpy are essential thermal efficiency parameters for PCMs, as they determine both the potential application areas and the heat storage capacity of the material. Thermodynamic principles and phase stability concepts serve as key frameworks for predicting the mass ratios and thermophysical properties of eutectic PCM mixtures. The Schroder-Van Laar equation is a thermodynamic model used to describe the phase behaviour in eutectic systems as follows (Kumar et al. 2017; Zhou et al. 2023).

$$\ln(x_i) = \frac{\Delta H_{mi}}{R}\left(\frac{1}{T_{mi}} - \frac{1}{T_i}\right) \tag{1}$$

Here, x_i is the mole fraction of the component i in liquid state. ΔH_{mi} denotes the molar latent heat of component i. The universal gas constant, R, has a value of 8.315 J/mol·K. T_{mi} represents the melting temperature of the pure ith component while T_i represents the melting temperature of the eutectic PCM.

For a BEPCM comprising components A and B, the Schroder van Laar equation can be represented as a set of Eqs. (2) and (3) (Kahwaji and White 2018).

$$T_{BA} = \frac{1}{\left(\frac{1}{T_{mA}} - \frac{R\ln x_A}{\Delta H_{mA}}\right)} \tag{2}$$

$$T_{AB} = \frac{1}{\left(\frac{1}{T_{mB}} - \frac{R\ln x_B}{\Delta H_{mB}}\right)} \tag{3}$$

Here, T_{AB} and T_{BA} indicates the temperature of the mixture when adding component A to B and vice versa, respectively. The intersection point of these curves in the phase diagram indicates the ideal composition for a particular eutectic composition.

The latent enthalpy of a eutectic system can be determined by Eq (4) (Kumar et al. 2017).

$$H_m = T_m \sum_{i=1}^{n}\left[\frac{x_i \Delta H_{mi}}{T_{mi}}\right] \tag{4}$$

3. RESULTS AND DISCUSSION

The selection of suitable PCMs is crucial for designing efficient TES systems. This study explores binary eutectic PCMs using camphene with various fatty alcohols. Thermal properties, including melting points and latent heat, were sourced from existing literature and summarized in Table 97.1. These values serve as input parameters for the Schroder-van Laar equation to predict phase behavior and eutectic compositions.

Table 97.1 Thermal properties of camphene and fatty alcohols from literature

PCM	Melting Temperature (°C)	Latent heat of fusion (J/g)	Reference
Camphene	50.00	238.00	(Sharma et al. 2009)
1-DD	22.96	213.81	(Zhang et al. 2021)
1-TD	37.02	229.55	(Zhang et al. 2021)
1-HD	48.95	237.42	(Zhang et al. 2021)
1-OD	56.36	248.81	(Zhang et al. 2021)

Using the Schroder-van Laar model, the melting temperature as a function of mole fraction was calculated for binary systems comprising camphene and each fatty alcohol. The model accounts for ideal mixing behavior and the suppression of melting points due to intermolecular interactions. The computed melting temperature profiles were plotted against the mole fraction of camphene to generate phase diagrams for each binary system.

Figure 97.1 presents a representative phase diagram of the camphene-fatty alcohol system, showing two distinct liquidus lines converging at a eutectic point.

The phase diagrams demonstrated a typical V-shaped behavior, with melting temperatures decreasing as compositions approached the eutectic point. The sharp transitions at these points indicated the co-crystallization of both components.

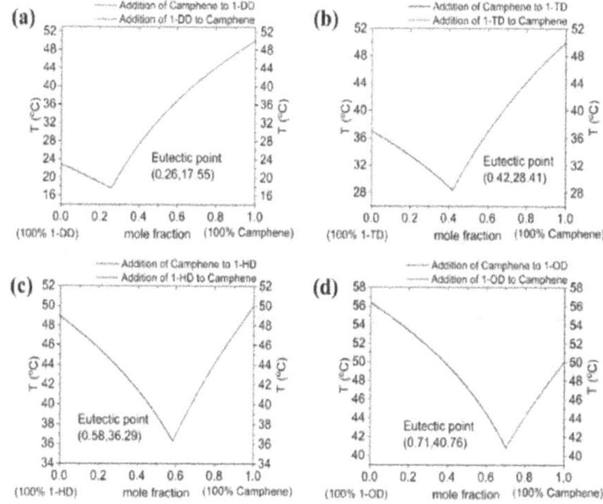

Fig. 97.1 Phase diagram of the (a) camphene-1-DD, (b) camphene-1-TD, (c) camphene-1-HD and (d) camphene-1-OD binary system

The eutectic composition, eutectic temperature, and latent heat were extracted from the phase diagrams for each binary system and are detailed in Table 97.2.

Table 97.2 Computed eutectic properties of camphene-fatty alcohol systems

BEPCMs	Eutectic molar composition	Eutectic melting temperature (°C)	Eutectic latent heat of fusion (J/g)
Camphene – 1-DD	26:74	17.55	211
Camphene – 1-TD	42:58	28.41	223
Camphene – 1-HD	58:42	36.29	228
Camphene – 1-OD	71:29	40.76	234

The results indicate that the eutectic temperatures for camphene-fatty alcohol systems were consistently lower than the melting points of the individual components, highlighting the potential for reducing operating temperatures in TES applications. The latent heat of fusion of the eutectic mixtures, although slightly lower than that of the pristine components, remains substantial, ensuring efficient energy storage and retrieval. These tailored thermophysical properties of camphene-fatty alcohol mixtures are due to intermolecular interactions between camphene's double bond and the hydroxyl (-OH) group of fatty alcohol. These interactions, including weak hydrogen bonding and van der Waals forces, disrupt the crystalline structures of both components, reducing the energy required for melting and lowering the eutectic temperature and latent heat. Theoretically predicting the eutectic compositions and thermal properties with minimal experimental effort represents a significant step forward in developing efficient TES materials.

4. CONCLUSIONS

This study theoretically evaluated the thermophysical properties of innovative eutectic PCMs formed from camphene and different fatty alcohols. The analysis was conducted using thermodynamic principles rooted in the Schroder Van Laar equation, providing valuable output for the design and performance of these PCMs. The results reveal eutectic mixtures with compositions ranging from 26-71 mol% camphene, melting temperatures between 17.55 °C and 40.76 °C, and high latent heat of fusion in the 211-234 J/g range. These findings highlight the efficiency of theoretical calculations for screening candidate materials, addressing the limitations of experimental trial-and-error methods by offering a cost-effective alternative for PCM discovery. The results provide a solid basis for utilizing camphene-fatty alcohol-based eutectic PCMs in TES applications, particularly for building heating and cooling.

REFERENCES

1. Alipour, A., Eslami, F., and Sadrameli, S. M. (2023). A novel bio-based phase change material of methyl palmitate and decanoic acid eutectic mixture: thermodynamic modeling and thermal performance. Chemical Thermodynamics and Thermal Analysis. 10: 100111.
2. Fan, Z., Zhao, Y., Liu, X., Shi, Y., and Jiang, D. (2022). Thermal properties and reliabilities of lauric acid-based binary eutectic fatty acid as a phase change material for building energy conservation. ACS omega. 7(18): 16097-16108.
3. Kahwaji, S., and White, M. A. (2018). Prediction of the properties of eutectic fatty acid phase change materials. Thermochimica Acta. 660:94-100.
4. Kumar, R., Kumar, R., and Dixit, A. (2016). Thermal phase diagram of acetamide-benzoic acid and benzoic acid-phthalimide binary systems for solar thermal applications. In AIP conference proceedings 1728 (1). AIP Publishing.
5. Kumar, R., Vyas, S., and Dixit, A. (2017). Fatty acids/1-dodecanol binary eutectic phase change materials for low temperature solar thermal applications: design, development and thermal analysis. Solar Energy. 155: 1373-1379.
6. Sharma, A., Tyagi, V. V., Chen, C. R. and Buddhi, D. (2009). Review on thermal energy storage with phase change materials and applications. Renewable and Sustainable energy reviews. 13(2):318-345.
7. Singh, P., Sharma, R. K., Ansu, A. K., Goyal, R., Sari, A., & Tyagi, V. V. (2021). A comprehensive review on development of eutectic organic phase change materials and their composites for low and medium range thermal energy storage applications. Solar Energy Materials and Solar Cells. 223:110955.
8. Zhang, Q., Yan, H., Zhang, Z., Luo, J., Yin, N., Tan, Z., and Shi, Q. (2021). Thermal analysis and heat capacity study of even-numbered fatty alcohol (C12H25OH-C18H37OH) phase-change materials for thermal energy storage applications. Materials Today Sustainability. 11-12:100064.
9. Zhou, D., Xiao, S., Xiao, X., & Liu, Y. (2023). Preparation, Phase Diagrams and Characterization of Fatty Acids Binary Eutectic Mixtures for Latent Heat Thermal Energy Storage. Separations. 10(1):49.

Note: All the figures and tables in this chapter were made by the authors.

Crystal Growth and Structural Characterization of Bi0.95Sb0.05 Topological Single Crystal

Rabia Sultana*

Department of Electronics, Asutosh College, University of Calcutta, Kolkata, India

Abstract: We report the growth and structural characterization of $Bi_{0.95}Sb_{0.05}$ topological insulator. $Bi_{0.95}Sb_{0.05}$ single crystal (SC) was synthesized by melting bismuth and antimony together using the facile self-flux method via solid state reaction route in an automated programmable tube furnace. The on surface X-ray diffraction (XRD) pattern on as grown crystal flake displayed 001 alignment and revealed the crystalline nature of the resultant $Bi_{0.95}Sb_{0.05}$ crystal. Rietveld refinement of powder XRD data confirmed the phase purity and rhombohedral crystal structure with R-3m space group. Further, the unit cell structure formed using the VESTA software verified the rhombohedral crystal structure. Scanning electron microscope (SEM) and energy dispersive X-ray analysis (EDXA) mapping images taken on freshly cleaved $Bi_{0.95}Sb_{0.05}$ SC flakes displayed the layered directional growth with homogeneous distribution of Bi and Sb elements. Elemental mapping of the resultant $Bi_{0.95}Sb_{0.05}$ SC showed that the constituent elements (Bi and Sb) are homogeneously distributed in accordance to their stoichiometric ratio.

Keywords: Topological Insulator, Single Crystal, structural

1. INTRODUCTION

Topological insulators (TIs) have grabbed significant attention owing to their unique physical properties and potential technological applications. Researchers and condensed matter physicists are continuously exploring this (TI) fascinating field of research in order to realize new topological materials with topological phenomena that could be useful for the development of robust photonic devices (optoelectronics), nanoelectronics, thermoelectrics, quantum computers and spintronics. TIs are quantum materials where the interior / bulk is insulating but the edges/surfaces can conduct electrons. The conducting states located along the edges/surfaces are topologically protected by time reversal symmetry (TRS) and have spin momentum locking property. Unlike, the other topological materials (Bi_2Se_3, Bi_2Te_3, Sb_2Te_3), bismuth antimony alloys ($Bi_{1-x}Sb_x$) are binary alloys of antimony (Sb) and bismuth (Bi) in various ratios. Among the TIs, BiSb alloy emerged as the most exciting field of research because of its exceptional high spin–orbit torque (SOT) efficiency and tunable electronic band structure. The electronic band structure of $Bi_{1-x}Sb_x$ can be regulated by varying the stoichiometric ratio, temperature and thickness. Already known for its thermoelectric properties, BiSb alloys act as promising material for spin-orbit torque magneto resistive random-access memory (SOT-MRAM) devices, spintronics and nanoelectronics applications.

2. EXPERIMENTAL DETAILS

$Bi_{0.95}Sb_{0.05}$ SC was synthesized from stoichiometric mixtures of high purity (99.99%) Sb and Bi powders. Briefly, stoichiometric ratio of Sb and Bi powders were precisely weighed, mixed well and thoroughly grounded inside a glove box under Argon atmosphere. Further, the homogeneously grounded powder was pelletized and then enclosed in a quartz ampoule (10^{-3} Torr).

The ampoule was loaded into a tube furnace and heated up to 650°C (42°C/hour) followed by a hold duration of 8hours. The ampoule was then gradually cooled from 650°C to 250°C (3°C/hour), hold for 95hours and then

*Corresponding author: rabiauit@gmail.com

DOI: 10.1201/9781003684718-98

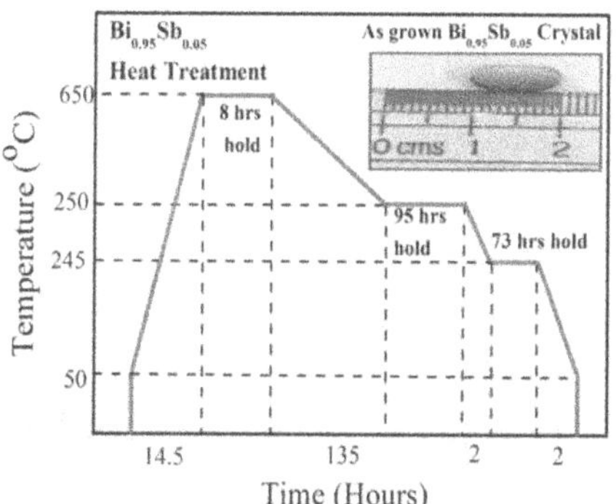

Fig. 98.1 Heat treatment schematic diagram for the growth of $Bi_{0.95}Sb_{0.05}$ SC. Inset shows the synthesized $Bi_{0.95}Sb_{0.05}$ SC

slowly cooled to 245 °C (3 °C/hour). The ampoule was maintained at 245 °C for 73 hours and subsequently permitted to cool down to room temperature (120 °C/hour). The diagram of heat treatment along with the synthesized crystal is displayed in Fig. 98.1. The synthesized crystals of $Bi_{0.95}Sb_{0.05}$ were mechanically cleaved along the basal plane and prepared for further characterizations.

3. RESULTS AND DISCUSSION

Figure 98.2 shows the Rietveld refinement of gently crushed powder X-ray diffraction (PXRD) pattern of synthesized $Bi_{0.95}Sb_{0.05}$ SC. The single crystal XRD (SCXRD) pattern obtained on mechanically cleaved crystal flake of $Bi_{0.95}Sb_{0.05}$ SC is shown in the Fig. 98.2 inset. Rigaku Miniflex II powder x-ray diffractometer with Cu-Kα radiation (λ = 1.5418 Å) was used for both the PXRD and SCXRD. The Rietveld fitted data verified the phase purity and confirmed the rhombohedral structure with R-3m space group. Figure 98.2's inset depicts sharp diffraction peaks (0, 0, 3n with n = 1, 2, 3...etc.) growing along the [001] direction and thus, confirms the good crystalline quality of as grown $Bi_{0.95}Sb_{0.05}$ SC.

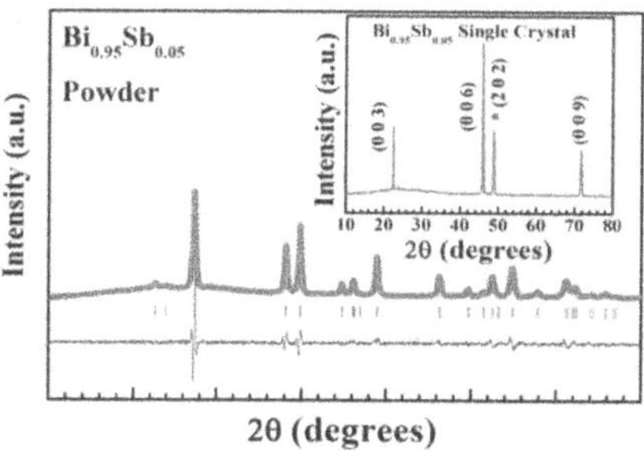

Fig. 98.2 PXRD pattern for $Bi_{0.95}Sb_{0.05}$ SC. SCXRD pattern of $Bi_{0.95}Sb_{0.05}$ SC is displayed in the inset

Unit cell structure of $Bi_{0.95}Sb_{0.05}$ SC is shown in Fig. 98.3. Similar to other TIs (Bi_2Se_3, Bi_2Te_3, Sb_2Te_3), $Bi_{0.95}Sb_{0.05}$ crystal also exhibits rhombohedral crystal structure. However, the atoms are arranged differently in the unit cell structure i.e., each of the unit cells in this case has three bi-layers of Bi and Sb layered one over the other (Fig. 98.3).

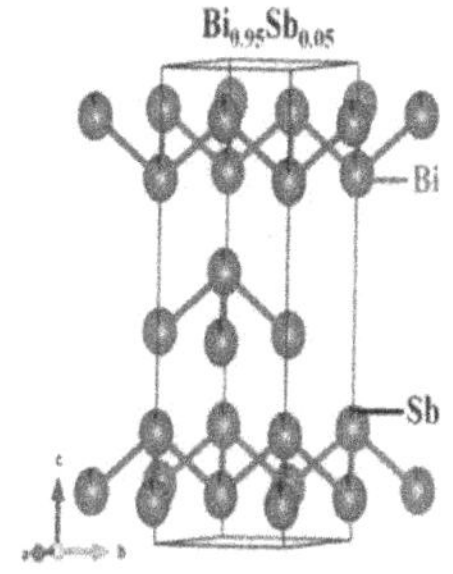

Fig. 98.3 Unit cell structure of $Bi_{0.95}Sb_{0.05}$ SC

To check the elemental composition and morphological characteristics of $Bi_{0.95}Sb_{0.05}$ SC, we performed the EDXA and SEM measurements (Fig. 98.4). EDXA analysis was done on a small portion of the SEM figure. Bi and Sb elements are present in a nearly stoichiometric ratio. i.e. composition is close to $Bi_{1-x}Sb_x$ (x = 0.05).The EDXA analysis further confirmed that the as grown $Bi_{0.95}Sb_{0.05}$ crystal are free from foreign contamination i.e., they are of pure form. The SEM image (inset of Fig. 98.4) of the $Bi_{0.95}Sb_{0.05}$ SC displayed a smooth layered structure without any kind of impurity, indicating the good quality feature.

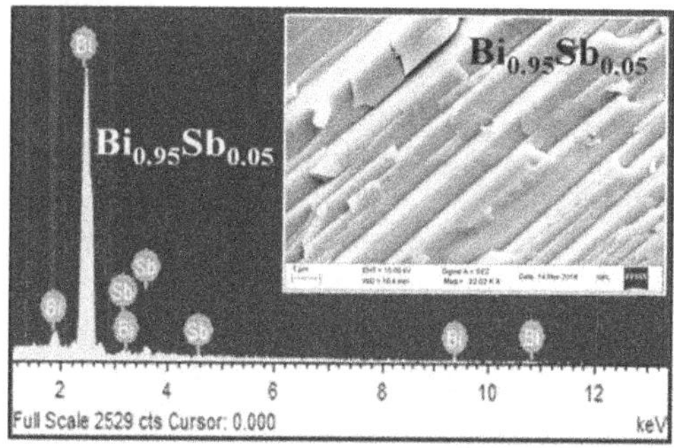

Fig. 98.4 EDXA mapping image of $Bi_{0.95}Sb_{0.05}$ SC. Inset shows the SEM image of $Bi_{0.95}Sb_{0.05}$ SC

4. CONCLUSION

In conclusion, crystal growth and structural characterization of $Bi_{0.95}Sb_{0.05}$ SC is discussed. The resultant crystal has 001 alignment and is highly crystalline. Further, the well depicted crystal structure, phase purity and morphological characterizations indicate that the synthesized $Bi_{0.95}Sb_{0.05}$ SC is of good quality.

REFERENCES

1. Hasan, M. Z. and Kane, C. (2010). Colloquium: Topological insulators. Rev. Mod. Phys. 82, 3045-3067.
2. Ando,Y. (2013). Topological Insulator Materials J. Phys. Soc. Jpn. 82, 102001-32.
3. Zhang, H., Liu, C. X., Qi, X. L., Dai, X., Fang, Z. and Zhang, S. C. (2009). Topological insulators in Bi_2Se_3, Bi_2Te_3 and Sb_2Te_3 with a single Dirac cone on the surface. Nat. Phys. 5, 438-442.
4. Sultana, R., Gurjar, G., Gahtori, B., Patnaik, S., and Awana, V.P.S. (2019). Flux free single crystal growth and detailed physical property characterization of $Bi_{1-x}Sb_x$ (x = 0.05, 0.1 and 0.15) topological insulator. Mater. Res. Express 6, 106102.

Note: All the figures in this chapter were made by the authors.

Recent Trends in Applied Physics and Material Science – Sudhir Bhardwaj et al. (eds)

Temperature Dependent Modulation of Textures and Switching Voltage in Cholesteric Liquid Crystals for Application in Optical and Sensor Devices

Pooja, Vandna Sharma[1] and Pankaj Kumar[2]

Centre for Liquid Crystal Research, Chitkara University Institute of Engineering and Technology, Chitkara University, Punjab, India

Abstract: The cholesteric liquid crystal (CLC) mixtures known for their stabilizing property in the planar (P) and focal conic (FC) states, are widely used in optical shutters and sensor devices. The temperature dependent pitch of the CLC mixtures directly affects the performance and sensitivity of CLC cells, impacting their different practical applications. In this work, the CLC mixtures were prepared using nematic liquid crystal (NLC) with 4.0 and 5.0wt% of chiral dopant (R1011) to observe the effect of temperature on their textural and electro-optical behavior. The measurements were taken at 25, 30, 35, 40, and 45 °C temperatures using the polarized optical microscope (POM), temperature controller and electro-optical setup. Textural changes under an applied electric field showed the initial rotation of the helical axis in FC state and gradual unwinding of the helices in homeotropic state. Further, as the temperature increases, these changes occur at lower voltages because of the reduced viscosity and increase in the pitch make it easier for the electric field to overcome the molecular interactions that maintain the helical structure. The electro-optical results for both CLC mixtures showed a decrease in initial transmittance with temperature, making the FC state more random and resulted in reduced switching voltage. At 45 °C, the switching voltage was found decreased from 46 V (at 25°C) to 38 V and 59 V (at 25°C) to 48 V for 4.0 and 5.0wt% chiral doped CLC mixtures, respectively. However, the better contrast was found at 40°C and 45°C for 4.0 and 5.0wt% samples, respectively.

Keywords: Cholesteric liquid crystal, Focal conic, Switching voltage, Contrast ratio

1. INTRODUCTION

The cholesteric liquid crystal (CLC) mixtures exhibited unique optical properties such as variable pitch length, stabilized planar (P) and focal conic (FC) states, that allow it to produce amazing results in various applications such as light shutters [Zhao et al. (2024), Gahrotra et al. (2022), Gahrotra et al. (2020), Kumar et al. (2012)], broadband mirrors [Yi et al. (2024)], sensors [Hussain et al. (2024)] and displays [Lin et al. (2024)]. The performance of CLC mixtures for varied applications depends on the different stimuli such as electric field, temperature, magnetic fields, concentration of chiral dopant etc. Figure 99.1 (a, b) shows the basic switching operation of CLC molecules modulating from FC to H state under the effect of externally applied field [Gahrotra et al. (2022), Sadigh et al. (2021)].

In the FC state, CLC molecules are in random orientation, while in H state, LC molecules are parallel to field direction. In the present work, effect of temperature on electro-optical properties has been studied as the change in temperature directly influences the pitch length of CLC mixtures which affects the switching voltage from one state to another. The CLC mixtures were prepared using 4.0

Corresponding Authors: [1]shandlyavandna90@gmail.com, [2]pankaj.kumar@chitkara.edu.in

DOI: 10.1201/9781003684718-99

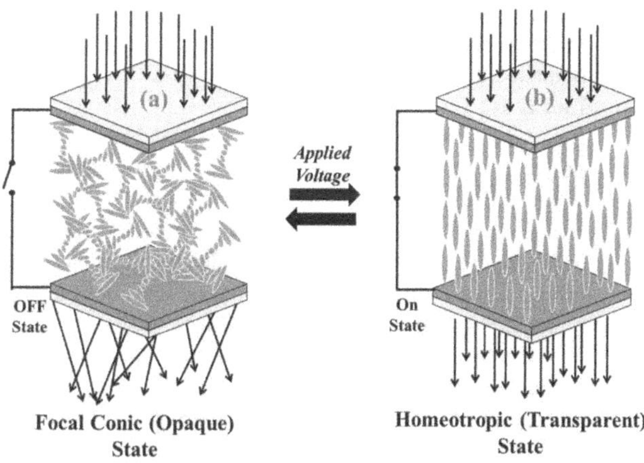

Fig. 99.1 Schematic showing switching of CLC molecules from (a) FC-OFF to (b) H-ON state

and 5.0wt% chiral (R1011, Merck, HTP 37.6 μm^{-1}) doped LC (E7, Merck). The thickness of CLC cells was 10µm. The temperature dependent textural and electro-optical studies of CLC cells were performed using polarizing optical microscope (POM, Nikon Eclipse LV100POL), connected with camera (Q28378) and electro-optical setup assembled with temperature controller (LINKAM THMS-600), respectively.

2. RESULT AND DISCUSSION

Figure 99.2(i, ii) shows the temperature dependent electro-optical cum textural behavior of 4.0 and 5.0 wt% chiral doped CLC mixtures. The points (a_1-a_5) in graphs (i, ii) of temperature dependent voltage-transmittance (VT) behavior correspond to the textures showing the FC state [Fig. 99.2 i(a_1-a_5) and 2 ii(a_1-a_5)] due to the random arrangement of CLC molecules under no applied voltage. Initial transmittances of the 4.0 and 5.0wt% CLCs reduced with the increase of temperature, resulting in the FC states with more randomness near the isotropic temperature. Further applied voltage twists the helices of CLC molecules towards the unwinding state (unstable state), thus, attaining the increase in transmittance value as represented in Fig. 99.2 with points i(b_1-b_5) and ii(b_1-b_5) for 4.0 and 5.0wt% CLC mixtures, respectively. It was observed that intermediate transmittance for unstable state was less in 4.0wt% CLC mixture than 5.0wt% CLC mixture. On further increasing the voltage, the unstable arrangement of CLC molecules obtained minimum transmittance positions at all temperatures as represented in Fig. 99.2 (i, ii) with points (c_1-c_5). Further increase of voltage caused the complete unwinding of the helices, thus, a mixture achieved H (maximum transmittance) state as shown in Fig. 99.2 i (d_1- d_5) and ii (d_1- d_5) corresponds to the black textures observed under crossed polarizers as shown in Fig. 99.2 (i(d_1-d_5), ii(d_1-d_5)) for 4.0 and 5.0wt% CLC mixtures at varied temperatures. The switching voltage reduced at higher temperatures due to decrease in the viscosity and increase in the pitch length of CLC mixture, thus CLC molecules reoriented easily at low voltages from FC to the H state. At 45°C, the switching voltage decreased from 46 V (at 25°C) to 38V and 59V (at 25°C) to 48 V for 4.0 and 5.0wt% CLC mixtures, respectively. However, the better contrast was found at 40°C and 45°C for 4.0 and 5.0wt% samples, respectively, as given in Table 99.1.

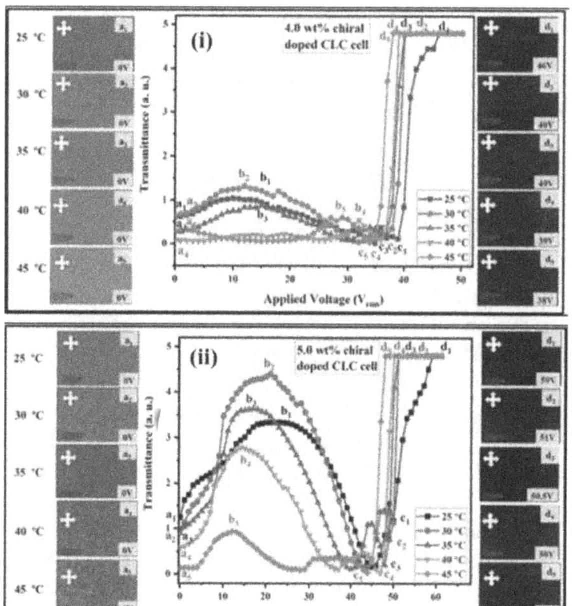

Fig. 99.2 POM textures and voltage *vs.* transmittance of (i) 4.0 and (ii) 5.0wt% CLCs at varied temperatures

3. CONCLUSION

From this work, it is concluded that with the increase of temperature, both the mixtures showed better contrast ratios and lower switching voltages. 4.0wt% chiral doped CLC mixture switched to H state at 39 V with 80.17 contrast ratio at 40 °C, while 5.0 wt% chiral doped

Table 99.1 Summarized values of contrast ratio and switching voltages for 4.0 and 5.0wt% CLCs at varied temperatures

Concentration of chiral	Parameters	T (25°C)	T (30°C)	T (35°C)	T (40°C)	T (45°C)
4.0 wt%	Contrast Ratio	8.02	7.88	17.81	80.17	15.52
	Switching Voltage (V_{rms})	46	40	40	39	38
5.0 wt%	Contrast Ratio	3.89	5.38	5.45	8.26	33.5
	Switching Voltage (V_{rms})	59	51	50.5	50	48

CLC cell showed better results at 45°C having switching voltage of 48V with 33.5 contrast ratio. Thus, rise in the temperature decreases the switching voltage of the CLC mixtures with a better contrast ratio.

REFERENCES

1. Gahrotra, R., Sharma, V., Dogra, A.R., Malik, P. and Kumar, P. (2022). Performance augmentation of bistable cholesteric liquid crystal light shutter-effect of dichroic dye on morphological and electro-optical characteristics. *Opt. Mater. 127:*112243.
2. Gahrotra, R., Sharma, V., Dogra, A.R., Malik, P. and Kumar, P. (2022). Temperature dependent morphological and electro-optical characteristics of dye doped cholesteric liquid crystal. *ECS Trans. 107*:5481.
3. Gahrotra, R., Sharma, V., Malik, P. and Kumar, P. (2020). Textural and electro-optical study of cholesteric liquid crystal for light shutter. *AIP Conf. Proc. 2220*:090021.
4. Hussain, S. and Zourob, M. (2024). Solid-state cholesteric liquid crystals as an emerging platform for the development of optical photonic sensors. *Small, 20*, 2304590.
5. Kumar, P., Kang, S.W. and Lee, S.H. (2012). Advanced bistable cholesteric light shutter with dual frequency nematic liquid crystal. *Opt. Mater. Express 2:*1134.
6. Lin, K.W., Lu, J.Y., Tseng, H.Y., Lin, C.Y., Huang, L.Y., Lin, Y.L., Chang, H.H. and Lin, T.H. (2024). Study of polymer-stabilized cholesteric liquid crystals for reflective displays in active matrix driving. *J. Inf. Disp.* 1.
7. Sadigh, M.K., Naziri, P., Ranjkesh, A. and Zakerhamidi, M.S. (2021). Relationship of pitch length of cholesteric liquid crystals with order parameter and normalized polarizability. *Opt. Mater. 119*:111373.
8. Yi, H., Zhao, Y., He, Z., Gao, H., Gao, J., Wang, D. and Luan, Y., (2024). Doped nanofibers achieve tunable broadband in cholesteric phase liquid crystals. *Liq. Cryst.* 1-15.
9. Zhao, Y., Yu, P., Su, Y., Guo, Z., Song, W., He, Z. and Miao, Z. (2024). Flexible bistable polymer stabilised cholesteric texture light shutter display. *Liq. Cryst.* 1.

Note: All the figures and tables in this chapter were made by the authors.

Recent Trends in Applied Physics and Material Science – Sudhir Bhardwaj et al. (eds)

Effect of Silica Nanoparticles on Polymer Dispersed Liquid Crystal and its Analysis for Smart Windows

Meenakshi
Centre for Liquid Crystal Research,
Chitkara University Institute of Engineering and Technology,
Chitkara University, Punjab, India
Department of Applied Physics, Mukand Lal National College, Yamunanagar, Haryana, India

Pooja, Vandna Sharma[1] and Pankaj Kumar[2]
Centre for Liquid Crystal Research,
Chitkara University Institute of Engineering and Technology,
Chitkara University, Punjab, India

Abstract: Polymer-dispersed liquid crystals (PDLCs) are the materials comprised of liquid crystal (LC) droplets dispersed within a polymer matrix, and are extensively employed in modern electro-optical (E-O) applications. This study explores the comparison of morphological and E-O properties of PDLC and 0.8 wt% of silica (SiO_2) nanoparticles (NPs) dispersed PDLC cured at 0.15mW/cm^2 UV light intensity. The morphological study of the samples was performed using Polarized Optical Microscope (POM), the average droplet size was increased from ~2.5 μm to ~4.6 μm with the dispersion of 0.8wt% SiO_2 in PDLC. In E-O study, the threshold voltages (V_{th}) were almost same in pure and 0.8wt% SiO_2 dispersed PDLCs, however, a significant decrease in the operating voltage (V_{op}) of PDLC from 15.11 V to 12.49V was achieved in 0.8wt% SiO_2 dispersed PDLC due to the bigger average size of droplets. Further, the maximum transmittance (T_{max}) value was found slightly increased in SiO_2 dispersed PDLC. Moreover, 0.8 wt% of SiO_2 NPs dispersed PDLC showed higher contrast ratio (14.12) as compare to pure PDLC (7.2). The improved results observed in 0.8wt% of SiO_2 NPs dispersed PDLC are substantially correlated with the uniformity of increased droplets' size. Thus, the primed PDLCs show up a prospective relevance with the sustainable energy efficient smart windows.

Keywords: Polymer-dispersed liquid crystal (PDLC), Nanoparticles, Electro-optical, Contrast ratio, Threshold Voltage and Operating Voltage

1. INTRODUCTION

Polymer dispersed liquid crystals (PDLCs) are a class of materials that comprises of micro- or nano-sizes' liquid crystal (LC) droplets embedded into a polymer matrix used for smart applications [Ellahi et al. (2024), Zhang et al. (2023), Jain et al. (2020) Zhang et al. (2023)]. Because of the anisotropic nature of LC, the PDLC films show different optical states such as opaque and transparent under the external applied electric field. They can be fabricated into different categories of systems depending on preparation conditions, and concentration of polymer which can be as large as 70% or as small as 2%. Various distinct techniques such as thermally induced, solvent induced, and polymerization induced phase separations, are also used to form LC droplets in polymer matrix [Sharma et al. (2021), Ji et al. (2017)]. Researchers executed the dispersion of nanoparticles (NPs) with minute concentrations in PDLCs for the further enhancement in the electro-optical properties [Katariya et al. (2024), Sharma et al. (2022), Kumar et al. (2018)]. Thus, in the present work, PDLC mixture was doped with a higher concentration (0.8 wt%) of SiO_2 NPs for comparative analysis of morphological and electro-optical (E-O) properties of pure and NPs dispersed PDLC. For the preparation of PDLC mixtures, the positive nematic

Corresponding Authors: [1]shandlyavandna90@gmail.com, [2]pankaj.kumar@chitkara.edu.in

DOI: 10.1201/9781003684718-100

LC E7 (Merck) and UV curable polymer NOA-65 were used in 1:1 ratio. Further, spherical SiO_2 NPs (0.8 wt%) having average particle size ~80 nm (Merck) were used to prepare NPs dispersed PDLC cell. The thickness of PDLC cells was maintained using 10μm mylar spacer and polymerization induced phase separation method was used to cure polymer in the PDLC cells, UV curing light intensity was 0.15mW/cm^2. For characterizations, polarizing optical microscope (POM, Nikon Eclipse LV100POL), connected with camera (Q28378) and assembled electro-optical setup were used for the morphological and E-O studies of prepared PDLC cells.

2. RESULTS AND DISCUSSION

Figure 100.1((a_1-e_1), (a_2-e_2)) denotes the POM morphology of the prepared PDLC cells under the effect of external applied field. At 0V, the LC droplets possess bipolar configuration (scattering) as represented in Fig. 100.1(a_1, c_1). On applying external field, the droplet configuration changed into radial configuration (transparent) as represented in Fig. 100.1(a_2, c_2) for pure and 0.8 wt% SiO_2 doped PDLC cell, respectively. At more higher voltages, dark black textures were observed, confirming the complete alignment of LC molecules toward the field direction. With doping of the 0.8 wt% SiO_2, the average droplet size was increased from ~2.5 μm to ~4.6 μm in PDLC. This change is clearly represented in Figure 1(e_1, e_2) taken in the parallel polarizer and analyzers. This is due to an increase in phase separation and polymerization process.

The voltage-transmittance behavior represented in Fig. 100.1(b), shows the variation in threshold (V_{th}) and operating voltages (V_{op}) of 0.8wt% SiO_2 NPs dispersed PDLC compared with pure PDLC. Table 100.1 compared the values of minimum transmittance, maximum transmittance, contrast ratio, and operating voltage for pure and 0.8wt% SiO_2 dispersed PDLC cells. As shown in the Table 100.1 significant decrease in the V_{op} of PDLC from 15.11 V to 12.49 V was achieved in 0.8wt% SiO_2 dispersed PDLC, whereas, a slight change in the V_{th} was observed. Moreover, 0.8 wt% of SiO_2 NPs dispersed PDLC showed a higher contrast ratio (14.12) as compared with pure PDLC (7.2). The maximum transmittance

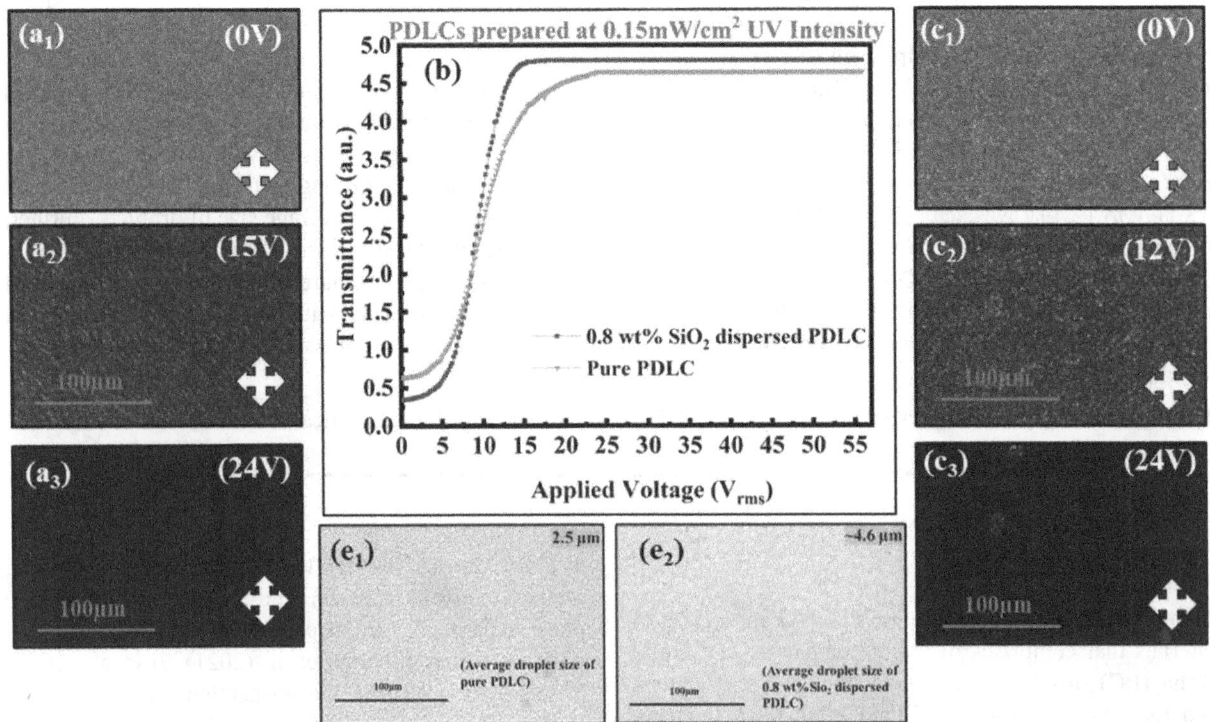

Fig. 100.1 POM textures (a_1-a_3) pure PDLC, (c_1-c_3) 0.8wt% SiO_2, (b) voltage – transmittance curve of pure, and 0.8 wt% SiO_2 dispersed PDLC cells. Average droplet size of (e_1) pure and (e_2) 0.8wt% SiO_2 dispersed PDLCs under parallel polarizers at 50x magnification

Table 100.1 Comparison of minimum transmittance, maximum transmittance, contrast ratio, operating voltage and average droplet sizes for pure and 0.8wt% SiO_2 dispersed PDLC cells

Fabricated PDLC Cells	T_{min}	T_{max}	Contrast Ratio	V_{op}	Average Droplet Size
Pure	0.64	4.65	7.2	15.11	~2.5 μm
0.8wt% SiO_2	0.34	4.8	14.12	12.5	~4.6 μm

(T_{max}) value was also found slightly higher in 0.8 wt% SiO_2 dispersed PDLC. These results indicate that using the varied concentration range of NPs, best results can be optimized to fabricate energy efficient smart windows.

3. CONCLUSION

This work concluded that with the dispersion of 0.8 wt% SiO_2 NPs in the PDLC system increased the average size of LC droplets from ~2.5 μm to ~4.6 μm. The 0.8 wt% SiO_2 NPs dispersed PDLC showed reduction in V_{op} from 15.11 V to 12.49V. In addition, contrast ratio was also found higher in 0.8 wt% SiO_2 NPs dispersed PDLC (14.12) as compared with pure PDLC (7.2). Thus, improved E-O results of PDLC, open the ways to make energy-efficient PDLC smart windows.

REFERENCES

1. Ellahi, M., Taimur, S., Baloch, N., Bhayo, A.M., Sarwar, A. and Qadar, F. (2024). Study of Polymer-Dispersed Liquid Crystal (PDLC) Thin Film Technology for Smart Electronic Devices. *Journal of Electronic Materials, 53:*1094.
2. Zhang, P., Tong, X., Gao, Y., Qian, Z., Ren, R., Bian, C., Wang, J. and Cai, G. (2023). A sensing and stretchable polymer-dispersed liquid crystal device based on spiderweb-inspired silver nanowires-micromesh transparent electrode. *Advanced Functional Materials, 33:*2303270.
3. Jain, A.K. and Deshmukh, R.R., (2020). An overview of polymer-dispersed liquid crystals composite films and their applications. *Liquid Crystal Display and Technology*, 1.
4. Zhang, R., Zhang, Z., Han, J., Yang, L., Li, J., Song, Z., Wang, T. and Zhu, J., (2023). Advanced liquid crystal-based switchable optical devices for light protection applications: principles and strategies. *Light: Science & Applications, 12*:11.
5. Ji, S.M., Huh, J.W., Kim, J.H., Choi, Y., Yu, B.H. and Yoon, T.H., (2017). Fabrication of flexible light shutter using liquid crystals with polymer structure. *Liquid Crystals*, 44:1429.
6. Sharma, V., Kumar, P. and Raina, K.K., (2021). Simultaneous effects of external stimuli on preparation and performance parameters of normally transparent reverse mode polymer-dispersed liquid crystals—a review. *Journal of Materials Science, 56*:18795.
7. Katariya-Jain, A., Mhatre, M.M., Dierking, I. and Deshmukh, R.R., (2024). Enhanced thermo-electro-optical and dielectric properties of carbon nanoparticle-doped polymer dispersed liquid crystal based switchable windows. *Journal of Molecular Liquids, 393*:123575.
8. Sharma, V. and Kumar, P. (2022). Electro-optically oriented Kerr and orientational phase study of normal mode polymer dispersed liquid crystals–Effect of dispersion of nanoparticles. *Journal of Molecular Liquids, 348*:118030.
9. Kumar, P., Sharma, V. and Raina, K.K. (2018). Studies on inter-dependency of electrooptic characteristics of orange azo and blue anthraquinone dichroic dye doped polymer dispersed liquid crystals. *Journal of Molecular Liquids, 251*:407.

Note: All the figures and tables in this chapter were made by the authors.

Recent Trends in Applied Physics and Material Science – Sudhir Bhardwaj et al. (eds)
© 2026 Taylor & Francis Group, London, ISBN 978-1-041-16452-4

Study of Dispersion Characteristics of SPPs at Metal/Lithium Niobate (Ferroelectric) Waveguide Structures

Gishamol Mathew*, Anusha P. P. and Anamika K
Department of Physics,Maharaja's College, Ernakulam, Kerala

Abstract: Surface Plasmon Polaritons are collective electron plasma excitations at the interface of two materials having opposite dielectric functions. The transfer of field energy of SPP modes is not possible into the bulk of the metal due to the screening effect of free electrons but can propagate along the interface with very high field confinement. In this paper, we aim to investigate the dispersion characteristics of SPPs along ($LiNbO_3$ (strip)/Ag/SiO_2) nano-waveguide structures. The presence of $LiNbO_3$ strip provides ferroelectric influences such as blue or red shift of plasmon resonance. Both symmetric and asymmetric modes were excited and asymmetric modes are of long ranging. We observed propagation length in μm and geometrical/ material dispersion which ensure a plasmonic waveguide operating at plasmon frequencies.

Keywords: Surface plasmon polaritons, Ferroelectric, Plasmon-waveguides

1. INTRODUCTION

Surface plasmon polaritons (SPPs) are electromagnetic waves coupled to the oscillations of conduction electrons propagating along the metal/dielectric interface with exponentially decaying field in the direction perpendicular to the interface (Barnes et al., 2003, Maier, 2007). These interface SPP modes are now considered as a promising candidate for transferring energy at subwavelength scale. It was reported that both subwavelength confinement and a reasonable propagation length can be achieved in plasmonic waveguides (Mathew et.al., 2022). In addition, SPP modes show a strong dependence on geometrical and material parameters and the frequency of mode excitation, thereby making it suitable for the design of integrated active plasmonic components.

In this paper, we propose an alternate material, a ferroelectric $LiNbO_3$, to excite SPP modes at the metal interface. Since ferroelectric crystals have a permanent polarization, there is a distinct separation of charges on the two sides or two opposite domains of the $LiNbO_3$ crystal. This will result in an increase or decrease in electron density at the $LiNbO_3$/metal interface at the side with positive or negative charges respectively, which will give rise to a blue or red-shift of surface plasmon frequency (Zhijie et.al., 2016). A strip plasmonic waveguide structure, $LiNbO_3$ (strip)/Ag on SiO_2 substrate is proposed for the effective propagation and field confinement. The coupling of interface plasmons leads to the existence of symmetric and asymmetric modes. The dispersion relation is then numerically studied for the mode propagation, field confinement and ferroelectric tunability.

2. THEORY

Figure 101.1 depicts the plasmonic slot-waveguide geometry under consideration.

The optical response of metals is defined by the classical or Drude electron theory, where the frequency dependent dielectric function $\varepsilon_m(\omega)$ is assumed for complex permittivity of the metallic region as:

$$\varepsilon_m = \left(1 - \frac{\omega_p{}^2}{\omega^2 + \upsilon^2}\right) - j\left(\frac{\omega_p{}^2 \upsilon}{\omega\left(\omega^2 + \upsilon^2\right)}\right) \quad (1)$$

*Corresponding Author: mathew.gishamol@maharajas.ac.in

DOI: 10.1201/9781003684718-101

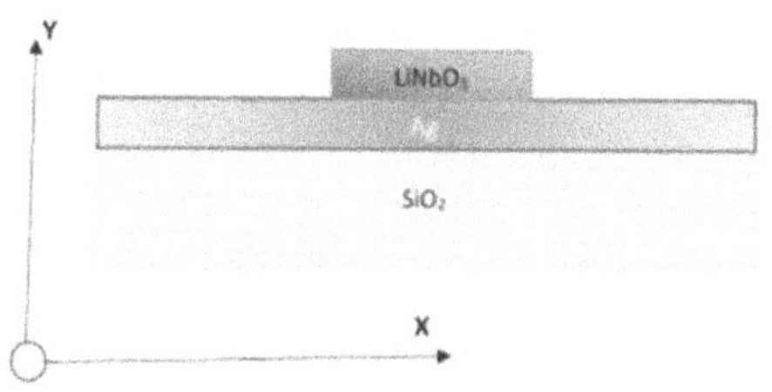

Fig. 101.1 Structure of a plasmonic strip-waveguide

where ω_p is plasmon-resonance frequency and υ is the effective collision frequency with the relaxation time of electrons (τ) in the metal. The relative permittivity tensor of $LiNbO_3$ medium under Voigt geometry is,

$$\varepsilon_d = \begin{bmatrix} 4.938 & 0 & 0 \\ 0 & 4.770 & 0.168j \\ 0 & -0.168j & 4.770 \end{bmatrix} \tag{2}$$

For SiO_2, $\varepsilon_d = 2.25$.

The dispersion relation of propagating modes can be obtained by solving Maxwell's uncoupled wave equations separately for each layer and applying the interface boundary conditions (Mathew et.al., 2022). The propagation is assumed to be along z-direction with propagation constant, $\gamma = \alpha + j\beta$, where α is the attenuation constant and β is the phase constant (Mathew et.al., 2012). The superposition of TE_x and TM_x modes will characterize any mode propagating through the structure including the hybrid ones. Due to the structural asymmetry and complex field coupling, the Method of Lines algorithm is used with appropriate boundary conditions applied at horizontal and vertical limits (Mathew et.al., 2022). The dispersion codes were developed in Matlab with the aid of root finding algorithms: Muller method, Nelder-Mead minimisation method, complex root finding algorithms, to get the non-trivial solution (Mathew et.al., 2022).

3. RESULT ANALYSIS

The waveguide geometry, $LiNbO_3$ (w=1 µm)/Ag/SiO_2 is studied for dispersion. The slot-geometry supports two fundamental bound modes symmetric and asymmetric modes in accordance with the symmetry of the electric field with the two symmetry planes of the ferroelectric-strip. For asymmetric modes, the value of phase constant increases when the thickness of the metal film increases and then attains an approximately fixed value. But the symmetric modes exhibit a decreasing trend with core metal-film and then attains a saturated value. Figure 101.2 shows the dispersion behaviour of normalised phase constant as a function of metal-core thickness for asymmetric modes. As the slot region is with a ferroelectric material, the SPPs can have some ferroelectric influences too. It was seriously noted that the ferroelectric material, LiNbO3, is capable of producing localized and propagating SPPs (Zhijie et.al., 2016).

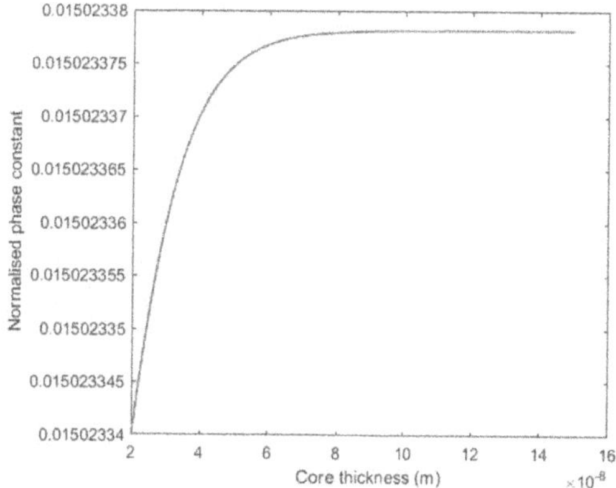

Fig. 101.2 The dispersion of normalised phase constant with core-film thickness for asymmetric modes at 4.741×10^{12} Hz

It was observed that the asymmetric modes are of long-ranging when compared to symmetric modes. The propagating SPP modes show (1/e) attenuation lengths due to low mode confinement for a nano-meter film thickness. Fig. 101.3 shows the variation of propagation length with the metal-core thickness, and it shows a decreasing trend. The proposed structure supports ferroelectric tunability on SPP modes and are able to sustain SPPs with micro-meter range propagation lengths.

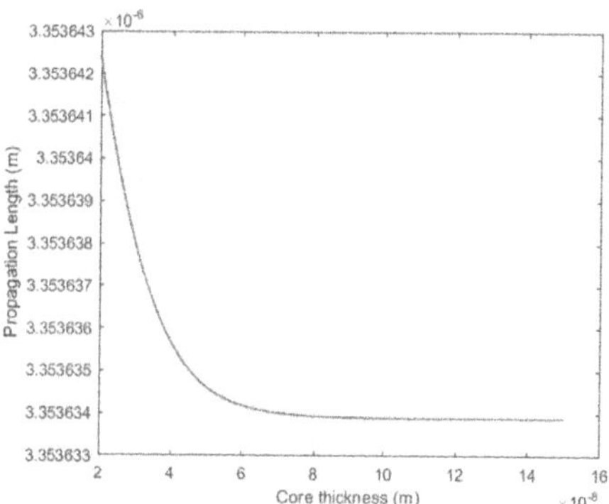

Fig. 101.3 The variation of propagation length with metal-film thickness for asymmetric modes at 4.741×10^{12} Hz

4. CONCLUSION

We presented a comprehensive theoretical and numerical analysis of the dispersion properties of $LiNbO_3$(strip)/Ag/SiO_2 plasmonic waveguides. The coupling of interface plasmons leads to the evolution of symmetric and asymmetric modes. It was observed that asymmetric modes are long ranging due to the balance between SPP mode localization and propagation loss. In addition, the presence of $LiNbO_3$ strip provides ferroelectric influences on SPPs. The change of plasmon resonance frequency demonstrates proper tuning of plasmon waveguides which ensure a promising alternative for the realization of advanced optoelectronic applications.

REFERENCES

1. Barnes, W. L., Dereux, A. and Ebbesen, T. W. (2003). Surface plasmon subwavelength optics. Nature 424: 824–830.
2. Maier, S. A. (2007) Plasmonics: Fundamentals and Applications, Springer Science + Business Media LLC, New York, NY, USA.
3. Mathew, G. and Louie Frobel, P. G. (2022) Nonreciprocal mode dispersion in a coupled metal-slot waveguide structure. Optik 270 (170019).
4. Zhijie, W. et.al. (2016) Manipulation of charge transfer and transport in plasmonic-ferroelectric hybrids for photoelectrochemical applications. Nature Communications 7, 10348.
5. Mathew, G. and Mathew, V. (2012) Tunable surface plasmon polaritons in metal strip waveguides with magnetised semiconductor substrates in Voigt configuration. Semiconductor Science and Technol. 27(5), 055010.

Note: All the figures in this chapter were made by the authors.

Recent Trends in Applied Physics and Material Science – Sudhir Bhardwaj et al. (eds)

102

Investigating the Impact of Calcium Carbonate and Water on Black soil Resistivity and Geotechnical Properties

Sujata R. Jadhav[1], Vaijanath V. Navarkhele[2]
Department of Physics, Dr. Babasaheb Ambedkar Marathwada University, Chhatrapati Sambhajinagar, India

Abstract: This study examines the impact of calcium carbonate ($CaCO_3$) and water on the resistivity of soil and its subsequent effects on friction angle, plasticity, and cohesion. Laboratory experiments were carried out with varying $CaCO_3$ contents and moisture content. A factorial design varying $CaCO_3$ concentrations (0-8%) and water content (0-50%) revealed significant correlations. Resistivity was measured with two-probe resistivity meter. Friction angle, plasticity and friction angle were measured using formulae depends on resistivity given in previous literatures. Results shows that increasing $CaCO_3$ content and water addition decreases soil resistivity, leading to reductions in friction angle and cohesion, while plasticity increases. These finding have significant implications for agricultural soil management, soil erosion control, and soil structure interaction.

Keywords: Resistivity, Plasticity, Friction angle, Cohesion and moisture content

1. INTRODUCTION

Agriculture is vital sector in Maharashtra, India. However, soil degradation, erosion, and nutrient depletion are concerns that needs attention. Soil resistivity is an important parameter in geotechnical engineering, agriculture, and environmental monitoring.

The hereditary physical, chemical, and biological characteristics of soil are excessively impacted by $CaCO_3$ forms (Mustafa Ismail Umer et. al 2020). In practice, the study of resistivity is important to examine the soil-cement quality (Song Yu Lu et. al 2008). Electrical resistivity if cost effective and time saving way to check soil- cement mixture quality (Narongchai wiwattchang et. al 2023).

Agricultural soil resistivity is crucial for understanding soil behaviour, water infiltration and root growth. Calcium carbonate ($CaCO_3$) is well known and common additive used to neutralize acidity and improve structure of soil. However, its impact on soil physical properties, resistivity and other resistivity dependent is not well understood.

This is first study to investigate $CaCO_3$ and water effects on agricultural soil resistivity and geotechnical properties. It reveals a previously unreported correlation between $CaCO_3$ content and resistivity of agricultural soils. Provides new insights into the impact of $CaCO_3$ and water on soil plasticity, challenging existing understanding.

2. MATERIALS AND METHODS

2.1 Sample Collection and Sample Preparation

Sample of soil were collected from agricultural areas close to Chhatrapati Sambhajinagar, Maharashtra, India. The collected soil had dried out, sieved with 2mm mesh size sieve to eliminate debris and large particles, and mixed to confirm uniformly. Then prepared soil samples by mixing $CaCO_3$ as (0-8%) and water content (0-50%) for experimental investigations.

2.2 Resistivity Measurement

The resistivity was measured using the two-probe resistivity meter. For use in the lab, the soil samples were

[1]sujatajadhav18@gmail.com, [2]vvn_bamu@yahoo.co.in

DOI: 10.1201/9781003684718-102

collected in a glass container. The resistivity meter was operated by inserting the two probes into the soil. The copper rods used to make the probes are connected at one end and tapered at the other. The probes' sensitivity to changes in soil characteristics is high. Probes create an electrical connection with the soil when they are implanted. Through the probes, the meter continuously introduces current into the soil. The soil current creates potential difference between two probes. V and I measurements, ohms law was used to compute the resistance,

$$R = V/I$$

After measurement of resistance resistivity calculated by using formula,

$$\varrho = RA/l$$

2.3 Soil Water Content

A moisture sensor SM-150 was used to measure soil water content. The manufacturer's water content measurement has an error of around ±3%.

2.4 Dielectric Constant

"The dielectric constant of a substance is a measure of its capability to store electrical energy."

Dsielectric constant (ε) was calculated by,

$$\varepsilon = 3.03 + 9.3\theta + 146.0\theta^2 + 76.7\theta^3$$

...(Muhammad et. al 2013)

Where θ is moisture content.

2.5 Friction Angle

Soil dryness is measured by friction angle. The correlation between resistivity(ϱ) and friction angle(fr) was given by Jusoh and Osman as,

$$fr = (4.7036 * \ln(\varrho)) + 6.6297$$

.....(Vivek Prashar et. al 2021)

2.6 Plasticity of Soil

The amount of the fine particles content in the soil is known as plasticity. Jusoh and Osman give correlation between plasticity index(pl) and resistivity as,

$$pl = (-2.7 * \ln(\varrho)) + 29.793$$

.....(Vivek Parashar et. al)

3. RESULT AND DISCUSSION

Figure 102.1 shows the change in resistivity of soil with increasing percentage of $CaCO_3$ and water. There is decrease in resistivity with addition of water in soil. This is because increase in moisture content leads to more conductive path due to water fills pore spaces, reducing the distance between particles and increasing electrical contact. For 0% water content the resistivity of soil rises with increase in $CaCO_3$ content (2-8%).

But when add mixture of $CaCO_3$ and water in soil for all 2-8% $CaCO_3$ and soil mixture resistivity decreases with increasing water content. Because $CaCO_3$ dissolves in water, releasing Ca^{2+} and CO_3^{2-} increasing electrical conductivity and ionic strength.

$$CaCO_3 + H_2O \longrightarrow Ca^{2+} + CO_3^{2-} + 2H^+ + 2e^-$$

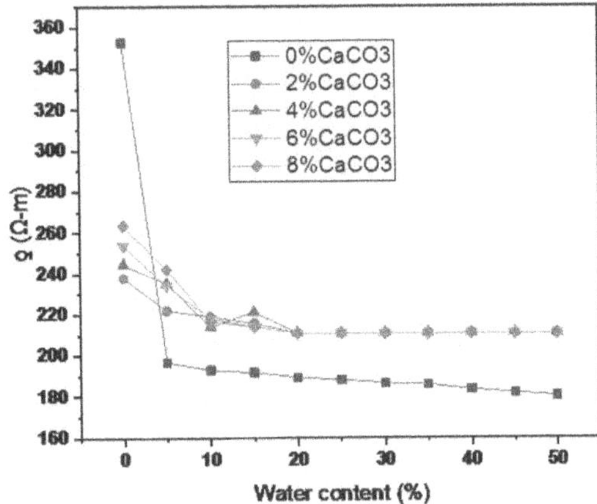

Fig. 102.1 Effect of $CaCO_3$ and water content on resistivity of soil

Figure 102.2 is variation of dielectric constant with increase in $CaCO_3$ and water content in soil. It is seen that dielectric constant continuously increases with increase in $CaCO_3$(0-8%) and water content in soil (0-50%). Because $CaCO_3$ dissolves in water enhances polarization, increasing dielectric constant.

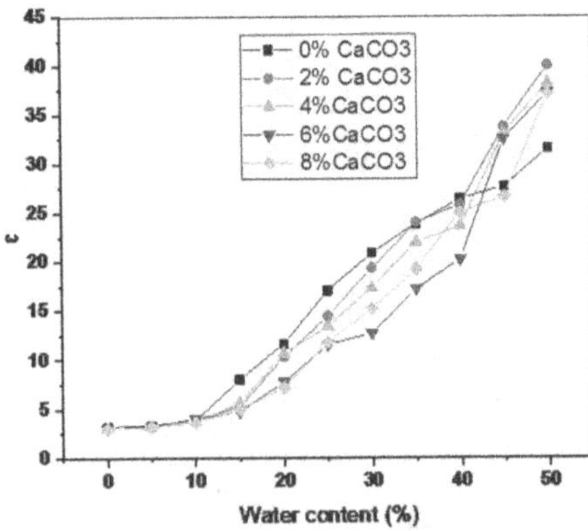

Fig. 102.2 Effect of $CaCO_3$ and water content on dielectric constant(ε) of soil

Figure 102.3 is change in friction angle(fr) of soil with change in $CaCO_3$(0-8%) and water content (0-50%) in soil. Up to 20% water content friction angle(fr) of soil decreases rapidly because $CaCO_3$ particles coat soil grains, reducing direct contact and alters soil structure. Water helps to reduce friction between soil particles. Therefore, combinedly water and $CaCO_3$ reduce soils internal friction results into decrease in friction angle(fr). After 20% water content friction angle(fr) of soil remains unchanged. Beyond 20% water content, soil pores become

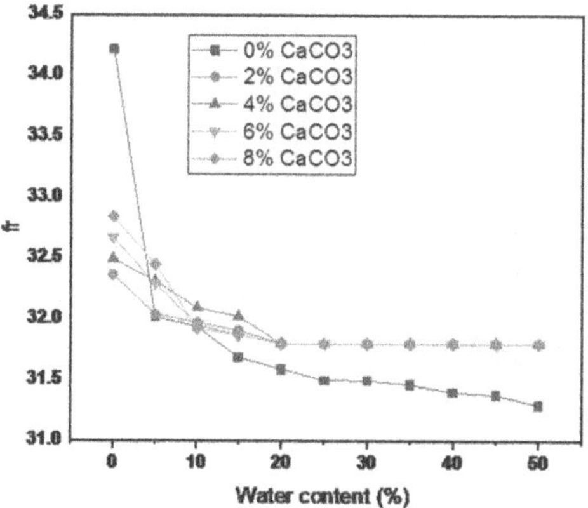

Fig. 102.3 Effect of $CaCO_3$ and water content on friction angle(fr) of soil

saturated. Excess water fills larger pores, so friction angle(fr) is unaffected.

For (0-8%) $CaCO_3$ and (0-20%) water content plasticity index(pI) of soil increases rapidly is shown in Fig. 102.4. Increase in $CaCO_3$ adds clay-like properties facillates particle bonding and water reduces interparticle friction in soil. Therefore, $CaCO_3$ and water enhances particle bonding in soil leading to increase in plasticity.

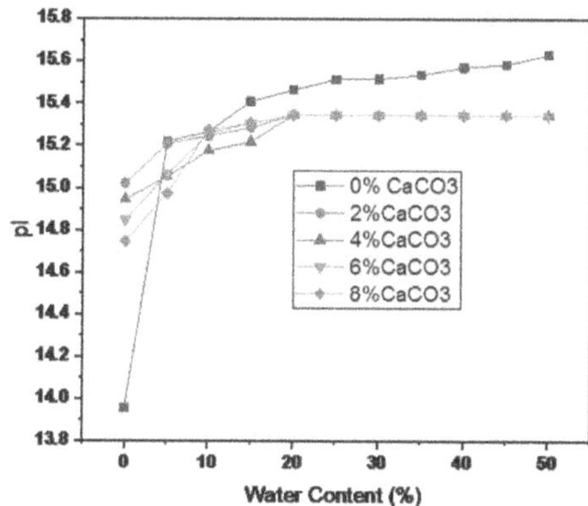

Fig. 102.4 Effect of $CaCO_3$ and water content on plasticity index(pI) of soil

After 20% water content plasticity of soil remains constant (Fig. 102.4). After 20% water content, soil pores become saturated. This excess water does not alter soils plasticity.

4. CONCLUSION

This experimental study investigated the impacts of $CaCO_3$ content (0-8%) and water content on the resistivity, dielectric constant, friction angle and plasticity of soil.

The results show significant correlations between $CaCO_3$ and water content and the studied soil properties. $CaCO_3$ content and water content significantly decreases resistivity and friction angle while increasing dielectric constant and plasticity index. Synergistic interaction between $CaCO_3$ and water enhances these effects.

This work contributes to the comprehension of the intricate relation between $CaCO_3$, water and soil properties, providing valuable implications for soil stabilization and foundation design, geotechnical and geological applications, soil erosion and sedimentation control and agricultural and environmental management.

REFERENCES

1. Andrew. B, Sionhan, H. P., Bem. S, (1996). Examination of solute transport in an undisturbed soil column using electrical resistance tomography. Water. Resour. Res 32(4), 763-769
2. Bryson, L. S., Bathe. A. (2009) Determination of selected geotechnical properties of soil using electrical conductivity testing. Geotech. Test. J. 32(3), 1-10
3. Hisyam Jusoh; Syed Baharom Syed Osman (2017) The correlation between resistivity and soil properties as an alternative to soil investigation. Indian J Sci Technol:10(6):1-5
4. Muhmmad Mukhalisin; Almushfi Saputra (2013). Performance evaluation of volumetric water content and relative permittivity models. The Scientific world Journal 421762
5. Mustafa Ismail Umer; Shayma Mohammad Rajab; Hajeen Khalil Ismail (2020). Effect of CaCO3 form on soil inherent quality properties of calcareous soils. Materials science forum, Vol. 1002, pp 459-467
6. Narongchai wiwattchang; Chanarop Vichalai; and Pham Huy Giao (2023). Influence of calcium carbonate sludge on cement stabilized subgrade quality as investigated by means of electrical resistivity measurement. Scientific report 13-19392
7. Song Yu Lu; Yan Jun Du; L.H Han; M.F Gu (2008). Experimental study on the electrical resistivity of soil-cement admixtures. Environ geol 54:1227-1233
8. Vivek Parashar; Bharat Mishra (2021). Designing efficient soil resistivity measurement technique for agricultural wireless sensor network. Int J Commun Syst.; 34: e4785.

Note: All the figures in this chapter were made by the authors.

Recent Trends in Applied Physics and Material Science – Sudhir Bhardwaj et al. (eds)
© 2026 Taylor & Francis Group, London, ISBN 978-1-041-16452-4

Optimized Photocatalytic Degradation of Xanthene Dyes in Wastewater Using Silver-Impregnated ZnO (Ag-ZnO)

Sivakumar Surabhi* and Jagannatha Rao Vepa
Anil Neerukonda Institute of Technology and Sciences,
Visakhapatnam, Andhra Pradesh, India

Abstract: Xanthene dyes are used to dye wool, cotton, silk and paper where brilliant shades are required. But the release of these complex dyes into waste water creates environmental pollution due to toxic and carcinogenic nature. Advanced oxidation processes, particularly semiconductor-assisted photocatalysis, have recently gained prominence as effective method for organic dye removal from wastewaters. The effectiveness of photocatalytic degradation lies in its potential to fully mineralize organic dyes into CO_2, water, and other benign mineral acids. Rhodamine 6G and Eosin Y dye photocatalytic degradation has been examined in this work using ZnO and silver impregnated ZnO (Ag-ZnO) catalysts under various conditions. Ag-ZnO containing silver at different concentration (0.5, 1.0 and 1.5 atom %) were prepared and the samples were calcined for 2 hours at various temperatures. The catalysts were analyzed using X-ray diffraction (XRD), UV-visible diffuse reflectance spectroscopy (UV-DRS), and scanning electron microscopy (SEM). Silver impregnation significantly enhances photocatalytic activity, with the 1.0 atom% Ag-ZnO catalyst achieving the highest degradation efficiency. The impact of operational parameters on degradation efficiency, including catalyst dose, dye quantity, and solution pH, was also investigated. The dye degradation followed pseudo fist order kinetics.

Keywords: Photocatalysis, Ag-ZnO, Xanthene dyes

1. INTRODUCTION

The textile industry produces substantial amounts of coloured dye effluents that are both toxic and non-biodegradable [Reif and Fremann, 1996]. Various treatment methods, including adsorption, precipitation, reverse osmosis, flocculation and ultrafiltration are commonly employed to remove colour from textile wastewater [Robinson et al, 2001]. However, these methods are non-destructive, as they merely transfer the pollutants into sludge, generating secondary pollution that requires further treatment [Chaudari and Sur, 2000].

Advanced oxidation processes (AOPs) have recently emerged as a promising substitute for water purification [Al- Ekabi and Ollis, 1993]. Among AOPs, heterogeneous photocatalysis has emerged as a highly effective approach for eliminating toxic organic pollutants from industrial wastewaters [Kusvuran et al., 2005]. Titanium dioxide (TiO_2) is widely used as a highly efficient photocatalyst, but its absorption is limited to the UV region of the solar spectrum. This drawback has urged interest in exploring alternative photocatalysts.

ZnO stands out due to its ability to absorb a larger portion of the solar spectrum compared to TiO_2, making it an ideal candidate for photodegradation under sunlight [S. Sakthivel et al., 2003].

This study investigates the photocatalytic degradation of Rhodamine 6G and Xanthene dyes from wastewaters using visible light and Ag-impregnated ZnO.

*Corresponding author: siva_ks9123@rediffmail.com

DOI: 10.1201/9781003684718-103

2. EXPERIMENTAL

2.1 Silver-impregnated ZnO (Ag-ZnO) preparation

Ag-ZnO photocatalyst with 1 atom% silver has been prepared as follows: 5 g of ZnO was added to silver nitrate solution (105 mg in 100 mL), and agitated constantly for 4 hours. The slurry was undisturbed for 24 hours. Afterward, the mixture subjected to heating at 100 °C for 16 hr in order to evaporate water. The solid produced was finely crushed into powder using an agate mortar and then calcined in muffle furnace at three different temperatures 200, 400 and 600 °C for 2 hours.

2.2 Characterisation

The structural and morphological features of the catalysts were evaluated using XRD and SEM. The powdered XRD of the prepared catalysts were obtained using X-ray diffractometer (Bruker D8 Advanced), and the SEM pictures were captured by scanning electron microscope (Philips, XL 30 ESEM).

2.3 UV-Visible Diffuse Reflectance Spectroscopy

Ag-ZnO and pure ZnO's light absorption behaviour was examined using UV-VIS spectroscopy. Shimadzu UV 3600 Spectrometer operated in diffuse reflectance mode (DRS) was chosen to record absorbance spectra from 200–800 nm wavelength range, with $BaSO_4$ serving as the reference material. Measurements were carried out at room temperature, and Kubelka-Munk function is used to convert the data.

2.4 Photocatalytic Measurements

The photocatalytic degradation was performed by by subjecting 100 mL of dye solution combined with 1 g/L of ZnO or Ag-ZnO catalyst to a 125 W Philips high pressure mercury vapor lamp. Aliquots were periodically collected, centrifuged, and analysed to determine dye degradation. The decolorization efficiency (%) was calculated as follows:

$$\% \text{ Degradation Efficiency} = \frac{C_0 - C}{C_0} X\ 100$$

where C_0 is the original dye concentration and C is the dye solution concentration after irradiation.

3. FINDINGS AND CONVERSATION

3.1 XRD and SEM Investigation

XRD peaks of ZnO and Ag-ZnO with silver loadings of 0.5, 1.0, and 1.5 at. % (dried at 100 °C) confirm the wurtzite structure matched with JCPDS file No. 36-1451 as presented in Fig. 103.1. The XRD peaks intensity increases with silver loading up to 1 at. %, indicating improved crystallinity, but decreases with further silver addition. Ag^+ ion incorporation does not alter the ZnO crystal structure, possibly due to low precent of silver addition and surface deposition. SEM pictures of Ag-ZnO (heated at 100 °C, Fig. 103.2) show diverse particle shapes and sizes, with silver impregnation having no significant effect on morphology.

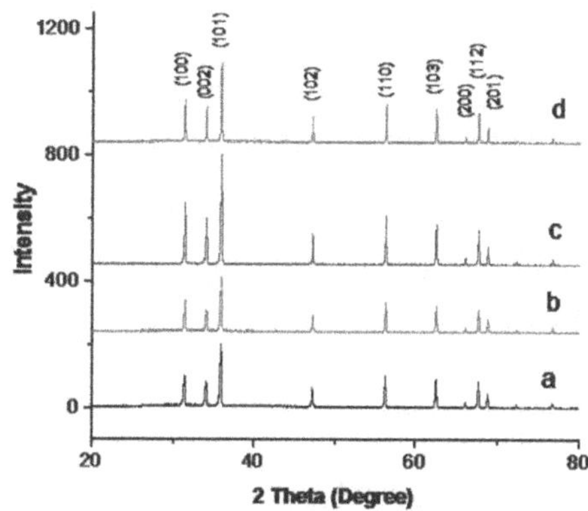

Fig. 103.1 Powdered XRD peaks of ZnO and different at.% of Ag–ZnO samples. (a) ZnO (b) 0.5 at.% Ag–ZnO (c) 10 at.% Ag-ZnO and (d) 1.5 at.% Ag-ZnO

Fig. 103.2 SEM images of 1 at. % Ag impregnated ZnO

3.2 Diffuse Reflectance Analysis

Pure ZnO and 1% Ag-ZnO both show significant absorption below 400 nm in their UV-Vis absorption spectra. Doping with silver causes a minor change in absorption to a higher wavelength. By extrapolating the linear region of the modified Kubelka–Munk plots $[F(R)\ h\nu]^2$ and photon energy (hv), band gap of semiconductor photocatalyst were calculated [Cimitan et al., 2009]. The observed red shift in absorbance was correlated with a modest decrease in the band gap, which was 3.28 eV for pure ZnO and 3.27 eV for 1at. % Ag-ZnO.

3.3 Photocatalytic Degradation

The efficiency of dye degradation changes with the amount of silver ion deposition. Table 103.1 clearly shows that 1 atom % silver loading exhibited greater photocatalytic performance for both dyes. The higher photocatalytic activity of Ag impregnation is accredited to its ability to trap electrons, reducing charge recombination

Table 103.1 Effect of silver impregnation on ZnO for degradation of R6G and Eosin Y; [Dye] = 50 mgL^{-1}, [Catalyst] = 1.00 mg L^{-1}

Time (min)	Pure ZnO	0.5 at. % Ag-ZnO	1.0 at. % Ag-ZnO	1.5 at. % Ag-ZnO	Pure ZnO	0.5 at. % Ag-ZnO	1.0 at. % Ag-ZnO	1.5 at. % Ag-ZnO
	% Degradation of R6G				% Degradation of Eosin Y			
0	0.00	0.00	0.00	0.00	0.00	0.00	0.00	0.00
15	33.66	29.06	44.77	28.79	31.32	29.89	42.26	30.15
30	52.18	49.43	64.26	48.07	47.47	48.12	60.23	44.76
45	63.26	67.18	75.69	61.60	58.94	63.72	74.95	59.18
60	72.18	80.46	85.24	78.87	68.87	79.14	88.34	73.60
90	83.67	90.00	94.51	86.64	86.38	97.20	100	96.44
120	90.07	97.00	100	94.67	99.02	100	100	100

and promoting substrate oxidation [Subbarao et al., 2003]. However, excessive silver impregnation can reduce efficiency, as doping below the optimal level facilitates electron-hole separation [Hermann et al., 1986], while higher concentrations may promote charge carrier recombination [Scalafani and Hermann, 1998]. Surface plasmon resonance of silver may also contributing to enhanced photocatalytic activity due to increased charge carrier generation [Toropov et al., 2016], [Snehal et al., 2023].

3.4 Effect of Catalyst Concentration

The catalyst dose influence on R6G and Eosin Y degradation was examined by changing the catalyst dosage from 0.25 to 2.0 g/L for 50 mgL^{-1} dye solution at its natural pH. Degradation efficacy increases with catalyst dose up to 1.5 g/L, beyond which no significant change was observed. This could be due to increased active surface area and availability of more sites with higher catalyst doses.

3.5 Impact of Dye Solution Concentration

The effect of starting dye concentration on photocatalytic degradation was investigated in the range of 5 mgL-1 to 200 mgL-1 dye concentration (Fig. 103.3). Results showed a gradual decrease in degradation efficiency with increasing dye concentration. Higher concentrations cause the catalyst to absorb less photons because the dye molecules block the photons before they reach the catalytic surface.

This reduces the prodcution of reactive species (OH· and O_2^-) on the catalyst surface, which lowers the efficiency of dye degradation [Sahoo et al., 2005].

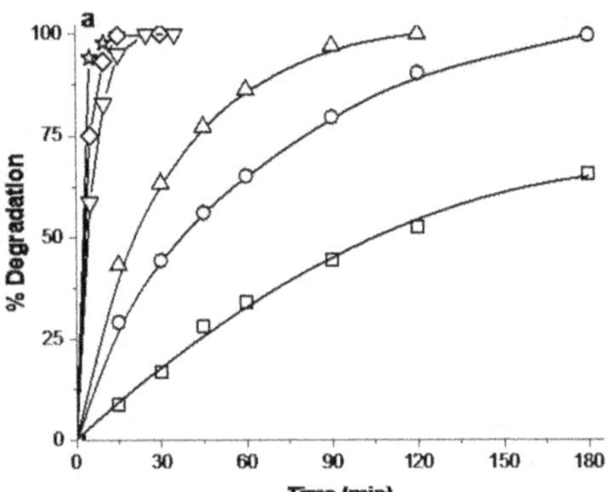

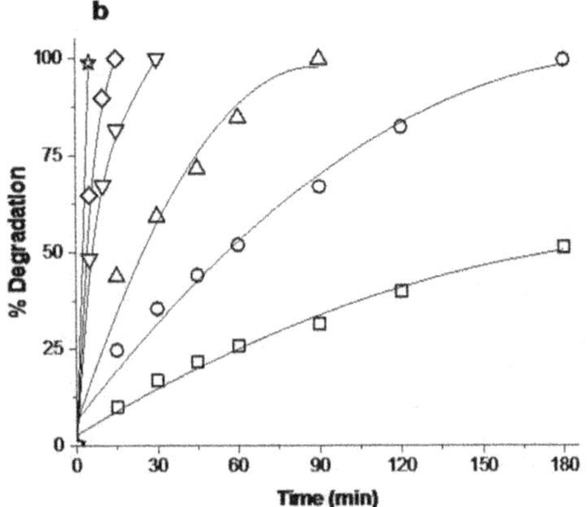

Fig. 103.3 Dye concentration effect on degradation of a) R6G and b) Eosin Y, [Catalyst] = 1.00 gL^{-1};1 atom% Ag-ZnO; (☆) 5 mgL^{-1}, (◇) 10 mgL^{-1}, (∇) 20 mgL^{-1}, (Δ) 50 mgL^{-1}, (O) 100 mgL^{-1}, (□) 200 mgL^{-1}

3.6 Effect of pH

The pH effect on decolorization efficiency was studied across a pH of 2 to 12 range, keeping dye content and catalyst amount constant. Decolorization efficiency significantly improved in the basic range and just modestly in the acidic pH range. Because of adsorbed OH-ions, the ZnO surface is negatively charged at pH 9.0 (ZnO's zero-point charge) and positively charged below. Numerous OH^- ions on the ZnO surface and in the reaction media encourage the creation of OH·, which increases photocatalytic activity [Akyol et al., 2004].

3.7 Kinetic Analysis

Figure 103.4 presents the kinetic graphs for R6G and Eosin Y degradation using the ideal dose of ZnO and

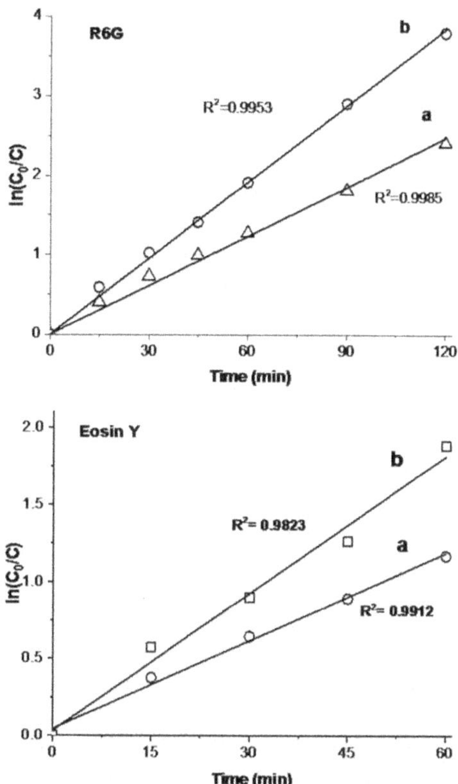

Fig. 103.4 Kinetic graphs for R6G and Eosin Y degradation on a) ZnO and b) 1 at. % Ag-ZnO; [Dye] = 50 mgL^{-1}, [Catalyst] = 1.00 gL^{-1}

Ag–ZnO catalyst. The data align well with the Langmuir–Hinshelwood kinetic model [Ozkan et al., 2004], with Ag–ZnO exhibiting higher rate constants, indicating superior photocatalytic activity.

4. CONCLUSIONS

Silver ion impregnation improved the photocatalytic performance of ZnO, with an optimal silver loading of 1 at.%. Degradation efficacy of dyes increased with dose of catalyst up to an optimum quantity, beyond which it is decreased. Degradation of both R6G and Eosin Y under photocatalytic conditions followed pseudo-first-order kinetics.

REFERENCES

1. Reife, A. and Fremann, H.S. (1996). Environmental chemistry of dyes and pigments. Wiley, New York.
2. Robinson. T.F., McMullan, G., Marchant, R. and Nigam, P. (2001). Remediation of dyes in textile effluents: a critical review on current treatment technologies with a proposed alternative. Bioresour. Technol. 77:247–255.
3. Chaudari, S.K. and Sur, B. (2000). Oxidative decolorization of reactive dye solution using fly ash as catalyst. J. Environ. Eng. 126:583–594.
4. Al- Ekabi, H.A. and Ollis, D. (2003). Photocatalytic Purification and Treatment of Water and Air, Elsevier, Amsterdam, edn. 2.
5. Kusvuran, E., Samil, A., Atanur, OM. and Erbatur, O. (2005). Photocatalytic degradation of di- and tri-substituted phenolic compounds in aqueous solution by TiO2/UV. Appl. Catal. B. 58:211–216.
6. Sakthivel, S., Neppolian, B., Shankar, B.V., Arabindoo, B., Palinich-amy, M. and Murugesan, V. (2003). Solar photocatalytic degradation of azo dye: comparison of photocatalytic efficiency of ZnO and TiO2. Sol. Energy Mater. Sol. Cells. 77:65–82.
7. Cimitan, S., Albonetti, S., Forni, L., Peri, F. and Lazzari, D. (2009). Solvothermal synthesis and properties control of doped ZnO nanoparticles. J. Colloid Interface Sci. 329:73–80.
8. Subbarao, KV., Lavendrine, B. and Boule, P. (2003). Influence of metallic species on TiO_2 for the photocatalytic degradation of dyes and dye intermediates. J. Photochem. Photobiol. A. 154:189–193.
9. Hermann, J.M., Disdier, J. and Pichat, P. (1986). Photoassisted platinum deposition on TiO_2 powder using various platinum complexes. J. Phys. Chem. 90:6028–6034.
10. Scalafani, A. and Hermann, JM. (1998). Influence of metallic silver and of platinum–silver bimetallic deposits on the photocatalytic activity of Titania (anatase and rutile) in organic and aqueous media. J. Photochem. Photobiol. A. 113:181–188.
11. Toropov, N. A., Kamalieva, A. N. and Vartanyan, T. A. (2016). Thin films of organic dyes with silver nanoparticles: enhancement and spectral shifting of fluorescence due to excitation of localised surface plasmons. Int. J. Nanotechnol. 13(8/9): 642-647.
12. Snehal. S Wagh., Vishal S. Kadam., Chaitali V. Jagtap., Dipak B. Salunkhe., Rajendra S. Patil., Habib M. Pathan. and Shashikant P. Patole. (2023). Comparative Studies on Synthesis, Characterization and Photocatalytic Activity of Ag Doped ZnO Nanoparticles. ACS Omega. 8 (8): 7779-7790.
13. Sahoo, C., Gupta, A.K. and Anjali Pal. (2005). Photocatalytic degradation of methyl red dye in aqueous solutions under UV irradiation using Ag doped TiO_2. Desalination. 181:91–100.
14. Ozkan, A., Ozkan, M.H., Gurkan, R., Akcay, M. and Sokmen, M. (2004). Photocatalytic degradation of textile azo dye, sirius gelb GC on TiO_2 or Ag-TiO_2 particles in the absence and presence of UV irradiation: the effect of some inorganic anions on the photocatalysis. J. Photochem. Photobiol. A. 163:29–35.

Note: All the figures and tables in this chapter were made by the authors.

Recent Trends in Applied Physics and Material Science – Sudhir Bhardwaj et al. (eds)
© 2026 Taylor & Francis Group, London, ISBN 978-1-041-16452-4

Impact of Environmental Factors on Human Health and Strategies for Mitigation

Shipra Gupta[1]
School of Management, Graphic Era Hill University, Dehradun, Uttarakhand, India

Sweta Bagdwal[2]
School of Management, Graphic Era Hill University, Dehradun, Uttarakhand, India

Abstract: Environmental factors play a significant role in the cause of human health, pollution is a public issue around world such as air pollution and water contamination caused serious illness, or emission substances affect global warming status., In the developing world around three million excess premature deaths are estimated to be associated with household air from use in non-hilly areas due to its major burden on respiratory diseases as well cardiovascular disease also lung cancer. Waterborne diseases caused by pathogenic and chemical pollutants in contaminated water sources can lead to long-term health implications and gastrointestinal infections Climate change exacerbates health dangers like extreme weather events, alterations in the spread of diseases and risks to food and water security. Many are associated with health issues, including lead, mercury and asbestos which can cause developmental delays in children, neurological problems for adults as well cancer. Similarly, deforestation and associated losses of biodiversity are important forms of environmental degradation that indirectly influence human health by disrupting ecosystem services. Better air and water quality, tackling climate change, stopping the river of toxic substance, in order to reduce emissions from industry, transport and homes as well as investing in clean water treatment facilities, negotiating household chemicals, clinical thresholds properly, disseminating high levels of greenhouse gases while returning man-made infrastructure riding toxics into controlling processes or land-use changes. These challenges need to be addressed through focused policies, public engagement and multi-disciplinary collaboration for human health as well as sustainable future.

Keywords: Sustainability, Human health, Environment, Air quality, Environment change

1. INTRODUCTION

Environmental factors are an important modulator for the important components of human health and well-being. These include air and water quality, chemical exposure, climate change issues such as temperature fluctuations or severe weather conditions and availability of resources. Environmental determinants are largely responsible for an estimated 24% of the global burden of disease, and this led to one quarter (Adelodun, et al. 2022). To understand the ways in which these factors influence human health is thereby core to organising public health and sustainable development (Agache et al 2023).

1.1 Influence of Environmental Factors on Human Health

Environmental health threats frequently manifest as a decline in air quality due to poor indoor and outdoor air pollution. Air pollution Air pollutants, including particulate matter (PM) nitrogen dioxide NO_2 & sulphur dioxides SO_2 can cause respiratory diseases like heart disease, hypertension stroke lung cancer and premature mortality (Babu et al. 2023). Indoor pollution resulting from the use of solid fuels for cooking and heating is another major public health concern. Clean drinking watery is important for good health, however contaminated water contributes

[1]shipragupta@gehu.ac.in, [2]swetabagdwal2000@gmail.com

DOI: 10.1201/9781003684718-104

to myriad illnesses like gastrointestinal infections, reproductive and neurological disorders. Pathogens, heavy metals and organic contaminants such as pesticides or industrial runoff these are all put in a category (Beneyto, & Vehi, 2024). Climate change poses a number of health risks, impacting extreme weather events disease patterns and food & water security. Heat waves, floods and droughts directly are threatening human Health while changes in Climate can facilitate a rise of Malaria or Dengue Fever mosquitoes. When people are uncovered to contaminants, they can get health conditions like the developing problem and nerve injury plus kinds of cancer (Brown et al. 2023). They are produced by industrial emissions, home products and polluted land. Environmental degradation processes such as Deforestation, loss of biodiversity and soil aggradations have the potential to harm human health by disrupting ecosystems that would provide valuable services like clean air, water or food (Garcia-Ayllon 2023).

1.2 Strategies for Mitigation

Reducing emissions from industrial sources, transportation, and household energy use can greatly enhance air quality. Strict air quality control standards and encouragement of the usage of clean fuels or adoption of newer technologies are very important (Gupta et al. 2022). Investment in improving water treatment facilities, protecting sources of water from contamination and attention to safe sanitation practices allows us the ability to ensure that everyone has access to clean drinking water. An equally important measure is monitoring water quality, as well enabling of the implementation and enforcement to regulate the sources through which these contaminants may enter groundwater (Jaiswal et al. 2022). The mitigation strategies should look at reducing greenhouse gas emissions by enhancing energy efficiency, use of renewable energies and afforestation. These communities are better able to cope with climate impacts and use adaptation measures such as building infrastructure robust enough so it can withstand a weather event or developing early warning systems for extreme events. This is implemented to restrict the use of toxic materials, promote safer alternatives, and govern proper disposal/quarantine for hazardous waste in order to minimize exposure risks. Action to raise awareness and screening programs specifically about the above can help in early detection of diseases ensuring preventive approach. Sustainable land use practices, biodiversity conservation and ecosystem restoration are critical facilitators of improved human health by maintaining the natural systems which generate essential resources and services (Kazancoglu et al. 2020). The mutually multifaceted relationship of the environmental factors are concerning the health today, water and air, makes continuity for sustainable development alongside preventative mitigation initiatives all important. By addressing the underlying determinants of environmental health risks and pursuing effective preventive policy, people can be better safeguarded from hazards that threaten their health while a more sustainable planet could co-exist with development. To accomplish these goals, it is necessary to work in solidarity between public engagement and various discipline sectors to innovate integrated solutions that hold mutual responsibility for this colonial environmental burden (Kumar et al. 2021).

2. OBSERVATION

Table 104.1 Differentiation for the impact of environmental factors on human health and corresponding strategies for mitigation

Environmental Factor	Health Outcomes	Vulnerable Populations	Mitigation Strategies
Air Pollution	Respiratory diseases, cardiovascular problems, and aggravated asthma.	Children, the elderly, and those with pre-existing health conditions.	Implement stricter emissions regulations, transition to clean energy, and promote public transport.
Water Pollution	Gastrointestinal diseases, neurological disorders, and reproductive issues.	Infants, pregnant women, and marginalized communities.	Enhance water treatment facilities, regulate industrial discharges, and ensure access to safe drinking water.
Climate Change	Heat-related illnesses, worsening air quality, and spread of vector-borne diseases.	Low-income communities and the elderly.	Reduce greenhouse gas emissions, invest in renewable energy, and develop climate-resilient infrastructure.
Chemical Exposure	Cancer, endocrine disruption, and developmental disorders.	Children and workers in hazardous industries.	Enforce regulations on hazardous substances, promote safer alternatives, and increase public awareness campaigns.
Urbanization	Mental health issues, respiratory problems, and decreased physical activity.	Urban residents, especially in low-income areas.	Create urban green spaces, improve public transport, and enhance community health programs.
Noise Pollution	Stress, sleep disturbances, and cardiovascular diseases.	Children, the elderly, and those with chronic illnesses.	Implement noise control regulations, enhance sound insulation, and develop quiet zones in urban areas.

Environmental Factor	Health Outcomes	Vulnerable Populations	Mitigation Strategies
Deforestation	Loss of biodiversity increased vector-borne diseases, and climate change effects.	Indigenous communities and rural populations.	Promote reforestation, enforce sustainable land-use policies, and educate communities about forest conservation.
Food Safety Issues	Foodborne illnesses, malnutrition, and exposure to harmful additives.	Children, the elderly and low-income families.	Strengthen food safety regulations, support local agriculture, and educate consumers on safe food handling practices.

3. LITERATURE REVIEW

The literature has repeatedly highlighted the significant effects of environmental conditions on health in humans thus making it compulsory to implement a holistic mitigation approach Kumar and Diwakar (2021). Ensuring that the average person can breathe clean air and drink safe water, is unexposed to dangerous chemicals or substances, and in general enjoys some level of environmental quality through regulatory mandates, technology development efforts as well as educational campaigns may be a key to improving public health outcomes while advancing sustainable development goals Meyer and Lutz (2023). Interdisciplinary research is needed to develop appropriate policies and practices for addressing these environmental health challenges Mishra and Badhotiya (2023).

4. RESULT AND DISCUSSION

The impact of environmental factors on human health and corresponding strategies for mitigation shows in Table 104.1. The extensive impact environmental factors can have on human health, providing details about individual risk-related outcomes for each factor as respiratory diseases and cardiovascular problems from air pollution that affect individuals like children/elderly. Water contamination risks diseases like gastroenteritis and affecting children in addition as marginalized group. Climate change contributes to heat-related illnesses, disproportionately afflicting lower-income residents. Exposure to chemicals is associated with cancers endocrine disrupting effects, and has clear evidence of harm primarily in children or high-risk occupational sectors. In urban regions, particularly in low-income slum locations where a disproportionately large number of the world's cities burgeoning population live and rapidly growing populations are concentrated. High rates of mental health problems are coupled with declining lung function ascribed to an inability to engage in beneficial outdoor exercise constitute on our diagnosis checklist. There are different effects accessible as well of noise pollution for instance it brings disruptive disorders or trigger stress & annoyance at the same time. The change in structure through aging is causing a gradual loss in sound compression level. Finally, food safety problems result in child malnutrition and illnesses from depraved families or poor groups.

Mitigation measures range from tougher emissions and pollutants standards to clean energy, sustainable land use as well as food safety practices. Similarly, the creation of urban green spaces supports public health infrastructure. In the end, these environmental issues need attention in order to protect our health and plan for a healthier tomorrow.

5. CONCLUSION

The complex relationships among environmental factors and human health highlight the imperativeness of combatting environmental degradation, pollution in this study. There is robust evidence that air and water pollution, climate change, and exposure to toxic substances are causally related to a range of health outcomes from respiratory diseases, cardiovascular conditions to cancer. Ending on an optimistic note, the platform is a call to action for dealing with some of these environmental factors that would not only worsen this pre-existing health issues takes them into another level which then generate new public health challenges.

There are many pathways by which environmental exposures can harm health and solutions must be cross-sectorial. Stricter environmental regulations governing pollution, as well as a limited exposure to toxic substances alone are of essential necessity. Promoting feasible agricultural, industrial and concrete growth practices to lower the environmental impact means encouraging renewable energy use, reducing waste and improving recycling. Educating people about the connections between the environment and health can help them to understand how better care for their own well-being. It can also improve lifestyles through education campaigns and promote environmental stewardship. Technological and scientific progress can provide novel technologies for environmental health surveillance, response and control. It is essential to deal with Trans boundary environmental health challenges. International initiatives and accords, such force the Paris Agreement on climate change, are also vital in that they address environmental determinants of health. Last but not least, when it comes to safeguarding public health from environmental dangers, varied stakeholders need further and sustained support. When combined with practice and public engagement, school policies can result in better generations now as well one the days to come. This is why the collaborative and interdisciplinary work done in concert with our global partners are important for securing a sustainable future.

REFERENCES

1. Adelodun, B., Kumar, P. and Odey, G. et al. (2022). A safe haven of SARS-CoV-2 in the environment: Prevalence and potential transmission risks in the effluent, sludge, and biosolids. Geoscience Frontiers. 13(6):101373.
2. Agache, I., et al. (2023). Climate Change-Associated Health Impacts: A Way Forward. *Frontiers in Public Health*. 11:993.
3. Babu, P. C., Kumar, D. G., Sireesha, N. V., Pushkarna, M., Prashanth, B. V. and Dsnmrao, B. V. et al. (2023). Modeling and Performance Analysis of a Grid-Connected Photovoltaic System with Advanced Controller considering Varying Environmental Conditions. International Journal of Energy Research. 1:1631605.
4. Beneyto, A., & Vehi, J. (2024). The Effects of Environmental Factors on General Human Health: A Scoping Review. *Healthcare*. 12(21), 2123.
5. Brown, P. et al. (2023). Synergies between Air Pollution Control and Health Benefits. *Environmental Research Letters*. 18(9):95012.
6. Garcia-Ayllon, S. (2023). Editorial: Interplay between Climate Change, Land Use Change, and Human Health. *Frontiers in Environmental Science*. 9:1258.
7. Gupta, M. et al. (2022). Use of biomass-derived biochar in wastewater treatment and power production: A promising solution for a sustainable environment. Sci. Tot. Environ. 825:153892.
8. Jaiswal, K. K. et al. (2022). Bio-flocculation of oleaginous microalgae integrated with municipal wastewater treatment and its hydrothermal liquefaction for biofuel production. Environ. Technol. Innov. 26: 1023402022.
9. Kazancoglu, Y., Ozkan-Ozen, Y. D., Mangla, S. K. and Ram, M. (2020). Risk assessment for sustainability in e-waste recycling in circular economy. Clean Technologies and Environmental Policy. 24(1):1-13.
10. Kumar, M., Dhangar, K., Thakur, A. K., Ram, B., Chaminda, T. and Sharma, P. et al. (2021). Antidrug resistance in the Indian ambient waters of Ahmedabad during the COVID-19 pandemic. Journal of Hazardous Materials. 416:126125.
11. Kumar, P. and Diwakar, M. (2021). A novel approach for multimodality medical image fusion over secure environment. Transactions on Emerging Telecommunications Technologies. 32(2):3985.
12. Meyer, T. and Lutz, W. (2023). Emerging Challenges in Health and Environmental Governance. *Global Health Journal*. 13(2):143-154.
13. Mishra, A., Badhotiya, G. K., Patil, A., Siddh, M. M. and Ram, M. (2023). Servitization in the circular supply chain: delineating current research and setting future research plan. Management of Environmental Quality: An International Journal. 34(4):1035-1056.
14. Sen Thapa, B. et al. (2022). Application of Microbial Fuel Cell (MFC) for Pharmaceutical Wastewater Treatment: An Overview and Future Perspectives. Sustainability. 14(14):8379.
15. Verma, M. et al. (2022). Capturing of inorganic and organic pollutants simultaneously from complex wastewater using recyclable magnetically chitosan functionalized with EDTA adsorbent. Process Saf. Environ. 167:56-66.

Note: The authors made the table in this chapter.

Field-induced Assembly of Anisotropic Microparticles for Applications in Micro-robotics

Indira Barros*
Department of Physics, Birla Institute of Technology and Science, Pilani-KK Birla Goa Campus, Zuarinagar, Goa, India

Sayanth Ramachandran
Max-Planck Institute for Polymer Research Mainz, Ackermannweg 10, 55128 Mainz, Germany

Zabana Azeem
Indian Institute of Science Education and Research Tirupati, Andhra Pradesh, India

Indrani Chakraborty
Department of Physics, Birla Institute of Technology and Science, Pilani-KK Birla Goa Campus, Zuarinagar, Goa, India

Abstract: Field-induced assembly and propulsion of colloidal particles is a cheap, convenient and versatile method for producing micro-robots for targeted drug delivery, diagnostics, and environmental remediation. Anisotropic particles offer a range of shapes, sizes and functionalities for building a micro-swimmer. Here we synthesize hematite (Fe_2O_3), a canted antiferromagnetic material, in pseudo-cubic and peanut shapes using a sol-gel method. Peanut-shaped particles exhibit controlled reorientation along their long axis under an AC electric field and function as micro-rotors when subjected to a rotating magnetic field. Cubic particles assemble into chain-like or 2D structures driven by electric and magnetic interactions. These field-activated assemblies demonstrate precise reconfigurability, enabling the development of active colloidal micro-rotors and complex functional structures.

Keywords: Micro-robotics, targeted drug delivery, colloids, field-induced assembly

1. INTRODUCTION

Colloidal particles have emerged as excellent building blocks for micro-robotics, owing to their diverse shapes, material properties, and responsiveness to various stimuli (Chakraborty et al. 2017). However, designing functional micro-bots remains challenging, particularly in achieving propulsion and control in low Reynolds number environments (Bechinger et al. 2016). Introducing anisotropy in colloidal systems has proven to be a straightforward and effective way to achieve propulsion and functionality in combination with activation methods, including chemical, thermal, acoustic, field-based and light-based techniques. Recently, electric and magnetic field induced assembly has gained the interest of researchers due to their versatility, precise tunability, and ease of implementation. Magnetic fields, in particular, have an added advantage of penetrating the human body, making them highly suitable for applications such as targeted drug delivery (Dasgupta et al. 2022). In this work, we produce anisotropic microparticles and demonstrate their promise in field activated assembly and propulsion. Our work opens up new avenues for developing controllable and functional micro-bots for applications in medicine and diagnostics, environmental remediation, and industrial manufacturing.

2. MATERIALS AND METHODS

Hematite cubes and peanuts (Fig. 105.1a-b) were synthesized by a sol-gel method outlined by Sugimoto *et al.* where the reactants are aged in an oven at 100° C for 8 days (Sugimoto et al. 1992, 1993). Hollow silica cubes were synthesized using hematite cubes as a precursor (Wang et al. 2013). For the electric field-based experiments, a capillary channel containing the dispersion of particles was fabricated by sandwiching two ITO coated coverslips with parafilm as a spacer (150 μm).

*Corresponding author: indranic@goa.bits-pilani.ac.in

DOI: 10.1201/9781003684718-105

Contacts were made using copper tape and silver paste and electric field was supplied using a function generator. For magnetic field experiments, a Helmholtz coil was used to supply magnetic field in situ. A bright field optical microscope was used for imaging along with a high-speed camera.

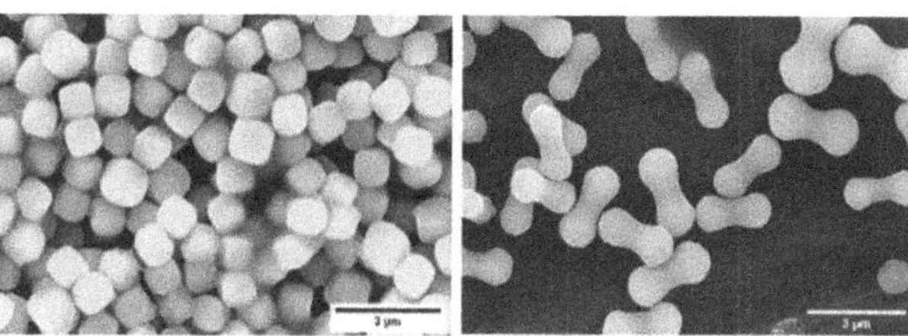

Fig. 105.1 SEM images of (a) Hematite cubes and (b) peanuts

3. RESULTS AND DISCUSSION

3.1 Hematite Cubes as Building Blocks for Cubic Crystals

Cubic hematite particles can be made to form cubic crystals by utilizing the inter-particle magnetic interactions arising from the canted antiferromagnetic nature of hematite at room temperature (Morrish 1995). When a magnetic field is applied to a dispersion of the cubic particles the particles form large chains and layers of chains in the direction of the applied magnetic field. Once the magnetic field is switched off, the chains collapse into smaller zones of crystalline structures having a coordination number of 4 as shown in Fig. 105.2a. Repeated annealing cycles can produce larger area cubic crystals and has potential for applications in colloidal lithography and photonic crystals.

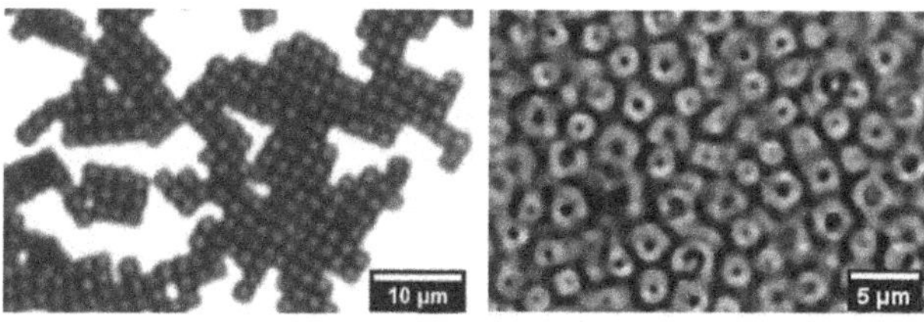

Fig. 105.2 (a) Cubic crystals assembled from hematite cubes under magnetic field cycling. (b) hollow silica particles form close-packed glassy structures when subject to an out of plane ac electric field of 5kHz and 20 Vpp.

3.2 Hollow Silica Cubes Under Electric Field

For applications where the intrinsic interparticle magnetic interactions might be detrimental, hollow silica cubes were synthesized. The assembly of these particles under the application of an AC electric field revealed close-packed glassy structures as shown in Fig. 105.2b (Barros et al. 2024). Although the behaviour is similar to isotropic particles, the hollow cubes provide a way to address a more fundamental problem of understanding field-based inter particle interactions, since the dielectric constant of the core of the particles is equal to that of the surrounding medium.

3.2 Hematite Peanuts as Micro Propellers

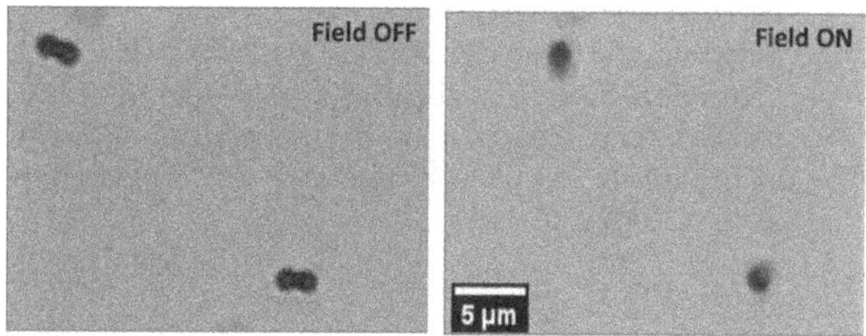

Fig. 105.3 (a) Long axis of hematite peanuts oriented parallel to the substrate without AC electric field and (b) oriented perpendicular to the substrate when subject to an out of plane ac electric field of 500 kHz and 20 Vpp

Micro-rotors were produced from hematite peanuts by using a combination of magnetic and AC electric fields. Under an out of plane AC electric field of frequency 500 kHz and peak to peak voltage (Vpp) of 20 V, hematite peanuts can be made to re-orient along their long axis in the direction of the field, thereby making a standing peanut. Since the magnetic moment of the peanut is along its waist, a rotating magnetic field in a plane perpendicular to the substrate makes the peanut rotate about an axis parallel to the substrate, making a propeller in the microscale Lee et al. (2009).

4. CONCLUSIONS

Our work highlights the potential of field-based assembly methods for creating functional components of micro-bots. Hematite particles in cubic and peanut shapes, as well as hollow silica cubes were synthesized and assembled into different structural configurations and functional components. Magnetic fields were used to assemble hematite cubes into cubic crystals, while AC electric fields created close-packed structures from hollow silica cubes. Combination of electric and magnetic fields was used to produce a micro-propeller from hematite peanuts. These results show that field-based activation is a promising approach for assembling micro-bots, with potential applications in targeted drug delivery, environmental remediation, and photonics.

REFERENCES

1. Chakraborty, I., Meester, V., van der Wel, C. and Kraft, D.J., 2017. Colloidal joints with designed motion range and tunable joint flexibility. *Nanoscale*, *9*(23),7814-7821.
2. Bechinger, C., Di Leonardo, R., Löwen, H., Reichhardt, C., Volpe, G. and Volpe, G., 2016. Active particles in

complex and crowded environments. *Reviews of modern physics*, *88*(4), 045006.

3. Dasgupta, D., Peddi, S., Saini, D.K. and Ghosh, A., 2022. Mobile nanobots for prevention of root canal treatment failure. *Advanced healthcare materials*, *11*(14), 2200232.
4. Sugimoto, T., & Sakata, K. (1992). Preparation of monodisperse pseudocubic α-Fe2O3 particles from condensed ferric hydroxide gel. *Journal of colloid and interface science*, *152*(2), 587-590.
5. Sugimoto, T., Khan, M.M. and Muramatsu, A., 1993. Preparation of monodisperse peanut-type α-Fe2O3 particles from condensed ferric hydroxide gel. *Colloids and Surfaces A: Physicochemical and Engineering Aspects*, *70*(2), 167-169.
6. Wang, Y., Su, X., Ding, P., Lu, S. and Yu, H., 2013. Shape-controlled synthesis of hollow silica colloids. *Langmuir*, *29*(37), 11575-11581.
7. Morrish, A.H., 1995. *Canted antiferromagnetism: hematite*. World Scientific.
8. Lee, S.H. and Liddell, C.M., 2009. Anisotropic magnetic colloids: a strategy to form complex structures using nonspherical building blocks. *Small*, *5*(17), 1957-1962.
9. Barros, I., Ramachandran, S. and Chakraborty, I., 2024. Structure, dynamics and phase transitions in electric field assembled colloidal crystals and glasses. *arXiv preprint arXiv:2410.16792*.

Note: All the figures in this chapter were made by the authors.

Recent Trends in Applied Physics and Material Science – Sudhir Bhardwaj et al. (eds)

106 New Trends in Skin Friendly and Comfortable Electronics

Himani Dahiya[1]
Graphic Era (Deemed to be University), Dehradun, Uttrakhand, India

Harindra Singh
CRA College, Sonipat, Haryana, India

Gautam Nain[2]
Graphic Era (Deemed to be University), Dehradun, Uttrakhand, India

Abstract: Day by day as electronic devices are being miniaturized, made more light weight, wearable and intelligent, we are entering into an era where electronics is ultra-thin, soft, flexible and stretchable form has no longer been a dream and is becoming a reality. A new branch of electronics known as skin friendly electronics has thus emerged in recent years and gaining attention of the researchers across the globe. As demand for human-machine interfaces is surging, developing materials having high-performance sensing capabilities is seeing a steep rise and to meet out this, researchers are making efforts to fabricate flexible, stretchable, comfortable and wearable material which responds to various stimuli such as strain, stress, pressure, temperature, humidity etc. and generate signals due to change in resistance, voltage, current, or capacitance etc. In the present review, we incorporate the recent trends and progress in skin friendly materials their fabrication aspects and prospective applications. Skin friendly electronic materials comprising of piezoresistive, piezoelectric, thermoelectric, conductive polymer, carbon nanomaterials etc. are analyzed here. In the end, practical applications of skin friendly electronic materials in fields of human healthcare, human motion detection, tactile perception and human-machine interface are discussed.

Keywords: Human-machine interface, Piezoresistive, Piezoelectric, Thermoelectric, Skin friendlyelectronics, Strain

1. INTRODUCTION

Conventional electronic devices usually incorporate silicon and rigid printed circuit boards (PCB), and thus these invariably fail to perform duty where movements of body parts viz. twisting, turning and folding are involved. The normal physiological processes of the skin like sweating and breathing are likely to be adversely affected because of the difference between flexibility, elastic properties and permeability of rigid electronic circuit boards and skin. Undesirability and unfeasibility of such devices led to the invention of skin friendly, flexible, thin and stretchable electronics. This gave a fresh platform for future research in this field and the basic target here is realizing high-fidelity signal/data from the body via interacting and establishing an intimate contact with the skin without hurting or damaging it. Thus, materials for fabricating skin friendly electronic devices should invariably be flexible, foldable and stretchable. Besides that, the parameters such as bio-compatibility, sweat permeability and an inbuilt powering source to ensure independent performance are prerequisites for developing such devices.

Taking into consideration various aspects viz. finding suitable materials, fabricating composites, developing supporting mechanical and electrical engineering parts and assembling these, researchers have made concerted efforts towards developing skin friendly devices and exploring their potential applications. The present review, is an attempt towards unravelling recent developments in skin friendly electronics. This is achieved by dividing the topic in three parts viz. materials involved in fabrication, recent designs and potential applications.

Initially a reporting of basic substrate materials is provided. Electrodes which establish a smooth contact between the flexible sensor and the signal measuring circuits are then explored. The designs that ensure comfort and turn rigid materials such as metals into flexible forms as well as others and different types of fabrication strategies to develop skin friendly electronics for wearing are then

Corresponding authors: [1] hsdahiya14@gmail.com, dahiya_himani@yahoo.com; [2]gautamnain60@gmail.com

DOI: 10.1201/9781003684718-106

briefly discussed. Finally, advanced research work on skin friendly electronics from applicability point of view is described.

2. MATERIALS INVOLVED IN FABRICATION

2.1 Matrix/Substrate Materials

Substrate materials for wearable electronics desirably needs to be flexible, biocompatible. Comparable elastic properties of substrate membranes and the skin are also equally important. Many commercially available polymers and elastomers are shown in Fig. 106.1. Polydimethylsiloxane (PDMS), Ecoflex, polyurethane (PU), Poly vinylidene fluoride (PVDF), and Poly vinyl alcohol (PVA) etc. have been found suitable for this purpose as reported by Yao et.al (2022).

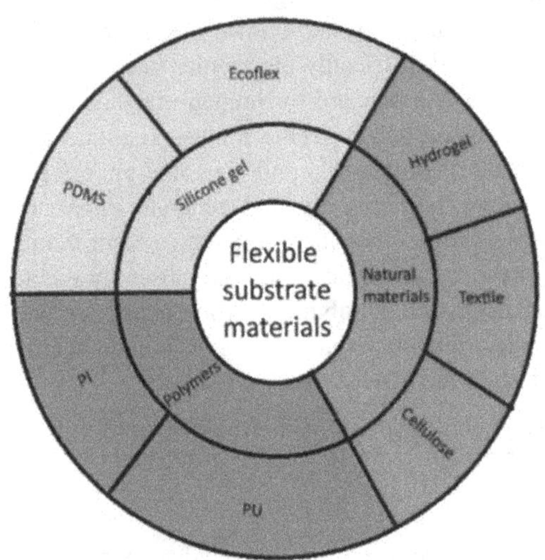

Fig. 106.1 Various substrate materials

2.2 Functional Component

Different substances convenient for skin friendly electronics have been fabricated by researchers to meet people's demand involving various mechanisms e.g. piezoresistive, piezoelectric, thermoelectric etc. Mukherjee et.al. (2023) developed stretchable and flexible piezoresistive strain sensors by depositing extremely thin nano-sized doped Poly Propylene yarn layers on PU membranes of various designs employing vapor deposition technique. Piezoelectric ceramic lead Pb $(Zr_{0.52} Ti_{0.48}) O_3$ films were fabricated by Yu et al. (2018) to determine elastic properties of human tissues. Peddigari et al. (2021) formulated a device where $Pb(Mg_{1/3} Nb_{2/3}) O_3–PbTiO_3$ being capacitor material functioned as storage material. Chun et. al. (2018) fabricated a soft and stretchable auto- powered mechano-receptor comprising of PVDF membrane for sensing pressure stimuli by skin. Flexible materials absorbing heat energy from body and generating thermoelectric signals are used for fabricating skin friendly devices. Ren et al. (2021) fabricated a self-healing, flexible thermoelectric generator comprising of composite thin layers of Bismuth and Antimony incorporated on polyimide membranes. Using BiTe-SbTe flexible thermocouples were developed through screen printing technique by Cao et al. (2016) for studying effect of various binders on the thermoelectric performance.

2.3 Conductive Materials

Conductive materials acting as electrodes are indispensable for inter-connection of power source with functional/ substrate materials. Metals such as gold and silver are extensively used. Graphene used by Huang et.al. (2022) and carbon nanotubes used by Sattar et.al. (2024) are also popular because of excellent conductivity and ease of processability besides high porosity.

3. RECENT DESIGN MECHANISMS FOR SENSING DEVICES

For better performance a proper design strategy is essential to make device efficient for a particular application. For skin, comfortability is extremely important. Consequently, several designs have been developed which can be broadly classified into two parts viz. mechanical and comfortability aspects as shown in Fig. 106.2.

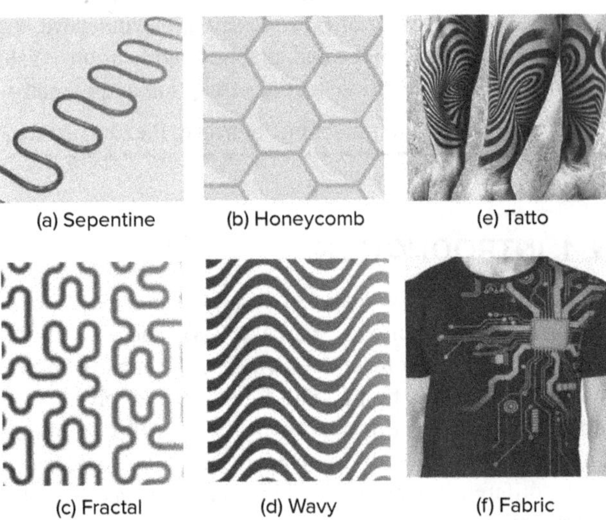

Fig. 106.2 Designs: Mechanics aspects (a) to (d) comfortability aspects (e) and (f)

3.1 Mechanical Aspects

Basically, four designs viz. serpentine, honeycomb, fractal and wavy are employed. Serpentine structure is mainly used for metallic materials. It is versatile in child health monitoring. Chung et.al. (2019) fabricated a device employing narrow serpentine metal strips inter-connect to form chip size integrated circuit which can be used for ECG and photoplethysmography observations. The developed component is capable of bending to a much smaller radius.

Inspired from the natural honeycomb of bees, Chen et.al. (2022) developed an organic electrochemical transistor (OECT) which exhibits stable electromechanical and mechanical properties over 1500 cycles. With stable ECG signal recording capability, it bears great significance in skin friendly electronics. Fractal patterns are infinitely complex and are self-similar across various scales such as branching pattern in trees, river networks and blood vessels etc. Multiple electrode arrays each comprising of seventeen fractal mesh electrodes in cross patterns, developed by Tian et.al (2019) are capable of elastic stretchability of 18% which is more than enough to handle straining during natural skin movements. Wavy pattern was employed by Kang et.al (2019) who developed a vibro-tactile actuator

3.2 Comfortability Aspects

To ensure comfortability, the sensing material should be lightweight, thin, porous, stretchable and biodegradable. Keeping these parameters in mind, broadly two designs viz. tattoo and fabric are adopted by researchers. A tattoo like foldable sensor easily mountable on human body fabricated by Nortan et.al. (2015) is capable of generating Hi-Fi prolonged lasting detection of EEG signals. It is suitable of being worn on skin for 14 days without adversely affecting human activities like walking, gyming, bathing etc. Wong et.al. (2022) also developed a thin tattoo like stable performance triboelectric nano generator exhibiting good potential in the field of human machine interface. Incorporating electronic sensors into garments is an innovative aspect to ensure comfortability. A triboelectric electric fabric device capable of excellent pressure sensitivity and exhibiting repeatability over 105 cycles was developed by Fan et.al. (2020). When mounted on human body, it is capable of counting pulse rate at wrist, neck and ankle etc.

4. APPLICATIONS

Skin friendly devices are capable of monitoring several physiological signals shown in Fig. 106.3.

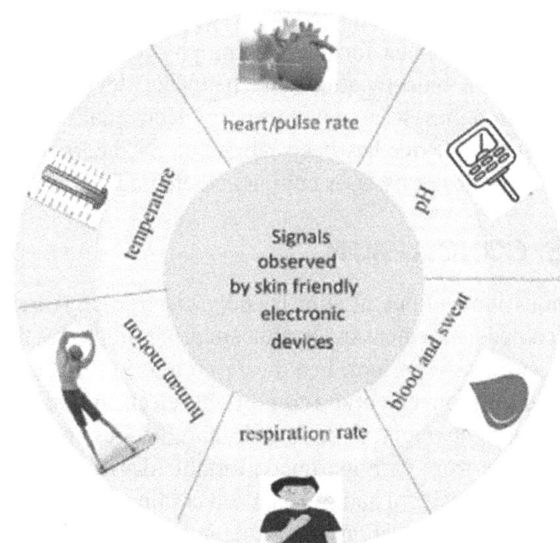

Fig. 106.3 Pictorial representation of various body signals observed by Skin friendly electronic devices

4.1 Body Temperature

Conventional devices can't do continuous monitoring although flexible and stretchable temperature sensor can do so. Recently, a specialized temperature sensor was developed Hong et.al. (2016) using polyaniline nanofiber and single-walled carbon nanotubes which quickly responds in less than 2 seconds and possesses low conductance sensitivity.

4.2 Blood Pressure

In medical field, continuous observation of heart beating rate and pulse are most essential for diagnosing cardilogical problems. A soft wearable pressure sensor, recently has been developed by Bijender and Ashok Kumar (2021) by inserting PDMS layer between indium tin oxide smeared PET electrodes. The developed sensor reportedly has low response of 110 ms and an extremely low pressure sensing capability of 1Pa.

4.3 Human Motion

A human movement sensing device was fabricated by He et.al. (2021) by involving a soft and bendable triboelectric nano-generator (TENG) fabric, which is washable and exhibits reasonably good skin compatibility. This sensor monitors movements on being attached to various movable parts of human beings. The innermost fold among the 5 folds of this device consists of a coil spring, both ends of which are hook shaped and joined to each other resulting in a multi array design for movement sensing.

4.4 Respiratory Rate

In the post COVID-19 era, people are more concerned towards observing respiratory rates. Miniature, low cost and reliable devices are thus the need of the day. Being small sized, flexible and free from EM interference, Fiber Bragg Grating sensors are extensively used for this purpose involving two methods. In one, the device monitors breathing by observing variation in volume of chest during inhaling and exhaling while in other respiration rate is monitored by observing variation in humidity as well as temperature while breathing as reported by Mihailov et.al. (2012).

4.5 Sweat Sensing

Sweat usually monitored by stretchable electrochemical sensors, provide vital data regarding blood glucose level. As sweat in copious amount is available only after body warmup, sweat sensors are put on duty after heavy exercise. Such a battery-less device was developed by Bandhokar et.al. ((2014) which is capable of providing real time sensing of various components in human sweat. Another important parameter for maintaining physiological health is pH and a biocompatible and bendable device for its observation has been developed by Hou et.al. (2021). The sensing device has been fabricated by incorporating MWCNTs in silk by laser cutting and etching technique.

5. CONCLUSION

Various innovations in skin friendly electronics is being welcomed by medical and healthcare professionals as they can ceaselessly perform healthcare monitoring and equip people with convenient and comfortable electronic devices. Wearing lightweight, flexible and smart devices upon skin relieves us from carrying uncomfortable devices, as well as makes it easier to harness data and continuously share it for storage. However, impediments in terms of efficiency, deterioration of the performance of sensor during repeated cycles, high manufacturing cost may be a hindrance for large scale use in the contemporary society. Besides this as novel materials are being used in fabricating these devices, some of these may prove allergic to sensitive human skin and harm users. Thus a systematic study in this direction needs to be carried out at the earliest.

REFERENCES

1. Bandodkar, A. J.,Gutruf, P., Choi, J. (2014) Battery-free, skin-interfaced microfluidic/electronic systems for simultaneous electrochemical, colorimetric, and volumetric analysis of sweat. Sci. Adv. 32:363–371.
2. Bijender and Kumar, A., (2021) Flexible and Wearable Capacitive Pressure Sensor for Blood Pressure Monitoring. Sensing and Bio-sensing Reserch. 33:100434.
3. Cao, Z., Koukharenko, E., Tudor, M. J., Torah, R. N. and Beeby, S. P. (2016). Flexible screen printed thermoelectric generator with enhanced processes and materials Sensors Actuators, A. Phys. 238:196–206.
4. Chen. J., Huang, W., Zheng, D. et al. (2022) Highly stretchable organic electrochemical transistors with strain-resistant performance," Nat. Mater., 21(5):564–571.
5. Chun, K. Y., Son, Y. J., Jeon, E. S., Lee, S. and Han, C. S. (2018). A Self-Powered Sensor Mimicking Slow- and Fast-Adapting Cutaneous Mechanoreceptors. Adv. Mater., 30:1706299.
6. Chung H. U., Kim, H. B., Lee J. Y. et.al. (2019). Binodal, wireless epidermal electronic systems with in-sensor analytics for neonatal intensive care," Science. 363(6430): eaau0780.
7. Fan, W., He, Q., Meng, K., et al. (2020) Machine-knitted washable sensor array textile for precise epidermal physiological signal monitoring. Sci. Adv. 6 (9) eaay2840.
8. He, M., Du, W., Feng, Y. (2021) Flexible and Stretchable Triboelectric Nanogenerator Fabric for Biomechanical Energy Harvesting and Self-Powered Dual-Mode Human Motion Monitoring. Nano Energy. 86:106058.
9. Hong, S.Y., Lee, Y.H., Park, H. et.al. (2016) Stretchable Active Matrix Temperature Sensor Array of Polyaniline Nanofibers for Electronic Skin. Adv. Mater.28:930–935.
10. Hou, C., Zhang, F., Chen, C. et.al. (2021), Wearable Hydration and PH Sensor Based on Protein Film for Healthcare Monitoring. Chem. Pap. 75:4927–4934.
11. Huang, X., Li, H., Li, J. et al. (2022) Transient, Implantable, Ultrathin Biofuel Cells Enabled by Laser-Induced Graphene and Gold Nanoparticles Composite. Nano Lett. 22:3447–3456.
12. Kang, J., Park, J. H., Choi, D. S., Kim, D. H., Kim, S. Y. and Cho, J. H. (2019) Design of Wavy Ag Microwire Array for Mechanically Stable, Multimodal Vibrational Haptic Interface. Adv. Funct. Mater.29:1902703.
13. Mihailov, S.J. (2012) Fiber Bragg Grating Sensors for Harsh Environments. Sensors. 12: 1898–1918.
14. Mukherjee, A., Dianatdar, A. A., Gładysz, M. Z. et.al. (2023). Electrically Conductive and Highly stretchable piezoresistive polymer nanocomposites via oxidative chemical vapor deposition. Appl. Mater. Interfaces, 15:31899–31916.
15. Norton J. J. S., Lee, D. S., Lee, J.W. and Rogers J.A. (2015). Soft, curved electrode systems capable of integration on the auricle as a persistent brain-computer interface. Proc. Natl. Acad. Sci. U. S. A.112(13):920–3925.
16. Peddigari, M., Ryu, J., Hwang, G., et al. (2021) Flexible Self-Charging, Ultrafast, High-Power-Density Ceramic Capacitor System," ACS Energy Lett., 6:1383–1391.
17. Ren, W., Sun, Y., Zhao, D. et. al. (2021) "High-performance wearable thermoelectric generator with self-healing, recycling, and Lego-like reconfiguring capabilities," Sci. Adv. 7:586.
18. Sattar, M., Lee, Y.J., Kim, H., et.al (2024) Flexible Thermoelectric Wearable Architecture for Wireless Continuous Physiological Monitoring. ACS Applied Mat. & Interfaces. 16:37401-17.
19. Tian, L., Zimmerman, B., Akhar, A., et al. (2019). Large-area MRI-compatible epidermal electronic interfaces for prosthetic control and cognitive monitoring. Nat. Biomed. Eng. 3(3):194–205.
20. Wong T. H., Liu, Y., Li. J., et. al. (2022) Triboelectric Nanogenerator Tattoos Enabled by Epidermal Electronic Technologies. Adv. Funct. Mater.32: 2111269.
21. Yao, K., Yang, Y., Wu, P., Zhao, G., Wang, L., Yu, X. (2022). Recent advances in Materials, designs and applications of skin electronics. IEEE Open Journal of Nanotechnology, 99:1-39.
22. Yu, X., Wang, H., Ning, X., et.al. (2018) Needle-shaped ultrathin piezoelectric microsystem for guided tissue targeting via mechanical sensing Nature biomedical engineering, 2:165-172.

Note: All the figures in this chapter were made by the authors.

Recent Trends in Applied Physics and Material Science – Sudhir Bhardwaj et al. (eds)

107

Future Applications of Junctionless Transistors Using Gallium Phosphide Channel for the Semiconductor Industry

Pooja Srivastava*, Aditi Upadhyaya, Shekhar Yadav, C. M. S. Negi
Department of Physical Sciences, Banasthali Vidyapith, Rajasthan, India

Abstract: The present study examines the junctionless field effect transistor (JLFET), characterized by no gradients in doping concentration and non-availability of junctions. According to the results, the proposed device offers a novel type of architecture that is less susceptible to short-channel effects (SCEs) and greatly improves transistor performance. The performance of the suggested JLFET has been examined using the Atlas 3D device simulator, specifically intended for low-power applications. This study uses Gallium Phosphide, a III-V compound semiconductor, to examine the electrical performance of a JLFET. The simulations' outcomes showed that, unlike other circuit topologies, the III-V JLFET has a favorable subthreshold slope and a lower DIBL. The suggested results show that the subthreshold slope is roughly 63.84 mV/V, with a low DIBL of about 22.50 mV/dec.

Keywords: Scaling, Short-channel effects, JLFET, Gallium phosphide material

1. INTRODUCTION

In the last four decades, metal-oxide-semiconductor field-effect transistor (MOSFET) devices have steadily decreased in size. New solutions to expand MOS scaling beyond the 100-nm node are being introduced more quickly due to the high pace of MOSFET downscaling. This acceleration necessitates thoroughly investigating the detrimental short-channel effects (SCEs) and their solutions to maintain the historical pace of downsizing and increase performance. This research details the latest developments in MOSFET device designs to identify the minimal allowable channel length and get over SCEs, which are considered the most difficult obstacles to sub-100nm MOSFET scaling. To overcome the carrier transport efficiency barrier and customize the SCEs of deep-submicron MOSFETs, the source/drain-engineering, substrate-engineering, and gate-engineering approaches show great promise. Continuous semiconductor scaling has made it possible to reduce the device's size, which has improved the VLSI circuits' operating speed and area requirements (Dhiman and Chandel, 2020)

Transistors with lower dimensions are superior to CMOS in terms of higher current drive and faster processing performance. A lot of digital devices rely entirely on Moore's law to function. These include packing density, memory size, and processing speed. The semiconductor industry's International Technology Road Map (ITRS) is updated using this law (Moore, 1975). Concerning the ongoing development of transistor technology. In current devices, the distance between junctions decreases to less than 10 nm due to device downsizing, necessitating extremely high doping concentration gradients. Researchers have proposed a junctionless field effect transistor (JLFET) to replace conventional transistors (CT). The entire planet is facing challenges regarding energy shortage. As a result, the whole world is searching for new technology for energy-sufficient products. The necessity for alternative MOSFET structures arises because CTs are getting so small that they are becoming harder to overcome SCEs. Since the advent of nano-engineering and nanowires, current research has focused on a novel JLFET technology (Colinge and Lee, 2010).

*Corresponding author: poojasrivastava@banasthali.in

DOI: 10.1201/9781003684718-107

2. METHODOLOGY AND DEVICE STRUCTURE OF JLFET

A junctionless transistor is a novel device that is a uniformly doped nanowire with no junction availability. Nowadays, the channel length in current devices is getting close to 10 nm, and making these junctions is crucial; JLFET offers a solution. In this work, the effects of gallium phosphide on the device performance have been extensively investigated. A device's enhancement may improve the VLSI circuits' size, power, and speed. In this work, the performance parameters of the GaP have been compared with the silicon by developing an analytical sub-threshold model based on pseudo-2D analysis. Gallium phosphide (GaP) is a semiconductor substance that has several uses in optical systems. For the preparation of GaP, several solution-based key reactions have been established over the past 20 years. These techniques include various solution-based pathways and thermolytic and metathetical processes (Fallahnejad and Vadizadeh, 2020).

Additionally, applications for III–V GaP-based quantum dots in bioimaging and gas sensing are established. Indirect band gap semiconductors, or GaPs, are utilized in microelectronic applications. A two-GaP nanowire optical X-coupler that enables signal spectrum separation can be created.

The suggested device arrangement is shown in Fig. 107.1, and the design parameters of the proposed device are noted in Table 107.1. The work's findings provide new avenues for using GaP nanowires in sophisticated photonic logic circuits and nanoscale interferometers. The Shockley-Read-Hall and AUGER recombination models were used to analyze recombination effects for high current densities. NEWTON and GUMMEL models were incorporated into equations with coupled and decoupled forms (Merad and Guen-Bouazza, 2020).

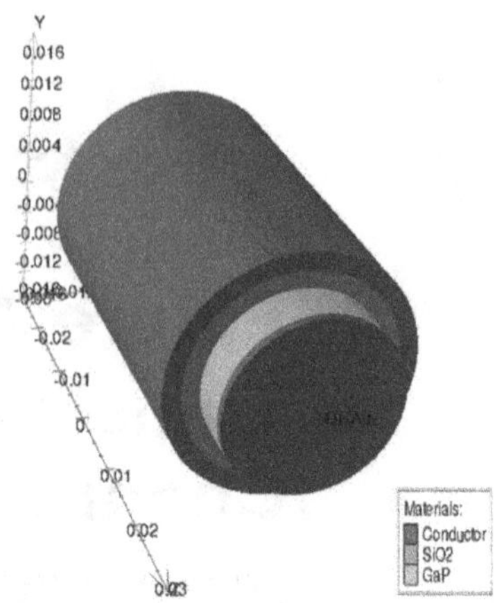

Fig. 107.1 20 nm GaP-based JLFET structure with the solid view

Table 107.1 Design parameters of the GaP-based JLFET

Parameters	GaP based JLFET
Channel Material	Gallium Phosphide (GaP)
Channel region	N-type – 10^{19} cm^{-3}
Source region	N-type – 10^{19} cm^{-3}
Drain region	N-type – 10^{19} cm^{-3}
Gate material	P^+ Polysilicon
Gate Work-function	5.4 eV
Oxide Permittivity	3.9
Channel Length	40 nm
Oxide Thickness	2 nm (radius)
Device Length	60 nm

The current equations for JLFET have been discussed in Table 107.2. This work thoroughly examines the electrical characteristics of a junctionless gate-all-around field effect transistor design with a silicon channel and gallium phosphide (GaP) as the III-V compound material, as shown in Table 107.3.

Table 107.2 Current equations for GaP-based JLFET (Colinge, 2012)

Bias	Drain Current
$V_{GS} > V_{po}$ $V_{GS} < V_{FB}$ $V_{DS} < V_{DSat1}$	$I_D = \frac{q\mu_{bN_D}}{L_{effb}}\left(\frac{1}{n+1}\frac{S_{max}-S_{min}}{(V_{FB}-V_{po})^n}\left((V_{GS}-V_{po})^{n+1}-(V_{GS}-V_{DS}-V_{po})^{n+1}\right)+S_{min}V_{DS}\right)$
$V_{GS} > V_{po}$ $V_{GS} < V_{FB}$ $V_{DS} > V_{DSat1}$	$I_D = \frac{q\mu_{bN_D}}{L_{effb}}\left(\frac{1}{n+1}\frac{S_{max}-S_{min}}{(V_{FB}-V_{po})^n}\left((V_{GS}-V_{po})^{n+1}\right)+S_{min}V_{DS}\right)$
$V_{GS} > V_{FB}$ $V_{DS} < V_{DSat2}$	$I_D = \frac{q\mu_{bN_D}}{L_{effb}}S_{max}C_{ox} + \frac{\mu_{acc}C_{ox}W_{eff}}{L_{effacc}}\left(V_{DS}(V_{GS}-V_{FB})-\frac{1}{2}V_{DS}^2\right)$

Table 107.3 Comparison of the proposed Gap-based JLFET performance with Si-based JLFET

Si and III-V Material	V_{th} (V)	Saturation Slope (A)	Max. Drain Current(A)	SS. (mV/dec)	DIBL (mV/V)	I_{on} (μA)	I_{off} (μA)	I_{on}/I_{off}
Si	0.501014	3.14567×10^{-6}	2.68907×10^{-5}	66.7865	47.7865	15.89876	1.3785×10^{-5}	1.34567×10^{6}
GaP	0.572443	8.47869×10^{-11}	5.78965×10^{-6}	63.84844	22.50066	5.42848	3.8905×10^{-4}	12631.8786

The simulations showed that GaP had a more favorable subthreshold slope and reduced drain-induced barrier lowering (DIBL) compared to other circuit topologies. Additionally, a strong association was found in the study between the values of drain-induced barrier lowering (DIBL) and subthreshold slope (SS) in gallium phosphide (GaP). The drain-induced barrier lowering (DIBL) of 22.50 mV/dec and the optimal value of 60 mV/V are fairly near the subthreshold slope of 63.84 mV/V (Ghosh and Jana, 2022).

3. RESULTS AND DISCUSSION

With V_{DS} ranging from 0 to 1 V, Fig. 107.2 shows the I_D-V_{DS} characteristics of the GaP-based JLFET at V_{GS} = 1.2 V. Having a high ON-state current and a low OFF-state current, this device has a maximum drain current and good electrostatic control. The threshold voltage in this study is positive because the JLFET is a typically ON-gated resistor, and the gate and silicon nanowires have different work functions.

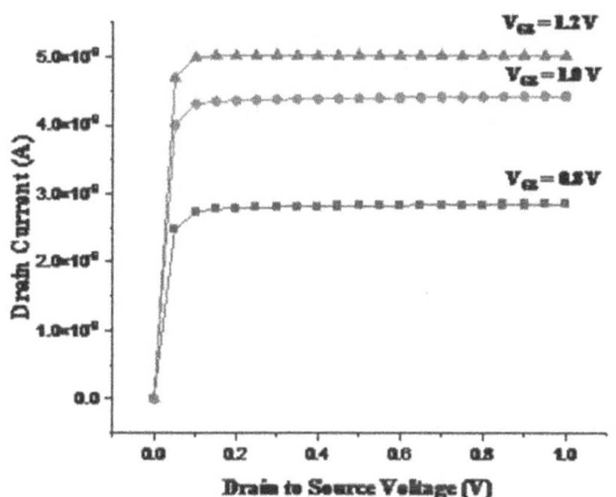

Fig. 107.2 Characteristics of the GaP-based JLFET output

The simulated transfer characteristics of a GaP-based JLFET for a 40 nm channel length are shown in Fig. 107.3, and P+ Polysilicon has been utilized to satisfy the need for

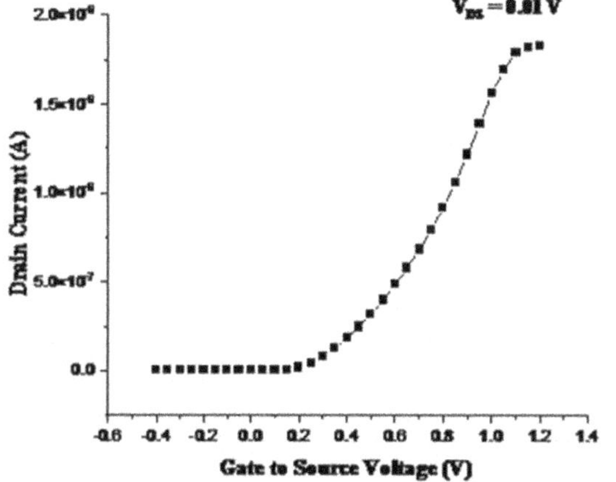

Fig. 107.3 Transfer characteristics of GaP-based JLFET

a high gate work function to obtain the proper VTH value. Here, it is clear that the drain current's speed against the gate voltage rises as the channel length decreases (Balasubramanian, 2006).

The transconductance of a cylindrical gate-all-around GaP-based JLFET for a channel length of 14 nm is shown to vary on the channel length in Fig. 107.4. The superior charge transport characteristics of the JLFET architecture over other CTs and NCTs are responsible for its superior transconductance. It can be observed here that the value of transconductance increases as channel length increases (Borli and Kolberg, 2008). Transconductance and MOSET threshold voltage (V_{TH}) are closely correlated and depend on gate channel size. The g_m curve climbs steeply for longer channel lengths, increasing amplification as the gate voltage increases (Zipperian and Chaffin,1982).

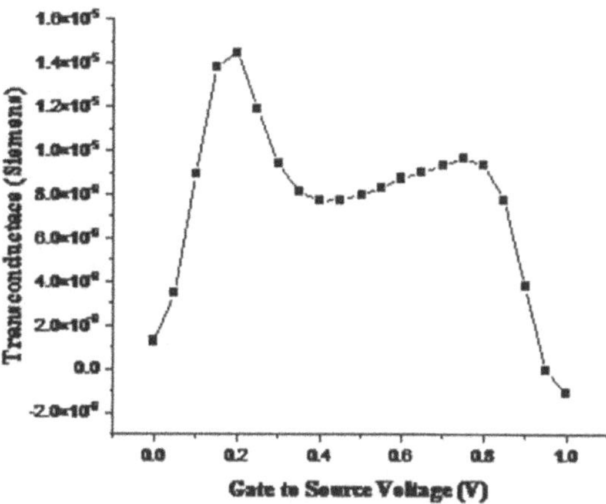

Fig. 107.4 Transconductance of GaP-based JLFET

4. CONCLUSION

Multigate FETs without PN, N+N, or P+P connections are called JLFETs. It essentially functions as a resistor in which the gate can adjust the mobile carrier density. Findings indicate that a JLFET with a GaP-based channel is less susceptible to SCEs than a device with a junction and has a virtually perfect sub-threshold slope, very low leakage current, and minimal drain-induced barrier lowering (DIBL). It also exhibits reduced mobility degradation with gate voltage and temperature.

REFERENCES

1. Balasubramanian, S. (2006). *Nanoscale thin-body and applications,* Doctoral Thesis MOSFET design dissertation, University of California, Berkeley).
2. Borli, H., Kolberg, S., Fjeldly, T., & Iñíguez, B. (2008). Precise modeling framework for short-channel double-gate and gate-all-around MOSFETs.Electron Devices, IEEE Transactions on, 55(10):2678-2686.
3. Colinge, J. P., Lee, C. W., Afzalian, A., Akhavan, N. D., Yan, R., Ferain, I., & Murphy, R. (2010). Nanowire transistors without junctions. Nature nanotechnology, 5(3):225-229.

4. Colinge, J. P. (2012). Junctionless transistors. In 2012 IEEE International Meeting for Future of Electron Devices, Kansai.
5. Dhiman R, Chandel R. (2020). Nanoscale VLSI: Devices, Circuits and Applications. 319. Springer Nature.
6. Fallahnejad, M., Vadizadeh, M., Salehi, A., Kashaniniya, A., & Razaghian, F. (2020). Impact of channel doping engineering on the high-frequency noise performance of junctionless In0. 3Ga0. 7As/GaAs FET: a numerical simulation study. Physica E: Low-dimensional Systems and Nanostructures, 115, 113715.
7. Ghosh, S., Jana, A., Agnihotri, A. K., Kundu, S., Das, D., & Sarkar, S. K. (2022). DC and analog/RF performance comparison of renovated GAA JLFET structures. In 2022 IEEE VLSI Device Circuit and System (VLSI DCS) IEEE. 80-84.
8. Merad, F., & Guen-Bouazza, A. (2020). DC performance analysis of a 20nm gate length n-type silicon GAA junctionless (Si JL-GAA) transistor. International Journal of Electrical and Computer Engineering, 10(4):4043.
9. Moore, G. E. (1975). Progress in digital integrated electronics. IEDM Tech. Digest, 21:11-13.
10. Zipperian, T. E., Chaffin, R. J., & Dawson, L. R. (1982). Recent advances in gallium phosphide junction devices for high-temperature electronic applications. IEEE Transactions on Industrial Electronics, 2:129-136.

Note: All the figures and tables in this chapter were made by the authors.

Recent Trends in Applied Physics and Material Science – Sudhir Bhardwaj et al. (eds)
 ISBN 978-1-041-16452-4

Synthesis and Characterization of Cu:Ni Co-doped ZnO Thin Films for Optoelectronics Applications

R.K. Pandey*, Anjali Vaishnaw, Dushyant Karpal
Department of Pure and Applied Physics, Guru Ghasidas Vishwavidalaya, Bilaspur (C.G.)

Avik Karmakar
Chakdaha College, Nadia, West Bengal

Koushik Ghosh
Department of Pure and Applied Physics, Guru Ghasidas Vishwavidalaya, Bilaspur (C.G.)

Abstract: The present paper reports the synthesis and characterization of Cu:Ni dual doped thin films of ZnO using different concentration of Cu:Ni. The as prepared thin films were characterized using XRD and UV-Vis spectrometer. It is apparent from XRD patterns that all the films are of crystalline nature and bear hexagonal structure. The energy band gap (E_g) of pristine ZnO thin films were found to be 3.23 eV and it slightly reduces with rise in Cu:Ni doping concentration.

Keywords: ZnO thin films, Optical materials, Semiconductor materials, Optoelectronic devices

1. INTRODUCTION

Over the past few decades, zinc oxide (ZnO) has gained prominence as a vital material because of its exceptional properties. It belongs to II–VI group and shows many significant properties such as piezoelectric, pyroelectric, high transparency, high electron mobility, high temperature sustainability, radiation hardening etc. It possesses 3.37 eV band gap (Fayaz Rouhi and Rozati, 2022) and 60 meV high exciton energy (Al-Naim, 2021) at room temperature. The significant properties, such as conductivity, energy band gap, chemical sensing, electro-optical and refractive index, effectively varies with donor and acceptor doping concentration. While, doping of rare earths offer limited solubility due to their larger ionic sizes, transition metals, with sizes similar to Zn^{2+}, are highly soluble (Singh et al. 2019).

2. EXPERIMENT

Un-doped and dual doped ZnO ($Zn_{0.99-x}Cu_xNi_{0.01}O$ (x=0.005, 0.015, 0.025) thin films were fabricated using solution processed method. Zinc acetate dehydrate (ZAD) [$Zn(CH_3CO_2)_2 2H_2O$], copper(II) acetate [$Cu(CO_2CH_3)_2$], and nickel(II) acetate tetrahydrate [$Ni(CH_3COO)_2 4H_2O$] were used as precursor materials for the synthesis of Cu:Ni:ZnO thin films. A pure ZnO solution was prepared by dissolving 0.3 Molar ZAD with 99.5% purity in 20 ml of 2-Methoxyethanol. The solution was stirred at 60 °C for approximately 2 hours to achieve a homogeneous mixture with the desired viscosity. When the solution turned milky, 1.8 ml of Monoethanolamine (MEA) was added to get a clear transparent homogeneous solution. The clear solutions were used for deposition after 24 hours of aging. Dual doped ZnO thin films were prepared using different concentrations of copper (II) acetate & nickel (II) acetate tetrahydrate in to pure ZnO solution. Prior to the coating the glass substrates were cleaned and rinsed. Films were deposited using spin coating technique operated at 3000 rpm for period of 30 seconds. Each layer was desiccated on a hot plate at 100°C for 5 minutes and performed again for 6 times to get the required thickness. Finally, all the films were annealed at 450 °C for 2 hours. Four samples, A1 (for un-doped) and A2, A3, A4 (for doped x=0.005, 0.015, 0.025) were prepared and characterized.

*Corresponding author: rkpandeyggv@gmail.com

DOI: 10.1201/9781003684718-108

3. RESULTS AND DISCUSSION

The XRD patterns of un-doped ZnO and dual doped $Zn_{0.99-x}Cu_xNi_{0.01}O$ (x = 0.005, 0.015, 0.025) thin films are illustrated in Fig. 108.1. It is evident from the figure that no other secondary phases have been found (Senol and Arda, 2022). The crystallite size was calculated using the Debye–Sherrer equation and got 7.964 nm, 7.961nm, 9.096nm, and 13.344nm for un-doped ZnO and dual doped (x=0.005, 0.015, 0.025) films, respectively (Senol et al. 2020).

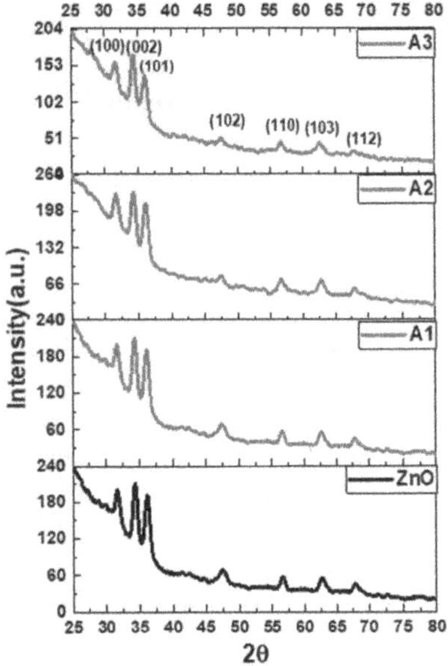

Fig. 108.1 XRD pattern of un-doped and dual doped ($Zn_{0.99-x}Cu_xNi_{0.01}O$, x = 0.005, 0.015, 0.025) ZnO thin films

The absorption spectra of thin films were plotted in Fig. 108.2. The Energy band gap (E_g) value using Tauc plot equation of un-doped ZnO film was found 3.23 eV and it decreases up to 3.08 eV with increasing concentration of Cu:Ni as shown in Figs 108.3 and 108.4. The reduction in the Eg value with higher Cu:Ni doping concentrations can be linked to the increased carrier concentration in the doped films (Alqadi et al. 2022).

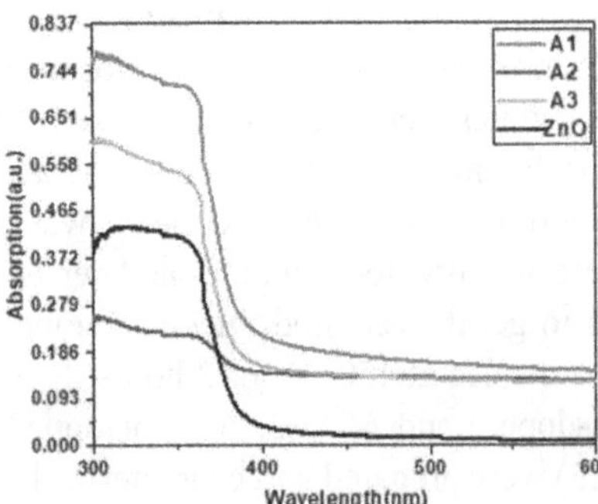

Fig. 108.2 Absorption spectra of un-doped and dual doped ($Zn_{0.99-x}Cu_xNi_{0.01}O$, x=0.005, 0.015, 0.025) ZnO films

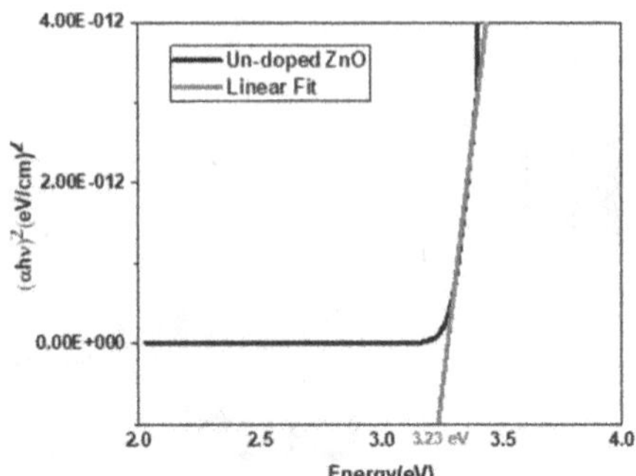

Fig. 108.3 E_g of un-doped ZnO thin film

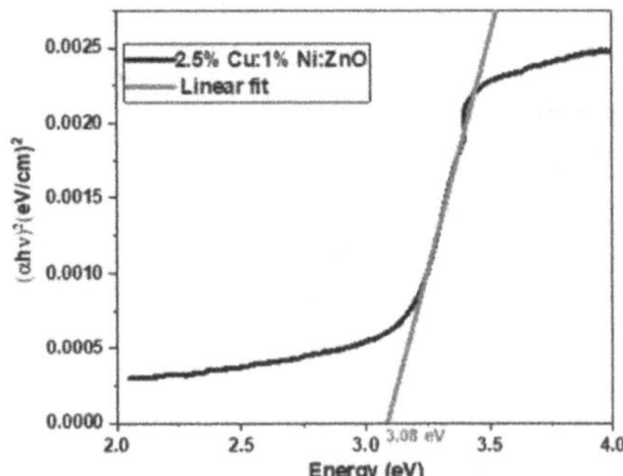

Fig. 108.4 E_g dual doped ZnO thin film

4. CONCLUSION

Thin films of un-doped and Cu:Ni dual doped ZnO were prepared using the solution process technique. It is observed that the E_g value slightly decreases with increasing doping concentration.

REFERENCES

1. Al-Naim A.F., Sedky A., Affy N. and Ibrahim S.S., Structural, FTIR spectra and optical properties of pure and dual doped Zn1-x-yFex MyO ceramics with (M = Cu, Ni) for plastic deformation and optoelectronic applications, 2021. Appl. Phys. A 127(11):840
2. Alqadi, M.K., Migdadi, A.B., Alzoubi, F.Y., Al-Khateeb, H.M. and Almasri, A.A., 2022. Influence of (Ag–Cu) co-doping on the optical, structural, electrical, and morphological properties of ZnO thin films. Journal of Sol-Gel Science and Technology, 103(2):319-3
3. Fayaz Rouhi, H., and Rozati S. M. Synthesis and investigating effect of tellurium-doping on physical properties of zinc oxide thin films by spray pyrolysis technique, 2022. Applied Physics A 128: 252.
4. S.D. Senol, E. Ozugurlu, L. Arda, (2020).Synthesis, structure and optical properties of (Mn/Cu) dual doped ZnO nanoparticles. J. Alloys Compd. 822(5): 153514
5. Senol, S.D. and Arda, L., 2022. The effects of Ni/Cu dual doped ZnO nanorods: Structural and optoelectronic study. Journal of Materials Science: Materials in Electronics, 33(26): pp.20740-20755.
6. Singh, P., Kumar R. and Singh R K. Progress on Transition Metal-Doped ZnO Nanoparticles and Its Application, 2019. Industrial & Engineering Chemistry Research 2019 58 (37): 17130-17163

Note: All the figures in this chapter were made by the authors.

Recent Trends in Applied Physics and Material Science – Sudhir Bhardwaj et al. (eds)

Dielectric Dispersion Analysis of Quaternary Ceramic BNT-BT-PMN-PT

Aparna Saxena
Hindu College, University of Delhi, Delhi, India

Abhilash J. Joseph
University of Delhi, Delhi, India

Raghvendra Sahai Saxena*
Solid State Physics Laboratory, Lucknow Road, Timarpur, Delhi, India

Abstract: A quaternary solid state solution with MPB composition, [$0.47(Bi_{0.5}Na_{0.5})TiO_3$-$0.04BaTiO_3$-$0.31Pb(Mg_{1/3}Nb_{2/3})O_3$-$0.18PbTiO_3$] has been synthesized using solid state reaction method by incorporating BNT-BT relaxor into the PMN-PT ferroelectric material to explore the combined advantages of relaxor and high Curie temperature ferroelectric material. This combination also eliminates the pyrochlore phase and reduces the lead content. The dielectric measurements were performed at temperatures 30 °C to 300 °C, in 200 Hz – 2 MHz frequency range. The material was found to have two overlapping Debye type relaxations in its dielectric behaviour. The corresponding two Debye peaks noticed in dielectric loss as function of frequency show a clear shift towards higher frequencies with increasing temperature of measurement.

Keywords: BNT-BT, Debye dispersion, PMN-PT, Quaternary ceramic

1. INTRODUCTION

Lead magnesium niobate - lead titanate (PMN-PT) is very well known and the most studied relaxor type ferroelctric, having excellent properties as a dielectric, piezoelectric, pyroelectric and ferroelectric material (Hussain et al. 2016). On the other hand, bismuth sodium titanate – barium titanate (BNT-BT) offers less toxicity and high Curie temperature, in addition to very good ferroelectric and piezoelectric properties (Li et al., 2022). It is also preferred due to the absence undesired pyroclore phase (Shieh et al., 2007 and Yao et al., 2017). The advantages of both the binary materials may be obtained by combining them. To explore this possibility, a quaternary solid-state solution of PMN-PT and BNT-BT with MPB composition, i.e., 0.47BNT-0.04BT-0.31PMN-0.18PT, has been prepared using solid state reaction technique (Joseph et al., 2019) and characterized for its ferroelectric, pyroelectric and dielectric properties. The enhanced properties of this material make it suitable for high field applications at elevated temperatures.

The investigation and analysis of dielectric performance of the synthesized 0.47BNT-0.04BT-0.31PMN-0.18PT material is presented here. It is shown that the material follows Debye model in both the ferroelectric and paraelectric phases. Multi-peak behaviour of the dielectric loss data as function of frequency was observed and analysed.

2. EXPERIMENTAL DETAILS

Sigma Aldrich chemicals with more than 99% purity, such as, $MgNb_2O_6$, $BaCO_3$, Bi_2O_3, Na_2CO_3, TiO_2, and PbO were used in powder form for synthesis of the BNTBT-PMNPT material discussed in this paper. These powders were mixed thoroughly as per the required stoichiometry and fired for 2 hours at 870 °C temperature for calcination. This calcinated powder was bound by adding Polyvinyl alcohol and pressed at 100 MPa pressure to form 1 mm thick pellets of 12 mm diameter. This was followed by the sintering process for 2 hours at 1030 °C temperature.

The material was subjected to microstructural analysis using X-ray diffraction. The Archimedes principle was applied for bulk density measurement in distilled water. Ferroelectric hysteresis analysis and extraction of displacement versus voltage curves have also been obtained and are reported in detail in our previous work (Saxena et al., 2022, Saxena et al., 2023 and Joseph et al., 2019).

*Corresponding author: rs_saxena@yahoo.com

DOI: 10.1201/9781003684718-109

The dielectric measurement results reported in this paper were obtained by performing dielectric measurements on the sample of PMNPT-BNTBT material pellet by mounting it in Agilent's 16048A sample holder. The measurements were performed using Agilent's LCR meter, E4890A in 200 Hz – 2 MHz frequency range and 30 °C – 300 °C temperature range. The analysis of these results was performed using Origin9Pro software tool.

3. RESULTS AND DISCUSSION

The dielectric parameters of this material were measured as function of frequency and temperature. These results are shown in the 3D surface plots in Fig. 109.1(a) and (b). The values of these parameters indicate that it is a promising material for a variety of applications, such as it has a low dielectric loss, high dielectric constant and very low DC conductivity.

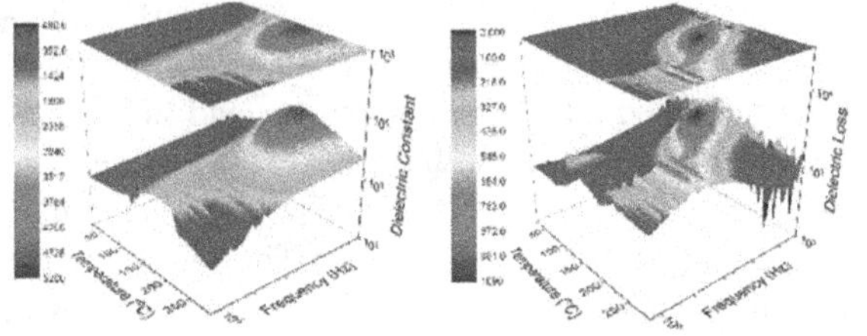

Fig. 109.1 Measured dielectric parameters, (a) Dielectric constant, (b) Dielectric loss shown as 3D surface plot

The material has a very high dielectric constant of ~5000 at 170 °C. It decreases when the temperature is changed to either side. Both ferroelectric and paraelectric phases show Debye behaviour of dielectric data as function of frequency in 125 °C to 175 °C temperature. An interesting phenomenon of the shift of peaks towards higher frequencies was observed when temperature increases.

The plot between the imaginary and real parts of the dielectric permittivity displayed in Fig. 109.2 show single semi-circular behaviour of the curves, as expected. The anomalies found in dielectric parameters due to frequency dependence are linked with polarized entities. On the other hand, the anomalies due to temperature are found to be due to polar species in the relaxor material.

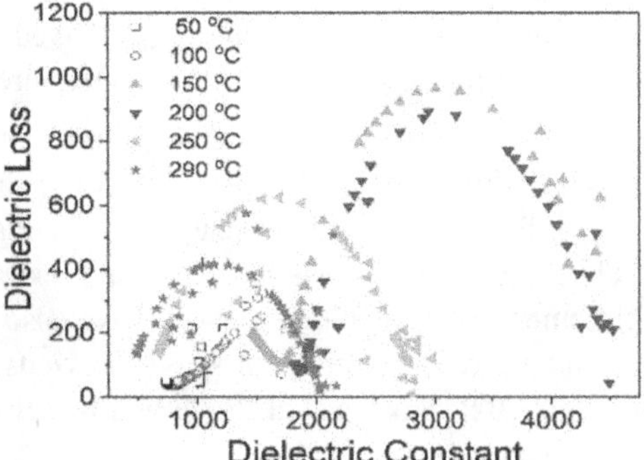

Fig. 109.2 Imaginary part of dielectric permittivity (dielectric loss) plotted as function of its real part (dielectric constant)

Two Debye peaks in dielectric loss were found in its frequency dependent curves, the first one at low temperature and the second one at high temperature. These peaks occur due to the two dispersion phenomena corresponding to the two binary components of this quaternary material. The relaxation times corresponding to the two peaks were extracted by fitting the dielectric data in Debye model (Saxena et al., 2023). We found that the two peaks have about three orders of magnitude different relaxation times. These relaxation times were further investigated to extract the activation energies by using the Arrhenius plot of the relaxation times versus 1000/T plots. Both the peaks are found to have a clear shift towards higher frequencies with increasing temperature.

4. CONCLUSION

A new quaternary ceramic BNTBT–PMNPT having MPB composition was investigated in detail for its dielectric properties. A very high dielectric constant of the order of 10^3 to 10^4 was measured with sufficiently low dielectric losses preferred in a variety of important applications over the large range of operating temperature and frequency.

The two peaks observed in the dielectric loss as function of frequency was investigated for temperature dependence. The material shows promising properties required for fabrication of highly efficient devices.

REFERENCES

1. Hussain A., Sinha N., Bhandari S., Yadav H., and Kumar B. (2016). Synthesis of $0.64Pb(Mg_{1/3}Nb_{2/3})O_3$ $-0.36PbTiO_3$ ceramic near morphotropic phase boundary for high performance piezoelectric, ferroelectric and pyroelectric applications. J. Asian Ceram. Soc. 4(3):337-343.
2. Joseph A. J., Sinha N., Goel S., Hussain A., Yadav H., Kumar B. (2019). New quaternary BNT–BT–PMN–PT ceramic: ferro/piezo/pyroelectric characterizations. Journal of Materials Science: Materials in Electronics. 30:12729-12738.
3. Li X., Zhang B., Cao X., Peng B., Ren K. (2022). Large strain response in $(Bi_{0.5}Na_{0.5})TiO_3$–$6BaTiO_3$ based lead-free ceramics at high temperature. Ceram. Intl. 48(7):9051-9058
4. Saxena A., Hussain A., Saxena A., Joseph A. J., Saxena R. S. (2022). Dielectric dispersion near the morphotropic phase boundary of 0.64PMN-0.36PT ceramics," Ceramics International. 48(5):26258–26263.
5. Saxena A., Joseph A. J., Saxena R. S. (2023). Investigation of Dielectric Properties of Quaternary Ceramic 0.47BNT-0.04BT-0.37PMN-0.18PT. Non-Metallic Material Science. 5(2):26-36.
6. Shieh J., Wu K. C., Chen C. S. (2007). Switching characteristics of MPB compositions of $(Bi_{0.5}Na_{0.5})TiO_3$–$BaTiO_3$–$(Bi_{0.5}K_{0.5})TiO_3$ lead-free ferroelectric ceramics. Acta Materialia. 55(9):3081-3087.
7. Yao K., Chen S., Guo K., Tan C. K. I., Mirshekarloo M S., Tay F. E. H. (2017). Lead-Free Piezoelectric Ceramic Coatings Fabricated by Thermal Spray Process. IEEE Trans. Ultrasonics, Ferroelectrics, and Freq. Control. 64(11):1758-1765.

Note: All the figures in this chapter were made by the authors.

Recent Trends in Applied Physics and Material Science – Sudhir Bhardwaj et al. (eds)
 ISBN 978-1-041-16452-4

Structural and Photoluminescence Properties of Er^{3+} Doped $SrBi_2Ta_2O_9$ Ceramics

Vyom Khare*, Devesh Garg, Megha Narwan, Surya Pratap Singh, Renuka Bokolia

Functional Materials Research Laboratory (FMRL),
Delhi Technological University (DTU), New Delhi, India

Abstract: This research investigated the structural and luminescent properties of Er^{3+}-doped $SrBi_2Ta_2O_9$ ceramic materials. The ceramic samples were fabricated using conventional solid-state reaction method. Variations in the dopant concentration (x = 0.00, 0.02, 0.04) were explored to assess their impact on its luminescence and structural characteristics. X-ray diffraction techniques were employedto confirm that the samples had a single-phase orthorhombic structure with space group A21am after heat treatment that included calcination at 900°C and sintering in the 1050°C temperature range. The SEM images shows a microstructure featuring grains of varying sizes that are randomly oriented, accompanied by noticeable porosity in the sintered pellets. Furthermore, the photoluminescence (PL) spectra were studied at specific excitation wavelengths of 488nm, highlighting the enhanced luminescent properties attributed to Er^{3+} doping. This research underscores the potential of rare-earth-doped ferroelectric ceramics in various energy-related applications, including optoelectronics and sensing technologies.

Keywords: Ceramic, SBT, Photoluminescence etc.

1. INTRODUCTION

Recent researches has highlighted the significance of photoluminescence in rare-earth-doped materials, as these materials have diverse applications in energy-related fields including temperature sensing, Lighting, display technologies, solar cells and biomedical applications.

Ferroelectric materials are characterized to display the spontaneous polarization that can be reversed under an external electric field. Their unique properties include high dielectric constants, fatigue resistance, high Curie temperatures, and notable piezoelectric effects, make them suitable for various energy-related applications, including optoelectronics and sensing technologies[1]. Among these materials, Bismuth-layered structure ferroelectrics (BLSFs), particularly from the Aurivillius family, are attracting significant interest. These materials are represented by the formula $(Bi_2O_2)^{2+}(A_{n-1}B_nO_{3n+1})^2$.A notable example is $SrBi_2Ta_2O_9$ (SBT), which is lead-free and features a high Curie temperature of 600 K along with low dielectric loss. Doping materials like $SrBi_2Ta_2O_9$ with rare-earth ions such as Er^{3+}, Eu^{3+}, and Pr^{3+} has been shown to improve both their electrical and luminescent properties. This study aims to synthesize Er^{3+}-doped $SrBi_2Ta_2O_9$ ceramics using traditional solid-state methods and investigate how this doping influences their crystal structures and luminescent characteristics.

2. SYNTHESIS OF MATERIAL

In this study, polycrystalline ceramics of Er^{3+}-doped $SrBi_{2-x}Ta_2Er_xO_9$with (x= 0.00, 0.02, 0.04) were prepared using a conventional solid-state reaction method. It began with the precise weighing of high-purity starting materials: Tantalum oxide (Ta_2O_5), Strontium carbonate ($SrCO_3$), Bismuth oxide (Bi_2O_3), and Erbium oxide (Er_2O_3). These materials were mixed in their stoichiometric ratios and ground together using a mortar and pestle with ethanol

*Corresponding author: vyomkhare18@gmail.com

DOI: 10.1201/9781003684718-110

added to aid the process, continuing until the mixture was dry; this grinding was performed for six hours to ensure uniformity. The resulting powders were calcined at 900 °C for three hours to facilitate solid-state reactions and improve material properties. After calcination, a binder solution consisting of5 wt% PVA was added to the powders and mixed thoroughly. Disk-shaped pellets were then formed using a hydraulic press operated manually at a pressure of 50 MPa. Finally, the pellets were sintered at 1050 °C for three hours to ensure that the final ceramic structure was free from organic residues, optimizing their electrical and luminescent properties.[2], [3]

3. CHARACTERIZATION DETAILS

X-ray diffraction techniques were employed to study the phase and crystal structure of $SrBi_{2-x}Ta_2Er_xO_9$sintered pallets. The XRD patterns were acquired using a Bruker D8 Discover X-ray diffractometer. This approach enabled the identification of the crystalline phases present in the samples and offered insights about their structural properties. To further examine the structural morphology of $SrBi2_{-x}Ta_2Er_xO_9$, unpolished sintered pellets were analyzed using scanning electron microscopy. The analysis was conducted using JEOL JSM 6610LV, enabling a comprehensive assessment of the surface morphology. Additionally, the photoluminescence (PL) properties of the Er^{3+}-doped $SrBi_2Ta_2O_9$ ceramics were assessed by recording PL spectra using an excitation wavelength of λ_{ex}=488 nm. This measurement provided valuable information regarding the luminescent behavior of the doped materials, highlighting their potential applications in optoelectronic devices.

4. RESULTS AND DISCUSSIONS

4.1 Structural Analysis

The XRD analysis of $SrBi_{2-x}Ta_2Er_xO_9$ ceramics, with varying concentrations of Er^{3+} (x = 0.00, 0.02, 0.04), are illustrated in Fig. 110.1. The XRD analysis was performed over a 2θ range of 10° to 80°, revealing Bragg reflections that closely correspond to the standard diffraction pattern of $SrBi_2Ta_2O_9$ as documented in JCPDS card #49-0609. This confirms the formation of a single-phasematerial without any detectable secondary or unreacted phases, indicating that Er^{3+} ions are fully incorporated into the SBT host lattice without significantly altering its structure. All samples exhibitedthe formation of orthorhombic unit cell, corresponding to the space group A21am. The most prominent peak was attributed to the (115) plane of $SrBi_2Ta_2O_9$, consistent with the bismuth layered structure characterized by n=2.[4]

The micro structural characteristics of the sintered ceramic surfaces for $SrBi_{2-x}Ta_2Er_xO_9$are depicted in Fig. 110.2. The morphology showed grains of varying sizes that are randomly oriented, accompanied by noticeable porosity, typical of solid-state synthesis methods. As the Er^{3+} content increased, no substantial alterations in grain size were detected.

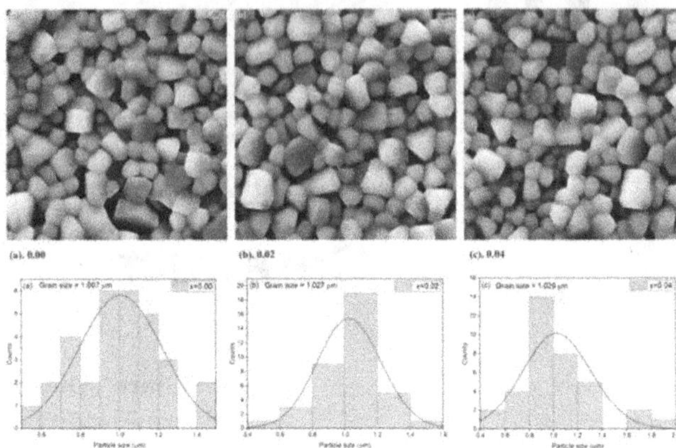

Fig. 110.1 SEM micrographs and histograms of $SrBi_{2-x}Ta_2Er_xO_9$ ceramic

4.2 Spectral Analysis of Photoluminescence

The Er^{3+}-doped $SrBi_2Ta_2O_9$ ceramics were analyzed for their photoluminescence (PL) properties using an excitation wavelength of 488 nm at room temperature. A prominent green emission was detected within the wavelength (λ) ranges from 536 nm to 556 nm. This strong emission reflects the$^4S_{3/2}$ to $^4I_{15/2}$and $^2H_{11/2}$ to $^4I_{15/2}$ electronic transitions of Er^{3+} ions.

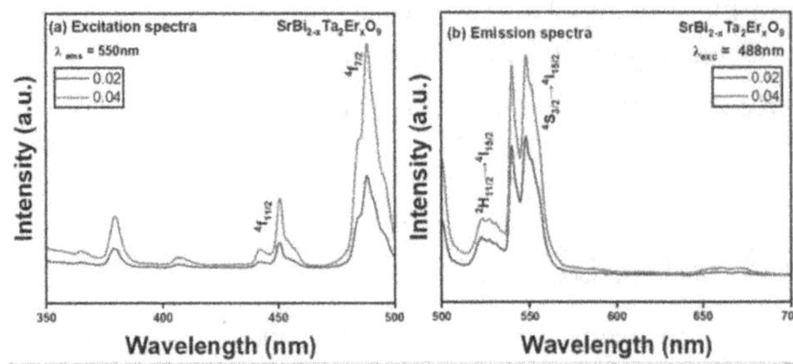

Fig. 110.2 Excitation and Emission Spectra of $SrBi_{2-x}Ta_2Er_xO_9$

5. CONCLUSION

The study successfully demonstrated the synthesis and characterization of Er^{3+} doped $SrBi_2Ta_2O_9$ ceramics, revealing their structural and photoluminescence properties. The orthorhombic structure was maintained with Er^{3+} doping, while exhibiting strong green emission under 488nm excitation. The optimal doping concentration of x=0.04 provides the most intense luminescence. These findings open up possibilities for further research into rare-earth doped ferroelectric materials for advanced optical applications. Future work could explore different rare-earth dopants, co-doping strategies, or alternative synthesis methods to further enhance the luminescent properties of SBT-based materials.

REFERENCES

1. Shaily, R. and Bokolia, R. (2021). Structural and photoluminescence properties of Er3+ doped SrBi2Nb2O9 ceramics, *Mater Today Proc*, 47:4657–4660.
2. Banwal, A. and Bokolia, R. (2022) Enhanced upconversion luminescence and optical temperature sensing performance in Er3+ doped BaBi2Nb2O9 ferroelectric ceramic," *Ceram Int*. 48(2):2230–2240.
3. Banwal, A. and Bokolia, R. (2021) Effect of Er3+ ion doping on structural, ferroelectric and up/down conversion luminescence in BaBi2Nb2O9 ceramic, *Mater Today Proc*, 47:4692–4695
4. Bokolia, R., Thakur, O. P., Rai, V. K., Sharma, S. K. and Sreenivas, K. (2015) Dielectric, ferroelectric and photoluminescence properties of Er3+ doped Bi4Ti3O12 ferroelectric ceramics, *Ceram Int*. 41(4):6055–6066.

Note: All the figures in this chapter were made by the authors.

Recent Trends in Applied Physics and Material Science – Sudhir Bhardwaj et al. (eds)
 ISBN 978-1-041-16452-4

Synthesis and Characterization of Magnesium Oxide(MgO) Nanoparticles via Wet Chemical Approach

H. B. Nayak
P.G. Department of Physics,
Fakir Mohan University, Balasore,
Odisha, India

Paramananda Jena
P.G. Department of Physics,
S.K.C.G. Autonomous College,
Paralakhemundi,
Odisha, India

R. Biswal*
P.G. Department of Physics,
Fakir Mohan University, Balasore,
Odisha, India
Centre of Excellence for BM & ECMD,
Fakir Mohan University, Balasore,
Odisha, India

Abstract: Magnesium oxide (MgO) was synthesized via a wet chemical method using $Mg(NO_3)_2 \cdot 6H_2O$ with 98% purity and deionized water. Thermogravimetric analysis showed the two steps weight loss processes of the synthesized powder before calcination. The first step observed below 190ºC is due to vaporization of water content and hydroxide groups and the next step occurs at 380-510ºC of about 57% is ascribed due to removal chemically absorbed hydroxide groups and oxidation of magnesium. The XRD pattern indicated a single-phase, cubic MgO structure. The Scherrer-equation was employed to estimate the crystallite size, which was ~16.5 nm. The optical energy forbidden gap is studied by UV-Visible spectroscopy and it is calculated 5.03 eV. The dielectric constant, loss factor, and AC conductivity were characterized across a broad frequency spectrum (90 Hz to 8 MHz) and a wide temperature range (room temperature to 550°C).

Keywords: MgO, TGA, XRD, Impedance and spectroscopy

1. INTRODUCTION

In recent years, inorganic materials have garnered significant attention due to their exceptional durability under demanding processing conditions. Among these, nanoscale magnesium-oxide (MgO) is a crucial material with unique properties (Nagappa and Chandrappa, 2007). Possessing a wide array of potential applications, MgO is a highly promising material. These applications encompass water purification, optoelectronics, and microelectronics. Moreover, MgO finds use as a bactericide and an essential component in industrial cables, crucibles, and refractory materials (Selvam et al., 2011). However, the development of novel nanostructures is anticipated to further enhance the already valuable properties of MgO.

This paper presents the characterization of MgO nano-powder synthesized via a wet chemical method (Hazmi et al., 2012). The following sections detail the comprehensive characterization of the synthesized samples, employing a range of analytical and spectroscopic techniques.

2. MATERIAL AND METHODS

2.1 Material Synthesis

Magnesium oxide nanoparticles were produced using a wet chemical method and magnesium nitrate hexahydrate $[Mg(NO_3)_2 \cdot 6H_2O]$ (CDH, 98% purity). The synthesis began with the preparation of 1M solution of the magnesium nitrate hexahydrate in deionized water. This

*Corresponding author: rajibbiswal@gmail.com

DOI: 10.1201/9781003684718-111

solution underwent continuous stirring with a magnetic-stirrer till complete solution was achieved. Stirring process lasted for 1 hour at 400 rpm and kept the solution for 24 hours, after that followed by heating at 60°C until the gel formation achieved. The ultimate precursor heated at 120°C in an oven over night for drying the gel. The dried powder is grinded by agate mortar pestle for one hour. Further the powder was calcined at temperature 600 °C for one hour in a furnace to yield magnesium oxide nanoparticles. A pellet was made by adding 3% PVA in MgO powder and sintered at 600 °C for impedance measurement.

2.2 Characterization

MgO nano powders were characterized using thermogravimetric analysis (TGA 4000 Perkin Elmer) to determine the decomposition temperature, X-ray diffraction (XRD) to gather crystallographic information, UV-Vis spectroscopic measurements were conducted using a Shimadzu UV-2600 spectrophotometer to assess the band gap, Fourier Transform Infrared (FTIR) spectroscopy (Shimadzu IRSpirit X- series) for presence of functional groups and Impedance measurement (IM 3536 LCR Meter) at different set temperatures was used for study of dielectric properties.

3. RESULTS AND DISCUSSION

3.1 TGA Analysis

The weight loss verses temperature of the as prepared sample before calcination is shown in Fig. 111.1. From this graph it is revealed that a decrease in weight as the temperature is increased which resulting a total weight loss of 60.2 % in this TGA measurement. TGA curve showed the two steps weight loss processes. The weight loss observed within 50-183 °C specifies the loss of water content and also hydroxide groups. The decomposition of the sample begins around 380°C, and complete after 510°C, which was well matched as reported in paper (Pradita et al., 2017) and is attributed to removal chemically absorbed hydroxide and nitride groups and oxidation of magnesium.

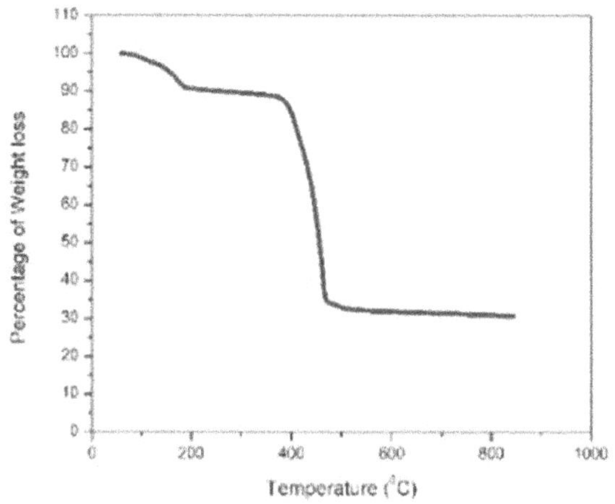

Fig. 111.1 TGA curve of the as prepared sample before calcination

3.2 Crystal Structure Analysis

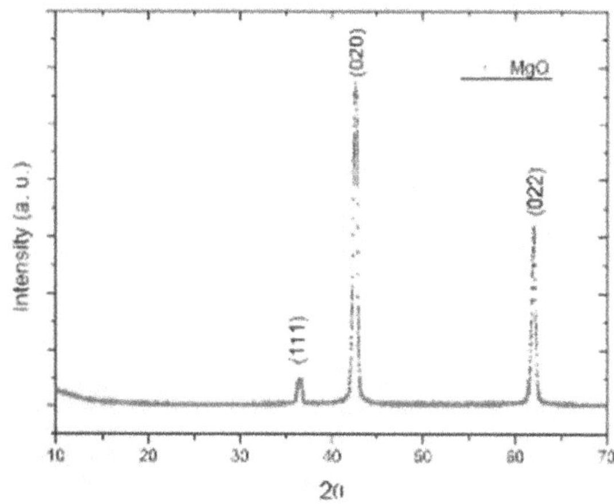

Fig. 111.2 Intensity vs 2θ plot of x-ray diffraction of prepared MgO powder

Figure 111.2 displays the XRD pattern, revealing prominent peaks at angle (2θ values) of 36.6°, 42.6°, and 61.95°. which are representing crystal planes having (hkl) values (111), (020), and (022), respectively (COD database code: 9000495). This analysis confirms the successful synthesis of single-phase MgO nano-powders with a cubic crystal structure from magnesium nitrate hexahydrate. The Scherrer equation, applied to the XRD peaks, revealed an average crystallite size of about 16.5 nm.

3.3 FT- IR Spectroscopy

The FT-IR spectrum of MgO is given in Fig. 111.3. Characteristic absorption peaks were observed at 669 cm^{-1}, 1490 cm^{-1}, and 2362 cm^{-1}. The spectral region between 410 and 665cm^{-1} is associated with metal-oxygen bending vibrations. A prominent peak at 418.3 cm^{-1} validates the existence of Mg-O vibrational bonds (Salman

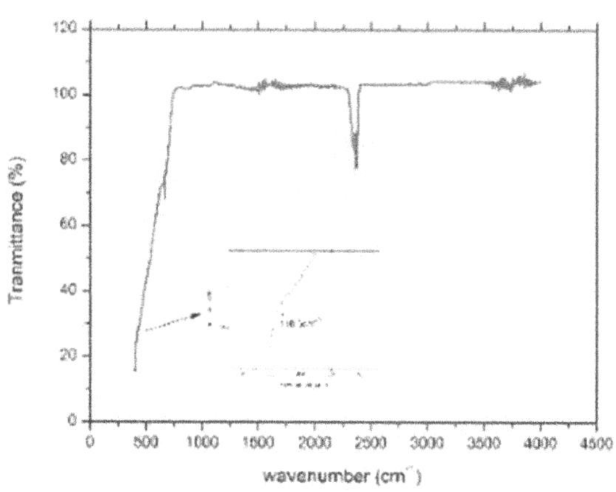

Fig. 111.3 Figure shows FT-IR spectra of MgO

et al., 2021). Notably, a broad band centred around 2342 cm^{-1} is observed in the spectrum, which is strongly indicative of the stretching vibrations within the H-O-H bond (Salman et al., 2021).

3.4 UV- Vis Spectroscopy

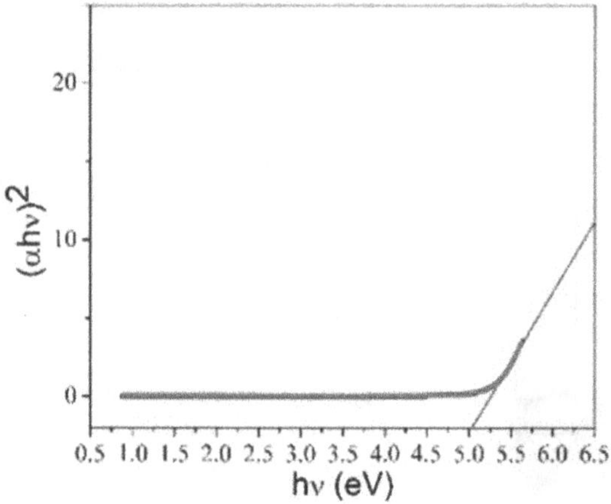

Fig. 111.4 Graph shows Tauc plot for MgO

The optical bandgap of synthesized nanoparticles was extracted according to the following relation (Kumar et al., 2017)

$$\alpha = \frac{C(h\nu - E_g)^n}{h\nu}$$

This equation relates the energy of a photon (hν) to the absorption coefficient (α), where α is influenced by material properties and is related to the optical bandgap (E_g) through a constant (C). Fig. 111.4 represents the Tauc plot of MgO. It is found the optical band gap of MgO ~ 5.03 eV. It is noted that the calculated MgO nanoparticles is consistent with reported literature (Almontasser et al., 2019) and (Farizan et al., 2023).

3.5 Impedance Spectroscopy

The impedance measurement results are shown in Fig. 111.5a-c. The dielectric constant (Fig. 111.5a) is increasing as the temperature is increasing. At room temperature (RT = 23°C), the dielectric constant (ε_r) and loss ($tan\delta$) were obtained 1.24 & 0.26 respectively. The dielectric constant & loss values measured from 90Hz-8MHz at room temperature and from 350 – 550 °C. The dielectric constant and dielectric loss value decreases continuously at lower frequency and remains constant at higher frequency (Fig. 5a-b) (Palanisamy et al., 2017) and (Suresh et al., 2014). The ac conductivity was found to be 1.79 $\times 10^{-9}$ Sm^{-1}, 2.84 $\times 10^{-7}$ Sm^{-1} for 8 MHz for RT and 550°C.

The ac conductivity (σ) was estimated using the equation from dielectric measurement data was given below (Mantry, 2020),

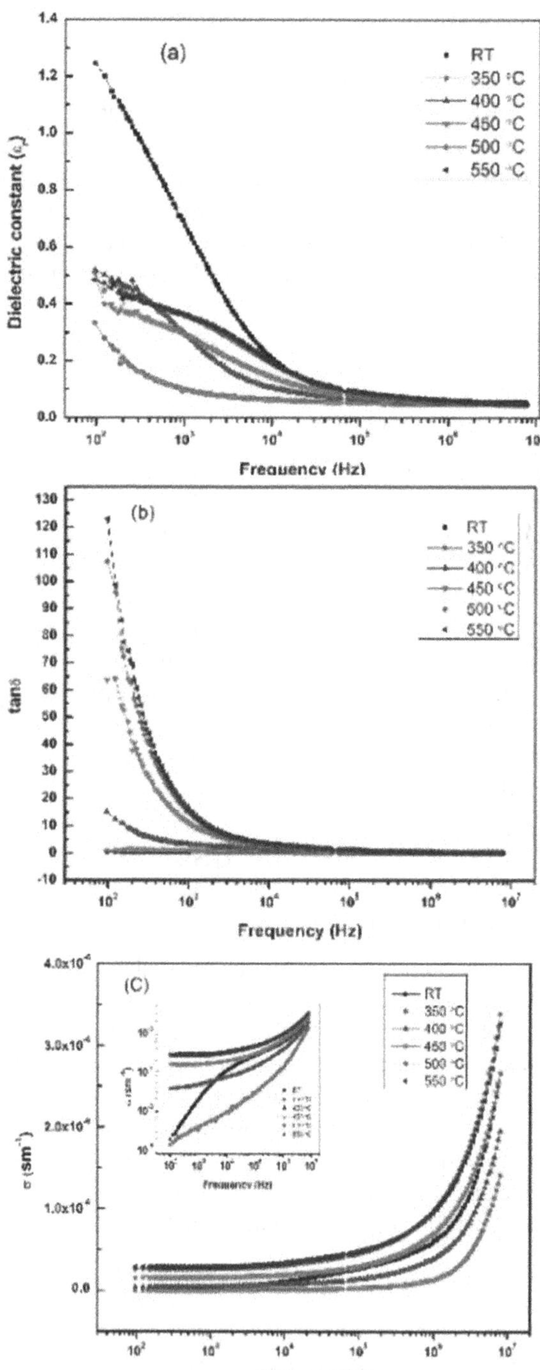

Fig. 111.5 Plot of (a) dielectric constant variation with frequency, (b) Dielectric-loss (tan δ) vs frequency and (c) ac conductivity variation with frequency

$$\sigma = 2\pi f \varepsilon_o \varepsilon_r \tan\delta$$

Conductivity rises with both rising temperature and converging at higher frequencies, particularly at elevated temperatures (inset of Fig. 5c). This surge is attributed to a significant increase in mobile charge carriers prompted by thermal agitation.

4. CONCLUSION

MgO nano-particles were successfully prepared via wet chemical method. In TGA analysis shows minimal weight loss after 510°C suggesting the complete oxidation of magnesium into MgO. XRD pattern revealed that the sample is a pure single phase with cubic structure. The crystallite size, determined using the Scherrer equation, was found to be approximately 16.5 nm. It is found from UV – visible spectroscopy data that the optical band gap of MgO was 5.03eV. The dielectric constant is increasing as the temperature is increasing. The dielectric constant and dielectric loss value decreases continuously at lower frequency and remains constant at higher frequency. AC conductivity rises with both rising temperature and converging at higher frequencies.

5. ACKNOWLEDGEMENTS

R. Biswal gratefully acknowledges the OHEPEE, a joint initiative of the DHE, Govt. of Odisha for the financial assistance. The authors extend their sincere appreciation to Dr. R. Naik, ICT-IOC, Bhubaneswar, for the invaluable experimental support and insightful guidance.

REFERENCES

1. Nagappa, B. and Chandrappa, G. T. (2007). Mesoporous nanocrystalline magnesium oxide for environmental remediation. Microporous and Mesoporous Materials. 106(1):212–218.
2. Selvam, N. C. S., Kumar, R. T., Kennedy, L. J. and Vijaya, J. J. (2011). Comparative study of microwave and conventional methods for the preparation and optical properties of novel MgO-micro and nano-structures. Journal of Alloys and Compounds. 509(41):9809-9815.
3. Hazmi, A. F., Alnowaiser, F., Al-Ghamdi, A. A., Aly, M. M., Al-Tuwirqi, R. M. and El- Tantawy, F. (2012). A new large - Scale synthesis of magnesium oxide nanowires: Structural and antibacterial properties. Superlattices Microstruct. 52(2):200–209.
4. Pradita, T., Shih, S. J., Aji, B. B. and Sudibyo, S. (2017). Synthesis of MgO Powder from Magnesium Nitrate Using Spray Pyrolysis. AIP Conf. Proc. 1823(1):020016-4.
5. Salman, Khansaa, D., Abbas, H. H. and Aljawad, H. A. (2021). Synthesis and characterization of MgO nanoparticle via microwave and sol-gel methods. Journal of Physics. Conference Series. 1973:012104.
6. Kumar, A. M. R., et al., (2017). UV - Sun light Photocatalytic and photoluminescence studies of rare-earth-doped (Sm3+) MgO nanopowders by Aloe Vera gel. Materials Today Proceedings. 4 (11):11737-11746.
7. Almontasser, A., Parveen, A. and Azam, A. (2019). Synthesis, characterization and antibacterial activity of magnesium oxide (MgO) nanoparticles, IOP Conference Series. Materials Science and Engineering. 577(1): 012051.
8. Farizan, A.F., Yusoff, H. M., Badar, N. et al. (2023). Green synthesis of magnesium oxide nanoparticles using mariposa christia vespertilionis leaves extract and its antimicrobial study toward s. aureus and e. coli. Arab J Sci Eng. 48:7373–7386.
9. Palanisamy, G., & Pazhanivel, T. (2017). Green synthesis of MgO nanoparticles for antibacterial activity. Int Res J Eng Technol. 4(9):137-141.
10. Suresh, S. (2014). Investigations on synthesis, structural and electrical properties of MgO nanoparticles by Sol-Gel Method. Journal of Ovonic Research. 10(6):205-210.
11. Mantry, S. P., Kumar, A. and Sarun, P. M. (2020). Temperature and frequency dependent electrical properties of $SrTi_{1-x}Ta_x O_3$ (x = 0.00 − 0.15) ceramics. Materials Chemistry and Physics. 242:0254-0584.

Note: All the figures in this chapter were made by the authors.

Recent Trends in Applied Physics and Material Science – Sudhir Bhardwaj et al. (eds)

112

Effect of Holmium Doping on Structural and Luminescence Characteristics in $BaBi_2Nb_2O_9$ Ferroelectric Ceramics

Lakshya Saini*, Megha Narwan, Surya Pratap Singh, Renuka Bokolia
FMRL Department of Applied Physics, Delhi Technological University, Delhi, India

Abstract: A comparative study was carried out between $BaBi_2Nb_2O_9$ ceramics and holmium-doped $BaBi_{2-x}Ho_xNb_2O_9$ ceramics. These samples were synthesized using solid state reaction technique. In this study, we investigated the structure and luminescence of pure $BaBi_2Nb_2O_9$ ceramics and Ho-doped $BaBi_{2-x}Ho_xNb_2O_9$ ceramics. The evaluations of the properties were carried out using X-ray diffraction (XRD), FTIR, Photoluminescence and Scanning electron microscopy (SEM). The ceramics were subjected to heat treatment including calcination at 950 °C followed by sintering at 1050 °C. X-ray diffractometry validated the formation of an orthorhombic structure with no secondary phase. Scanning electron microscopy (SEM) reveals a randomly oriented plate like microstructure with non-homogeneous grains. Furthermore, PL spectra were examined at 548 nm emission wavelengths and 454 nm excitation wavelength to emphasize the improved luminous property caused by doping of Ho^{3+}. The results indicate significant modifications in the material's properties, which can be tailored for specific applications.

Keywords: Ceramics, Ferroelectrics, Photoluminescence, Solid state reaction method

1. INTRODUCTION

Layered perovskite-like oxides are characterized with mostly unique structural and functional properties, which open them for a vast array of functions in the area of modern technology; thus, their intense and effective research applications. Among such materials, (BBN), as a member compound of the Aurivillius family, has drew significant interest for its photoluminescence, ferroelectric and dielectric properties by Banwal et al. (2022). These properties are important for application in high performance nonvolatile-memories, low loss capacitors and energy harvesting piezoelectric sensors. Chemical modifications including rare-earth doped BBN have the potential to tailor the structural and luminescence properties of BBN, making them functional and application relevant.

Doping of host materials with rare-earth ions, such as holmium (Ho^{3+}), is a common practice for tuning the physical properties of the material. The particular electronic configurations and magnetic behavior of holmium ions are prone to substantial modification on crystal structure, microstructure and photoluminescence in BBN.

2. RESULTS AND DISCUSSION

2.1 XRD Analysis

X-ray diffraction (XRD) patterns of the synthesized samples with compositions x=0.00 (pure) and x=0.02 (doped) are shown in Fig. 112.1. Both patterns exhibit well-defined diffraction peaks, indicating good crystallinity and a phase formation consistent with the target crystal structure. JCPDS file number 00-012-0403 confirmed orthorhombic structure with Fmmm space group. The peak narrowing observed in the doped sample suggests increase in crystallite size by Narwan et al. (2024).

*Corresponding author: lakshyasaini95@gmail.com

DOI: 10.1201/9781003684718-112

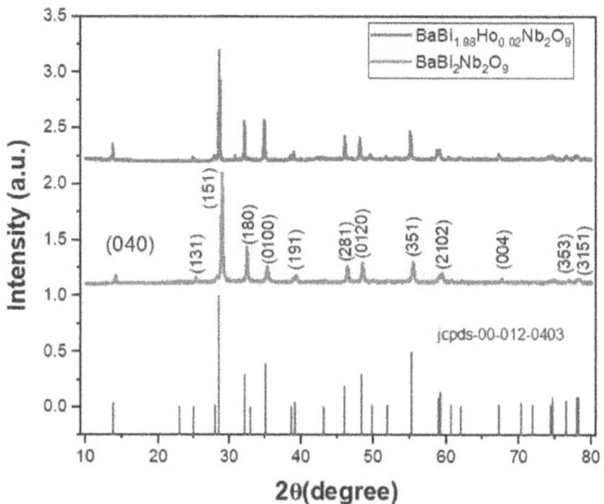

Fig. 112.1 XRD spectra of $BaBi_2Nb_2O_9$ and $BaBi_{1.98}Ho_{0.02}Nb_2O_9$

2.2 SEM Analysis

The SEM micrographs of the pure (x= 0.00) and doped (x=0.02) samples provide insights into the microstructural evolution with doping. For the pure sample, smaller and more uniformly distributed particles are observed, indicating controlled nucleation and growth. In contrast, the doped sample shows a noticeable increase in particle size along with slight agglomeration. This increase in grain size with doping suggests the enhanced mobility of ions during synthesis, leading to accelerated grain coalescence and growth.

Fig. 112.2 SEM micrograph of $BaBi_2Nb_2O_9$

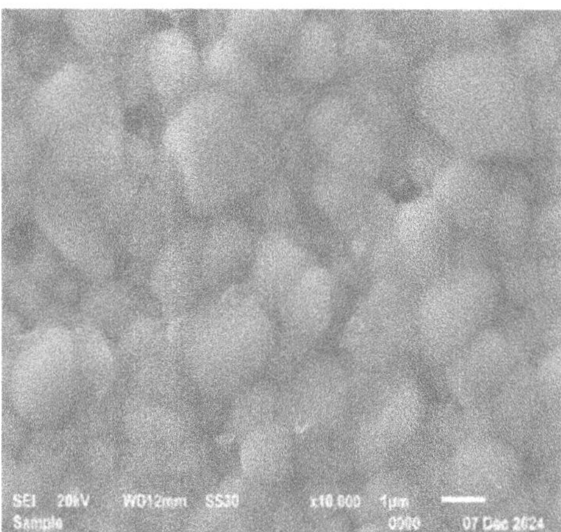

Fig. 112.3 SEM micrograph of $BaBi_{1.98}Ho_{0.02}Nb_2O_9$

The morphological changes observed in the SEM images correlate well with the XRD results, where peak narrowing and lattice expansion were noted. The larger particle size in the doped sample may also contribute to the improved crystallinity observed. Overall, the SEM analysis confirms that doping influences not only the structural but also the microstructural properties, with a clear trend of particle size growth upon doping.

2.3 FTIR Analysis

The two spectra shown in Fig. 112.4 exhibit peaks in the wave number span of 400–1000 cm^{-1}, which are likely related with NbO_6 octahedral vibrations (stretching and bending modes). The peak at 422 cm^{-1} is related to the bending vibrations of Nb–O bond, while the peak at 632 cm^{-1} ascribed to the stretching vibration of Nb–O bond. Also, the peak at 824 cm^{-1} is linked to the asymmetric stretching modes of NbO_6 units.

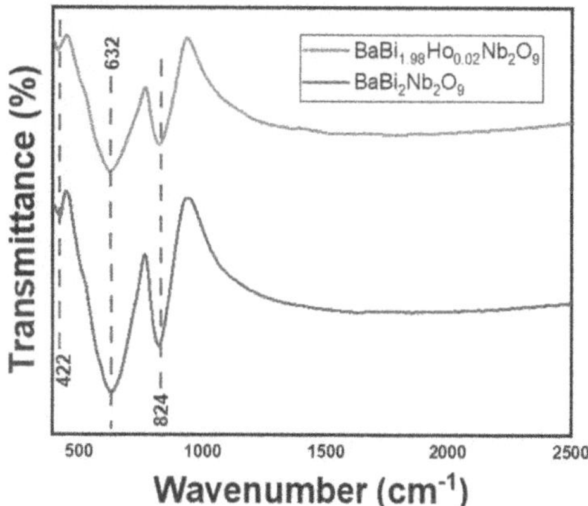

Fig. 112.4 FT-IR spectra of $BaBi_2Nb_2O_9$ and $BaBi_{1.98}Ho_{0.02}Nb_2O_9$

The red spectrum has lower transmittance in the 400–1000 cm^{-1} range, which could be attributed to the modified chemical environment and changes in bond polarizability.

2.4 Down Conversion PL Properties

The optical properties of Ho^{3+} doped BBN ceramics were characterised using photoluminescence spectra. The emission spectrum was recorded under an excitation wavelength of 454 nm a significant peak observed at 548 nm was ascribed to the $^5F_4/^5S_2 \rightarrow {}^5I_8$ transition, accompanied by a smaller peak at 660 nm ascribed to the $^5F_5 \rightarrow {}^5I_8$ transition. The intensity distribution of these peaks highlighted the efficient green emission at 548 nm which is significant for potential application in photonic devices.

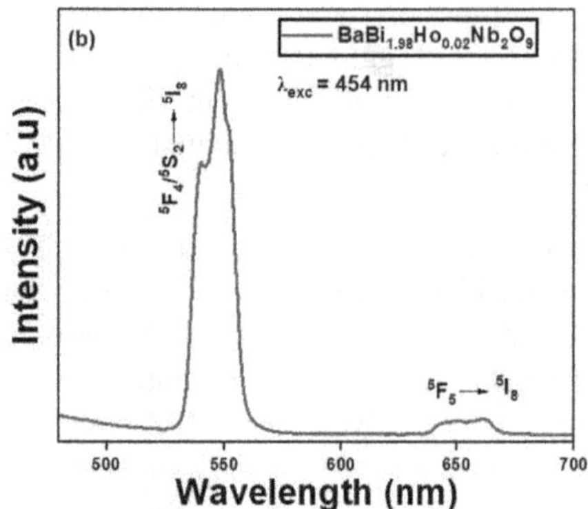

Fig. 112.5 PL spectra of $BaBi_{1.98}Ho_{0.02}Nb_2O_9$ under the excitation of 454 nm

3. CONCLUSION

Investigation of the impact of holmium (Ho^{3+}) substitution on structure and luminescence of $BaBi_2Nb_2O_9$ (BBN) ferroelectric ceramics. XRD analysis confirmed that all samples retained the orthorhombic Aurivillius-type layered perovskite structure, with no secondary phases observed. SEM analysis revealed that holmium doping promoted grain growth, with the average grain size increasing with introduction of doping. The doped ceramics showed enhanced densification and reduced porosity. FTIR spectra showed the presence of characteristic vibrational modes related to Nb–O, Bi–O, and Ho–O bonds, confirming the structural integrity of the doped ceramics. Photoluminescence (PL) analysis exhibited enhanced emission intensities in Ho^{3+}doped samples. Holmium doping improved the microstructure and luminescence properties of BBN ceramics, making them suitable for potential vibrations applications in optoelectronic devices, sensors, and photonic materials.

REFERENCES

1. Banwal, A., & Bokolia, R. (2022). Enhanced upconversion luminescence and optical temperature sensing performance in Er3+ doped BaBi2Nb2O9 ferroelectric ceramic. *Ceramics International*, *48*(2), 2230–2240.
2. Narwan, M., Sharma, R., & Bokolia, R. (2024). Optical temperature sensing and upconversion luminescence in Er3+/Yb3+ co-doped BNT ferroelectric ceramic. *Applied Physics A: Materials Science and Processing*, *130*(12).

Note: All the figures in this chapter were made by the authors.

Recent Trends in Applied Physics and Material Science – Sudhir Bhardwaj et al. (eds)
 ISBN 978-1-041-16452-4

A Solar Energy based Dryer for Environmental Sustainability, for Accelerated Drying of Vegetables

Alok Singh*, Tilak V. Chavda, S. H. Sengar
Department of Renewable Energy Engg, College of Agril. Engg. and Tech., Navsari Agril. University Navsari, Gujarat, India

Alok Singh
Department of Physics, JSS Academy of Technical Education, Noida, India

Ektaray Vasava, Riddhi Lakhtariya, Arti Luhar
Department of Renewable Energy Engg, College of Agril. Engg. and Tech., Navsari Agril. University Navsari, Gujarat, India

Abstract: The use of solar energy in drying application is becoming an important and viable alternative since it decreases consumption of conventional energy by 27-80 % at an average solar collector efficiency of 40 % (Arata et al., 1993). In this view, a mixed mode solar cabinet dryer for drying tomato and onion was developed at Department of Renewable Energy Engineering, CAET, NAU, Dediapada. Mixed mode solar dryer is the combination of direct and indirect type of solar dryers. The product to be dried is dried by subjecting it directly to the sun's radiation and also by supplying hot air from the bottom of the cabinet. Here, air is heated in a flat plate solar collector and then supplied to the drying chamber. Studies on load test, no load test and drying characteristics of tomato and onion in the developed solar dryer were performed. The experiments were conducted on the dryer without any load to find out the temperature profile at different locations in the dryer. Design parameter of the dryer such as temperature of dryer cabinet was recorded. Climatic parameters such as ambient temperature (T_a), solar radiation (I), wind velocity (V_w) and relative humidity (RH) were recorded at an interval of one hour. The temperature of dryer tray varied from 40°C to 64°C during No load test. Tomato and onion were dried in the solar dryer till their moisture content reached 8 % during Load test. Heat collection efficiency and system drying efficiency of the drying system was obtained as 64 % and 25 % respectively.

Keywords: Solar dryer, Mixed mode cabinet solar dryer, Design of solar dryer

1. INTRODUCTION

A solar dryer is most suitable device for drying of agricultural products as it can easily provide low temperature requirement of heat for drying (Mahapatra and Imre, 1990). In recent past, a considerable interest among researchers has been noticed in the design, development and testing of various types of solar dryer like direct (Singh et al., 2006; Saleh and Badran, 2009), indirect (El-Sebaii et al., 2002; Sreekumar et al., 2008) and mixed mode (Tripathi and Kumar, 2009). These dryer systems can be broadly grouped into three major types as direct, indirect and mixed mode, depending on arrangement of system components and mode of solar heat utilization (Leon et al., 2002). In fact, the operation of these dryers is primarily based on the principle of natural or forced air circulation mode. In this view a mixed mode cabinet solar dryer for drying tomato and onion was designed and developed and study was performed to evaluate its performance.

2. MATERIALS AND METHODS

Table 113.1 shows the parameters considered for designing the solar dryer and their computation.

*Corresponding author: salok@nau.in

DOI: 10.1201/9781003684718-113

Table 113.1 Design detail of mixed mode cabinet solar dryer

S. No.	Particulars	Symbol	Design parameter of dryer
1	Amount of moisture to be removed	W_w	W_w = 20 × (95-10)/(100-10) = 18.89 kg
2	Average drying rate	W_{dr}	W_{dr} = 18.89/16 = 1.18 kg/hr
3	Volume of air required for drying	V_a	V_a = (18.89 × 2320)/ (0.2426 × 1.18 × 16) = 5874 m^3
4	Volume flow rate of air required	Q_a	Q_a = 5874/16 = 368 m^3/hr
5	Useful heat energy required, (kJ)	E_u	E_u = 368 × 0.2426 × 1.02 × 30 × 8 = 21912 kJ
6	Total collector area, (m^2)	A_c	A_c = 21912/(500 × 3.6 × 8 × 0.50) = 3 m^2
7	Dryer dimensions	Area	Flat plate collector Area for auxiliary heating = 2 × 1 = 2 m^2 Area of base of absorber plate of drying cabinet = 1 × 1 = 1 m^2
8	Total area of tray for drying	A_t	Assuming avg. size of tomato/onion = 4cm × 5 cm
9	Number of trays	Tray	A_t = 60 × 20 × π × $(0.025)^2$ = 2.3 m^2 = Assumed as 3 m^2 3/1 = 3 trays are required

3. RESULTS AND DISCUSSION

The mixed mode cabinet solar dryer performance was analyzed during October to December (Winter season) and March to May (Summer season). The variations of different climatic and design parameters were studied throughout the day.

No load test of mixed mode cabinet solar dryer: The variation of temperatures and intensity of solar radiation inside mixed mode cabinet solar dryer was studied. It is observed that the temperature of dryer tray varied from 40°C to 80°C. A significant rise in the dryer tray temperature with ambient temperature was observed varying from 11° C to 43°C.

Load test of mixed mode cabinet solar dryer for drying of *Tomato* and *Onion*: Load test on tomato and onion was performed in the month of March for two days to reach the safe moisture content level (8 %) for storage. Both tomatoes and onion sliced into thickness of 8 mm were loaded in the dryer for the test. Fig. 113.1 and Fig. 113.2 show the variation of moisture content (wb) and drying rate against drying time duration for both the days for drying of tomato and onion respectively. Drying of both products was completed at about 01:00 PM on the day-2 (total 12 hours) using mixed mode cabinet solar dryer.

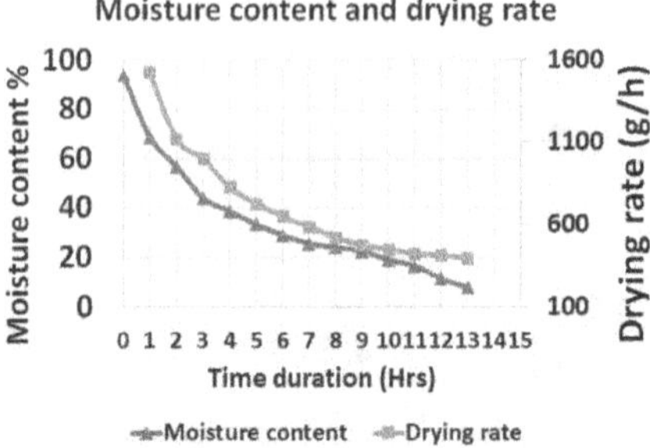

Fig. 113.1 Variation of moisture content and drying rate for drying of tomato

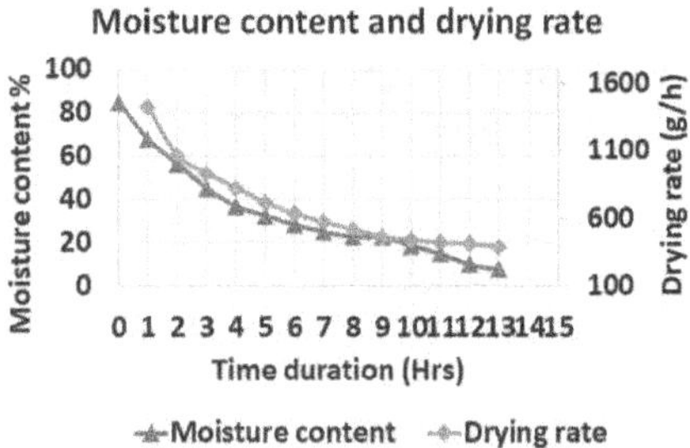

Fig. 113.2 Variation of moisture content and drying rate for drying of onion

Dryer efficiencies: The efficiency of the mixed mode cabinet solar dryer for the drying of different products were computed as mentioned in Table 113.2.

Table 113.2 Computation of drying efficiency of mixed mode cabinet solar dryer obtained while drying the products

S. No	Efficiency	Computation
1	Heat collection efficiency	[(365/3600) X 0.2426 X 30 X 1.02]/ [3 X 0.600 X 0.65 = **64%**
2.	System drying efficiency	(4.9 X 2320)/(0.80 X 0.75 X 0.600 X 3 X 12 X 3600) = **25 %**

4. CONCLUSIONS

Designed and developed Mixed mode cabinet solar dryer of 20 Kg capacity can be used for drying of tomato and onion up to 8% moisture content (safe for storage). Heat collection efficiency and system drying efficiency oh the dryer was found to be 64 % and 25 % respectively.

REFERENCES

1. Arata, A., Sharma, V.K., and Spagna, G. (1993). Performance evaluation of solar assisted dryers for low temperature drying application-II: experimental results, Energy Conversion Management, vol. 34, no. 5, pp. 417–426. Mahapatra, A. K. and Imre, L. (1990). Role of solar agricultural drying in developing countries, Int. J. Ambient Energy, vol. 2, pp. 205–210.

2. Singh, P. P., Singh, S., and Dhaliwal, S. S. (2006). Multi-shelf domestic solar dryer, Energy Conversion Management, vol. 47, pp. 1799–1815.
3. Saleh, A. and Badran, I. (2009). Modeling and experimental studies on a domestic solar dryer, Renewable Energy, vol. 34, pp. 2239–2245.
4. El-Sebaii, A. A., Aboul-Enein, S., Ramadan, M. R. I., and El-Gohary, H. G. (2002). Experimental investigation of an indirect type natural convection solar dryer, Energy Conversion Management, vol. 43, no. 16, pp. 2251–2266.
5. Sreekumar, A., Manikantan, P. E., and Vijayakumar, K. P. (2008) Performance of indirect solar cabinet dryer, Energy Conversion Management, vol. 49, pp. 1388–1395.
6. Tripathy, P. P. and Kumar, S. (2009). Influence of sample geometry and rehydration temperature on quality attributes of potato dried under open sun and mixed mode solar drying, Int. J. Green Energy, vol. 6, no.2, pp. 143–156.
7. Leon, M. A., Kumar, S., and Bhattacharya, S. C. (2002). A comprehensive procedure for performance evaluation of solar food dryers, Renewable Sustainable Energy, vol. 6, pp. 367–393.

Note: All the figures and tables in this chapter were made by the authors.

Recent Trends in Applied Physics and Material Science – Sudhir Bhardwaj et al. (eds)
© 2026 Taylor & Francis Group, London, ISBN 978-1-041-16452-4

Self-Organized Liquid Crystals for High-Efficiency Photovoltaics

Bhumika[1], Rohit Verma[2]
Physics Department, Amity Institute of Applied Sciences,
Amity University Uttar Pradesh, Noida, India

Abstract: In the field of photovoltaics, liquid crystals (LCs) have gained significant attention as a promising invention that could lead to the development of new solar energy conversion technologies. In this study, we have explored the integration of liquid crystal materials into photovoltaic systems, which also highlights the materials' variable phases, unique optical features, and capacity to improve light absorption. Because LCs can self-organize into ordered structures, they can control light in ways that conventional solid-state materials cannot. Scientists can enhance light trapping and boost solar cell efficiency by fine-tuning the orientation and alignment of liquid crystal molecules. As recent research has shown, liquid crystals can boost the spectrum sensitivity of photovoltaic systems, especially in the visible and near-infrared range. Furthermore, the built-in adaptability of liquid crystals' natural flexibility also makes it possible to create flexible and lightweight solar panels that may be incorporated into various surfaces, such as wearable technology and building materials. Controlling the molecular structure of liquid crystals also makes it possible to design dynamic photovoltaic systems that can optimize energy harvesting by changing their characteristics in reaction to external factors. This study investigates the fundamental mechanisms of light harvesting in liquid crystal photovoltaics and explores the engineering challenges hindering their practical implementation. Liquid crystal-based photovoltaics and traditional solar technologies are compared and discussed, with an emphasis on the latter's possible benefits—such as more adaptability and lower production costs. Moreover, we have also emphasized how crucial it is to maintain multidisciplinary study to fully utilize liquid crystals' potential in the development of sustainable energy solutions, opening the door for the next generation of environmentally friendly, multifunctional, and highly efficient photovoltaic systems.

Keywords: Photovoltaics systems, Solar energy

1. INTRODUCTION

The global demand for sustainable and renewable energy sources has driven the need for innovations in photovoltaic (PV) technology. Solar cells, which convert sunlight into electricity, are at the forefront of this effort. Over the past few decades, significant advancements have been made in photovoltaic materials, with organic photovoltaics (OPVs) emerging as a promising alternative to conventional silicon-based solar cells. Organic materials offer potential advantages such as flexibility, low-cost fabrication, and ease of processing. A particularly exciting class of materials that has shown great promise in OPV research is self-organized liquid crystals (LCs) (L. Schmidt-Mende *et al.* 2001). Liquid crystals, known for their unique structural properties and versatility in electronic and optical applications, are being increasingly explored as active layers in PV devices due to their ability to self-assemble into highly ordered structures that can enhance the efficiency of light absorption and charge transport (Kumar and Kumar 2017). This review explores the role of self-organized liquid crystals in high-efficiency photovoltaics, addressing their properties, fabrication methods, challenges, and future directions.

2. A BRIEF OVERVIEW OF LIQUID CRYSTALS

Liquid crystals (LCs) are substances that exhibit properties intermediate between those of conventional

[1]bhumika16@s.amity.edu, [2]rverma85@amity.edu

DOI: 10.1201/9781003684718-114

liquids and solid crystals. They possess both fluidity and long-range order, which makes them highly suitable for a wide range of applications, from display technologies to optical systems. LCs can be broadly classified into three main categories: nematic, smectic, and cholesteric, each with distinct structural characteristics. The ability of LCs to self-organize into ordered structures is of particular interest for photovoltaic applications.

In a nematic LC phase, molecules align parallel to each other in a directional arrangement, but they lack positional order. In contrast, smectic LCs exhibit both positional and orientational order, where molecules are organized into layers. Cholesteric LCs form a helical structure, which can be particularly useful in applications where optical activity is important. The dynamic and tunable nature of LCs allows them to adapt to external stimuli such as electric fields, temperature changes, and light, making them ideal candidates for advanced applications, including organic photovoltaics.

3. MECHANISMS BEHIND PHOTOVOLTAIC FUNCTIONALITY

Photovoltaic devices rely on the efficient conversion of sunlight into electricity through the photovoltaic effect. In OPVs, this process typically involves the absorption of photons by the active layer, followed by the generation of excitons (electron-hole pairs) that must be dissociated into free charges. These charges then travel through the material to the electrodes, where they are collected as an electrical current.

In organic photovoltaics, the active layer is often composed of a blend of donor and acceptor materials (Tang *et al.* 1986). The donor material absorbs light and generates excitons, while the acceptor material provides a pathway for charge separation. The efficiency of this process depends heavily on the morphology of the active layer, as well as the electronic properties of the materials used. Self-organization, particularly in the context of liquid crystals, plays a key role in enhancing both the light absorption and charge transport properties of the active material, ultimately improving the overall efficiency of the device.

4. SELF-ORGANIZED LIQUID CRYSTALS FOR PHOTOVOLTAICS

Self-organized liquid crystals offer several advantages when used as materials in photovoltaic devices. The ability of LCs to spontaneously form ordered structures at the nanoscale can improve the performance of OPVs by facilitating charge transport, enhancing light absorption, and minimizing recombination losses (Varotto *et al.* 2010). Furthermore, the use of liquid crystals in photovoltaics can lead to devices that are flexible, lightweight, and potentially more cost-effective than traditional rigid solar panels.

Table 114.1 Development in LC based photovoltaics

Year	Development	Researchers
2010	First generation of liquid crystal based organic solar cells with 1% efficiency	Varotto *et al.* 2010
2014	Improved efficiency to 2.5% using liquid crystal phases	J. E. Lyons *et al.*
2017	Introduction to liquid crystal molecules with high charge carrier mobility	H. W. Lee *et al.*
2019	Over 16% Efficiency Organic Photovoltaic Cells Enabled by a Chlorinated Acceptor with increased open-circuit voltages	Y. Cui *et al.*
2022	Record efficiency of 8.2% using LC-based ternary blend OSCs	J.M. Wang *et al.*

4.1 Light Absorption and Optoelectronic Properties

One of the most significant benefits of using liquid crystals in OPVs is their ability to enhance light absorption. The self-organized structures of LCs can form nanoscale domains that act as efficient light-trapping systems, increasing the optical path length of light within the material (Ma *et al.* 2005). This leads to improved absorption of incident photons across a wide range of wavelengths. Additionally, the optical properties of LCs can be tuned through external factors such as temperature and electric fields, offering a high degree of flexibility in tailoring the absorption spectrum of the device to match the solar spectrum.

The alignment of molecules within the LC phase also has a direct impact on the electronic properties of the material. Well-ordered structures can facilitate better charge transport by providing continuous pathways for electron and hole movement. The combination of high light absorption and efficient charge transport is crucial for improving the power conversion efficiency of OPVs.

4.2 Charge Transport and Recombination Suppression

Charge transport is one of the critical factors determining the efficiency of photovoltaic devices. In conventional organic solar cells, the disordered morphology of the active layer can lead to poor charge transport, increased recombination losses, and lower overall efficiency (Shirota *et. al.* 2007). Self-organized LCs can mitigate this issue by forming highly ordered nanostructures, which allow for better mobility of charge carriers (Green *et al.* 2015).

Moreover, the formation of ordered domains in LCs can suppress charge recombination, a major loss mechanism in photovoltaic devices. In conventional OPVs, the random morphology of the donor-acceptor blend can create traps that hinder the dissociation of excitons into free charges, leading to recombination. Self-organized liquid crystal phases can reduce the number of such recombination sites by providing a more uniform and well-structured environment for charge separation and transport.

4.3 Flexibility and Processability

Another key advantage of liquid crystals in photovoltaic applications is their inherent flexibility. Liquid crystal-based materials can be processed using low-cost techniques such as roll-to-roll printing, making them ideal for large-scale manufacturing of flexible solar cells. The self-assembly process of LCs allows for the creation of nanoscale structures without the need for complex lithographic techniques, further reducing the cost and complexity of fabrication (Hains *et al.* 2010).

The tunable nature of liquid crystals also means that their properties can be adapted to different processing conditions. By adjusting parameters such as temperature or the application of an external electric field, researchers can influence the alignment and phase behavior of the LCs, offering a high degree of control over the final morphology of the active layer.

5. CHALLENGES AND LIMITATIONS

Despite the significant promise of self-organized liquid crystals for high-efficiency photovoltaics, several challenges remain that must be addressed for widespread adoption.

5.2 Stability

One of the major concerns with using liquid crystals in photovoltaics is their long-term stability. LCs are generally sensitive to temperature fluctuations and environmental factors such as humidity and oxygen, which can degrade their performance over time. To ensure that self-organized liquid crystal-based PV devices remain reliable and efficient in real-world applications, researchers must focus on enhancing the stability of these materials. This may involve developing new liquid crystal formulations or incorporating stabilizing agents that can protect the material from degradation.

5.2 Scalability of Fabrication

While self-organized liquid crystals offer exciting potential for low-cost, flexible solar cells, scaling up the fabrication process remains a challenge. The ability to produce large-area, uniform films of liquid crystals with consistent ordering is crucial for the commercial viability of liquid crystal-based OPVs. Current fabrication techniques may not always yield the desired level of uniformity or reproducibility, which could hinder the development of large-scale production methods. Overcoming these challenges will require innovations in processing technologies and better understanding of the self-assembly mechanisms at play.

5.3 Material Design

Liquid crystals are a highly diverse class of materials, and not all LCs are equally suitable for use in photovoltaic applications. Researchers need to develop liquid crystal materials that have the right electronic and optoelectronic properties for efficient charge transport and light absorption. This includes optimizing the alignment, molecular structure, and intermolecular interactions within the LC phase. Furthermore, incorporating functional groups that can enhance the interaction between donor and acceptor materials within the LC matrix is an area of ongoing research.

6. RECENT ADVANCES AND FUTURE DIRECTIONS

Recent advances in the field of self-organized liquid crystals for photovoltaics have demonstrated promising results. For instance, researchers have successfully developed liquid crystal-based organic photovoltaics with high power conversion efficiencies, often outperforming traditional OPVs in certain areas. Many of these advancements have focused on optimizing the morphology of the active layer, enhancing charge transport, and increasing the stability of the materials.

Future research will likely focus on overcoming the remaining challenges, such as improving material stability and scalability. Additionally, the integration of self-organized liquid crystals with other emerging technologies, such as perovskite solar cells or tandem devices, could further boost the efficiency of photovoltaic systems.

Furthermore, advances in computational modeling and material design could help researchers better understand the complex interactions within self-organized liquid crystal phases, enabling the development of new materials with even higher performance. Exploring new LC structures, hybrid materials, and novel processing techniques will be key to realizing the full potential of liquid crystals in high-efficiency photovoltaics.

7. CONCLUSION

Self-organized liquid crystals represent a promising avenue for the development of high-efficiency photovoltaics. Their unique structural properties enable enhanced light absorption, improved charge transport, and greater flexibility in device fabrication. While challenges related to stability, scalability, and material design remain, recent advances in the field suggest that liquid crystal-based photovoltaics have the potential to become a major player in the future of renewable energy. With continued research and innovation, self-organized liquid crystals could play a critical role in the development of next-generation solar cells that are not only efficient but also cost-effective and flexible enough for a wide range of applications.

REFERENCES

1. Dimitrakopoulos, C. D., Purushothaman, S., Kymissis, J., Callegari, A. & Shaw, J. M. (1999). Low-voltage organic transistors on plastic comprising high-dielectric constant gate insulators. Science. 283:822–824.
2. Garnier, F., Hajlaoui, R., Yassar, A. & Srivastava, P. (1994). All-polymer field-effect transistor realized by printing techniques. Science. 265:1864–1866
3. Green, M. A., Emery, K., Hishikawa, Y., Warta, W. & Dunlop, E. D. (2015). Solar cell efficiency tables (Version 45). Prog. Photovoltaics. 23:1–9.
4. Hains, A. W., Liang, Z., Woodhouse, M. A. & Gregg, B. A. (2010). Molecular semiconductors in organic photovoltaic cells. Chem. Rev. 110:6689.
5. Ma, W., Yang, C., Gong, X., Lee, K. & Heeger, A. J. (2005). Thermally stable, efficient polymer solar cells with nanoscale control of the interpenetrating network morphology. Adv. Funct. Mater. 15:1617–1622
6. Schmidt-Mende, L., Fechtenkötter, A., Müllen, K., Moons, E., Friend, R. H. & MacKenzie, J. D. (2001). Self-organized discotic liquid crystals for high-efficiency organic photovoltaics. Science. 293:1119–1122.
7. Shirota, Y. & Kageyama, H. (2007). Charge carrier transporting molecular materials and their applications in devices. Chem. Rev. 107:953–1010.
8. Tang, C. W. & VanSlyke, S. A. (1987). Organic electroluminescent diodes. Appl. Phys. Lett. 51:913–915.
9. Tang, C. W. (1986). Two-layer organic photovoltaic cell. Appl. Phys. Lett. 48:183–185.

Note: The authors made the table in this chapter.

Recent Trends in Applied Physics and Material Science – Sudhir Bhardwaj et al. (eds)
© 2026 Taylor & Francis Group, London, ISBN 978-1-041-16452-4

A Detailed Overview of CSP and Thermal Energy Storage Technologies

Priyanka Devi
Department of Physics,
Shri Khushal Das University,
Hanumangarh, India

Renu Sharma*
Department of Physics,
JECRC University,
Jaipur, India

Robin Gupta
Department of Physics, Shri
Khushal Das University,
Hanumangarh, India

Pawan Kumar Jain
Department of Physics,
Swami Keshvanand Institute
of Technology Management
& Gramothan, Jaipur, India

Abstract: The integration of thermal energy storage (TES) systems in renewable energy technologies, particularly in plants that use concentrated solar power (CSP), a crucial step in the direction of attaining sustainable and is efficient energy solutions. This review explores the latest materials and technologies employed in TES, addressing associated limitations such as high-temperature corrosion and proposing innovative solutions. The renewable energy sector's future is closely tied to the development of CSP and TES technologies, which promise enhanced energy conversion efficiency and reliability. Additionally, Phase change materials (PCM) are incorporated into TES systems for nearly zero energy buildings demonstrates significant improvements in energy conservation. Various forms of TES system integration are examined, highlighting their potential to support sustainable energy infrastructures. Furthermore, the strategic integration of renewable sources into smart city energy systems is discussed as a forward-thinking approach to fostering clean and sustainable urban environments. The significance of TES to promote a more sustainable future and enhance renewable energy technology future.

Keywords: TES, CSP, PCM, Thermal fluid

1. INTRODUCTION

The most convincing and different thing about the technology for Collection of Solar Power and its Storage is that it may be paired with already existing Heat Storage Systems, this advantage could help it to be more flexible to implement. The technology is constantly being evolved and is expected to become more easier to implement even with minimum of the funds, some researchers are focused on subject such as improving the materials for the large mirrors in collectors, improving its designs, technologies used for mobility of heat energy across the thermal energy storage systems etc (Zhiwen Ma et al. 2023). The capacity for storage is anticipated to significantly encourage electrical utilities to adopt CSP facilities (F. Alnaimat, et al. 2019). By integrating TES (Thermal Energy Storage) with CSP (Concentrated Solar Plants thermal energy can be stored for use in powering a heat engine later. to produce electricity for prolonged periods from traditional chemical or mechanical technologies. TES can be significantly cost effective from traditional technologies as well as comes with increased efficiencies (D.S. Codd et al. 2020). TES systems fall into three categories: visible heat, latent heat, and thermochemical heat [M. Barrasso et al. 2023]. This chapter provides requirements for each technology and application as well as clues for each TES system. Specific and innovative applications are described together with a summary of system kinds [D. Jayathunga et al. 2023]. The three main parts of a mechanism for storing thermal energy is the containment system, the store medium, and the heat transmission mechanism. The storage medium holds onto thermal energy by reversible chemical reactions, perceptible heat, or latent heat of vaporization or fusion. Melted salt and synthetic oil are now the most often used sensible heat storage materials in large-scale CSP systems. Systems that use thermochemical reactions, Meanwhile, research is being done on latent heat and other sensible heat materials (Kurup et al. 2022). Delivering

*Corresponding author: renu.sharma@jecrcu.edu.in

DOI: 10.1201/9781003684718-115

or removing heat transport is a function of the storage medium mechanism. The energy transmission apparatus and storage medium are housed in the containment system, which also acts as insulation against the outside world. To guarantee optimum storage performance and lifespan, several parameters must be fulfilled, based on the kind of storage (S. Pascual et al. 2022). These requirements remain the same for each kind of TES technology. These technologies include: The storage material should have a high energy density, efficient heat transfer between the storage medium and the heat transfer fluid (HTF) is essential. The substance used for storage needs to be stable both chemically and mechanically (G. Ferruzzi et al. 2023).

The storage medium, heat exchanger, and HTF should all be chemically compatible, complete reversibility throughout multiple cycles of charging and draining should be possible with this approach. It reduces the amount of heat loss. The system should be cost-effective. The environmental impact should be low. With taking in note of the above conditions, these criteria should also be considered when designing TES systems. Specific enthalpy loss and nominal temperature during conversion and discharge Maximum capacity for load. Strategy for operations integration in the plant.

2. TYPES OF CSP COLLECTOR

The first commercial Concentrated Solar Power (CSP) plant was introduced in New Mexico in 1979 by the Sandia National Laboratory (Ali O M Maka et al. 2023). Since then, With the bulk of the installed CSP capacity worldwide, the USA and Spain have been in the forefront of CSP research and development. Globally, more than 700 MW of CSP capacity was built between 2010 and March 2012 (M. Papaelias et al. 2018). Over 20,000 MW of solar thermal electricity has been installed globally as of right now. The outcomes of the 2013 and 2020 CSP configurations, Notable examples include size of the solar field and the ability to store thermal energy. The anticipated significant increase in renewable energy sources' incorporation into power systems worldwide is the driving force behind the increasing interest in CSP with thermal energy storage. Traditional wind and solar plants generate energy on a variable basis and contribute less to asset sufficiency compared to fossil-fuel generation (M. Papaelias et al. 2018). The thermal energy storage (TES) system, power block, and solar field are the three primary parts of CSP systems. These systems consist of modules like the power block, thermal energy storage, and concentrating solar collector field (M. Papaelias et al. 2018).

The application of thermodynamic cycles with high-temperature input via CSP technology can result in high efficiency. Diffused and reflected portions of incoming solar radiation are lost since CSP technology only uses the direct component. Greater Direct Normal Irradiation (DNI) value is necessary for CSP technology.

Due to their higher initial costs, Concentrating Solar Power (CSP) systems are Small-scale solar power facilities are often not a good fit for. Concentrating Devices called solar collectors are made to absorb solar radiation and transform it into thermal energy, which can then be utilized to generate electricity. A solar collector's primary function is to convert incoming solar radiation into heat. A Heat Transfer Fluid (HTF) receives this heat and flows through the collector. After that, HTF transports the thermal energy from the collectors to a thermal storage system or central steam generator enabling the power generation process. Concentrating solar collectors come in two varieties: stationary, non-concentrating collectors and sun-tracking collectors.

2.1 Stationary or Non-Concentrated Collectors

The stationary non-concentrating solar collector functions by using a similar area for both absorbing and intercepting incoming radiation (M. Papaelias et al. 2018).

Sun-tracking concentrating solar collectors track the sun during its daily path to maintain maximum solar flux at their location by using optical components to focus a lot of radiation onto a small receiving area focal point.

Typically, CSP technologies come in four varieties: dish/engine, central receiver, linear Fresnel, and parabolic trough.

2.2 Parabolic Trough Collectors

The most sophisticated and extensively utilized technique is the Parabolic Trough Collector (PTC). Both parabolic trough and linear Fresnel collectors are designed for similar thermal power output and have comparable input and outlet oil temperatures. Solar troughs heat a transfer fluid, typically oil. A PTC collects solar radiation and transforms it into thermal energy that can be used in the Heat Transfer Fluid (HTF) that passes across the solar field. (M. Papaelias et al. 2018). The parabolic trough collector is illustrated in Fig. 115.1.

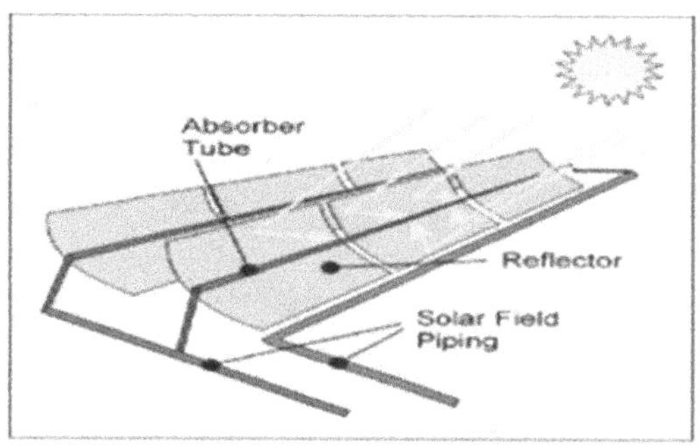

Fig. 115.1 Parabolic trough collector

The heated fluid generates steam, which drives a turbine to generate electricity. Maintaining the field's outlet temperature as near to a predetermined point as feasible is the major goal of a control system for a parabolic trough field. The oil flow rate (q) serves as the control signal because solar energy, the main energy source, is uncontrollable. Globally, CSP technologies are presently controlled by 90% of CSP power plants globally, which are parabolic trough collectors.

Three components make up a parabolic trough power plant with heat storage: the solar field equipped with a heat exchange circuit mechanism for heat storage, and a power block that includes a cooling system, a generator, and a turbine. This method directs sunlight onto a high-performance absorber pipe that runs along the concentrators' center axis using a curved mirror system, through which a heat transfer fluid (HTF) flow. The fluid, typically a mineral oil, absorbs heat from the sun and is sent to a power plant via receiver pipes. In this case, the heat is transformed into steam by a heat exchanger and sent to a turbine to produce energy. Through the receiver, HTF gathers and transfers thermal energy to storage facilities or power generation devices, usually a boiler and turbine generator. Water or oil are commonly used as the HTF in PTC systems; oil is usually used because of its greater boiling point and comparatively reduced volatility. Thomas has proposed a number of water boiler ideas.

2.3 Linear Fresnel Collectors

Despite this, from an annual performance perspective, linear Fresnel collectors tend to achieve lower efficiency compared to parabolic trough collectors, as indicated in technical literature. Typically, a linear Fresnel collector is made up of a collection of mirror panels, each of which might have different sizes and arrangements. The linear Fresnel collector has less expensive components, frequently necessary to make up for this optical disadvantage. Despite being parabolic trough plants, all current commercial Solar Thermal Power Plants (STPP) Fresnel collectors that are linear are becoming a more appealing alternative for generating power from solar radiation (M. Papaelias et al. 2018).

An improved version of the parabolic trough, linear Fresnel technology has the benefit of using flat reflectors, which are far less expensive than parabolic mirrors. More reflectors can be installed and utilized in compact plants. Fig. 115.2 illustrates the performance of linear Fresnel collectors.

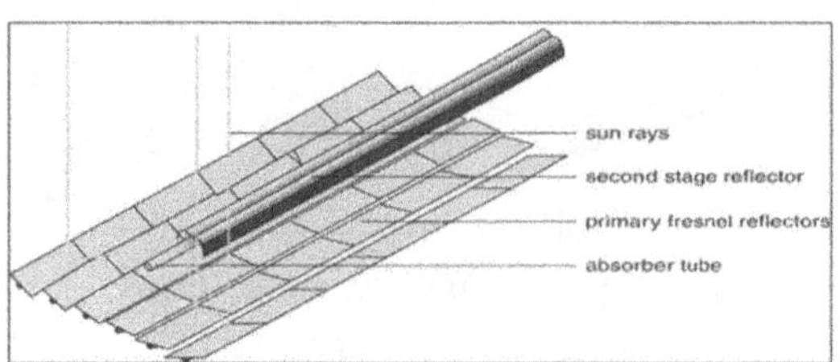

Fig. 115.2 Linear Fresnel Collectors

Long arrays of flat mirrors make up linear Fresnel reflectors, which focus light onto a linear receiver. The receiver is positioned above and along the reflector arrays on a tower that is typically 10 to 15 meters high. One or two axes can be used to mount the mirrors. Because the mirrors are flat and bendable, the Linear Fresnel reflectors are considerably less costly in design than Parabolic Trough Collectors (PTC).

Shadowing between adjacent reflectors is a major problem with linear Fresnel reflector systems. This problem can be solved by raising the reception tower's height or increasing the distance between mirrors, which requires more land, which raises costs. The advantages of linear Fresnel collectors include simplicity, robustness, and lower capital costs.

2.4 Central Receiver Systems

Compared to linear parabolic trough collectors, solar power tower technology is less developed economically. Nevertheless, various test sites have been established worldwide over the past 15 years. A field of heliostats that follow the sun and reflect its rays to a receiver perched atop a tall tower is a component of central receiver systems. When necessary, the energy is used to power a steam turbine. Heat is transferred from the receiver to a steam generator, which powers a turbine, via fluid. Air, liquid sodium, molten salts, or steam/water can all be used as the heat transfer medium. Extremely high temperatures can be reached by using compressed air or gas resulting in a significantly higher level of efficiency. The central receiver system is illustrated in Fig. 115.3.

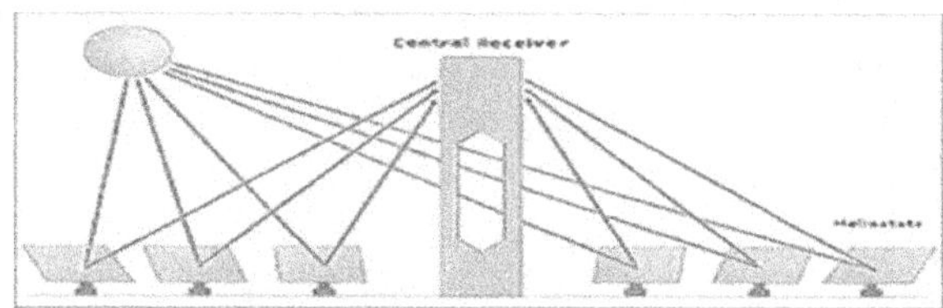

Fig. 115.3 Central receiver systems

The storage system ensures that energy is available even when solar energy is not directly accessible. Cloud cover can have an impact on the energy produced by thermal solar arrays, which are not always reliant on sunshine hours. Molten salts or other fluids can be heated by the central receiver, which also stores energy.

2.5 Parabolic Dish Engine

Parabolic dish concentrators are compact devices with a motor generator at the reflector's focal point. A stirling engine or a gas turbine with a power output of 5 to 25 kW powers the motor generating unit. The stirling engine uses hydrogen or helium, which is heated up to 750°C, utilizing solar operations for base loads with greater collector fields and thermal energy storage. Fig. 115.4 illustrates the parabolic dish engine under discussion.

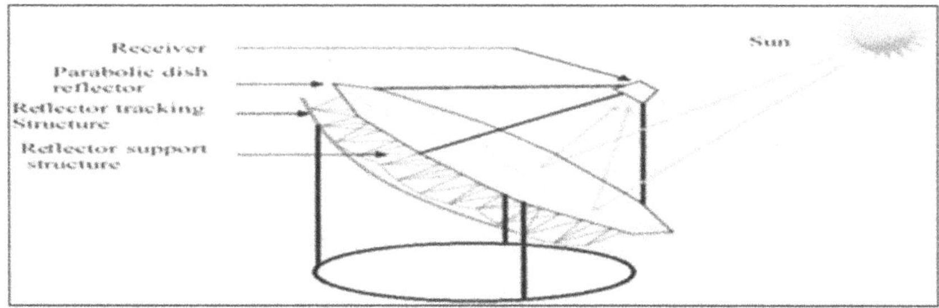

Fig. 115.4 Parabolic dish engine

The Stirling engine transforms the solar heat into mechanical energy, which is then transformed into electrical power by a generator that is attached to the engine. The concentrator must be correctly pointed toward the sun in order to function at its best. As a result, it is installed on a two-axis tracking system that enables both vertical and horizontal concentrator adjustment. Sensors that follow the sun or specialized software that continuously determines the sun's location regulate the alignment with the sun. In Fig. 115.4 demonstrates that the parabolic dish engine that has been examined.

3. CONCLUSION

In this paper, we have discussed and compare various types of CSP and thermal energy storage (TES) techniques. The significance of TES to promote a more sustainable renewable energy technology.

REFERENCES

1. Ma, Zhiwen, and Martinek J. (2024). Integration of Concentrating Solar Power with High Temperature Electrolysis for Hydrogen Production. International Conference on Concentrating Solar Power, Thermal, and Chemical Energy Systems. Solar Fuels and Chemical Commodities. National Renewable Energy Laboratory. 2 (2023)
2. Pascual, S., Lisbona, P. and Romeo, L.M. (2022). Thermal Energy Storage in Concentrating Solar Power Plants: A Review of European and North American R&D Projects. Energies 2022, 15 (22).
3. Alnaimat, F. and Rashid, Y. (2019). Thermal Energy Storage in Solar Power Plants: A Review of the Materials, Associated Limitations, and Proposed Solutions. Energies 2019, 12 (21), 4164.
4. Gentile, G., Manzolini, G. (2022). The Value of CSP with Thermal Energy Storage in Providing Flexible Electric Power. In: Sayigh, A. (eds) Sustainable Energy Development and Innovation. Innovative Renewable Energy. Springer, Cham.
5. Codd, D.S., Gil, A., Manzoor, M.T. et al. (2020). Concentrating Solar Power (CSP) Thermal Energy Storage (TES) Advanced Concept Development and Demonstrations. Curr Sustainable Renewable Energy Rep 7, 17–27.
6. Barrasso, M., Langella, G., Amoresano, A. and Iodice, P. (2023). Latest Advances in Thermal Energy Storage for Solar Plants. Processes. 11 (6), 1832.
7. Solar Thermal Energy Storage and Heat Transfer Media | Department of Energy
8. Jayathunga, D., Weliwita, J.A., Karunathilake, H., and Witharana, S. (2023). Economic Feasibility of Thermal Energy Storage-Integrated Concentrating Solar Power Plants. Solar 2023, 3, 132-160.
9. Kurup, Parthiv, Akar S., Glynn S., Augustine C., and Davenport P. (2022). Cost Update: Commercial and Advanced Heliostat Collectors. Golden, CO: National Renewable Energy Laboratory, 2022a.
10. Imran Khan, M., Asfand, F., & Al-Ghamdi, S. G. (2022). Progress in research and technological advancements of commercial concentrated solar thermal power plants. Solar Energy, 249, 183-226.
11. Xu, Pei, J., Yuan and J. et al. (2022). Concentrated solar power: technology, economy analysis, and policy implications in China. Environ Sci Pollut Res 29, 1324-1337.
12. Ferruzzi, G., Delcea, C., Barberi, A., Di Dio, V., Di Somma, M., Catrini, P., Guarino, S., Rossi, F., Parisi, M.L. and Sinicropi, A. et al. (2023). Concentrating Solar Power: The State of the Art, Research Gaps and Future Perspectives. Energies 2023, 16, 8082.
13. Ali O M Maka and Jamal M Alabid (2022). Solar energy technology and its roles in sustainable development, Clean Energy. 6(3) 476-483.
14. Papaelias, M., García Márquez, F.P., Ramirez, I.S. (2018). Concentrated Solar Power: Present and Future. Renewable Energies. Springer, Cham.

Note: All the figures in this chapter were made by the authors.

Recent Trends in Applied Physics and Material Science – Sudhir Bhardwaj et al. (eds)
© 2026 Taylor & Francis Group, London, ISBN 978-1-041-16452-4

Comparative Analysis of Different Versions of 'Area Method' in Thermoluminescence Glow Curve Analysis Considering Temperature Dependent Frequency Factor

Tusar Subhra Sarkar[1]
Department of Physics, Jadavpur University, Kolkata, India
Department of Physics, Sonarpur Mahavidyalaya, Rajpur, Kolkata, India

Soumya Sarkar[2]
Department of Physics, Jadavpur University, Kolkata, India

Mohan Kundu[3]
Department of Physics, Jadavpur University, Kolkata, India
Sadhanpur Uludanga Tulsiram High School (H.S), 24 Parganas (N), West Bengal, India

Sukhamoy Bhattacharyya[4]
Department of Physics, Jadavpur University, Kolkata, India

Abstract: 'Area Method' is one of the widely used techniques in thermoluminescence (TL) glow curve analysis to extract different trapping parameters of phosphors. The 'Area method' utilizes different portions of the glow curve to estimate the values of trapping parameters. In this study, we have used different versions of 'Area Method' on TL glow curves simulated in General Order Kinetics (GOK) and One Trap One Recombination center (OTOR) model considering temperature-dependent frequency factor (TDFF) under quasi-equilibrium approximations in linear heating scheme. Activation energy (E) is estimated for different values of temperature exponent (*a*) using the above mentioned models and a comparative study across different versions of 'Area Method' for simulated as well as experimental curves are also reported.

Keywords: Thermoluminescence, Temperature-dependent frequency factor, GOK model, OTOR model, Area methods

1. INTRODUCTION

Thermoluminescence (TL) occurs in insulating materials with localized energy levels within the band gap. When the materials exposed to ionizing radiation like X-rays or gamma rays, electrons and holes are generated in the conduction and valence bands respectively. The electrons become trapped in defect centres and holes migrate to luminescence centres (McKeever, 1985). Luminescence is emitted when trapped charges are thermally released and recombine at luminescence or hole centres. This TL process is analysed using models like the GOK model, OTOR model etc. A generalized temperature-dependent frequency factor (TDFF) is expressed as $s(T) = s_0 T^a$, where s_0 is the temperature independent component and a is a temperature exponent ranging between -2 and 2 (Keating, 1961, Chen, 1969, Fleming, 1990). Traditional methods usually assume $a = 0$ which will cause errors to estimate different parameters like activation energy. To address this, studies (Keating, 1961, Fleming, 1990) have incorporated TDFF into models like GOK, which still has limitations, prompting interest in band theory based models like OTOR model.

Conventional methods like peak shape method (Chen, 1969), initial rise method (Garlick and Gibson, 1948),

[1]thetusar@gmail.com, [2]smyasarkar@gmail.com, [3]mohanelsc@gmail.com, [4]sukhamoyb.physics@jadavpuruniversity.in

DOI: 10.1201/9781003684718-116

various heating rate method (Chen and Winer, 1970) etc., analyse limited sections of the TL glow curve which will often cause inaccuracies. In the other hand 'Area method' examines a larger portion of the glow curve, gives precise estimation of parameters like activation energy (E), frequency factor (s), and order of kinetics (b) (Chen and McKeever, 1997). Kirsh (1992) introduced a three-point method using coordinates and partial areas to estimate E and b simultaneously. Based on Moharil's approach (1982), Rasheedy (1993) developed a method for determining parameters of complex TL curves. The Three Point Area (TPA) method (Kundu et al., 2021) which is basically the extension of Kirsh's method, uses areas under TL peaks for selected points, providing more flexibility in analysis.

This study aims to employ different versions of 'Area Method' for analysing TL glow curves generated using GOK and OTOR model. By applying this method to both simulated and experimental data, we demonstrate its usefulness in determining various trapping parameters. Our work highlights the applicability of different versions of 'Area Method' and carries out a comparative study of the outcomes of these methods for different portions of the glow peak.

2. METHODOLOGY

The methods used for thermoluminescence (TL) analysis primarily rely on the General Order Kinetics (GOK) framework. The 'Area Method' calculates initial filled trap density (n_0) from the total area under the TL curve.

According to Fig. 116.1(a), $n_0 \propto A_0 = \int_{T_i}^{T_f} I(T)\,dT$ (shaded region) and trap density (n_1) at some temperature T_1 can be estimated using $n_1 \propto A_1 = \int_{T_1}^{T_f} I(T)\,dT$ (striped region). The simulated glow curve under the GOK model considering TDFF is generated from Ref. (Chen, 1969, May and Partridge, 1964). In OTOR model a single type of trap and recombination center are assumed and the simulated curves are generated using intensity expression given in literature (Sunta, 2014). The value of s_0 is calculated using maxima condition by setting $\frac{dI(T)}{dT} = 0$ for all TL models, where $I(T)$ represents TL intensity of the glow peak at any temperature T.

2.1 Method Proposed by Kirsh (Kirsh Method)

Using the natural logarithmic form of TL intensity (Chen, 1969), we can write

$$\ln I = -\frac{E}{kT} + b\ln\left(\frac{n}{n_0}\right) + \ln(sn_0) \tag{1}$$

Considering two different points (T_1, I_1) and (T_2, I_2) on the glow peak we obtain

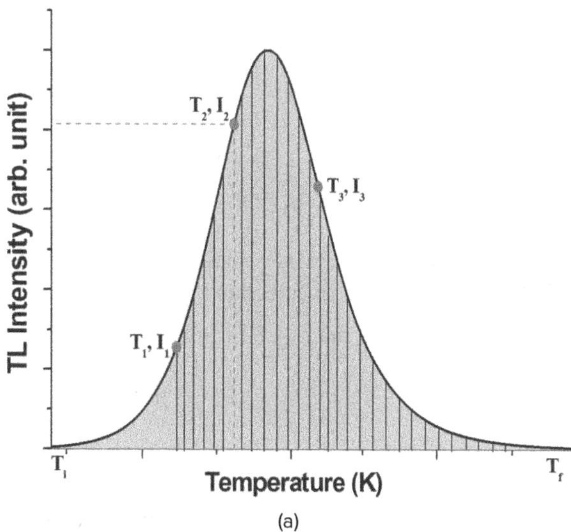

(a)

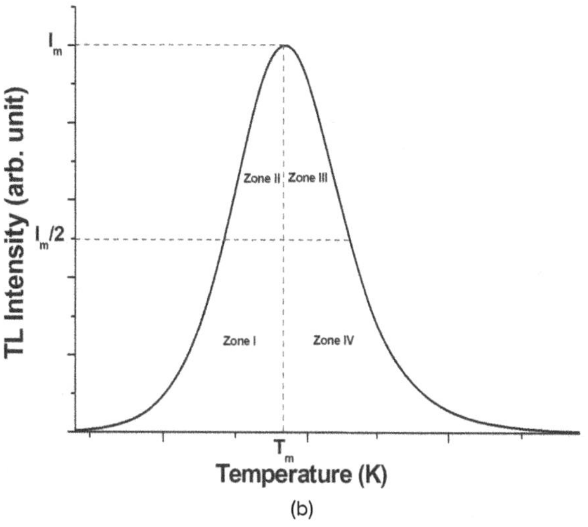

(b)

Fig. 116.1 (a) Schematic diagram of a TL glow peak, where the shaded area shows the area under the entire TL curve and the striped region shows the partial area. (b) Different zones in glow curve

$$\ln I_2 - \ln I_1 = -\frac{E}{k}\left(\frac{1}{T_2} - \frac{1}{T_1}\right) + b\ln\left(\frac{n_2}{n_1}\right) \Rightarrow$$

$$\Delta\ln I - b\Delta\ln\left(\frac{n}{n_0}\right) = -\frac{E}{k}\Delta\left(\frac{1}{T}\right) \tag{2}$$

Δ represents the difference between the corresponding values at T_1 and T_2. A plot of $\Delta\ln\left(\frac{n}{n_0}\right)$ against $\Delta(1/T)$ provides the values of E and b.

2.2 Method Proposed by Rasheedy (Rasheedy Method)

By analyzing areas under the glow curve and using relationships between intensities at different points considering Equation (1), the activation energy E is derived as

$$E = \left\{(\ln y) - b\ln\left(\frac{A_1}{A_2}\right)\right\}\frac{kT_1T_2}{T_1 - T_2} \quad \text{and}$$

$$E = \left\{(\ln z) - b\ln\left(\frac{A_1}{A_3}\right)\right\}\frac{kT_1T_3}{T_1 - T_3} \tag{3}$$

Where $I_2 = I_1/y$ and $I_3 = I_1/z$. By equating these expressions for E, the value of b can be calculated by

$$b = \frac{T_2(T_1 - T_3)\ln(y) - T_3(T_1 - T_2)\ln(z)}{T_2(T_1 - T_3)\ln\left(\frac{A_1}{A_2}\right) - T_3(T_1 - T_2)\ln\left(\frac{A_1}{A_3}\right)} \tag{4}$$

2.3 TPA Method

An extension of the Kirsh method, this technique involves deriving E and b using intensity values at different points and given by

$$E = -k\frac{Y_{12} - Y_{23}}{X_{12} - X_{23}}, \quad b = \frac{X_{23}Y_{12} - X_{12}Y_{23}}{X_{12} - X_{23}} \tag{5}$$

where $Y_{12} = b - \frac{E}{k}X_{12}$, $Y_{23} = b - \frac{E}{k}X_{23}$

with $X_{12} = \dfrac{\frac{1}{T_2} - \frac{1}{T_1}}{\ln\frac{n_2}{n_0} - \ln\frac{n_1}{n_0}}$, $X_{23} = \dfrac{\frac{1}{T_3} - \frac{1}{T_2}}{\ln\frac{n_3}{n_0} - \ln\frac{n_2}{n_0}}$.

3. RESULTS AND DISCUSSION

Activation energies are evaluated using different versions of 'Area Method' and compared across all the models. A large number of glow curves are simulated by varying system parameters, with activation energies from 0.5 to 2.0 eV. The temperature exponent a is varied between –2 and +2, with linear heating rate 1 K/s. Identical TL curves are reproduced for different a values by incorporating a correction term akT_m in the activation energy (E_0) and expressed as $E_{in} = E_0 - akT_m$ (Kundu, 2022).

For the GOK model, glow curves are generated by varying b from 1 to 2. The peak is divided into four distinct regions and the zones are shown in Fig. 116.1(b). Mixed zones, formed by combining these four regions, are analysed for detailed comparison. Zones I and II yield more precise values of activation energies compared to Zones III and IV. Mixed zones incorporating Zones I and II provide

Table 116.1 Comparison of different versions of 'Area Method' for glow peaks simulated in OTOR model. Here $E_{in} = E_0 - akT_m$, $E_0 = 1$ eV, $n_0/N = 0.5$ and heating rate $\beta = 1.0$ K/s

Input Parameters					Output values of activation energies (E_{out} in eV)		
T_m(K)	a	$s_0(s^{-1}K^{-1})$	E_{in}(eV)	Region	Kirsh Method	Rasheedy Method	TPA Method
				Zone I	0.9936	0.9935	0.9935
				Zone II	0.9966	0.9966	0.9966
	2	9.20×10^5	0.9337	Zone III	0.9684	0.9685	0.9685
				Zone IV	0.9702	0.9702	0.9702
				Mixed Zone	0.9952	0.9954	0.9954
				Zone I	0.9968	0.9968	0.9968
				Zone II	0.9983	0.9983	0.9983
	1	9.57×10^8	0.9669	Zone III	0.9842	0.9843	0.9843
				Zone IV	0.9851	0.9851	0.9851
				Mixed Zone	0.9976	0.9977	0.9977
				Zone I	1.0000	1.0000	1.0000
				Zone II	1.0000	1.0000	1.0000
384.56	0	1.00×10^{12}	1	Zone III	1.0000	1.0000	1.0000
				Zone IV	1.0000	1.0000	1.0000
				Mixed Zone	1.0000	1.0000	1.0000
				Zone I	1.0032	1.0032	1.0032
				Zone II	1.0017	1.0017	1.0017
	-1	1.05×10^{15}	1.0331	Zone III	1.0158	1.0157	1.0157
				Zone IV	1.0149	1.0149	1.0149
				Mixed Zone	1.0024	1.0023	1.0023
				Zone I	1.0064	1.0064	1.0064
				Zone II	1.0033	1.0034	1.0034
	-2	1.09×10^{18}	1.0663	Zone III	1.0316	1.0315	1.0315
				Zone IV	1.0298	1.0297	1.0297
				Mixed Zone	1.0048	1.0046	1.0046

Table 116.2 Comparison of output values of activation energies (with errors) estimated using different versions of 'Area Method' of experimental glow peak reproduced for material ($Sr_2P_2O_7$:Cu,Pr) (Yazici et al., 2010) using GOK model

Input Parameters for simulation						Output values of activation energies (in)		
Peak	T_m(K)	*a*	E_{in}(eV)	*b*	$s_0(s^{-1}K^{-1})$	Kirsh Method	Rasheedy Method	TPA Method
		2	0.8695		4.57×10^5	0.9280 (6.7%)	0.9282 (6.7%)	0.9282 (6.7%)
		1	0.9013		4.58×10^8	0.9305 (3.2%)	0.9306 (3.2%)	0.9306 (3.2%)
1	368.52	0	0.9330	1.1	4.59×10^{11}	0.9329 (0.0%)	0.9329 (0.0%)	0.9329 (0.0%)
		-1	0.9647		4.59×10^{14}	0.9354 (-3.0%)	0.9353 (-3.1%)	0.9353 (-3.1%)
		-2	0.9965		4.6×10^{17}	0.9378 (-5.9%)	0.9376 (-5.9%)	0.9376 (-5.9%)

even greater accuracy than other combinations. Results are consistent across all the methods for fixed *a* values.

Although the OTOR model does not explicitly involve *b*, a similar approach is applied. First-order ($b \rightarrow 1$) and second-order kinetics ($b = 2$) were simulated in the OTOR model by setting $R = 0$ and $R = 1$ respectively, with intermediate *R* values for general-order kinetics. Where *R* is retraping to recombination ratio. Table 116.1 presents sample output values for OTOR model. Zones I and II consistently yield mostly accurate activation energies for fixed value of *a*, aligning with findings from the GOK model results.

Accurate activation energy estimation depends on selecting appropriate data point set. We have studied different data point sets from a particular zone including mixed zone and we have noted the change in the output values from different versions of 'Area Method'. The observed difference is minimal, measuring less than 1%.

The methods were also applied to experimental curves of $Sr_2P_2O_7$: Cu, Pr reported by Yazici et al. (2010). For peak 1 ($b = 1.1$) the reported value of activation energy was 0.993 eV. We have reproduced the glow peak considering the reported data using GOK intensity expression. The output values of activation energies for $a = 0$ give quite satisfactory values with the reported data for all area methods and the fractional errors estimated for all area methods are within 7% for different values of a, as reported in Table 116.2.

4. CONCLUSION

In this study, we have analysed thermoluminescence glow curves generated using the OTOR and GOK models incorporating Temperature Dependent Frequency Factor. Activation energies are evaluated through various 'Area Method', revealing consistent accuracy across methods, particularly in zones I and II. Mixed zone analysis further enhanced precision, and results are validated with experimental data, showing minimal fractional error. This study also highlights the importance of selecting appropriate data point set for estimating activation energies. The methods applied are effective in both simulated and experimental contexts, demonstrating the robustness and applicability of the proposed approach for TL analysis.

5. ACKNOWLEDGEMENTS

The authors acknowledge the financial support from UGC-DAE CSR through a collaborative Research Scheme (CRS) project number CRS/2022-23/02/833.

REFERENCES

1. Chen, R. (1969). Glow curves with general order kinetics. J. Electrochem. Soc. 116(9):1254.
2. Chen, R., and Winer, S. (1970). Effects of various heating rates on glow curves. J. Appl. Phys. 41(13):5227-5232.
3. Chen, R., and McKeever, S. W. (1997). *Theory of thermoluminescence and related phenomena.* World Scientific Publishing Company.
4. Fleming, R. (1990). Activation energies and temperature-dependent frequency factors in thermally stimulated luminescence. J. Phys. D: Appl. Phys. 23(7):950.
5. Garlick, G., and Gibson, A. (1948). The electron trap mechanism of luminescence in sulphide and silicate phosphors. Proc. Phys. Soc. 60(6):574.
6. Keating, P. (1961). Thermally stimulated emission and conductivity peaks in the case of temperature-dependent trapping cross sections. Proc. Phys. Soc. 78(6):1408.
7. Kirsh, Y. (1992). Kinetic analysis of thermoluminescence. Phys. Status Solidi A. 129(1):15-48.
8. Kundu, M., Bhattacharyya, S., Karmakar, M., and Majumdar, P. (2021). Three-point area method for thermoluminescence glow curve analysis and its application to the glow peak of $K_2SrP_2O_7$:Pr. Radiat. Prot. Dosim. 193(3-4):247-258.
9. Kundu, M., Chakrabarty, S., Bhattacharyya, S., and Majumdar, P. (2022). Thermoluminescence glow curve analysis using temperature-dependent frequency factor in OTOR model. Radiat. Meas. 156:106820.
10. May, C., and Partridge, J. (1964). Thermoluminescent kinetics of alpha-irradiated alkali halides. J. Chem. Phys. 40(5):1401-1409.
11. McKeever, S. W. (1985). *Thermoluminescence of solids.* Cambridge University Press.
12. Moharil, S. (1982). On the general-order kinetics in thermoluminescence. Phys. Status Solidi A. 73(2):509-514.
13. Rasheedy, M. S. (1993). On the general-order kinetics of the thermoluminescence glow peak. J. Phys.: Condens. Matter. 5(5):633.
14. Sunta, C. (2014). *Unraveling thermoluminescence.* Springer India.
15. Yazici, A. N., Seyyidoğlu, S., Toktamiş, H., and Yilmaz, A. (2010). Thermoluminescent properties of $Sr_2P_2O_7$ doped with copper and some rare earth elements. J. Lumin. 130(10):1744-1749.

Note: All the figures and tables in this chapter were made by the authors.

Unveiling the Potential of Tungsten-Lithium-Strontium-Bismuth-Borate Glasses

Shwetha M., Madhu A.*
Department of Physics, Dayananda Sagar College of Engineering, Kumaraswamy layout, Bengaluru, India

Srinatha N
Department of Physics, R V Institute of Technology and Management, Bengaluru, India

Abstract: Glasses based on lithium-strontium-bismuth-borate incorporating tungsten ions exhibit distinctive synthesis and thermal, structural, and physical properties. These glasses are effectively fabricated utilizing melt-quenching techniques, facilitating the integration of tungsten ions. Including tungsten substantially impacts the thermal properties and physical characteristics, enhancing the overall glass performance. XRD studies confirmed the amorphous characteristics of the prepared glasses. Additionally, physical and differential thermal analyses indicated increasing density, Refractive index, and glass transition temperature following the incorporation of tungsten. The prepared glasses exhibit an increase in basicity when there is a low level of electron donation, leading to higher chemical durability. Furthermore, the polarizability of the oxide ion is enhanced as the concentration of tungsten increases. This enhancement results in a greater formation of non-bridging oxygen (NBO) bonds, contributing to the overall properties of the glasses. The FTIR analysis indicates that the glass composition's OH bond is absent at 3300 cm^{-1}. This absence is expected to enhance the luminescence features of the fabricated present glass by preventing quenching effects. Furthermore, the FTIR results also disclose the existence of major borate and metal ions bonds in the composition. The differential scanning calorimetry (DSC) analysis indicates that the glass transition temperature falls between 410 to 430 degrees Celsius.

Keywords: HMO, Borate glass, FTIR, DSC

1. INTRODUCTION

The synthesis and properties of bismuth-based glasses, especially those doped with tungsten oxide (WO_3), have gained significant attention(Ali et al., 2023)(Wang et al., 2023). This review focuses on five key areas: thermal properties, optical behaviour, structural analysis, electrical characteristics, and their implications for various applications. Recent research shows that incorporating WO_3 improves thermal stability and increases glass transition temperatures (T_g), with higher WO_3 concentrations leading to enhanced network connectivity. These changes are crucial for applications in thermal insulation and high-temperature environments(Balueva et al., 2021). The optical properties of these glasses have also been extensively examined. WO_3 increases the refractive index while lowering optical basicity, which is important for optics and photonics. Additionally, tungsten enhances the emission spectra of Er^{3+} ions, benefiting fibre optic technologies. Structural analyses through X-ray diffraction (XRD) and Fourier Transform Infrared (FTIR) spectroscopy indicate that WO_3 modifies the glass's structural framework, improving thermal and mechanical properties(Çamiçi et al., 2023) 10, 15, 20, and 25 mol% (Iliyasu et al., 2023). Moreover, the electrical characteristics of WO_3-doped bismuth glasses show promise, with increased ionic conductivity and reduced activation energy making them suitable for sensors and capacitors(Pershina, 2023). This study explores how tungsten ions influence the physical, structural, thermal, and optical properties of lithium-strontium-bismuth-borate glass(El-Shamy et al., 2024). We examine changes

*Corresponding author: mmathi.33@gmail.com

DOI: 10.1201/9781003684718-117

in density, refractive index, and thermal stability to deepen our understanding of tungsten's interaction with the glass matrix to develop advanced materials with tailored properties.

2. FABRICATION AND EXPERIMENTAL TECHNIQUES USED

2.1 Fabrication

The study utilized H_3BO_3, Bi_2O_3, $SrCO_3$, Li2CO$_3$, and WO_3 from Merck, all with 99.9% purity. These materials were combined in stoichiometric ratios for a 12 g batch using https://batchcalculator.in/ and mixed thoroughly in an alumina crucible capable of withstanding temperatures up to 1300°C. The crucible was heated in a programmable furnace from room temperature to 1100°Cover 120 minutes, with stirring intervals at 10 minutes to ensure a uniform melt. The melt was then rapidly quenched into preheated brass molds and underwent annealing at 250°C for 3 hours. Finally, circular glass discs measuring 3 mm thick were polished for optical studies and finely ground for structural and thermal analyses. Glass composition: (60-x) B_2O_3 + 20Bi_2O_3 + 10SrO + 10Li_2O + xWO_3; where: x = 0, 1, 3 and 5 mole %.

2.2 Characterization Techniques

Before the characterization, we determined the density and refractive index of the glass samples using Archimedes' principle with toluene as the immersion liquid, measured by a digital ATAGO Abbe refractometer at 589.3 nm. To explore the samples' amorphous or crystalline nature, we performed X-ray diffraction (XRD) analysis with a RIGAKU ULTIMA IV instrument, utilizing Cu Kα radiation in the 10° to 80° 2θ range. Fourier Transform Infrared (FTIR) spectroscopy was conducted on the Thermo-Nicolet Avatar 370 to obtain spectra from 400 cm^{-1} to 4000 cm^{-1}. Additionally, differential scanning calorimetry (DSC) measurements were executed from room temperature to 500 °C using a Mettler Toledo DSC-1 instrument in a nitrogen atmosphere at a heating rate of 10 °C/min.

3. RESULTS AND DISCUSSION

3.1 Physical Parameters

The analysis showed that adding tungsten significantly increased the density and refractive index of the glass materials, with these enhancements directly related to the amount of tungsten incorporated. The glass samples also exhibited improved basicity when electron donation levels were kept low, suggesting that lower levels enhance chemical durability. Additionally, higher tungsten concentrations increased the polarizability of the oxide ion, indicating that tungsten influences the electronic structure of the glass matrix, affecting its optical and chemical properties. These findings highlight tungsten's role as an effective additive in glass formulation. A summary of the estimated using the formulas mentioned in our previous paper (A et al., 2023)and estimated physical parameters for the W series is provided in Table 117.1.

Table 117.1 Physical parameter

Parameters	W_0	W_1	W_3	W_5
Molecular weight (g/mol)	148.31	149.93	153.17	156.42
Density (g/cm^3)	4.47	4.71	4.52	4.58
Molecular volume (cm^3/mol)	33.17	31.79	33.88	34.10
Refractive Index	1.6545	1.655	1.6555	1.656
Optical Basicity	0.77	0.74	0.82	0.84
Electro Negativity	1.27	1.24	1.34	1.37
Electronic Oxide polarizability (Å^3)	1.85	1.80	1.96	2.02

3.2 Nature of Samples

The two distinct humps are evident in Fig. 117.1, particularly between 20 and 60 degrees. As the concentration of tungsten increases, these humps exhibit noticeable broadening, which may clarify the determination of the material's amorphous characteristics.

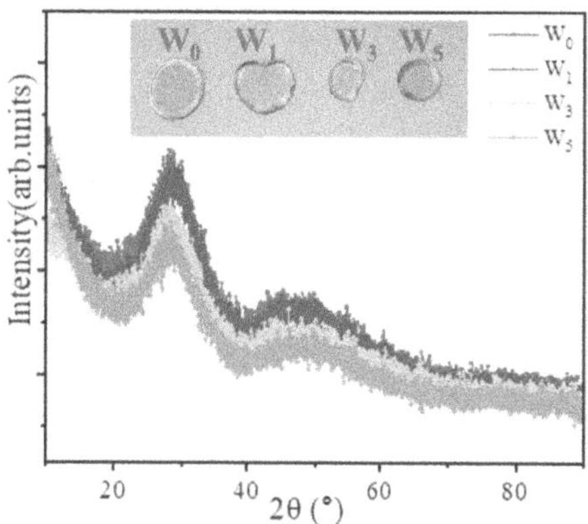

Fig. 117.1 X-ray diffraction profile of W series

3.3 Fourier-transform Infrared Spectrum Analysis

Figure 117.2 illustrates the variations in vibration modes of the sample with added Tungsten, spanning a spectral range of 400 to 4000 cm^{-1}. The Fourier-transform infrared (FTIR) spectrum shows three significant bands: one below 800 cm^{-1}, another between 800 and 1200 cm^{-1}, and the third from 1200 to 1600 cm^{-1}. The absence of the hydroxyl (OH) band around 3300 cm^{-1} indicates no water-related vibrations. The band below 800 cm^{-1} is linked to bending vibrations of borate segments influenced by Tungsten. The second band, below 1200 cm^{-1}, signifies the stretching of B to O bonds in tetrahedral (BO_4) structural units, while the

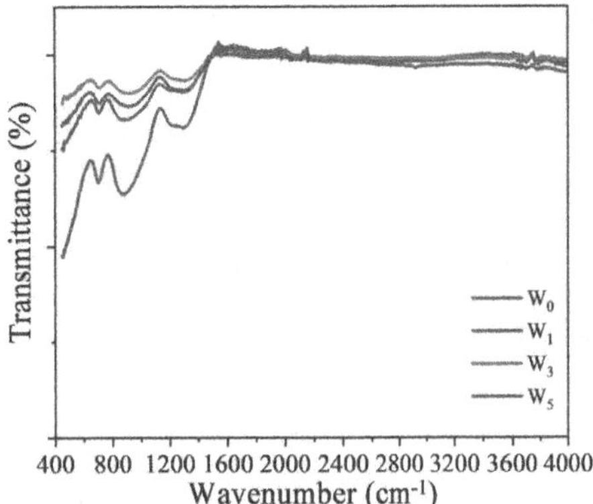

Fig. 117.2 FTIR profile of W series

band above 1200 cm^{-1} corresponds to stretching vibrations of B to O links in trigonal borate units, highlighting the sample's diverse interactions.

3.4 Glass Transition Temperature

The differential scanning calorimetric (DSC) profile for the W series, illustrated in Fig. 117.3, demonstrates a clear increase in the glass transition temperature (T_g) corresponding to higher concentrations of tungsten oxide (WO_3). The recorded T_g values of 411 °C, 424 °C, 428 °C, and 434 °C highlight a significant upward trend as the concentration of WO_3 rises.

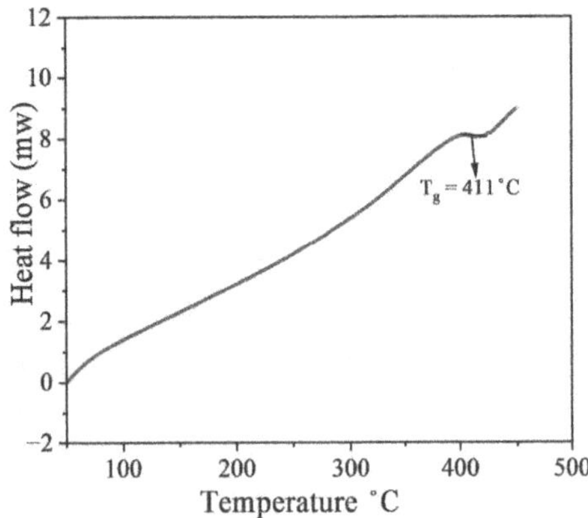

Fig. 117.3 Glass transition temperature of W_0 sample

4. CONCLUSION

Research into lithium-strontium-bismuth-borate glasses infused with tungsten ions has led to key advancements in their synthesis, structure, and physical properties. The melt-quenching techniques employed allow for effective tungsten incorporation, resulting in glasses with improved thermal stability and basicity. Notable enhancements in density, refractive index, and glass transition temperature highlight tungsten's transformative role as an additive. The polarizability of the oxide ion is enhanced with increased tungsten concentration, influencing optical and chemical properties. The absence of hydroxyl (OH) bonds, confirmed by FTIR analysis, improves luminescence by reducing quenching effects. Additionally, tungsten modifies the glass network, as seen in X-ray diffraction and FTIR analysis, leading to the formation of non-bridging oxygen and improved structural integrity. In summary, tungsten ions optimize the properties of the glasses, providing exciting opportunities for future research and applications in advanced materials science.

REFERENCES

1. A, M., Al-Dossari, M., EL-Gawaad, N. S. A., Gelija, D., Ganesha, K. N., & Srinatha, N. (2023). Structural, optical and luminescence properties of Nd3+ ions in B2O3+SiO2+TeO2+Na2O glasses. *Optical Materials*, *136*, 113436. https://doi.org/10.1016/j.optmat.2023.113436
2. Ali, A. A., Fathi, A. M., & Ibrahim, S. (2023). Material characteristics of WO3/Bi2O3 substitution on the thermal, structural, and electrical properties of lithium calcium borate glasses. *Applied Physics A*, *129*(4), 299. https://doi.org/10.1007/s00339-023-06537-w
3. Balueva, K. V., Kut'in, A. M., Plekhovich, A. D., Motorin, S. E., & Dorofeev, V. V. (2021). Thermophysical characterization of TeO2-WO3-Bi2O3 glasses for optical applications. *Journal of Non-Crystalline Solids*, *553*, 120465. https://doi.org/10.1016/j.jnoncrysol.2020.120465
4. Çamiçi, H. C., Guérineau, T., Rivera, V. A. G., Falci, R. F., LaRochelle, S., & Messaddeq, Y. (2023). The role of tungsten oxide in Er3+-doped bismuth-germanate glasses for optical amplification in L-band. *Scientific Reports*, *13*(1), 8835. https://doi.org/10.1038/s41598-023-35995-8
5. El-Shamy, N. T., Mahrous, E. M., Alghamdi, S. K., Tommalieh, M. J., Abomostafa, H. M., Abulyazied, D. E., Rabiea, E. A., Abouhaswa, A. S., Ismael, A., & Mohaymen Taha, T. A. (2024). Synergistic effect of WO3 on structural, physical, optical, and dielectric characteristics of multicomponent borate glasses for optoelectronic applications. *Journal of Materials Research and Technology*, *30*, 7927–7937. https://doi.org/10.1016/j.jmrt.2024.05.164
6. Iliyasu, U., Mohd Sanusi, M. S., & Ahmad, N. E. (2023). The effect of WO3 on the optical and radiation shielding properties of zinc-lead-borate glass. *Radiation Physics and Chemistry*, *209*, 111007. https://doi.org/10.1016/j.radphyschem.2023.111007
7. Pershina, S. V. (2023). Structural, thermal, and DC conductivity properties of WO 3 –P 2 O 5 glasses doped with B 2 O 3. *International Journal of Applied Glass Science*, *14*(2), 247–255. https://doi.org/10.1111/ijag.16620
8. Wang, B., Qiu, T., Yuan, L., Fang, Q., Wang, X., Guo, X., Zhang, D., Lai, C., Wang, Q., & Liu, Y. (2023). A comparative study between pure bismuth/tungsten and the bismuth tungsten oxide for flexible shielding of gamma/X rays. *Radiation Physics and Chemistry*, *208*, 110906.

Note: All the figures and tables in this chapter were made by the authors.

Recent Trends in Applied Physics and Material Science – Sudhir Bhardwaj et al. (eds)
 ISBN 978-1-041-16452-4

118 Solvent-Dependent Optical Properties and Photoluminescence Behaviour of 1,3,5-Triphenylbenzene

Aryan, Balwan Singh, Banit Kumar
Department of EEE, Guru Jambheshwar University of Science & Technology, Hisar, India

Satya Dev
Department of Physics, Chaudhary Ranbir Singh Univeristy, Jind, India

Vinita
Department of EEE, Guru Jambheshwar University of Science & Technology, Hisar, India

Pardeep Kumar
Haryana Education Department, Chandigarh, Haryana, India

Vinod Kumar, Sardul Singh Dhayal*
Department of EEE, Guru Jambheshwar University of Science & Technology, Hisar, India

Abstract: This study examines the effect of some solvent environments as well as solvent interactions on the absorption as well as the photoluminescence (PL) properties of the 1,3,5-Triphenylbenzene (TPB) using UV-Vis spectroscopy and PL analysis. Certain solvents which include the polar aprotic DMF and acetonitrile protic solvent like ethanol and non-polar solvent toluene were evaluated. The UV-Vis absorption spectra show absorption peaks in the range 300 nm to 316 nm as well as optical bandgaps attributable to TPB also lies around 3.65 eV. It was observed that the PL spectra was solvent-dependent in character, with two peaks of emissions recorded in the ethanol and only one emission peak in the toluene solvent. The excitation wavelength dependence of PL intensity was explored, showing that TPB demonstrates an excitation dependent PL behaviour in polar solvents and a stable emission in non-polar toluene. The single and stable emission for toluene was observed at 368 nm, eliminating the excitation dependent behaviour. The CIE color coordinates also enabled the confirmation of blue emission of TPB which certainly enables its application for both OLEDs and white LEDs. The results of the findings from this study further suggest that TPB may be used in the manufacture of optoelectronic devices and thus expounds on the stability, solvent interactions and the applicability of the material in modern light emitting technologies.

Keywords: Organic semiconductors, Photoluminescence (PL), Solvent effects, 1,3,5-Triphenylbenzene (TPB)

1. INTRODUCTION

After years of development, several outstanding performance organic semiconductors have been discovered and found that certain organic electronic materials' overall qualities have even surpassed those of polycrystalline silicon ("Organic Semiconductor - an Overview | ScienceDirect Topics" 2024). Organic semiconductors have distinct advantages over their inorganic counterparts, such as inherent flexibility, low weight, solution processing, and good biocompatibility. These advantages are demonstrated by organic flexible displays, organic circuits, organic sensors, and organic batteries, among other optoelectronic device applications (Ravikumar and Dangate 2024; Kaliramna et al. 2024; Singh Dhayal et al. 2023).

In recent years, conjugated oligomers and polymers have emerged as one of the most significant types of innovative materials. Organic materials offer low-cost processing applications, leading to substantial research in this field compared to inorganic materials.

Soluble conjugated polymers have the potential to serve as the foundation for a new large area electronics technology

*Corresponding author: sardulsingh@gmail.com

DOI: 10.1201/9781003684718-118

since they are easily processed over vast areas by spin coating, printing, and other techniques (Brandão et al. 2014). These materials offer unique shapes and features that may be tailored through molecular design, doping, makes them highly advantageous ("Doping (Materials) - an Overview | ScienceDirect Topics" 2024; Dev et al. 2020; Kumar et al. 2021). Significant advancements have recently been made in fields including nonlinear optical devices, thin-film transistors, light-emitting diodes, photovoltaic cells, and other optoelectronic devices (Singh Dhayal et al. 2023).

Small molecule fluorophores are employed in a broad variety of applications, including biophysics, molecular biology, medicine, and material sciences (Islam et al. 2025). Researchers are increasingly interested in organic materials because of its widespread applications in optoelectronic devices like as OFET and OLED (Kaliramna et al. 2024; Dhayal et al. 2023; Xu et al. 2016; Singh, Nain, and Kumar 2022), and studying their behaviour in various solvents is critical to understanding their structural, optical as well as electrical properties. The investigation of a solvent's effects on a solute's structure and spectroscopic behaviour is also crucial for the progress of solution chemistry (Mabesoone, Palmans, and Meijer 2020). The optical bandgap and Stokes shift are key characteristics to investigate. The energy difference between the absorption maximum and the lower-energy emission is known as Stokes shift.

In this work, the influence of several polar and non-polar solvents namely; acetonitrile, ethanol, methanol, toluene and DMF are investigated. The absorption spectra of Triphenylbenzene (TPB) in different solvents and its corresponding photoluminescence emission is detected. The Stokes' shift is seen in various solvents and CIE colour coordinates reveal the emission color for solvents.

2. MATERIALS AND METHOD

1,3,5-Triphenylbenzene (TPB) and all solvents were procured from Sigma Aldrich. A 5 mg sample of TPB was mixed with 10 mL of solvent and sonicated for 5 minutes to ensure a uniform solution. The mixture was observed to be completely mixed, resulting in a mostly transparent solution. For absorption and PL spectroscopy, the solution was diluted with an additional 10% of the prepared solvent. To eliminate the effect due to solvent the same solvent it taken as reference and baseline is created. The structure of TPB is shown in Fig. 118.1. The UV-visible was done using UV-1800/Shimadzu and for Photoluminiscence spectroscopy spectra Horiba PL was used.

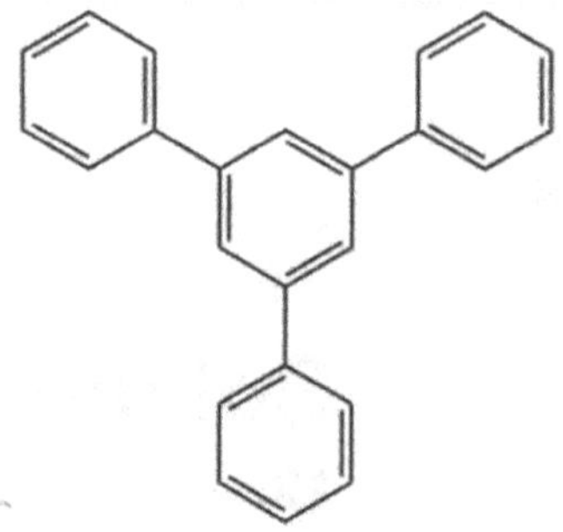

Fig. 118.1 Structure of 1,3,5-Triphenylbenzene (TPB) molecule

3. RESULT AND DISCUSSION

The influence of various solvent environments on the absorption properties of TPB was investigated using UV-Vis spectroscopy. This study included both polar solvents comprising aprotic solvents such as DMF and acetonitrile, along with protic solvents like ethanol as well as non-polar solvents, specifically toluene. The Tauc's model was used to calculate the optical bandgap (Eg) of molecule given by equation

$$(\alpha h\upsilon)n = k(h\upsilon - Eg) \tag{1}$$

Where α is absorption coefficient, hυ is incident photon energy, k is energy independent constant, Eg is band gap and n are nature of transition. In this case n is equal to 2 for direct band gap. The absorption spectrum of TBP in ethanol is shown in Fig. 118.2(a), and it show characteristic peaks at 300 nm with a shoulder peak at 250 nm, the two absorptions are mostly due to availability of two possible transitions which are dominated by the TPB molecule and another due to the molecule-solvent interaction (TPB-ethanol (Li and Binnemans 2021). The corresponding bandgap for TPB in ethanol is 3.64 eV as shown in inset of Figure 2(a). The absorption spectrum of TBP in DMF is shown in Figure 2(b), and it show characteristic peaks at 312 nm while the inset show corresponding the optical bandgap of 3.72 eV. The aprotic solvent acetonitrile also show adjacent peaks at 316 nm and 230 nm, while for toluene the single absorption peak is observed at 306 nm, as shown in Fig. 118.2(c, d). Inset show the bandgap of 3.65 eV and 3.62 eV respectively. It can be observed from the Fig. 118.2(e) that characteristic peak due to TPB does not show significant change but the effect of solvent is dominant for ethanol and acetonitrile, while the DMF and toluene has no such effect.

The absorption spectrum of TPB was observed around 300 nm to 316 nm corresponding to the $\pi \rightarrow \pi^*$ and $n \rightarrow \pi^*$ electronic transitions from the ground state to a higher state. The effect of solvent can cause the slight shift in the absorption spectra indicating that the observed bands correspond to a more charged excited state (Amos and Burrows 1973). In general, the stability of the $n \rightarrow \pi^*$ states decrease as the π-system size increases. However, in ethanol solution, the $n \rightarrow \pi^*$ states are less stabilized compared to the ground state, resulting in a blueshift in transitions. Conversely, the $\pi \rightarrow \pi^*$ excited states exhibit the opposite trend, becoming less energetic as the π-system size increases. The solvent-induced shift in these

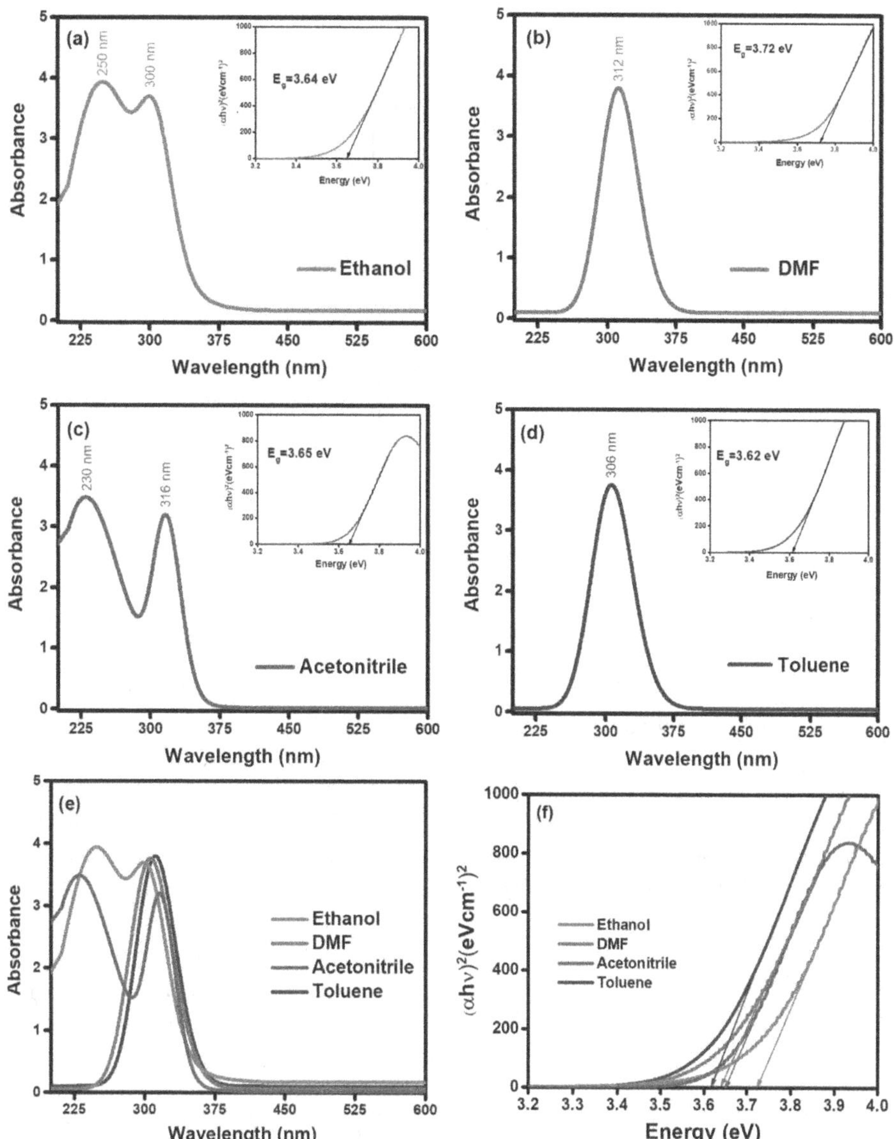

Fig. 118.2 Absorption spectra of TPB (inset shows Tauc's plot) in Ethanol (a), DMF (b), Acetonitrile (c), Toluene (d), overlap of absorption spectra (e), and Tauc's plot for various solvents (f)

states strongly depends on the size of the π-system and the presence of intramolecular hydrogen bonding, with the shift diminishing as the system size increases (Rančić et al. 2019). For bandgap there is no significant change and the bandgap lies in the range 3.64 eV to 3.72 eV, as shown in Fig. 118.2 (f). This range is most suitable for the optoelectronic devices mostly OLEDS and organic solar cells which is further studied using PL.

The photoluminescence (PL) spectra of TPB in various solvents were recorded using an excitation wavelength of 320 nm (3.87 eV). The photon energy at this wavelength is sufficient to excite the system effectively. The PL spectrum of TPB in ethanol shows two emission peaks at 363 nm and 459 nm and the corresponding bandgap is shown in the inset of Fig. 118.3(a). The maximum emission of DMF is observed at 368 nm, while for acetonitrile and toluene it is at 368 nm and 364 nm, as shown in Fig. 118.3(b-d). The loss of energy (or the difference in bandgap) is due to non-radiative transitions, shown in the inset of Fig. 118.3(b-d). Non-radiative decay processes in large molecules involve mechanisms such as electronic relaxation between two electronic states or unimolecular rearrangement reactions occurring in the excited electronic states. These processes enable the dissipation of energy without the emission of photons, contributing to the deactivation of the excited state.

The excitation wavelength plays a crucial role in the performance of practical optical devices. Therefore, the effect of excitation wavelength was studied to evaluate the stability of TPB in solvents across a broad range of excitation energies. The emission was observed for excitation wavelengths 300 nm to 330 nm, with increase in excitation wavelength the PL intensity increases, as shown in Fig. 118.4(a). In ethanol, the PL intensity increases for

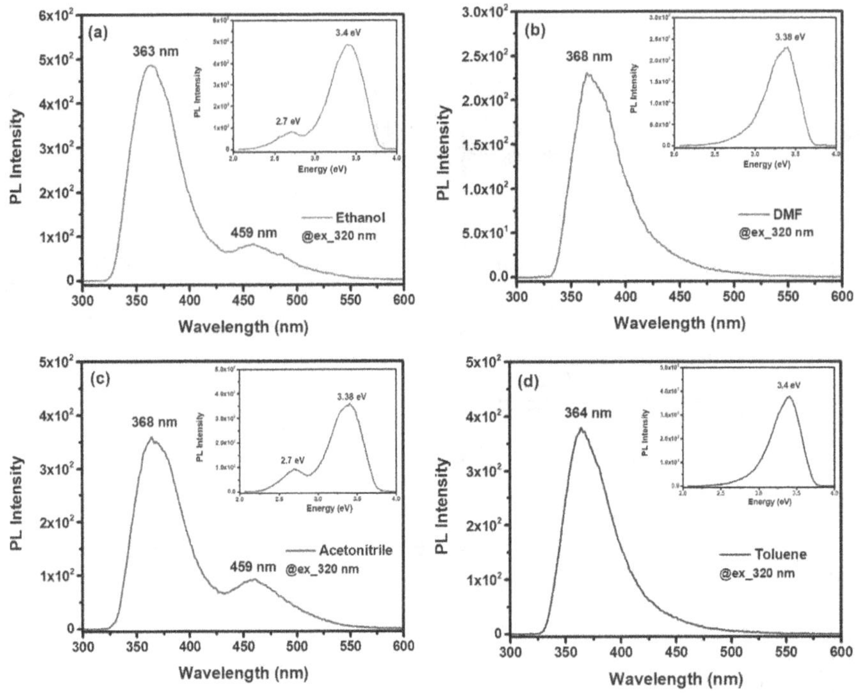

Fig. 118.3 PL of TPB (inset shows eV vs PL) in Ethanol (a), DMF (b), Acetonitrile (c), and Toluene (d)

an excitation wavelength of 310 nm, accompanied by a shift in wavelength and the loss of the characteristic peak. For DMF the gradual increase in the PL intensity was observed as shown in Fig. 118.4(b) along with slight shift after excited with 320 nm. For acetonitrile the same behaviour was observed as shown in Fig. 118.4(c), while for toluene almost no increase in PL intensity was observed from 300 nm to 310 nm. The TPB molecule doesn't show the excitation dependent PL for toluene and gives single emission at 364 nm (Kumsampao et al. 2022) as shown in Fig. 118.4(d), the single emission for range of excitement wavelength is highly useful for optoelectronic devices.

The PL spectra were overlapped to compare the PL intensity, as shown in Fig. 118.5(a). It was observed that, despite the effect of the solvent, the characteristic peak does not show significant differences. The normalized PL of TPB in various solvents is shown in Fig. 118.5(b) confirms the same trend. To further investigate its

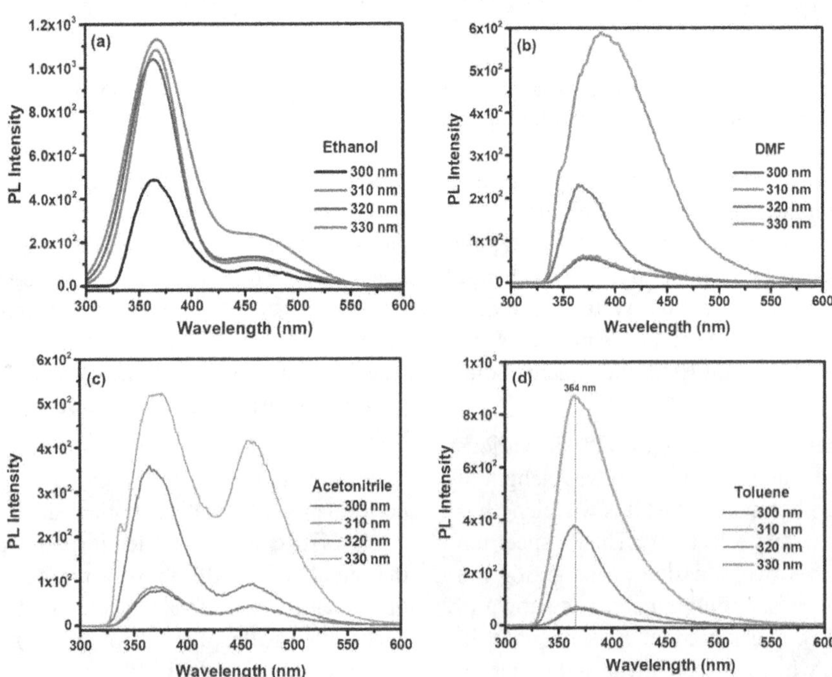

Fig. 118.4 PL spectra of TPB at different excitation wavelength for Ethanol (a), DMF (b), Acetonitrile (c), and Toluene (d)

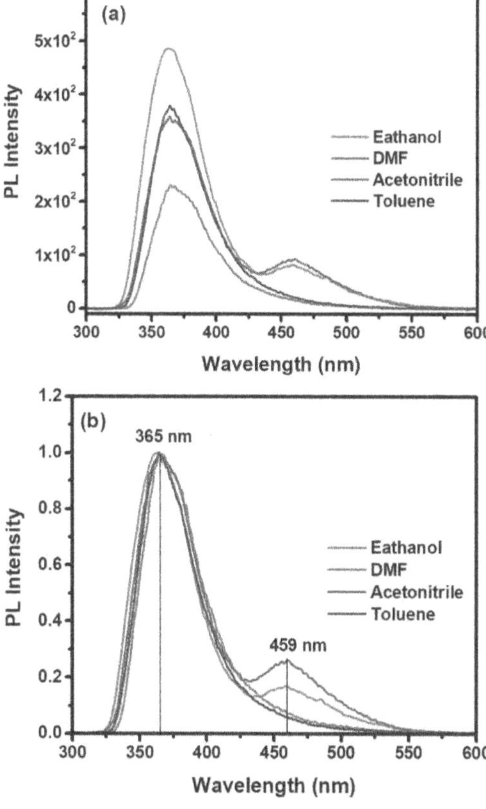

Fig. 118.5 Overlap of PL (a), Normalised PL (b) of TPB in various solvents, CIE (c) for TPB

potential application in OLEDs, the CIE (Commission Internationale de l'éclairage 1931) color coordinates were plotted, as shown in Fig. 118.5(c). The color coordinate diagram revealed that the TPB film emits blue light, with a dark blue hue observed at an excitation wavelength of 320 nm. White LEDs (WLEDs) typically incorporate blue or ultraviolet diodes, which are surrounded by phosphor converters that emit luminescence when stimulated by the diode. This combination of luminescence and the partial transmission of chip radiation results in the generation of white light (Chenna et al. 2023).

Stokes shift refers to the difference in wavelength (or energy) between the positions of the peak of the absorption spectrum and the peak of the emission spectrum of a molecule or atom (Subuddhi et al. 2006). This is also illustrated as an intra-molecular charge-transfer excitation of an electron from the HOMO to the LUMO is responsible for the observed high Stokes shifts (Horváth et al. 2015). Figure 118.6(a-d) illustrates the overlap of normalised absorption and PL for stokes shift. The maximum Stokes shift observed is 5785 cm-1 which is for ethanol and least Stokes' shifts for acetonitrile (Stopel, Blum, and Subramaniam 2014). The detailed comparison of absorption, Pl and the Stokes shifts are shown in Table 118.1. The Stokes shift is calculated in term of wavenumber ($\Delta\bar{\upsilon}$ (cm-1)) and also in term of wavelength ($\Delta\lambda$ (nm)).

Table 118.1 Optical parameters of TPB in various solvents

	Absorption		Photo luminescence		Stokes Shift	
	Abs (nm)	E_g (eV)	PL (nm)	E_g (eV)	$\Delta\bar{\upsilon}$ (cm^{-1})	$\Delta\lambda$ (nm)
Ethanol	250	3.72	363	3.4	5785	63
	300		459	2.7		
DMF	312	3.64	368	3.38	4877	56
Aceto nitrile	230	3.65	368	3.38	4471	52
	316		459	2.7		
Toluene	306	3.62	364	3.4	5207	58

4. CONCLUSION

The investigation was conducted to understand the solvent effects on the absorption and PL characteristics of the 1,3,5-Triphenylbenzene dissolved in a series of polar and non-polar solvents like ethanol, DMF, acetonitrile and Toluene. The UV-Vis spectra absorption showed TPB having features at certain wavelengths of each solvent and optical bandgaps and most of the solvents did not experience any change in the absorption behaviour except for the ethanol and acetonitrile. PL results showed that the production of emission was dependent on the solvent with two emissions being produced in the presence of ethanol, while non-polar solvents like Toluene maintained a constant emission. In addition, the photoluminescence intensity of TPB was relatively constant even when the excitation wavelength was changed indicating that TPB has good compatibility with different optical materials and excitation-emission behaviours of TPB were quite different in ethanol and DMF. The analysis through the stokes shift revealed the differences in energy losses for absorption and emission for different solvents with the most significant stoke shift being in the ethanol solvent. The CIE color coordinates ascertained the ability of TPB to act as a blue-emitting component that can be used when fabricating OLEDs and whites LEDs.

REFERENCES

1. Amos, A. T., and B. L. Burrows. 1973. "Solvent-Shift Effects on Electronic Spectra and Excited-State Dipole Moments and Polarizabilities." In Advances in Quantum Chemistry, edited by Per-Olov Löwdin, 7:289–313. Academic Press. doi:10.1016/S0065-3276(08)60566-3.
2. Brandão, Lúcia, Júlio Viana, David G. Bucknall, and Gabriel Bernardo. 2014. "Solventless Processing of Conjugated Polymers—A Review." Synthetic Metals 197 (November): 23–33. doi:10.1016/j.synthmet.2014.08.003.
3. Chenna, Praveen, Suman Gandi, Sujith Pookatt, and Saidi Reddy Parne. 2023. "Perovskite White Light Emitting Diodes: A Review." Materials Today Electronics 5 (September): 100057. doi:10.1016/j.mtelec.2023.100057.

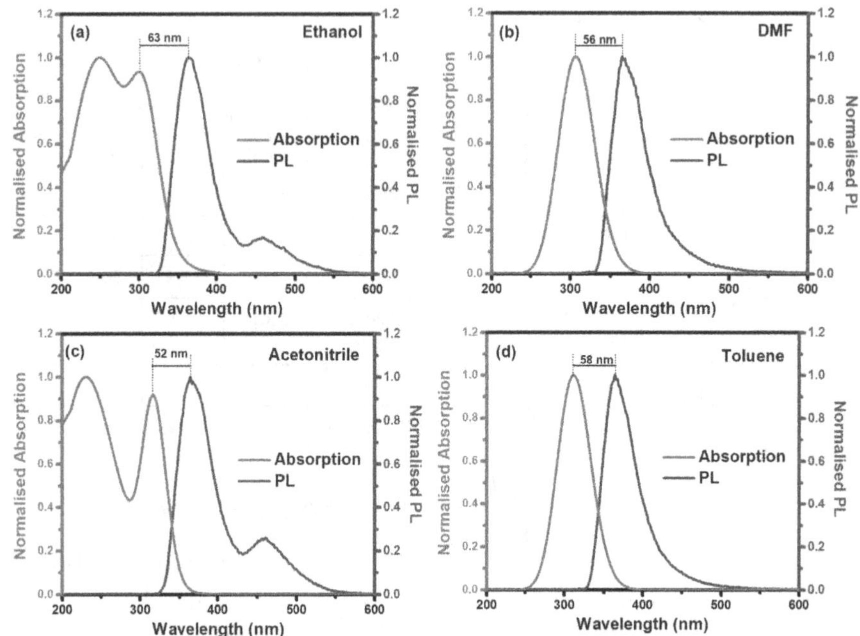

Fig. 118.6 Overlap of normalised absorption and PL for stokes shift for Ethanol (a), DMF (b), Acetonitrile (c), and Toluene (d)

4. Dev, Satya, Pardeep Kumar, Asha Rani, Ajay Agarwal, and Rakesh Dhar. 2020. "Development of Indium Doped ZnO Thin Films for Highly Sensitive Acetylene (C2H2) Gas Sensing." Superlattices and Microstructures 145 (September): 106638. doi:10.1016/j.spmi.2020.106638.
5. Dhayal, Sardul Singh, Abhimanyu Nain, Rajesh Punia, Ashutosh Shrivastava, and Amit Kumar. 2023. "Charge Transport Studies of Tris[4-(Diethylamino)Phenyl]Amine and OFET Application." Journal of Materials Science: Materials in Electronics 34 (20): 1556. doi:10.1007/s10854-023-10926-5.
6. "Doping (Materials) - an Overview | ScienceDirect Topics." 2024. Accessed December 31. https://www.sciencedirect.com/topics/physics-and-astronomy/doping-materials.
7. Horváth, Peter, Peter Šebej, Tomáš Šolomek, and Petr Klán. 2015. "Small-Molecule Fluorophores with Large Stokes Shifts: 9-Iminopyronin Analogues as Clickable Tags." Journal of Organic Chemistry 80 (3). American Chemical Society: 1299–1311. doi:10.1021/JO502213T/SUPPL_FILE/JO502213T_SI_001.PDF.
8. Islam, Md Mustahidul, Sakshi Priya, Shivani Kasana, Balak Das Kurmi, and Preeti Patel. 2025. "Small Molecule N-Heteroatomic Fluorophores: Bridging Chemistry, Principles and Photophysical Properties for Biomedical Applications." Journal of Molecular Structure 1326 (April): 141146. doi:10.1016/j.molstruc.2024.141146.
9. Kaliramna, Sonu, Aryan, Sardul Singh Dhayal, and Narendra Kumar. 2024. "Fabrication and Investigation of PMMA-Doped 1, 3, 5-Triphenylbenzene (TPB) Thin Film's Structural, Optical, and Electrical Properties for Optoelectronic Devices." Optical Materials 151 (May): 115381. doi:10.1016/j.optmat.2024.115381.
10. Kumar, Pardeep, Satya Dev, Sardul Singh Dhayal, Vishwas Acharya, Sanjeet Kumar, Sanjay Kumar, Namita Singh, and Rakesh Dhar. 2021. "Synergistic Effect of Mg and Se Co-Doping on the Structural, Optical and Anti-Bacterial Activity of ZnO Thin Films." Inorganic Chemistry Communications 131 (September): 108801. doi:10.1016/j.inoche.2021.108801.
11. Kumsampao, Jakkapan, Chaiyon Chaiwai, Chattarika Sukpattanacharoen, Phattananawee Nalaoh, Thanyarat Chawanpunyawat, Pongsakorn Chasing, Supawadee Namuangruk, Nawee Kungwan, Taweesak Sudyoadsuk, and Vinich Promarak. 2022. "Solid-State Fluorophores with Combined Excited-State Intramolecular Proton Transfer-Aggregation-Induced Emission as Efficient Emitters for Electroluminescent Devices." Advanced Photonics Research 3 (3): 2100141. doi:10.1002/adpr.202100141.
12. Li, Zheng, and Koen Binnemans. 2021. "Mechanism of Ferric Chloride Facilitating Efficient Lithium Extraction from Magnesium-Rich Brine with Tri-n-Butyl Phosphate." Industrial & Engineering Chemistry Research 60 (23). American Chemical Society: 8538–47. doi:10.1021/acs.iecr.1c01003.
13. Mabesoone, Mathijs F. J., Anja R. A. Palmans, and E. W. Meijer. 2020. "Solute–Solvent Interactions in Modern Physical Organic Chemistry: Supramolecular Polymers as a Muse." Journal of the American Chemical Society 142 (47). American Chemical Society: 19781–98. doi:10.1021/jacs.0c09293.
14. "Organic Semiconductor - an Overview | ScienceDirect Topics." 2024. Accessed December 31. https://www.sciencedirect.com/topics/engineering/organic-semiconductor.
15. Rančić, Milica P., Ivana Stojiljković, Milena Milošević, Nevena Prlainović, Maja Jovanović, Miloš K. Milčić, and Aleksandar D. Marinković. 2019. "Solvent and Substituent Effect on Intramolecular Charge Transfer in 5-Arylidene-3-Substituted-2,4-Thiazolidinediones: Experimental and Theoretical Study." Arabian Journal of Chemistry 12 (8): 5142–61. doi:10.1016/j.arabjc.2016.12.013.
16. Ravikumar, Kavinkumar, and Milind Shrinivas Dangate. 2024. "Advancements in Stretchable Organic Optoelectronic Devices and Flexible Transparent

Conducting Electrodes: Current Progress and Future Prospects." Heliyon 10 (13): e33002. doi:10.1016/j.heliyon.2024.e33002.

17. Singh Dhayal, Sardul, Abhimanyu Nain, Amit Kumar, and Atul Kumar. 2023. "Recent Trends in Selection of Small Molecules for OFET Applications: A Mini Review." Materials Today: Proceedings, Innovative Advancements in Engineering & Technology (IAET-2022), 79 (January): 34–38. doi:10.1016/j.matpr.2022.08.205.
18. Singh, Sardul, Abhimanyu Singh Nain, and Amit Kumar. 2022. "Solvent Effects on the UV-Visible Absorption and Emission of Tris[4-Diethylamino)Phenyl]Amine." Key Engineering Materials 934. Trans Tech Publications Ltd: 37–46. doi:10.4028/p-eft062.
19. Stopel, Martijn H.W., Christian Blum, and Vinod Subramaniam. 2014. "Excitation Spectra and Stokes Shift Measurements of Single Organic Dyes at Room Temperature." Journal of Physical Chemistry Letters 5 (18). American Chemical Society: 3259–64. doi:10.1021/JZ501536A/SUPPL_FILE/JZ501536A_SI_001.PDF.
20. Subuddhi, Usharani, Sourav Haldar, S. Sankararaman, and Ashok K. Mishra. 2006. "Photophysical Behaviour of 1-(4-N, N-Dimethylaminophenylethynyl)Pyrene (DMAPEPy) in Homogeneous Media." Photochemical and Photobiological Sciences 5 (5). Springer: 459–66. doi:10.1039/B600009F/METRICS.
21. Xu, Xiaomin, Yifan Yao, Bowen Shan, Xiao Gu, Danqing Liu, Jinyu Liu, Jianbin Xu, Ni Zhao, Wenping Hu, and Qian Miao. 2016. "Electron Mobility Exceeding 10 Cm2 V−1 S−1 and Band-Like Charge Transport in Solution-Processed n-Channel Organic Thin-Film Transistors." Advanced Materials 28 (26): 5276–83. doi:10.1002/adma.201601171.

Note: All the figures and tables in this chapter were made by the authors.

Recent Trends in Applied Physics and Material Science – Sudhir Bhardwaj et al. (eds)

Tuning the Structural and Optical Properties of the Mo and W Oxide Nanostructured Thin Films

Megha Singh*
CSIR-National Physical Laboratory, Dr KS Krishnan Marg, Pusa, New Delhi, India
Academy of Scientific and Innovative Research (AcSIR), Ghaziabad, India

Sujit K. Saini
Department of Physics and Materials Sciences and Engineering, Jaypee Institute of Information and Technology, Noida, India

Rabindar K. Sharma
Department of Physics D. D. U. Govt. P. G. College, Sitapur, Lucknow, India

Prabhat Kumar, G. B. Reddy
Formerly-Thin film laboratory, Department of Physics, Indian Institute of Technology Delhi, Delhi, India

Abstract: In this work, we explore the tuning of structural and optical properties of molybdenum (Mo) and tungsten (W) oxide nanostructured thin films (NTFs) through controlled variation of process parameters. By systematically adjusting these parameters, we observed significant changes in the morphology of the films, as characterized by Scanning Electron Microscopy (SEM). The films displayed distinct structural modifications. In addition to the structural analysis, UV-VIS spectroscopy was employed to investigate the optical properties of the thin films, revealing notable shifts in the diffused reflectance. These shifts were strongly correlated with the composition and morphology of the films, providing insights into the relationship between structural tuning and optical behaviour. Our findings demonstrate that precise control over the process parameters can effectively tailor the structural and optical characteristics of Mo and W oxide nanostructured thin films, offering potential for their use in advanced optical applications. This study highlights the significance of material composition and fabrication techniques in optimizing the functional properties of oxide-based nanostructures.

Keywords: Composite MoO_3, WO_3, Nanostructures, Diffuse reflectance

1. INTRODUCTION

The tunability of properties of transition metal oxides (TMOs) is an important characteristic for development of the next-gen applications and opening avenues that were previously unexplored. Earlier many methods were reported for this purpose. In this report, we present a relatively unique method (Kumar, 2018, 2019) for mixing two different TMOs and thus obtain the desired structural and optical properties.

2. EXPERIMENTAL DETAILS

The nanostructured films were deposited using plasma Assisted Sublimation Process, wherein metal foils of pure tungsten and molybdenum were oxidized in oxygen plasma ambient and these foils act as source of oxide for deposition of (a-c) MoO_3, WO_3 and composite film of MoO_3-WO_3. For individual pure oxide films, their respective power supplies (to the metal foil) were used. For composite film, both power supplies were used simultaneously. Other deposition parameters were kept constant throughout the deposition process and are listed in Table 119.1. The schematic of deposition set up is shown in Fig. 119.1. The samples thus obtained were analysed using XRD for structural analysis, SEM for morphological analysis and UV-VIS spectroscopy for analysis of optical properties.

3. RESULTS AND DISCUSSION

The morphology of thin films was studies using SEM and the results are shown in Fig. 119.2. Thin films containing

*Corresponding author: meghasingh.life@gmail.com

DOI: 10.1201/9781003684718-119

Table 119.1 Deposition parameters for oxide and composite films

Parameters	Value
Substrate Temperatures	350 °C
Electrode separation	7.5 cm
Deposition Period (t)	30 min
Plasma voltage	2500 V
Base pressure	7.5×10^{-6} Torr
Oxygen partial pressure	1.2×10^{-1} Torr

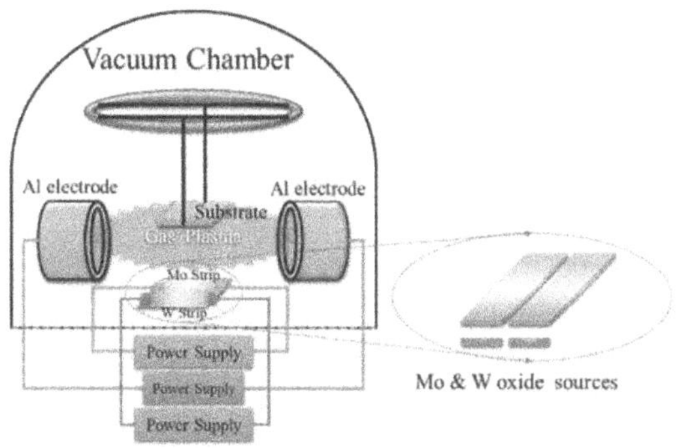

Fig. 119.1 Schematic diagram of experimental set up

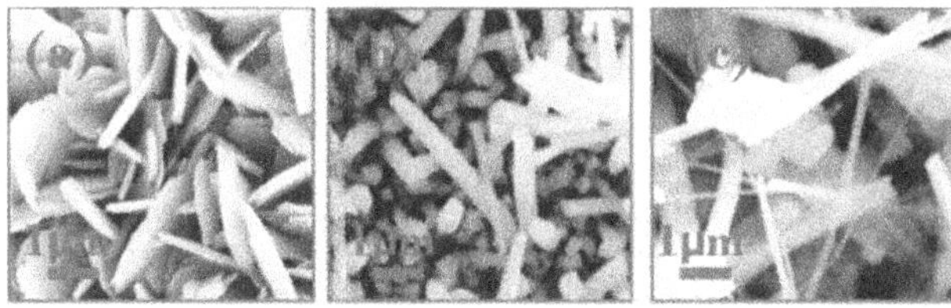

Fig. 119.2 SEM images of nanostructure thin films of (a) MoO_3, (b) WO_3 and (c) MoO_3-WO_3

only MoO_3 (Sharma, 2016) are seen as having nanoflakes-like structure (Fig. 119.2a), whereas thin films containing pure WO_3 (Fig. 119.2b) have nanorods-like structures. The films that contain both oxides of Mo and W (Fig. 119.2c) have a morphology that is a combination of both nanoflakes and nanorods with clear geometrical boundaries between structures of two types. The pure phases of MoO_3 and WO_3 show uniform shape and size, the combination of both oxides show structures that deviation in morphology of pure-phase films indicating that while films might be composed of both the pure tungsten and molybdenum oxides, there's another phase present which can be further analysed in XRD.

The figures 119.3(a), (b) and (c) show XRD diffractograms of MoO_3, WO_3 and MoO_3-WO_3. The peaks in Fig. 119.3(a) correspond to MoO_3 [JCPDS ref: 89-5108], and Fig. 119.3(b) correspond to WO_3 [JCPDS ref: 43-1035]. On observing Fig. 119.3(c), it can be inferred that not only the peaks corresponding to MoO_3 and WO_3 are present. The peaks indicated by ^ in Fig. 119.3(c) show presence of a mixed or composite phase in the thin

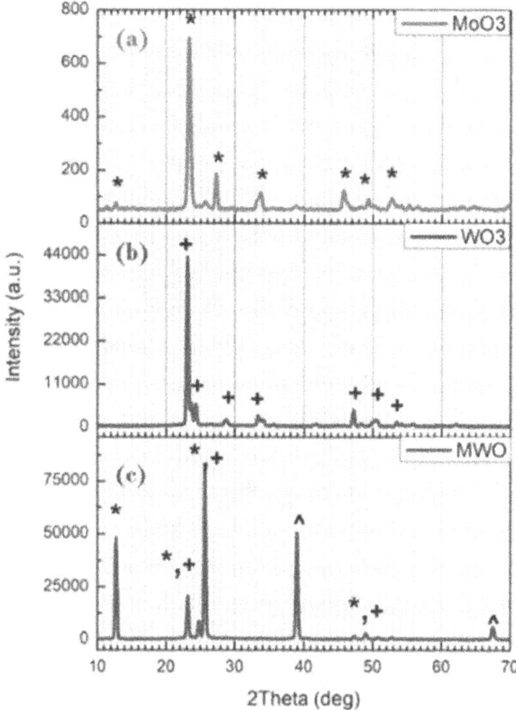

Fig. 119.3 X-Ray Diffraction patterns of (a) MoO_3 (marked by *), (b) WO_3 (marked by +) and (c) MoO_3-WO_3 NTFs

films. This implies that all three samples are different in composition.

As the thin films are nanostructured in morphology, the optical properties of samples can be most appropriately studied by diffuse reflectance (DR) using integrating sphere of UV-VIS spectrophotometer.

The transmittance of tungsten and molybdenum oxides had been studied by Wei, et.al. (2023). The results of spectroscopy are shown in Fig. 119.4(a-c). Figure 119.4(d) is depicting DR spectra of all three samples shown in Figs. 119.4(a-c). On analysing peak reflectance, the sample containing MoO_3 (red) shows peak reflectance at 420 nm, whereas sample containing only WO_3 (blue) shows peak reflectance at 500 nm. However, an interesting phenomenon is observed when both oxides are combined. The peak reflectance is observed at 470 nm. The band gap of these films can be obtained by applying Kubelka-Munk equation to the data obtained from diffuse reflectance spectra. However, it is clear from the spectra that mixing two oxides i.e. MoO_3 and WO_3 has resulted in tuning of optical properties away from either of the oxides and hence, by changing and controlling composition of mixed oxide films, optical properties can be similarly altered.

4. CONCLUSION

In present work, molybdenum oxide, tungsten oxide and composite of both these TMDs were deposited using

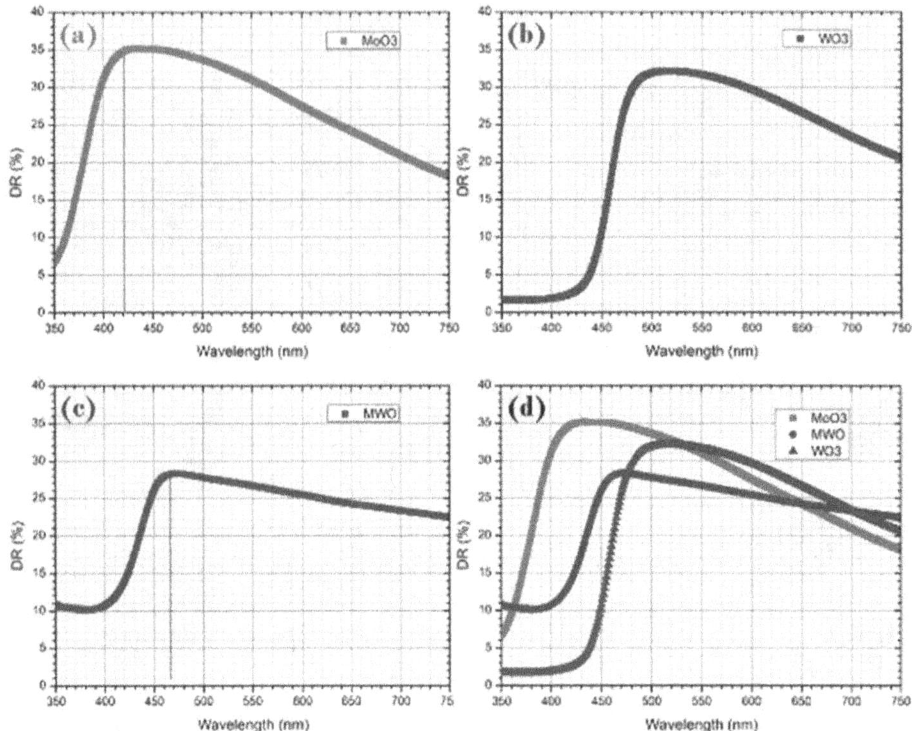

Fig. 119.4 Diffuse reflectance spectra of (a) MoO_3, (b) WO_3 (c) MoO_3-WO_3 NTFs and (d) all three spectra in one graph

Plasma Assisted Sublimation Process. The structural analysis of samples after both the growth steps explores the presence of WO_3 and MoO_3 with excellent crystallinity. The morphological analyses reveal MoO_3 film has nanoflakes, WO_3 film has nanorods and mixed film has combination of both the morphologies. Optical properties explored via diffuse reflectance (UV-VIS) spectroscopy reveal that diffuse reflectance of composite/mixed film was altered from either of its constituting components strongly indicating that properties are alterable and tuneable.

5. ACKNOWLEDGEMENT

The work presented here had been done under supervision of Prof. G.B. Reddy, formerly-Department of Physics, IIT Delhi. We thankfully acknowledge the use of facilities at Nanoscale Research Facility, at I.I.T. Delhi.

REFERENCES

1. Kumar, P., Singh, M., Gopal, P., Sharma, R.K., Reddy, G.B. (2018). Influence of substrate on growth of WO_3 nanostructured thin films. AIP Conference Proceedings. 1953
2. Kumar, P., Singh, M., Reddy, G.B. (2019). Core-shell WO_3-WS_2 nanostructured thin films via plasma assisted sublimation and sulfurization. ACS Applied Nano Materials (2)D; 1691-1703.
3. Sharma, R.K., Saini, S.K., Kumar, P., Singh, M., Reddy, G.B. (2016). Plasma Assisted Growth of MoO_3 Films on Different Substrate Locations Relative to Sublimation Source. International Conference on Condensed Matter and Applied Physics (ICC 2015) Proceedings (1728): 020116
4. Wei, C. C., Wu, T.H., Huang, J.W., Young, B.L., Jian, W.B., Lin, Y.L., Chen, J.T., Hsu, C.S., Ma, Y.R., Tsukagoshi K., (2023) Nanoparticulate films of WO_3 and MoO_3 composites for enhancing UV light electrochromic transmittance variation and energy storage applications. Electrochimica Acta (442):141897

Note: All the figures and tables in this chapter were made by the authors.

Recent Trends in Applied Physics and Material Science – Sudhir Bhardwaj et al. (eds)

A Review on Role of Image Processing Algorithms in Dynamic Speckle Imaging Analysis and their Applications

Shubhashri Kumari*

Department of Physics, SoS, O P Jindal University, Raigarh, Chattisgarh, India

Abstract: The dynamic speckle technique has been considered as a new growing emerging technique which can be applied in various field including agricultural, medical and industrial applications. In this technique, speckle image is captured and then analysed with different image analysis algorithms namely history of speckle pattern with respect to time (THSP), Inertia moment (IM), Generalized difference (GD), Fujii and many existing methods. In the present paper, it has been shown that how a change in parameters can change the result of existing methods with more significant result, however for some cases existing method can be better but proposed can be used for same applications. New or proposed algorithms can be beneficial for such application also where many of the existing algorithms does not work. Results of verified and proposed algorithm in different application has been shown and compared. Furthermore, it is concluded that proposed algorithms can be applied for more significant results in many applications.

Keywords: Dynamic speckle technique, Image processing algorithms, THSP, IM and GD method

1. INTRODUCTION

Dynamic speckle technique is considered as one of newly growing technique applicable in various applications. The technique being harmless, non-contact, inexpensive and smooth handled is used to evaluate bio-speckle activity. A diffuse surface scatters the coherent light beam and the granulated shape is obtained in free-space and known as a speckle. When the optical path of interfering rays shows changes due to scattering surface, the resulting pattern grows with time, and this temporal variation gives information about dynamic speckle activity of the sample. The phenomenon occurring in biological sample is known as bio-speckle. Here, in the first part paper deals with speckle image capturing using dynamic image analysis.

And the another part deals with Image analysis using algorithms important for presented work.

2. LITERATURE REVIEW

Laser speckle technique has been used broadly in various applications. Bruise on fruits has been detected (Pajuelo et al., 2003). Viability as well as non-viability of seeds with different specific humidity levels has also been detected by Braga and Fabbro et al. (2003). Bio speckle activity for root tissues grown in a gel substrate has been evaluated quantitatively (Braga et al. 2009). Lu et al. (2010) detected bruise using Pressure-sensitive film with laser speckle technique. Dynamic speckle technique has been used to assess localized paint - coat drying (Sierra-Sosa et al., 2018); Ansari and Nirala (2015) monitored blood flow using modified form of Lasca method. (Ansari et al., 2016) detected effect of drug in Trypanosoma cruzi parasites using Motion history of image. Due to the perfusion in hand surface, the vascular network, blood flow and

*Corresponding author: shubhashri.kumari@opju.ac.in

DOI: 10.1201/9781003684718-120

pressure can be detected (Zhang et al., 2017). Another study by Toderi et al. (2020) suggests that Biospeckle laser is useful in studying the red blood cell dynamics in a capillary tube. Black rot development on apple surface has been evaluated (Koley et al., 2023).

2.1 Image analysis using Existing and Proposed Algorithms

The bio-speckle activity of samples can be evaluated using algorithms based on Intensity i. e. Co-occurrence matrix (COM) method (Arizaga et al., 1999; Minz et al., 2015) and Generalized Difference method (GD) (Arizaga et al., 2002). Details of Fujii algorithm can be found in (Fujii et al., 1987) and Parameterized Fujii (PF) in (Saude et al., 2012).

Many algorithms have been proposed such as Alternative Fujii (Fal) by (Minz and Nirala, 2014), Parametrized Generalised Difference (PGD), Parameterized Global Average Fujii (PGAF) and Alternative GD(AGD) by (Kumari and Nirala, 2016); Kumari and Nirala (2019) proposed Parametrized geometrical mean of GD (GD_{PGM}), Image Sequence Mean of Parameterized GD (PGD_{ISM}) and Squared temporal difference (STD)

3. EXPERIMENTAL SET UP

The details of the schematic diagram can be found in (Kumari and Nirala, 2019). To get a good contrast image, the distance and angle is adjusted between the camera and the sample. It may vary according to the sample (biological/non-biological). Image size may be also different for different applications.

4. RESULTS AND DISCUSSION

4.1 Comparative Results Obtained using Existing and Proposed Algorithms

P.D. Minz and A. K Nirala proposed, an alternative of Fujii for betterment of the result obtained using Fujii method (Existing) for uneven illumination. In addition, Temporal difference method with some parameters (PTD) has also been introduced in this research article. The method evaluates temporal variation between two consecutive image frames and is different from GD method because GD method is dependent on order of image sequence whereas PTD method does not require the order of image sequence. With the help of parameters better results has been achieved. Qualitative and Quantitative results have been shown and compared in the paper. (Minz & Nirala, 2014).

A new statistic in GD method has been proposed by (Ansari & Nirala, 2016) and with this proposed algorithm, it has been observed that visibility of functional blood vessels can be enhanced using appropriate choice of q values.

P. S. Thakur et al. (2022) proposed FTHSP which is full-field time history of speckle pattern (FTHSP) and can be useful in numerical quantification of bio-speckle activity (BA). There are many more research where the important of algorithms has been shown.

Two regions of Apple have been differentiated using verified and proposed numerical methods and some of the result has been shown to compare the result and discuss the importance of the proposed algorithms (Kumari & Nirala, 2016).

4.2 Qualitative Analysis

Figures 120.1 and 120.2 show the spectral maps of bruised region of apple obtained using existing GD and proposed Alternative GD for p = –2. Spectral maps are used to evaluate bio-speckle activity (BA). In spectral maps highest activity is represented with red colour whereas blue colour shows the vice versa. Similar activity decay can be shown using both the methods with increasing shelf life but if we compare both the result, in Fig. 120.2 the bruised region can be noticed denoted with arrow marks in spectral map also. Hence, it is concluded that more visibility is possible with proposed algorithm. Though all the results have been obtained for all the proposed algorithms but shown only the best to compare with the existing one.

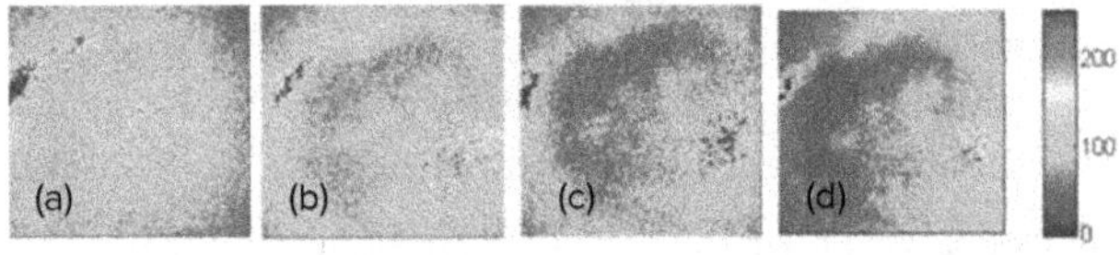

Fig. 120.1 Bruised region's spectral images with colour bar activity map obtained for the apple using GD for (a) Day1, (b) Day 3, (c) Day 6 and (d) Day 9 respectively

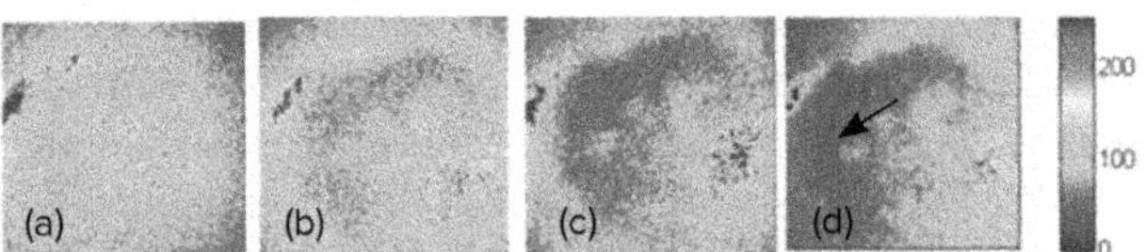

Fig. 120.2 Bruised region's spectral images with colour bar activity map obtained for the apple using AGD method with parameter = -2 for (a) Day1, (b) Day 3, (c) Day 6 and (d) Day 9 respectively

4.3 Quantitative Analysis

(Kumari & Nirala, 2016) concluded that the overall difference calculated from the mean values of plot for bruised and fresh region has been found maximum for the existing Parameterized Fujii than the existing GD as well as the proposed PGD and PGAF algorithms with a higher overall BA difference (39). So, it can be concluded that sometimes it is not necessary that all the proposed algorithms will give better result but can be used for the same purpose as obtained by existing one. But when the result among existing and proposed algorithms is compared, Alternative GD gives the best result with highest difference (43).

So, proposed AGD algorithm is the best among all the existing and proposed algorithms. In addition, one more conclusion can be obtained that change in parameter plays a very important role to find the better result.

Bio speckle activity evaluated from different image processing algorithms has been used to decide maturity and ripe stages of Mango, Guava and Banana and hence their ripening period has also been evaluated. Here some of the important and best result has been shown to discuss the importance of proposed algorithms (Kumari & Nirala, 2016).

4.4 Comparison between the Result Obtained using Existing Fujii and Proposed STD

It can be observed in Fig. 120.3 that the result of mean activity plot is not predictable. It is random in nature so the result obtained using Fujii method is not acceptable.

To overcome the inaccuracy of result obtained using Fujii method, three new algorithms has been proposed and result obtained using the best algorithms has been shown and compared. The result obtained using STD method gives the best among the proposed as shown in 3 and is in comparable with the existing result. In addition, computation time is very less for this method and it does not need any additional parameters.

In blood flow monitoring (Kumari & Nirala, 2019) shown that many of the proposed algorithms have been tried for blood flow monitoring but only PGAF can be applied to monitor blood flow, however the resut is found better for existing GD but the result obtained using PGAF plays an important role to monitor blood flow with similar significance with existing GD. So, It is not necessary to find better result, sometimes it is better for different applications, where existing ones fails.

5. CONCLUSION

1. Proposed algorithms have been proven better than existing algorithms and may be used in several applications.
2. Though, sometimes result can be found better for existing algorithms but still proposed algorithms may be useful in same evaluation and can be significant with the existing and can cause nobility in work. In addition, it may be useful in several applications, where other existing one fails.
3. Overall, it can be concluded that change in parameters plays an important role to find the best result. In addition, there are some limitations in parameter changes. All the parameters may be different for different applications.

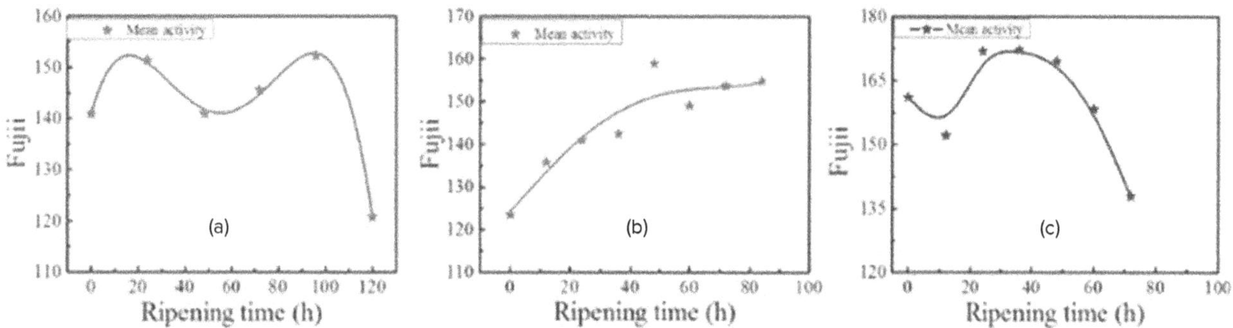

Fig. 120.3 Mean BA activity plots versus ripening time in hour for (a) Mango (b) Guava and (c) Banana respectively; evaluated using the Existing Fujii method

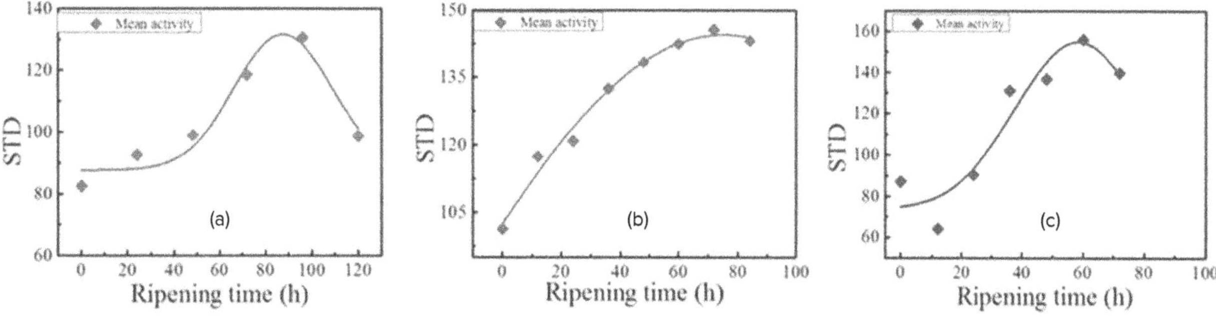

Fig. 120.4 Mean BA activity plots versus ripening time in hour for Mango (b) Guava and (c) Banana respectively; evaluated using the Proposed STD method

REFERENCES

1. Pajuelo, M., Baldwin, G., Rabal, H., Cap, N., Arizaga, R., Trivi, M. (2003). Bio-speckle assessment of bruising in fruits. Optics and Lasers in Eng. 40(1-2):13-24.
2. Braga Jr., R. A., Fabbro, I. M. D., Borem, F. M., Rabelo G., Arizaga, R., Rabal, H. J. and Trivi, M. (2003). Assessment of seed viability by laser speckle techniques. Biosys. Eng. 86(3):287-294.
3. Braga, R. A., Dupuy, L., Pasqual, M. and Cardoso, R. R. (2009). Live biospeckle laser imaging of root tissues. Eur. Biophys. J. 38(5):679-686.
4. Lu F., Ishikawa, Y., Kitazawa, H. and Satake, T. (2010) Measurement of impact pressure and bruising of apple fruit using pressure - sensitive film technique. Journal of Food Eng., 96(4):614-620.
5. Sierra-Sosa, D. S., Tebaldi, M., Grumel, E., Rabal, H. and Elmaghraby, A. (2018). Localized analysis of paint-coat rying using dynamic speckle interferometry. Optics and Lasers in Eng.106:61-67.
6. Ansari, M. Z. and Nirala, A. K. (2015). Monitoring capillary blood flow using laser speckle contrast analysis with spatial and temporal statistics. Optik. 126(24):5224-5229.
7. Ansari, M. Z., Grassi, H. C., Cabrera, H., Velasquez, A. and Andrades, E. D. J. (2016). Online fast Biospeckle monitoring of drug action in Trypanosoma cruzi parasites by motion history image. Lasers in Med. Sci. 31(7):1447-54.
8. Zhang, L., Ding, L., Li, M., Zhang, X., Su, D., Jia J., and Miao, P. (2017). Dual-Wavelength laser speckle contrast imaging (dwLSCI) improves chronic measurement of superficial blood flow in hands. Sensors. 17(12):2811(1-13).
9. Toderi, M.A. Riquelme, B. D. and Galizzi, G. E. (2020). An experimental approach to study the red blood cell dynamics in a capillary tube by biospeckle laser. Opt. and Lasers in Eng. 127:105943.
10. Koley, C., Das, R., and Nirala, A. K. (2023). Assessment of black rot development on apple surface using laser biospeckle technique. Phys. Scr. 98(3):035501.
11. Arizaga, R., Trivi, M. and Rabal, H. (1999). Speckle time evolution characterization by the co-occurrence matrix analysis. Opt. & Laser Technol. 31(2):163-169.
12. Minz, P. D., Ansari, M. Z., and Nirala, A. K. (2015) Effect of antibrowning agents on fresh-cut potato tubers using frequency filtering of biospeckle images). Laser Phys. 25(5):055601-055608.
13. Arizaga, R. A., Cap, N. L., Rabel, H. J., and Trivi M. (2002). Display of local activity using dynamical speckle patterns. Opt. Eng. 41(2):287-294.
14. Fujii, H., Nohira, K., Yamamoto, Y., Ikawa, H., and Ohura, T. (1987). Evaluation of blood flow by laser speckle image sensing. Applied Opt. 26(24):5321-5325.
15. Saude, A. V., Menezes, F. S., Freitas, P. L. S., Rabelo, G. F. and Braga Jr., R. A. (2012). Alternative measurement for biospeckle image analysis. J. Opt. Soc. Am. A. 29(8):1648-1658.
16. Minz, P. D. and Nirala, A. K. (2014). Intensity based algorithms for biospeckle analysis. Optik. 125(14):3633-3636.
17. Kumari, S. and Nirala, A. K. (2016). Biospeckle technique for the non-destructive differentiation of bruised and fresh regions of an Indian apple using intensity-based algorithms. Laser Phys. 26(11):115601-115611.
18. Kumari, S. and Nirala, A. K. (2019) Biospeckle image processing algorithms for non-destructive differentiation between maturity and ripe stages of Indian climacteric fruits and evaluation of their ripening period. Laser Phys. 29(7):075601-075613.
19. Thakur P. S., Chatterjee, A., Rajput, L.S., Rana, S., Bhatia V. and Prakash S. (2022). Laser biospeckle technique for characterizing the impact of temperature and initial moisture content on seed germination. Optics and Lasers in Eng. 153:106999.

Note: All the figures in this chapter were made by the authors.

Recent Trends in Applied Physics and Material Science – Sudhir Bhardwaj et al. (eds)
© 2026 Taylor & Francis Group, London, ISBN 978-1-041-16452-4

121

Assessing Water Quality Using the Normalized Difference Water Index (NDWI): A Remote Sensing Approach with Satellite Imagery

Balaji Yadav*
Department of Physics, Shri Chhatrapati Shivaji College, Omerga, MS India

Sandipan Sawant, Shafiyoddin Sayyad
Microwave and Imaging Spectroscopy Laboratory, Miliiya College, Beed, MS India

Abstract: The Normalized Difference Water Index (NDWI) is a widely utilized remote sensing technique for detecting and monitoring surface water bodies, as well as assessing various water quality parameters such as turbidity, suspended sediments, chlorophyll-a, and total dissolved solids. This study explores the application of NDWI across diverse hydrological settings, emphasizing its ability to delineate water features by leveraging the differential reflectance in the green and NIR (near-infrared) spectral bands. NDWI's strength lies in its capacity to enhance the contrast between water and non-water elements.

Utilizing high-resolution satellite imagery, including data from sensors like Landsat and Sentinel this paper demonstrates NDWI's versatility in monitoring both small and large water bodies. Through case studies ranging from urban reservoirs to natural lakes, NDWI's efficacy in tracking water extent changes, seasonal variations, and the impacts of anthropogenic activities is highlighted. Environments like turbid or mixed-use water systems.

The study also evaluates the correlation between NDWI and critical water quality indicators, validated through in-situ measurements and ground-truth data. Results show a strong relationship between NDWI-derived data and water quality parameters, confirming its utility in large-scale, real-time water monitoring. The integration of NDWI with other spectral indices and machine learning techniques is explored to improve water detection and quality predictions. This paper touches on the broader applications of NDWI, such as its role in supporting sustainable water resource management in the face of global water scarcity challenges.

Keywords: NDWI, Remote sensing, Turbidity, Sentinel-2, Image pre-processing

1. INTRODUCTION

Water is a vital natural resource, crucial for sustaining life and maintaining ecological balance. With increasing concerns over water scarcity and quality degradation, effective monitoring tools are essential. Remote sensing, particularly multispectral satellite imagery, has become a powerful method for water resource management. Among these, NDWI has gained prominence for detecting surface water bodies and evaluating quality indicators like turbidity, suspended sediments, and total dissolved solids,to the work of Kumar, S., et al.

This study advances NDWI applications by calibrating thresholds for saline-affected reservoirs, correlating NDWI-derived turbidity with groundwater salinity, and analyzing turbidity hotspots linked to agricultural runoff in semi-arid regions. NDWI leverages reflectance differences in the green and NIR spectral bands, enhancing water feature detection across diverse hydrological settings. The availability of high-resolution satellite imagery (Landsat, Sentinel) has improved NDWI's effectiveness, enabling it to monitor water bodies of varying scales and track changes in extent, seasonal variations, and anthropogenic impacts, (Smith, J. A., & Johnson, R. B.).

*Corresponding author: balajiby3555@gmail.com

DOI: 10.1201/9781003684718-121

Recent advancements have expanded NDWI's role beyond water detection. Its integration with other spectral indices and machine learning has enhanced water quality predictions. Strong correlations between NDWI-derived data and water quality indicators, validated through in-situ measurements, confirm its utility for large-scale, real-time monitoring. (Bid, Sumanta; Siddique El.)

This study focuses on Sentinel-2 imagery for turbidity assessment, detailing image acquisition, preprocessing, band selection, and formulation of turbidity indices. Calibration and validation procedures integrate field measurements to establish empirical relationships between NDWI and actual turbidity levels. It proposes region-specific NDWI thresholds, validated against in-situ samples, supporting improved water monitoring in saline-impacted semi-arid reservoirs. These findings highlight NDWI's potential for sustainable water resource management amidst global challenges. (Potes, M., Costa M.J. Salgado, R).

2. STUDY AREA

The research site, located at 19°08'03"N, 76°05'54"E in Maharashtra's Beed district near Majalgaon, features a dam 31.9 meters high and 6,488 meters long, creating a reservoir with a capacity of 5.759 million cubic meters. The reservoir experiences significant salinity gradients (TDS: 1,200–4,500 mg/L) and seasonal turbidity spikes during monsoon runoff (June–September), making it ideal for testing NDWI in semi-arid, saline conditions. As a key drinking water source, the reservoir faces increasing salinity and pollution, affecting its suitability for consumption and irrigation. Satellite data (400–1000 m resolution) was collected on January 16, 2023, alongside on-site sampling to obtain real-time surface water quality data and spectral measurements.

2.1 Data Acquisition

The Sentinel-2 images for the study area were acquired from the Copernicus Open Access Hub. The ESA SNAP (Sentinel Application Platform) software was utilized for image correction and preprocessing due to its comprehensive tools for atmospheric correction, radiometric calibration, and geometric corrections. The powerful features and intuitive interface of SNAP made it the perfect tool for processing Sentinel-2 data, ensuring high-quality preprocessing essential for accurate index calculations to the work of Y. Jiao, S. Wang.

2.2 Band Selection

In-Depth Overview of Band Selection for NDWI

The process of selecting bands for the Normalized Difference Water Index (NDWI) is carefully crafted to leverage the distinct spectral characteristics of water bodies Bands B3 (560 nm) and B8 (842 nm) were selected for NDWI based on ANOVA comparing reflectance variance across turbidity gradients ($p < 0.05$). B3 maximizes water reflectance, while B8 minimizes interference from vegetation and saline groundwater (see Table 121.1)."

Table 121.1 Sentinel-2 multispectral imager bands and wavelength

Band	wavelength region	wavelength (nm)	Resolution (m)
Band – 1	Coastal aerosol	443	60
Band – 2	Blue	490	10
Band – 3	Green	560	10
Band – 4	Red	665	10
Band – 5	Vegetation red edge	705	20
Band – 6	Vegetation red edge	740	20
Band – 7	Vegetation red edge	783	20
Band – 8	NIR	842	10
Band – 8A	Vegetation red edge	865	20
Band – 9	Water Vapor	945	60
Band – 10	SWIR - Cirrus	1375	60
Band – 11	SWIR	1610	20
Band – 12	SWIR	2190	20

The Green Band (B3) shows high reflectance from water, making it distinct from other land covers like vegetation or soil, which aids in accurate NDWI calculations. In contrast, the NIR Band (B8) is strongly absorbed by water, leading to low reflectance values, while vegetation and soil reflect more NIR radiation. This contrast between water and non-water surfaces in the NIR band is key for NDWI calculation. (McFEETER S. K.).

3. METHODOLOGY

Pre-process Sentinel-2 Data: The Sentinel-2 data was pre-process using ESA SNAP, which involved atmospheric correction using the Sen2Cor plugin to remove atmospheric distortions and radiometric calibration to convert raw data into surface reflectance values.

3.1 Cloud Detection and Masking

Shi Qiu [a c], Zhe Zhu by Fmask 4.0 was implemented using spectral thresholds ($B1 < 0.09$, $B2/B1 > 1.5$) and thermal tests ($B10 < 295$ K) to classify clouds/shadows. Contaminated pixels were replaced with temporally adjacent scenes using SNAP's interpolation tool.

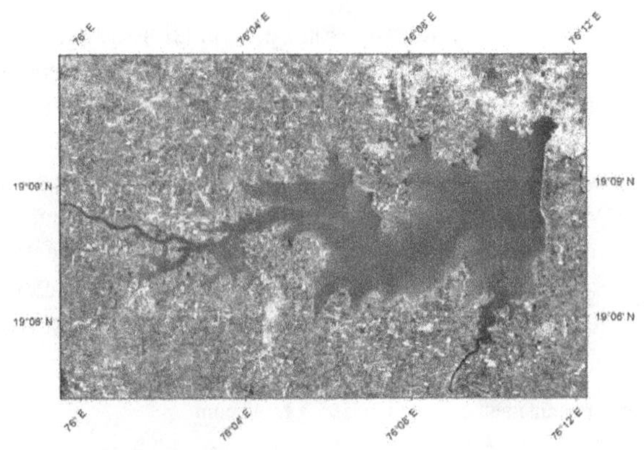

Fig. 121.1 Pre-processing image

3.2 NDWI Thresholding

"Optimal NDWI > 0.2 was determined via ROC curves (AUC = 0.89) using 30 in-situ sampling points, minimizing false positives from saline groundwater."

3.3 Spatio-Temporal Analysis

Monthly Sentinel-2 data (Jan–Dec 2023) were analyzed to quantify turbidity trends. Hotspots were correlated with monsoon rainfall (IMD data) and groundwater salinity maps (Central Groundwater Board) using Pearson's correlation (r = 0.72)." NDWI Calculation: Utilizing the NIR and B3 spectral bands from Sentinel-2 imagery, we calculate the NDWI.

$$NDWI = \frac{(XGreen - XNear\text{-}Infrared)}{(XGreen + XNear\text{-}Infrared)}$$

NDWI thresholds were optimized using ROC curves (AUC = 0.91), with NDWI > 0.18 achieving 89% accuracy for water detection in saline conditions. Turbidity was estimated via NDWI using the regression model: Turbidity (NTU) = 112 × (1 – NDWI) + 15, validated against 35 field samples ($R^2 = 0.82$). Water pixel mask: The NDWI image is used to mask out non-water pixels. Optimal NDWI > 0.2 was determined via ROC curves (AUC = 0.89) using 30 in-situ sampling points, minimizing false positives from saline groundwater

3.4 Converted Infrared Spectra

NDWI adjusts infrared reflectance based on water presence, varying inversely with its value. It is a remote sensing index used to detect and monitor water bodies by quantifying reflectance differences between water-sensitive and non-sensitive wavelengths. Widely applied in environmental and hydrological studies, NDWI helps track changes in water bodies, wetlands, and aquatic features using satellite or aerial imagery. The NDWI spectral index is computed as to the work of Xiucheng Yang, Shanshan Zhao El.

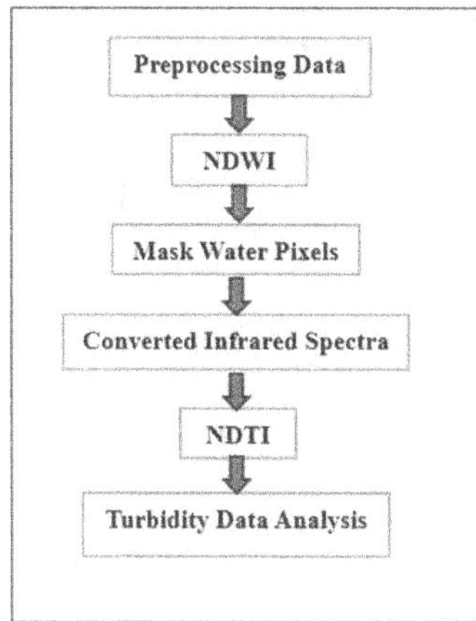

Fig. 121.2 Flow chart

NDWI is calculated using reflectance from the green and NIR bands, specifically B3 (green) and B8 (NIR) from Sentinel-2 imagery, leveraging their distinct properties

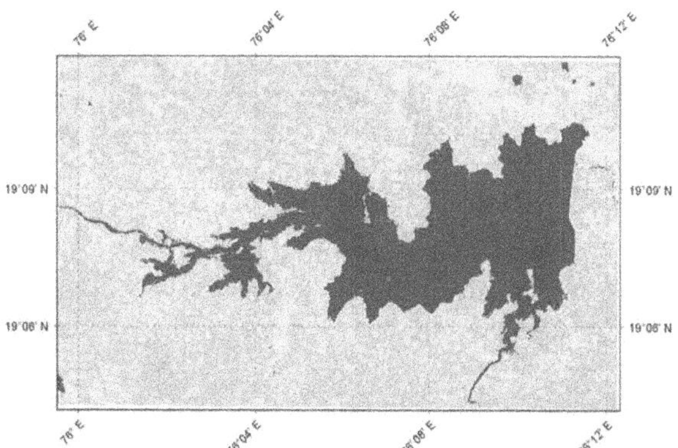

Fig. 121.3 Process image of NDWI

for water detection. The index effectively detects water bodies, even in complex hydrological conditions. High turbidity and sedimentation increase NIR absorption, lowering NDWI values. This relationship enables the identification of areas with high turbidity, offering a valuable method for assessing water clarity and sediment dynamics in aquatic systems.

4. RESULT AND DISCUSSION

4.1 NDWI Results for Water Detection

The **NDWI** derive from Sentinel-2 imagery using bands B3 (green) and B8 (NIR), the value of NDWI index ranges from -1 to 1. Water bodies have an NDWI index value greater than 0. An NDWI index value is 0.27 indicates the presence of water in the image. However, never the less, it is crucial to acknowledge that the NDWI index is not flawless and can be influenced by additional elements like the existence of vegetation and sediment (McFEETER S. K., Ujwala Bhangale, El) Confirming the capability of NDWI to enhance. The distinction between water and non-water features is achieved by utilizing the varying reflectance characteristics of the green and NIR spectral bands.

NDWI maps revealed clear spatial distributions of water bodies, capturing both permanent and seasonal features. Additionally, the findings showcased NDWI's capability to identify small and dispersed water bodies effectively. Highlighting its utility for applications in areas with diverse hydrological conditions.

4.2 Assessment of Turbidity Using NDWI

While primarily used for water detection, the NDWI also proved useful in identifying variations in water turbidity. Spatio-temporal analysis (2018–2023) revealed post-monsoon turbidity peaks (NDWI < 0.12, NTU > 85) near agricultural inflows, correlating with groundwater salinity (r = 0.68). NDWI's accuracy in saline zones improved by 24% compared to McFeeters (1996) after calibration. These findings were validated against field measurements, showing a strong correlation between NDWI values and turbidity indicators, as demonstrated in previous works of Dogliotti, A.I. Ruddick, Vaibhav Garg El

This study builds on existing research by integrating NDWI with calibration techniques to quantify turbidity levels. The NDWI-derived turbidity indices provided an effective means of monitoring spatial and temporal some change in water quality, particularly in response to seasonal or anthropogenic factors.

Seasonal Turbidity Trends: Monthly NDWI-derived turbidity maps (Fig. 121.4) showed a 45% increase in turbidity during monsoon months (June–September), linked to sediment runoff from adjacent farmlands. Groundwater salinity (TDS > 3,000 mg/L) amplified turbidity misclassification, necessitating a correction factor (1.2×) for saline pixels."

4.3 Comparative Analysis and Accuracy

The results confirm that NDWI is a reliable index for water detection and turbidity assessment, outperforming other remote sensing indices in terms of simplicity and applicability with Sentinel-2 data ([8]). By integrating NDWI with salinity-adjusted thresholds and spatio-temporal trend analysis, this study addresses NDWI's limitations in saline, semi-arid reservoirs. Future work will combine NDWI with machine learning to disentangle salinity-turbidity spectral overlaps where additional indices or more complex algorithms may be required for improved accuracy, as highlighted studied by Dogliotti, A.I. Ruddick Domenico Miglino El.

4.4 Implications for Water Resource Management

The NDWI results offer critical insights into water extent, distribution, and quality, supporting sustainable water resource management. By providing high-resolution spatial data, NDWI can guide policy decisions on irrigation, urban planning, and ecological conservation. The strong correlation between NDWI and turbidity metrics further demonstrates its potential as a real-time monitoring tool for assessing water quality.

5. CONCLUSION

This research demonstrates the significant utility of the Normalized Difference Water Index (NDWI) in remote sensing for identifying water bodies and assessing water quality parameters like turbidity, suspended sediments, and water clarity. Leveraging high-resolution Sentinel-2 satellite imagery, the study showcased NDWI's effectiveness in distinguishing water from non-water features, even in hydrologically complex and challenging environments. Furthermore, NDWI proved useful in detecting changes in water quality, particularly turbidity, making it a valuable tool for large-scale, real-time water monitoring. With thorough pre-processing and the incorporation of additional spectral indices, NDWI demonstrated its capability to analyse both small and extensive water bodies in diverse settings, including urban and natural landscapes. The strong relationship observed between NDWI-derived values and in-situ turbidity measurements further validated its effectiveness in monitoring seasonal and anthropogenic influences on water quality

REFERENCES

1. Kumar, S., et al. (2021). "Integrating Groundwater Salinity and Satellite Indices for Turbidity Assessment in Semi-Arid Reservoirs." Journal of Hydrology, 603, 127012.
2. Smith, J. A., & Johnson, R. B. "Remote sensing of turbidity in coastal waters using multispectral satellite imagery." *Journal of Environmental Monitoring*, 10(3), 456–470. 2022.
3. Bid, Sumanta; Siddique, Giyasuddin "Identification of seasonal variation of water turbidity using NDTI method in Panchet Hill Dam, India". *Modeling Earth Systems and Environment*. 2019.
4. Potes, M.; Costa, M. J.; Salgado, R. "Satellite remote sensing of water turbidity in Alqueva reservoir and implications on lake modelling." *Hydrology and Earth System Sciences* volume 16, issue 6 2012.
5. Y. Jiao, S. Wang, Y. Zhou, F. Yan, W. Zhou, L. Zhu. "Using Unmixing Method to Retrieve the Concentration of Chl-a in Lake Tai", *IEEE International Symposium on Geoscience and Remote Sensing*, 2006.
6. McFEETER S. K. "The use of the Normalized Difference Water Index (NDWI) in the delineation of open water features." *International Journal of Remote Sensing*, 17(7), 1425–1432. 1996.
7. Xiucheng Yang , Shanshan Zhao, "Mapping of Urban Surface Water Bodies from Sentinel-2 MSI Imagery at 10 m Resolution via NDWI-Based Image Sharpening." Remote Sens. 2017, 9, 596.
8. Ujwala Bhangale, Swapnil More, Tanishq Shaikh, Suchitra Patil, Nilkamal More. "Analysis of Surface Water Resources Using Sentinel-2 Imagery." *Procedia Computer Science* 171, 2645–2654, 2020.
9. S. K. McFEETERS , "The use of the Normalized Difference Water Index (NDWI) in the delineation of open water features" International Journal of Remote Sensing Received 25 May 1995, Accepted 04 Jan 1996
10. Vaibhav Garg, Shiv Prasad Aggarwal, Prakash Chauhan, "Changes in turbidity along Ganga River using Sentinel 2 satellite data during lockdown associated with Covid 19." *GEOMATICS, NATURAL HAZARDS AND RISK 2020,* VOL. 11, NO. 1, 1175–1195. 2020
11. Dogliotti, A.I. Ruddick, K.G, Nechad, B. Doxaran, D. Knaeps, E. "A single algorithm to retrieve turbidity from remotely-sensed data in all coastal and estuarine waters." *Remote Sensing of Environment.* Volume 156 2015.
12. Domenico Miglino 1, Seifeddine Jomaa 2, Michael Rode 2, Francesco Isgro 3 and Salvatore Manfreda "Monitoring Water Turbidity Using Remote Sensing Techniques." *Environ. Sci. Proc.* 21, 63, 2022.
13. MOORE, GERALD K. "Satellite remote sensing of water turbidity" *Hydrological Sciences Bulletin* volume 25, issue 4 1980.
14. Shi Qiu [a c], Zhe Zhu "Fmask 4.0: Improved cloud and cloud shadow detection in Landsats 4–8 and Sentinel-2 imagery" Remote Sensing of Environment Volume 231, 15 September 2019.

Note: All the figures and tables in this chapter were made by the authors.

Recent Trends in Applied Physics and Material Science – Sudhir Bhardwaj et al. (eds)
 ISBN 978-1-041-16452-4

Coherent Raman Scattering in n-InSb Semiconductor Quantum Plasma: Influence of Szigeti Effective Charge

Ravi Vanshpal*, Gopal Chand Dangi
Department of Physics,
Shri Vaishnav Vidyapeeth Vishwavidyalaya,
Indore (M.P.) India

Ratna Agrawal
School of Studies in Physics, Vikram University,
Ujjain (M.P.) India

Abstract: An analytical study examines the impact of Szigeti's effective charge on coherent Raman scattering (CRS) of laser radiation in semiconductors. The CRS arising from electron density perturbations and molecular vibrations of the medium are produced at the transverse optical phonon frequency. The Szigeti effective charge is important in determining the strength of the coherent Raman scattering signal, influencing the interaction between the laser radiation and the medium's vibrational modes. The quantum hydrodynamic model is considered to obtain the effective Raman susceptibility arising due to induced nonlinear current density and molecular vibrations. Numerical estimations are made for n-type InSb semiconductor shining by pulsed 10.6 μm CO2 laser. The results reveal a strong influence of Szigeti effective charge on Raman gain coefficient and conceptually modify the way Raman amplification. The quantum effects, particularly the Bohm potential and Fermi pressure lead to a more accurate prediction of resonance conditions in the Raman scattering process.

Keywords: Quantum plasma, Coherent raman scattering, Laser-plasma interaction

1. INTRODUCTION

The Coherent Raman Scattering (CRS) in semiconductor plasmas is attributed to the interaction of a strong electromagnetic field, that produces new photons at shifted frequencies, corresponding to vibrational excitations of the system. However, in semiconductors, the coupling between molecular vibrations and pump waves depends on Szigeti's effective charge, which cannot be ignored in the infrared regime. The Szigeti effective charge corresponds to the effective interaction of ions with the surrounding electronic environment in a medium, particularly in the context of polar vibrations and dielectric properties. It originates from the concept of dynamic polarization, where the displacement of ions under an external field induces a corresponding electronic charge distribution that modifies the overall interaction.

The theoretical framework for CRS coherent Raman scattering has traditionally been rooted in classical electrodynamics and nonlinear optics. Bloembergen and Shen (1964) laid the foundation for understanding nonlinear optical processes, by developing a comprehensive formalism for nonlinear susceptibilities in materials. Sen et al. (1980) have reported significant research on CRS and its consequent instabilities in isotropic and magnetized semiconductors. Neogi et al. (1993) observed an enhancement in Raman gain coefficient in the presence of an external static magnetic field. In the previously reported works, Ghosh et al. (1982) established the fact that the origin of CRS in semiconductor plasmas has been taken into finiteness of differential polarizability only.

The concept of quantum plasma refers to the robust framework for analysing quantum effects in plasmas. The Quantum Hydrodynamic (QHD) model is appropriate for incorporating quantum corrections such as the Bohm potential and Fermi pressure into classical fluid dynamics (Manfredi, 2006). The basic QHD equations consist

*Corresponding author: ravivanshpal@gmail.com

DOI: 10.1201/9781003684718-122

of the continuity, momentum, and Poisson equations, modified to include quantum effects. By applying these equations, the authors derived expressions for nonlinear susceptibility and gain coefficient in the CRS process, allowing a detailed investigation of Raman amplification in highly doped n-type InSb semiconductors in quantum plasma as a medium.

Advanced theoretical modelling with rigorous numerical analysis, we attempt to shed light on the crucial role of Szigeti's effective charge in quantum plasma in determining the response of optical nonlinear polarization due to the coupling of the molecular vibration of CRS process. In section 2, we develop the theoretical formulation for the gain coefficient for CRS processes, considering the quantum mechanical nature of the system. Additionally, we provide a detailed analysis of the influence of various parameters such as number density, electric field intensity, and wave number in the results and discussion section 3 followed by section 4 with the conclusions.

2. THEORETICAL FORMULATIONS

The scattering of a large amplitude electromagnetic pump wave is enhanced due to the excitation of a molecular vibration consisting of N harmonic oscillators in a Raman active medium. The equation of motion for an oscillator

$$\frac{\partial^2 u(x,t)}{\partial t^2} - \omega_t^2 u(x,t) + 2\Gamma_a \frac{\partial u(x,t)}{\partial t} = \frac{F(x,t)}{M} \tag{1}$$

Where $F(x,t) = q_s E + \frac{1}{2}\left(\frac{d}{du}\right)_0 E^2(x,t)$ the driving force emerging from Szigeti's effective charge q_s and differential polarizability $\left(\frac{d}{du}\right)_0$. The basic equations employed are

$$\frac{\partial \upsilon_0}{\partial t} + \nu \upsilon_0 = \frac{e}{m} E_0 \tag{2}$$

$$\frac{\partial \upsilon_1}{\partial t} + \nu \upsilon_1 + \left(\upsilon_0 \cdot \frac{\partial}{\partial x}\right)\upsilon_1 = \frac{e}{m} E_1 - \frac{1}{m n_0}\frac{\partial P}{\partial x} + \frac{\hbar^2}{4m^2 n_0}\frac{\partial^3 n_1}{\partial x^3} \tag{3}$$

$$\upsilon_0 \frac{\partial n_1}{\partial x} + n_0 \frac{\partial \upsilon_1}{\partial x} = -\frac{\partial n_1}{\partial t} \tag{4}$$

$$\frac{\partial E_1}{\partial x} + \frac{n_{1\alpha} e}{\varepsilon} = \frac{\beta}{\varepsilon}\frac{\partial^2 u}{\partial x^2} \tag{5}$$

$$P = q_s E + \varepsilon \mathrm{N}\left(\frac{d\alpha}{du}\right)_0 u E \tag{6}$$

The meaning of symbols is prevalent and defined in the paper published by Guha et al. (1979). Further explanation of quantum corrections is found in subsequent research studies developed by Manfredi and Haas (2001). The high-frequency pump field gives rise to a carrier density perturbation, which in turn derives an electron–plasma wave and induces current density in the Raman active medium. Now following the standard approach adopted by Sen and Sen (1985), the perturbed electron density (n_T) of the Raman active medium due to molecular vibrations:

$$n_T = \frac{i\varepsilon k}{e}\left[\frac{\delta_2^2 + i\Omega\Gamma_a - \frac{\varepsilon \mathrm{N}}{2M}\left[\frac{2q_s\left(\frac{d\alpha}{du}\right)_0}{\varepsilon} - \left(\frac{d\alpha}{du}\right)^2 E_0\right]}{\frac{q_s}{M} + \frac{\varepsilon}{2M}\left(\frac{d\alpha}{du}\right)_0 E_0}\right] \tag{7}$$

The density perturbation associated with fast components combines with the pump at frequency ω_0 and produces the Stokes component of ($\omega_0 - \omega_s$) of the scattered electromagnetic waves. We can obtain the Stokes component of density perturbation using the procedure adopted by Ghosh et al. (2010), which incorporates the effect of quantum correction through a modified plasma frequency $\varpi_p^2 = \omega_p^2 + k^2 V_F'^2$ in term $\delta_1^2 = \varpi_p^2 - \omega_s^2$.

$$n_s = \frac{ik\bar{E}}{\left(\delta_1^2 - i\omega_1 \nu\right)} n_T \tag{8}$$

The resonant Stokes component of the current density becomes

$$J_1(\omega_1) = \frac{\varepsilon k^2 E_1 E_0^2}{\omega_1} \left[\frac{\omega_p^2}{k^2 E_0^2} - \frac{i\omega_1 e^2 k^2}{m\left(\delta_1^2 - i\omega_1 \nu\right)} \frac{\delta_2^2 + i\Omega\Gamma_a - \frac{\varepsilon \mathrm{N}}{2M}\left[\frac{2q_s\left(\frac{d\alpha}{du}\right)_0}{\varepsilon} - \left(\frac{d\alpha}{du}\right)^2 E_0\right]}{\frac{q_s}{M} + \frac{\varepsilon}{2M}\left(\frac{d\alpha}{du}\right)_0 E_0}\right] \tag{9}$$

Where

$$\bar{E} = \left(\frac{e}{m}E_0\right), \varpi_p^2 = \omega_p^2 + k^2 V_F'^2$$

$$V_F' = V_F\sqrt{1+\gamma_e}, \gamma_e = \frac{\hbar^2 k^2}{8mK_B T_F}$$

In deriving Eq. (9) the component of oscillatory electron fluid velocities in the combination of the pump and perturbed fields are obtained from Eqs. (2) and (3) include linear and nonlinear current density induced by the coupling amongst the three interacting waves. Here, we ignored the Doppler shift under the assumption that $\omega_0 >> \nu > k\upsilon_0$; $\omega_p = \left(\frac{n_0 e^2}{m\varepsilon}\right)^{1/2}$ is the plasma frequency.

Now treating the induced polarization P_{cd} as the time integral of induced nonlinear current density $J_{nl}(\omega_1)$, using eq. (9) we may obtain:

$$P_{cd}(\omega_1) = \frac{-\varepsilon\,\omega_0 k^2 E_1 E_0^2}{\omega_1}\left[\frac{\omega_p^2}{k^2E_0^2} - \frac{i\omega_1 e^2 k^2}{m\left(\delta_1^2 - i\omega_1 \nu\right)}\,\frac{\delta_2^2 + i\Omega\Gamma_a - \frac{\varepsilon \mathrm{N}}{2M}\left[\frac{2q_s\left(\frac{d\alpha}{du}\right)_0}{\varepsilon} - \left(\frac{d\alpha}{du}\right)^2 E_0\right]}{\frac{q_s}{M} + \frac{\varepsilon}{2M}\left(\frac{d\alpha}{du}\right)_0 E_0}\right] \quad (10)$$

The polarization created by the interaction of the pump field with the molecular vibration is obtained by

$$P_{mv}(\omega_1) = \frac{\frac{\varepsilon \mathrm{N}}{2M}\left(\frac{d\alpha}{du}\right)_0 E_0^2 E_1(\omega_1)}{\delta_1^2 + i\Omega\Gamma_a} = \varepsilon_0\left(\chi_B^{(3)}\right)_{es} E_0^2 E_1(\omega_1) \quad (11)$$

In a doped semiconductor, the total induced nonlinear polarization is proportional to $E_0^2 E_1(\omega_1)$ with finite electrostrictive coupling is given by

$$P_{nl}(\omega_1) = P_{cd}(\omega_1) + P_{mv}(\omega_1) = \varepsilon_0 \chi_R E_0^2 E_1(\omega_1) \quad (12)$$

Hence effective Raman susceptibility χ_R can be obtained from Eqs. (10) and (11) by using Eq. (12) as

$$\chi_R = \frac{\varepsilon\,\omega_0 k^2}{\omega_1}\left[\frac{i\omega_1 e^2 k^2\, \delta_2^2 + i\Omega\Gamma_a - \frac{\varepsilon \mathrm{N}}{2M}\left[\frac{2q_s\left(\frac{d\alpha}{du}\right)_0}{\varepsilon} - \left(\frac{d\alpha}{du}\right)^2 E_0\right]}{\left(\frac{q_s}{M} + \frac{\varepsilon}{2M}\left(\frac{d\alpha}{du}\right)_0 E_0\right) m\left(\delta_1^2 - i\omega_1\nu\right)} - \frac{\omega_p^2}{k^2E_0^2} + \frac{\frac{\varepsilon \mathrm{N}}{2M}\left(\frac{d\alpha}{du}\right)_0}{\delta_1^2 + i\Omega\Gamma_a}\right] \quad (13)$$

Equation (13) has real and imaginary parts, including the influence of Szigeti effective charge q_s. This allows for a more comprehensive understanding of the material's response to varying electric fields, particularly at high frequencies. The principal aim of the present research is to study the impact of the Szigeti effective charge through susceptibility on the CRS gain coefficient $(g_R)_{QE}$ of backward scattered mode. In doing so, the following expression is:

$$(g_R)_{QE} = -\frac{k}{2\varepsilon_1}\mathrm{Im}(\chi_R)|E_0|^2 \quad (14)$$

It is evident from the above equations that the Szigeti effective charge and quantum effect considerably influence the third-order nonlinearity of the medium.

3. RESULTS AND DISCUSSION

To examine the dependence of $(g_R)_{QE}$, we chose an n-Insb semiconductor as the Raman medium. The influence of Szigeti's effective charge on CRS we have plotted $(g_R)_{QE}$ as a function of different parameters with $q_s = 0$ and $q_s \neq 0$. The following material parameters are taken as representative values to establish the analytical discussion (Vanshpal et al., 2013): $m = 0.015m_0$, m_0 being the free electron mass, $\varepsilon_1 = 15.8$, $\gamma = 5 \times 10^{-10} Fm^{-1}$, $\rho = 5.8 \times 10^3\ kgm^{-3}$, $\omega_1 = 2 \times 10^{11} s^{-1}$, $\omega_0 = 1.78 \times 10^{14} s^{-1}$, $v = 4 \times 10^{11} s^{-1}$, $N = 1.48 \times 10^{28} m^{-3}$, $\left(\frac{d}{du}\right)_0 = 1.68 \times 10^{-16}$ MKS Unit, $M = 236.47$, $q_s = 1.2 \times 10^{20}\ C$.

Figure 122.1 illustrates the variation of gain constant $(g_R)_{QE}$ with pump electric field E_0. It is observed that in both cases the Raman gain increases with input pump amplitude. It is interesting to note that for $E_0 < E_0 = 4.5 \times 10^8$ the contribution of q_s is more as compared to $\left(\frac{d\alpha}{du}\right)_0$. The inclusion of Szigeti's effective charge in the analysis modifies the way Raman amplification is achieved, providing a more reliable solution for enhancing signal gain in optical communication systems. By incorporating this concept, the efficiency and performance of Raman amplifiers can be significantly improved, leading to enhanced signal quality and reduced transmission losses.

Figure 122.2 illustrates the Variation of gain constant $(g_R)_{QE}$ with wave vector k. The gain increases as the value of the wave vector rises in both curves, but the change intensity

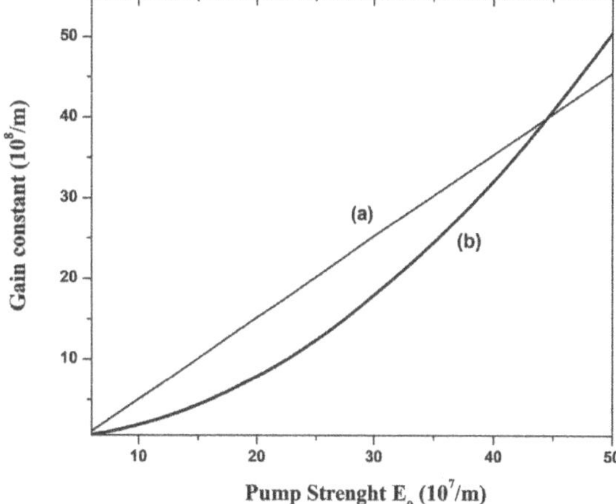

Fig. 122.1 Variation of gain constant $(g_R)_{QE}$ ($q_s \neq 0$, curve (a) and $q_s = 0$, curve (b) with pump electric field E_0 at $k = 3 \times 10^8\ m^{-1}$ and $n_0 = 3 \times 10^{24}\ m^{-3}$

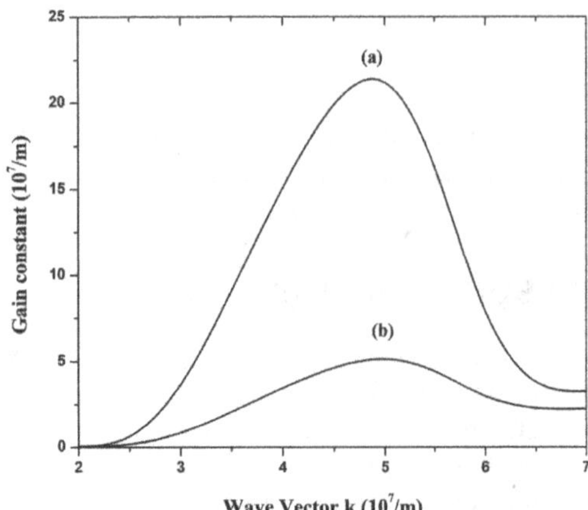

Fig. 122.2 Variation of gain constant $(g_R)_{QE}$ ($q_s \neq 0$, curve-a and $q_s = 0$, curve-b) with wave vector k at $n_0 = 3 \times 10^{24}\ m^{-3}$ and $E_0 = 8 \times 10^7\ Vm^{-1}$

is high when we incorporate the Szigeti effective charge. The characteristics of the two curves are identical, curve-a demonstrated a rapid change in nature, with four times the covered area of curve-b. This effective charge takes into account the screening effects of the surrounding medium, resulting in a more accurate representation of the actual intensity. Therefore, incorporating the Szigeti effective charge can provide a more comprehensive understanding of the gain in relation to wave vector values.

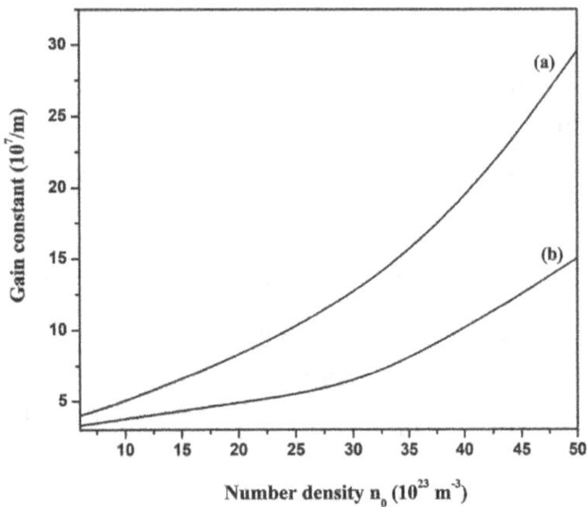

Fig. 122.3 Variation of gain constant $(g_R)_{QE}$ ($q_s \neq 0$, curve-a and $q_s = 0$, curve-b) with number density n_0 at $k = 3 \times 10^8\ m^{-1}$ and $E_0 = 8 \times 10^7\ Vm^{-1}$

Figure 122.3 illustrates the variation of gain constant $(g_R)_{QE}$ with number density n_0. In this illustration, the gain increases as the number density rises and both curves are identical, with a positive relationship between gain and number density. The number density measures the material's doping level, demonstrating a clear relationship between gain and the material's ability to interact with intense electromagnetic radiation. The quantum plasma improvised plasma frequency ϖ_p^2 $\left(\varpi_p^2 = \omega_p^2 + k^2 V_F'^2\right)$ and modified the characteristics of coherent Raman scattering in the material, resulting in a more accurate prediction of the scattering process.

4. CONCLUSIONS

1. The Szigeti effective charge conceptually modifies the way Raman amplification is achieved, providing a more reliable solution for enhancing signal gain in optical communication systems.
2. The QHD model has been effectively used to investigate the Influence of Szigeti effective charge *n*-type direct band gap semiconductor and replaces the existing concept of using high field power to obtain noticeable Raman gain.
3. The results reveal a strong dependence of the Raman gain coefficient on plasma density and laser intensity, highlighting the potential for tuning the gain by controlling these parameters, which is significant for practical applications in semiconductor-based photonics and diagnostics.
4. The quantum effects, particularly the Bohm potential and Fermi pressure, play a crucial role in modifying the absorption characteristics of plasma waves, leading to a more accurate prediction of resonance conditions in the Raman scattering process.

ACKNOWLEDGEMENT

The authors sincerely thank Prof. S. K. Ghosh and Prof. Swati Dubey for their invaluable guidance, insightful suggestions, and unwavering support throughout this research.

REFERENCES

1. Bloembergen, N. and Shen, Y. R. (1964). Nonlinear Optics and Raman Scattering. Physical Review, 133):37–49.
2. Ghosh, S. and Agrawal, V. K. (1982). Effect of magnetostatic field on the stimulated Raman scattering of transversely polarized electromagnetic waves and Raman instability in semiconductor plasmas. Physica Status Solidi b (112):119–126.
3. Ghosh, S., Dubey, S. and Vanshpal, R. (2010). Modulational instability of ion-acoustic waves in quantum dusty plasmas. Physics Letters A (375): 43–47.
4. Guha, S., Sen, P. K. and Ghosh, S. (1979). Parametric instability of acoustic waves in transversely magnetised piezoelectric semiconductors, Physica Status Solidi a (52):407–414.
5. Manfredi, G. (2006). How to Model Quantum Plasmas. Fields Institute Communications (46):263–287.
6. Manfredi, G. and Haas, F. (2001). Self-consistent fluid model for a quantum electron gas. Physical Review B (64): 075316: 1-9.
7. Neogi, A., Ghosh, S. and Sinha, D. K. (1993). Stimulated scattering in magnetoactive semiconductors. Physical Review B (47):16590–16597.
8. Sen, P. and Sen, P. K. (1985). Theory of stimulated Raman and Brillouin scattering in noncentrosymmetric crystals. Physical Review B (31):1034–1040.
9. Sen, P. K., Apte, N. and Guha, S. (1980). Raman instability in n-type piezoelectric semiconductors. Physical Review B (22):6340–6346.
10. Vanshpal, R., Dubey, S. and Ghosh, S. (2013). Stimulated Brillouin Scattering in Semiconductors: Quantum Effects. Chinese Journal of Physics, 51(6):1251–1269

Note: All the figures in this chapter were made by the authors.

Recent Trends in Applied Physics and Material Science – Sudhir Bhardwaj et al. (eds)
 ISBN 978-1-041-16452-4

Enhancing Second Harmonic Generation through Zinc Nitrate Mixed Oxalic Acid Crystals: Growth, Characterization, and Optical Properties Investigation

Sujata B. Bade*
Milliya Arts, Science and Management Science College, Beed, Maharashtra, India

R. V. Late
Ahmednagar College Ahmednagar, Ahmednagar, Maharashtra, India

Y. B. Rasal
S. K. Gandhi Arts, Amolak Science and P. H. Gandhi Commerce College Kada, Beed, Maharashtra, India

V. P. Shirsat, S. S. Hussaini
Milliya Arts, Science and Management Science College, Beed, Maharashtra, India

Abstract: In this study, oxalic acid crystals were grown with zinc nitrate at 20°C using the slow evaporation technique. Crystallographic parameters were analyzed via single-crystal X-ray diffraction. FTIR spectroscopy confirmed the presence of functional groups. UV-visible studies (200–900 nm) revealed optical transparency, with a band gap of 4.10 eV. SHG efficiency was 1.2 times higher than KDP. Photoluminescence and shock wave spectroscopy were also conducted. Dielectric studies examined the thermal effect of zinc nitrate on the crystals. Thermal stability was assessed using TGA/DTA analysis.

Keywords: Semi-organic crystal, Nonlinear optics, UV-visible, TGA/DTA

1. INTRODUCTION

NLO crystals are widely used in optoelectronic devices, while metal-doped organic crystals are useful for optical switching, modulation, and integrated telecommunication bistability [Dinakaran S. et al., 2010]. Organometallic crystals offer advantages like high laser damage thresholds, low dielectric constants, and high SHG efficiency. Oxalic acid, with two directly bonded carboxylic groups, is highly effective for optical applications. Over recent decades, researchers have focused on the linear and nonlinear properties of organic, inorganic, semi-organic, and metal-doped compounds for optical, laser, and photonic uses. This study involves the growth of zinc nitrate-doped oxalic acid crystals via the slow evaporation technique for optical applications.

2. LITERATURE REVIEW

The metals have advantageous properties such as high conductivity, shiny appearance, toughness, malleability, and ductility. And organic materials having unique features optical clarity, thermal stability and presence of central carbon atom with an amine group forming covalent bonds serving as nucleation substrates. Currently, there is a demand of metal doped organic crystals that exhibits promising optical (linear as well as nonlinear) and photonic properties. The non-linearity of the organic crystals enhances due to metal doping found in the literature. J. G. Oliveira Neto et al explored high-temperature effects on bis (β-alanine) nickel (II) dihydrates through structural, thermal and vibrational mode assignments [J. G. Oliveira Neto et al, 2020].

*Corresponding author: sujatabade02091995@gmail.com

DOI: 10.1201/9781003684718-123

3. EXPERIMENTAL PROCEDURE

AR-grade oxalic acid was dissolved in 200 ml of distilled water to prepare a saturated solution at room temperature. Then, 0.2 mol of zinc nitrate was added to the homogeneous solution. The mixture was filtered using Whatman filter paper, and the filtrate was left at room temperature. Oxalic acid and zinc nitrate crystals were grown via slow solvent evaporation over 21 days. After recrystallization, transparent crystals with dimensions of 7 mm × 15 mm × 7 mm were obtained, as shown in Fig. 123.1.

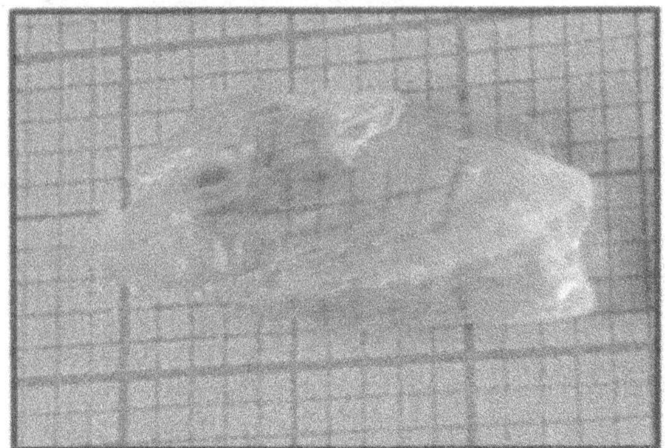

Fig. 123.1 Photograph of grown crystal

4. CHARACTERIZATIONS

4.1 XRD Analysis

A crystal measuring 0.2 mm × 0.4 mm × 10 mm was analyzed using the Enraf Nonius CAD4-MV31 Bruker

Table 123.1 Functional groups

Sr. No	Absorption peak/ bands (cm^{-1})	Assignments
1	3739	NH asymmetric stretching
2	1517	NO_2 stretching of out phase
3	1675	-C=N stretching
4	1085	-N-C-N stretching

Kappa X-ray diffractometer to determine its crystal system and lattice parameters. The results confirmed a monoclinic crystal system, with cell parameters. a = 6.13, b = 3.714, c = 11.93, α = 90°, β = 103.20°, Y = 90°.

4.2 FTIR

FTIR spectroscopy (500–4000 cm^{-1}) using Bruker ALPHA-T was used to identify functional groups in the crystals (Fig. 123.2). Peaks at 675 cm^{-1} indicate C=N stretching, 1085 cm^{-1} corresponds to N-CN stretching, and 1517 cm^{-1} shows NO_2 bending vibrations.

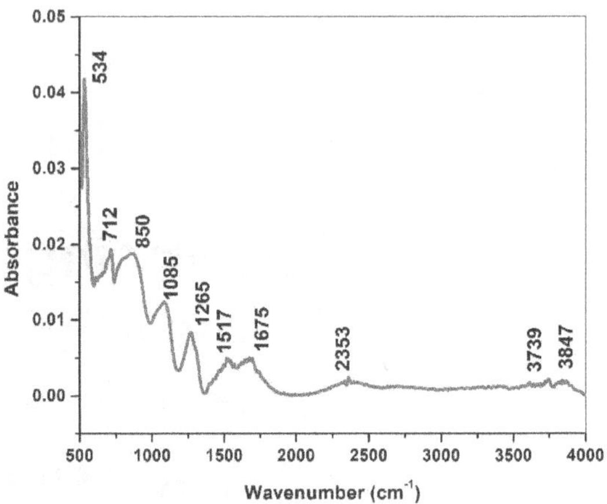

Fig. 123.2 FTIR assignments

Optical Studies

The optical study covered a wavelength range of 100–1000 nm, with the crystal's lower cut-off at 120 nm. The material showed 96% transmittance, meeting the key requirement for non-linear optical effects [Balamurugan S. et al., 2008]. This high transmittance makes it suitable for optical applications, and relevant quantities were calculated using the given equation. The transmittance, calculated as 99%, highlights the crystal's suitability for optical devices. The optical band gap (Eg), shown in Fig. 123.3, was determined using the relation ahn=A(hn-E_g)$^{1/2}$. With high optical conductivity (2.6 × 10202.6 \times 10^{20}2.6 × 1020) and minimal extinction coefficient, indicating low energy loss, the crystal is ideal for optical applications [Mohd A. et al., 2016; ; Mohd A. et al., 2017].

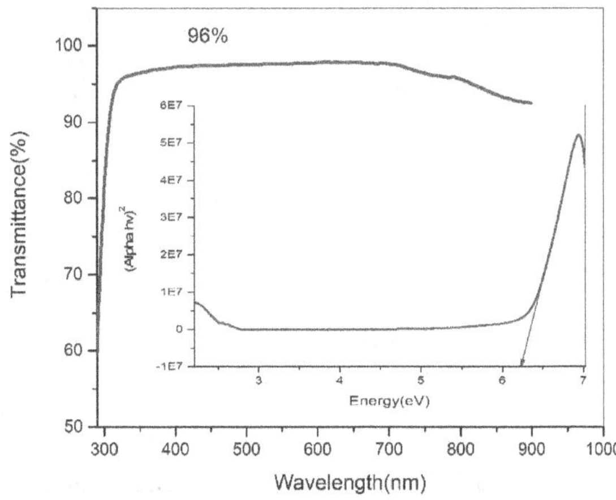

Fig. 123.3 Transmittance and B. G. spectrum

Second Harmonic Generation

Figure 123.4 shows the SHG efficiency of zinc nitrate-doped oxalic acid crystals, which is 1.54 times higher than KDP and greater than pure oxalic acid. The strong non-linear optical response makes these crystals suitable for laser frequency conversion and optical modulation applications [Rasal Y. B. et al., 2015].

4.3 TGA/DTA

Figure 123.5 displays the TGA and DTA plots of the material. TGA shows a slight weight loss at 120°C, indicating the start of decomposition. The DTA curve reveals endothermic transitions at 150°C and 200°C, corresponding to water and other content decomposition, in agreement with the TGA results.

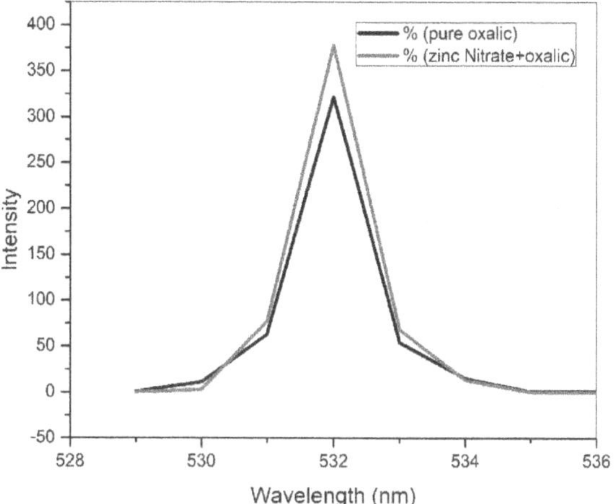

Fig. 123.4 SHG spectrum

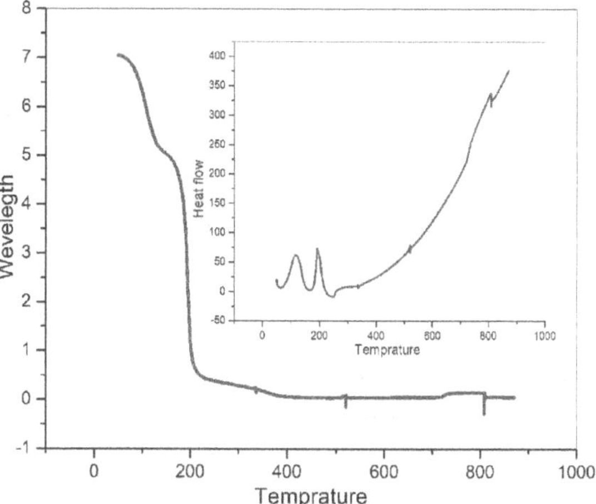

Fig. 123.5 Thermal studies

Both plots confirm the material's stability up to 180°C [Shaikh R. N. et al., 2015]

5. CONCLUSIONS

This A high-quality zinc nitrate-doped oxalic acid crystal was synthesized using the slow evaporation method at 20°C. X-ray diffraction and FTIR analysis confirmed its crystallographic and vibrational properties. UV spectroscopy showed a 3.9 eV band gap. The crystal offers excellent transmittance, high SHG efficiency (1.54 times that of KDP), and thermal stability up to 180°C, making it ideal for nonlinear optical and optoelectronic applications.

REFERENCES

1. Dinakaran S.(2010). Influence of forced convec tion on unidirectional growth of crystals. Physica B 405(3): 3919–3923.
2. Oliveira Neto J. G. (2020) Growth, structural, vibrational, DFT and thermal studies of bis(β-alanine) nickel(II) dihydrate crystals Phys. Chem. Solids. 141: 109435
3. Mohd A., (2017). Bulk growth of undoped and Nd_3 doped zinc thiourea chloride (ZTC) monocrystal: Exploring the remarkably enhanced structural, optical, electrical and mechanical performance of Nd_3 doped ZTC crystal for NLO device applications, Opt. Laser Technol. 90:190.
4. Mohd A (2016) Synthesis growth and optical studies of novel organometallic NLO crystal Calcium bis thiourea chloride optic 127:2137–2142.
5. Rasal Y. B. Investigation on thiourea crystal grown in presence of mmonium acetate. Indian Journal of Pure & Applied Physics. 56 (7):522–528.
6. Shaikh R. N. et al., (2015) Study on optical properties of L-valine doped ADP crystal Spectrochim. Acta A 136: 1243.

Note: All the figures and tables in this chapter were made by the authors.

Recent Trends in Applied Physics and Material Science – Sudhir Bhardwaj et al. (eds)
© 2026 Taylor & Francis Group, London, ISBN 978-1-041-16452-4

Investigation of Vacancy Formation in Muscovite Mica by Different Projectile Ions

Joy Mukherjee
Department of Physics, Indian Institute of Technology, Kanpur, India

D. Bhowmik
Division Micro-robotics and Control Engineering, University of Oldenburg, Oldenberg, Germany

S. Bhowmick
Variable Energy Cyclotron Centre, HBNI, Kolkata, India

Paramita Patra
Institute of Engineering and Management, University of Engineering and Management, Kolkata, India

Abstract: The study illustrates the influence of the different mass of the projectile ion on formation of defect/vacancy in muscovite mica, using two ions (Ar^+ and O^+) with low energy (10 keV) irradiation for normal and grazing incident conditions. Through theoretical analysis, we demonstrate that the heavier projectile ion significantly influences the vacancy production within the constituent elements of mica. It is observed that being heavier, Ar^+ ions induce more vacancy formation than the lighter O^+ ions due to their greater available damage energy for the defect generation. Although both ions transferring similar amounts of energy, the resultant vacancy formation differ, underscoring the vital role of projectile mass in the damage process. In addition, the ion distribution profile shifts towards deeper penetration depths, contrasting with the Bragg's peak of the energy distribution profile, which varies the incidence angle. The mechanism of vacancy formation due to the projectile mass are studied with the help of Monte Carlo simulation implemented code in SRIM and the NRT displacement model. Present work sheds light on the complex relationship between energy transfer from different projectile ion species and the resulting defects production in muscovite mica.

Keywords: NRT displacement model, Vacancy/defect formation, Projectile mass, Transfer of energy by projectiles

1. INTRODUCTION

Naturally occurring 2D muscovite mica is a wide band-gap insulting material, which make it a significant substrate for various applications, including solid-state device, AFM calibration, biomolecule deposition device deposition, and most importantly in optoelectronic device fabrication (Bhowmik et al., 2019 and Kim et al., 2015). A major change challenge in utilizing muscovite mica is reducing its bandgap by inducing defects at different layers. Several studies have explored methods to achieve this, including thermoluminescence, swift heavy ion irradiation, and impurity incorporation to enhance the electronic properties of material by introducing controlled defects (Ma et al., 2020), exfoliation (Kim et al., 2015), and introduction of defect (Pan et al., 2018). Hence, defect creation in muscovite mica has become a topic of considerable interest. In this context introduction of defect by low-energy bombardment is an established technique as it facilitates the sputtering of constituent elements, thereby enabling precise controlled defect formation (Bhowmik et al., 2021 and Mukherjee et al., 2022). The formation of vacancies in muscovite mica can be influenced by several parameters, such as ion energy, ion species, incidence angle (Bhowmik et al., 2021). By tailoring these parameters, the formation of defect/vacancy can be controlled various layers of the material. Nevertheless, the precise contribution of incident ion mass to vacancy formation remains unclear and requires further study.

*Corresponding author: patra.paro369@gmail.com

DOI: 10.1201/9781003684718-124

This chapter investigates the influence of different energetic (10 KeV) projectile ions on defect production in muscovite mica, employing Monte Carlo simulations with Ar^+ and O^+ ions at same energy (10 keV) under varying incidence angles. The result reveals that Ar^+ ions create more vacancies due to their mass similarities to K atoms in mica and also highlights the dependence of ion incidence angle on ion distribution and energy transfer within the material.

2. METHODOLOGY

A single layer of muscovite mica with formula $KAl_2(Si_3Al)O_{10}(OH)_2$ and density ~2.8g/cm^3 was demonstrated for the SRIM -TRIM calculation (Ziegler et al., 2010) based on Monte Carlo Simulation. Ar^+ and O^+ ions with two different masses of ions were chosen for the simulation for their easy accessibility from ion sources and their mass similarities of constituent elements, K and O of the mica, respectively, making them ideal for the objectives of this study.

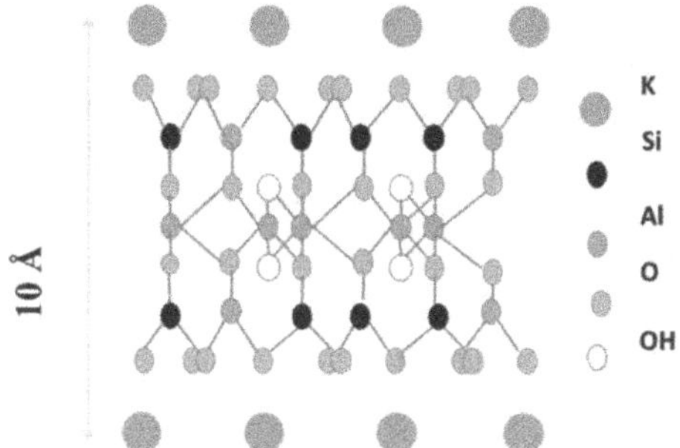

Fig. 124.1 Schematic of muscovite mica layer

3. RESULTS AND DISCUSSION

Figure 124.2 (a-d) and Fig. 124.2 (e-h) illustrate the intricate details of energy deposition by two projectile ions (10 keV Ar-ions and O-ions bombardment at normal incidence) and subsequent formation of vacancy on muscovite mica, resulting from, respectively. Figure 124.2(a) illustrates that the argon-ions in penetrates nearly 12 nm into muscovite mica, signifying the damage of 12 layers of mica. The highest energy deposition by Ar-ion is at a depth of 10 nm insight the muscovite mica shown in Fig. 124.2(b). The peak position of argon-ion distribution and its transfer of energy differ by approximately 2 nm, which can be interpreted as the slowing-down distance of the Ar-ion in mica. Detailed investigation in Figs. 124.2(c) and 124.2(d) highlights the critical role of oxygen atoms in the Ar^+ ions and muscovite mica interaction, where they absorb a significant portion of the transferred energy. Although Si atoms have a higher surface binding energy compared to other atoms, oxygen vacancies are more prominent. Notably, formation of vacancy is greatest at shorter distances compared to ion energy deposition. Additionally, oxygen atoms absorb the most energy, resulting in maximum formation of vacancy in oxygen atoms.

In contrast, from the Figs. 124.2(e) and 124.2(f), it is observed that a shift of approximately 5 nm between the ion distributions peak and energy transfer when O^+ ions bombarded in mica. The O-ions penetrates approximately 26.92 nm in the muscovite mica, signifying that defect formation extends across nearly 27 layers of mica. As with Ar-ions, oxygen atoms predominantly absorb the transferred energy, while hydrogen atoms retain significantly less energy, as illustrated in Fig. 124.2(g). However, formation of vacancy due to oxygen ion bombardment is lower compared to Ar^+ ion bombardment, as shown in Fig. 124.2(h). Although, the impact of incidence angle of projectile ions does not influence the energy transfer by the incident ion, however, the projectile ion range or the ion penetration depth in the material depends on the implantation angle. To confirm this, the projectile ion (Ar^+

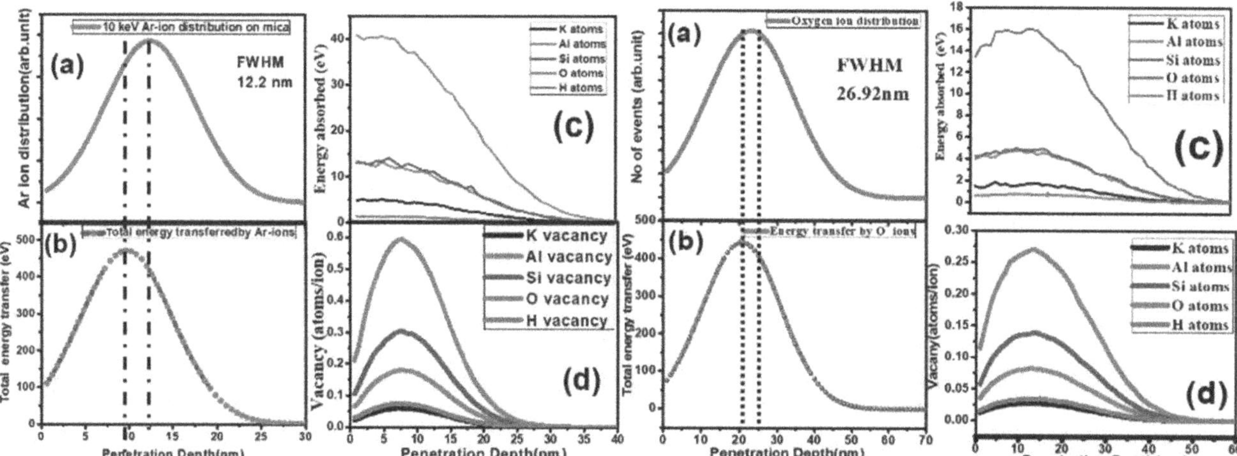

Fig. 124.2 For normal incidence of 10 KeV ions: Distribution of ion with penetration depth on mica on (a) Ar-ion (e) O-ion; Energy transfer by ions (b) Ar-ion (f) O-ion; Energy transferred by (c) Ar-ions (g) O-ions; Formation of vacancy of constituent elements by (d) Ar-ions (h) O-ions

and O^+) distribution and their energy were analyzed at both normal and grazing incidence angles, as presented in Figs. 124.3(a-d) and Figs. 124.3(e-h), respectively. Figure 124.3(a) illustrates the distribution of 10 keV Ar^+ ions bombardment at an 85° angle. The concentration of argon ions is highest within 8 layers, 5 layers fewer than at incidence normally on the material. The distribution in Fig. 124.3(a) peaks at 3 nm, notably lower than the peak observed for normal incident of Ar^+ ion shown in Fig. 124.2(a). At grazing incidence, the maximum energy transferred by Argon ions is about 0.8 nm, revealing a peak shift of two distributions, which is smaller than the shift observed at normal incidence (Fig. 124.3(b)). Similar to normal incidence projectile ions, the maximum formation of vacancies occurs in oxygen atoms as they absorb the most of the energy depicted in Fig. 124.3(c) and Fig. 124.3(d)). Although, the overall formation of vacancy is not as much of ion irradiated at the normal incidence as shown in Fig. 124.2(c) and Fig. 124.2(d). In additional information from Fig. 124.3(e), the distribution of 10 keV O^+ are primarily distributed over 18 layers of the muscovite mica as depicted by the full width half maximum (FWHM). In contrast, the maximum energy of O^+ transferred up to 0.4 nm of the mica shown in Fig. 124.3(f). As in the other cases, the O vacancy is higher due to the significant absorption of energy by O-atoms, depicted in Figs. 124.3(g) and Fig. 124.3(h). However, for O^+ ion deposition, the energy absorbed by each constituent elements of mica and the resulting formation of vacancy are relatively smaller than those observed with Ar^+ incidence, as shown in Fig. 124.3(c) and Fig. 124.3(d). The theoretical explanation of vacancy formation in the material surface by incidence of energetic ions is described by the well-established Norgett-Robinson-Torrens (NRT) displacement model (Norgett et al., 1975). The NRT model states that the formation of vacancies does not occur below the damage threshold (i.e., $T_{dam} = E_d$), here T_{dam} represents the energy transfer afterward the primary knock-on atom (PKA) event, and E_d is the displacement energy needed to form basically a vacancy-interstitial atom pair known as a Frenkel pair. When the energy exceeds the threshold, i.e., above 2.5 E_d, the damage and vacancy formation are expected to formed. According to NRT model, the number of stable Frenkel pair formed by a primary knock-on atom at damage energy T_{dam} is determined by

$$\vartheta_{NRT} = 0.8T_{dam}/2E_d \tag{1}$$

In this work, the displacement energy values are 25 eV for Al, K, and Si, 28 eV for O atoms, and 10 eV for H atoms. The Ar^+ ion, being heavier than the O^+ ion, has a mass identical to that of K atoms in the constituent atoms of muscovite mica. Additionally, the Ar ion projectile range is shallower than that of the O^+ ion, meaning the total energy transferred (E_{tot}) is concentrated in fewer layers. Due to the higher mass of Ar^+ ion, the recoil ionization energy is lower, allowing more energy to be available for defect formation. This results in greater damage and vacancy creation for Ar^+ ions compared to other ions, as the displacement energies of the K, Al, and Si atoms in mica are similar. Furthermore, the O atom has a higher displacement energy, leading to a greater formation of O vacancies in the mica compared to other constituent elements.

4. CONCLUSION

In conclusion, this study provides a qualitative understanding of the formation of vacancy in muscovite mica when subjected to ions of varying mass and projectile incidence angles. The findings confirm that formation of vacancy in the material is influenced by the mass of the incident ion, heavier ions being effective in including vacancies. The energy of the displacement of O atoms is

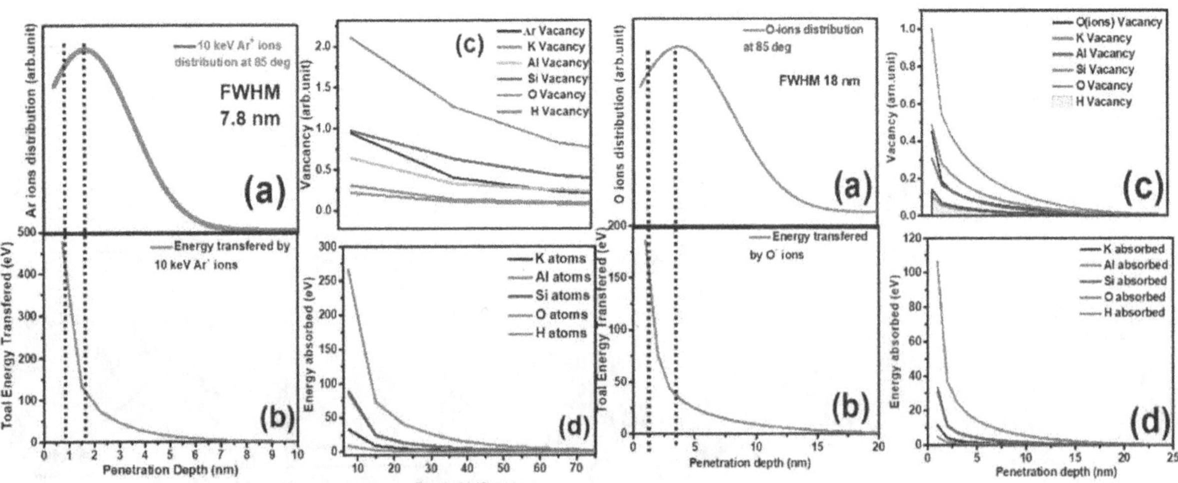

Fig. 124.3 For grazing incidence of 10 KeV ions: Distribution of ion with penetration depth on mica on (a) Ar-ion (e) O-ion; Energy transfer by ions (b) Ar-ion (f) O-ion; Energy transferred by (c) Ar-ions (g) O-ions; Formation of vacancy of constituent elements by (d) Ar-ions (h) O-ion

highest and is lowest for H atoms of all constituent atoms of the Muscovite mica, which leads to a larger oxygen vacancy compared to other elements. This emphasizes the critical role of ion mass and energy transfer in determining the extent of the formation of vacancy in muscovite mica.

REFERENCES

1. Bhowmik, D., and Karmakar, P. (2019). Tailoring and investigation of surface chemical nature of virgin and ion beam modified muscovite mica. Surf. Interf. Analysis 51(6):667–673.
2. Kim, S. S., Van Khai, T., Kulish, V., Kim, Y.H., Na, H. G., Katoch, A., Osada, M., Wu, P. and Kim, H. W. (2015). Tunable Bandgap Narrowing Induced by Controlled Molecular Thickness in 2D Mica Nanosheets. Chem. Mat. 27(12):4222–4228.
3. Ma, Z., Skumryev, V. and Gich. M. (2020). Magnetic properties of synthetic fluorophlogopite mica crystals. Mater. Adv. 1(5):1464–1471.
4. Pan, X. F., Gao, H. L., Lu, Y., Wu, C. Y., Wu, Y.D., Wang, X. Y., Pan, Z. Q., Dong, L., Song, Y. H., Cong, H. P. and Yu, S. H. (2018). Transforming ground mica into high-performance biomimetic polymeric mica film. Nat. Commun. 9(1):2974.
5. Bhowmik, D. and Mukherjee, J. (2021). Study of low energy ion beam induced sputtering parameters of muscovite mica $[KAl_2(Si_3Al)O_{10}(OH)_2]$ using Monte Carlo simulation. Mater. Today. Proc.47(8):1515–1519.
6. Mukherjee, J. Dalsaniya, M. H., Bhowmick, S., Bhowmik, D., Jha, P. K. and Karmakar, P. (2022). Band gap engineering of the top layer of mica by organized defect formation. Surf. Interf. 33:102283.
7. Bhowmik, D., Mukherjee, J. and Karmakar, P. (2021). Projectile mass dependent nano patterning and optical band gap tailoring of muscovite mica. Rad. Phy. and Chem. 187:1029568.
8. Ziegler, J. F., Ziegler, M. D. and Biersack, J. P. (2010). SRIM - The stopping and range of ions in matter (2010). Nucl. Instrum. Meth. Phys. Res. B 268(11–12): 1818–1823.
9. Norgett, M. I., Robinson, M. T. and Torrens, I. M. (1975). A Proposed Method of Calculating Displacement Dose Rates. Nucl. Eng. Degn. 33:50–54. North-Holand Publishing Company.

Note: All the figures in this chapter were made by the authors.

Recent Trends in Applied Physics and Material Science – Sudhir Bhardwaj et al. (eds)
© 2026 Taylor & Francis Group, London, ISBN 978-1-041-16452-4

Investigating the Salt Concentration and Temperature on the Mean Squared Displacement of Ionic Species in EC-LiTFSI and PEO-$NaPF_6$ Electrolytes

Manisha Kumari Meena,
Rakhi Katyal, Akash Kumar Meel,
Hema Teherpuria, and Santosh Mogurampelly*
Polymer Electrolytes and Materials Group (PEMG),
Department of Physics, Indian Institute of Technology Jodhpur, Rajasthan, India

Abstract: Understanding the impact of salt concentration and temperature on ion dynamics and the structural properties of electrolytes provides critical insights useful for designing advanced rechargeable battery electrolytes. In this work, we employ atomistic molecular dynamics (MD) simulations to study two distinct classes of battery electrolyte systems: a liquid electrolyte (LE: EC-LiTFSI) and a polymer electrolyte (PE: PEO-$NaPF_6$). We analyze the mean squared displacement (MSD) of ionic species and density as a function of salt concentration and temperature. The chosen salt concentrations include regimes of maximum ionic conductivity, and the temperature range spans below and above the glass transition temperature (T_g). Our results reveal a significant increase in MSD above a certain temperature, accompanied by corresponding changes in density that qualitatively correlate with ion mobility. This transition temperature is found to increase with salt concentration, saturating at higher concentrations for both systems. These findings provide critical insights into the ion transport mechanisms and structural dynamics, offering guidelines for optimizing electrolyte performance in practical battery applications.

Keywords: Glass transition, Liquid electrolyte, Polymer electrolyte, MD simulation, Density, MSD

1. INTRODUCTION

Electrolytes are vital for rechargeable batteries, enabling efficient ion transport between electrodes. Their performance, dictated by ionic conductivity, depends on ion mobility, which is influenced by the structure, salt concentration, and temperature. The mean squared displacement (MSD) quantifies ion motion and links directly to conductivity, while density changes reveal ion-solvent organization critical for ion dynamics. Electrolytes operate across temperatures, from rigid glassy states below the glass transition (T_g) to flexible, fluid states above it. Ion mobility increases above T_g, making the study of MSD and density across this transition key for optimizing performance. Salt concentration further complicates dynamics; low concentrations enhance mobility, while higher concentrations promote ion clustering and reduced mobility. This work employs molecular dynamics simulations to analyze MSD and density in ethylene carbonate-lithium bis(trifluoromethanesulfonyl) imide (EC-LiTFSI) and polyethylene oxide-sodium hexafluorophosphate (PEO-$NaPF_6$) electrolytes across varying concentrations and temperatures. By examining behavior across T_g, we uncover mechanisms influencing

*Corresponding author: santosh@iitj.ac.in

DOI: 10.1201/9781003684718-125

ion transport and structural organization, providing insights for advanced battery electrolytes with enhanced ionic conductivity and mechanical stability.

2. METHODOLOGY

All-atom molecular dynamics (MD) simulations examined ion transport and structural properties of EC-LiTFSI and PEO-$NaPF_6$ electrolytes using established force fields. For PEO-$NaPF_6$, a single PEO chain ($n = 50$) was constructed and combined with $NaPF_6$ salts in PACKMOL to achieve specific ethylene oxide-to-sodium ratios (EO:Na). For EC-LiTFSI, ethylene carbonate (EC) and LiTFSI salts were similarly placed at predefined concentrations. The potential included bonded (bond, angle, dihedral) and nonbonded (LJ, Coulombic) terms:

$$U(r) = \frac{1}{2}k_r\left(r - r_0\right)^2 + \frac{1}{2}k_\theta\left(\theta - \theta_0\right)^2 + \frac{1}{2}\sum_{n=1}^{4} C_n\left[1 + (-1)^{n+1}\cos(n\phi)\right] + \sum 4\epsilon\left[\left(\frac{\sigma}{r_{ij}}\right)^{12} - \left(\frac{\sigma}{r_{ij}}\right)^{6}\right]$$

The first three terms contain bonded potential containing bond, angle, and dihedral information, and the nonbonded terms contain standard LJ and Coulombic electrostatic potential. Simulations were performed using GROMACS 2020.2 with the OPLSAA force field for PEO developed by Acevedo et al. After energy minimization and equilibration (1 ns NVT, 5 ns NPT), 50 ns production runs analyzed MSD, density, and ion dynamics.

3. RESULTS AND DISCUSSION

Figure 125.1 shows the mean squared displacement (MSD) of ionic species at 250 K with a lag time of 5 ns for EC-LiTFSI and PEO-$NaPF_6$. In EC-LiTFSI, the MSD of Li^+ and $TFSI^-$ decreases with increasing salt concentration due to enhanced ion-ion correlations and clustering, which hinder mobility. Similarly, in PEO-$NaPF_6$, Na^+ and PF_6^- exhibit reduced MSD at higher concentrations. In polymer electrolytes, ion mobility is influenced by ion-polymer interactions and polymer segmental motion, both of which are restricted at elevated salt concentrations due to stronger ionic interactions.

The MSD temperature dependence at fixed salt concentrations is also shown in Fig. 125.1. A marked increase in MSD occurs as the temperature exceeds T_g. Below T_g, ion motion is constrained in the glassy state, while above T_g, the system transitions to a more flexible state, enabling higher ion mobility. For EC-LiTFSI at 1.25 M, Li^+ and $TFSI^-$ show a sharp MSD increase above 250 K. In PEO-$NaPF_6$ at 1.43 M, Na^+ and PF_6^- demonstrate enhanced mobility at a higher predicted T_g, emphasizing

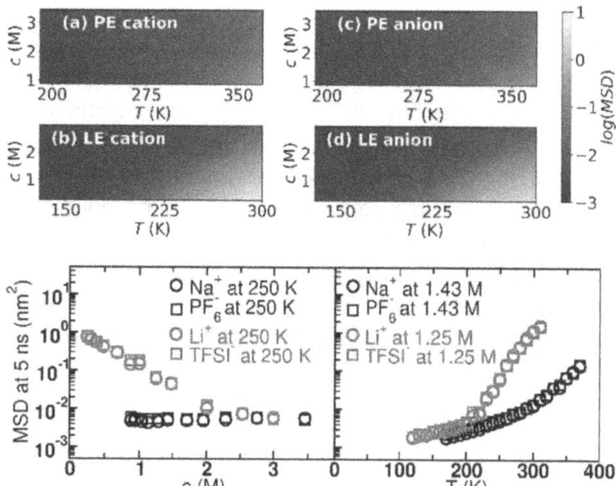

Fig. 125.1 MSD of the ionic species at different temperatures and salt concentrations in both the liquid and polymer electrolyte systems

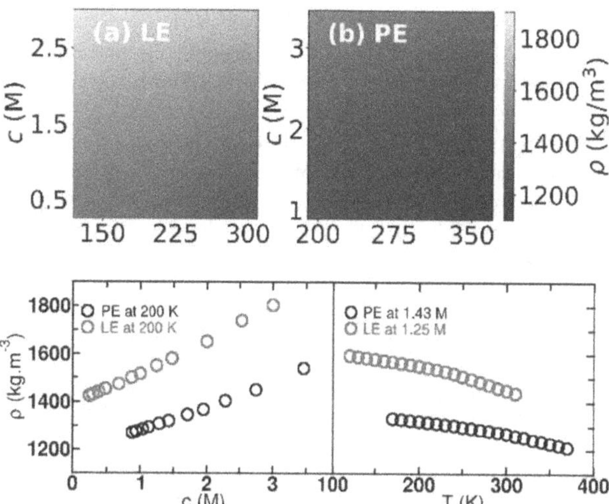

Fig. 125.2 Density of the liquid and polymer electrolyte systems at different temperatures and salt concentrations

the role of T_g in ion transport in both systems. Figure 125.2 depicts the density (ρ) of EC-LiTFSI and PEO-$NaPF_6$ as a function of salt concentration at 200 K. In both systems, ρ increases with salt concentration due to added salt ions compacting the structure. Conversely, at constant salt concentration, ρ decreases with rising temperature, reflecting thermal expansion. This effect intensifies above T_g, where the system becomes more disordered, reducing density. The correlation between ρ and MSD reveals that higher densities at lower temperatures limit ion mobility, while lower densities above T_g promote faster transport.

4. CONCLUSIONS

In this study, atomistic molecular dynamics simulations were used to examine how salt concentration and temperature influence ion dynamics and structural properties in EC-LiTFSI and PEO-$NaPF_6$ electrolytes. The mean squared displacement (MSD) of ionic species decreases with higher salt concentrations, reflecting reduced mobility due to stronger ion-ion correlations. Above the glass transition temperature (T_g), a sharp increase in MSD underscores the importance of T_g in facilitating ion transport. Similarly, density increases with salt concentration and decreases with temperature, corresponding to structural compaction and thermal expansion effects. These findings reveal the role of T_g and salt concentration in shaping ion mobility and electrolyte performance.

REFERENCES

1. Teherpuria, H., Paul Chowdhury, S. S., Kannam, S. K., Jaiswal, P. K., and Mogurampelly, S., (2024). Ionic Conductivity Mechanisms in EC-LiTFSI Electrolytes: Interplay of Ion-ion Correlations and Viscosity. arXiv:2401.11182
2. Mohapatra, S., Halder, S., Chaudhari, S. R., Netz, R. R. and Mogurampelly, S., (2023). Insights into the Structure and Ion Transport of Pectin-[BMIM][PF6] Electrolytes, *J. Chem. Phys. 159*, 154902
3. Mogurampelly, S.; Borodin, O.; Ganesan, V. (2016). Computer Simulations of Ion Transport in Polymer Electrolyte Membranes. *Annu Rev Chem Biomol Eng, 7* (Volume 7, 2016):349–371.
4. Maitra, A.; Heuer, A. (2007). Cation Transport in Polymer Electrolytes: A Microscopic Approach. *Phys Rev Lett, 98* (22), 227802.
5. Kauzmann, Walter. (1948). The Nature of the Glassy State and the Behavior of Liquids at Low Temperatures. *Chem Rev , 43* (2):219–256.
6. Haghkhah, H.; Ghalami Choobar, B.; Amjad-Iranagh, S. (2020). Effect of Salt Concentration on Properties of Mixed Carbonate-Based Electrolyte for Li-Ion Batteries: A Molecular Dynamics Simulation Study. *J Mol Model, 26* (8), 220.
7. Gudla, H.; Zhang, C. (2024). How to Determine Glass Transition Temperature of Polymer Electrolytes from Molecular Dynamics Simulations. *J Phys Chem B, 128* (43):10537–10540.
8. Martínez, L.; Andrade, R.; Birgin, E. G.; Martínez, J. M. (2009). PACKMOL: A Package for Building Initial Configurations for Molecular Dynamics Simulations. *J Comput Chem, 30* (13):2157–2164.
9. Van Der Spoel,D.; Lindahl, E.; Hess, B.; Groenhof, G.; Mark, A. E.; Berendsen,H. J. C.(2005)GROMACS: Fast, Flexible, and Free. *J Comput Chem,26* (16):1701–1718.
10. Jorgensen, W. L.; Maxwell, D. S.; Tirado-Rives, J. (1996). Development and Testing of the OPLS All-Atom Force Field on Conformational Energetics and Properties of Organic Liquids. *J Am Chem Soc, 118* (45), 11225–11236.

Note: All the figures in this chapter were made by the authors.

Recent Trends in Applied Physics and Material Science – Sudhir Bhardwaj et al. (eds)

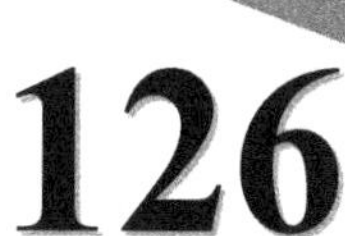

Computational Algorithms for Biospeckles Analysis and Evaluation: Review

Sadhana Tiwari*
Electronics & Telecommunication, IET-DAVV, India
Electronics & Communication, PIEMR, India

Shivangi Bande
Electronics & Telecommunication,
IET-DAVV, India

Abstract: The biospeckle technique optical technique and non-destructive in nature. This is used for the activity analysis of the biological samples based on the pattern received on screen via a backscattered laser beam from the sample. This technique has been successfully applied in various domains like medicine, agriculture, and microbiology. In this paper, we perform a survey focused on the computational models and algorithms used for the analysis of speckle signals. We examine the biospeckle acquisition process and methods used for analysis. Our findings indicate that computational models based biospeckle analysis getting an attention for different application domain in recent years.

Keywords: Biospeckle, Computational models, Biological processes

1. INTRODUCTION

Biospeckle laser (BSL) technique is applicable to various fields agriculture, food processing, medical and biometrics(Zdunek et al. 2014). The technique includes acquisition of signals and images, analysis, interpretation and prediction of activity undergoing inside the sample. Biospeckle pattern is produced by stationary and moving scatterers of the sample when illuminated by a laser beam. There are intensity fluctuations in the patterns received via scattered beams. These variations in intensity characterize the process or activity. In order to analyze the overall activity from bio speckle pattern, two methods are reported namely numerical and visual methods. Visual methods such as generalized difference (GD) (Saúde et al. 2012), Fuji's method (Chatterjee, Disawal, and Prakash, n.d.) used to derive qualitative information of the activity. Saúde et al presented the generalized difference (GD), and the Fujii's method for the analysis of video sequence. Chatterjee et al compared, Fuji's and alternative Fujii method for bread spoilage detection. Numerical methods as inertia-moment (IM) (Braga et al. 2011), absolute value of difference (AVD) (Ansari and Nirala 2013) used to derive quantitative information about sample. Braga et al applied Co-occurrence matrix (COM) based AVD algorithm on biospeckles of bean seed to analyse the drying process.

Traditional visual methods do not offer real-time monitoring and numerical methods are suited only for homogeneous sample based applications. Inability to handle heterogeneous samples, low sensitivity, high standard deviation, manual procedure, and manual region of interest (ROI) selection, are the limitations of conventional methods for bio speckle analysis. However, real-time automation and monitoring are now feasible because to developments in intelligent systems. Use of intelligent systems in image processing facilitate enhancement, analysis, and interpretation of images by leveraging advanced algorithms and computational techniques. These systems incorporate computational techniques to solve complex tasks which are challenging with traditional methods. In this paper, Section II describes the basic set up and different hardware used for acquisition of biospeckles. Section III includes different models and algorithms for analysis.

*Corresponding author: sadhana.khandekar3@gmail.com

DOI: 10.1201/9781003684718-126

2. ACQUISITION OF BIOSPECKLE SIGNAL AND IMAGES

The acquisition of bio speckle signal/image involves capturing of the dynamic light interference caused due to biological activity within a biological sample. Acquisition of Biospeckle signal & images is demonstrated in Fig. 126.1. It comprised of laser source, CCD camera, filter, beam expander, and computing device/algorithms (Tiwari and Bande 2024).

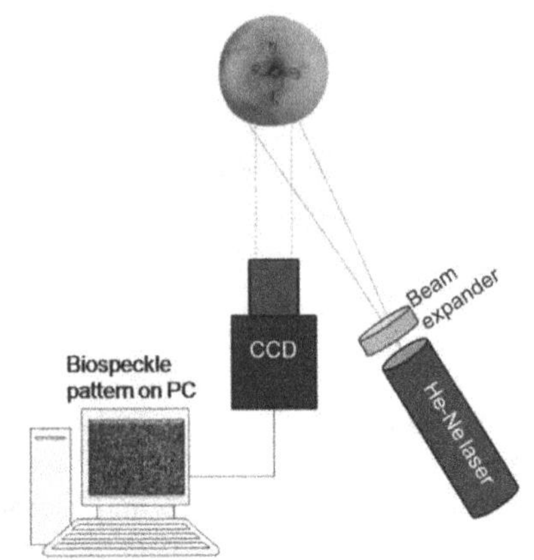

Fig. 126.1 Acquisition of biospeckle signal & images

Table 126.1 depict the hardware specification of the experimental set up used by researchers for their research problem:

The unavailability of standard and commercial equipment limits the accessibility and usability of this technology in the field despite having the capability to measure biological activity. As an alternative, Diego et al (Diego, Pujaico, and Braga 2019) used a diode device with 635 nm, a smartphone camera with a macro zoom feature and computing facilities, and Android 5.0 for biospeckle data acquisition.

3. ANALYSIS OF BIOSPECKLES

In order to estimate dynamicity from bio speckle pattern, statistical, graphical and numerical method are suggested by the researchers. Graphical methods perform qualitative analysis based on activity map generated from bio speckles. These activity maps obtained by various methods: Fujii's method, Generalized Difference (GD), and Subtraction Average (SA). Fujii's method (Fujii and Asakura 1985) was applied to observe blood flow using speckles. For this weighted sums of the differences between two consecutive pixels' intensities was calculated. Braga et al analyzed the activity in the seed tissue using GD method. GD evaluates intensity changes with time instances and assign new value to every pixel of processed image. Graphical methods perform qualitative evaluation. Whereas, numerical methods assign a numeric value to each dynamic level. Numerical methods are based on a temporal history of speckle pattern (THSP) matrix. Which is further used to calculate COM, IM and AVD.Traditional analysis methods demand high computational costs and huge set of images. Laser Speckle Contrast Analysis (LASCA) needs low computational cost offers the advantage of real-time monitoring and processing of bio specimen (Briers et al. 2012).

This method is based on computation of contrast for each pixel across a series of images. High contrast indicates static regions and low contrast indicates high dynamic regions (e.g., blood flow).The Motion history image (MHI) method was also used for the analysis of real-time phenomena and reduces loss of resolution while monitoring of images (Godinho et al. 2012). MHI visualize areas of intense movement over a sequence of frames and recent motion is indicated by change in intensity. The use of automation techniques reduces computational and experimental overheads and enhance adaptability for practical applications. The inclusion of intelligent and computational algorithms (Salehi et al. 2023) make the classification, prediction and analysis fast and more reliable. This fact has motivated the researchers to use computational models and algorithms for the analysis of bio speckles in different domains. Table 126.2 and 126.3 Methods/algorithms for biospeckle analysis and use case on computational algorithms.

Table 126.1 Hardware specification used in BSL applications

Laser	Camera	Computing devices/algorithms	Distance between camera & sample	Number of frames acquired	Reference
GaAs laser	Near-infrared camera	Laser Speckle Network (LSNet) /Personal computer	20 cm	263	(Lu and Yang 2021)
He-Ne laser	Camera with CMOS image sensor	Quantitative and Qualitative algorithms / Biospeckle Laser Tool Library, MATLAB	140 and 200 mm	256	(Wing et al. 2022)
Green laser	BASLER digital camera	MATLAB	25 mm	300	(Tang et al. 2021)
Solid state laser	Basler Digital camera (CMOS)	STATISTICA V12 (StatSoft, used to build model	focal length of 8mm	41	(Rahmanian et al. 2020)

Table 126.2 Methods/algorithms for biospeckle analysis

Method/Algorithm	Description
Mean	First order statistics that indicate level of temporal variation of speckle. (Dai et al., n.d.)
Autocorrelation	Second-Order Statistics gives average correlation between the intensities at two successive sampling instants in speckle pattern, function (Xu 2015)
Spectral density	Second-Order Statistics giving velocity information of time-varying speckles. (Xu 2015)
Wavelet transform	Frequency domain methods and gives spatial information of speckles. (Kleber Mariano Ribeiro et al. 2013)
Entropy	Measures the information produced by the amplitude of each sample. (Dai et al., n.d.)
THSP	Indicate the evolution of biospeckle with respect to time. (Zdunek et al. 2014)
Normalized COM	Number of pixels in THSP where intensity increases from value I(k)=i to I(k+1)=j, where k is time. (Zdunek et al. 2014)
Inertia moment	It gives mean value of intensity jumps that occurred at any pixel at any time. (Braga et al. 2011)
AVD	Modified version of IM and more sensitive to intensity jump. (Md Zaheer Ansari and Nirala 2013

Table 126.3 Use case based on computational algorithms

Methods/Algorithms	Description
Tunable algorithm	Tunable filters based on addition and subtraction operations used to process biospeckels tuning parameter is determined on basis of the desired result.(Buffarini et al. 2020)
Morphological and geo-statistical operators	Morphological and geo-statistical operators with alpha variogram and numerical indexing techniques used for automated analysis of biospeckles. (Chatterjee et al. 2020)
Principal component analysis (PCA)	PCA technique used to filter the data from the biospeckles and then signals were analyzed using graphical and numerical methods for analysis. (Ribeiro 2014)
Support vector machine (SVM), Artificial neural networks (ANN)	SVM and ANN algorithms were used to classify oranges into different categories using biospeckle descriptors and machine learning algorithms..(Rahmanian et al. 2020)
Convolutional neural network (CNN)	The CNN method is applied to identify and classify apple defects (Wu et al 2020)

4. CONCLUSION

In presented work, systematic review the state- of- art of biospeckle signal/image analysis was performed to identify and describes different analysis methods of biospeckles signal/images. The results showed a continuous advancement in the analysis technique. BSL usages is maximum in agriculture and medical applications. Most of the analysis methods were based on numerical methods, using IM and AVD algorithms. In recent years' computational methods are getting attention for the analysis and interpretation of activity using bio speckle images. Maximum research is carried out using images captured in the laboratory set up. This limits the BSL based research for field applications. Further research and investigations needs is to be done for improvement in image analysis methods and field application of BSL. Presented review is helpful for the researchers and entrepreneurs willing to work in BSL domain.

REFERENCES

1. Ansari, Md Zaheer, and A. K. Nirala. (2013). Assessment of Bio-Activity Using the Methods of Inertia Moment and Absolute Value of the Differences." *Optik* 124 (6): 512–16.
2. Braga, R. A., C. M.B. Nobre, A. G. Costa, T. Sáfadi, and F. M. Da Costa. (2011). Evaluation of Activity through Dynamic Laser Speckle Using the Absolute Value of the Differences. *Optics Communications* 284 (2): 646–50.
3. Briers, J. David, Paul M. Mcnamara, Marie Louise O'Connell, and Martin J. Leahy. 2012. "Laser Speckle Contrast Analysis (LASCA) for Measuring Blood Flow." *Microcirculation Imaging*, no. April: 147–63. https://doi.org/10.1002/9783527651238.ch8.
4. Buffarini, L, H Rabal, N Cap, E Grumel, M Trivi, and P Finger. 2020. "Tuneable Algorithms for Tracking Activity Images in Dynamic Speckle Applications." *Optics and Lasers in Engineering* 129 (September 2019): 106084. https://doi.org/10.1016/j.optlaseng.2020.106084.
5. Chatterjee, Amit, Reena Disawal, and Shashi Prakash. n.d. "Biospeckle Assessment of Bread Spoilage by Fungus Contamination Using Alternative Fujii Technique," 395–401. https://doi.org/10.1007/978-981-10-3908-9.
6. Chatterjee, Amit, Puneet Singh, Vimal Bhatia, and Shashi Prakash. 2020. "An Efficient Automated Biospeckle Indexing Strategy Using Morphological and Geo-Statistical Descriptors." *Optics and Lasers in Engineering* 134 (May): 106217.
7. Dai, Ana L, Isabel L Passoni, G Hernan Sendra, Marcelo Trivi, Hector J Rabal, Grupo Inteligencia Artificial, Departamento De Matemática, Juan B Justo Av, Mar Plata, and Bs As. n.d. "Biospeckle Signal Descriptors : A Performance Comparison," 1–14.

8. Diego, Marcos, Fernando Pujaico, and Roberto A Braga. 2019. "Optik Viability of Biospeckle Laser in Mobile Devices." *Optik - International Journal for Light and Electron Optics* 183 (July 2018):897–905.
9. Fujii, Hitoshi, and Toshimitsu Asakura. 1985. "Blood Flow Observed by Time-Varying Laser Speckle" 10 (3): 104–6.
10. Godinho, R. P., M. M. Silva, J. R. Nozela, and R. A. Braga. 2012. "Online Biospeckle Assessment without Loss of Definition and Resolution by Motion History Image." *Optics and Lasers in Engineering* 50 (3): 366–72. https://doi.org/10.1016/j.optlaseng.2011.10.023.
11. Lu, Yuri, and Simon X Yang. 2021. "LSNet : IDENTIFICATION OF COPPER AND STAINLESS STEEL USING LASER SPECKLE IMAGING IN," no. April. https://doi.org/10.2316/J.2021.206-0568.
12. Rahmanian, Alireza, Seyed Ahmad, Saeid Sadri, Mahdiyeh Gholami, and Majid Nazeri. 2020. "Postharvest Biology and Technology Application of Biospeckle Laser Imaging for Early Detection of Chilling and Freezing Disorders in Orange." *Postharvest Biology and Technology* 162 (July 2019): 111118. https://doi.org/10.1016/j.postharvbio.2020.111118.
13. Ribeiro, K M. 2014. "Principal Component Analysis in the Spectral Analysis of the Dynamic Laser Speckle Patterns" 14009.
14. Salehi, Ahmad Waleed, Shakir Khan, Gaurav Gupta, Bayan Ibrahimm Alabduallah, and Abrar Almjally. 2023. "A Study of CNN and Transfer Learning in Medical Imaging : Advantages , Challenges , Future Scope."
15. Saúde, André V., Fortunato S. de Menezes, Patricia L. S. Freitas, Giovanni F. Rabelo, and Roberto A. Braga. 2012. "Alternative Measures for Biospeckle Image Analysis." *Journal of the Optical Society of America A* 29 (8): 1648. https://doi.org/10.1364/josaa.29.001648.
16. Tang, Xin, Ping Zhong, Zhisong Li, Yinrui Gao, and Haowei Hu. 2021. "Optik Evaluation of Biological Speckle Activity : Using Variational Mode Decomposition." *Optik* 243 (December 2020): 167475. https://doi.org/10.1016/j.ijleo.2021.167475.
17. Tiwari, Sadhana, and Shivangi Bande. 2024. "Intelligent Systems And Applications In Engineering LASCA-Based Monitoring of Drug Impact and Classification Using Machine Learning for Biospeckle Images of Melanoma Cells" 12: 506–12.
18. Wing, Spotted, Drosophila Drosophila, Arkadiusz Ratajski, and Monika Janaszek-ma. 2022. "Applied Sciences Biospeckle Activity of Highbush Blueberry Fruits Infested By."
19. Wu, Ang, Juanhua Zhu, and Taiyong Ren. 2020. "Detection of Apple Defect Using Laser-Induced Light Backscattering Imaging and Convolutional Neural Network ☆" 81. https://doi.org/10.1016/j.compeleceng.2019.106454.
20. Xu, Zijie. 2015. "Temporal and Spatial Properties of the Time- Varying Speckles of Botanical Specimens" 34 (5): 1487–1502.
21. Zdunek, Artur, Anna Adamiak, Piotr M. Pieczywek, and Andrzej Kurenda. 2014. The Biospeckle Method for the Investigation of Agricultural Crops: A Review. *Optics and Lasers in Engineering*. https://doi.org/10.1016/j.optlaseng.2013.06.017.

Note: All the figures and tables in this chapter were made by the authors.

Recent Trends in Applied Physics and Material Science – Sudhir Bhardwaj et al. (eds)

127 Buckling Analysis of Porous FGM Plate using First Order Shear Deformation Theory

Kanishk Sharma[1], Vikash Choudhary[2]
Department of Mechanical Engineering Department, JECRC University, Jaipur, India

Abstract: The buckling analysis of a porous rectangular FGM plate is presented using first-order shear deformation plate theory based finite element method. A laminate with fully bonded isotropic layers is used to mimic an FGM plate having even-type porosity. The material properties at the middle of each layer are calculated using power law and the rule of mixture. Numerous parametric experiments are carried out to ascertain how various characteristics, such as porosity, geometry variables, or material gradation exponent, affect the buckling reaction in the FGM plate after the current formulation has been validated with prior literature. The buckling strength in the FGM the material gradation exponent is shown to have a considerable effect on the plate. Additionally, the rise in porosity results in a monotonic decrement in the buckling of the FGM plate's strength. It is also shown that the aspect ratio of rectangular FGM plates considerably impacts FGM plates' buckling behavior.

Keywords: FGM (Functionally graded material), Buckling behavior, Porous FGM plate, Finite element analysis, FSDT

1. INTRODUCTION

Advanced composites known as FGMs were created to mitigate the drawbacks i.e., stress concentration, delamination, and debonding associated with traditional composite materials by providing a smooth and continuous variation of volume fractions of parent constitute materials, generally metal (good mechanical strength) and ceramic (good thermal resistance) in the prespecified directions. Owing to its enhanced structural performance the FGM is finding increasing applications in multiple engineering fields, especially under high-temperature conditions, and gained worldwide research interests (Dash et al., 2023). It is well known that under the axial compressive load, the thin-walled structure can fail catastrophically due to buckling. Many researchers have conducted the buckling analysis of FGM structures since it is a crucial design requirement. (Belounar et al., 2023). Moreover, porosity refers to the existence of gaps or vacant areas in the substance that can be induced due to manufacturing and /or processing defects. Also. by inducing porosity, the FGMs can exhibit desired specific mechanical, thermal, or electrical properties. (Ramteke & Panda, 2023). However, porosity can affect the response of FGM structures significantly as shown in numerous studies ((Chaabani et al., 2023). It is Observed from the preceding literature study indicates research in buckling examination of porous FGM plates with mechanical loadings is rather restricted. This work aims to investigate buckling analysis of buckled porous FGM plates beneath mechanical stress loading.

2. MATHEMATICAL MODEL OF POROUS FGM PLATE

This work considers a porous FGM plate shown in Fig. 127.1, which has length a, breadth b, and thickness h. Along the thickness of FGM plate's the material properties change by a power law using power law index n, with metal on the bottom and pure ceramic on the top. With porosity volume fraction α and with the metal and ceramic

[1]kanishksharma2009@gmail.com, [2]vjvikash106@gmail.com

DOI: 10.1201/9781003684718-127

volume fractions, *Vc & Vm,* respectively, the FGM plate's Young's modulus can be expressed as follows:

$$E(z) = (E_c - E_m)\left(\frac{2z+h}{2h}\right)^n + E_m - \frac{\alpha}{2}(E_c + E_m) \quad (1)$$

where, the subscripts '*c*' & '*m*' stand, respectively, for ceramic and metal; and *n* is power law exponent.

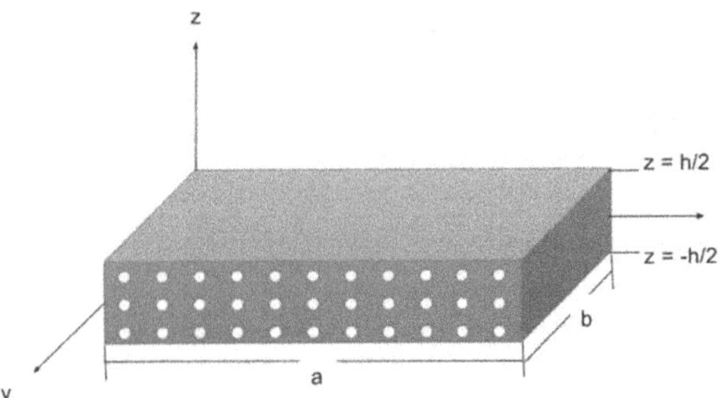

Fig. 127.1 Porous FGM plate with porosity volume fraction $0 \leq \alpha < 1$

A macro written in ANSYS® Parametric Design Language is used to execute the investigation of the porous FGM plate's buckling wherein this analysis is performed in two phases in ANSYS®.

3. NUMERICAL RESULTS

The number of elements to mesh the porous Al/Al_2O_3 FGM plate with simple support has been selected by carrying out the convergence study for buckling analysis using 10 layers along with 4 × 4, 8 × 8, and 16 × 16 elements. A non-dimensional critical buckling force, i.e., $(N_x a^2)/(E_m h^3)$ of FGM plate are tabulated in Table 127.1. The value of Young's modulus for parent material i.e., for Alumina and Aluminum are taken as 380 GPa & 70 GPa respectively, while the poison's ratio for both materials is considered as 0.30. Table 127.1 shows that the outcomes are in agreement with a mesh of 8 × 8 and are in good accordance with that produced by (Zenkour & Aljadani, 2022). Accordingly, the number of elements to mesh the FGM plate for carrying out the study is kept. 8 × 8.

4. CONCLUSION

Several parametric experiments are carried out to reveal the impacts of gradation profile, porosity, boundary conditions, slenderness ratio, & aspect ratio of the FGM plate's buckling load following the current FEM model has been validated with the literature. The buckling strength of the material gradation exponent has a major effect on the FGM plate, and the reduced volume % of ceramic content results in a significant drop in the minimum load required for buckling the porous FGM plate. Further, the buckling strength gradually decreases as the porosity increases of the FGM plate. Also, the rectangular FGM plate's aspect ratio raises the plate's critical buckling load and has a significant impact on its buckling behavior.

Table 127.1 Validation and convergence analysis with critical buckling load ($\lambda = N_{cr}a^2/E_m h^3$) of simply-supported FGM square plate compressed in a single direction, for *a/h* = 10 and *n* = 1

n	No. of elements (in *x* × in *y*)	Present Study		Zenkour & Aljadani (2022))	
		α=0	α=0.1	α=0	α=0.1
0	4 × 4	20.25	19.08	18.90	17.44
	8 × 8	18.15	17.10		
	16 × 16	17.46	16.45		
1	4 × 4	9.87	9.35	9.33	8.22
	8 × 8	8.85	8.40		
	16 × 16	8.52	8.08		
2	4 × 4	8.01	6.43	7.26	6.04
	8 × 8	7.19	5.77		
	16 × 16	6.92	5.56		

Table 127.2 Material inhomogeneity's influence on the critical load for buckling ($\lambda = N_{cr}a^2/E_m h^3$) of Al/$Al_2O_3$ FGM square plate with simple support Uniaxial compression is applied, for *a/h* = 100

n	α=0	α=0.1	α=0.2
n=0	21.10	19.90	18.60
n=1	10.30	8.86	7.30
n=2	8.27	6.62	4.82
n=5	7.16	5.44	3.45
n=10	6.73	5.11	3.21

The effect of clamped edges is to increase the porous FGM plate's buckling strength, regardless of the power law index or porosity index values with all edges clamped exhibiting maximum buckling load whereas simply supported plate possesses minimum buckling strength.

Table 127.3 Aspect ratio's impact on the critical buckling load ($\lambda = N_{cr}a^2/E_m h^3$) in simply-supported Al/Al_2O_3 FGM rectangular plate that is compressed uniaxially, for *a/h* = 100 and *n* = 1.

b/a	α=0	α=0.1	α=0.2
b/a=1	10.30	8.86	7.30
b/a=2	16.20	13.80	11.40
b/a=3	28.80	24.60	20.30
b/a=4	46.80	40.00	33.00
b/a=5	70.00	59.90	49.40

Table 127.4 Effect of slenderness ratio on critical buckling load ($\lambda = N_{cr}a^2/E_mh^3$) of simply-supported Al/Al_2O_3 FGM plate under uni-axial compression, for $n = 1$

a/h	α=0	α=0.1	α=0.2
a/h =100	10.30	8.86	7.30
a/h =80	10.30	8.85	7.30
a/h =60	10.30	8.84	7.29
a/h =40	10.30	8.82	7.28
a/h =20	10.10	8.69	7.18

Table 127.5 Boundary conditions' (BC) influence on the critical load for buckling ($\lambda = N_{cr}a^2/E_mh^3$) of Al/$Al_2O_3$ FGM plate under uni-axial compression, for $n = 1$ and $a/h = 100$

n	BC*	α=0	α=0.1	α=0.2
n=0	BC1	21.10	19.90	18.60
	BC2	65.10	61.30	57.40
	BC3	41.40	39.00	36.50
n=1	BC1	10.30	8.86	7.30
	BC2	32.00	27.40	22.60
	BC3	20.30	17.40	14.30
n=2	BC1	8.27	6.62	4.82
	BC2	25.60	20.50	14.90
	BC3	16.20	13.00	9.47
n=5	BC1	7.16	5.44	3.45
	BC2	22.10	16.80	10.70
	BC3	14.00	10.70	6.78
n=10	BC1	6.73	5.11	3.21
	BC2	20.70	15.70	9.90
	BC3	13.20	10.00	6.29

*BC1, BC2, and BC3 refer to the plate with any clamped edges for x=0 and x=a, edges with simple support, & every single edge clamped boundary conditions.

REFERENCES

1. Belounar, A., Boussem, F., & Tati, A. (2023). A Novel C0 Strain-Based Finite Element for Free Vibration and Buckling Analyses of Functionally Graded Plates. *Journal of Vibration Engineering & Technologies*, *11*(1), 281–300.
2. Chaabani, H., Mesmoudi, S., Boutahar, L., & Bikri, K. El. (2023). A high-order finite element continuation for buckling analysis of porous FGM plates. *Engineering Structures*, *279*, 115597.
3. Dash, A., Kishore Joshi, K., Jha, P., Behera, A., Kumar Mohapatra, S., & Rahul. (2023). Various plate theory used in the analysis of FGMs-A review. *Materials Today: Proceedings*, *78*, 565–569.
4. Ramteke, P. M., & Panda, S. K. (2023). Computational Modelling and Experimental Challenges of Linear and Nonlinear Analysis of Porous Graded Structure: A Comprehensive Review. *Archives of Computational Methods in Engineering*.
5. Zenkour, A. M., & Aljadani, M. H. (2022). Buckling response of functionally graded porous plates due to a quasi-3D refined theory. *Mathematics*, *10*(4), 565.

Note: All the tables in this chapter were made by the authors.

First-Principles Study of Br-Functionalized Zigzag Gallium Nitride Nanoribbons: Structural and Electronic Properties

Ankita Nemu
2-D Materials Research Laboratory, Discipline of Physics, Indian Institute of Information Technology, Design and Manufacturing, Jabalpur, India

Neha Tyagi*, Asita Kulshreshtha
Amity School of Applied Sciences, Amity University Lucknow, Uttar Pradesh, India

Neeraj K. Jaiswal
2-D Materials Research Laboratory, Discipline of Physics, Indian Institute of Information Technology, Design and Manufacturing, Jabalpur, India

Abstract: This study explores the structural and electronic impacts of Br functionalized zigzag GaN nanoribbons (ZGaNNR). Our findings reveal that Br functionalization causes slight distortions in the optimized geometries compared to H functionalization, yet it results in more stable structures. Moreover, functionalizing ZGaNNR with Br reduces the band gap from 0.74 eV to 0.19 eV, indicating its potential to enhance band gap engineering for various technological applications.

Keywords: Gallium nitride, Functionalization, Band structures, Electronic properties

1. INTRODUCTION

In 2004, a low-dimensional material known as graphene was discovered by researchers (Geim and Novoselov, 2004). Despite its semimetal characteristics, graphene is not well-suited for use in nanoelectronic devices (Lu et. al., (2013). Group III nitrides, such as gallium nitride nanoribbons (GNNRs) (Yang et. al., 2004) offer potential applications. Various modifications can alter the properties of Gallium nitride (GaN), a wide bandgap semiconductor. Studies have shown that edge functionalization significantly alters the electronic characteristics of nanoribbons (Inge et. al.,2017). Previous research indicates that zigzag GaN nanoribbons (ZGaNNRs) exhibit metallic behavior, while armchair GaN nanoribbons (AGaNNRs) are semiconducting. This study focuses on investigating the electronic properties and stability of bromine-functionalized zigzag GaN nanoribbons (Br-ZGaNNRs) compared to both bare and hydrogen-functionalized (pristine) ZGaNNR structures.

2. COMPUTATIONAL DETAILS

We present results for bare, pristine, and Br-ZGaNNR with N_z = 4, 6, and 8, respectively. The first-principles computations were done with the generalized gradient approximation and the density functional theory of the plane-wave kinetic energy cut-off (Perdew et. al., 1996). To explore the structural stability, we compute the binding energies (E_b). Using the relationship below to calculate E_b:

$$E_b = 1/t\,[E_T - (n_{Ga}E_{Ga} + n_N E_N + n_X E_X)] \qquad (1)$$

The total number of atoms and energy of edge-functionalized nanoribbons are denoted by the atom numbers t, E_T, E_{Ga}, E_N, and E_X.

3. RESULTS AND DISCUSSION

Figure 128.1 illustrates the optimized geometries for different ZGaNNR configurations. Regardless of width size, Br-ZGaNNR structures have a higher binding

*Corresponding author: ntyagi@lko.amity.edu

DOI: 10.1201/9781003684718-128

energy (E_b) than bare and pristine ones. The higher electronegativity of Br-ZGaNNR makes it more stable by letting more charge move from edge atoms to Br atoms, which makes the bonding stronger. To be even more sure of this, we calculated the electron localization function (ELF) of Br-ZGaNNR and compared it with structures that were completely clean, as shown in Fig. 128.2.

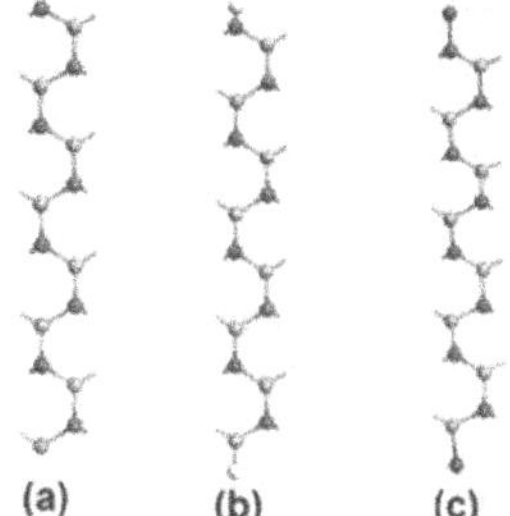

Fig. 128.1 The optimized geometries of : (a) Bare (b) Pristine (c) Br-ZGaNNR at N_z=8

The ELF demonstrates the bonding properties of functionalized Br with Ga and N, revealing greater ionic Ga-Br interactions and more covalent Br-N bonds near the N-edge. Upon examining the band structures and density of states (DOS) of these nanoribbons, we noticed that the

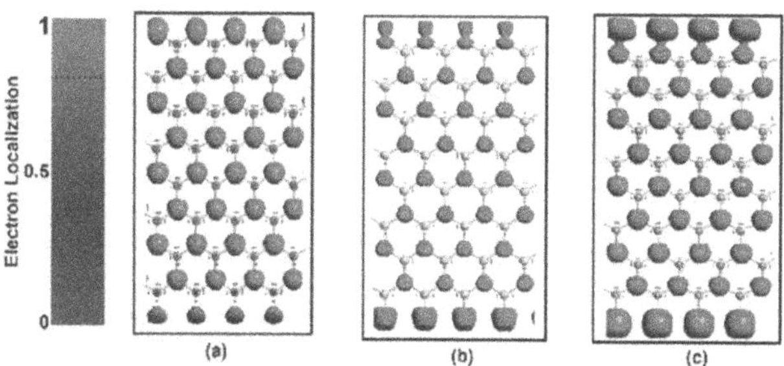

Fig. 128.2 Electron Localization Function plots of 8-ZGaNNR at isovalue 0.8. (a) Bare (b) Pristine (c) Bromine functionalized

bare ZGaNNR exhibits metallic characteristics, while the pristine ZGaNNR displays band gaps that indicates indirect semiconducting behavior, ranging from 2.71 eV to 2.32 eV, as illustrated in [Figs. 128.3, 128.4]. However, the functionalization of Br results in reduced direct band gaps at the Γ point, signifying a quantum confinement effect.

Table 128.1 represents the binding energy (E_b) and the band gap (E_g) for all the configurations. Notably, irrespective of width the Br-ZGaNNR leads to highest E_b and narrow E_g compared to others.

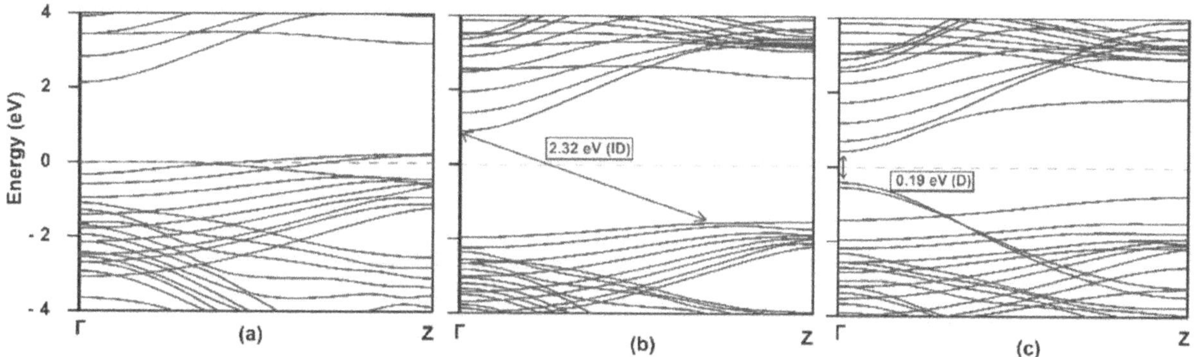

Fig. 128.3 Band structure representations for (a) bare, (b) pristine, (c) Br-ZGaNNR are shown at N_z=8. The red dash line indicates the Fermi level

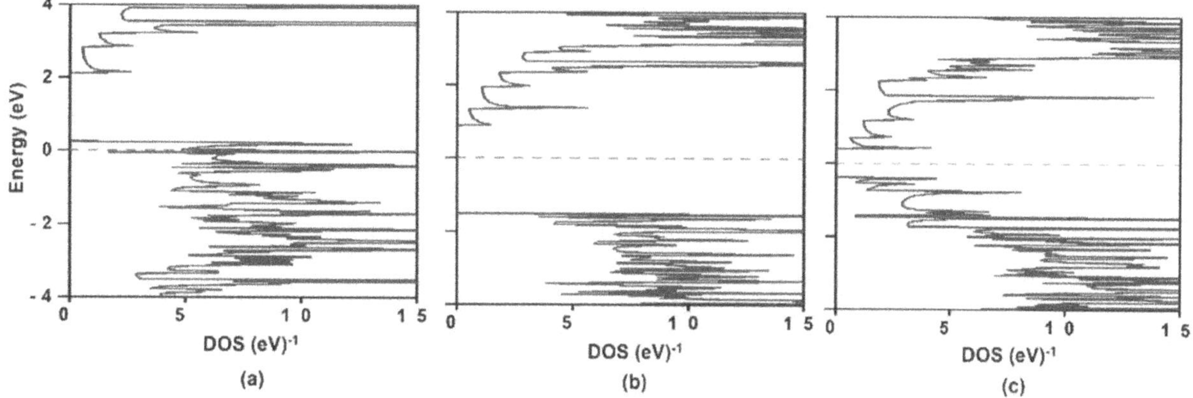

Fig. 128.4 DOS representations for (a) bare, (b) pristine (c) Br-ZGaNNR are shown at N_z=8. The red dash line indicates the Fermi level

Table 128.1 E_b and E_g comparison of ZGaNNR geometries

ZGaNNR geometries	E_b (eV)			E_g (eV)		
Width (N_z)	4	6	8	4	6	8
Bare	-4.96	-5.13	-5.22	M	M	M
Pristine	-5.03	-5.16	-5.24	2.71	2.52	2.32
Br-ZGaNNR	-5.11	-5.43	-5.57	0.74	0.56	0.19

4. CONCLUSION

Conclusively, the research explores the impact of Br-induced edge functionalization on the stability and electronic structure of ZGaNNR. The findings demonstrate that, in contrast to the bare and pristine, the bromine functionalization significantly reduces the band gap and makes the structure more stable for all widths. It suggests that bromine functionalization may be beneficial for applications demanding small band gaps, such as low-power electronic devices, perhaps opening the way for future advances in nanoelectronics.

REFERENCES

1. Geim, A.K. and Novoselov, K.S., (2007). The rise of graphene. Nat. Mater., 6(3), 183–191.
2. Lu, G., Yu, K., Wen, Z. and Chen, J., (2013). Semiconducting graphene: converting graphene from semimetal to semiconductor. Nanoscale, 5(4), 1353–1368.
3. Yang, L., Zhang, X., Huang, R., Zhang, G. and An, X., (2004). Synthesis of single crystalline GaN nanoribbons on sapphire (0001) substrates. Solid State Commun., 130(11), 769–772.
4. Inge, S.V., Jaiswal, N.K. and Kondekar, P.N., (2017). Realizing negative differential resistance/switching phenomena in zigzag GaN nanoribbons by edge fluorination: A DFT investigation. Adv. Mater. Interfaces, 4(19), 1700400.
5. Perdew, J.P., Burke, K. and Ernzerhof, M., (1996). Generalized gradient approximation made simple. Phys. Rev. Lett., 77(18), 3865.

Note: All the figures and tables in this chapter were made by the authors.

Recent Trends in Applied Physics and Material Science – Sudhir Bhardwaj et al. (eds)

Radiation Therapy Set-Up Error Study and its Impact on Treatment Delivery Accuracy

Laishram Amarjit Singh, Rohit Verma*
Department of Applied Physics,
Amity Institute of Applied Sciences, Amity University,
Uttar Pradesh

Manindra Bhushan
Division of Medical Physics, Department of Radiation Oncology, Rajiv Gandhi Cancer Institute and Research Centre, Rohini, Delhi

Abstract: The importance of this paper is to analyze the set-up error or deviation in radiation treatment of cancer diseases. We do take the help of imaging technology to correct and reduce the errors of the patient's position so that a quality treatment is provided to the patients. The deviations or mismatch in daily set-up are studied by using image registration with planning CT. The QAs are generated in the phantom (Octavius II, PTW Germany) and errors and deviations are introduced to create or mimic the set-up error. The results are analyzed in the form of Gamma Passing Criteria.

Keywords: Computed tomography, Cone beam computed tomography, Intensity modulated radiotherapy, Image-guided radiotherapy

1. INTRODUCTION

In the era of IGRT, online image review and offline image review have become a regular routine for all the radiotherapy setup. The set-up errors are frequently found in several forms in the radiation treatment delivery process (Gupta, 2012). Delivering the appropriate amount of radiation to the target cells by protecting vulnerable organs is the critical goal of radiotherapy (Khan et al., 2010). Due to technological advancement, image-guided radiotherapy (IGRT) could achieve a far better accuracy for set-up as well as delivery (Rudat, 2023). Still, the external body mismatch between the planned computed tomography (CT) image and the XVI cone beam computed tomography (CBCT) image can be caused by a number of variables, including body weight loss and mechanical or physical set-up errors. Since there is variation in the voxel matrix patient model, a discretized patient representation, this can alter the actual dosage calculation and its dosimetric impact on the illness site and other organs at risk. Additionally, the patient's body exhibits non-homogeneity and abnormalities in several sites, including the chest, abdomen, and lungs. This discrepancy could lead to a greater inaccuracy in the dosage that the doctors recommend (Khan et al., 2010). The purpose of this study is to determine how much the patient's body changed in dose distribution and how much the dose varied. The Gamma Analysis pass-fail criterion for patient-specific Intensity Modulated Radiotherapy (IMRT) quality assurance (QA) (Das et al., 2022) is being used in this study and we found that the gamma passing rate gets reduced depending on the variation in the patient's body. The study would find out the quantitative evaluation of dose distribution variations due to a mismatch in the patient's body or external between the planning CT image and XVI (CBCT) image acquired before treatment. This research will help identify the best ways to prevent these setup mistakes so that patients can receive more precise therapy going forward.

2. METHODOLOGY

This method is a retrospective analysis to keep the patient's data confidential and look for a better workflow to reduce error during treatment delivery. The patient's body is prone

*Corresponding author: rverma85@amity.edu

DOI: 10.1201/9781003684718-129

to variation from the initial treatment fraction. A close watch during the plan implementation is required if any major changes is there in the set-up for patient's positions. Mainly, in this study we look for changes in the external surface or body of the patients and analyze the image registration in all the fractions. We select those patients with the maximum changes in the external body due to weight loss or mold compression etc. We generate QA plans for those patients and expose those QA plans on slab phantom. We vary the thickness of the phantom anteriorly from 0 to 2-3 cm and same QA plans are exposed and see the variations in the dose deposition through the Gamma Analysis pass-fail criterion. The variation in Gamma passing criterion is seen which may indicate further dose reduction to the target cells and OARs.

3. RESULTS

The data collected from the QA plan's delivery were analyzed using Gamma Analysis and were presented for 11 selected patients.

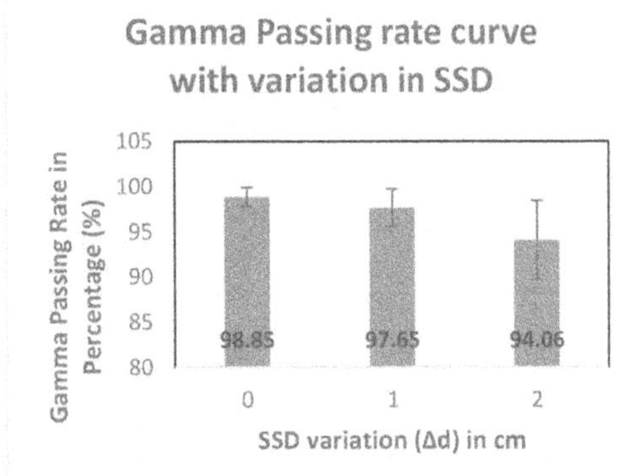

Fig. 129.1 Graphical presentation of gamma passing rate

4. CONCLUSION

There is significant dose variation observed with the change in SSD. This analysis gives plan verification and comparison of measured dose and TPS calculated dose. There is increase in failing rate of Gamma analysis as SSD variation increases. We need to search ways to quantify the deviation in Prescribed dose to tumor and dose to OARs. This is a quantitative comparison of two dose distribution (Evaluated and Reference dose). Only 11 patients were analyzed. More data (Patients) is required to get the actual information about the gamma passing rate.

Table 129.1 Gamma passing rate (data analysis)

Patients	Gamma Passing rate with SSD Variation of 0 cm, 1cm & 2 cm		
	0 cm	1 cm	2 cm
24049	98.9	96.2	88.8
24054	98.4	97.1	94.9
24073	99.2	98.4	94.7
24075	98.9	99	97
24081	100	100	99.5
24083	99.7	99.5	98.8
24099	99.7	100	97.9
24065	99	95.5	86
23002	99.5	98.8	95.6
23007	97.7	95.9	90
24121	96.4	93.8	91.5
Average	98.85455	97.65455	94.06364
STDV.s	1.042462	2.075747	4.411411

REFERENCES

1. Das, S., Kharade, V., Pandey, V.P., Kv, A., Pasricha, R.K., Gupta, M. (2022). Gamma Index Analysis as a Patient-Specific Quality Assurance Tool for High-Precision Radiotherapy: A Clinical Perspective of Single Institute Experience. Cureus. 14(10):e30885. doi: 10.7759/cureus.30885.
2. Gupta, T., Narayan, C., Anand. (2012). Image-guided radiation therapy: Physician's perspectives. Journal of Medical Physics 37(4):p 174–182 doi: 10.4103/0971-6203.103602.
3. Khan, F.M., Gibbons, J.P. (2010). Khan's the physics of radiation therapy. Fifth. Lippincott Williams and Wilkins. ISBN-10. 9781451182453. ISBN-13. 978-1451182453
4. Rudat, V., Shi, Y., Zhao, R., Xu, S., Yu, W. (2023). Setup accuracy and margins for surface-guided radiotherapy (SGRT) of head, thorax, abdomen, and pelvic target volumes. Sci Rep. 13(1):17018. doi: 10.1038/s41598-023-44320-2.

Note: All the figures and tables in this chapter were made by the authors.

Recent Trends in Applied Physics and Material Science – Sudhir Bhardwaj et al. (eds)
© 2026 Taylor & Francis Group, London, ISBN 978-1-041-16452-4

Anti-Fungal Activity of Different Phytochemicals Against Malassezia sp.

Prashasti Sinha*, Anil Kumar Yadav
Department of Physics, Babasaheb Bhimrao Ambedkar University, Lucknow, Uttar Pradesh, India

Abstract: A frequent ailment that affects the state of the scalp is dandruff. Dandruff can only be adequately managed; it cannot be entirely eradicated. Products that are currently on the market perform poorly or cause symptoms to return. Plant extracts contain a variety of active substances that might aid in preventing the establishment of resistance in the causing organism. Therefore, the goal of the current study is to develop a poly-herbal blend that may be added to hair oil to improve its anti-dandruff effectiveness. This mixture will be rich in alkaloids, flavonoids, phenolic compounds, terpenoids, and tannins. Therefore, the present study deals with the computational analysis of the interaction of the phytochemicals present within margosa, rosemary, chamomile and tea tree against Malassezia globosa.

Keywords: Malassezia globosa, Molecular docking, Molecular dynamics, Phytochemicals

1. INTRODUCTION

Malassezia globosa is a scalp condition affecting half of the population at the prepubertal stage, affecting men aged 20-30. It is a significant aesthetic issue and raises public health concerns. The condition is characterized by mild to severe scalp scaling, erythema, and dryness. Hair loss is common in dandruff patients. Due to the ineffectiveness of existing medications, the search for novel anti-fungal substances has increased. Plants have their own defence mechanisms against fungi, leading to an interest in plant-derived antifungal substances. Plants like Azadirachta indica, Rosmarinus officinalis, Matricaria chamomilla, and Melaleuca alternifolia have anti-fungal properties and have been proven effective in treating skin allergies, burns, and rashes. The current study aims to study the anti-fungal activity of selected plant extracts individually and find new possibilities for designing novel combinations of poly-herbal anti-dandruff.

2. METHODOLOGY

The structure of the phytochemicals of the selected plants was obtained from PubChem database. Nibolide (CID 12313376), Carnosol (CID 442009), Apigenin (CID 5280443) and ZINC; diheptoxy-sulfanylidene-sulfido-lambda-5-phosphane (CID 22833361) are the major constituent of margosa, rosemary, chamomile and tea tree. Therefore, these structures were utilized throughout the study. The protein structure of Malassezia globosa was downloaded from RCSB PDB site. PDB ID: 3UUF which is the crystal structure obtained from mono- and diacylglycerol lipase of Malassezia globosa. The docking simulation was carried out via AutoDock 4.2 using Lamarckian Genetic Algorithm and empirical and further, MD simulation was carried out using GROMACS 2022.3. The RMSD plots were obtained using OPLS-AA force field at TIP3P water model at 100ns time steps. The analysis of the results of docking simulation and MD simulation was performed using Biovia Discovery Studio and xmgrace, respectively. MD simulation was followed by MM/PBSA to study the free energy of the components using following equations.

$$\Delta G_{bind} = \Delta E_{MM} + \Delta G_{SOLV} - T\Delta S \quad (1)$$

$$\Delta E_{MM} = \Delta E_{int} + \Delta E_{ele} + \Delta E_{vdw} \quad (2)$$

$$\Delta G_{SOLV} = \Delta G_{GB} + \Delta G_{SA} \quad (3)$$

*Corresponding author: prashasti.bbau@gmail.com

DOI: 10.1201/9781003684718-130

3. RESULTS AND DISCUSSION

3.1 Molecular Docking

The molecular structure of CID 12313376, CID 442009, CID 5280443 and CID 4367869 were docked with PDB ID: 3UUF. **Table 130.1** shows the energy parameters obtained during docking process. From the table, it can be noted that CID 12313376 has comparatively higher binding affinity as compared to the other phytochemicals. On comparing the total energies of the selected compounds, it can be seen that CID 442009 and CID 12313376 possess relatively higher total energy than the other compounds. The higher vdW and lower electrostatic energy of CID 12313376 and CID 4367869 signifies the greater strength with which ligand binds with the binding site. Since, experimentally all these phytochemicals have shown efficiency in acting as anti-fungal activity. Also, from the table given below, binding affinity lies in the order CID 12313376 > CID 442009 > CID 4367869 > CID 5280443, which is in accordance with the experimental data. The docking simulation was carried out within a predefined binding site, which was obtained from the parameterization of the native ligand (alpha-D-mannopyranose). In this cavity, Val 233, Val 266, Arg 236, Ile 269, Glu 264, Phe 276 and Trp 229 plays an active role during the formation of different types of interactions such as hydrogen bond, van der Waals interaction, pi-alkyl etc.

Table 130.1 Energy parameters of the plant extracts

Parameters	CID 12313376	CID 442009	CID 5280443	CID 4367869
Binding Affinity (kcal/mol)	-7.908	-7.352	-6.506	-7.127
Total Energy (kcal/mol)	-50.753	-53.093	-9.938	-41.807
vdW Energy (kcal/mol)	-18.396	-10.052	-0.822	-18.750
Electrostatic Energy (kcal/mol)	-8.557	-21.143	-28.171	-5.035

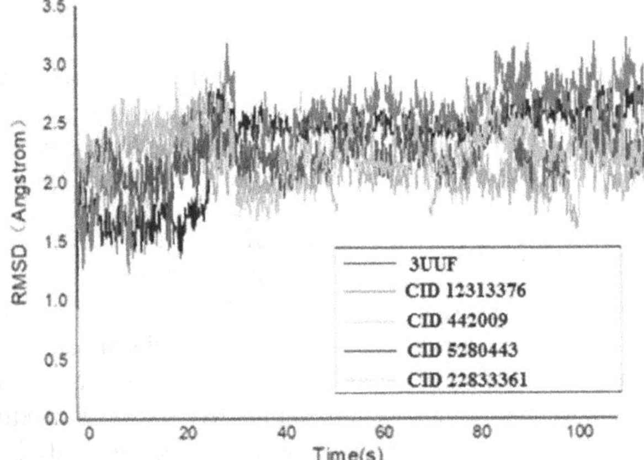

Fig. 130.1 RMSD plot of each docked complex and protein backbone

3.2 Molecular Dynamics Simulation

Figure 130.1 shows the RMSD plot of each complex and the protein backbone (3UUF). Different colours are used to illustrate the trajectories of the docked complexes of CID 12313376, CID 442009, CID 5280443 and CID 4367869 with 3UUF and the protein backbone itself. It was revealed that the RMSD was found to be 1.5 ± 0.02 Å, 1.78 ± 0.04 Å, 2.10 ± 0.05 Å and 2.15 ± 0.07 Å for the complexes of CID 12313376, CID 442009, CID 5280443 and CID 4367869, respectively. The RMSD of the backbone was found to be 1.95 ± 0.06 Å. Therefore, on comparing the individual RMSD of each complex it was found that results of MD simulation validated the docking results.

3.3 MM/PBSA Analysis

Table 130.2 highlights the parameters involved in binding free energy such as ΔE_{ele}, ΔE_{vdw} and ΔG_{SASA}; moreover, ΔG_{GB} was not favored. After analyzing the tabulated data, it was found that the CID 12313376 has higher ΔG_{bind} as compared to the other phytochemicals. Moreover, the trend of ΔGbind was observed as CID 12313376 > CID 442009 > CID 4367869 > CID 5280443 which is very similar to the trend observed in the docking simulation.

4. CONCLUSION

Selected plants have their own defence mechanisms against fungus diseases; there has recently been a lot of interest in plant-derived antifungal chemicals. Our present research describes the utilization of herbal active ingredients to treat dandruff. The phytochemicals of the selected plants were studied using the tools of molecular docking, molecular dynamics and MM/PBSA. It was found that Azadirachta indica (known as margosa) has better binding affinity (-7.908 kcal/mol), shows lower

Table 130.2 MM/PBSA parameters

Compound	3UUF				
	ΔE_{ele}	ΔE_{vdw}	ΔG_{GB}	ΔG_{SASA}	ΔG_{bind}
CID 12313376	-13.18 ± 0.10	-82.37 ± 0.15	31.84 ± 0.16	-2.25 ± 0.13	-63.96 ± 0.02
CID 442009	-10.33 ± 0.10	-73.24 ± 0.05	26.75 ± 0.05	-4.34 ± 0.10	-59.16 ± 0.20
CID 5280443	-12.96 ± 0.07	-55.84 ± 0.12	38.41 ± 0.02	-1.98 ± 0.30	-30.37 ± 0.45
CID 4367869	-12.18 ± 0.05	-69.34 ± 0.15	57.81 ± 0.08	-3.55 ± 0.55	-31.26 ± 0.67

RMSD (1.5 ± 0.02 Å) and possess higher ΔG_{bind} (-63.96 ± 0.02 kcal/mol) as compared to the other phytochemicals of rosemary, chamomile and tea tree. Therefore, our computational analysis is parallel to the experimental evidences. Hence, the extract of Azadirachta indica is a better suited in treating dandruff problems.

REFERENCES

1. Kim S., Thiessen P.A., Bolton E.E., Chen J., Fu G., Gindulyte A., Han L., He J., Shoemaker B.A., Wang J. (2016). Nucleic acids research, 44(D1):D1202–13
2. Morris G.M., Huey R., Lindstrom W., Sanner M.F., Belew R.K., Goodsell D.S., Olson A.J. (2009) AutoDock4 and AutoDockTools4: Journal of computational chemistry, 30(16):2785–91
3. Bauer P., Hess B., Lindahl E. (2022) GROMACS 2022.1 Manual. Zenodo:1–673
4. Srinivasan J., Cheatham T.E., Cieplak P., Kollman P.A., Case D.A. (1998) Journal of the American Chemical Society; 120(37):9401–9.
5. Bhowmik, D., Chiranjib, Y.J., Tripathi, K.K., Kumar, K.S. (2010) J Chem Pharm Res; 2(1):62–72
6. Ambareen, Z., Chinappa, A. (2014) Int J Dent Sci Res; 2(6B):21–25
7. Etewa, S.E., Abaza, S.M. (2011) Parasitol United J; 4(1): 3–14
8. Khanpara, P., Jadeja, Y. (2022) Journal of Medicinal Plants; 10(5):131–140
9. Bernardes, W.A., Lucarini, R., Tozatti, M.G., Souza, M.G., Andrade Silva, M.L., da Silva Filho, A.A., Cunha, W.R. (2010) Chemistry & biodiversity; 7(7):1835–1840

Note: All the figures and tables in this chapter were made by the authors.

Recent Trends in Applied Physics and Material Science – Sudhir Bhardwaj et al. (eds)
© 2026 Taylor & Francis Group, London, ISBN 978-1-041-16452-4

Assessment of Normal Tissue Objective Function using Intensity Modulated Radiosurgery of Single Brain Metastasis

Shabbir Ahamed*
Department of Physics,
Jawaharlal Nehru Technological University Anantapur,
Ananthapuramu, Andhra Pradesh, India
Department of Radiation Physics,
MNJ Institute of Oncology and Regional Cancer Center,
Hyderabad, Telangana, India

R. Padma Suvarna
Department of Physics,
Jawaharlal Nehru Technological University Anantapur,
Ananthapuramu, Andhra Pradesh, India

Abstract: Efficacy of radiosurgery planning in terms of normal tissue sparing and tumor dose conformation is essential for favourable treatment outcomes. This study examined normal tissue objective (NTO) based radiosurgery planning for single brain metastasis (SBM), with volumetric-modulated arc radiosurgery (VMAT) technique and intensity-modulated radiosurgery (IMRT) technique. The NTO function is an exponential function based on of dose and distance parameters such as start-dose (Ds), end-dose (De), start-distance (x_i), slope (k) and priority. Twenty plans were retrospectively planned using each technique. The applied NTO input parameters were Ds = 99%, De = 10%, x_i = 1 mm k = 1.0 mm^{-1}, and priority = 100. All targets were planned using 20 Gy margin dose while maximum dose was constrained to 25 Gy. The plans were analysed using 12 Gy normal brain dose-volume (V12- NBD), gradient index (PGI), conformity index (CI), the prescription isodose-level (PIDL), and MU/Gy. VMAT plans had substantially less V12 compared to IMRT plans. V12 was lower by 1.5 cm^3 for the VMAT plans, $P < 0.001$. PGI was better for VMAT (2.9 ± 0.3) than for IMRT (3.5 ± 1) for all planned cases, $P = 0.0025$. CI was similar between both plan techniques. Mean PIDL was also similar, at about 70%, for both techniques. MU/Gy was significantly lower for IMRT compared to VMAT, $P < 0.001$. The NTO based modulated radiosurgery techniques exhibited varied dosimetry. VMAT plans showed overall betterment in dosimetry compared to IMRT plans.

Keywords: Metastasis, Radiosurgery, Efficacy

1. INTRODUCTION

Normal tissue objective function (NTO) is a distance dependant exponentially decaying function commonly used to minimize non-target tissue dose surrounding the irradiated target and is described in Eclipse photon and electron algorithms reference guide, 2015. Advancements in planning radiosurgery facilitated tailoring conformal target dose while reducing normal brain dose. The NTO parameters can be chosen as per requirement to define dose fall-off. Few attempts have been made to report the use of NTO for radiosurgery [Indrayani, 2022, Wang, 2020, and Lobb, 2020]. Therefore, we attempted to explore NTO function with IMRT and VMAT techniques applied for radiosurgery of single brain metastasis (SBM). The dosimetric parameters for evaluation of the NTO included the V12-NBD i.e., 12 Gy dose volume in normal brain, gradient index (PGI), conformity index (CI), prescription isodose-level (PIDL), and monitor units per Gray (MU/Gy).

*Corresponding author: shabbirahmedp@gmail.com

DOI: 10.1201/9781003684718-131

2. MATERIALS AND METHODS

Twenty computed-tomography (CT) sets of SBM cases were used for this retrospective dosimetric planning study. Planning target volume (PTV) structure outline was obtained by 1 mm expansion of GTV and normal brain outline was obtained by subtracting GTV from total brain outlined. A 20 Gy marginal dose was prescribed to each case. All plans were planned using 6 MV flattening-filter-free photon beams having 1400 MU/min dose rate provided by TrueBeam linac (linear accelerator) having high definition (HD 120) multileaf-collimator with 0.25cm leaf projection width. The Photon Optimizer (Ver 13.6) and Anisotropic Analytical Algorithm (Ver 13.6) were used for dose optimization and computation with a 1.25 mm grid size. Twelve to thirteen non-coplanar beams in three couch planes were used for generating IMRT plans. Whereas for VMAT planning three non-coplanar arcs were utilized to mimic IMRT beam arrangement. Both techniques employed NTO parameter values priority = 100, slope, k = 1.0 mm^{-1}, and the end-dose, D_e = 10%, intended to produce steep dose-fall around the target. The start dose D_s was set to 99% of the PD at start distance x_i = 1 mm, for adequate dose coverage. The PTV dose minimum and dose maximum were assigned 20 Gy and 25 Gy with corresponding priority values of 100 and 75. All plans ensured normalization of prescribed dose encompassing 99% of target volume.

Dosimetric parameters such as V12-NBD [Milano et al., 2021], Paddick gradient index (PGI) [Paddick and Lippitz, 2006], conformity index (PCI) [Paddick, 2000], PIDL, and MU/Gy were analyzed. PGI is the ratio of 50% and 100% prescription isodose volumes. PCI is expressed as $(PIV \cap PTV)^2/(PIV \times PTV)$, where PTV∩PIV is overlap of prescription isodose-volume and PTV. PIDL is quotient of maximum and prescription doses. MU per Gy indicates monitor units required to deposit unit PTV dose. For data comparison, two tailed t-test with P<0.05 was used.

3. RESULTS AND DISCUSSION

Comparison metrics for NTO based techniques are summarized in Table 131.1. PGI and V12-NBD were computed to determine the dose gradient beyond target boundary. The mean V12-NBD volume recorded for VMAT plans was lesser than for IMRT plans. A substantial reduction of 1.5 cm^3 resulted for the VMAT plans, P < 0.001. The mean PGI was lesser for VMAT compared to IMRT, P = 0.0025. The average PCI was similar between the planned techniques. Concerning PIDL, both plan techniques showed comparable values. MU/Gy values were lower for significantly lower for IMRT versus VMAT, P < 0.001. This study results indicate VMAT superiority for all the studied metrics, except MU/Gy (Table 131.1). The differences in plan metrics arise due to the delivery mode of each technique. IMRT technique uses static fields from fixed directions while VMAT technique uses arc spanning large volume. Further for the IMRT technique the fluence is created first by the optimizer and the relevant MLC patterns later. This fluence created by the optimizer determines the collimator jaw extents of each static IMRT field. Whereas for VMAT the optimizer utilizes user supplied collimator jaw extents and determined dynamic parameters such as MLC patterns, dose rate and gantry speed during optimization. The use of NTO parameters, De = 10% and k = 1.0 mm^{-1} produced better plan optimization for VMAT than IMRT.

Table 131.1 Plan metric summary of IMRT Vs VMAT

Plan metric	IMRT	VMAT	P-value
			IMRT Vs VMAT
V12-NBD (cm^3)	7.3 ± 2.7	5.8 ± 2.2	< 0.001
PGI	3.5 ± 1	2.9 ± 0.3	0.0025
CI	0.7 ± 0.1	0.7 ± 0.1	0.2119
PIDL (%)	70.8 ± 2.7	73.5 ± 2.7	<0.0001
MU/Gy	248.4 ± 18	354.7 ± 22.9	< 0.001

This study shows that the VMAT plans are always superior in terms of mean PIDL, which is <74% and is in line with literature. Existing research suggests that PIDL of 50% to 75% is ideal/favourable for the linac based radiosurgery for single lesions using dynamic conformal technique [Zhao et al., 2014] and 60% to 70% using VMAT technique [Xu et al., 2019]. In summary, the VMAT technique offers better V12, GI with comparable CI which would make it a choice of radiosurgery of solitary brain metastasis targets.

4. CONCLUSION

Comparison of the NTO based radiosurgery techniques showed varied dosimetry. The VMAT technique produced lesser normal brain dose and also overall superior metrics compared to IMRT technique.

REFERENCES

1. Indrayani, L., Anam, C., Sutanto, H., Subroto, R., Dougherty, G. (2022). Normal tissue objective (NTO) tool in Eclipse treatment planning system for dose distribution optimization. Pol. J. Med. Phys. Eng. 28(2):99–106.
2. Lobb, E,C., and Degnan, M. (2020). Comparison of VMAT complexity-reduction strategies for single-target cranial radiosurgery with the Eclipse treatment planning system. J. Appl. Clin. Med. Phys. 21(10):97–108.
3. Milano, M.T., Grimm, J., Niemierko, A., Soltys, S,G., Moiseenko, V., Redmond, K.J., *et al.* (2021). Single- and Multifraction Stereotactic Radiosurgery Dose/Volume Tolerances of the Brain. Int. J. Radiat. Oncol. Biol. Phys. 110(1):68–86.
4. Paddick, I., and Lippitz, B. (2006). A simple dose gradient measurement tool to complement the conformity index. J. Neurosurg, 105:194–201.

5. Paddick, I. (2000). A simple scoring ratio to index the conformity of radiosurgical treatment plans. J. Neurosurg. 93:219–22.
6. Varian Medical Systems (2015). *Eclipse photon and electron algorithms reference guide*. Palo Alto, CA: Varian; https://jpneylon.github.io/ABR/PDFs/Add_052418/EclipseAlgorithms13.6_RefGuide.pdf.
7. Wang, D., DeNittis, A.S., and Hu, Y. (2020). Strategies to optimize stereotactic radiosurgery plans for brain tumors with volumetric-modulated arc therapy. J. Appl. Clin. Med. Phys. 21(3):45–51.
8. Xu, Y., Ma. P., Xu, Y., Dai, J. (2019). Selection of prescription isodose line for brain metastases treated with volumetric modulated arc radiotherapy. J. Appl. Clin. Med. Phys. 20:63–9.
9. Zhao, B, Jin, J.Y., Wen, N., Huang, Y., Siddiqui, M.S., Chetty, I,J., *et al.* (2014). Prescription to 50-75% isodose line may be optimum for linear accelerator based radiosurgery of cranial lesions. J. Radiosurg. SBRT. 3:139–47.

Note: The authors made the table in this chapter.

132

Linear Calibration-Dose Tuning Adjustment of X-Band Linear Accelerator in Accuray™ Cyberknife M6

Vaibhav Soni*, Rushikesh Talole, Gouse Nidamanuri, Annex E. H., Ashitha M. K., Debnarayan Dutta

Department of Radiation Oncology, Amrita Institute of Medical Sciences, Ponnekara P.O, Kochi, Kerala

Abstract: Cyberknife M6 radiotherapy treatment machine having an X-band LINAC which is not a conventional gantry based system instead it is bolted on a robotic arm. It uses 9.3 GHz X Band accelerator which provides 6 MV FFF photon beam energy with dose rate of 800 MU/min which enables to precisely irradiate small tumor area with high dose-rate non-coplanar beams. This beam tuning procedure allows you to calibrate and adjust the LINAC dosimetry system, nominal dose amounts are delivered for MU value ranging from 10 to 200 using a fixed collimator opening and measuring the dose detected. X-Band LINAC is calibrated for a range of MU's instead of calibrating with a single larger MU value of conventional S-Band LINAC. Since it is used for SRT and SBRT hypofractionated treatments where high dose per fraction are delivered with high precision and every small dose counts, thus it is important to maintain the MU-Dose linearity and MU stability especially for smaller MU values for a beam delivering with high dose rate. When delivering smaller MU with high dose rate having high dose per pulse, the chamber should be sensitive enough so that the before the subsequent pulse comes, the charge has accumulated and recombined, resulting in instantaneous complete ionization per pulse. It's important to tune the monitor chamber to have linear output throughout the MU range of small to high while delivering every pulse in high dose rate.

Keywords: Cyberknife, Tuning, Monitor chamber

1. INTRODUCTION

Clinical radiation dosimetry requires a high level of accuracy because of a relatively small therapeutic dose window for tumor target and surrounding normal organ at risk (OAR), which can induce side effects, so controlling the tumor and inducing side effects have to be optimally balanced since having direct consequences for patients by either delivering too little or too much dose to a tumor, thereby reducing the chances for treatment success

In a Linear accelerator (LINAC) accurate dose delivery is assured by dose monitoring system based on transmission ionization chambers for dose measurement and calibration of radiation beam, it basically measure the ionization produced by the charged particles in order to calculate the energy transferred by the radiation beam in the medium.

Instead of gantry to carry an x-ray source Accuray™ cyberknife M6 is a radiotherapy treatment machine using an X-band LINAC where a small packed LINAC is bolted on a robotic arm, here LINAC operates at a radiofrequency (RF) of 9.3 GHz, providing 6 MV photon beam energy with a high dose rate of 800 MU/min. (Anon n.d.)

In radiotherapy the radiation beam is calibrated to ensure the accuracy of LINAC by adjusting the output of radiation beam. This output is a measurement of the absorbed dose

*Corresponding author: sonivaibhan44@gmail.com

by beam and calibrated to deliver a specific amount of dose per monitor unit (MU) at the specific point.(IAEA 2024) This is done to ensure accurate and consistent dose delivery to patient so in ionization chamber based beam monitor each ion pair got collected and transformed into a pulse, used to signal the presence of the particle and signify individual ionization in detector.

The intent of this work is to verify the stability and linearity of dose with MU variation for X-band LINAC delivered for small to high MU range which is achieved after the linear beam calibration thereby gives a comparative analysis with S-band LINAC on the basis of the linearity of the output by collected charge readings using radiation detector for both the monitor chambers.

2. LITERATURE REVIEW

In cyberknife dose is given in multiple smaller MU segments, MU dose linearity be maintained in every region.(Sudahar et al. 2012) Since dose rate plays an important role in MU dose linearity especially in lower MU range.(Mohr et al. 2007)

3. DATA AND VARIABLES

Linear calibration adjustment of X-band LINAC of 6 MV flattening filter free (FFF) photon beam with nominal dose rate of 800 MU/min. Charge is measured using a 0.6cc cylindrical ionization chamber (PTW TN 30013) connected with PTW WeblineT10023 electrometer. Ionization chamber positioned at effective point of measurement, from the depth of 10 cm having source to chamber-axis distance (SAD) 80 cm for a fixed collimator size of 60 mm. Uncalibrated nominal MU values ranges from 10 to 200 is delivered and calibration gain curve is plotted to study MU linearity and stability by measuring the output in SNC 3D scanner water phantom on minilift.

For comparison between X-band and S-band LINAC for MU linearity, same MU values delivered with C-Arm Elekta Synergy S-band LINAC having photon beam of energy 6 MV flattening filter (FF) photon beam with dose rate of 600 MU/min and charge is measured using a 0.6cc cylindrical ionization chamber (PTW TN 30013) positioned at a depth of 10 cm, field size of 10×10 cm^2 in a setup of SAD 100 cm connected with SNC PC-electrometer, in solid water slab phantom.

4. EMPIRICAL RESULTS

This graph represents the linear beam calibration adjustment plotting uncalibrated MU values of 10 to 200 on y-axis and dose measured in centigray (cGy) on x-axis.

For output comparison of both the LINAC's same MU is delivered and measured the charge collected in nano coulomb (nC) and thus plotting the graph taking MU values in x-axis and collected charge readings in y-axis are being normalized with maximum reading.

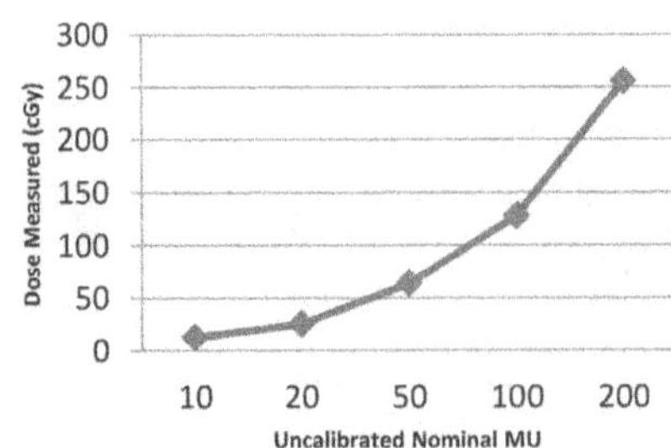

Fig. 132.1 Nominal uncalibrated MU vs. dose

Table 132.1 Normalized meter readings for X-band and S-band LINAC

MU Delivered	Normalized Meter Reading (nC)- X Band LINAC	Normalized Meter Reading (nC)- S Band LINAC
3	0.0153	0.015
5	0.0253	0.025
7	0.0352	0.035
10	0.0504	0.050
15	0.0753	0.075
20	0.1002	0.100
30	0.1500	0.150
40	0.2000	0.200
50	0.2497	0.251
80	0.3998	0.401
100	0.4999	0.501
125	0.6250	0.626
150	0.7500	0.751
175	0.9000	0.876
200	1.0000	1.000

5. CONCLUSION

In the comparative study since The reading for both X-Band and S-band will follow same trendline, it proves the aim of doing linear calibration for X-band LINAC to archive the fine beam tuning for the lower range of MU rather calibrating with a single larger MU like in S-Band LINAC it suggest the special utility of cyberknife where radiation beam is delivered in multiple smaller MU sets having high dose per pulse. It's important to tune the monitor chamber in X-band LINAC to have linear output throughout the MU range of small to high while delivering every pulse in high dose rate. So each pulse produces ionization instantly, and the charge is generated and collected prior to the subsequent pulse.

In Cyberknife X-band LINAC, delivering high doses per fraction demands to tune LINAC for maintaining MU dose linearity in all MU ranges to guarantee the accuracy and precision in dose delivery to the patient.

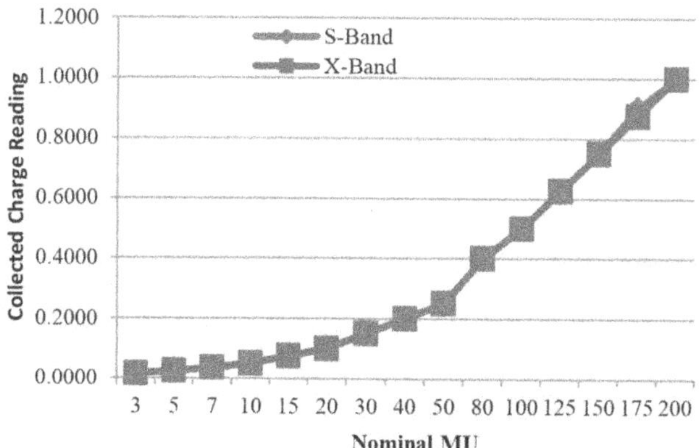

Fig. 132.2 Comparative curves for X-band and S-band LINAC for MU delivered v/s normalized collected charge readings

REFERENCES

1. Anon. n.d. Technical-Specifications-CyberKnife-M6-Series.Pdf.
2. INTERNATIONAL ATOMIC ENERGY AGENCY. 2024. *Absorbed Dose Determination in External Beam Radiotherapy*. Rev. 1. INTERNATIONAL ATOMIC ENERGY AGENCY.
3. Mohr, Peter, Sabrina Brieger, Jürgen Stahl, and Gerlo Witucki. 2007. "Linearity of the Dose Monitor System at Low Monitor Units." *Strahlentherapie Und Onkologie: Organ Der Deutschen Rontgengesellschaft ... [et Al]* 183(6):327–31. doi: 10.1007/s00066-007-1596-2.
4. Sudahar, H., P. G. G. Kurup, V. Murali, and J. Velmurugan. 2012. "Dose Linearity and Monitor Unit Stability of a G4 Type Cyberknife Robotic Stereotactic Radiosurgery System. *Journal of Medical Physics / Association of Medical Physicists of India* 37(1):4. doi: 10.4103/0971-6203.92714.

Note: All the figures and tables in this chapter were made by the authors.

Recent Trends in Applied Physics and Material Science – Sudhir Bhardwaj et al. (eds)

133

Comparison of Accuracy of Ray Tracing and Monte Carlo in Regions of Tissue Heterogeneity

Megha Philip*, Annex E. H., Ashitha M. K., Alma Peter, Debnarayan Dutta

Department of Radiation Oncology, Amrita Institute of Medical Sciences, Kochi, Kerala, India

Abstract: Ensuring accurate dose calculation is essential for an effective radiation therapy, especially in heterogeneous areas, such as the lungs. The dose optimization and calculation algorithms play a crucial role in achieving this. The CyberKnife Robotic Radiosurgery system utilizes two such algorithms- Ray Tracing (RT) and Monte Carlo (MC). The RT algorithm, being a correction-based algorithm, takes into account the effective path length of the beam traversed through the lung volume and the related attenuation along the path. The MC algorithm, on the other hand, is capable of simulating the individual particles' trajectories across the heterogeneous regions. The treatment planning was performed in the Accuray Precision® 3.3.1.3[2] treatment planning system (TPS) for the CyberKnife M6 system. Treatment plans were created using RT algorithm for water-equivalent slab phantoms, both with and without Styrofoam (air equivalent material), the former indicating the presence of inhomogeneity. The same plans were recalculated with MC, maintaining the original beam parameters. The setup is reproduced, the treatment plans are executed using CyberKnife M6 and point doses are measured at reference point for each plan using calibrated 0.6cm^3 farmer chamber. Measured doses are tabulated and compared with the planned dose obtained from the TPS. In homogenous medium, the behaviour of RT and MC are comparable with a minimal difference of less than 0.5%. In heterogenous region, a deviation of more than 2% is observed when RT algorithm is used. The result highlights the better performance of MC algorithm in heterogenous as well as the homogenous medium.

Keywords: Heterogeneity, CyberKnife, Monte Carlo

1. INTRODUCTION

Stereotactic Body Radiation Therapy (SBRT) is a specialized treatment approach that delivers extremely high radiation doses to a very small volume of tissue in a very few numbers of treatment sessions. This unique feature of SBRT requires precise and accurate treatment planning and delivery. The accuracy of a radiation treatment plan depends significantly on the dose optimization and calculation algorithms. This requirement is quite challenging in regions of inhomogeneity.

The CyberKnife M6 robotic radiosurgery system delivers 6MV photon beams with sub-millimetre accuracy. It uses two algorithms for dose optimization and calculation- Ray Tracing (RT) and Monte Carlo (MC). The Ray Tracing algorithm, uses the concept of effective depth of the beam to account for density variations when calculating dose. It does not accurately consider the electron and photon scatter in heterogeneous regions. The Monte Carlo algorithm, regarded as the gold standard in calculating the doses in heterogeneous regions, takes into account individual particles' interactions while traversing the medium.

This present study investigates the performance of Ray Tracing and Monte Carlo in regions of heterogeneity. An inhomogeneous setup was simulated and planned accordingly. The plan results were verified dosimetrically by reproducing the setup.

2. LITERATURE REVIEW

Sharma S et al., in 2010, conducted a comparison of the CyberKnife Monte Carlo algorithm with other dose calculation techniques, such as the Analytical Anisotropic

*Corresponding author: meghaphilip1510@gmail.com

DOI: 10.1201/9781003684718-133

Algorithm, Pencil Beam Convolution, and Convolution-Superposition algorithms, using a heterogeneous slab phantom. Planned and measured dose distributions using radiochromic films were compared. The MC algorithm gave more accurate results compared with the other algorithms.

Wilcox et al., in 2008, conducted a similar study where the dose measurements in heterogeneous phantom using an EBT film were compared with the dose calculations done with Ray Tracing and Monte Carlo algorithms of the CyberKnife treatment planning system. The study concluded that the dose calculated using Monte Carlo agree with the EBT film measurements.

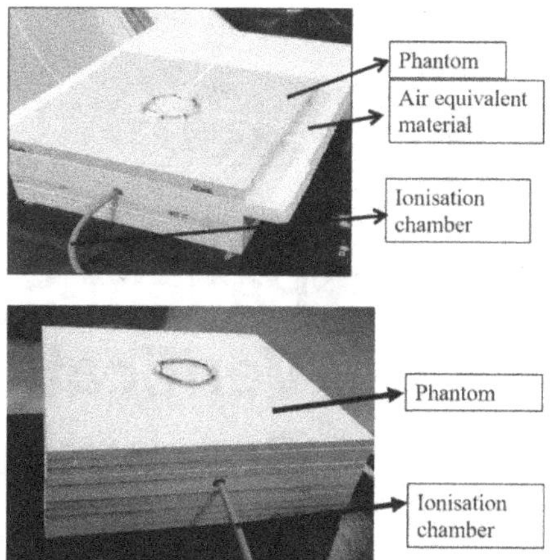

Fig. 133.1 Setup with and without inhomogeneity

3. MATERIALS AND METHODS

The tissue inhomogeneity was simulated by introducing a low-density material into a high-density medium. An air-equivalent material like Styrofoam (density= 0.001293 g/cm^3) was used to represent the former and solid water phantom (density = 1.032 g/cm^3) representing the latter. The solid water phantom is made up of an epoxy-based resin mixture having radiation absorption and scattering properties similar to that of tissue.

For setup with inhomogeneity, the Styrofoam was placed in between the phantoms and for setups without inhomogeneity, Styrofoam was removed and phantoms kept instead.

The CT images for the two setups, with and without inhomogeneity were taken in the institute's CT scanner; GE Optima 580 W with a slice thickness of 1.25 cm. Plans were created using the Accuray Precision® 3.3.1.3[2] treatment planning system (TPS) with Fiducial tracking chosen as the tracking method. This was achieved with the help of a "ring model" where four fiducials were attached to a ring, placed on top of the phantom setup. Sequential optimisation was performed. Four plans were created for each of the setups, with both Raytracing and Monte Carlo dose calculation algorithms.

All plans were executed and dose measured using a calibrated 0.6 cm^3 ionisation chamber and compared with the TPS calculated doses.

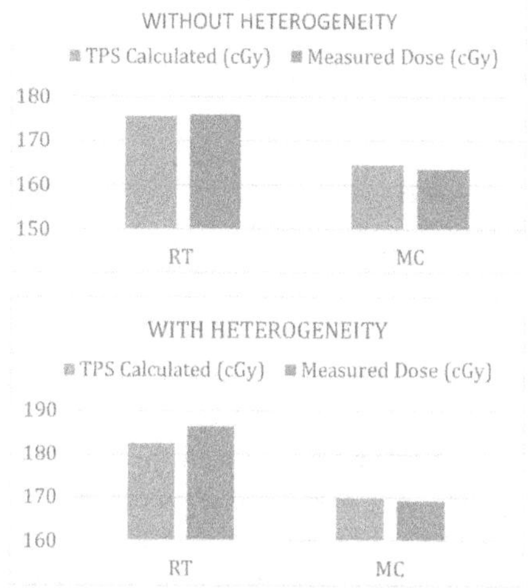

Fig. 133.2 TPS calculated and measured doses for the setups with and without heterogeneity

4. RESULTS

The plans with heterogeneity calculated using RT algorithm showed difference of 2% between TPS calculated and measured doses. In the setup without heterogeneity, the measured doses were in agreement with the calculated doses. The MC algorithm provided comparable results between measured and calculated doses, irrespective of heterogeneity.

5. CONCLUSIONS

The RT algorithm does not fully take into account the photon and electron transport in an inhomogeneous medium during planning resulting in a significant difference between planned and measured doses. The MC algorithm equally takes into account both the particles' interactions in the medium and provides a better performance in predicting doses irrespective of heterogeneity.

ACKNOWLEDGEMENT

I would like to express my gratitude to the faculties of Amrita Institute of Medical Sciences, Kochi for their valuable support.

REFERENCES

1. Sharma S., Ott J., Williams J. and Dickow D. (2010) Dose Calculation Accuracy of the Monte Carlo Algorithm for CyberKnife Compared with Other Commercially Available Dose Calculation Algorithms. Medical Dosimetry 36(4): 347–350.
2. Wilcox E. E., Daskalov G. M. (2008) Accuracy of dose measurements and calculations within and beyond heterogeneous tissues for 6 MV photon fields smaller than 4 cm produced by CyberKnife. Medical Physics 35(6): 2259–2266.

Note: All the figures in this chapter were made by the authors.

Patient-Specific Dosimetric Verification Using TLD in HDR Brachytherapy for Cervical Cancer with an In-House Phantom

Prabhat Krishna Sharma*
Department of Physical Sciences,
Banasthali Vidhyapith,
Newai Tonk, Rajasthan, India
Department of Radiological Physics,
Dr. S.N. Medical College & AGH,
Jodhpur, Rajasthan, India

Saral Kumar Gupta
Department of Physical Sciences,
Banasthali Vidhyapith,
Newai Tonk, Rajasthan, India

Devesh Gupta,
Sushil Kumar Shukla,
Subhash Verma
Department of Radiological Physics,
Dr. S.N. Medical College & AGH,
Jodhpur, Rajasthan, India

Abstract: To ensure the correct distribution of the dosage to the tumor volume and the avoidance to the normal tissues, patient-specific quality assurance (QA) is crucial in high-dose-rate (HDR) brachytherapy for cervical cancer. The subject of this research is QA for individual patients in HDR brachytherapy to confirm the accurate delivery of treatment plan. We designed a phantom for QA with a stable structure, which can accommodate applicators to treat cervical cancer. TLDs were placed at strategic positions on the phantom including the applicator surface, virtual rectum, and bladder. Treatment plans were created and TLD measurements were compared to TPS calculations post dose delivery for varying vaginal cylinder diameters (2 cm, 2.5 cm, 3 cm). The dose deviations between TLD and TPS values closely agree with clinical measurement for LiF: Mg, Ti. The study established that the developed phantom is useful in patient-specific QA that improves dosimetric accuracy in HDR brachytherapy. This method offers a reliable solution for patient treatment delivery verification and dosimetric accuracy.

Keywords: HDR brachytherapy, Cervical cancer, Patient-specific QA, Thermoluminescent dosimeters (TLDs), CVS applicator, Dosimetric accuracy, Treatment planning systems (TPS)

1. INTRODUCTION

When it comes to malignancies of the female genital tract, cervical cancer is by far the most frequent and one of the top causes of cancer-related death in underdeveloped nations (Gultekin et al., 2020). Surgery, radiation, chemotherapy, and targeted therapy are among the most prevalent methods used to treat gynaecological malignancies. Concurrent chemo-radiation, including external beam irradiation, is followed by brachytherapy, as the majority of patients in underdeveloped nations arrive with large or locally advanced cancers (Eter et al., 1999). Cervical, endometrial, breast, esophageal, head and neck, and other malignancies are often treated using HDR brachytherapy. With the advent of HDR after-loading equipment, brachytherapy has progressed to a more sophisticated level, further demonstrating its efficacy in radical and palliative treatments (Chargari et al., 2019). It is possible to administer cervical cancer brachytherapy using either intracavitary or interstitial methods, or a mix of the two. To target the upper vagina, cervix, and uterus, intracavitary brachytherapy places a radioactive source within an applicator that is introduced into the vaginal canal (Banerjee & Kamrava, 2014). Cylindrical tube applicators are used for individuals undergoing treatment for vaginal cancer or cervical cancer after surgery; the

*Corresponding author: prabhat91krishna@gmail.com

DOI: 10.1201/9781003684718-134

diameter of the cylinder is selected according to the patient's anatomy. Depending on institutional guidelines, patients will undergo a CT scan after having the Central Vaginal Cylinders (CVS) applicator inserted. The CT images are utilized by the Treatment Planning System (TPS) to develop a personalised plan for administering the prescribed dose.

This plan will target the upper 6 cm of the vagina, either at the vaginal surface or at a depth of 0.5 cm (Nag et al., 2000). The use of surrogate points in the calculation of dosages to potentially harmful organs is outlined in the ICRU38 recommendations. As a reference, the bladder point is defined at the surface of the Foley catheter balloon that represents the maximal bladder dosage, which is 7 cc of contrast media. The rectal point is defined as 0.5 cm posterior to the vaginal wall (Bethesda, 1985; Pelloski et al., 2005). It is well acknowledged that QA is crucial in HDR brachytherapy (Kubo et al., 1998). Making a QA system for HDR brachytherapy that is patient-specific was the primary goal of this work. Present QA practices include verifying treatment plan delivery and organ-at-risk (OAR) doses using TLDs. The main objective of our research is to guarantee the correctness of treatment planning by performing patient-specific QA prior to the first therapy session. Our facility makes use of the Sagiplan (Bebig, Brachytherapy) program for treatment planning (Kohr & Siebert, 2007; Sharma & Jursinic, 2013). A phantom developed for patient-specific QA is required to carry this out(Ochoa et al., 2007). The TLDs were selected from among the available dosimetry tools for the phantom's setup in order to facilitate absolute dosage comparisons. An important consideration was making sure the applicator was securely fastened to minimize the possibility of inaccurate dwell position, stepwise movement, and dwell duration measurements. To improve the applicator's positional precision, agar gel was shaped to suit the solid phase of the phantom perfectly.

2. MATERIALS AND METHODS

2.1 Fabrication of Phantom

Figure 134.1 shows the custom-fabricated phantom, which is made from tissue-equivalent materials with dimensions 30.5 cm x 30.5 cm x 7 cm. The phantom base was designed to hold individual applicators using custom-shaped foam inserts. Agar gels, including purified agarose, are widely used in phantoms due to their tissue-mimicking properties (Kim et al., 2011).

The bladder and rectum points were created within the phantom to measure absolute doses to the corresponding organs. A key focus during fabrication was to ensure the applicator insertion component was designed to keep the applicator securely fixed throughout and after insertion. To ensure the applicators would remain undamaged, the hot agar gel solution was cooled slowly to about 50°C

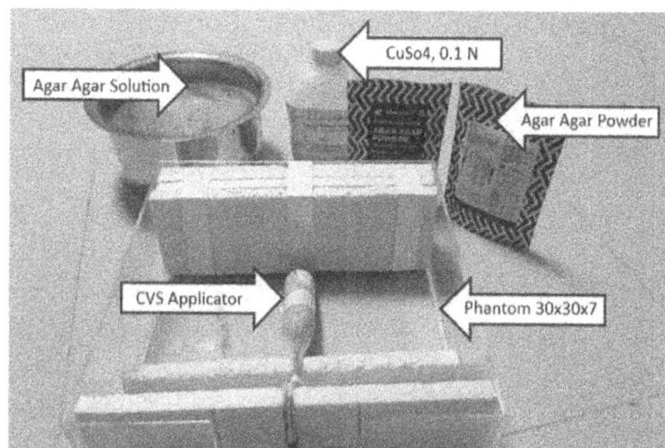

Fig. 134.1 Materials used for phantom fabrication

while stirring intermittently until it reached a pourable and uniform consistency (Fagerstrom & Kaur, 2020; Haack et al., 2009). The heated gel was carefully poured into the prepared phantom to fix the applicators and left to set after it had cooled enough, but before it had fully solidified.

2.2 Dose Measurement and Comparison

We simulated 9 CVS applicator plans for various diameters of the cylinders of 2, 2.5, and 3 cm and treatment lengths of 6.0 cm. The dose delivery point is kept at the surface of the applicator. The HDR brachytherapy plan was developed using a treatment planning system (TPS) on DICOM images of the phantom (Bansal et al., 2013). The treatment plans were created on the phantom consistent with clinical practice, and a dose of 4 Gy, 5 Gy, and 6 Gy was prescribed at the surface of the CVS applicator of different diameters 2, 2.5, and 3 cm to the vaginal vault irradiation. The MTS 100 (LiF: Mg, Ti) of dimension 0.32 cm x 0.32 cm x 0.09 cm thick TLD was placed at the surface of the cylinder, bladder, and rectum points. After the placement of TLD, the phantom was Exposed as per the treatment plan. The delivered doses at the respective positions calculated by TPS were recorded. After irradiation, all TLDs were read using a TLD reader to determine the absorbed dose at respective points in the phantom. After then, the TPS dosages computed for those sites were compared to them. We evaluated the percentage difference between the measured and estimated dosages to find out how accurate the dose administration was.

3. RESULTS

The Dosimetric study showed that the dose measured using TLDs changed with prescription techniques and the size of the applicator used in the treatment. The percentage error among the calculated dose by TPS and the dose delivered as measured by TLDs was calculated for the vaginal cylinder assembly.

The dose distribution at the applicator surface, bladder, and rectum for different vaginal cylinder diameters is shown in Table 134.1. The percentage deviation of the dose reported by the TPS when compared to TLD measurements is obtained as -13.17% to -5.22% at the applicator surface,

Table 134.1 Comparison of TPS calculated dose with TLD measured doses

Vaginal cylinder diameter (mm)	Pre-scribed dose (Gy)	Mean dose at the surface of the applicator (Gy)			Mean dose at Bladder Point (Gy)			Mean dose at Rectum Point (Gy)		
		TPS	TLD	% Variation	TPS	TLD	% Variation	TPS	TLD	% Variation
30	6	5.82	5.13	-11.86	2.66	2.52	-5.26	3.21	3.04	-5.30
30	5	4.6	4.13	-10.22	2.54	2.31	-9.12	3.33	2.99	-10.21
30	4	3.84	3.50	-8.85	2.2	2.03	-7.58	2.64	2.34	-11.36
25	6	5.39	4.68	-13.17	2.48	2.36	-4.91	3.06	2.81	-8.17
25	5	4.72	4.24	-10.17	2.62	2.32	-11.58	3.19	2.91	-8.78
25	4	3.45	3.27	-5.22	2.09	1.89	-9.49	2.52	2.34	-7.14
20	6	5.53	4.87	-11.93	2.77	2.52	-9.15	3.47	3.09	-10.95
20	5	4.46	4.22	-5.38	2.35	2.1	-10.64	3.00	2.77	-7.67
20	4	3.50	3.11	-11.14	1.85	1.72	-6.76	2.30	2.09	-9.13
		Mean % Variation		-9.77	Mean % Variation		-8.28	Mean % Variation		-8.75

-11.58% to -5.26% at the bladder point and -11.36% to -5.30% at the rectum point. From the above tables, the mean percentage deviation of dose by TPS for the surface of the applicator, Bladder, and rectum is -9.77%, -8.28%, and -8.75% respectively.

4. DISCUSSION AND CONCLUSION

In the present work, TLDs were used to verify the brachytherapy treatment for patients treated with vaginal cylinders. The dosimetric consequences of CVS applicators were assessed using TLD. A new method was presented, using an in-house phantom developed for patient QA in brachytherapy to evaluate the treatment reproducibility for various applicators. Results obtained using this method closely matched those measured with TLDs. The TG-43 method, which is used by TPS for dosage estimates, fails to take medium heterogeneity into consideration. The overestimation of dose by the TPS was caused in target and OAR dosages, since the TG-43 formalism fails to account for the attenuation induced by vaginal cylinders and metallic applicators. The deviation strongly depends on the inherent uncertainties of dose measurement by TLDs can be minimized by selecting the uniform response TLD for the measurement. The integration of TLDs for dose verification in HDR brachytherapy, along with the in-house phantom, provides a precise and reliable approach for patient-specific QA. This study underscores the potential of this method to enhance treatment accuracy and delivery.

REFERENCES

1. Gultekin, M., Ramirez, P. T., Broutet, N., & Hutubessy, R. (2020). World Health Organization call for action to eliminate cervical cancer globally. In International Journal of Gynecological Cancer (Vol. 30, Issue 4, pp. 426–427).
2. Eter, P., Ose, G. R., Undy, R. N. B., Dwin, E., Atkins, Aiman, I. A. M., & Larke -P Earson, A. L. C. (1999). Concurrent cisplatin-based radiotherapy and chemotherapy for locally advanced cervical cancer.
3. Chargari, C., Deutsch, E., Blanchard, P., Viswanathan, A. N., & Haie-Meder, C. (2019). Brachytherapy: An overview for clinicians. CA: A Cancer Journal for Clinicians, 69(5), 386–401.
4. Banerjee, R., & Kamrava, M. (2014). Brachytherapy in the treatment of cervical cancer: A review. In International Journal of Women's Health (Vol. 6, Issue 1, pp. 555–564).
5. Nag, S., Erickson, B., Thomadsen, B., Orton, C., Demanes, J. D., & Petereit, D. (2000). The American Brachytherapy Society recommendations for high-dose-rate brachytherapy for carcinoma of the cervix.
6. Bethesda, M. (1985). Dose and Volume Specification for Reporting Intracavitary Therapy in Gynaecology: ICRU Report No. 38.
7. Pelloski, C. E., Palmer &Eifel, P. J. (2005). Comparison between CT-based volumetric calculations and ICRU reference-point estimates of radiation doses delivered to bladder and rectum during intracavitary radiotherapy for cervical cancer. International Journal of Radiation Oncology Biology Physics, 62(1), 131–137.
8. Kubo, H. D., Glasgow & Williamson, J. F. (1998). High dose-rate brachytherapy treatment delivery: Report of the AAPM Radiation Therapy Committee Task Group No. 59.
9. Kohr, P., & Siebert, F. A. (2007). Quality assurance of Brachytherapy Afterloaders using a multi-slit phantom. Physics in Medicine and Biology, 52(17).
10. Sharma, R., & Jursinic, P. A. (2013). In vivo measurements for high dose rate brachytherapy with optically stimulated luminescent dosimeters. Medical Physics, 40(7).
11. Ochoa, R., Gómez, F., Ferreira, I. H., Gutt, F., & de Almeida, C. E. (2007). Design of a phantom for the quality control of high dose rate 192Ir source used in brachytherapy. Radiotherapy and Oncology, 82(2), 222–228.
12. Kim, Y., Muruganandham, M., Modrick, J. M., & Bayouth, J. E. (2011). Evaluation of artifacts and distortions of titanium applicators on 3.0-tesla MRI: Feasibility of titanium applicators in MRI-guided brachytherapy for gynecological cancer. International Journal of Radiation Oncology Biology Physics, 80(3), 947–955.
13. Fagerstrom, J. M., & Kaur, S. (2020). Simple phantom fabrication for MRI-based HDR brachytherapy applicator commissioning. Journal of Applied Clinical Medical Physics, 21(11), 283–287.
14. Haack, S., Nielsen, S. K., Lindegaard, J. C., Gelineck, J., & Tanderup, K. (2009). Applicator reconstruction in MRI 3D image-based dose planning of brachytherapy for cervical cancer. Radiotherapy and Oncology, 91(2), 187–193.
15. Bansal, A. K., Semwal, (2013). A phantom study on bladder and rectum dose measurements in brachytherapy of cervix cancer using FBX aqueous chemical dosimeter. Physica Medica, 29(4), 368–373.

Note: All the figures and tables in this chapter were made by the authors.

Recent Trends in Applied Physics and Material Science – Sudhir Bhardwaj et al. (eds)

135

Impact of Aperture Size Selection in Oligometastatic of Brain—Dosimetric Evaluation

Sruthi G. S.*, Megha Philip, Annex E. H., Ashitha M. K., Debnarayan Dutta
Department of Radiation Oncology, Amrita Institute of Medical Sciences,
Ponekkara P.O, Kochi, Kerala

Abstract: The approach for collimator selection is vital in controlling the plan quality criteria of Stereotactic radiosurgery (SRS). CyberKnife® Robotic Radiosurgery System(M6) delivers beam in 3D workspace unlike 360-degree rotation of the conventional Linear accelerator-based SRS treatment system. Different approach for aperture size selection for the treatment of brain metastasis is evaluated. Accuray Precision® 3.3.1.3[2] Treatment planning system provides multiple options for the collimator selection. Plans created for 3 PTVs which are separated more than 50mm apart. First approach used individual cone selection (7.5 mm and 10 mm) for each PTV, second one used combined PTV volume for cone size selection (12.5 mm, 30 mm, 50 mm) and in the third approach combined PTV with minimal cone opening (7.5 mm and 10 mm) which is derived from first approach. Plan created using Volo optimization algorithm. Plan evaluation criteria include delivery system parameters Total MUs, Number of Nodes, Beams, and dosimetric parameters like Homogeneity Index (HI), Conformity Index (CI), Volume receiving 2% and 98% dose (D2% and D98%), volume of 30% isodose line for spillage, percentage of isodose prescription and PTV maximum and minimum (D_{max},D_{min}) dose. The treatment time is considerably reduced when using the higher aperture size and spillage volume increased. Plans with optimized collimator size gives CI closer to unity and lower spillage volume ($V_{30\%}$ Isodose line). Lower prescribed isodose line in larger collimator opening and higher prescribed isodose line in smaller collimator opening is observed. The plan selection criteria are derived from As Low As Reasonably Achievable (ALARA) and the total treatment delivery time in accordance with the clinical goals. Wise selection of the cone size will help in the effective planning of SRS.

Keywords: SRS, Optimization, Collimator aperture, ALARA

1. INTRODUCTION

Stereotactic radiosurgery (SRS) and hyper-fractionated stereotactic radiotherapy (HFSRT) are recognized as highly effective and well-tolerated treatment modalities for brain metastases, offering substantial clinical benefits, particularly in patients with multiple brain metastases (MBM). SRS involves a single-fraction radiation therapy, using narrow, non-coplanar beams delivered through isocentric arcs and a stereotactic system to target intracranial lesions. When the treatment involves multiple fractions. It is termed stereotactic radiotherapy (SRT). Both methods rely on three-dimensional imaging to accurately localize lesions and deliver focused radiation while minimizing exposure to surrounding healthy brain tissue. A key feature of SRS is its high dose conformity, achieved by using circular beams tailored to the lesion shape, optimizing arc angles, and utilizing dynamic treatment field shaping with mini or micro multileaf collimators (MLCs). A specialized stereotactic apparatus ensures precise beam delivery throughout the process, including imaging, target localization, head immobilization, and treatment setup.

The selection of aperture size plays a crucial role in CyberKnife® radiotherapy, impacting treatment precision, dose distribution, and healthy tissue sparing.

*Corresponding author: srayamgeetha@gmail.com

DOI: 10.1201/9781003684718-135

Smaller apertures offer improved tumor targeting and dose conformity but can lead to longer treatment times and increased risk of inaccuracies due to patient movement. Larger apertures, while reducing treatment time, may compromise precision and increase radiation exposure to surrounding healthy tissues. Therefore, optimizing aperture size based on tumor characteristics, location, and proximity to critical organs is essential for maximizing treatment efficacy and minimizing potential adverse effects.

2. LITERATURE REVIEW

CyberKnife treatment planning for multiple brain metastases Tianlong Ji et al (2022), where they compared two approaches: one in which each Planning Target Volume (PTV) was treated individually, and another where a single, combined PTV encompassing all lesions was targeted. In the separate planning method, each lesion was targeted individually using the collimator auto-selection technique, while in the combined approach, a single collimator was employed to target the entire PTV. The study found that the separate planning approach improved treatment efficiency and provided better dosimetry to normal tissues.

In a separate study, Si Young Jang (2016) compared the dosimetric outcomes of cone/iris-based versus InCise MLC-based CyberKnife plans for treating single and multiple brain metastases. They concluded that intracranial SRS using the InCise MLC was dosimetrically feasible and offered the advantage of reducing beam delivery time compared to the cone/iris method. However, they also observed that the MLC-based planning might not be ideal for very small targets.

2.1 Methodology and Model Specifications

Patient with multiple lesions is taken for the study and contour delineation was based on 0.625mm slice thick CT and MRI images registered. Plans created for three planning target volumes (PTVs) that were spaced more than 50 mm apart with prescribed dose of 20Gy per fraction. Plans were created for Three different approaches to collimator selection were created and evaluated as follows:

- **Plan 1:** Individual cone selection (7.5 mm and 10 mm) for each PTV.
- **Plan 2:** Combined PTV volume used to select larger cone sizes (12.5 mm, 30 mm, 50 mm).
- **Plan 3:** Combined PTV with minimal cone openings (7.5 mm and 10 mm), derived from the first approach.

The plans were developed using the Accuray Precision® 3.3.1.3 treatment planning system

3. EMPIRICAL RESULTS

In Plan 2 the treatment time is considerably reduced when using the higher aperture size and spillage volume increased. While in Plan1 with optimized collimator size gives CI closer to unity and lower spillage volume ($V_{30\%}$ Isodose line).

Lower prescribed isodose line in larger collimator opening and higher prescribed isodose line in smaller collimator opening is observed.

4. CONCLUSION

The planning criteria are derived as the dose to critical structures As Low as Reasonably Achievable, optimum coverage to the target volume and reduced total treatment delivery time in accordance with the clinical goals. Wise selection of the aperture size will help in the effective planning of SRS.

Table 135.1 Evaluation of Plan 1, Plan 2 and Plan 3

Parameters	Plan 1	Plan 2	Plan 3
Number of nodes	110	60	115
Number of beams	282	90	267
Treatment time (min)	97	44	94
Prescription Isodose (%)	90	84	90
Conformity Index (CI)	1.14	1.36	1.14
Homogeneity Index (HI)	1.11	1.19	1.11
D 2%	2188	2340	2186
D 98%	1988	1965	1978
Total MU	36387.1	12259.7	35496.9
ISO 30% (cm^3)	23.2	155.78	23.86
Brain stem 0.01 cm^3	504	467	535

REFERENCES

1. Jang Si Y, Lalonde Ron, Ozhasoglu C, Burton S., Dwight H., and Huq M. S (2016) Dosimetric comparison between cone/Iris-based and InCise MLC-based CyberKnife plans for single and multiple brain metastases. J Appl Clin Med Phys.2016 Sep 8;17(5):184–199.
2. Ji T, Song Y, Zhao X, Wang Yand Li G (2022) Comparison of Two Cyberknife Planning Approaches for Multiple Brain Metastases.Front. Oncol. 12:797250.

Note: The authors made the table in this chapter.

Recent Trends in Applied Physics and Material Science – Sudhir Bhardwaj et al. (eds)

A Comparative Study of Technological Advancements from 5G to 6G

Shipra Gupta[1], Vijay Kumar[2] and Amit Verma[3]
Graphic Era Hill University, Dehradun, Uttrakhand, India

Abstract: In this comparative study, the main technological evolutions are reviewed between 5G to 6G technologies in terms of frequency, spectrum and MIMO (Multiple Input, Multiple Output). The previous generation (5G) of mobile technology, brought about the capability for high-speed data transfer, low latency and even improved connectivity. The research describes the anticipated enhancements such as data rate, network capacity, energy efficiency and security. It also looks at the emerging use cases for 6G, encompassing immersive XR, holographic-type communications in real-time and end-to-end autonomous systems. This study also analyses the potential of 6G systems, but it also speaks to other critical issues around 6G that includes requirements for sustainable architecture, an advanced level of security and global standardization. The comparative analysis is expected to give a full picture how differently 6G can be from its predecessor, with highlighting the revolutionary capability of 6G in building up highly connected and intelligent world.

Keywords: Smart technologies, Wireless communication, 5G, 6G, AI integration

1. INTRODUCTION

The transformation process which starts from 5G to 6G has a tremendous change in wireless communication technology and connectivity, user experience along with wide range of new applications. This comparative survey discusses the salient characteristics of 6G. Its technical innovations against current status (5G), by considering new frequency band usage (≥100 GHz), advanced MIMO technologies, such as hybrid analogue-digital beam forming methodologies. 5G networks deployment has not only delivered faster speeds, lower latency but the new network capacity as a significant breakthrough in mobile communication. Such advancements have resulted in applications like high-definition video streaming, augmented reality (AR), virtual reality (VR), and the explosion of Internet of Things (IoT) devices (Davuluri and Hasan et al., 2021). In order to process higher data rates low latency connectivity and limitless experience reality, the 5G has also some of the boundaries. This gap opens up an urgent demand for wireless technology of the next generation 6G. 6G represents a step further from 5G, transforming existing infrastructure for the next era of cooperation while creating an even more connected and intelligent world (Dhiman and Nagar, 2022). Therefore, a comparative analysis between 5G and the upcoming Era of 6G is important as to what shall change these innovations will bring & how it would waive off few limitations of today. 5G networks use different types of spectrum ranging from sub-1 GHz to above 24 GHz, including mid band frequencies between them, i. e., roughly between 1–6 GHZ for capacity-minded deployments and underutilized regulatory landscapes as in developed nations (Dhiman and Rho et al., 2022). With the introduction of these frequency bands, data rates will improve significantly when compared to previous generations. However, they are plagued with issues of signal attenuation due to high frequencies and low power levels, restricted range limitation band width utilization inefficiency fragmentation leading a thousand points networking model (Gupta et al., 2021). One of the likely advances in 6G that many are looking at is using 100 GHz. This spectrum offers significant bandwidth and data rate improvements, up to terabit-per-second speeds. Featuring blazing fast data throughput, the THz spectrum supports 6G and beyond with data rates to enable UHD content as well as next-generation applications such holographic communication (Mirza et al., 2023).

[1]drshipragupta16@gmail.com, [2]drvijaykumar.geu@gmail.com, [3]av726091@gmail.com

DOI: 10.1201/9781003684718-136

2. LITERATURE REVIEW

This literature review investigates the fundamental enablers and explores features from 5G to 6G for performance enhancements, use cases, and technical innovations.

2.1 The Modernization of 5G Technology

5G, or fifth-generation wireless, offers faster speeds and thousands of time more capacity than 4G. 5G offers a mindboggling download speed ranging up to 10 Gbps or 100 times faster than today (Mishra et al., 2023). In 5G networks, the latency can be as low as 1 ms, useful for applications that require real-time communication like self-driving cars or telemedicine. Massive Connectivity means 5G can connect up to 1 million connected devices per square kilometre, which will help power the explosion of growth in IoT based solutions (Wazid et al., 2021). 5G enables operators to provision multiple virtual networks within a single physical infrastructure, optimizing network resources according to specific use-cases, like high-speed internet or low-power IoT-connected devices. Despite the better connectivity 5G provides, it is fundamentally developed for better mobile broadband services such as smart cities, autonomous driving or industrial automation (Wazid et al., 2021).

3. OBSERVATION

Table 136.1 Comparative analysis of key performance metrics of 5G and 6G wireless technologies

Performance Metric	5G	6G
Peak Data Transfer Rates	Up to 10 Gbps	Up to 1 Tbps (100 times faster than 5G)
Typical User Data Rates	100 Mbps to 1 Gbps	10 Gbps to 100 Gbps
Latency	~1 millisecond (ms)	Below 1 ms; potentially as low as 0.1 ms (100 microseconds)
Network Capacity	Supports up to 1 million devices per km^2	Supports up to 10 million devices per km^2 (10 times more than 5G)
Device Density	High, but performance may degrade in dense environments	Extremely high, designed for dense environments without performance degradation
Spectrum Efficiency	3-4 times more efficient than 4G	Further improved with AI-driven management and advanced modulation
Bandwidth	Operates in sub-1 GHz, mid-band (1-6 GHz), and high-band (above 24 GHz, mmWave)	Operates in terahertz (THz) frequencies (100 GHz to 10 THz) with much larger bandwidth
Energy Efficiency	More energy-efficient than 4G but with higher overall energy consumption	Up to 90% more energy-efficient than 5G, with AI-driven energy management and sustainable practices
Reliability	High reliability with 99.999% availability	Ultra-high reliability with 99.9999% availability
Security and Privacy	Enhanced security features, but increased attack surface due to more connected devices	Advanced security with quantum-resistant cryptography and AI-driven threat detection

4. DISCUSSION

Table 136.1 represents that 6G has the potential to reduce its energy consumption by as much as 90% over that of 5G, with an ultimate target for sustainability thanks in part to AI and green technologies. The 6G is aimed to provide ultra-reliability with less than one millisecond latency, a high reliability of up to 99.99% assurance availability which goes far beyond human intuition as well. The introduction of this generation system would boast advanced security features that includes quantum-resistant cryptography in order prevent emerging cyber threats from exploiting its evolvable vulnerabilities. Table 136.2 summarizes from 5G to 6G in terms of the types of technological advancements. 5G deployment is already a very capital intensive including small cells, but 6G with ultra-dense networks and THz frequencies with advances hardware employ. 5G is already high in its power consumption with the advent of higher data rates but balancing ultra-high data rate plus sustainability and energy efficiency for 6G are critical tasks. 5G technologies is achieving low latency across diverse environments, ensuring consistent reliability. 5G technologies are with massive MIMO, network slicing and edge computing but handling even greater complexity with extremely large-scale MIMO, AI-driven networks, and new use cases with 6G. The deployment and maintenance cost of 5G is already high for consumers and businesses. In 6G technologies, it is higher costs for developing, deploying, and maintaining 6G technologies, potential affordability issues. 6G must managing the environmental footprint of 6G, ensuring sustainable practices and green technologies while 5G Concerns over increased energy consumption and electronic waste. Regulatory challenges for 5G, especially regarding spectrum allocation will be even more complicated with the new use of frequency bands required by 6G as well as the integration of AI into them and a global coordination that is increasingly needed. 5G has to balance innovation with performance & security.

5. CONCLUSION

From the discussion of the above four tables' data the conclusion of this manuscript shows that we are facing a technological revolution compared to 5G. This is for the first time so much markedly with previous generation. Although 5G has provided a solid framework of fast data transfer, increased connectivity and low latency, the

Table 136.2 Key technological aspects of 5G and 6G wireless technologies

Technology Aspect	5G	6G
Frequency Bands	Sub-1 GHz, 1-6 GHz (mid-band), and above 24 GHz (mm Wave)	Terahertz (THz) frequencies (100 GHz to 10 THz)
MIMO (Multiple Input, Multiple Output)	Massive MIMO with hundreds of antennas for enhanced beam forming	Extremely large-scale MIMO with thousands of antennas for higher capacity and resolution
Network Architecture	Cloud-based architecture, centralized and edge computing	AI-driven decentralized and self-optimizing networks, with more advanced edge computing
Network Slicing	Supports network slicing for dedicated virtual networks	Enhanced slicing with more dynamic and fine-grained control for specialized services
AI and Machine Learning Integration	AI for network optimization, resource management, and predictive maintenance	Deep integration of AI for autonomous networks, real-time analytics, and intelligent decision-making
Security Mechanisms	Enhanced encryption, secure boot, and network slicing	Quantum-resistant cryptography, AI-driven threat detection, and real-time security analytics
Latency Reduction Techniques	Edge computing and network slicing	Ultra-low latency techniques with AI optimization and advanced edge intelligence
Connectivity and Interoperability	Designed for IoT, smart cities, AR/VR, and autonomous vehicles	Seamless integration of AI, XR (Extended Reality), tactile internet, and holographic communication
Energy Efficiency Techniques	Power-saving techniques and energy-efficient hardware	AI-driven energy management, green technologies, and sustainable infrastructure
Service and Application Support	High-definition streaming, IoT, smart cities, AR/VR	Advanced AR/VR, tactile internet, holographic communication, real-time AI processing, and immersive experiences

speed at which these functions can be performed is going to increase significantly with 6G. It will also eliminate some problems that were not possible in its predecessor bringing a host full new applications. Prominent features of 6G technology are its operation in the terahertz (THz) bands. It would make possible data rates and network capacity that far exceed those offered by 5G, as well as a potentially limitless supply of new spectrum open only at THz frequencies, more advanced massive MIMO systems with very large antenna arrays to form extremely high directivity beams even at much higher frequency ranges. AI-driven intelligent networks are enforcing its applications efficiently while maintaining performance gains over existing technologies. These advancements will provide the necessary foundations for developing next-gen applications like immersive extended reality (XR), real-time holographic communication, and completely autonomous systems which go far beyond 5G. Still, other problems may arise as we proceed from 5G to 6G. This will complicate spectrum management, infrastructure deployment and energy consumption by the communications system of various regions which needs new solutions with considerable investment. The depth of encryption and threat detection will only get deeper as AI becomes embedded in various processes leading to a rise in both securities, privacy risks. Furthermore, 6G will require consistent global standards to be set and for it to be offered at an affordable price in order for everyone on the planet to respond rapidly. In conclusion, 6G is a revolutionary wireless technology which provides significant enhancements on speed, capacity and scalability as well as green communication via energy efficiency over security issues for applications towards a more connected world.

REFERENCES

1. Davuluri, S. K., Alvi, S. A. M., Aeri, M., Agarwal, A., Serajuddin, M. and Hasan, Z. (2021). A Security Model for Perceptive 5G-Powered BC IoT Associated Deep Learning. IEEE Transactions on Industrial Informatics. 17(10): 6754–6764.
2. Dhiman, G. and Nagar, A. K. (2022). Editorial: Block chain-based 6G and industrial internet of things systems for industry 4.0/5.0. Expert Systems. 39(10):e13162.
3. Dhiman, G., Nagar, A. K., Vimal, S. and Rho, S. (2022). Guest Editorial: Cyber twin-Driven 6G for Internet of Everything: Architectures, Challenges, and Industrial Applications. IEEE Transactions on Industrial Informatics. 18(7):4846–4849.
4. Gupta, A., Ghanshala, K. and Joshi, R. C. (2021). Machine Learning Classifier Approach with Gaussian Process, Ensemble boosted Trees, SVM, and Linear Regression for 5G Signal Coverage Mapping. International Journal of Interactive Multimedia And Artificial Intelligence. 6(6):156–163.
5. Mirza, M. A., Junsheng, Y., Raza, S., Ahmed, M., Asif, M., Irshad, A. and Kumar, N. (2023). MCLA Task Offloading Framework for 5G-NR-V2X-Based Heterogeneous VECNs. IEEE Transactions on Intelligent Transportation Systems. 24(12): 14329–14346.
6. Mishra, A. K., Wazid, M., Singh, Devesh P., Das, A. K., Singh, Jaskaran, V. and Athanasios, V. (2023). Secure Block chain-Enabled Authentication Key Management Framework with Big Data Analytics for Drones in Networks Beyond 5G Applications. Drones. 7(8):508–514.
7. Wazid, M., Das, A. K., Shetty, S., Gope, P. and Rodrigues, J. P. C. (2021). Security in 5G-Enabled Internet of Things Communication: Issues, Challenges, and Future Research Roadmap. IEEE Access. 9: 4466–4489.
8. Wazid, M., Das, A. K., Kumar, N. and Alazab, M. (2021). Designing Authenticated Key Management Scheme in 6G-Enabled Network in a Box Deployed for Industrial Applications. IEEE Transactions on Industrial Informatics. 17(10): 7174–7184.

Note: All the tables in this chapter were made by the authors.

Recent Trends in Applied Physics and Material Science – Sudhir Bhardwaj et al. (eds)
© 2026 Taylor & Francis Group, London, ISBN 978-1-041-16452-4

Cation Distribution in the Spinel Oxide $MnFe_2O_4$: Magnetization and Raman Spectroscopy Study

P. Suchismita Behera*
Department of Physics, Indian Institute of Technology Madras, Chennai, India

Abstract: $MnFe_2O_4$ is a magnetic spinel oxide whose crystal structure and magnetization can be tuned by the sample preparation conditions and chemical composition. In this study, a single phase polycrystalline $MnFe_2O_4$ spinel oxide (cubic, Fd-3m) was synthesized via solid-state reaction. The spinel structure (Space group Fd-3m) of $MnFe_2O_4$ has been confirmed through Rietveld refinement analysis and Raman spectroscopy. Magnetization vs temperature obtained in a field of 1 kOe reveal a paramagnetic state to a ferrimagnetic phase at high-temperature regime with $T_C = 561$ K, $\theta_P = 580$ K, an $\mu_{eff} = 8.3\ \mu_B$/(f.u.) Magnetic susceptibility is well explained by Néel two-sublattice model in the paramagnetic region.

Keywords: Spinel ferrite, X-ray diffraction, Raman spectroscopy, Ferrimagnetism

1. INTRODUCTION

Among the Fe-based spinel oxides, the $MnFe_2O_4$ system has high chemical stability for room-temperature applications that can be prepared easily. It exhibits soft magnetic behaviour, moderate saturation magnetization at room temperature, and high initial permeability [1, 2]. Interestingly, bulk $MnFe_2O_4$ predominantly crystallizes in a conventional spinel structure, where Mn and Fe ions hold the positions of tetrahedral(A)-site and octahedral(B)-site respectively, exhibiting ferrimagnetic properties. In contrast, nanocrystalline $MnFe_2O_4$ adopts a partially inverse spinel structure, with Mn^{2+} (20%) ions at the B-sites and Fe^{3+} (80%) at the A-sites, resulting in unique behaviours such as metastable cation distribution, superparamagnetic, coercivity enhancement, and boosted electrical resistivity [2 - 5]. In the present study, the cation distribution in polycrystalline $MnFe_2O_4$ has been estimated and DC magnetization and Raman spectroscopy data are obtained.

2. EXPERIMENTAL DETAILS

Solid State technique was used to prepare polycrystalline $MnFe_2O_4$ sample. High purity (≥99.9%) stoichiometric quantities of MnO and Fe_2O_3 were mixed well, ground, and calcined at 1000 °C for 45h. Final sintering was carried out at 1200 °C for 5h. Powder X-ray diffraction (XRD) at 300 K (Rigaku) with Cu Kα (λ = 1.541Å) radiation was applied to characterize the sample structure. The Rietveld refinement technique examined the resulting pattern using FULLPROF software [6]. Based on the anticipated cation arrangement, the crystallographic spinel ferrite structural model of $MnFe_2O_4$ was adjusted using VESTA software. The sample's elemental composition was determined by energy dispersive x-ray analysis (EDXA) employing a FEI Inspect scanning electron microscope. The He-Ne laser (633 nm) was used as an excitation source in micro-Raman spectrometer Horiba Jobin-Yvon HR800 UV to conduct Raman spectroscopy. Utilizing a commercial magnetometer (PPMS-VSM, Quantum Design), DC magnetization (M) was performed up to 70 kOe of applied magnetic field and temperatures between 3 to 350 K. A commercial VSM magnetometer [Lakeshore Inc.] recorded high-temperature DC magnetization in magnetic fields of 1 kOe between 300 and 900 K.

3. RESULTS AND DISCUSSIONS

XRD pattern of $MnFe_2O_4$ at 300 K shows that the sample

*Corresponding author: psuchi06@gmail.com

DOI: 10.1201/9781003684718-137

exists in a single phase with a cubic structure that belongs to *Fd-3m* space group [Fig. 137.1(a)]. The lattice constant (a) and volume of the unit cell (V) were determined to be 8.5121(4) Å and 616.741(3) ($Å^3$) respectively and these values are similar to those obtained in the previous study [7]. Figure 137.1(b) illustrates the crystallographic structure of $MnFe_2O_4$ spinel ferrite according to the cation distribution at A and B sites as obtained from the Rietveld refinement method. The cationic distributions depict that the $MnFe_2O_4$ compound is a mixed spinel ferrite system because 85 % of Mn^{2+} divalent and 15% of Fe^{3+} trivalent cations occupied at A-sites with Wyckoff position 8a (1/8, 1/8, 1/8) and the B-sites are occupied by 85% of Fe^{3+} trivalent and 15% of Mn^{2+} divalent cations with 16d (1/2, 1/2, 1/2). The anionic distribution of Oxygen ions O^{2-} attributed at 32e (0.26, 0.26, 0.26) as shown in Table 137.1. The cationic distribution in $MnFe_2O_4$ can be written as $(Mn^{2+}_{0.85}\,Fe^{3+}_{0.15})_{Tetra}\,(Mn^{2+}_{0.15}\,Fe^{3+}_{1.85})_{Octa}\,O_4^{2-}$.

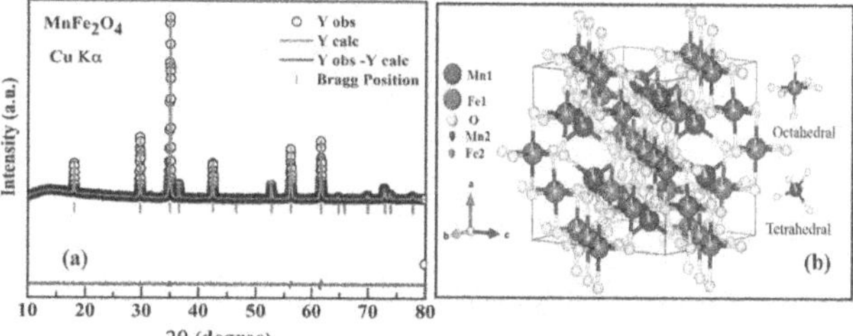

Fig. 137.1 (a) Rietveld refinement fit for the powder XRD pattern of $MnFe_2O_4$ at 300 K and (b) The crystallographic structure of $MnFe_2O_4$ spinel ferrite using VESTA

Table 137.1 Structural parameters derived from rietveld refined XRD data of $MnFe_2O_4$ (Cubic, Fd-3m) at 300 K. Lattice parameters: a=b=c= 8.5121(4) Å and interaxial angle (α=β=γ=90°), Factors; R_p = 9.47, R_{wp} = 5.52, χ^2 = 2.05, R_{Bragg} = 2.25, R_F = 1.92. Bond lengths (A-O) = 2.016(3) Å, (B-O) = 2.033(3) Å and Bond Angles ∠AOB = 121.3° (3), ∠BOB=95.5° (3)

Atoms	Wyckoff Position	x	y	z	Occupancy
Mn1	8a	0.125	0.125	0.125	0.03921
Fe1	8a	0.125	0.125	0125	0.00354
Mn2	16d	0.500	0.500	0.500	0.00832
Fe2	16d	0.500	0.500	0.500	0.07512
O	32e	0.2617	0.2617	0.2617	0.17213

The average cation radii are estimated using equations (1) and (2) [8], as r_A = 0.6345 Å for A-site and r_B = 0.6911 Å for B-site respectively. Here, C^A and C^B, denoted as ionic concentration at these sites and ionic radii of Mn^{2+} and Fe^{3+} are denoted as $r_{Mn^{2+}}$ and $r_{Fe^{3+}}$ respectively. The results suggest that the value of r_A is smaller than r_B which can be due to the larger ionic radii of Mn^{2+}(0.83Å) and Fe^{3+}(0.645 Å) distribution at octahedral B-site.

$$r_A = \left(C^A_{Mn^{2+}}\right)\cdot\left(r_{Mn^{2+}}\right)+\left(C^A_{Fe^{3+}}\right)\cdot\left(r_{Fe^{3+}}\right) \quad (1)$$

$$r_B = \frac{\left(C^B_{Mn^{2+}}\right)\cdot\left(r_{Mn^{2+}}\right)+\left(C^B_{Fe^{3+}}\right)\cdot\left(r_{Fe^{3+}}\right)}{2} \quad (2)$$

EDXA determines the elemental composition of the sample. The atomic % of Mn, Fe and O are 16, 29, and 55 respectively and it closely matches with the initial values ascertaining stochiometric composition. In addition, the elemental mapping confirms the uniform distribution of elements in the sample [Fig. 137.2(a)].

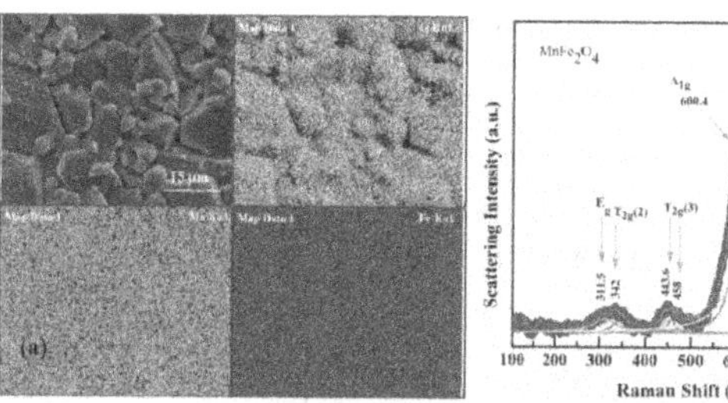

Fig. 137.2 (a) Elemental mapping of $MnFe_2O_4$ and (b) Deconvoluted Raman spectrum of $MnFe_2O_4$ at 300 K

The XRD result reveals 15% inversion at tetrahedral sites of the $MnFe_2O_4$ sample. However, Raman spectroscopy is considered as highly effective and non-destructive method for distinguishing normal, inverse and mixed spinel phases. In the present study, the observation of prominent vibrational modes ($A_{1g} + E_g + 3T_{2g}$) in the Raman spectra of the $MnFe_2O_4$ oxide sample, recorded at 300 K within the 100-1000 cm^{-1} range, confirmed the formation of a spinel cubic structure. For this, the spectrum was deconvoluted as shown in the Fig. 137.2 (b). It is commonly accepted that the highest frequency A_{1g} mode correlates to the symmetric stretching mode of the O^{2-} atoms with in the spinel lattice of Mn/FeO_4 tetrahedral unit. The low-frequency modes are ascribed to vibrations within the octahedral FeO_6 sites [9]. The observed vibrational mode (E_g) at 311.5 cm^{-1}, T_{2g} (2) mode at 342 cm^{-1}, and T_{2g} (3) mode at 443.6 cm^{-1} and 458 cm^{-1} are less intense, weaker, and shoulder-like features suggest the reattribution of Mn^{2+} and Fe^{3+} cations within A and B sites of the spinel lattice. However, in the Raman spectra of cubic spinel ferrites, the highly intense A_{1g} active mode at 633.5 cm^{-1} shows an additional shoulder-like peak at a lower wavenumber of 600.4 cm^{-1}, which is a characteristic feature of mixed spinels, similar explanation also suggested by few reports [10, 11].

Initially, the magnetization vs temperature (M-T) of $MnFe_2O_4$ in the zero-field cooled (ZFC) mode was obtained from 3 K to 350 K in applied fields of 5 kOe [Fig. 137.3(a)]. Then the high-temperature M(T) under 1kOe applied field was recorded from 300 to 900 K in ZFC mode [Fig. 137.3(b)]. As the T decreases, M increases, showing a sharp rise below transition temperature

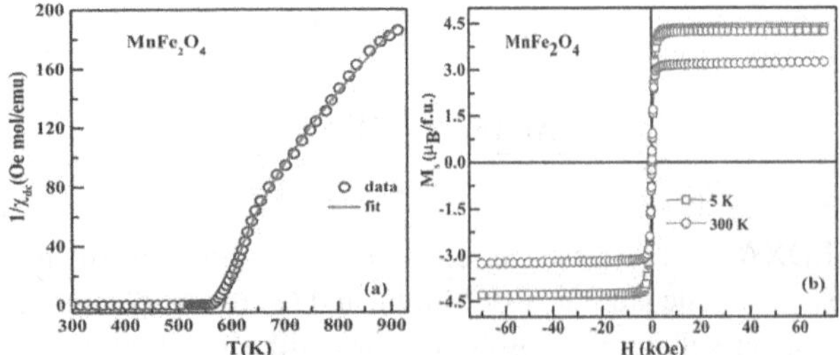

Fig. 137.3 (a) Inverse magnetic dc susceptibility vs temperature plot measured under H=1 kOe. The red solid line denotes the fit to Eq. (3) as described in the text. (b) Field-dependent magnetization data at 5 K and 300 K

$T_C \sim 561$ K [Inset of Fig. 137.3(b)], manifesting a long-range ordered state. The shift from a paramagnetic state to a ferrimagnetic phase is responsible for the transition temperature, which agrees with the previously published value [12]. Fig. 137.4(a) represents the inverse magnetic dc susceptibility $\left(\frac{1}{\chi_{dc}}\right)$ vs temperature in the paramagnetic zone. According to the Néel's ferrimagnetic theory for two-sublattices, the below formula represents the paramagnetic inverse dc susceptibility [4, 12].

$$\frac{1}{\chi_{dc}} = \frac{T-\theta_a}{C} - \frac{\xi}{T-\theta'} \tag{3}$$

High-temperature asymptotic behaviour is represented in the first term of the equation where θ_a is the asymptotic Curie temperature. The robustness of the antiferromagnetic interaction between the two sublattices can be revealed by the size of this θ_a value [4, 13, 14] and C is the combined Curie constant of both sublattices, expressed as $C = C_A + C_B$ [4, 12, 13]. The hyperbolic nature close to the transition temperature is explained by the second term where ξ and θ' are the molecular field constant parameters [4, 12]. The solid curve in Fig. 137.4(a) shows that Eq. (3) fits very well with the experimental data and its fitting parameters are provided in Table 137.2. Fitting parameters of single crystal $MnFe_2O_4$ [4] are also listed in Table 137.2 for comparison. The value of θ_a for polycrystalline $MnFe_2O_4$ is similar to that in single crystal $MnFe_2O_4$, indicating that the antiferromagnetic coupling in both polycrystalline and single crystal $MnFe_2O_4$ is weaker, resulting in higher ferrimagnetic transition temperatures. The overall effective paramagnetic moment is evaluated to be $\mu_{eff} = 8.3\ \mu_B/(f.u.)$ by using fitted value $C = \frac{N\mu_{eff}^2}{3k_B}$ = 8.6 emu-K/mol and extrapolated fitted line gives the paramagnetic temperature ($\theta_p \approx 580$ K). The expression, $\mu_{theor} = \sqrt{(\mu_T^2 + \mu_O^2)}$ can be determine the theoretical value of the magnetic moment, where μ_T and μ_O are the magnetic moments of the tetrahedral and octahedral sublattices, respectively [4, 14]. Hence, assuming the high-spin state of both Mn and Fe, the theoretical magnetic moment (μ_{theo}) of $MnFe_2O_4$ is calculated as 10.2 μ_B [15].

The field-dependent magnetization M(H) data values at 5 K and 300 K are found to be 4.4 μ_B/f.u.. and 3.3 μ_B/f.u. in 70 kOe respectively with soft ferrimagnetic behaviour, which is comparable with the single crystal $MnFe_2O_4$ [Fig. 137.3 (b)].

4. CONCLUSION

A single-phase polycrystalline mixed spinel oxide $MnFe_2O_4$ (cubic, Fd-3m) has been synthesized and its structure and cationic distribution are studied through XRD and Raman spectroscopy. EDX analysis confirms the sample composition. Magnetization versus temperature measurements show that the sample exhibits long-range ferrimagnetic ordering around 561 K. The high-temperature paramagnetic region is best described by Néel's two-sublattice model.

ACKNOWLEDGEMENTS

P.S.B thanks DST-SERB, New Delhi for fellowship through the NPDF scheme, sanction order no. PDF/2022/001808. P. S. B thanks R. Nirmala for support. The PPMS (Quantum Design) facility used in this study is supported by DST-FIST project no. SR/FST/PSII-038/2016.

Table 137.2 Comparison of the magnetic parameters and characteristic temperatures acquired from Néel's ferrimagnetic expression for polycrystalline and single-crystal $MnFe_2O_4$

Sample	μ_{theor} (μ_B)	High Temp μ_{eff} (μ_B)	Low Temp μ_s (μ_B)	Curie Temp T_C (K)	θ_a	θ'	Ref.
Polycrystalline $MnFe_2O_4$	10.2	8.3	4.4 at 5 K	561	-295	511	This work
Single crystalline $MnFe_2O_4$	10.2	8.8	4.5 at 2 K	575	-400	524	[4]

REFERENCES

1. Dippong T., Andrea Levei E. and Cadar O. (2021). Recent Advances in Synthesis and Applications of MFe_2O_4 (M = Co, Cu, Mn, Ni, Zn) Nanoparticles. Nanomaterials 11: 1560.
2. Tang Z. X., Sorensen C. M., Klabunde K.J., et al. (1991). Size-Dependent Curie Temperature in Nanoscale $MnFe_2O_4$ Particles. Phys. Rev. Lett. 67: 3602–3605.
3. Aslibeiki B., Kameli P., Ehsani M. H. (2016). $MnFe_2O_4$ bulk, nanoparticles and film: A comparative study of structural and magnetic properties. Ceram. Int. 42: 11.
4. Nepal R., Saghayezhian M., et. al. (2020). Observation of three magnetic states in spinel $MnFe_2O_4$ single crystals. J. Magn. Magn. Mater. 497: 165955.
5. Almessiere M. A., Güner S., et. al. (2021). Structural, Magnetic, and Mossbauer Parameters' Evaluation of Sonochemically Synthesized Rare Earth Er^{3+} and Y^{3+} Ions Substituted Manganese–Zinc Nanospinel Ferrites. ACS Omega,6: 22429–22438.
6. Rodríguez-Carvajal J. (2001). Recent Developments of the Program FULLPROF, in Commission on Powder Diffraction (IUCr). Newsletter, 26: 12–19.
7. Anwar M. S., Bon Heun Koo (2022). Observation of the magnetic entropy change in Zn doped $MnFe_2O_4$ common ceramic: Be cool being environmental friendly. Curr. Appl. Phys. 39: 77–83.
8. Diab R. S., El-Deen L. M. S., Nasr M. H. (2024). Structural, cation distribution, Raman spectroscopy, and magnetic features of Co-doped Cu–Eu nanocrystalline spinel ferrites. Mater Sci: Mater Electron 35:290.
9. Yadav P., Choudhary P., Saxena P., et al. (2019) Spectroscopic analysis and temperature-dependent dielectric properties of bulk Ni–Zn ceramics. J. Adv. Dielect. 9: 1950014.
10. Simon C., Blosser A., Eckardtet M., et al. (2021). Magnetic properties and structural analysis on spinel $MnFe_2O_4$ nanoparticles prepared *via* non-aqueous microwave synthesis. Z. Anorg. Allg. Chem. 647: 2061–2072.
11. Ansari S. M., Ghosh K. C., Devan R. S., et al. (2020). Eco-Friendly Synthesis, Crystal Chemistry, and Magnetic Properties of Manganese-Substituted $CoFe_2O_4$ Nanoparticles. ACS Omega 5: 19315–19330.
12. Morrish A. H. (2001). "Ferrimagnetism" in The Physical Principles of Magnetism, The Institute of Electrical and Electronics Engineers, Inc., John Wiley & Sons, Inc. 486–538.
13. Srinivasan G., Seehra Mohindar S. (1983). Magnetic properties of Mn_3O_4 and a solution of the canted-spin problem. Phys. Rev. B. 28: 1–7.
14. Thota S., Seehra M. S. (2013). Co-existence of ferrimagnetism and spin-glass state in the spinel Co_2SnO_4. J. Appl. Phys. 113: 203905.
15. Hastings J. M., Corliss L. M. (1956). Neutron diffraction study of manganese ferrite, Phys. Rev. 104 (2): 328–331.

Note: All the figures and tables in this chapter were made by the authors.

138

Terahertz Radiation Generation through Cylindrical Nanoparticles

Anuj Dandain[1]
Department of Physics, Lovely Professional University, Phagwara, Punjab, India
K. M. Government College, Narwana, Jind, Haryana, India

Shivani Vij[2]
Department of Applied Sciences, DAV Institute of Engineering & Technology, Jalandhar, India

Niti Kant
Department of Physics, University of Allahabad, Prayagraj, Uttar Pradesh, India

Oriza Kamboj[3]
Department of Physics, Lovely Professional University, Phagwara, Punjab, India

Abstract: This paper presents an analytical model for terahertz (THz) radiation generation from cylindrical nanoparticles (CNPs) under laser excitation. Cylindrical nanoparticles exhibit radiation generation due to their distinct electronic and optical properties. Their strong interaction with laser fields significantly enhances nonlinear optical effects, which are crucial for efficient THz radiation generation. Additionally, this study focuses on the influence of their geometric parameters on the amplitude of THz radiation produced by these nanoparticles. The results provide valuable insights into optimizing the design and geometry of cylindrical nanoparticles for improved THz radiation generation, contributing to the development of compact and efficient THz sources.

Keywords: THz radiation, Cylindrical nanoparticles, Laser excitation, Nonlinear optical effects

1. INTRODUCTION

Terahertz (THz) radiation, which ranges from 0.1 to 10 THz, bridges the microwave and infrared regions (Siegel, 2002; Singh et al., 2017). It offers non-ionizing properties and unique interactions with materials (Guerboukha et al., 2018; Sizov, 2019; Yang et al., 2016; Zeitler and Gladden, 2009). These characteristics make THz radiation valuable for applications such as non-destructive testing, biomedical imaging (Hansson et al., 2019), security scanning (Kemp, 2011), and communication (Akyildiz et al., 2014). Current THz generation techniques, such as solid-state electronic devices, nonlinear crystals, optical rectification, and photoconductive antennas (Siles and Grajal, 2010; Blanchard et al., 2007; Vodopyanov, 2008; Bhasin and Tripathi, 2009; Gholami and Bahari, 2022) are mostly constrained by damage thresholds, narrow bandwidths, and high costs. Laser-plasma interactions, especially using nanostructures, provide a promising alternative (Punia et al., 2022; Thakur et al., 2023; Zhang et al., 2015). The nanostructure's plasmon frequencies align with the THz range, and their interaction with intense femtosecond laser pulses generates dense plasmas that serve as broadband THz sources (Polyushkin et al., 2011; Sepehri and Erdi, 2017). This study focuses on cylindrical nanoparticles (CNPs), emphasizing their geometric parameters and how they influence THz amplitude, contributing to the development of efficient THz sources.

2. NON-LINEAR CURRENT DENSITY ($\vec{J}^{NL}$)

We analyze the interaction of two co-propagating linearly polarized laser beams in an argon gas medium embedded with cylindrical nanoparticles, with their electric field profiles defined as:

$$\vec{E}_j = \hat{x} A_j \exp(-i(\omega_j t - k_j z)), \tag{1}$$

where $A_j = A_{j0} e^{-y^2/a_0^2}$ and j = 1,2 represents the laser index, μ_j is the laser beam's modulation index, q is the

Corresponding author: [1]anuj.dandian@gmail.com, [2]svij25@yahoo.co.in.com, [3]kambojoriza1964@gmail.com

periodicity parameter, ω_j is the angular frequency, and k_j is the wavenumber. We analyze cylindrical nanoparticles (CNPs) with two basal plane orientations relative to the electric field: one perpendicular $(\vec{E}_j \| \hat{c}_i)$ and one parallel $(\vec{E}_j \perp \hat{c}_i)$, varying with the crystallographic axis (c_i) normal to the basal plane. The CNPs in the host medium contribute to the macroscopic nanoparticle density, n_0 is given as (Sharma et al.,2020)

$$n_0 = \sum n_{0,k_i}^0 (f_{ck_i}). \tag{2}$$

In this equation, n_c represents the density of cylindrical nanoparticles, and n_{0,k_i}^0 denotes the density of conduction electrons for both basal plane orientations. The volume fraction, $f_{ck_i} = \pi r_{ck_i}^2 h / d_{ck_i}^3$, denotes the ratio of the nanoparticle volume to the unit cell volume, where, r_{cki} is the nanoparticle radius, and d_{cki} s the average inter-nanoparticle distance for a given orientation. Assuming an even charge distribution, the electron cloud surrounding the nanoparticles, influenced by the laser's electric field, behaves like a single fluid while the heavy positive ions remain stationary. The restoring force, exerted by immobile ions, plays a crucial role in returning the electronic cloud to its equilibrium state when subjected to a laser electric field. Lasers impart the oscillatory velocity $(\vec{v}_j)$ to electrons, which can be obtained after solving the equation of motion,

$$\vec{v}_j = -\frac{\iota \omega_j e \vec{E}_j}{m\left(\omega_j^2 - \frac{\omega_p 2}{2} + \iota \Gamma \omega_j\right)}. \tag{3}$$

Lasers beat together and exert a beat-frequency ($\omega = \omega_1 - \omega_2$) ponderomotive force $F_{p\omega}$, i.e,

$$F_{p\omega} = -\frac{e^2}{2m}\left[\frac{\omega_1 \omega_2 A_{10} A_{20}^* \left(-\frac{4x}{a_0^2}\hat{x} + \iota k \hat{z}\right)}{\left(\omega_1^2 - \frac{\omega_p 2}{2} + \iota \Gamma \omega_1\right)\left(\omega_2^2 - \frac{\omega_p 2}{2} - \iota \Gamma \omega_2\right)}\right] e^{-2y^2/a_0^2} e^{-i(\omega t - kz)}. \tag{4}$$

Where $\vec{k} = (\vec{k}_1 - \vec{k}_2)$. The beat frequency ponderomotive force imparts nonlinear velocity to plasma electrons at $(\omega, \vec{k})$, which can be derived as:

$$\vec{v}_{\omega,k}^{NL} = -\frac{e^2}{2m^2}\left[\frac{\omega_1 \omega_2 A_{10} A_{20}^*}{(\omega^2 - \frac{\omega_p 2}{2} + \iota\text{“}\omega)}\right] e^{-2y^2/a_0^2} \left[\frac{\left(-\frac{4x}{a_0^2}\hat{x} + \iota k \hat{z}\right)}{\left(\omega_1^2 - \frac{\omega_p 2}{2} + \iota \Gamma \omega_1\right)\left(\omega_2^2 - \frac{\omega_p 2}{2} - \iota \Gamma \omega_2\right)}\right] e^{-i(\omega t - kz)}. \tag{5}$$

This nonlinear velocity is responsible for the origination of nonlinear current and the x component of nonlinear current density $\vec{J}_{NL}$, can be expressed as:

$$\vec{J}_{\omega,k}^{NL} = \frac{n_0 e^3}{2m^2}\left[\frac{\omega \omega_1 \omega_2 A_{10} A_{20}^* \left(\frac{4x}{a_0^2}\hat{x}\right)}{\left(\omega_1^2 - \frac{\omega_p 2}{2} + \iota \Gamma \omega_1\right)}\right] e^{-2y^2/a_0^2} \left[\frac{1}{\left(\omega_2^2 - \frac{\omega_p 2}{2} + \iota \Gamma \omega_2\right)\left(\omega^2 - \frac{\omega_p 2}{2} + \iota \Gamma \omega\right)}\right] e^{-i(\omega t - kz)}. \tag{6}$$

3. TERAHERTZ GENERATION

Wave equation for propagation of terahertz can be derived from the Maxwell's equations, which can be written as:

$$\nabla^2 \vec{E}_{THz} - \vec{\nabla}(\vec{\nabla}.\vec{E}_{THz}) + \frac{\omega^2}{c^2}\epsilon_{eff}\vec{E}_{THz} = -\frac{4\pi\iota\omega \vec{J}^{NL}}{c^2}, \tag{7}$$

where ϵ_{eff} represents the effective permittivity of the medium with cylindrical graphite nanoparticles. It accounts for both conducting and non-conducting bound electrons and can be expressed as (Sihvola, 1999),

$$\epsilon_{eff} = \epsilon_i^{bound} + \epsilon^{free}. \tag{8}$$

where,

$$\epsilon^{free} = 1 - \frac{\omega_p^2}{\left(\omega^2 - \frac{\omega_p 2}{2} + \iota \Gamma \omega\right)}. \tag{9}$$

where, $\frac{\omega_p 2}{2}$ term is considered because of restoring force in the case of CNPs. As this medium contains cylindrical graphite nanoparticles oriented differently, the effective permittivity of the host medium is modified. The contribution of bound electrons to the effective permittivity, due to the presence of CNPs ($\epsilon(c)$), can be represented as:

$$\epsilon_i^{bound} = \epsilon(c), \tag{10}$$

where,

$$\epsilon(c) = \epsilon_h + \sum_{ki} f_{cki} \frac{(\epsilon_{ki} - \epsilon_h)(\epsilon_{ki} + 5\epsilon_h)}{(3 - 2f_{cki})\epsilon_{ki} + (3 + 2f_{cki})\epsilon_h}. \tag{11}$$

In the above equation, ϵ_h represents the permittivity of the host medium, while ϵ_{ki} pertains solely to the permittivity of bound electrons and these values vary with different nanoparticle orientations.

Terahertz field can be found by taking the div. of equation (7) and solving for $\vec{E}_{THz}$, we can obtain the expression for the normalized electric field of terahertz, ($|A_{THz}/A_{10}|$) as follows:

$$\vec{E}_{THz} = \vec{A}_{THz}\, e^{-i(\omega t - kz)} \tag{12}$$

$$\left|\frac{E_{THz}}{A_{10}}\right| = \frac{4\pi\iota}{\epsilon_{eff}}\left(\frac{e^3}{2m^2}\right)\left(\sum n^0_{0,k_i}\left(\frac{\pi r^2_{ck_i} h}{d^3_{ck_i}}\right)\right)\left(\frac{4x}{a_0^2}\right)\left[\frac{\omega_1\omega_2 A^*_{20}}{\left(\omega^2 - \frac{\omega_p 2}{2} + \iota\Gamma\omega\right)}\right]e^{-2y^2/a_0^2}\left[\frac{1}{\left(\omega_1^2 - \frac{\omega_p 2}{2} + \iota\Gamma\omega_1\right)\left(\omega_2^2 - \frac{\omega_p 2}{2} + \iota\Gamma\omega_2\right)}\right]. \tag{13}$$

4. RESULTS AND DISCUSSION

This study investigates the dependence of THz radiation generation on the geometric parameters of cylindrical nanoparticles (CNPs). In this study, we consider the CO_2 lasers with frequencies $\omega_1 = 1.85 \times 10^{14}$ rad/s, $\omega_2 = 2 \times 10^{14}$ rad/s, and intensity $I \approx 10^{16}$ W/cm^2. The plasma frequency is set at $\omega_p = 1.2 \times 10^{13}$ rad/s, corresponding to a plasma density of $n_0^0 = 4.5 \times 10^{16}$ cm^{-3}.

The plots in Figs. 138.1 and 138.2 demonstrate the relationship between the normalized THz amplitude (A_{THz}/A_{10}) and normalized frequency (ω/ω_p) for varying nanoparticle radii r_c and heights h. As the nanoparticle radius increases, the amplitude of THz radiation also increases due to the higher conduction electron density per unit volume, which enhances the nonlinear current density, resulting in stronger THz emission at the resonance frequency ($\omega/\omega_p = 0.97$). Similarly, Figure 138.2 shows that increasing the height of the cylindrical nanoparticles leads to a significant increase in normalized THz amplitude as the larger nanoparticle volume allows for a greater number of conduction electrons to interact with the laser field. These results align with previous studies (Punia et al., 2022; Sepehri and Erdi, 2017; Sharma et al., 2020).

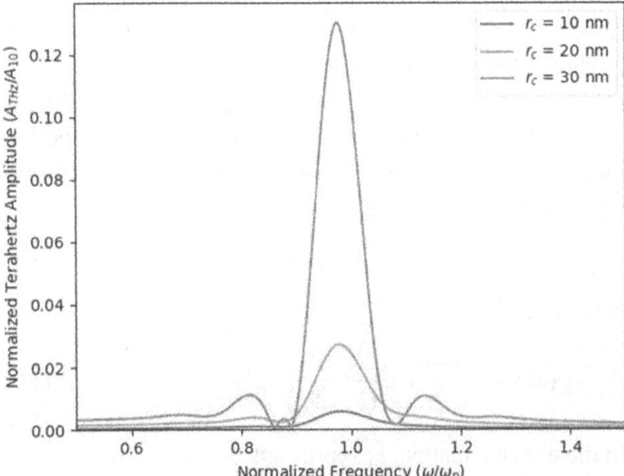

Fig. 138.1 Variation of normalized terahertz amplitude (A_{THz}/A_{10}) with normalized frequency (ω/ω_p) at different radii (r_c) of CNPs

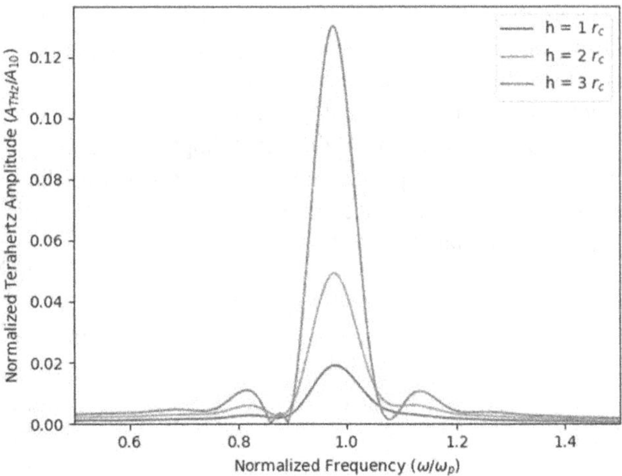

Fig. 138.2 Variation of normalized terahertz amplitude (A_{THz}/A_{10}) with normalized frequency (ω/ω_p) at the height of CNPs (h)

Figure 138.3 shows the relationship between the normalized THz amplitude (A_{THz}/A_{10}) and normalized THz frequency (ω/ω_p) for varying inter-particle distances d_c. As the inter-particle distance decreases, the THz amplitude increases, which is attributed to the enhanced effective conduction electron density. This increase in conduction electrons strengthens the nonlinear current, resulting in a higher THz amplitude. These findings are consistent with previous research (Sharma et al., 2020; Vij, 2024).

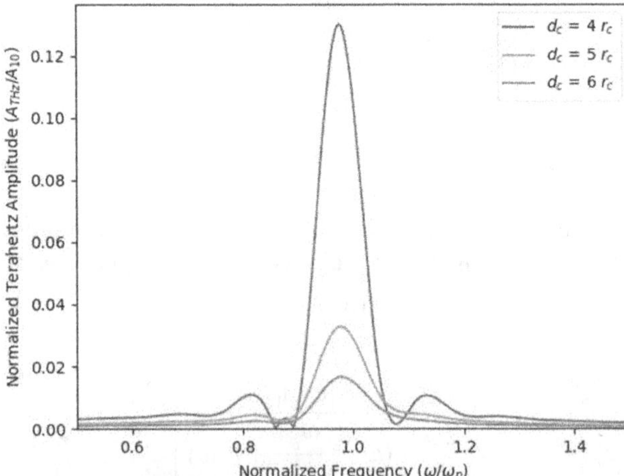

Fig. 138.3 Variation of normalized terahertz amplitude (A_{THz}/A_{10}) with normalized frequency (ω/ω_p) at different interparticle distances (d_c)

5. CONCLUSION

This study provides a theoretical framework for THz generation from cylindrical nanoparticles under laser excitation. The results highlight the significant role of geometric parameters, such as nanoparticle size, height, and inter-particle distance, in influencing THz radiation output. By optimizing these parameters, the THz amplitude can be significantly enhanced, offering valuable insights for designing efficient and compact THz sources.

REFERENCES

1. Akyildiz, I. F., Jornet, J. M., & Han, C. (2014). Terahertz communication: A comprehensive survey. *Physical Communication*, 12, 16.
2. Bhasin, L., & Tripathi, V. K. (2009). Plasma-based THz radiation generation and applications. *Physics of Plasmas*, 16.
3. Blanchard, F., Razzari, L., Bandulet, H.-C., Sharma, G., Morandotti, R., Kieffer, J.-C., Ozaki, T., Reid, M., Tiedje, H., Haugen, H., et al. (2007). Generation of terahertz radiation using a graphene-based structure. *Optics Express*, 15, 13212.
4. Gholami, S., & Bahari, A. (2022). Modeling of THz radiation from plasmonic nanostructures. *Optical and Quantum Electronics*, 54, 147.
5. Guerboukha, H., Nallappan, K., & Skorobogatiy, M. (2018). Advances in terahertz science and technology: Optical approaches. *Advances in Optics and Photonics*, 10, 843.
6. Hansson Mild, K., Lundstrom, R., & Wilen, J. (2019). Terahertz research in environmental science. *International Journal of Environmental Research and Public Health*, 16, 1186.
7. Kemp, M. C. (2011). Terahertz radiation for communications. *IEEE Transactions on Terahertz Science and Technology*, 1, 282.
8. Polyushkin, D., Hendry, E., Stone, E., & Barnes, W. (2011). THz emission from nanostructures: Role of nanoparticle geometry. *Nano Letters*, 11, 4718.
9. Punia, T., Sharma, D., & Malik, H. K. (2022). Modeling of terahertz radiation from nanoparticle configurations. *IEEE Transactions on Plasma Science*, 50, 1087.
10. Sepehri Javan, N., & Rouhi Erdi, F. (2017). Theoretical study of the generation of terahertz radiation by the interaction of two laser beams with graphite nanoparticles. *Journal of Applied Physics*, *122*(22).
11. Sharma, D., Singh, D., & Malik, H. K. (2020). Terahertz radiation generation through nanoparticle interaction with laser fields. *Plasmonics*, 15, 177.
12. Siegel, P. H. (2002). Terahertz radiation: An introduction and overview. *IEEE Transactions on Microwave Theory and Techniques*, 50, 910.
13. Sihvola, A. H. (1999). *Electromagnetic Mixing Formulas and Applications*. Elsevier.
14. Siles, J. V., & Grajal, J. (2010). Physics-based design and optimization of Schottky diode frequency multipliers for terahertz applications. *IEEE Transactions on Microwave Theory and Techniques*, *58*(7), 1933–1942.
15. Singh, R. K., Singh, M., Rajouria, S. K., & Sharma, R. (2017). Investigating plasma-based THz generation mechanisms. *Physics of Plasmas*, 24.
16. Sizov, F. F. (2019). Brief history of THz and IR technologies. *Semiconductor physics, quantum electronics & optoelectronics*, 67–79.
17. Thakur, V., Vij, S., Kant, N., & Kumar, S. (2023). THz generation by propagating lasers through magnetized SWCNTs. *Indian Journal of Physics*, 97(7), 2191–2196.
18. Vij, S. (2024). Terahertz radiation from laser interactions with nanostructures. *Journal of Applied Physics*, 136.
19. Vodopyanov, K. L. (2008). Terahertz generation in nonlinear optical materials. *Laser & Photonics Reviews*, 2, 11.
20. Yang, X., Zhao, X., Yang, K., Liu, Y., Liu, Y., Fu, W., & Luo, Y. (2016). Biomedical applications of terahertz spectroscopy and imaging. *Trends in biotechnology*, *34*(10), 810–824.
21. Zeitler, J. A., & Gladden, L. F. (2009). Terahertz time-domain spectroscopy for pharmaceutical analysis. *European Journal of Pharmaceutics and Biopharmaceutics*, 71(2), 2.
22. Zhang, L., Mu, K., Zhou, Y., Wang, H., Zhang, C., & Zhang, X.-C. (2015). Compact terahertz radiation sources for imaging. *Scientific Reports*, 5, 12536.

Note: All the figures in this chapter were made by the authors.

Design Optimization of Compact Microstrip Patch Antenna for 5G Networks

Renu Sharma*
Department of Physics,
JECRC University,
Jaipur, India

Raghvendra Patidar
Department of Electronics
& Communications
Engineering,
Arya College of Engineering
& IT, Jaipur, India

Pawan Kumar Jain
Department of Physics,
Swami Keshvanand Institute
of Technology Management
& Gramothan, Jaipur, India

Anna Varughese
Department of Information &
Technology,
University of Rajasthan,
Jaipur, India

Abstract: Since the commencement of the progression toward the next generation of communication systems, an explosive exploitation of the millimetre wave band/millimetre band has been observed. Since 5G systems are currently of utmost relevance in the present telecommunication scenario. This paper is penned with the goal to introduce a rectangular microstrip patch antenna capable of operating in the millimetre band frequency, with the operational frequency of 42.6 GHz with an appreciable gain of 6.41 dB, Voltage Standing Wave Ratio (VSWR) of 0.6 and a return loss of -28.71 dB, along with a radiation efficiency of 97%. The paper contains the design parameters as well as the results obtained from the proposed antennas simulation conducted utilising the High-Frequency Structure Simulator (HFSS) software.

Keywords: VSWR, HFSS, Communication

1. INTRODUCTION

Wired communication was widely used and implemented despite its various challenges, however since the discovery of wireless communication in the late 19th century a steady but consistent preference and implementation of wireless communication can be observed in all sectors. With advancements in technology, various concepts and ideas can only be implemented with wireless systems. In the present world, wireless communication is omnipresent. Wireless communications rely on radio frequency signals for transmitting and receiving information between devices. Currently the 5G communication systems are being tested before their widespread utilization (Ramli, Nurulazlina et al. (2020)). The millimeter band, although it has various shortcomings, is widely being employed for the new generation 5G communication systems. They also have applications in other niches such as in the military, airport security and so on to list a few. Antennas play a pivotal role in wireless communications as they are capable of both transmitting and receiving radio signals, which is essential in realizing wireless communication (Sura K, Sainab T. (2021)). Since antennas are responsible for such an important task in the whole schema of wireless communication, meticulous attention and precision are needed in the construction of antennas. Designing of antennas precedes its construction and hence antennas are designed using simulation software before they are constructed and employed in real world systems. Since the conception of the antennas a myriad of antenna options are available (Naresh Kumar Darimireddy et al. (2015)). Despite this, the microstrip patch antenna (MPA) since its inception in 1955 is consistently and widely being preferred and utilized. Modern day devices need to accommodate a large number of components in a very limited space hence miniaturization of components is a necessity. The Microstrip Patch Antenna marked a significant milestone in the never-ending attempts at miniaturization. The appeal

*Corresponding author: renu.sharma@jecrcu.edu.in

DOI: 10.1201/9781003684718-139

of Microstrip Patch Antennas lies in their multifaceted advantages including their small size, light weight, ease of fabrication, cost-effectiveness and so on. Apart from this they show high adaptability and scope of customization for diverse applications. Microstrip Patch Antenna are also highly compatible with integrated circuits pertaining to their planar geometry.

The main components of a patch antenna are substrate, ground plane, patch and transmission line. The ground plane serves as a reflector and improves the radiation efficiency of the antenna. A dielectric substance with a specific dielectric constant and specific height are the characteristic features of the substrate and it is responsible for providing mechanical support and electrical insulation. Substrates usually have a dielectric constant in the range 2.2 to 12. Substrates that have lower dielectric constants and which are thick are more favoured as they maximize efficiency and bandwidth. However, like two sides of a coin, the thickness of the substrate leads to antennas which are bulkier and account for the dielectric loss hence a balance needs to be struck while choosing the substrate height to optimize the antennas performance. The fringing fields increase as the value of the dielectric decreases, as a result of which the radiated fields increase. The patch is embedded on top of the substrate and will have a specific geometric shape. Most commonly used geometric shapes are square, rectangle, circular, ellipse, triangular etc. Among these rectangular and circular are popularly employed, since they are simple to design, have great feed line flexibility, support multiple frequency operation and have great compatibility with array configurations. The patch is fed using either contact or non-contact mode.

The microstrip patch antenna proposed in this paper optimally operates at the resonant frequency of 42.6 GHz and has a bandwidth of 3 GHz. The suggested antenna features a rectangular patch with slots carved into it, to further reduce the size of the patch and the antenna's substrate material was chosen to be Arlon AD300A with a dielectric constant ε_r of 3. The various simulation results and substrate parameters for the proposed design have been provided in the sections below.

2. DESIGN CRITERIA

The design process of a microstrip patch involves selecting the patch antenna's primary components in order to maximize the antenna's performance within whichever device it is being incorporated into. One of the key factors that plays a significant impact on the performance of the antenna, is the material of the substrate over which the patch is mounted. The dielectric material, Arlon AD300A with dielectric constant 3 is being employed in the antenna proposed in this paper. Arlon AD300A was chosen since it provides a higher degree of impedance control and helps to obtain higher antenna gains and has good dimensional and thermal stability. The substrate height was set as 0.5mm.

The proposed microstrip patch antenna was fed with a feed line of length $\lambda/4$ to ensure impedance matching with the 50Ω transmission line. The $\lambda/4$ feeding technique was chosen since this feeding technique is highly versatile, simple to design and has superior impedance matching capabilities, it also aids in the bandwidth enhancement and in lowering the return loss, this feeding technique due to its ease in designing also proves to be a cost-effective technique. The dimensions of the transmission line are 2.0532mm x 1.257mm. The $\lambda/4$ feedline has dimensions of 1.0266mm x 0.314mm. Impedance matching is essential to ensure two major things, one being to lower the reflection loss and the other being to enhance the antenna's radiation efficiency. The operational frequency of the proposed antenna was chosen as 42 GHz.

The rectangular patch was further miniaturized using slots cut into it. The slots, as can be seen in Fig. 139.1 (a) and Fig. 139.1 (b), have two hexagons positioned at either side of the feedline towards the center of the patch. The hexagon has a side of dimension 0.17mm. The hexagons are connected by means of two thin rectangular strips each of width 0.02 mm and 0.505 mm.

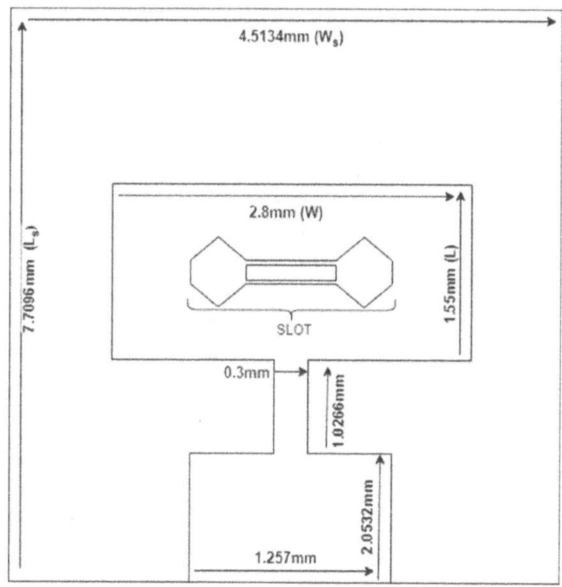

(a) 2D representation of proposed antenna

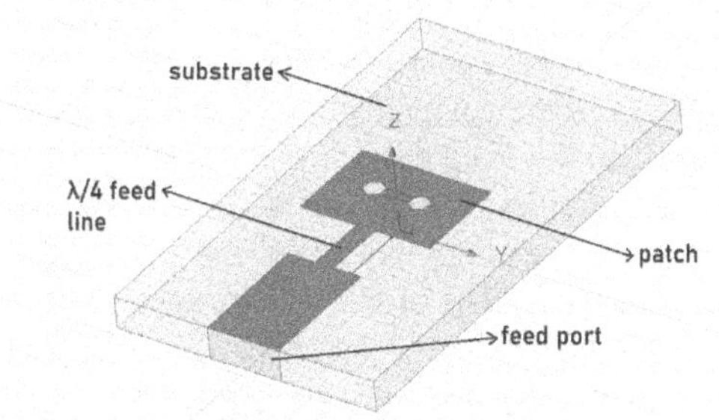

(b) 3D representation of proposed antenna

Fig. 139.1 Representation of proposed antenna in (a) 2D and (b) 3D dimensions with parameters

The calculations for the dimensions of the proposed rectangular microstrip patch antenna, namely the length and width, was carried out using the equations stated below:

$$W = \frac{c}{2f_o\sqrt{\frac{(\varepsilon_r + 1)}{2}}} \tag{1}$$

$$\varepsilon_r = \frac{\varepsilon_r + 1}{2} + \frac{\varepsilon_r - 1}{2}\left(1 + 12\frac{h}{w}\right)^{\frac{-1}{2}} \tag{2}$$

$$L_{eff} = \frac{c}{2f_o\sqrt{\epsilon_{eff}}} \tag{3}$$

$$\Delta L = 0.412h\frac{\left(\epsilon_{eff} + 0.3\right)\left(\frac{w}{h} + 0.265\right)}{\left(\epsilon_{eff} - 0.258\right)\left(\frac{w}{h} + 0.8\right)} \tag{4}$$

$$L = L_{eff} - 2\Delta L \tag{5}$$

The patch's width is given by w and the velocity of light in vacuum is denoted by c, ε_{eff} is the effective dielectric constant. Height of the substrate is shown by *h*. ΔL denotes the extension in length caused as a result of the fringing effect, thereby L_{eff} denotes the effective length of the patch and L is the actual patch's length. Computation was done employing the equations listed above and the results obtained pertaining to all the vital dimensions to model the proposed microstrip patch antenna are as stated in the Table 139.1 given below:

Table 139.1 Dimensions of proposed antenna

Parameters	Value(mm)
Width of substrate (W_s)	4.5
Length of substrate (L_s)	7.7
Height of substrate (h)	0.5
Breadth of patch (W)	1.55
Length of patch (L)	2.8

3. SIMULATED RESULTS

The Ansys HFSS (High-Frequency Structure Simulator) software allows for the design and simulation of high frequency electronic components with high accuracy and helps to reduce the need of multiple physical prototypes. HFSS also provides various tools for analysis of the components and has high integration capacity with other simulation software. The proposed antenna was designed and simulated in HFSS, and the results obtained from the simulation analysis is as described below:

3.1 VSWR

The proposed antenna when analyzed using the software, a peak was obtained at 42.65 GHz with a VSWR value of 0.6370 as can be seen from Fig. 139.2. VSWR provides knowledge regarding an antenna's power reflection capabilities. The performance of an antenna can be gauged directly from its VSWR value, which should be minimal and positive. The extent of the impedance matching the antenna's patch with that of the transmission line can be ascertained from the VSWR value. The value of VSWR is directly proportional to the degree of mismatch.

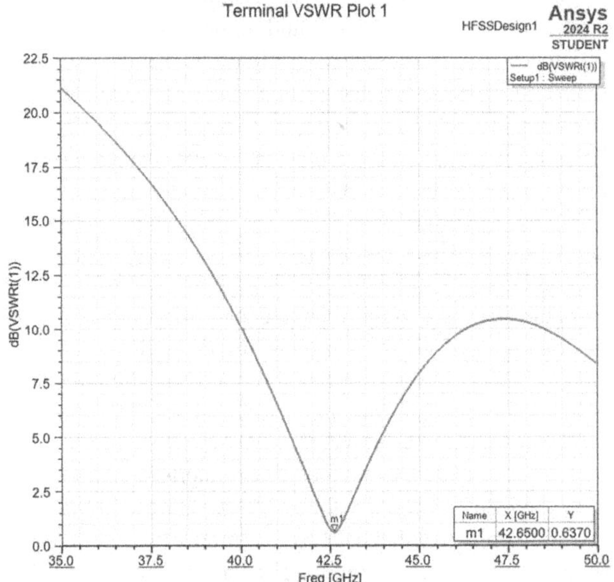

Fig. 139.2 VSWR plot

3.2 Return Loss *(RL)*

When a signal is transmitted through a transmission line, if an impedance mismatch is present then some/all of the incident power will undergo reflection back to the source depending on the amount of mismatch, which will result in only a fractional/no power reaching the patch for radiation. If the mismatch is very great, then the patch will not radiate any power which renders the antenna useless. Return loss is defined as the measure of the ratio of incident power to that of reflected power. Thus, it is the efficiency measure of the power transmission capabilities from the source to the antenna. For instance, a return loss of -10 dB signifies that 10% of the incident power is reflected meaning that 90% of power is successfully delivered to the antenna. The S11 parameter is a measure of the return loss in the software. The S11 of the proposed antenna showed a peak at 42.65 GHz with a return loss value of -28.718 dB as can be seen from the graph (Fig. 139.3) shown below.

3.3 Gain

Gain is the antenna performance parameter that measures how much power is transmitted in the peak's direction to that of an isotropic source. On the other hand, directivity indicates the antenna's capability to concentrate the radiation in a particular direction. Both serve as crucial parameters in determining the effectiveness of the

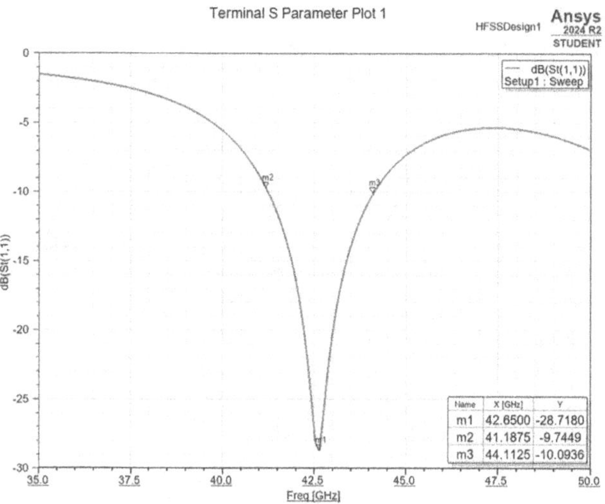

Fig. 139.3 S parameter/Return loss plot

antenna. At the operational frequency for the antenna proposed in this paper i.e. 42.65 GHz, the analysis of the gain and directivity was performed, and the gain was noted to be 6.41dB which is considered to be relatively high, and the visual illustration of the same in the form of a 3-dimensional polar plot is provided below (Fig. 139.4). The directivity for the antenna introduced was obtained as 6.55 dB, the 3D polar plot for which is as shown in Fig. 139.5.

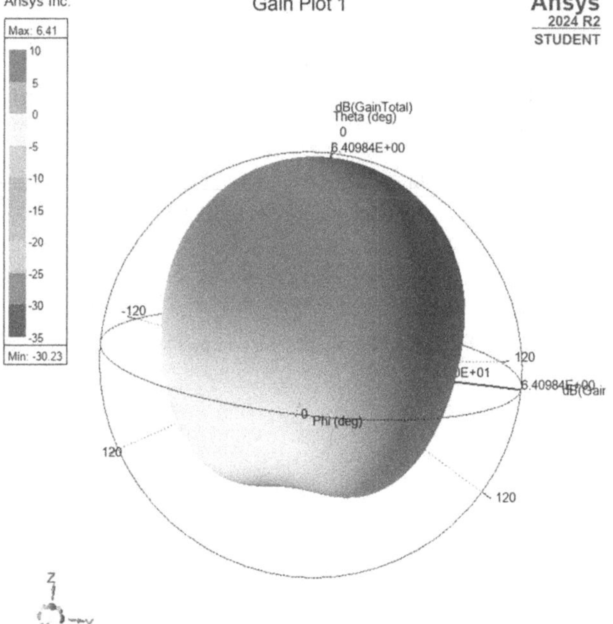

Fig. 139.4 Gain plot

3.4 Radiation Efficiency

Radiation efficiency refers to the ratio of the total power radiated by an antenna to the power accepted by the antenna through its excitation port. It is, thus, a way to measure the antenna's efficiency in converting the electrical power

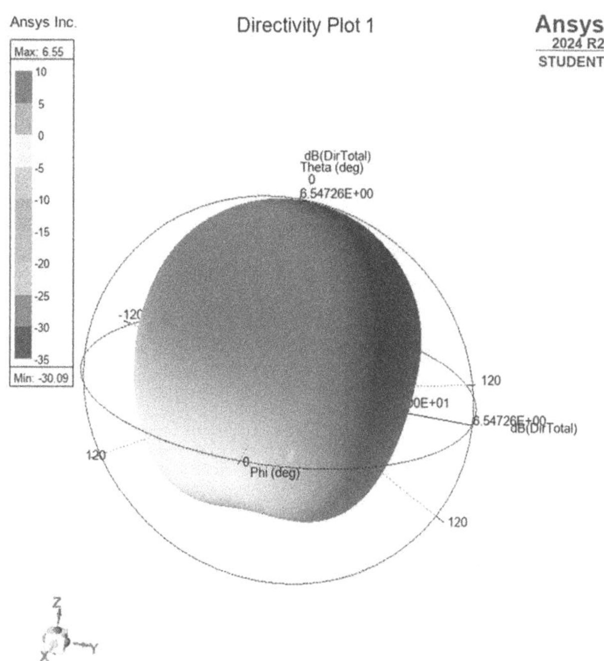

Fig. 139.5 Directive plot

input into electromagnetic energy that can be radiated. For the proposed antenna, a total radiation efficiency of 97.2% was measured. Radiation efficiency is a fundamental parameter in measuring the efficiency of the antenna and is mainly influenced by the frequency, antenna design, impedance mismatch and the characteristics of the materials used in the fabrication of the antenna.

In a microstrip patch antenna, the performance parameter radiation pattern depicts how the electromagnetic energy is distributed in space as it is radiated from or received by the antenna. The Radiation Pattern is portrayed graphically as a function of direction. The perusal of the radiation pattern of the antenna introduced in this paper was conducted at the resonant frequency 42 GHz at both values of phi namely 0° and 90°. Figure 139.6(a) shows the graph obtained from the analysis at phi=0° and Fig. 139.6(b) depicts the graph acquired as result of the analysis at phi=90°.

4. CONCLUSION

Conforming with the objectives of the paper, a microstrip patch antenna that can be employed in the new 5G communication systems by operating in the millimeter band with an operational frequency of 42.65 GHz was introduced. The proposed antenna has a rectangular patch into which integration of slots was done so as to further miniaturize the antenna. The proposed antenna adopted the $\lambda/4$ feeding technique for more effective impedance matching with the transmission line so as to enhance the antenna's performance. The proposed antenna was effectively crafted and simulated using Ansys HFSS software and the antenna was analyzed for different parameters as stated in the results. According

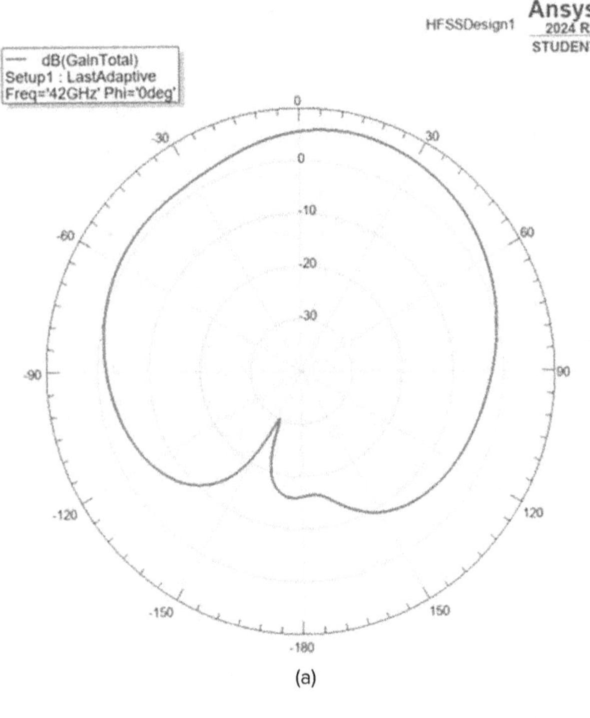

(a)

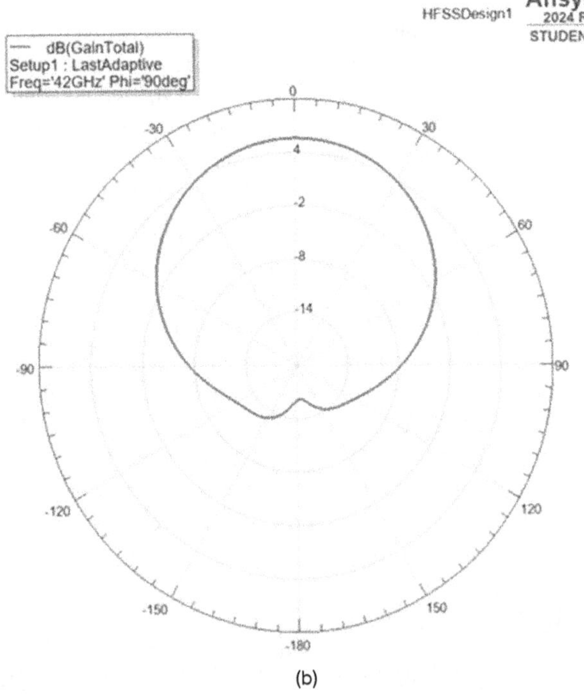

(b)

Fig. 139.6 (a) and 6(b) depicts the results of the radiation pattern obtained at Phi=0° and at Phi=90° respectively

to the findings, at the resonant frequency of 42.65 GHz, the proposed antenna boasted a commendable gain of 6.41dB, with a VSWR of 0.637. A satisfactory return loss of -28.718 dB was obtained at the peak. Furthermore, the proposed antenna has a commendable radiation efficiency of approx. 97.2%. Radiation efficiency is a critical parameter in antenna design, as higher efficiency antennas can transmit and receive signals more effectively thereby enhancing the overall communication quality and improving the systems' performance. Future scope of the proposed antenna would involve tweaking the antenna parameters to further improve the results obtained and thus enhance the performance of the proposed antenna when employed in real life systems.

REFERENCES

1. Md.Sohel Rana, Sifat Hossain, Shuvashish Biswas Rana, Md. Mostafizur Rahman (2022). Microstrip Patch antenna for various applications: a review.
2. J. Colaco and R. Lohani (2020). Design and Implementation of Microstrip Patch Antenna for 5G applications.
3. Omar Darboe, Franklin Manene, Dominic Konditi (2019). A 28 GHz Rectangular Microstrip Patch Antenna for 5G Applications.
4. Ramli, Nurulazlina et al. (2020). Design and Performance Analysis of Different Dielectric Substrate based Microstrip Patch Antenna for 5G Applications.
5. Sura K, Sainab T. (2021). A High Gain Compact Rectangular Patch Antenna For 5G Applications.
6. SN Nafea, Nasser N. Khamiss (2023). For 5G applications, high gain patch antenna in Ka-Band.
7. Khalil H. Sayidmarie, Neil J. McEwan, Peter S. Excel, Raed A. Abd-Alhameed (2019). Antennas for the emerging 5G systems.
8. Zeeshan Siddiqui, Marko Sonkki et al. Dual-band Dual Polarised Planar Antenna for 5G millimeter-wave antenna-in-package Applications.
9. Naresh Kumar Darimireddy et al. (2015) Design of triple-layer double U-slot patch antenna for wireless applications.

Note: All the figures and tables in this chapter were made by the authors.

Recent Trends in Applied Physics and Material Science – Sudhir Bhardwaj et al. (eds)

Design and Performance of a Dual-Band Planar Dipole Antenna for Enhanced Connectivity

Renu Sharma*
Department of Physics, JECRC University, Jaipur, India

Raghvendra Patidar
Department of Electronics & Communications Engineering, Arya College of Engineering & IT, Jaipur, India

Pawan Kumar Jain
Department of Physics, Swami Keshvanand Institute of Technology Management & Gramothan, Jaipur, India

Jaidev Bairwa
Department of Information & Technology, University of Rajasthan, Jaipur, India

Abstract: The suggested antenna design features a microstrip patch antenna with dual-band characteristics was designed using a combination of meander lines. The planar dipole antenna (PDA) has a gain of 3.55 dB and a return loss of -20.69 dB while operating at 17.6 GHz. At 8 GHz, it has a realized gain of 1.63 dB and a return loss of -22.65 dB. This versatile high-performance antenna maintains reliable communication performance by switching to a lower frequency when necessary, ensuring connectivity, effective spectrum management, and resilience against interference. Its capability to provide backup links and alternative frequencies enhances service availability for broadcasting, telecommunication, data transmission, and remote sensing applications.

Keywords: PDA, Telecommunication, Broadcasting

1. INTRODUCTION

Dipole antennas are typically employed for radio and satellite transmission because of their omnidirectional characteristics. In contemporary communication, there's a preference for planar dipole antennas (PDAs) because to their small size, simplicity in production, and capacity for integration, and suitability for array configurations (G. A. Deschamps (1953)). As technology advances and the demand for processing large data volumes grows, the importance of higher frequency antennas in satellites has become more pronounced. These antennas enable more efficient data transmission within limited bandwidth, facilitating tasks like high-resolution imaging, real-time video streaming, and rapid internet services (A. Colin et.al 2010).

The frequency flexibility offered by dual-band antennas in satellite communication systems allows for optimized coverage, bandwidth allocation, and adaptability to diverse operational environments (Zhang, H., & Xin, H. (2009)). By supporting multiple frequency bands like C-band and Ku-band, these antennas can tailor communication services based on specific requirements, ensuring reliable connectivity and effective spectrum management. This flexibility enables operators to mitigate interference issues by switching between frequency bands, maintaining uninterrupted communication even in congested spectrum conditions. Additionally, dual-band antennas enhance system resilience by providing backup links and alternative frequency options, ensuring continuous service availability and minimizing downtime. Moreover, the reliable performance of dual-band antennas is highlighted by their ability to mitigate rain attenuation using lower frequency bands like C-band during adverse weather conditions, while leveraging higher frequency bands like Ku-band for high-speed data applications (A. Mallikarjuna Prasad et.al (2015). This reliability is essential for delivering dependable satellite communication services across broadcasting, telecommunication, data transmission, and remote sensing applications.

*Corresponding author: renu.sharma@jecrcu.edu.in

DOI: 10.1201/9781003684718-140

2. ANTENNA GEOMETRY DESIGN

There are various designs of microstrip patch available for multi-band antennas either incorporating multiple layers and/or U-slots. The proposed antenna has simple structure overall to keep fabrication and manufacturing simple and easy operating at a frequency of 17 GHz. Changes have been made in the parameters of the antenna to make it a dual band. The Dielectric material chosen is RT/duroid 5880 with thickness 0.15 mm as it is widely used, cheap to fabricate, and compatible with PCB boards. The size of the substrate is 15 mm in length and 7 mm in width with a dielectric constant (ε_r) of 2.5. To get a higher bandwidth and better efficiency low dielectric constant is used. The arms of the dipole patch are identical in geometry with each other with rotational symmetry. The transmission line is a simple direct coaxial feed that has been tuned to match the antenna's impedance, which is 50 Ω with a 1 mm feed gap. The antenna uses meander lines with an extra strip of patch between meander lines as shown in Fig. 140.1 with measurements.

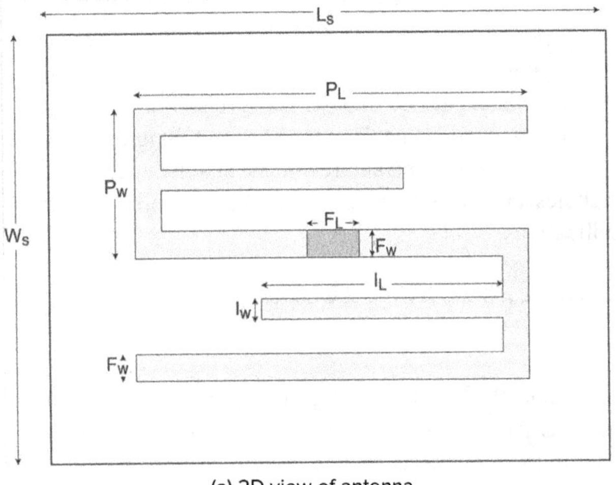

(a) 2D view of antenna

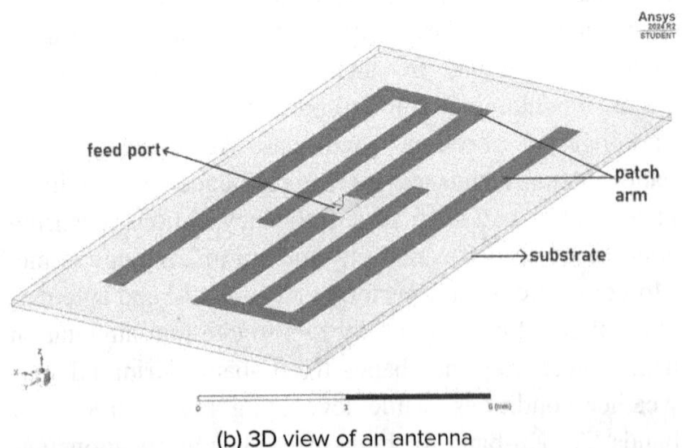

(b) 3D view of an antenna

Fig. 140.1 Structure of antenna with (a) 2D and (b) 3D view. All dimensions are in mm

The antenna's structural design has been kept simple to facilitate easy manufacturing and fabrication, ensuring minimal production costs. The geometrical values of the antenna have been mentioned below in Table 140.1.

Table 140.1 Measurements of antenna

Parameters	Dimension
Length of the substrate (Ls)	15 mm
Width of the substrate (Ws)	7 mm
Length of total patch (PL)	11.6 mm
Width of total patch (Pw)	4.7 mm
Length of inside arm (IL)	6.9 mm
Width of the inside arm (Iw)	0.4 mm
Length of feed port (FL)	1 mm
Width of feed port (Fw)	0.5 mm

It should be noted, a normal half-wave 8 GHz dipole antenna has overall dimensions of 17mm x 9mm. By incorporating meander lines, we have managed to successfully reduce the antenna's overall size by 31% without compromising its performance. This modification also allows for a dual-band operation while maintaining a simple shape that is easy to manufacture.

3. SIMULATED RESULTS

The High-Frequency Structure Simulator (HFSS) simulation results are obtained using Ansoft Ansys software at a frequency of 17 GHz. The resonance frequency of the antenna was found to be at 17.34 GHz with maximum and an additional resonating frequency of 8.13 GHz was also observed. The S11 plot of wideband from 6 GHz to 19 GHz is showcased below with markings for return loss on y-axis and frequency on x-axis in Fig. 140.2.

3.1 Return Loss

The power reflected from an antenna's impedance mismatch is known as return loss, or S11. It shows how well the transmission line and antenna are matched. The return loss ought to be at its lowest at the resonance frequency. The following formula is used to determine the amount of power lost to reflection using the reflection coefficient (Γ):

$$Return\ Loss\ (RL) = -20\ log_{10}|\Gamma|$$

A high return loss indicates a low reflection coefficient, meaning that most of the power is being transferred to the antenna, with very little being reflected back. This suggests good impedance matching. A return loss of at least -10 dB is desirable, as it indicates that 90% of the power is being transmitted or received. As shown in Fig. 140.2, a resonance frequency of 17.6 GHz was achieved with a minimum return loss of -20.69 dB, indicating that 99% of the power is being transmitted or received. Another resonance frequency was observed at 8.16 GHz, with a return loss of -22.65 dB.

It is important to note that, apart from the resonance frequency, the return loss is close to 0 dB, suggesting that

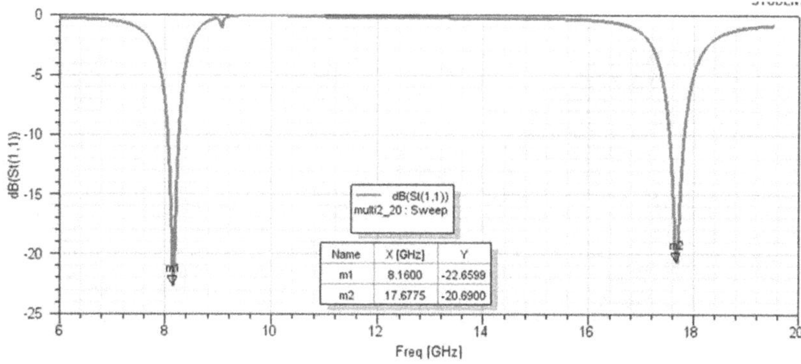

Fig. 140.2 S11 graph with wide band from 6 GHz to 19 GHz

nearly 100% of the power is reflected back. This indicates that the antenna is highly selective and effectively minimizes interference and noise by not interacting with other frequencies.

3.2 VSWR

VSWR stands for Voltage Standing Wave Ratio. It measures how efficiently power is transmitted from a power source and how much is reflected due to load mismatch. A portion of the power is reflected from the antenna when the transmission line impedances and the antenna are out of alignment. Standing waves are produced along the transmission line by the combination of this reflected wave and the forward wave, which is the wave that travels from the source to the load. It is the antenna's maximum voltage divided by its minimum voltage. The ideal value for VSWR is '1' indicating proper matching with no load mismatch. The acceptable range for VSWR is below 2.

The PDA VSWR obtained was 1.60 dB for 17 GHz suggesting approximate 94.6% of the power is being transmitted, and approximate 5.4% of the power is reflected. For 8 GHz 1.82 dB was obtained, indicating 91.8% of power transmitted and 8.2% reflected, presenting a good VSWR as the standard 1.2 dB. The graph representing VSWR has been shown in Fig. 140.3.

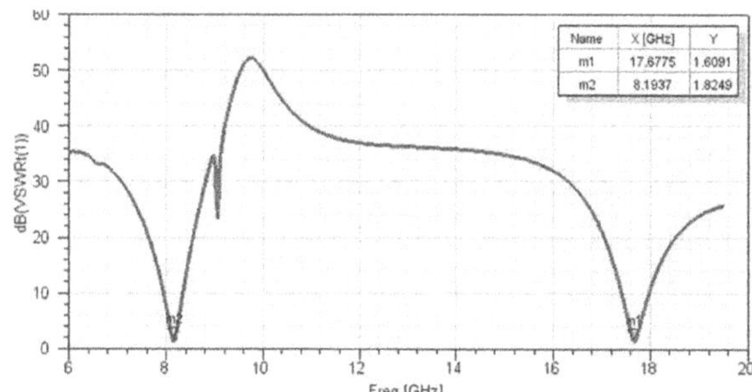

Fig. 140.3 VSWR of proposed PDA

3.3 Gain

The gain is the maximum effectiveness with which the antenna can radiate the power delivered to it. The acceptable gain for a dipole antenna is 1.5-1.7 dB. At 17.6 GHz the gain of the antenna was 3.55 dB and at 8 GHz the gain was 1.63 dB, 3D polar plot for both 8 GHz (Fig. 140.4(a)) and 17.6 GHz (Fig. 140.4(b)) has been mentioned below. Notably, even though return loss of the antenna was arguably the same, gain at 17.6 GHz is much better than 8 GHz.

3.4 Radiation Pattern

With phi=0 and phi=90, the antenna's radiation pattern has been simulated at 8 GHz and 17 GHz. The maximum gain of 3.55 GHz was found to be at 17 GHz with phi=0. For 8 GHz uniform radiation pattern like donut was observed depicted in Fig. 140.5 (a) while for 17 GHz four clover leaf like pattern was observed in phi=90 depicted in Fig. 140.5(b).

4. CONCLUSION

A dual-band planar dipole antenna has been simulated, demonstrating several notable characteristics. The measured resonance is at 17.6 GHz and 8 GHz. The antenna efficiently radiates most of its energy, as indicated by a low return loss of -22.6 dB. The low reflections across most of the frequency spectrum, close to 0 dB return loss, suggest that the antenna maintains signal integrity and minimizes loss, ensuring minimal interference and efficient transmission and reception of signals. The measured values of peak gains at lower and upper frequency are 3.5 dB and 1.5 dB for 17.6 GHz and 8 GHz respectively.

The antenna features two distinct peaks in different bands, specifically the C-band and the K-band. If the K-band faces challenges in maintaining a stable connection due to its limited range, the antenna can seamlessly switch to the C-band without sacrificing performance. This ensures reliable connectivity, optimizing communication based on distance and data rate requirements. The ability to operate efficiently at both frequencies makes it ideal for systems that need to adapt to varying environmental conditions or communication requirements such as satellite communication, radar system, wireless communication, 5G network etc.

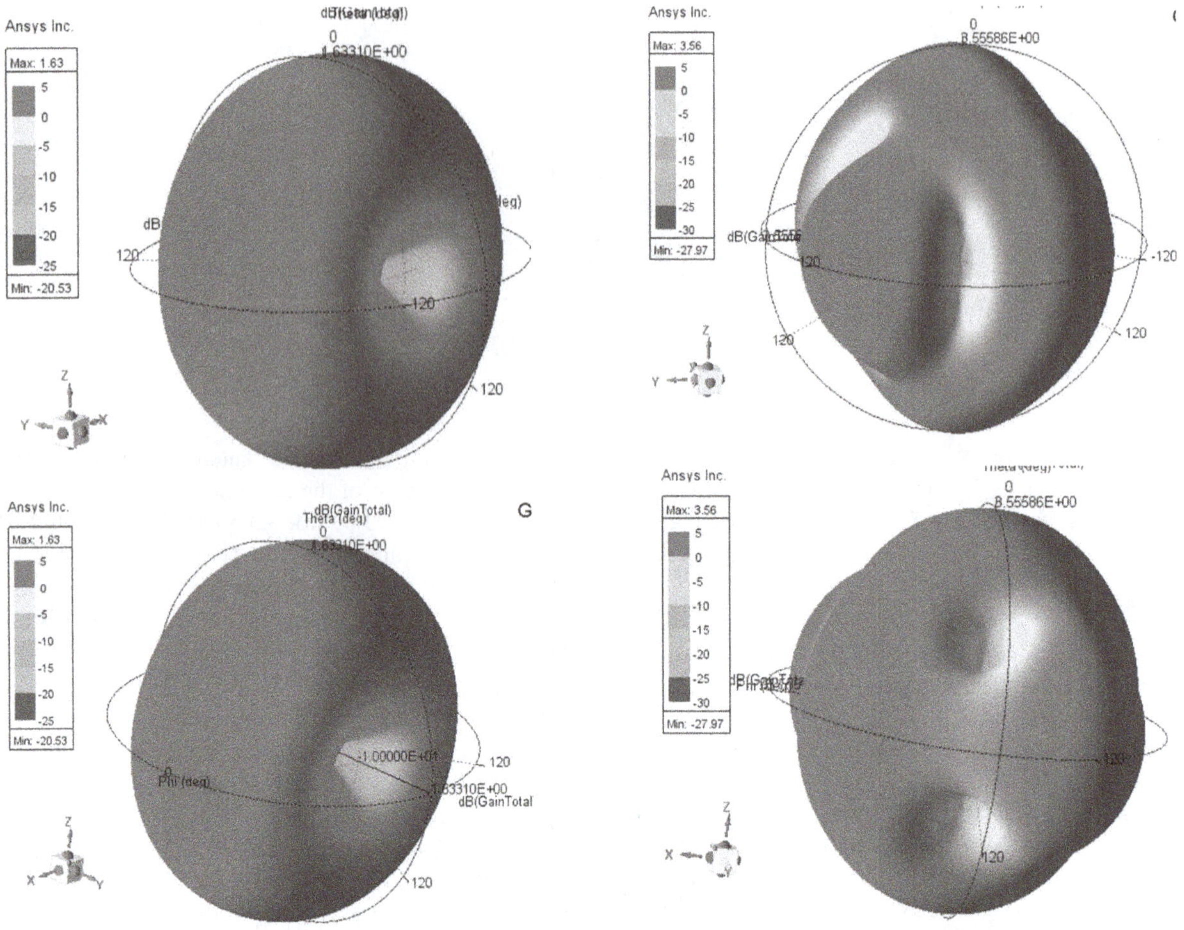

(a) for 8 GHz

(b) for 17 GHz

Fig. 140.4 3D plot of gain (a) for 8 GHz and (b) for 17 GHz

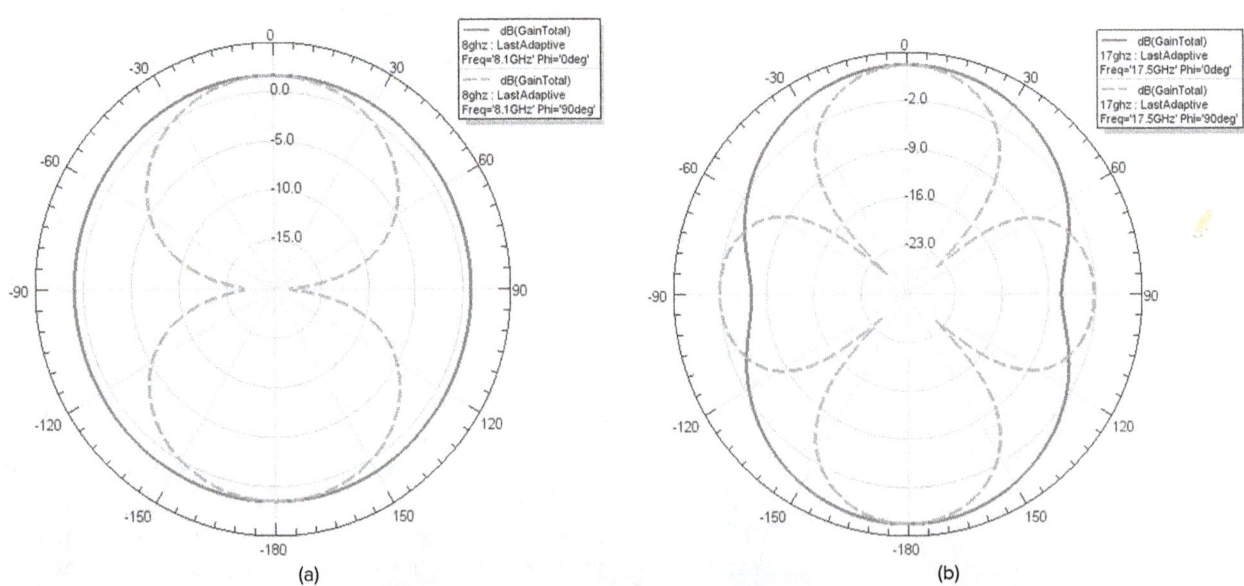

(a)

(b)

Fig. 140.5 Antenna Radiation pattern at (a) 8 GHz and (b) 17 GHz

In summary, this dual-band planar dipole antenna exhibits low return loss across a wide frequency range, coupled with band-switching capability. These characteristics contribute to enhanced efficiency, reliability, and flexibility for communication systems operating in multiple frequency bands, ensuring optimal performance under diverse operating.

REFERENCES

1. G. A. Deschamps (1953). Microstrip microwave antennas, presented at the Third USAF Symp. On Antennas.
2. Colin, A., E. Artal, E. Villa, D. Ortiz, and E. Mart´ınez-Gonz´alez (2010). Bow-tie slot antenna for 29 to 45 GHz band, Far East Journal of Electronics and Communications, Vol. 4: 69–76.
3. Z. D. Liu, P. S. Hall, and D. Wake (1997). Dual-frequency planar in-verted-F antenna, IEEE Trans. Antennas Propag., Vol. 45(10): 1451–1458.
4. Zhang, H., & Xin, H. (2009). A Dual-Band Dipole Antenna with Integrated-Balun. IEEE Transactions on Antennas and Propagation, 57(3): 786–789.
5. Shuchita Saxena & Kanaujia, B.K. (2011). Design And Simulation of A Dual Band Gap Coupled Annular Ring Microstrip Antenna, International Journal of Advances in Engineering & Technology Vol. 1(2):151–158. ISSN: 2231–1963.
6. K.L. Wong (2003). Planar antennas for wireless communications, Wiley, New York.
7. C.M. Su, H.T. Chen, and K.L. Wong (2002). Printed dual-band dipole antenna with U-slotted arms for 2.4/5.2-GHz WLAN band operation, Electron Lett 38:1308–1309.
8. N. Kumar Darimireddy, R. Ramana Reddy, and A. Mallikarjuna Prasad (2015). Design of triple-layer double U-slot patch antenna for wireless applications, JART, 13(5).
9. Gunaram, & Deegwal, J.K. & Sharma, Vijay. (2020). Dual Band Circular Polarized Printed Dipole Antenna for S and C Band Wireless Applications. Progress In Electromagnetics Research C. 105. 129–146. 10.2528/PIERC20050301.

Note: All the figures and tables in this chapter were made by the authors.

Recent Trends in Applied Physics and Material Science – Sudhir Bhardwaj et al. (eds)

Design and Performance Analysis of a U-Shaped Microstrip Patch Antenna for Multiband Wireless Applications

Renu Sharma*, Suraj Samal
Department of Physics, JECRC University,
Jaipur, India

Pawan Kumar Jain
Department of Physics,
Swami Keshvanand Institute of Technology Management
and Gramothan, Jaipur, India

Abstract: The design and analysis of a U-shaped microstrip patch antenna employing Rogers RT/duroid 5880 dielectric materials, designed for a resonant frequency of 25 GHz, are described in detail. The antenna, which has a substrate thickness of 1 mm and dimensions of 1 mm in width by 3.15 mm in length, is made to reduce return loss for better wireless communications performance. HFSS software simulations that concentrate on important parameters including gain, directivity, return loss, and VSWR produce a maximum gain of 8.04 dB and a return loss of -23.7726 dB. The antenna uses a coaxial probe for effective power transmission and impedance matching, and it has good broadside radiation patterns, which make it appropriate for multiband applications including 5G, Wi-Max, and Bluetooth.

Keywords: Microstrip, HFSS, Bluetooth

1. INTRODUCTION

The flat, low-profile design of a microstrip patch antenna makes it a popular antenna type that is perfect for a variety of microwave and radio frequency (RF) communications applications. With a ground plane on the other side, the antenna is made up of a thin metallic patch—which can be round or rectangular—mounted on a dielectric substrate. When a microwave signal is delivered to the patch, this arrangement enables effective electromagnetic wave radiation (G. Kaur et al. 2016).

A microstrip patch antenna's design is essential since the patch's resonance frequency is determined by its dimensions. By changing the patch's dimensions and form, the antenna can be tuned to operate effectively at specific frequency ranges, making it versatile for different applications. The typical operating frequency range for microstrip patch antennas spans from a few hundred megahertz (MHz) to several gigahertz (GHz).

Microstrip patch antennas' small size and light weight, which make it simple to integrate them with other microwave components, are among its main benefits. They are affordable and appropriate for mass manufacturing since they can be produced using conventional printed circuit board (PCB) procedures (R. Bhavani, at al. 2019). In order to improve gain and directivity, microstrip antennas can also be grouped in arrays and tailored for different polarization types, such as linear or circular.

Microstrip patch antennas are used in many different fields, such as satellite communications, GPS devices, and wireless communication systems (including cell phones and Wi-Fi). They do, however, have certain drawbacks, such as constrained bandwidth and possible problems with radiation efficiency and surface wave losses, especially at higher frequencies (M. H. Kabir et al. 2018).

In summary, a popular option for many applications, microstrip patch antennas are a flexible and crucial part of contemporary communication technology, providing a balance of performance, size, and manufacturing ease.

*Corresponding author: renu.sharma@jecrcu.edu.in

DOI: 10.1201/9781003684718-141

2. ANTENNA DESIGN AND SPECIFICATIONS

The Microstrip patch antenna's design (shown in Fig. 141.1), specifically the U-shaped variant discussed in the document, involves several key considerations and steps:

Substrate Selection: A dielectric substrate with a particular thickness and dielectric constant—like Rogers RT/duroid 5880—is used to design the antenna. The substrate thickness in this instance is 1 mm, while the dielectric constant is 2.2.

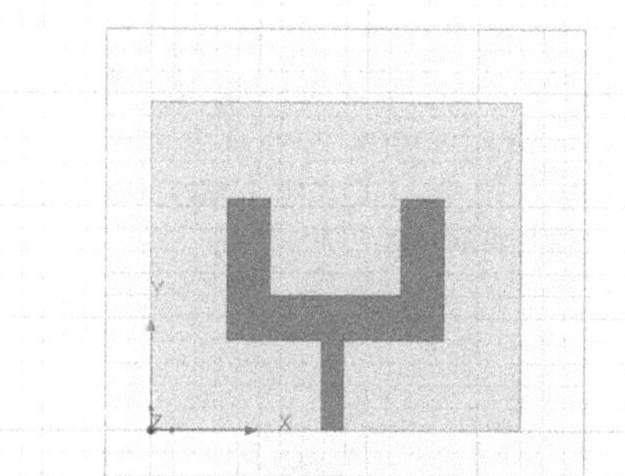

Fig. 141.1 U-shaped patch antenna top view

Patch Geometry: The antenna's overall performance and resonant frequencies are largely determined by the patch's shape. The U-shaped patch is created by modifying a rectangular patch design, introducing slots at the top and bottom to achieve desired resonant characteristics. This design frequency for multiband operation was 25 GHz.

Feeding Mechanism: A coaxial probe is used to feed the antenna, a standard technique for microstrip antennas. This feeding technique helps in achieving a good match of impedance and facilitates efficient power transfer.

Simulation and Optimization: The design process involves using HFSS software to simulate the antenna's performance. This includes analysing parameters such as Voltage Standing Wave Ratio (VSWR), directivity, gain, and return loss. The model helps in optimizing the design to enhance bandwidth and minimize losses.

Performance Metrics: The design is evaluated based on several performance metrics:

Return Loss: Shows the amount of power reflected back from the antenna; smaller numbers are better.

Gain: Indicates how well an antenna performs by measuring its capacity to focus radio frequency energy in a certain direction.

Directivity: Shows the degree of focus of the radiation pattern from the antenna, which is important for effective communication.

VSWR: A measure of impedance matching; lower values indicate better matching and efficiency.

Bandwidth: Using the U-shaped antenna achieving a bandwidth which is within the frequency range.

The patch's size radiator is determined from numerical analysis using these equations: -

For width calculation (W): -

$$W = \frac{1}{2f_o\sqrt{\mu_o\varepsilon_o}}\sqrt{\frac{2}{\varepsilon_r+1}} \tag{1}$$

Where ε_r is the substrate's dielectric constant, fo is the resonant frequency, and c is the free-space velocity of the particle given by

$$c = \frac{1}{2f_o\sqrt{\mu_o\varepsilon_o}}$$

Calculation of length (l): -

$$l = \frac{1}{2f_o\sqrt{\varepsilon_{reff}}\sqrt{\mu_o\varepsilon_o}} - 2\Delta L \tag{2}$$

Where ΔL is the length extension given by

Calculation of length extension is

$$\Delta L = 0.412(h)\frac{(\varepsilon_{reff}-0.3)\left(\frac{W}{h}+0.264\right)}{(\varepsilon_{reff}-0.258)\left(\frac{W}{h}-0.8\right)} \tag{3}$$

Here, h is the height of dielectric substrate.

Effective dielectric constant calculation (ε_{reff}): -

$$\varepsilon_{reff} = \frac{\varepsilon_r+1}{2} + \frac{\varepsilon_r-1}{2}\left[1+12\frac{h}{w}\right]^{\frac{1}{2}} \tag{4}$$

Table 141.1 Design parameter and corresponding values

Subject	Dimensions
Ground Size	8.5mm × 7.3mm
Patch Size	1mm × 3.15mm
Substrate Used-(Rogers RT/ duroid 5880(tm))	8.5mm × 7.3mm × 1mm
Thickness	1mm
Inset feed	0.5mm × 2mm
Radiation box	11mm × 10mm × 5mm

3. ANALYSIS OF SIMULATED RESULT

The Rogers RT/duroid 5800 dielectric medium, which has a dielectric constant of 2.2 and a loss tangent of 0.0009, was utilized to construct the U-shaped microstrip patch antenna for a resonance frequency of 25 GHz using HFSS software. The patch's measurements are 1 mm for width, 3.15 mm for length, and 1 mm for substrate thickness. In order to improve performance, the design seeks to reduce return loss at the feeding point. The antenna is appropriate for a range of wireless communication applications because it runs at a frequency of 25 GHz.

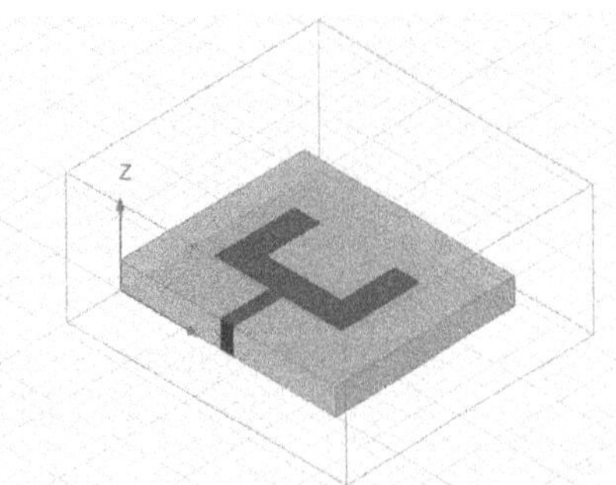

Fig. 141.2 3D view of U-shaped antenna design

The S parameter, specifically the S11 parameter, represents the return loss of an antenna, showing the amount of power that is reflected back from the antenna as opposed to being transmitted. Better performance is indicated by a lower S11 value (more negative), which shows that more power is being efficiently delivered or received and less is being reflected.

In the Fig. 141.3 the return loss is -23.7726 dB at resonance frequency of 24.6875 GHz.

Figure 141.4 and Fig. 141.5 show the VSWR plot for U-shaped patch antenna. Figure 141.6 displays the far field region where the radiation and overall efficiency are mentioned, frequency of 25 GHz. Its design is optimized for efficient radiation and reception at this frequency. Utilizing a lower dielectric constant substrate (ε_r = 2.2) allows for a compact size while enhancing bandwidth, enabling the antenna to perform effectively across a broader frequency range. The gain plot of an antenna indicates its efficiency in radiating energy in different directions compared to

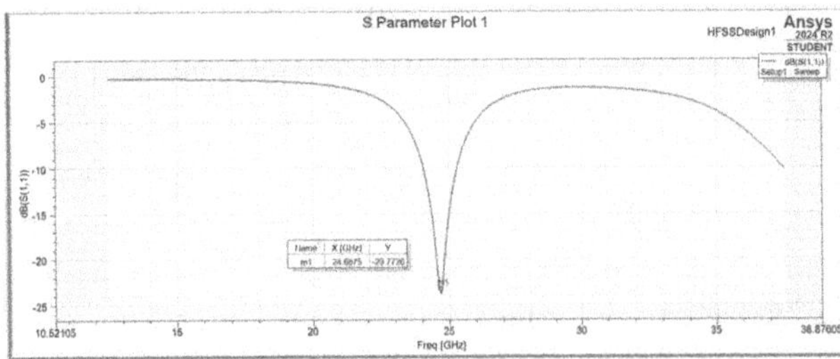

Fig. 141.3 Return loss graph for U-shaped antenna

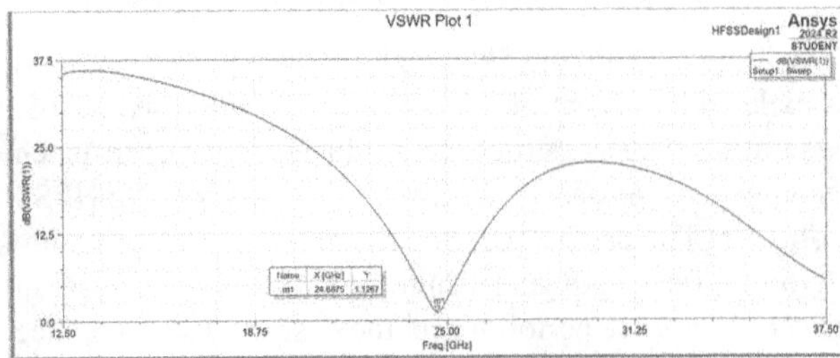

Fig. 141.4 A U-shaped patch antenna's VSWR plot (in dB)

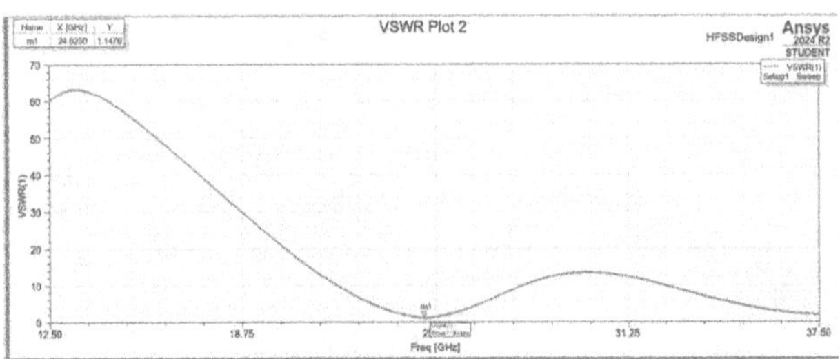

Fig. 141.5 VSWR plot for U-shaped patch antenna

an isotropic radiator. For the U-shape microstrip patch antenna, gain is typically measured in decibels (dB) and ranges from 6 to 9 dB, depending on design and frequency. The plot shows a gain total of max value of 8.04 dB.

In a 3D polar plot, the U-shaped Microsoft patch antenna's radiation pattern, gain (Fig. 141.6 and 141.7), and directivity (Fig. 141.8) are better than those of the rectangular patch antenna.

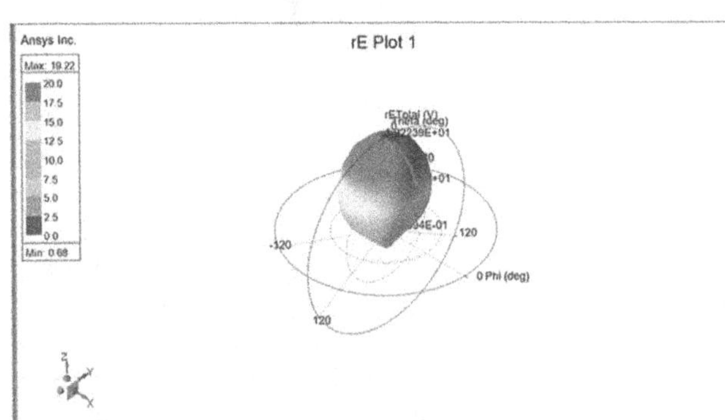

Fig. 141.6 Far field rE plot

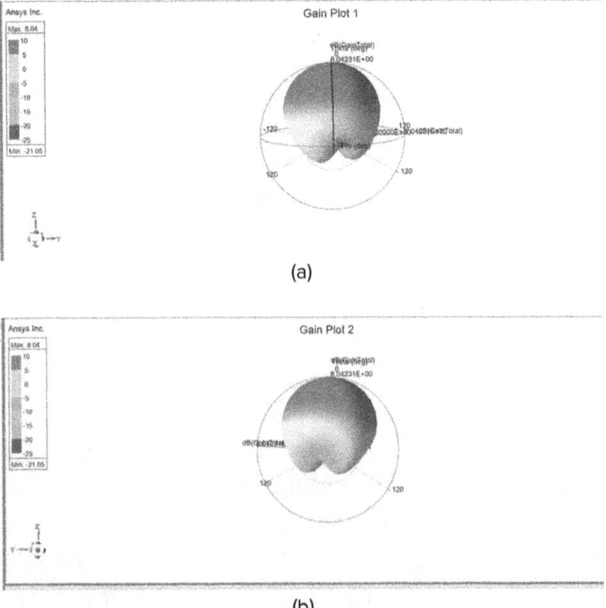

(a)

(b)

Fig. 141.7 (a) Gain plot 1, (b) Gain plot 2

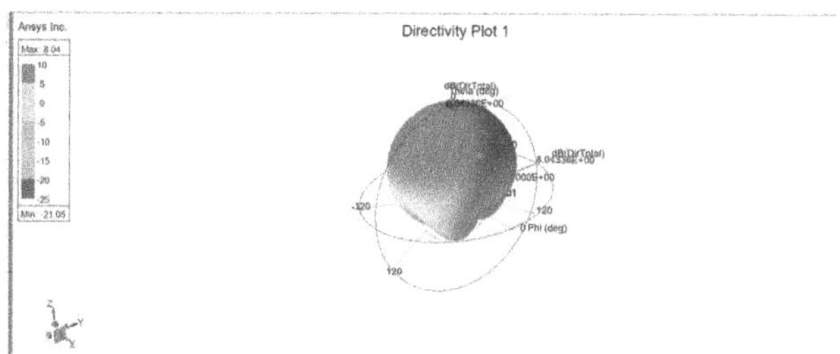

Fig. 141.8 Directivity of U-shaped antenna

3.1 Radiation Pattern

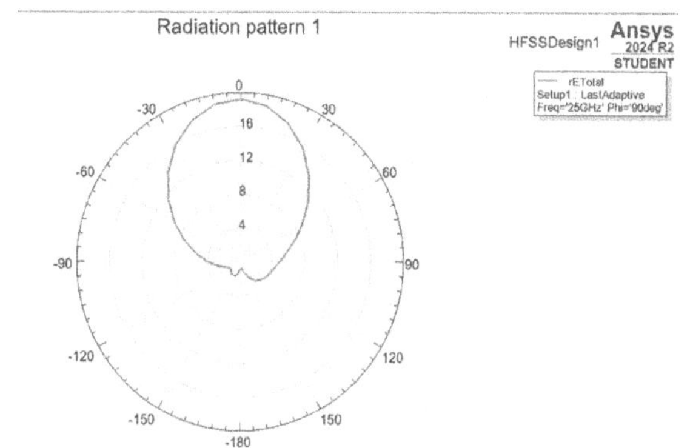

Fig. 141.9 The U-shaped antenna's radiation pattern in H-field

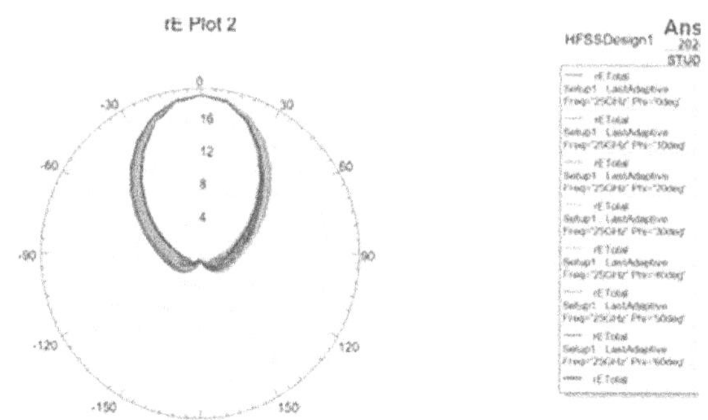

Fig. 141.10 The U-shaped antenna's radiation pattern in E-field

3.2 Field in a Microstrip U-shaped Patch Antenna

H-field: The H-field plan for the microstrip patch antenna in the U form shows the distribution of magnetic field intensity (Fig. 141.9), highlighting stronger orientations that enhance radiation effectiveness. This information is vital for understanding the antenna's radiation pattern and optimizing its design for improved gain and directivity in wireless communication applications.

E-field: Electric field distribution is seen (Fig. 141.10) by the U-shaped microstrip patch antenna's E-field plot with higher intensity in the broadside direction, indicating effective radiation. It helps to analyze the interaction with

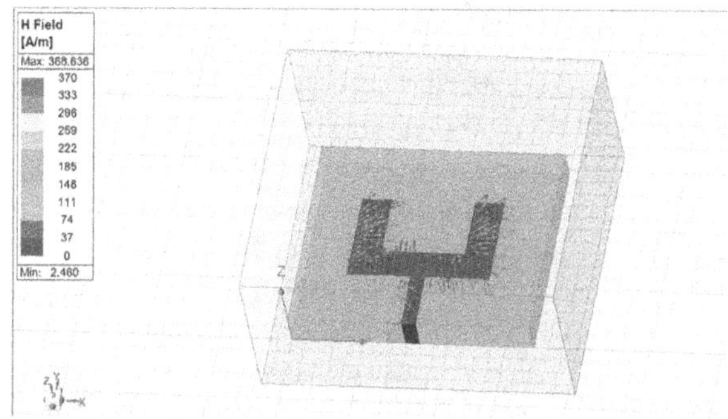

Fig. 141.11 H-field for Microstrip patch antenna with a U shape

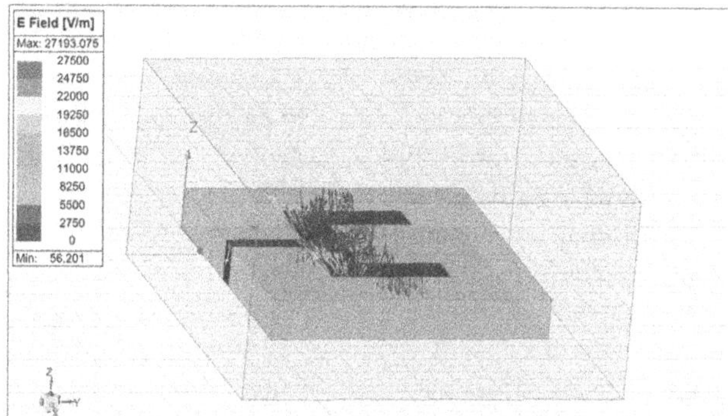

Fig. 141.12 E-field for Microstrip patch antenna with a U shape

the H-field, informing design optimizations for enhanced gain and directivity in communication applications.

The interaction between the H-field and E-field is key to the antenna's effective radiation.

3.3 Total Gain

Here's a brief summary of the total gain graph:

Maximum Gain: reaches an 8.04 dB maximum gain.

Radiation Efficiency: Indicates powerful performance in directing radio frequency energy.

Directional Focus: Higher gain values are observed broadside to the patch.

Application Suitability: Effective for wireless communication applications like 5G, Wi-Max, and Bluetooth.

Signal Strength: Ensures better signal strength and reception.

Loss Minimization: Designed to minimize losses, enhancing overall performance.

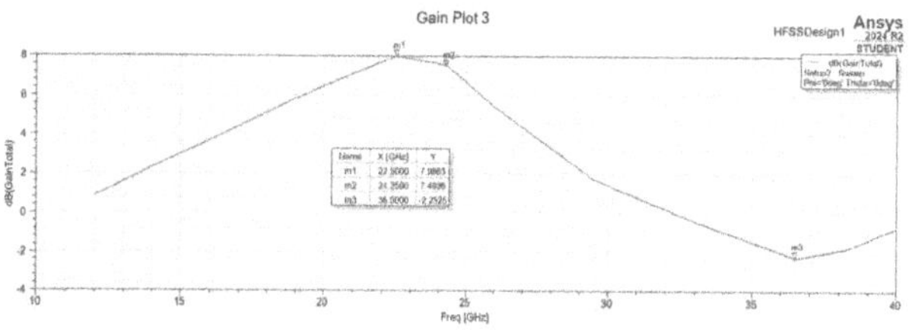

Fig. 141.13 Gain plot 3

4. CONCLUSION

The conclusion of the document underscores the successful design and performance of the microstrip patch antenna with a U shape that is tuned for a 25 GHz resonant frequency. The antenna is 1 mm wide, 3.15 mm long, and has a substrate thickness of 1 mm. It uses Rogers RT/duroid 5880 as the dielectric material. The design effectively minimizes return loss, achieving a value of -23.7726 dB, which indicates efficient power transmission and reception. Additionally, the antenna demonstrates a maximum gain of 8.04 dB, reflecting its capability to direct radio frequency energy effectively.

The document highlights the antenna's suitability for various wireless communication applications, including 5G technology, Wi-Max, and Bluetooth, due to its multiband capabilities and enhanced performance metrics. The U-shaped configuration is particularly advantageous, offering superior bandwidth and reduced losses compared to traditional rectangular patch designs. Overall, the analysis confirms that the U-shaped microstrip patch antenna is a popular option for a variety of wireless applications because it combines excellent performance with a small size, making it a flexible and effective solution for contemporary communication needs.

REFERENCES

1. Kaur, G., Goyal, S. (2016). To Study the Effect of Substrate Material for Microstrip Patch Antenna. International Journal of Engineering Trends and Technology (IJETT). 36 (9).
2. Microwave Engineering, David M. Pozar, John Wiley & Sons, 2011.
3. Zaidi, S.A. and Tripathy M.R. (2014). Design & Simulation Based Study of Microstrip E-Shaped Patch Antenna Using Different Substrate Material. Advance in Electronic & Electric Engineering, 4:611–616.
4. Ahmed, B., Saleem, I., Zahara, H., Khurshid, H. and Abbas, S. (2012). Analytical Study on Substrate Properties on the Performance of Microstrip Patch Antenna", International Journal of Future Generation Communication & Networking, 5(4):113–122.
5. Rathi, V., Rawat, S., Pokhariya, H. S. (2011). Study the Effects of Substrate Thickness and Permittivity on Patch Antenna. IEEE Conf. Signal Processing, Communications and Computing (ICSPCC).
6. Bhavani, R. Durga, Narayana, J.L. (2019). Microstrip H-Shaped Patch Antenna's Analysis and Layout.
7. International Journal of Innovative Technology and Exploring Engineering (IJITEE). 8(10).
8. Kabir, M. H., Uddin, M. A. and Hossain, M. Z. (2018). Bandwidth Enhancement of U-Shape Microstrip Patch Antenna. International Journal of Computer Science and Information Security. 16 (12): 118–122.
9. Islam, M. M., Hasan, R. R., Rahman M.M., Islam, K. S. and Al-Amin S.M. (2016). Design & Analysis of Microstrip Patch Antenna Using Different Materials for WiMAX Communication system. International journal of Recent Contribution from Engineering, Science & IT. 4 (1)

Note: All the figures and tables in this chapter were made by the authors.

For Product Safety Concerns and Information please contact our EU
representative GPSR@taylorandfrancis.com
Taylor & Francis Verlag GmbH, Kaufingerstraße 24, 80331 München, Germany

www.ingramcontent.com/pod-product-compliance
Lightning Source LLC
Chambersburg PA
CBHW080126220326
41598CB00032B/4975
9781041164524